Trapping and the Detection, Control, and Regulation of Tephritid Fruit Flies

Some Important Tephritid Species

Fly species (photo credit):

Top row (left to right): *Rhagoletis pomonella* (Juan Rull), *Ceratitis capitata* (Giovanni Benelli), *Ceratitis rosa* (Robert Copeland)

Middle row: *Anastrepha ludens* (Ana Rodriguez), *Anastrepha fraterculus* (M.Teresa Vera), *Bactrocera oleae* (Giovanni Benelli)

Bottom row: *Bactrocera tryoni* (Jaye Newman), *Bactrocera cucurbitae* (Ana Rodriguez), *Bactrocera dorsalis* (Ana Rodriguez)

Todd Shelly • Nancy Epsky • Eric B. Jang •
Jesus Reyes-Flores • Roger Vargas
Editors

Trapping and the Detection, Control, and Regulation of Tephritid Fruit Flies

Lures, Area-Wide Programs, and Trade Implications

Springer

Editors
Todd Shelly
APHIS
United States Department of Agriculture
Waimanalo, Hawaii, USA

Nancy Epsky
ARS
United States Department of Agriculture
Miami, Florida, USA

Eric B. Jang
ARS
United States Department of Agriculture
Hilo, Hawaii, USA

Jesus Reyes-Flores
Insect Pest Control Section
International Atomic Energy Agency
Food and Agriculture Organization
Vienna, Austria

Roger Vargas
ARS
United States Department of Agriculture
Hilo, Hawaii, USA

ISBN 978-94-017-9192-2 ISBN 978-94-017-9193-9 (eBook)
DOI 10.1007/978-94-017-9193-9
Springer Dordrecht Heidelberg New York London

Library of Congress Control Number: 2014947215

Printed on acid-free paper

Springer is part of Springer Science+Business Media (www.springer.com)

Dedicated to the Memory of Robert R. Heath [1945–2011] and Donald A. Lindquist [1930–2011]

Bob Heath's expertise in the development and application of insect semiochemicals along with Don Lindquist's vision and coordination of full scale multinational programs to test novel approaches have resulted in substantial advances in the use of trapping for tephritid fruit fly detection and control.

Preface

This project emerged from three simple facts: (i) Certain species of tephritid fruit flies are among the world's most notorious pests of commercially important fruits and vegetables; (ii) trapping these flies is vital to identifying infestations, controlling detected populations, and establishing guidelines for international transport of agricultural commodities; and (iii) despite its central role, there exists no comprehensive repository of factual or theoretical material relating specifically to trapping issues for economically important Tephritidae. While the editors (and we assume many of the authors) would admit to a scientific fascination with this group of insects, production of a volume devoted strictly to trapping of a relatively small number of pest species reflects, not just this scientific curiosity, but also the serious impact these pests have on global commerce. As Aldo Malavasi notes in his Introductory Remarks, every major fruit and vegetable growing county in the world maintains some program relating to surveillance and control of tephritid fruit fly pests. Thus, trapping issues concern scientists, regulatory agencies, and trade organizations in countries of every continent, from Australia and Brazil through the alphabet to Yemen and Zimbabwe.

We thank all the authors for their contributions, which were produced without financial compensation. Collectively, they exhibited a spirit of industry, cooperation, and patience that smoothed the task of editing. We extend special thanks to A. Malavasi, who graciously provided introductory remarks. TS also thanks J.C. Stewart, who allowed him time to initiate and complete this project.

Each chapter was reviewed by at least one editor and at least one external reviewer. We extend deep appreciation and gratitude to the following individuals, who served as reviewers: R. Dowell, J. Duan, R. Duthie, W. Enkerlin, Y. Gazit, S. Geib, T. Holler, P. Kendra, L. Leblanc, A. Liebhold, N. Manoukis, A. Manrakhan, D. McInnis, M. De Meyer, D. Midgarden, S. Myers, A. Norrbom, J. Piñero, J. Rojas, D. Rubinoff, M. San Jose, D. M. Suckling, S. Thornsbury, M. Virgilio, T. Yamanaka, B. Yuval, and J. L. Zavala Lopez.

We also thank those who graciously provided the photos appearing in the preceding gallery.

Our goal was to produce a comprehensive synthesis of tephritid-centric trapping issues, and accordingly the topics included are far-ranging and address lures and traps, population ecology and detection, suppression and eradication strategies, and regulatory issues. We hope we have achieved this goal and that this volume proves useful for years to come.

Waimanalo, HI, USA — Todd Shelly
Miami, FL, USA — Nancy Epsky
Hilo, HI, USA — Eric B. Jang
Vienna, Austria — Jesus Reyes-Flores
Hilo, HI, USA — Roger Vargas

Introductory Remarks

From an economic point of view, true fruit flies are, by far, the most important insect family attacking horticultural crops. Tephritid flies cause both direct losses and also indirect losses as their presence can result in major international trading constraints. Total damage caused in all production, harvesting, packing, and marketing worldwide is estimated to amount to more than 2 billion dollars annually. Their economic and trading importance is so high that in every fruit growing country there is at least one unit dedicated to fruit fly detection and control under the National Plant Protection Organization.

In this context, an essential issue is to determine the density and distribution of fruit fly populations in the field. In all cases, fly populations vary from zero to high numbers, depending on many factors, but mainly host availability and climate conditions. All this critical information, obtained mainly through trapping, is required to design the most effective strategies in order to suppress or eliminate the population.

The big challenge for researchers and managers of action programs is to choose the best trapping system available for a particular growing area or region and for a specific fruit fly species or group of species. Four critical parameters are involved: trap type, fly attractant, trap density, and service interval. Once such parameters are defined, the operation and logistics of the surveillance network need to be planned to provide the most accurate possible estimates of the actual fruit fly populations in the field – whether an orchard or vegetable field, natural vegetation or an urban area, or an area-wide landscape that includes a mosaic of these different types of areas.

Defining the optimal trap type and fly attractant is an endless task. Both by chance or by active search, many researchers in all countries are deeply involved in developing more effective, selective, inexpensive, and easier to handle combinations of trap and attractant. A huge number of solutions can be found in the literature or in local/regional fruit fly manuals. However, there is a worldwide effort to harmonize the solutions in order to have comparable data that can be internationally recognized.

The fruit fly trapping system selected affects a wide range of stakeholders and interests, from the government officer in charge of a detection program, to the grower that needs to know the population density in his orchard to start control measures, and up to the packers and trading partners who import or export horticultural products.

With the ever increasing invasive process linked to globalization, resulting in the movement of exotic fruit flies to all corners of the world, reliable detection programs are essential to plant protection services with the responsibility to safeguard their countries from unwanted new fruit fly pests.

Furthermore, many exporting programs must have in place an efficient trapping system to help both growers and inspectors make the right decisions regarding the fresh fruit to be exported. Also, in cases of a systems approach, where a low resident adult population is acceptable, the monitoring of fruit flies is a critical issue to guarantee the quality of the commodity. In countries or regions considered fruit fly free, an essential component is a surveillance system to demonstrate to trading partners the absence of the target species.

In conclusion, the establishment of a trapping system should take into consideration many elements from natural history to genetics and modeling, from design to cost and logistics, from international plant protection standards to international trade, and this exhaustive book will be an extremely valuable source of information for all readers in this respect.

Many experts with deep knowledge and actual field experience on fruit fly trapping contributed to this book. Here, for the first time, very valuable information often not found in the refereed literature is consolidated, reviewed and synthesized, not only for the fruit fly community – fruit fly technical officers, plant protection inspectors, trappers in charge of surveillance and managers that need to update their trapping program – but also for common growers and academic researchers with interest on fruit fly biology. The editors of this book are commended for their comprehensive effort.

Biofabrica Moscamed Brazil, Juazeiro, BA, Brazil Aldo Malavasi

Contents

Part I Introduction

1 **Fruit Fly Alphabets** 3
Todd E. Shelly

Part II Lures and Traps

2 **Pheromones, Male Lures, and Trapping of Tephritid Fruit Flies** 15
Keng Hong Tan, Ritsuo Nishida, Eric B. Jang, and Todd E. Shelly

3 **History and Development of Food-Based Attractants** 75
Nancy D. Epsky, Paul E. Kendra, and Elena Q. Schnell

4 **Plant Odors as Fruit Fly Attractants** 119
Serge Quilici, Toulassi Atiama-Nurbel, and Thierry Brévault

5 **Interactions Between Tephritid Fruit Fly Physiological State and Stimuli from Baits and Traps: Looking for the Pied Piper of Hamelin to Lure Pestiferous Fruit Flies** 145
Francisco Díaz-Fleischer, Jaime C. Piñero, and Todd E. Shelly

Part III Ecology and Detection

6 **Trapping to Monitor Tephritid Movement: Results, Best Practice, and Assessment of Alternatives** 175
Christopher W. Weldon, Mark K. Schutze, and Minette Karsten

7 **Fruit Fly Invasion: Historical, Biological, Economic Aspects and Management** 219
Nikos T. Papadopoulos

8 Fruit Fly Detection Programs: The Potentials and Limitations of Trap Arrays 253
A. Meats

9 Spatial Analysis of Tephritid Fruit Fly Traps 277
David Midgarden, Estuardo Lira, and Micha Silver

10 Using Molecules to Identify the Source of Fruit Fly Invasions 321
Norman Barr, Raul Ruiz-Arce, and Karen Armstrong

11 Modeling Trapping of Fruit Flies for Detection, Suppression, or Eradication 379
Hugh J. Barclay and Jorge Hendrichs

Part IV Attract and Kill

12 Priorities in Formulation and Activity of Adulticidal Insecticide Bait Sprays for Fruit Flies 423
Robert L. Mangan

13 Recent Developments and Applications of Bait Stations for Integrated Pest Management of Tephritid Fruit Flies 457
Jaime C. Piñero, Walther Enkerlin, and Nancy D. Epsky

14 Male Annihilation, Past, Present, and Future 493
Roger I. Vargas, Luc Leblanc, Jaime C. Piñero, and Kevin M. Hoffman

15 Mass Trapping for Fruit Fly Control 513
Vicente Navarro-Llopis and Sandra Vacas

Part V Phytosanitary Programs and Regulations

16 Integrating Tephritid Trapping into Phytosanitary Programs 559
D.R. Lance

17 Trapping Related to Phytosanitary Status and Trade 589
Eric B. Jang, Walther Enkerlin, Charles "Ed" Miller, and Jesus Reyes-Flores

Part VI Coda

18 The Complexities of Knowing What It Is You Are Trapping 611
Anthony R. Clarke and Mark K. Schutze

Index 633

Contributors

Karen Armstrong Bio-Protection Research Centre, Lincoln University, Lincoln, Christchurch, New Zealand

Toulassi Atiama-Nurbel CIRAD, UMR PVBMT CIRAD/Université de La Réunion, Saint-Pierre, La Réunion, France

Hugh J. Barclay Victoria, BC, Canada

Norman B. Barr Center for Plant Health Science and Technology, Mission Laboratory, USDA-APHIS-PPQ-CPHST, Edinburg, TX, USA

Thierry Brévault CIRAD, UR AIDA, Montpellier, France

Anthony R. Clarke School of Earth, Environmental and Biological Sciences, Science and Engineering Faculty, Queensland University of Technology, Brisbane, QLD, Australia

Francisco Díaz-Fleischer Instituto de Biotecnologia Aplicada (INBIOTECA), Universidad Veracruzana, Xalapa, Veracruz, México

Walther Enkerlin Programa Regional Moscamed, Ciudad de Guatemala, Guatemala

Nancy D. Epsky Subtropical Horticultural Research Station, USDA-ARS, Miami, FL, USA

Jorge Hendrichs Insect Pest Control Section, Joint FAO/IAEA Division of Nuclear Techniques in Food and Agriculture, International Atomic Energy Agency, Vienna, Austria

Kevin M. Hoffman California Department of Food and Agriculture, Pest Detection/Emergency Projects, Sacramento, CA, USA

Eric B. Jang USDA-ARS, Daniel K. Inouye, U.S. Pacific Basin Agricultural Research Center, Hilo, HI, USA

Minette Karsten Department of Conservation Ecology and Entomology, University of Stellenbosch, Matieland, South Africa

Paul E. Kendra Subtropical Horticultural Research Station, USDA-ARS, Miami, FL, USA

David R. Lance USDA-APHIS-PPQ-CPHST, Otis Laboratory, Buzzards Bay, MA, USA

Luc Leblanc Department of Plant and Environmental Protection Sciences, College of Tropical Agriculture and Human Resources, University of Hawaii, Honolulu, HI, USA

Estuardo Lira USDA-APHIS-IS Medfly Program, Guatemala City, Guatemala

Aldo Malavasi Biofabrica Moscamed Brazil, Juazeiro, BA, Brazil

Robert L. Mangan Subtropical Horticultural Research Station, USDA-ARS, Miami, FL, USA

Alan Meats School of Biological Sciences A12, University of Sydney, Sydney, NSW, Australia

David Midgarden USDA-APHIS-IS Medfly Program, Guatemala City, Guatemala

Charles "Ed" Miller Frederick, MD, USA

Vicente Navarro-Llopis Instituto Agroforestal del Mediterráneo-Universidad Politécnica de Valencia, Valencia, Spain

Ritsuo Nishida Graduate School of Agriculture, Laboratory of Chemical Ecology, Pesticide Research Institute, Kyoto University, Sakyo-ku, Kyoto, Japan

Nikos T. Papadopoulos Department of Agriculture, Laboratory of Entomology and Agricultural Zoology, Crop Production and Rural Environment, University of Thessaly, Magnisias, Greece

Jaime C. Piñero Cooperative Research and Extension, Lincoln University, Jefferson City, MO, USA

Serge Quilici CIRAD, UMR PVBMT CIRAD/Université de La Réunion, Saint-Pierre, La Réunion, France

Jesus Reyes-Flores Insect Pest Control Section, International Atomic Energy Agency, Vienna, Austria

Raul Ruiz-Arce Center for Plant Health Science and Technology, Mission Laboratory, USDA-APHIS-PPQ-CPHST, Edinburg, TX, USA

Elena Q. Schnell Subtropical Horticultural Research Station, USDA-ARS, Miami, FL, USA

Mark K. Schutze School of Earth, Environmental and Biological Sciences, Queensland University of Technology, Brisbane, QLD, Australia

Todd E. Shelly USDA-APHIS-PPQ-CPHST, Waimanalo, HI, USA

Micha Silver Arava Development Company, Sapir, Israel

Keng Hong Tan Tan Hak Heng, Tanjong Bungah, Penang, Malaysia

Sandra Vacas Instituto Agroforestal del Mediterráneo-Universidad Politécnica de Valencia, Valencia, Spain

Roger I. Vargas United States Department of Agriculture, Agricultural Research Service, U.S. Pacific Basin Agricultural Research Center, Hilo, HI, USA

Christopher W. Weldon Department of Zoology and Entomology, University of Pretoria, Hatfield, South Africa

Part I
Introduction

Chapter 1
Fruit Fly Alphabets

Todd E. Shelly

Abstract The routine operations associated with tephritid fruit fly programs are divorced from the workings of the underlying science on traps and lures, which has developed more haphazardly through the work of individual researchers. Additionally, all trapping outcomes are probabilistic, rendering data interpretation problematic. Mark-release-recapture studies have proven valuable in providing estimates of minimum detectable population sizes for invasive fruit fly species. Both intra- and interspecific variation in lure/trap responsiveness demand further investigation, as the notion that "one trap/bait combination fits all" is probably not maximally effective.

Keywords Trapping program • Detection probability • Incipient populations • Trimedlure • *Ceratitis capitata* • Mark-release-recapture • Trapping sensitivity • Male lures • Food baits • Fly responsiveness

In early morning, the working space of the fruit fly surveillance program resembles a war room. A dozen workers are organizing and entering data on trap captures from yesterday's work. A few more are organizing supplies for today's routes. Another is on the phone trying to locate a replacement for a sick employee. Encapsulating this whole process – its magnitude, its importance, its military feel – is a large wall map of the surveyed region divided into regular grids, each bordered with thick red lines and prominently numbered as a distinct sampling unit. Thousands of grids, tens of thousands of traps, each checked every 2 weeks, each with bait replenished every 6 weeks, all year long, year after year. The daily movement starts: trappers check their supplies, grab their lunch, and drive away singly in their trucks to run their daily routes. The room is empty. Data collection has begun.

T.E. Shelly (✉)
USDA-APHIS, 41-650 Ahiki Street, Waimanalo, HI 96795, USA
e-mail: todd.e.shelly@aphis.usda.gov

T. Shelly et al. (eds.), *Trapping and the Detection, Control, and Regulation of Tephritid Fruit Flies*, DOI 10.1007/978-94-017-9193-9_1,

The orderliness of the survey operation speaks to a dedication to routine. The industrial nature of the process is unmistakable – a complex problem (sampling fruit flies over a large area) is broken down into regular, well-defined units (grids) whose uniform oversight and sampling are assigned to trained managers (trappers). The logic and regularity of the process are strikingly clear and overwhelmingly mechanistic. At its core, of course, the entire process is a sampling effort, but one in which the underlying science exists separately from its implementation. Isolated, the underlying science lacks the tenor of the trapping operations. Instead of constant routine, it proceeds erratically via the trial-and-error approach, which, of course, is the driver of progress in empirical research. Instead of certain routine, it faces complex questions and generates partial answers, which then pose novel complex questions for which partial answers are obtained and so on. In other words, until some high level of reliability or predictability is achieved, the science of trapping moves in a manner typical of empirical science in general: hypotheses are tested, some are falsified, and new hypotheses emerge for further testing. In addition, whereas survey operations constitute a unified group with a shared goal, the science of tephritid trapping proceeds largely through the unconcerted efforts of single individuals or laboratory teams working on specific projects chosen for any number of reasons, including academic interests and research experience, available tephritid species, local agricultural concerns, and international economic and trade issues as well as more practical factors, such as funding opportunities, availability of equipment and manpower for research, and travel possibilities.

This approach, which is hardly unique to trapping research, results in a mosaic of knowledge and understanding of trapping-related issues. Some tephritid species are well-studied, others not. Some trap types and baits have been examined extensively, others not. And so on.

Compounding the matter is the obvious fact that, owing to the large number of uncontrolled variables, field research on tephritid trapping typically produces probabilistic conclusions, not absolute ones. Changes in weather (particularly, temperature, rainfall and humidity, and wind speed and direction), inter-site differences (e.g., in climate, host plant availability, and predation risk), and spatiotemporal variability in population size and physiological profile (e.g., age structure and mating status) are key factors that may render true replication problematic and so promote variability in test results. This is neither a novel nor a particularly insightful comment, but it does describe accurately the context in which both data collection and data interpretation occur. A cynical view of statistics has no place in field research on trapping. Nobel Laureate Ernest Rutherford's quote "If your experiment needs statistics, you ought to have done a better experiment" may suit the laboratory setting in atomic research but is largely inapplicable to outdoor trapping studies (catchy, but also equally irrelevant, is the statement "Torture numbers, and they'll confess to anything" attributed to the science journalist Gregg Easterbrook).

If the topic being investigated addresses specific problems of limited generality, then the situation of evolving research generating probabilistic outcomes has little consequence. However, in broader questions with substantial academic as well as commercial implications, this situation becomes extremely important. Perhaps the best illustration of this involves the current debate regarding the establishment of the Mediterranean fruit fly (medfly), *Ceratitis capitata* (Wiedemann), in California. Kaneshiro (1993) cited the low attractiveness of trimedlure (a male-specific lure widely used in medfly surveillance programs) as a key factor responsible for – in his opinion – the undetected invasion and establishment of *C. capitata* in California. As evidence, he reported that, during a field study in Hawaii, 140 males were captured, marked, and released within 10–20 m of a US Department of Agriculture maintained trimedlure-baited trap, and none was captured in the trap over the succeeding 3 days. Carey (1996) later cited this observation in proposing the idea of 'early stage subdetectability', whereby incipient or small populations escape discovery by virtue of their small size and, by implication, the poor attractancy of sentinel lures.

Relative to the lures used for detection of certain lepidopterans (e.g., gypsy moth) or even other tephritid fruit flies (e.g., *Bactrocera* spp.), trimedlure is clearly a weak attractant. But, are we to conclude from Kaneshiro's (1993) observation that it is completely ineffective and that the state of California is simply wasting money by purchasing trimedlure for use in tens of thousands of traps? Of course not. We will never know why none of the 140 marked Hawaiian flies was trapped, but other, more rigorous studies have recorded capture probabilities >0 for male medflies in trimedlure-baited traps (a point made by Lance and McInnis (1993) in their reply to Kaneshiro (1993)). Ultimately, the Hawaiian observation sheds more heat than light, and the challenge is quantifying the effective capture rate of trimedlure-baited traps.

Attraction of male medflies to trimedlure was described over 50 years ago (Beroza et al. 1961), and since then trimedlure has been used in detection programs worldwide. Given this rather long history, it is surprising that data on capture rates remain scant, i.e., few empirical studies have attempted to measure capture probabilities of male medflies at varying distances from a trimedlure-baited traps. Seminal papers by Cunningham and Couey (1986) and Lance and Gates (1994) provided initial estimates, and the latter authors used distance-dependent capture probabilities to estimate the detection sensitivity of the California trapping program. Considering the area covered by an individual trap, Lance and Gates (1994) weighted capture probabilities for specific release distances by the relative amount (%) of area corresponding to these distances, summed these adjusted values across distance zones, and calculated the probability that at least 1 fly would be trapped for populations of varying sizes. The central question was: what is the minimum population size certain (defined operationally as >99.9 % probability) to be detected by the California trapping program? Assuming point occurrence of the flies, the estimate provided by Lance and Gates (1994) was approximately 2,200 males. This value pertains to a single generation, however, and given a stable size over 5 generations, populations with approximately 300 males were certain to be detected within this time interval.

What is the importance of this estimate? Alone, of course, it does not answer the question of whether the medfly is established in California, a complex, arguably unresolvable issue whose debate draws, not only from trapping data, but from diverse sources, including interception records and molecular genetic analyses among others (Papadopoulos et al. 2013 and references therein). However, it does provide a key piece of information, namely a numerical estimate of the upper limit of subdetectable populations, and given the possibility of multiple generations per year in southern California, it sharpens attention on the postulated checks to population growth (Carey 1991) that purportedly suppress populations to very low (and subdetectable) levels in a region with a favorable climate and abundant host plants. This has been a key argument of skeptics of medfly establishment in California: How do we reconcile the r-selected, high reproductive capacity of the medfly with its obvious scarcity? Again, an estimate of detectable population size does not provide the answer, but it does generate an abundance limit below which populations must exist through time in order to escape discovery through trapping. Knowledge of this limit may also be useful in various modeling efforts, particularly those involving the occurrence and impact of the Allee effect on the extinction of small populations of medfly.

In logic, argument is a technical term (lacking emotional overtones) for the process of convincing others to believe a certain statement or claim and consists of one or more premises (statements proposed as true) and a conclusion (a statement whose acceptance as true derives from the demonstrated validity of the premises). For heuristic purposes, Kaneshiro's (1993) observation (carried to its extreme) produces the argument:

1. Even when placed in suitable locations, trimedlure-baited traps capture no male medflies (premise).
2. Given this finding, male medflies are obviously not attracted at all to trimedlure (premise).
3. Therefore, trimedlure-baited traps provide no useful information on medfly presence or abundance (conclusion).

This caricature is plainly false as trimedlure-baited traps do, of course, capture male medflies, and a more realistic argument is:

1. When placed in suitable locations, trimedlure-baited traps capture male medflies (premise).
2. However, only a portion of the male population is attracted to and captured in trimedlure-baited traps (premise).
3. Therefore, trimedlure-baited traps provide useful information, albeit couched in probabilistic terms, on medfly presence and abundance (conclusion).

As stated above, a key challenge is the measurement, preferably through mark-recapture studies, of trap capture rate and subsequent estimation of the sensitivity of medfly trapping programs. Recognizing that trimedlure is a relatively weak lure should not preclude efforts to obtain more robust estimates of detectable population sizes; these estimates have inherent value regardless of the level of trap efficiency. In fact, somewhat surprisingly, although the data set is small, existing studies for the medfly suggest some uniformity across regions in estimates of minimum detectable population sizes for the medfly. Following the computational methods of Lance and Gates (1994) for the trap density used in California, Shelly et al. (unpublished data) found that data on distance-dependent capture probabilities (single generation, point source) from Hawaii (Cunningham and Couey 1986) and Australia (Meats and Smallridge 2007) yielded estimates of minimum detectable population sizes of 2,000–6,000, which are similar to Lance and Gates' (1994) aforementioned estimate. Whether this conformity is an outcome of small sample size or actually a robust finding awaits additional data.

The chapters of this book deal with a diversity of topics relating to trapping tephritid fruit flies, including, not only detection, but also dispersion and invasiveness, suppression, and regulatory issues in phytosanitation. Still, the focus in this introductory essay on detection was deliberate as the response of fruit flies to trap stimuli is at the core of all trapping issues. The foremost student of tephritid foraging behavior, Ron Prokopy (1995), emphasized this point in his contribution to a medfly symposium held nearly 20 years ago:

> It seems unlikely that truly robust progress can be made toward developing more sensitive approaches to detecting medflies and safer or more effective approaches to controlling medflies without first developing (a) a much firmer understanding of how medfly behavior is organized in space and time in natural habitats, and (b) a more complete appreciation of how variation in environmental factors and fly physiological, informational, or genetic state factors affects patterns of behavioral organization.

Prokopy appears to be advocating for the quantification of capture probability for each trap-lure combination and for each fly state in a population (e.g., sex, age, mating status, hunger level, etc.) under different combinations of relevant environmental factors (e.g., temperature, resource availability, predation risk, etc.). Implicitly, such quantification would allow the development of traps that maximize captures for a particular subset(s) of the population (e.g., virgin females) or over the entire population (including all fly states). While this goal may be largely unachievable, it has value in identifying factors potentially important in trap design and bait development.

By highlighting variable response to trap/lure parameters, Prokopy sends the tacit warning that a "one trap/bait combination" fits-all-approach may not be effective. Conceivably, differential attraction to a specific trap-bait combination could represent interspecific or intraspecific variation. The latter could reflect variation between different populations of the same species occurring in different regions or different seasons in the same location or (as Prokopy emphasized) between different sub-groups (based on gender, age, mating status, etc.) existing within the same population. Working with two *Anastrepha* species, Díaz-Fleischer et al. (2009) document both between- and within-species variation in response to food baits. They conclude by acknowledging the appeal of a "generic, 'magic' trap" that attracts flies of all physiological states of all species equally but suggesting, more realistically, that effective trapping of multiple, syntopic tephritid species may require species-specific trap/lure combinations.

While scant data exist regarding interspecific variability in response to food baits, even fewer data exist regarding potential between-species differences in response to male lures. For example, although methyl eugenol is well known as a powerful lure for males of several economically important *Bactrocera* species, we know virtually nothing about interspecific variation in attraction to this lure. Yet, several studies suggest that such variation exists and may have practical implications for control efforts. Wee et al. (2002) offered different dilutions of methyl eugenol to mature males of *Bactrcera dorsalis* (Hendel) and *Bactrocera carambolae* Drew and Hancock and found that the dose required to elicit a response (landing and feeding on a methyl eugenol source) was 17 times higher for *B. carambolae* than for *B. dorsalis* males. In additional feeding trials, *B. carambolae* males consumed significantly smaller amounts of the lure than *B. dorsalis* males. Based on these findings, the authors describe *B. carambolae* as having a "lower sensitivity to methyl eugenol". Given this finding, it is noteworthy that the Male Annihilation Technique (MAT), which involves the distribution of poisoned methyl eugenol-coated blocks in the environment, was unsuccessful in eradicating *B. carambolae* in Suriname and French Guyana (Vargas et al., Chap. 14, this volume), whereas the MAT has successfully eliminated populations of *B. dorsalis* in various locations (e.g., Steiner et al. 1965; Ushio et al. 1982; Seewooruthun et al. 2000). In a parallel example, Jang and Siderhurst (unpublished data) investigated possible alternatives to cue-lure, another male lure for certain *Bactrocera* species, and found that an analogue of cue-lure is attractive to males of *Bactrocera tryoni* (Froggatt) but not those of *Bactrocera cucurbitae* (Coquillett). Thus, interspecific variation in male responsiveness appears to exist among cue-lure-responding species as well.

The preceding examples validate Prokopy's suggestion that the "one trap/bait combination" fits-all-approach is perhaps not the most effective strategy when surveying multiple species or multiple physiological states within a species. For trapping aimed primarily at a single target species, it might be expected that the most effective trap would simply combine male- and female-preferred odors. However, the few studies that have tested multiple odors in a single trap have generally not demonstrated enhanced trap performance. Several researchers (Hill 1986; Tóth et al. 2004; Reboulakis et al. 2004) compared captures between traps containing both a food odor and a male lure versus traps containing each of these odors alone and found no improvement or even decreased capture in the combination traps. Likewise, the combination of food/host odor plus male pheromone has not proven particularly effective. In *Anastrepha ludens* (Loew), for example, the pairing of chapote fruit (both a larval and adult food source) odor plus male pheromone was never more attractive than fruit odor alone, and, in certain tests, was actually less attractive than either odor presented singly (Robacker and Garcia 1990). Similar studies on the olive fruit fly (*Bactrocera oleae* (Rossi)) have generated inconsistent results (e.g., Haniotakis and Vassiliou-Waite 1987).

So, where does all this leave us? As noted above, large-scale surveillance programs, particularly in the USA, function as industrial processes, where globally accepted trap/bait combinations (methyl eugenol, cue-lure, and trimedlure used in Jackson traps and torula yeast solution used in McPhail traps) are deployed, and have been deployed for the decades, according to international guidelines. Based on the science of fruit trapping, however, it seems apparent that (i) fruit fly baits and lures are – as a group – relatively weak attractants that attract only a proportion of the flies in an area and (ii) particular bait/trap combinations do not sample all individuals equally. Moreover, it seems likely that sampling biases – even for the same bait/trap pair – are themselves not constant but vary temporally and spatially with climate, habitat, and/or population size and structure. Thus, there is an apparent disconnect between the surveillance programs that rely on a standard set of baits/lures and supporting science that shows these substances are weak and, to some degree, selective attractants.

A straightforward explanation can account for this disconnect, i.e., while diversified trapping might be preferable, there are simply not sufficient resources to expand the trapping protocol to include deployment of more and, in some cases, novel bait/trap combinations. Financial restrictions on detection programs are clearly an important element, but I suggest that the perceived success of the currently used food baits and male lures has served to de-emphasize research and development of alternative attractants. This opinion is admittedly coarse and heavy-handed, as research on new substances continues. Jang et al. (2001), for example, investigated ceralure and Mwatawala et al. (2013) studied enriched ginger root oil as possible alternatives to trimedlure. The discovery by Tan and Nishida (2007) that the compound zingerone attracts both methyl eugenol- and cue-lure-responding *Bactrocera* species is noteworthy as well. Likewise, various researchers (e.g., Robacker et al. 2011) have examined heretofore unstudied plant odors as possible trap baits; ongoing work by Epsky and her colleagues (Niogret et al. 2011) is exploring the potential of essential plant oils as fruit fly attractants. Still, despite these examples, there is little doubt that, compared to the intensive screening of potential fruit fly attractants in the mid-twentieth century (Beroza and Green 1963), contemporary efforts to identify new or improve existing attractants reflect more the work of independent researchers and less the shared objective of a large-scale, coordinated research project.

The following poem by W.S. Merwin serves as succinct conclusion to this essay. Although he emphasizes the acoustic "language" of insect song (what else would a poet do?), his thesis – much of insect biology, including communication, remains unknown – applies equally well to olfactory communication, a process central to the success of tephritid trapping. Thus, while this book describes important and substantive progress in the efficacy of tephritid trapping, there is clearly much more work to be done.

After the Alphabets

I am trying to decipher the language of insects
they are the tongues of the future
their vocabularies describe buildings as food
they can depict dark water and the veins of trees
they can convey what they do not know
and what is known at a distance
and what nobody knows
they have terms for making music with the legs
they can recount changing in a sleep like death
they can sing with wings
the speakers are their own meaning in a grammar without horizons
they are wholly articulate
they are never important they are everything

Acknowledgments Conversations with N. Epsky, E. Jang, and L. Light and comments by T. Allen, F. Díaz-Fleischer, D. Lance, D. McInnis, and P. Gomes improved the paper, and to all I express a warm mahalo. "After the Alphabets" from THE RAIN IN THE TREES by W.S. Merwin copyright © 1988 by W.S. Merwin. Used by permission of Alfred A. Knopf, an imprint of the Knopf Doubleday Publishing Group, a division of Random House LLC. All rights reserved.

References

Beroza M, Green N (1963) Materials tested as insect attractants. USDA-ARS, Agricultural handbook no 239. Washington, DC

Beroza M, Green N, Gertler SI, Steiner LF, Miyashita DH (1961) New attractants for the Mediterranean fruit fly. Agric Food Chem 9:361–365

Carey JR (1991) Establishment of the Mediterranean fruit fly in California. Science 253:1369–1373

Carey JR (1996) The incipient Mediterranean fruit fly population in California: implications for invasion biology. Ecology 77:1690–1697

Cunningham RT, Couey HM (1986) Mediterranean fruit fly (Diptera: Tephritidae): distance/response curves to trimedlure to measure trapping efficiency. Environ Entomol 15:71–74

Díaz-Fleischer F, Arredondo J, Flores S, Montoya P, Aluja M (2009) There is no magic fruit fly trap: multiple biological factors influence the response of adult *Anastrepha ludens* and *Anastrepha obliqua* (Diptera: Tephritidae) individuals to MultiLure traps baited with BioLure or NuLure. J Econ Entomol 102:86–94

Haniotakis GE, Vassiliou-Waite A (1987) Effect of combining food and sex attractants on the capture of *Dacus oleae* flies. Entomol Hellenica 5:27–33

Hill AR (1986) Reduction in trap captures of female fruit flies (Diptera: Tephritidae) when synthetic male lures are added. J Aust Entomol Soc 25:211–214

Jang EB, Raw AS, Carvalho LA (2001) Field attraction of Mediterranean fruit fly, *Ceratitis capitata* (Wiedemann), to synthetic stereoselective enantiomers of the ceralure B1 isomer. J Chem Ecol 27:235–242

Kaneshiro KY (1993) Introduction, colonization, and establishment of exotic insect populations: fruit flies in Hawaii and California. Am Entomol 39:23–29

Lance DR, Gates D (1994) Sensitivity of detection trapping systems for Mediterranean fruit flies (Diptera: Tephritidae) in southern California. J Econ Entomol 87:1377–1383

Lance DR, McInnis DO (1993) Misrepresented reality? Am Entomol 39:196

Meats A, Smallridge CJ (2007) Short- and long-range dispersal of medfly, *Ceratitis capitata* (Dipt., Tephritidae), and its invasive potential. J Appl Entomol 131:518–523

Mwatawala M, Virgilio M, Quilici S, Dominic M, de Meyer M (2013) Field evaluation of the relative attractiveness of enriched ginger root oil (EGO)lure and trimedlure for African *Ceratitis* species (Diptera: Tephritidae). J Appl Entomol 137:392–397

Niogret J, Montgomery WS, Kendra PE, Heath RR, Epsky ND (2011) Attraction and electroantennogram responses of male Mediterranean fruit fly to volatile chemicals from *Persea*, *Litchi*, and *Ficus* wood. J Chem Ecol 37:483–491

Papadopoulos NT, Plant RE, Carey JR (2013) From trickle to flood: the large-scale, cryptic invasion of California by tropical fruit flies. Proc Roy Soc B 280: 20131466. http://dx.doi.org/10.1098/rspb.2013.1466

Prokopy RJ (1995) Impact of Medfly foraging behavior on trapping and eradication. In: Morse JG, Metcalf RL, Carey JR, Dowell RV (eds) The Mediterranean fruit fly in California: defining critical research. University of California, Riverside, pp 109–122

Reboulakis C, Mavrikakis PG, Economopoulos AP, Ragoussis N (2004) Orange fruit volatiles are Medfly species- and male-specific attractants. In: Barnes BN (ed) Proceedings of the 6th international symposium on fruit flies of economic importance. Isteg Scientific Publications, Irene, pp 291–294

Robacker DC, Garcia JA (1990) Responses of laboratory-strain Mexican fruit flies, *Anastrepha ludens*, to combinations of fermenting fruit odor and male-produced pheromone in laboratory bioassays. J Chem Ecol 16:2027–2038

Robacker DC, Massa MJ, Sacchetti P, Bartelt RJ (2011) A novel attractant for *Anastrepha ludens* (Diptera: Tephritidae) from a Concord grape product. J Econ Entomol 104:1195–1203

Seewoorthunun SI, Permalloo S, Gungah B, Soonnoo AR, Alleck M (2000) Eradication of an exotic fruit fly from Mauritius. In: Tan KH (ed) Area-wide control of fruit flies and other insect pests. Penerbit Universiti Sains Malaysia, Pulau Pinang, pp 389–394

Steiner LF, Mitchell WC, Harris EJ, Kozuma TT, Fujimoto MS (1965) oriental fruit fly eradication by male annihilation. J Econ Entomol 58:961–964

Tan HK, Nishida R (2007) Zingerone in the floral synomone of *Bulbophyllum baileyi* (Orchidaceae) attracts *Bactrocera* fruit flies during pollination. Biochem Syst Ecol 35:334–341

Tóth M, Nobili P, Tabilio R, Ujváry I (2004) Interference between male-targeted and female-targeted lures of the Mediterranean fruit fly *Ceratitis capitata* (Dipt., Tephritidae) in Italy. J Appl Entomol 128:64–69

Ushio S, Yoshioka K, Nakasu K, Waki K (1982) Eradication of the oriental fruit fly from Amami Islands by male annihilation (Diptera: Tephritidae). Jpn J Appl Entomol Zool 26:1–9

Wee SL, Hee AKW, Tan KH (2002) Comparative sensitivity to and consumption of methyl eugenol in three *Bactrocera dorsalis* (Diptera: Tephritidae) complex sibling species. Chemoecology 12:193–197

Part II
Lures and Traps

Chapter 2
Pheromones, Male Lures, and Trapping of Tephritid Fruit Flies

Keng Hong Tan, Ritsuo Nishida, Eric B. Jang, and Todd E. Shelly

Abstract Both sex pheromones and male lures appear to play an important role in the mating systems of many species of economically important tephritid species. Typically, stationary males emit pheromone attractive to searching females, and recent evidence indicates that naturally occurring male lures may function as precursors in pheromone synthesis. Here, we review (i) the basic biology of sex pheromones and the importance of naturally occurring male lures as pheromone components or precursors and (ii) the use of sex pheromones and male lures as trap baits, primarily in fruit fly detection programs, for the major genera of *Anastrepha*, *Bactrocera*, *Ceratitis*, *Dacus*, *Rhagoletis*, and *Toxotrypana*. Relatively few studies have examined the effectiveness of pheromone-based trapping, and most of these have involved only three species, the Mediterranean fruit fly, *Ceratitis capitata* (Wiedemann), the Mexican fruit fly, *Anastrepha ludens* (Loew), and the Caribbean fruit fly, *A. suspensa* (Loew). In general, the results have been inconsistent, with traps baited with live males or male pheromone extracts or components attracting more females than blanks or food-baited traps in some studies but not in others. This inconsistency, along with the chemical complexity of pheromones and the

K.H. Tan (✉)
Tan Hak Heng, 20, Jalan Tan Jit Seng, 11200 Tanjong Bungah, Penang, Malaysia
e-mail: tan.kenghong@yahoo.com

R. Nishida
Graduate School of Agriculture, Laboratory of Chemical Ecology, Pesticide Research Institute, Kyoto University, Sakyo-ku, Kyoto 606-8502, Japan
e-mail: tan.kenghong@yahoo.com

E.B. Jang
USDA-ARS, U.S. Pacific Basin Agricultural Research Center, 64 Nowelo Street, Hilo, HI 96720, USA
e-mail: eric.jang@ars.usda.gov

T.E. Shelly
USDA-APHIS, 41-650 Ahiki Street, Waimanalo, HI 96795, USA
e-mail: todd.e.shelly@aphis.usda.gov

T. Shelly et al. (eds.), *Trapping and the Detection, Control, and Regulation of Tephritid Fruit Flies*, DOI 10.1007/978-94-017-9193-9_2,

multimodal nature of sexual communication (where olfaction is but one of several sensory channels used in male signaling and courtship), has limited research on the development of pheromone baits. Male lures, on the other hand, have proven incredibly useful and consistently effective trap baits. The major male lures – methyl eugenol, cue-lure/raspberry ketone, and trimedlure – are discussed as are possible replacements/modifications, such as fluorinated analogues of methyl eugenol, raspberry ketone formate, zingerone, ceralure, and enriched ginger root oil. In addition, we discuss various factors influencing the efficacy of male lures, including fly age, prior lure ingestion, selection for non-responsiveness, interspecific differences in responsiveness, and the use of liquid versus solid dispensers.

Keywords Aggregation • Anisylacetone • Attractant • α-copaene • Cuelure • Electroanntenogram • Floral volatile • Kairomone • Male lure • Methyl eugenol • Orchid • Phenylbutanoid • Phenylpropanoid • Pheromone • Raspberry ketone • Rectal gland • Sesquiterpene • Synomone • Trimedlure

1 Introduction

Chemical cues and signals influence the behavior, physiology, and ecology of insects in a remarkably large number of ways. It is hardly surprising, then, that strategies designed to protect agricultural systems are often based on chemical stimuli and cues important to pestiferous insects. These strategies are themselves diverse and may involve the elimination, modification, disruption, imitation, or circumvention of chemical information important to the target insect. Tephritid fruit flies are trapped for a variety of reasons – surveillance, suppression, and ecological study among others – and chemical baits have played a central role in these efforts. The existence of male lures was reported approximately 100 years ago (Howlett 1912, 1915), and such lures have been among the most widely used in programs to detect and manage tephritid fruit fly pests. Likewise, the presence of sex pheromones in economically important Tephritidae has been recognized for over 50 years (Féron 1959), and though not yet as effective as male lures, they have received considerable attention as possible tools in fruit fly surveillance and control.

This chapter provides an overview of the use of pheromones and male lures in trapping economically important fruit flies of the genera *Anastrepha*, *Bactrocera*, *Ceratitis*, *Dacus*, *Rhagoletis*, and *Toxotrypana*. Given the broad scope of this topic and the accompanying rich body of literature, our review is not exhaustive. Though somewhat idiosyncratic, reflecting invariably our own research experiences, we nonetheless believe we have highlighted main themes and introduced some new ideas or perspectives as well.

As evidenced by the chapter title, we have decided to describe compounds, such as methyl eugenol, cue-lure, raspberry ketone, trimedlure, and others, as male lures or male attractants and to avoid the oft-used term ‘parapheromone’. We do so for three main reasons: (i) Payne et al. (1973) originally defined parapheromones as "compounds which are not naturally used in intraspecific insect communication". However, several studies (Nishida et al. 1988a, b, 1993; Tan and Nishida 1995, 2007; Tan et al. 2011) have demonstrated that certain male lures (e.g., methyl eugenol, raspberry ketone, and zingerone) are used in synthesizing male sex pheromones, and so the original definition of parapheromone does not apply to tephritids; (ii) in a recent review of insect parapheromones, Renou and Guerrero (2000) restrict parapaheromones to "chemical compounds of anthropogenic origin not known to exist in nature". Once again, this criterion does not apply to methyl eugenol and raspberry ketone, which occur in many different plant species (Tan and Nishida 1995, 2012), and so excludes these two important tephritid male attractants (indeed, Renou and Guerrero's review does not even include discussion of the Tephritidae), and (iii) the very use of the root 'pheromone' implies that male lures produce behavioral and/or physiological effects that resemble those of natural pheromones. There is evidence that male-produced sex pheromones may attract conspecific males and so act as aggregation pheromones (Nishida et al. 1988b; Tan and Nishida 1996; Hee and Tan 1998; Khoo and Tan 2000; Wee and Tan 2005a; Wee et al. 2007). Because male lures may (upon ingestion) be used in pheromone synthesis (references above), the idea that the male lures mimic the male sex pheromone appears reasonable and may eventually be shown to be valid. However, the available data regarding male-male olfactory attraction derive exclusively from laboratory studies (with a single exception, Nishida et al. 1988b). With few field data available, we consider it premature to conclude that male lures resemble pheromones in function. That said, we also recognize that the term male lure is not completely accurate, since the lures are known to occasionally attract females (Steiner et al. 1965; Nakagawa et al. 1970; Fitt 1981a; Verghese 1998). While not dismissing the importance of these observations, our collective field experience (except for a female *Bactrocera umbrosa* (Fabricius) captured by Tan in 2014) is that males comprise the vast majority of all individuals observed at point sources (traps, flowers, etc.) of known male lures, and hence the terms male lure or male attractant are generally, if not always, appropriate.

2 Tephritid Pheromones and Trapping

The family Tephritidae contains several genera, namely *Anastrepha, Bactrocera, Ceratitis, Dacus, Rhagoletis* and *Toxotrypana*, with species that are major agricultural pests of fruits and vegetables. Information on the identification of pheromones and their possible use in trapping programs is summarized below for each of these genera.

2.1 Anastrepha *Pheromones*

The genus *Anastrepha* contains approximately 200 species distributed in tropical and subtropical areas of the New World (Norrbom et al. 2000) of which eight (*A. distincta* Greene – Inga fruit fly, *A. fraterculus* (Wiedemann) – South American fruit fly, *A. grandis* (Macquart) – South American cucurbit fruit fly, *A. ludens* (Loew) – Mexican fruit fly, *A. obliqua* (Macquart) – West Indian fruit fly, *A. serpentina* (Macquart) – sapote fruit fly, *A. striata* Schiner – guava fruit fly, and *A. suspensa* (Loew) – Carribean fruit fly) are major agricultural pests, attacking a wide variety of fruits and vegetables (Aluja 1994; Norrbom et al. 2012). Although field observations are incomplete, many of the polyphagous and economically important species appear to display a lek mating system in which males occupy mating territories on leaves and attract females to the territory via a complex suite of visual, acoustic, and olfactory signals (Aluja et al. 2000). Regarding the latter, pheromone-calling males emit volatiles from everted pleural pouches and anal membranes, with aerial dispersion aided by intense bouts of rapid wing vibrations (Nation 1972). Volatile components are also released via the mouth (Nation 1990), and abdominal dipping of the evaginated anal membranes to the leaf surface may amplify pheromone attractiveness by increasing the evaporative surface area of the volatile components (Sivinski et al. 1994). Pheromone calling has been observed for a small number of *Anastrepha* species, and chemical analysis and identification of pheromonal components has been undertaken for only a subset of these species (Fig. 2.1 and Table 2.1).

Measurements of female attraction to male sex pheromone, or components thereof, have been made for an even smaller subset of all species, with nearly all of the research conducted on *A. suspensa* (Table 2.2) or *A. ludens* (Table 2.3). The biological activity of male pheromones has been studied in only four other species, with only one study undertaken for each. In both *A. fraterculus* and *A. obliqua,* freshly dissected salivary glands of males were found to attract mature and virgin females in laboratory cage tests (Lima et al. 2001; Ibañez-López and Cruz-López 2001). However, in *A. serpentina*, three putative pheromonal components were examined, with no strong female response observed for any of them (Robacker et al. 2009a, b). In *Anastrepha sororcula* Zucchi, field tests found no difference in female captures in traps baited with live calling males versus blank control traps (Santos Felix et al. 2009).

Despite the large amount of research conducted on *A. suspensa* and *A. ludens*, trapping and detection efforts for these two species still rely primarily on food-based lures (e.g., Robacker and Thomas 2007; Epsky et al. 2011), and an effective pheromone-based trap has not been developed. Several authors (Landolt and Heath 1996; Landolt and Averill 1999) have enumerated the reasons for this, and these generally include:

First, the long-range attractiveness of the male sex pheromone has weak empirical support, since the majority of research has been conducted in the small, laboratory cages and thus measures only short-range attractiveness or arrestant

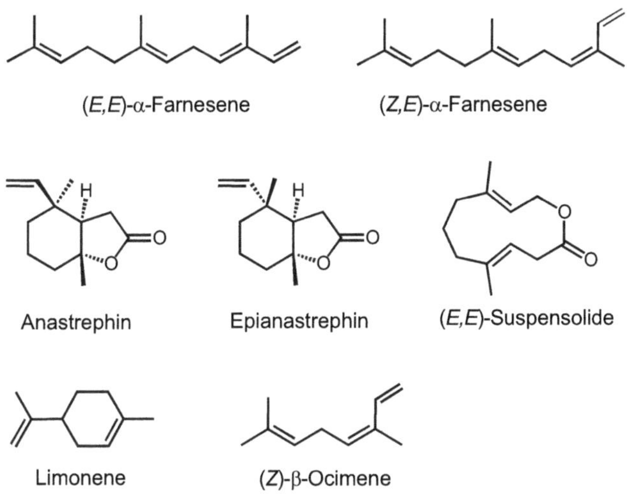

Fig. 2.1 Some sex pheromonal components of *Anastrepha* (Adapted from Rocca et al. 1992)

properties of the test chemicals. The widely used cage-top bioassay, for example, has generally been used in cubical cages only (0.2–0.3 m per side). Moreover, although the cage-top assay has generally indicated female response to male-derived chemicals, other laboratory tests involving slightly larger cages have failed to demonstrate long-range attraction of females to male pheromone. For example, arrivals of *A. ludens* females did not differ significantly between citrus trees having chemically-treated (with male pheromone extract or pheromonal components) or control (blank) leaves (Robacker and Hart 1986; Robacker 1988). However, using a 1.2 m long wind tunnel and a videotape system, Heath et al. (1993) found that *A. suspensa* females land more frequently on male-baited traps than control traps but spent equal amounts of time on the two trap types after landing. These data clearly indicate that the male volatiles are an attractant and not a simple arrestant. In sum, then, use of small cages does not allow rigorous identification of long-range attraction of *Anastrepha* females, and other laboratory results from slightly larger cages are inconsistent in this regard.

Second, field tests have yielded inconsistent results regarding female attraction to male-produced odors. In an early study, sticky traps baited with aggregations of 20 or 40 males captured significantly more released virgin females of *A. suspensa* than McPhail traps baited with an aqueous solution of yeast hydrolysate (Perdomo et al. 1975). A follow-up study on the same species (Perdomo et al. 1976) generated the same result and also documented attraction of released males to the male-baited traps. In contrast, although the difference was not statistically significant, Robacker and Wolfenbarger (1988) found that food-baited McPhail traps captured three times as many *A. ludens* females as pheromonal traps (baited with extracts of male abdomens). Similarly, and as noted previously, field tests involving *A. sororcula* found no difference in female captures in traps baited with live calling males versus

Table 2.1 *Anastrepha* species for which male pheromone calling has been observed and the incidence of chemical analyses of pheromonal components in these species

Species	Pheromone calling	References	Pheromone Chemistry	References
A. bistrigata Bezzi	+	Da Silva et al. (1985)	–	–
A. fraterculus (Wiedemann)	+	Malavasi et al. (1983), Morgante et al. (1983), Lima et al. (1994), Segura et al. (2007)	+	Cáceres et al. (2009), Lima et al. (1996), (2001)
A. ludens (Loew)	+	Aluja et al. (1983), (1989), Moreno et al. (1991), Robacker et al. (1991), (2003), Robacker and Hart (1985a), Aluja et al. (2008)	+	Battiste et al. (1983), Stokes et al. (1983), Robacker and Hart (1985b), Rocca et al. (1992), Baker and Heath (1993)
A. obliqua (Macquart)	+	Aluja et al. (1983), (1989) Da Silva et al. (1985), Meza-Hernández et al. (2002), López Guillén et al. (2008), Henning and Matioli (2006)	+	Meza-Hernández et al. (2002), Ibañez-López and Cruz-López (2001), López-Guillén et al. (2008)
A. pseudoparallela (Loew)	+	Da Silva et al. (1985), Polloni and Da Silva (1986)	–	–
A. robusta Greene	+	Aluja (1993)	–	–
A. serpentina (Macquart)	+	Aluja et al. (1989), Castrejón -Gómez et al. (2007), Robacker et al. (2009a)	+	Robacker et al. (2009a)
A. sororcula Zucchi	+	Da Silva et al. (1985)	–	–
A. striata Schiner	+	Aluja et al. (1993), (2008)	–	–
A. suspensa (Loew)	+	Nation (1972), (1989), (1990), Dodson (1982), Burk (1983), (1984), Landolt and Sivinski (1992),	+	Nation (1975), (1989), (1990), (1991)), Battiste et al. (1983), Chuman et al. (1988), Mori and Nakazono (1988), Tumlinson

Table 2.1 (continued)

Species	Pheromone calling	References	Pheromone Chemistry	References
		Epsky and Heath (1993a, b), Sivinski et al. (1994)		(1988), Rocca et al. (1992), Epsky and Heath (1993a, b), Baker and Heath (1993), Heath et al. (1993), Lu and Teal (2001)
A. zenildae (Zucchi)	+	De Almeida et al. (2011)	–	–

Table 2.2 Published accounts for *Anastrepha suspensa* regarding the behavioral response of females to the sex pheromones of conspecific males

		Positive response		
Odor source	Bioassay arena	Identified	Observed	Reference
Live males	Wind tunnel (0.45 m long)	Move 25 cm toward source	Yes	Nation (1972)
Pheromonal components[a]	Screen cage (0.45 m per side)	Enter trap	Yes	Nation (1975)
Live males	Avocado grove	Capture in male-baited sticky traps[b]	Yes	Webb et al. (1983)
Pheromone extract	Field cage	Capture in sticky traps	Yes	Webb et al. (1983)
Filter paper treated with volatiles	Screen cage (0.20 m per side)	Aggregation near treated paper[c]	Yes	Sivinski and Heath (1988)
Filter paper treated with major pheromone components[d]	Screen cage (0.2 × 0.12 × 0.10 m)	Aggregation near treated paper[c]	Yes	Nation (1991)
Filter paper treated with minor pheromonal components[e]	Screen cage (0.2 × 0.12 × 0.10 m)	Aggregation near treated paper[c]	No[f]	Nation (1991)
Live males	Flight tunnel (0.3 × 0.3 × 1.22 m)	Enter trap	Yes	Heath et al. (1993)

[a]Later identified (Nation 1983) as (Z)-3-nonenol, (Z,Z)-3; 6-nonadienol; anastrephin; epianastrephin; attraction was observed for individual compounds as well as pairs and trios, with greatest attraction observed for a combination of all four components

[b]Released 13 m from any trap

[c]A so-called cage-top test, where control and treated papers were placed, one per quadrant, placed on top of cage, and distribution of females in four quadrants was measured

[d]Same four as in footnote a

[e]Bisabolene, ocimine, suspensolide

[f]No response was observed when these compounds tested individually, but each increased female response when added to blend containing the major components (listed in footnote a)

Table 2.3 Published accounts for *Anastrepha ludens* regarding the behavioral response of females to the sex pheromones of conspecific males

Odor source	Bioassay Arena	Positive response Identified	Observed?	Reference
Filter paper treated with pheromone extract[a]	Screen cage (0.3 m per side)	Aggregation near treated paper[b]	Yes[c]	Robacker and Hart (1984), Robacker et al. (1990), Moreno et al. (1991)
Filter paper treated with pheromonal components[d]	Screen cage (0.3 m per side)	Aggregation near treated paper[b]	Y[e]	Robacker and Hart (1985b)
Citrus leaf treated with pheromonal components[f]	Wind tunnel (2 × 0.7 × 1.3 m)	Arrivals to treated leaves	No[g]	Robacker (1988)
Citrus leaf treated with combinations of pheromonal components[h]	Wind tunnel (2 × 0.7 × 1.3 m)	Arrivals to treated leaves	Yes	Robacker (1988)
Citrus leaf treated With pheromone extract[a]	Wind tunnel (2 × 0.7 × 1.3 m)	Arrivals to treated leaves	Yes	Robacker (1988)
Citrus leaf treated with pheromone extract[a]	Screen cage (0.7 × 1.6 × 1.0 m)	Arrivals to treated leaves	Yes[i]	Robacker and Hart (1986)
Cigarette filter treated with pheromone extract[a]	Citrus grove	Capture in treated McPhail traps	Yes[j]	Robacker and Wolfenbarger (1988)
Filter paper treated with pheromone extract[a]	Hallway (30.0 × 2.5 × 2.0 m)	Upwind movement; flight[k]	Yes	Robacker and Moreno (1988)

Table 2.3 (continued)

Odor source	Bioassay Arena	Positive response		Reference
		Identified	Observed?	
Pheromone extract[l]	Screen cage (0.3 m per side)	Aggregation near treated paper[b]	Yes[m]	Robacker and Garcia (1990)
Live males	Flight tunnel (30 × 30 × 122 cm)	Enter trap	Yes	Heath et al. (1993)

[a]Obtained from filtering and concentrating extract obtained from grinding abdomens of adult males
[b]A so-called cage-top test, where control and treated papers were placed, one per quadrant, placed on top of cage, and distribution of females in four quadrants was measure
[c]Attraction much stronger for mature virgin females than immature or recently mated females
[d]Six components were tested individually and in various combinations: (Z)-3-nonenol, (Z,Z)-3,6-nonadienol, (R,R)-(+)-anastrephin, (S,S)-(−)-anastrephin, (R,R)-(+)-epianastrephin, (S,S)-(−)-epianastrephin
[e]Only three components (the two alcohols plus epianastrephin) elicited female responses individually. Both synergistic and inhibitory effects were reported among the 15 combinations of paired components
[f]The six components listed in footnote d were tested individually
[g]With a single exception: (Z,Z)-3,6-nonadienol attracted more females than control (untreated) leaves
[h]Three combinations involving pheromonal components listed in footnote d were tested: (Z)-3-nonenol + (S,S)-(−)-epianastrephin; (Z,Z)-3,6-nonadienol + (S,S)-(−)-epianastrephin; all 6 components
[i]Females did not distinguish between treated and control trees but within trees were more attracted to treated than control leaves
[j]Highest dose did not attract more females than control suggesting an overdose effect; male attraction also observed
[k]Females behavior monitored in screen cages placed at different distances from odor source, with upwind movement scored as the number of females on upwind versus downwind sides of cages
[l]Obtained from crushing whole males
[m]Host fruit odor inhibited attraction of mature virgin females

blank control traps (Santos Felix et al. 2009). A field trial on the attractiveness of a pheromonal blend likewise yielded negative results. In laboratory tests conducted in small cages, *A. suspensa* females responded to four major components of the male pheromone as well as various mixtures of these chemicals (Nation 1975; see also Robacker and Hart 1985b for similar findings for *A. ludens*). However, a synthetic blend of these same four components placed in the field failed to attract flies of either sex over a 5-day period in Florida (Nation 1989). The uneven performance of pheromone-baited traps in the field, coupled with data showing that host fruit odors are equally or more attractive to females than male odors alone (Robacker and Garcia 1990), has been an important constraint on further research on the development of pheromone-based traps for *Anastrepha*.

Third, the pheromones of different *Anastrepha* males are complex and contain multiple chemicals with different isomers (Heath et al. 2000). This complexity has several important implications. It appears, for example, that the component ratios affect the attractiveness of the blend. Differences in the attractiveness of two dispensers to *A. ludens* females, for example, apparently reflected differential release of pheromone components, which resulted in the emission of abnormal component ratios for one of the dispensers (Robacker and Wolfenbarger 1988). In general, data support the conclusion that individual pheromonal components stimulate little behavioral response but instead function as an integrated unit to elicit behavior (Robacker 1988), and identifying the specific nature of this complex signal is seen as a daunting challenge. Moreover, the composition of male pheromone may vary with time of day (Tumlinson 1988; Nation 1990), social context (calling singly or in a group, Nation 1990), and food availability (Epsky and Heath 1993a), making it even more difficult to identify the particular blend most attractive to females. Analogously, variability in pheromone release rate (Epsky and Heath 1993b; Meza-Hernández et al. 2002) confounds identification of those rates that may be maximally attractive to females. In addition, the different components have different volatilities (Landolt and Averill 1999) and liabilities (Robacker et al. 2009b), which render production of synthetic pheromones difficult from a methodological perspective and imprecise from a biological one.

Finally, the importance of male pheromone to female mate searching remains uncertain, and it appears likely that a combination of visual, auditory, and olfactory cues may be involved. The pheromone appears to attract females to the vicinity of calling males but not to point sources (Robacker 1988), and after approach, females may rely on acoustic and/or visual signals to locate males (Webb et al. 1983; Sivinski and Calkins 1986). As with the complex pheromonal blend, the multifaceted nature of mate location appears to have lessened the impetus to develop pheromone-based traps for *Anastrepha*.

2.2 Bactrocera *Pheromones*

The genus *Bactrocera* consists of over 500 species distributed in the tropical and subtropical regions of Asia (Smith et al. 2003) and includes many serious and/or highly invasive polyphagus pest species, namely *B. correcta* (Bezzi) – guava fruit fly, *B. cucurbitae* (Coquillett) – melon fly, *B. carambolae* Drew & Hancock – carambola fruit fly, *B. dorsalis sensu stricto* (Hendel) – oriental fruit fly, *B. invadens* Drew, Tsuruta & White, *B. papayae* Drew & Hancock – Asian papaya fruit fly, *B. philippinensis* Drew & Hancock – Philippines fruit fly, *B. latifrons* (Hendel) – solanaeous fruit fly, *B. tryoni* (Froggatt) – Queensland fruit fly, *B. umbrosa* (Fabricius) – *Artocarpus* or jack-fruit fly, and *B. zonata* (Saunders) – peach fruit fly. Males of these species, with the exception of *B. cucurbitae* and *B. tryoni* (both attracted to cue-lure (CL)/raspberry ketone (RK)) and *B. latifrons* (not attracted to either CL/RK or methyl eugenol (ME), are attracted to ME, a compound found in a wide diversity of plant species (Tan and Nishida 2012) and now known to be a pheromonal precursor. As discussed below, the strong attraction of males to ME has, to some degree, limited impetus to explore sex pheromones as a trapping tool for *Bactrocera* species. Here, we summarize the chemistry of *Bactrocera* pheromones and note studies that have monitored male or female attraction to pheromonal emissions.

As true for most of the economically important tephritid species examined thus far, sexual signaling in *Bactrocera* typically involves the production and broadcast of sex pheromone by males (a behavior termed "calling") while resting on vegetation and detection and subsequent mate searching by receptive females. Most accounts of male calling and mating derive from laboratory or field cage observations (e.g., Tychsen 1977; Ohinata et al. 1982; Arakaki et al. 1984; Kuba and Koyama 1985), and the few field studies conducted – all on *B. dorsalis*– indicate plasticity in that species' mating system. Working in Hawaii, Shelly and Kaneshiro (1991) observed calling males and matings in the canopy of a single fruiting tree within a citrus orchard, suggestive of a lek mating system. In contrast, Stark (1995), also working in Hawaii, observed *B. dorsalis* females moving from papaya trees to non-host (*Panax*) trees in the late afternoon followed by males 30–60 min later. Although their incidence was not quantified, Stark (1995) observed matings on this nonhost plant. Finally, working in Thailand, Prokopy et al. (1996) released *B. dorsalis* within a non-fruiting orchard and experimentally added food, water, and host fruits to the trees. In this case, and in contrast to the aforementioned studies, all sexual behavior and all matings were recorded on trees with fruits and on the fruit itself, leading the authors to suggest that the importance of host fruits as foci for sexual activity may vary with microclimatic conditions. The behavioral variability described for *B. dorsalis*, along with the lack of field studies on *Bactrocera* species in general, serves as a cautionary prelude to the following discussion: little is known about the importance of male pheromones in sexual selection in the genus, and consequently evaluation of male pheromones as potential trap attractants is necessarily preliminary and inconclusive.

2.2.1 Sex Pheromone of *B. dorsalis* Complex Species

The *B. dorsalis* species complex comprises over 70 recognized species (White and Elson-Harris 1992), several of which, namely *B. dorsalis*, *B. invadens*, *B. papayae*, *B. philippinensis*, and *B. carambolae*, are serious agricultural pests. Recent molecular (Tan et al. 2011, 2013; Schutze et al. 2012; Krosch et al. 2013), morphological (Mahmood 1999, 2004; Schutze et al. 2012; Krosch et al. 2013), behavioral (i.e., mating compatibility; McInnis et al. 1999; Tan 2000, 2003; Wee and Tan 2005b; Schutze et al. 2013), and pheromone chemistry (Tan and Nishida 1996, 1998; Tan et al. 2011, 2013) data have raised doubts regarding the validity of species status for these sibling taxa (except *B. carambolae* – see below). Below, we retain the names as originally used but recognize that results obtained for one species may, if taxonomic synonymies are eventually recognized (Schutze et al. 2014), apply to other currently recognized species in the complex.

In the first published description on the pheromone chemistry of male *Bactrocera*, Ohinata et al. (1982) analyzed "smoke" produced by male *B. dorsalis* and found that trisodium phosphate was the major component (90 %) with much smaller amounts of N-(2-methylbutyl)propanamide and heptacosane. Perkins et al. (1990a) examined an acetate extract of rectal glands of sexually mature male *B. dorsalis* from a colony maintained in Hawaii and detected the trimethyl ester of citric acid (a major component), the trimethyl ester of phosphoric acid, and dimethyl succinate along with methyl esters of fatty acids and two spiroacetals. The males sampled in this study had not fed on ME, and no biological activity was demonstrated for the compounds identified. However, Nishida et al. (1988a, b) and Tan and Nishida (1996) demonstrated that males of *B. dorsalis* and *B. papayae* transformed consumed synthetic ME to two major pheromonal components – *E*-coniferyl alcohol (ECF) and 2-allyl-4,5-dimethoxy phenol (DMP), with trace quantity of *Z*-3.4-dimethoxycinnamyl alcohol (detected in some males). Nishida et al. (1988a) also detected these compounds in wild *B. papayae* males, indicating the males had fed on ME-bearing plants in the field, and a later study (Tan et al. 2002) showed that *B. papayae* males that fed on an ME-bearing orchid flower contained ECF and DMP in the rectal gland (Fig. 2.2). In laboratory tests, males deprived of ME did not have ECF or DMP in the rectal gland. As an aside, *B. papayae* males visiting an orchid whose floral fragrance contained zingerone (a compound structurally similar to ME) were found to have zingerol in the rectal gland, suggesting a role in pheromone synthesis for this compound as well (Tan and Nishida 2000, 2007). More recently, Tan et al. (2011, 2013) found ECF and DMP in the rectal sac of ME-fed males of *B. invadens* and *B. philippinensis*. Males of *B. carambolae* differ from the aforementioned species in that they produce only ECF after ingesting ME (Tan and Nishida 1996; Wee and Tan 2005a). Moreover, the sex pheromone of *B. carambolae* contains larger amounts of endogenously produced compounds, including 6-oxo-1-nonanol (a major component that is also detected in a closely related sibling species, *Bactrocera occipitalis* (Bezzi) and a distant species, *B. umbrosa* (Perkins et al. 1990b)) and three minor components, *N*-3-methylbutyl acetamide, ethyl benzoate, and 1,6-nonanediol (Wee and Tan 2005a).

Since Nishida et al.'s reporting, a number of studies have demonstrated that ME consumption increases male mating success in several species in the *B. dorsalis* complex (Shelly and Dewire 1994; Shelly et al. 1996; Shelly 2010a; Tan and Nishida 1996, 1998; Wee et al. 2007; Orankanok et al. 2013; Obra and Resilva 2013). However, the role of pheromone composition in determining this outcome is not known with certainty. In laboratory cage assays, Kobayashi et al. (1978) demonstrated attraction of *B. dorsalis* females to both live males and male rectal gland extract even when males were not previously fed ME. Wee and Tan (2005a) likewise reported zigzag anemotaxis by *B.carambolae* females to live males and endogenously produced rectal gland substances. Thus, the breakdown products of ME are not necessary to elicit female response. Nonetheless, using a wind tunnel or laboratory cages, several studies on *B. dorsalis* complex species (Shelly and Dewire 1994; Hee and Tan 1998; Wee et al. 2007) have reported greater female attraction to males that had previously fed on ME than to unfed males, and Hee and Tan (1998) and Khoo et al. (2000) showed female attraction to ECF and DMP individually (with greater attraction to ECF than DMP in these tests) and in combination (Fig. 2.3). Importantly, greater female response to ME-fed males has been

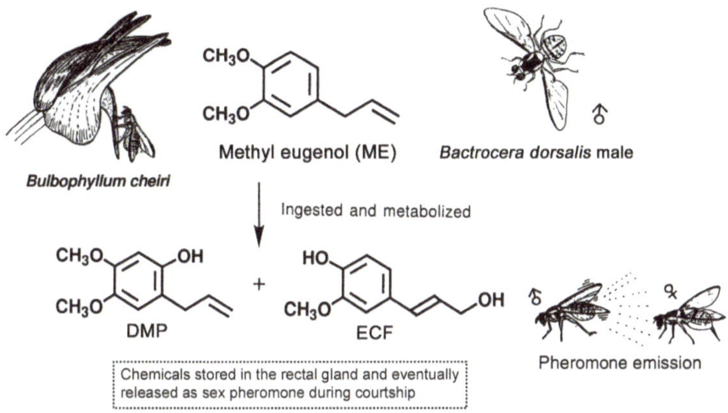

Fig. 2.2 Acquisition and biotransformation of methyl eugenol to sex pheromone by *Bactrocera dorsalis* males

Fig. 2.3 *Bactrocera dorsalis* females and males attracted to *E*-coniferyl alcohol. *A* An attracted female with ovipositor extruded (*arrow*), *B* Aggregation of males and an attracted female with extruded ovipositor at bottom left

documented, not only using synthetic ME, but also after male feeding on natural floral (Shelly 2000a, 2001a) or fruit (Shelly and Edu 2007) sources of ME. Several studies (Hee and Tan 1998; Wee and Tan 2005a; Wee et al. 2007) have documented maximum female attraction to male sex pheromone at dusk, the time of peak sexual activity in *B. dorsalis* species complex.

To our knowledge, only two studies have examined the long-range attractiveness of male pheromone to females in the field. In a study examining female attraction to groups (leks) of varying size, Shelly (2001b) placed *B. dorsalis* males (none of which had fed on ME) in screen-covered cups, which were in turn placed on trees situated in a circular (10 m radius) array around a central female release point. Approximately 10 % of released females were sighted near male-containing cups over all groups. In a second study also conducted on *B. dorsalis* in Hawaii, Shelly (2001c) performed two experiments in which groups of (i) ME-fed or ME-deprived males or (ii) flower-fed or flower deprived males (where the flower used [puakenikeni, *Fagraea berteriana* A. Gray ex Benth] was known to contain ME-like compounds [Nishida et al. 1997]) were placed in cups suspended in host trees (one male type [i.e., fed or non-fed] per tree) situated in a circle (12 m radius), and females were released from the center. Compared to non-fed males, both ME- and flower-fed males were found to signal more frequently and attract greater total numbers of females as well as greater numbers of females per signaling male. These studies were not designed to test explicitly the function of pheromone signaling (since blank controls were not run in either study), but they nevertheless hint at long-range attraction mediated by male pheromone and thus suggest the potential for male pheromone as a trap bait for species in the *B. dorsalis* complex.

Data on pheromonally-mediated male-male attraction are inconsistent. In laboratory cages, *B. dorsalis* males showed no attraction to conspecific males (non-ME-fed, Kobayashi et al. 1978). In contrast, Nishida et al. (1988b) found that traps baited with DMP captured as many wild males as traps baited with ME. In wind tunnel tests, Hee and Tan (1998) found that *B. papayae* males were attracted to both ME-fed and control (unfed) conspecific males but showed greater attraction to the treated males. Also using a wind tunnel, Wee et al. (2007) found non-ME-fed males of *B. carambolae* were attracted to ME-fed conspecific males at a much higher level than observed in the converse situation (i.e., ME-fed males responding to non-ME-fed males). Moreover, field cage observations showed that unfed males aggregated around ME-fed males and fed on anal secretions of ME-fed males (see also Tan and Nishida 1996). Results for *B. papayae* and *B. carmabolae* thus suggest that male sex pheromone may also serve as an aggregation pheromone. However, this function implies an evolutionary advantage to aggregation per se (e.g., increased mating success), whereas the possibility remains that male-male attraction simply represents a special case of male attraction to ME (or ME-like compounds), where the ME source is a male rather than a plant.

2.2.2 Presumed Sex Pheromone of Two Sibling Species of *B. zonata* Complex

ME also acts as a pheromone precursor for both *B. correcta* and *B. zonata*. In *B. zonata*, it is transformed to two male sex pheromonal components, DMP and Z-coniferyl alcohol (ZCF), although final confirmation awaits tests of biological activity on female response (Tan et al. 2011). In *B. correcta*, however, ME is converted to ZCF and (Z)-3,4-dimethoxycinnamyl alcohol (ZDMC) (Tokushima et al. 2010). Furthermore, wild *B. correcta* males also accumulate large quantities of sesquiterpene hydrocarbons, namely β-caryophyllene, α-humulene, and alloaromadendrene, in the rectal gland in addition to, or instead of, ZCF and ZDMC (Tokushima et al. 2010). The distinct difference in sex pheromonal profiles, albeit having a common ZCF component, between the two sibling species, most likely, plays an important role in maintaining reproductive isolation.

Interestingly, recent comparative field tests conducted in Thailand during 2012–2013 and based on average flies/trap/day using a similar lure dosage per trap showed that β-caryophyllene caught on average 7 (range 3–16) times more *B. correcta* wild males than ME during the first 3 days of trapping (Tan, Chinvinijkul, Wee & Nishida, unpublished data). This is the first case of a lure being more attractive than the very potent ME to a ME-sensitive *Bactrocera* species. Therefore, further behavioral/ecological studies, especially related to the role of the sequiterpene and its possible replacement of ME in the trapping of *B. correcta,* are warranted.

2.2.3 Sex Peromone of *B. umbrosa*

Rectal gland extracts showed the presence of (*E*)- and (*Z*)-2-methyl-1,6-dioxaspiro [4.5]decanes, 3-methylbutanol, 1,7-dioxaspiro [5.5]undecane, and 6-oxononan-ol (Perkins et al. 1990b). In addition, some unidentified ME metabolites (identities currently being evaluated) were detected in the rectal gland after consumption of ME by males (Nishida and Tan, unpublished data). In Malaysia, *B. umbrosa* and *B. papayae* are endemic and sympatric species as well as serious pests of jackfruit, *Artocarpus heterophyllus* Lam., but they do not interbreed. Apparent reproductive isolation was observed between the two species even when both males and females of both the species were kept together in a cage for approximately 2 months; intraspecific but no interspecific matings were observed (Tan, unpublished observations).

2.2.4 Sex Pheromone of *B. cucurbitae* and *B. tryoni*

Males of these species are attracted to RK/CL. Rectal gland secretions of *B. cucurbitae* contain *N*-3-methylbutyl acetamide, two spiroacetals, and three pyrazines (Baker et al. 1982a; Baker and Bacon 1985). Later, ethyl 4-hydroxybenzoate (a major component) and propyl 4-hydroxybenzoate (a minor component) were also detected in the rectal gland of the melon fly (Perkins et al. 1990b). Nishida et al. (1990) showed that sexually mature male melon flies produce, endogenously in the rectal gland, relatively small quantities of *N*-3-methylbutyl acetamide, methoxy-acetamide, methyl, ethyl, and propyl 4-hydroxybenzoate, and a large quantity of 1,3-nonanediol, which was not detected in the previous studies. The amounts of 1,3-nonanediol and ethyl 4-hydroxybenzoate stored in the rectal gland increased with age, starting 2 weeks after adult eclosion, thus coinciding with attainment of sexual maturity (Nishida et al. 1990). Additionally, at sexual maturity males of *B. cucurbitae* consume and sequester RK from anthropogenic (Nishida et al. 1990) and natural (Nishida et al. 1993; Tan and Nishida 2005) sources into the rectal gland. As noted above for *B. papayae*, males of *B. cucurbitae* are also attracted to and feed on zingerone, an orchid floral volatile, and store it unmodified in the rectal gland (Tan and Nishida 2000).

Males of *B. tryoni* produce endogenously six amides as major sex pheromonal components, and three of the six, namely, *N*-3-methylbutyl acetamide (MBA), *N*-3-methylbutyl propanamide (MBP), and *N*-3-methylbutyl-2-methyylpropanamide, are frequently detected in the rectal gland (Bellas and Fletcher 1979). Furthermore, MBA and MBP increase significantly from 14 to 17 day-old males corresponding with attaining sexual maturity (Tan and Nishida 1995). This suggests that the two chemicals may act as close range sex pheromone. Males consume plant-borne RK or RK from spontaneous hydrolysis of CL and sequester it in the rectal gland as a major pheromonal component (Tan and Nishida 1995).

Analogous to the *B. dorsalis* complex, ingestion of CL/RK has been shown to enhance male mating success, though the effect appears short-lasting both for *B. cucurbitae* (1 day after feeding, Shelly and Villalobos 1995; Shelly 2000b) and *B. tryoni* (1–3 days after feeding, Kumaran et al. 2013). More recently, *B. tryoni* males fed zingerone were also found to have a mating advantage over control males deprived this compound (Kumaran et al. 2013). The role of the sex pheromone in influencing male mating success is unknown. Kobayashi et al. (1978) found that *B. cucurbitae* females were attracted to male rectal glands as well as live males (in neither case were males fed CL/RK) but that the attraction was far weaker than that observed for *B. dorsalis* females to conspecific males. In wind tunnel trials, Khoo and Tan (2000) demonstrated that CL-fed and zingerone-fed males of *B. cucurbitae* attracted more females compared to males deprived these compounds, which strongly suggests a sex pheromonal role for these exogenous phenylbutanoids. To our knowledge, there are no laboratory or field data available investigating the effect of the male sex pheromone on female attraction or male mating success in *B. tryoni*.

2.2.5 Sex Pheromone of *Bactrocera oleae*

Bactrocera oleae (Rossi) [formerly *Dacus oleae* (Gmelin)], the olive fruit fly, unlike the other major pest *Bactrocera* species mentioned above, is a monophagous pest species. Additionally, the species differs from other *Bactrocera* in that the *B. oleae* females attract males for mating and not vice versa (Haniotakis 1974; but see below). Baker et al. (1980) identified the major component of the female sex pheromone as (1,7-dioxaspiro[5.5]undecane (also known as olean; as noted above, this compound was also identified from the pheromone of *B. umbrosa*). Additional studies (Mazomenos and Haniotakis 1981) confirmed this finding and also identified three minor components, o-pinene, n-nonanal, and ethyl dodecanoate, in the female pheromone (see also Baker et al. 1982b, who identified two hydroxyspiroacetals from *B. oleae* females). Other components of the female sex pheromone were reported (Gariboldi et al. 1982), but their isolation and biological activity (tested with synthetic products) was not corroborated (Jones et al. 1983; Mazomenos 1989). Interestingly, olean was also isolated from the rectal gland of male *B. oleae* along with other components (Mazomenos and Pomonis 1983). Canale et al. (2012) reported that, among males, olean production is greatest among young males (5–8 days old) and then ceases by 11 day of age. Also, in a recent finding, Carpita et al. (2012) identified (Z)-9-tricosene from male rectal gland extracts and reported female attraction to this compound in synthetic form.

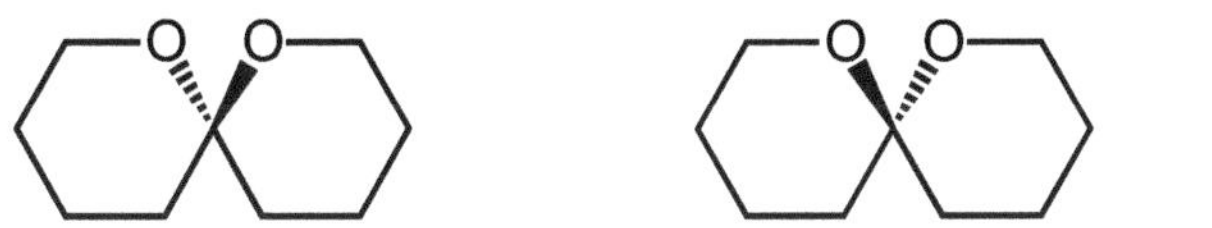

Fig. 2.4 Stereo enantiomers of (R)- and (S)-olean found in *B. oleae* sex pheromone

Several studies (Haniotakis 1974; Mazomenos and Haniotakis 1981, 1985) have demonstrated male attraction to natural or synthetic components or whole blends of the female pheromone in *B. oleae*. Laboratory and field experiments demonstrated that olean was more attractive than the remaining three components but that the combination of all four components was more attractive than olean alone. More detailed chemical analysis (Haniotakis et al. 1986a) revealed that olean exists as (*R*) and (*S*) mirror (stereo) image enantiomers, (*R*)-olean and (*S*)-olean (Fig. 2.4) and that (i) males are more strongly attracted to (*R*)-(−)-oleanthan(S)-(+)-olean, (ii) the converse was true for females, and (iii) overall, males showed greater attraction to response to the compound than did females. Haniotakis et al. (1986a) suggest olean may serve an aggregation or aphrodisiac function for females. Relative to the strong evidence gathered for male attraction to the female sex pheromone, data regarding female attraction to male olfactory signals are less conclusive. Mazomenos and Pomonis (1983) reported negligible female response in laboratory tests to extracts of rectal glands of mature males. More recently, however, Mavraganis et al. (2010) demonstrated that whole body extracts of *B. oleae* males were highly attractive to females and suggest that the previous negative results may have reflected low pheromone concentrations in the rectal gland extracts compared to those of whole body. Benelli et al. (2013) found that young males, which, as noted above, produce olean at higher levels than old males, did not have a mating advantage over older individuals.

In contrast to the other economically important species discussed here, several studies have demonstrated the usefulness of olean in baiting traps. In general, because olean is primarily a male attractant, the most effective traps appear to be those that combine the pheromone with ammonium or some other food bait that targets females (Haniotakis and Vassiliou-Waite 1987; Broumas and Haniotakis 1994). Traps baited with this combination have been used both in detection (Rice et al. 2003; Yokoyama et al. 2006) and in mass-trapping efforts to lower olive infestation (Haniotakis et al. 1986b, 1991; Iannotta et al. 1994; Petacchi et al. 2003; Noce et al. 2009; see also Navarro-Llopis and Vacas, Chap. 15, this volume).

2.2.6 Synthesis

Because males of many economically important *Bactrocera* species are strongly attracted to male lures (see below), the development of pheromone-based trapping for this genus would target females primarily. The important finding that ME and CL/RK are used in pheromone synthesis and that their incorporation in this process enhances the attractiveness of the male olfactory signal could facilitate the production of effective pheromone baits. Nonetheless, many of the obstacles noted above for *Anastrepha* apply to *Bactrocera* as well, namely the lack of data on (i) the long-range attractiveness of the male pheromone, (ii) the optimal blend (relative amounts) and release rate of pheromonal components that produce maximal attractiveness, and (iii) the importance of olfactory signals relative to other modalities (i.e., visual, acoustic) in the mating behavior of *Bactrocera* species

2.3 Ceratitis *Pheromones*

The genus *Ceratitis* contains approximately 80 species, most of which are found in tropical Africa, although C. *rosa* Karsch (Natal fruit fly) has invaded Mauritius and Réunion and *C. capitata* (Wiedemann) (Mediterranean fruit fly or medfly) has spread globally (South and Central America, Western Australia, and Hawaii) (De Meyer 2000). The medfly is, of course, one of the most harmful agricultural pests worldwide, with females ovipositing in soft fruits of more than 300 plant species (Liquido et al. 1990). Other major economic pests in the genus include C. *rosa*, *C. cosyra* (Walker) – mango fruit fly, and *C. catoirii* Guérin-Méneville (White and Elson-Harris 1992). Because of its economic importance, the medfly has been studied far more intensively than its congeners, and this review will necessarily focus on this species.

Féron (1959, 1962) provided the first detailed description of calling behavior in *C. capitata* males, which he associated with the emission of volatiles attractive to females. While the notion of male-produced olfactory stimuli had been proposed decades earlier (Martelli 1910; Back and Pemberton 1918, both cited in Jones 1989), Féron supplied empirical evidence by reporting female attraction to a cotton wick previously exposed to calling males. Quilici et al. (2002) and Briceño et al. (2005) report similar pheromone-calling behavior in *C. rosa* and *C. catoirii*, but data showing female attraction to calling males are not yet available for these species. For the medfly, Ohinata et al. (1977) and Nakagawa et al. (1981a) provided the first quantitative demonstration of the long-range, female attraction to calling male in the field by recording female captures in male-baited traps. Female attraction to live, calling medfly males (or their odor) was reported in further laboratory (Landolt et al. 1992a; Jang 1995; Jang and Light 1996; Jang et al. 1994, 1998) and field (Shelly 2000c) studies. The importance of olfaction to females has been demonstrated conclusively via antennal ablation: in existing studies, females with antennae removed either mated not at all (Nakagawa et al. 1973; Levinson et al. 1987) or at very low levels (Shelly et al. 2007).

Research aimed at characterizing the chemical composition of the male sex pheromone in *C. capitata* and identifying the biologically active components was undertaken even before female attraction to calling males was demonstrated in the field. Jacobson et al. (1973) identified two putative pheromonal components – methyl (*E*)-6-nonenoate and (*E*)-6-nonen-1-ol – and indicated that females were attracted to both compounds in assays performed in small cages. Ohinata et al. (1977, 1979) identified the same two components as well as 15 carboxylic acids, which were presumed to 'activate' the two main components. However, contrary to Jacobson et al. (1973), various blends of these different chemicals were found to attract males but not females. Delrio and Ortu (1987, cited in Millar 1995) likewise reported no female attraction to methyl (*E*)-6-nonenoate. Jacobson and Ohinata (1980) also reported (−)-β-fenchol in the medfly sex pheromone, but subsequent analyses failed to detect this compound.

In fact, it appears that, due to inadequate analytical methods, initial efforts to identify pheromonal components led to inaccurate results, which could not be confirmed by later studies. Despite the potential usefulness of pheromone-based lures in medfly surveillance programs, relatively few studies have further investigated pheromonal composition and/or the role of particular components as female attractants. Baker et al. (1985) identified nine components in male medfly emissions, with the three most abundant being ethyl (*E*)-3-octenoate, geranyl acetate, and (*E*, *E*)-α-farnesene. They further proposed that another component, 3,4-dihydro-2-H-pyrrole (1-pyrroline), functioned as the key attractive element to females (although no data on its purported biological activity were provided). In a follow-up study, Baker et al. (1990) tested the attractiveness of four compounds (linalool, two pyrazines, and geranyl acetate) in field trials in Mexico. Both individually and in various blends, these chemicals attracted both sexes of *C. capitata*, although the olfactory stimuli used bore little resemblance to the emissions of calling males.

More thorough chemical analyses (Jang et al. 1989a; Flath et al. 1993) confirmed the presence of the nine components reported by Baker et al. (1985), with one exception, and revealed a pheromonal complexity far greater than previously documented. Jang et al. (1989a) identified a total of 56 compounds from the odor of calling males, and Flath et al. (1993) identified four additional compounds, thus revealing that the sex pheromone of *C. capitata* males consists of approximately 60 different compounds. Jang et al. (1989b) established four abundance categories for the pheromonal constituents, with five considered major components (ethyl acetate, 1-pyrroline, ethyl (*E*)-3-octenoate, geranyl acetate, and (*E*,*E*)-α-farnesene). Based on electroantennogram (EAG) recordings, ethyl acetate, 1-pyrroline, and (*E*, *E*)-α-farnesene elicited low EAG responses (relative to a standard), geranyl acetate elicited a moderate response, and ethyl (*E*)-3-octenoate elicited a high response. Overall, the sexes displayed similar EAG responses to the different pheromonal components. Additional studies (Casaña-Giner and Primo-Millo 1999; Gonçalves et al. 2006) have identified additional components from the volatiles of calling

C. capitata males, and Cossé et al. (1995) confirmed strong female EAG responses to some of the previously identified compounds of the male pheromone.

Since Jang et al.'s (1989b) seminal paper, few studies have examined behavioral responses to the sex pheromone of *C. capitata* males. These investigations have employed simplified blends, including only the major components or a subset thereof, owing to (i) the difficulty of creating close mimics to the naturally complex male odor and (ii) the assumption that only a small portion of the pheromonal chemicals identified actually have biological activity (Landolt et al. 1992a). Thus, working in a coffee field in Guatemala, Heath et al. (1991) demonstrated attraction of medfly females to a synthetic blend containing three of the major components (ethyl (*E*)-3-octenoate, geranyl acetate, and (*E, E*)-α-farnesene). In two testing periods (lasting 6 and 8 days, respectively), 259 and 368 wild females, respectively, were captured in traps baited with the synthetic blend. However, the effectiveness of the blend as a trap-bait, could not be ascertained, because (i) no estimates of the size of the wild population were made and (ii) no traps baited with live males were operated, thus precluding assessment of the relative competitiveness of the simplified blend. Working with the same 3-component blend, Landolt et al. (1992a) reported an oriented response (i.e., upwind movement coupled with course-correcting, zigzagging flight) of female medflies to the stimulus in a wind tunnel. However, only a small proportion (3 %) of the females actually contacted the odor source, a level not significantly different from the contact rate observed in the absence of an olfactory stimulus. Jang et al. (1994) studied female response to live males, each of the five major components, and a mixture of these five compounds in a wind tunnel. Although females showed greatest attraction to live males, the five component blend was more attractive than the individual compounds and appeared to elicit a much greater female response than the 3-component blend used by Landolt et al. (1992a). While the above studies reveal the attractiveness of simple pheromone blends, Casaña-Giner et al. (2001) reported very low catch of female medflies in traps baited with a 6-component mixture and questioned the long-range effectiveness of male pheromone-baited traps.

Adopting a different approach, Mavraganis et al. (2008) obtained whole body extracts of medfly males and monitored female attraction to complete extracts of laboratory vs. wild males as well as the major components of the extracts either individually or in different combinations. Interestingly, females showed greater attraction in laboratory assays to the extracts from wild males than laboratory males. Samples from wild males contained larger amounts of the compound α-copaene than those from laboratory males, and this compound was found to have greatest attractancy to females in comparisons of the individual components. Field trials further revealed that total male extracts as well as synthetic blends of major components were highly attractive to wild females, the majority of which were virgin. Thus, in contrast to other studies (Heath et al. 1991; Casaña-Giner et al. 2001), the total extracts and blends tested by Mavraganis et al. (2008) appear to be highly attractive to female medflies and clearly merit additional field testing.

The chemical complexity of the male sex pheromone has, it appears, discouraged efforts to develop or improve the attractiveness of synthetic sexual lures to medfly females. Not only might some trace components await identification, but knowledge of the relative amounts of the constituent compounds is imprecise. In addition, the blend containing the five major components was far less attractive to females than the odor emitted by live males, suggesting that simplified formulations will not be able to compete against calling males in the wild (Jang et al. 1994). Similarly, whole body extracts were far more attractive to wild females than simplified blends with relatively few components (Mavraganis et al. 2008). In this regard, Howse and Knapp (1996), noting similarity in the volatiles released by host fruit, citrus in particular, and calling male medflies, suggest that competition with host fruit odors may further limit the effectiveness of male pheromonal traps in orchards (but see Mavraganis et al. 2008). In addition to the chemical composition, the importance of release rate in the development of a potent pheromone-based lure is uncertain as relevant data are inconsistent. In particular, results from Ohinata et al. (1977) and Jang et al. (1994) suggest that the amount of male emission does not have a marked effect on its attractiveness, whereas Heath et al. (1991) found that an intermediate release rate resulted in higher female captures than lower or higher rates. Finally, while the identification of a female lure for *Ceratitis* species is recognized as a worthy research objective, the wide usage of a male-lure (trimedlure) may lessen the impetus to achieve this goal.

2.4 **Dacus** ***Pheromones***

To date, little effort has been devoted to identifying possible sex or aggregation pheromones of *Dacus* species. This lack of interest is probably due to a combination of several factors, such as insufficient funding, the small number of pest species in the genus, which are generally moderate pests relative to highly invasive *Bactrocera* species, and the availability of male lures for surveillance purposes.

2.5 **Rhagoletis** ***Pheromones***

Male sex pheromones have been demonstrated in several *Rhagoletis* species. Using caged host trees, Prokopy (1975) and Katsoyannos (1976) furnished evidence for a male sex pheromone in *R. pomonella* (Walsh), the apple maggot fly, and *R. cerasi* Loew, the cherry fruit fly, respectively, by reporting attraction of mature, virgin females to cages of live males as well as to empty cages that had housed males. No male-to-female or male-to-male attraction was observed in either of these species. Also, males of these two species did not display any behavior typically associated with pheromone release in other tephritids (e.g., wing fanning), and consequently the manner of pheromone release was unclear. Additional tests on *R. cerasi* further showed that immature and mated females do not respond to male pheromone and that mature virgin females responded to an extract obtained from whole body preparations of mature males (Katsoyannos 1982). In addition to the male pheromone, several studies (Prokopy and Bush 1972; Katsoyannos 1975) have shown that female host marking pheromone acts as a male arrestant, possibly to increase the frequency of intersexual encounters. Research on sexually-oriented, pheromonal communication has not continued beyond these few studies, and the potential use of male pheromones in *Rhagoletis* trapping or detection has not been investigated.

2.6 Toxotrypana curvicauda *Pheromone*

Landolt and Hendrichs (1983) reported "puffing" of the pleural areas of abdomen in male *Toxotrypana curvicauda* Gerstaecker, the papaya fruit fly, a behavior associated with pheromome release in other tephritids, and Landolt et al. (1985) later demonstrated female attraction to male pheromone in laboratory assays, including wind tunnel trials. The pheromone has a single chemical component, which was identified as 2-methyl-6-vinylpyrazine (Chuman et al. 1987). Additional observations (Landolt and Heath 1988; Landolt et al. 1991) showed that the pheromone attracts, not only virgin females, but also mated females and males and that female response is increased when green papaya fruit or an extract thereof was presented with the male pheromone (Landolt et al. 1992b).

The male pheromone with sticky-coated green spheres was tested in Florida and resulted in high captures of *T. curvicauda* females (Landolt et al. 1988; Landolt and Heath 1990). To facilitate field use, Heath et al. (1996) developed a membrane-based formulation system and showed that release rates, which were dependent on the amount of pheromone loaded into the system, were relatively constant over trials lasting 23 days. They also showed that green opaque cylindrical traps yielded higher captures than spherical traps. In field tests conducted in a papaya orchard in Guatemala, greater numbers of females were, as expected, captured in pheromone-baited than blank cylinders. Surprisingly, however, similar tests run in a Mexican papaya plantation detected no influence of pheromone presence/absence on female captures in green cylindrical traps. Reasons for this discrepancy are unknown, and additional field trials are required to evaluate the efficacy of pheromone-baited traps in detecting and/or suppressing populations of *T. curvicauda*.

3 Male Lures

There are two types of male lures: anthropogenic (e.g., CL, trimedlure [TML], fluorinated methyl eugenol analogs, raspberry ketone-formate (RKF)) and plant-borne (e.g., α-copaene, ME, RK, and zingerone). For certain species, male lures are relatively cheap to synthesize due to the simplicity of the chemical structures, which are often not stereoisomeric. In addition, they are very potent attractants in most cases and thus appear a more robust option for trapping programs than the development of the male sex pheromone as baits. As a result, they are frequently used as baits in trapping for surveys and detection of invasive species, delimitation of an infestation, and control or eradication via the male annihilation technique. Because of their importance in trapping, numerous reviews (Chambers 1977; Cunningham 1989; Metcalf 1990; Millar 1995; Jang and Light 1996; Sivinski and Calkins 1986; Oliver et al. 2004; Vargas et al. 2010a) of male lures already exist, and rather than re-hashing information, our present aim was to address a few selected topics.

3.1 Anastrepha

At present, there are no identified male lures for any *Anastrepha* species. This fact does not reflect a lack of effort. Approximately 8,000 compounds were screened as possible attractants for *A. ludens* (Chambers 1977) and 1,320 compounds were tested for *A. suspensa* (Burditt and McGovern 1979; cited in Cunningham 1989). Despite intense screening, no male attractant was identified in both the investigations.

3.2 Bactrocera

While most species in the genus remain untested, *Bactrocera* has been broadly categorized into three groups of species based on male response to two very potent attractants (Drew 1974; Hardy 1979; Drew and Hooper 1981; White and Elson-Harris 1992; Fig. 2.5). Nearly 200 species have been identified as CL/RK responders and 81 species as ME-responders (IAEA 2003). A third group includes approximately 15 species (limited to an Australian survey) that do not respond to either CL/RK or ME as evidenced by their absence in traps from areas where species were known to be present (Drew and Hooper 1981). No species has been identified that responds to both CL/RK and ME. Response to these compounds correlates with morphologically-based taxonomic classification (Drew 1974; Drew and Hooper 1981), and several authors (Metcalf et al. 1979, 1981, 1983, 1986; Metcalf 1990; Raghu 2004) suggest that the existence of distinct CL/RK- and ME-responding species groups reflects evolutionary divergence from a common saprophytic relationship with rotting fruits. Broadly, coevolution between plants, specifically the appearance of novel metabolic pathways and the subsequent integration of those products into essential oils, and dacine tephritids, specifically adaptation of olfactory receptors to chemically diversified plant essential oils, is considered to underlie the present-day CL/RK or ME distinction in species responsiveness.

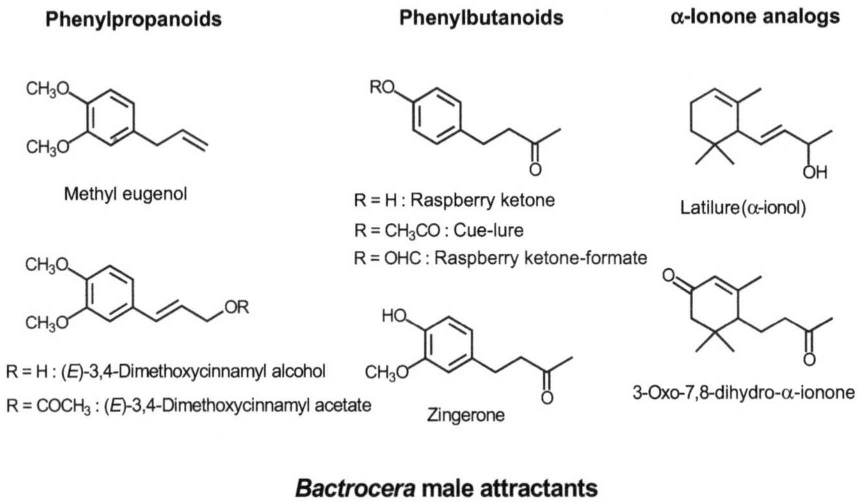

Fig. 2.5 Chemical structures of lures for *Bactrocera* males

3.2.1 Methyl Eugenol

ME is widely recognized as the most powerful male lure currently in use for detection, control, and eradication of any pestiferous tephritid species. The chemical occurs naturally in more than 450 plant species representing 80 families spanning 38 different orders in varying amounts, from a trace quantity to over 90 %, in essential oils extracted from flowers, leaves, roots, stems, or whole plant extracts (see review by Tan and Nishida 2012). It was first discovered as a fruit fly attractant by Howlett (1912, 1915), who observed males of *B. dorsalis* and *B. zonata* responding to ME-containing citronella grass, *Cymbopogon nardus* (L.) Rendle. Steiner (1952) further documented the strong attraction of *B. dorsalis* male to ME and noted their vigorous feeding on the chemical. Metcalf et al. (1975) exposed *B. dorsalis* males to ME as well as 34 analogs and found that ME elicited the greatest feeding response. The powerful attraction of this lure was further demonstrated by (a) a simple test in which approximately 1 nanogram (10^{-9} g) of ME spotted on a silica gel TLC plate placed in the field attracted *B. papayae* males, which readily consumed the minute amount of attractant (Tan and Nishida 2000); and (b) trap placement for trapping *B. dorsalis* and *B. umbrosa* hung at different heights – ground level (0.3–0.5 m), below (1.5–2 m), middle (5–7 m) and above (10–12 m) tree canopy – using traps baited with 0.5 ml ME/trap and set up using a 4 × 4 Latin square design in a 5 ha Penang village (a 4-day experiment conducted in two fruiting seasons), showed no significant difference in daily fly captured (flies/trap/day) (Tan 1984).

Because of its potency, ME-baited traps have been used in a variety of ways including (i) detection and surveillance of invasive species (Drew et al. 2005; McQuate et al. 2008a; Jessup et al. 2007), (ii) quarantine surveys and delimitation

(Allwood 2000; Sookar et al. 2008; OEPP/EPPO 2010), (iii) suppression and eradication (Steiner et al. 1965; Hancock et al. 2000; Hsu et al. 2010; Vargas et al. 2008), and (iv) ecological studies, including faunal surveys (Tan and Lee 1982), population dynamics and phenology (Tan 1985; Tan and Serit 1988, 1994; Ye and Liu 2005; Han et al. 2011; Kamala Jayanthi and Verghese 2011), adult survivorship (Tan and Jaal 1986), and dispersal (Iwahashi 1972; Tan and Serit 1988; Chen et al. 2006; Froerer et al. 2010). As noted above, existing reviews address many of these topics (see, in particular, Vargas et al. 2010a, b), and here we briefly address three topics, namely (i) the effect of ME feeding on subsequent attraction to ME-baited traps , (ii) age-dependent variation in response to ME, and (iii) interspecific differences in attraction to ME.

There is some evidence that feeding on ME reduces the likelihood of future ME feeding. Males of *B. dorsalis* given access to ME for 0.5–24 h prior to release were captured in ME-baited traps at much lower rates (1–4 %) than control, unfed males (22 %; Shelly 1994). In a second test, *B. dorsalis* males exposed to ME for 2 h and then held for 7–35 days prior to release were also captured at lower rates (11–18 %) than control, unfed males (34 %). Other studies, however, indicate that these results may be an artifact of the experimental design. When treated males were provided ME-bearing flowers instead of commercial ME, there was no difference in capture rate in ME-baited traps between floral-exposed and control males, indicating that the unnaturally high purity and availability of synthetic ME in the previous study may have accounted for the diminished capture rate of treated males (Shelly 2000d). Moreover, dissection of wild males of several *Bactrocera* species attracted to ME-baited traps (designed to prevent feeding) revealed the presence of ME metabolites in the rectal gland of nearly all individuals. Wee and Tan (2001) extracted the rectal gland from 76 wild *B. papayae* males in Penang, Malaysia, and found that all individuals possessed some ME-metabolites, ranging from trace quantities to approximately 103 μg per male. Similarly, Tan et al. (2011) reported that nearly all wild-caught *B. invadens* and *B. zonata* males dissected contained at least trace amounts of ME metabolites (maximum observed: 10 μg per gland for both species). Interestingly, after ad libitum feeding on synthetic ME, laboratory measurements regarding accumulation of ME breakdown products showed that males of *B. papayae*, *B. invadens*, and *B. zonata* sequester, on average, 20, 170, and 25 μg/gland 1–2 days after feeding (Wee and Tan 2001; Tan et al. 2011; see also Tokushima et al. 2010 for comparable data on *B. correcta*). As the quantity of ME derivatives detected in wild males attracted to ME traps was often less than these averages, it appears that, in general, males are unable to "tank up" at any one ME source and therefore must visit multiple ME sources to gather a sufficient amount of the chemical, a result consistent with the aforementioned result regarding the high capture rate at ME traps of *B. dorsalis* males experimentally fed ME-bearing flowers in the laboratory (Shelly 2000d).

Interestingly, when *B. papayae* males were exposed to commercial ME (isolated to prevent feeding) for various time periods within a trap, they became habituated after an hour and would not respond to subsequent ME exposure for a week (unpublished data, discussed in Tan et al. 2002). Because of this, the trapped males were removed from ME-traps every 0.5 h to avoid possible habituation to ME when using the mark-release-recapture technique to estimate population size (Tan 1985). While access to commercial ME of high purity may induce habituation depending on the length of exposure and underestimate the incidence of repeat ME feeding in nature, it nonetheless suggests a means of improving the effectiveness of control programs against *B. dorsalis* (or related species), namely the simultaneous application of the male annihilation and sterile insect techniques. Providing sterile males access to ME before release may both increase their mating competitiveness (McInnis et al. 2011) and reduce their attraction to insecticidal-laden ME sources deployed for male annihilation. Barclay and Hendrichs (Chap. 11, this volume) examine the improved control afforded by this strategy through a modeling approach.

As another caveat, to the extent that male responsiveness to ME has a heritable component, it seems possible that prolonged application of a male annihilation program might select for males showing low attraction to ME, thus eclipsing the effectiveness of the program. Faced with persistent populations of *B. dorsalis* on several Japanese Islands despite prolonged attempts at male annihilation, Itô and Iwashashi (1974) and Habu et al. (1980, 1984) suggested that selection for ME-insensitive males was responsible, and, as a result, SIT was implemented and finally achieved eradication. In support of their claim, Itô and Iwahasi (1974) exposed *B. dorsalis* males to ME in the laboratory and selected non-responding males as sires. Within only two generations of such selection, they produced a strain with lower ME responsiveness than a control line. Working with *B. dorsalis* in Hawaii, Shelly (1997) likewise reported a consistent reduction in ME responsiveness over 8 generations for several lines sired by non-responding males. These studies indicate that, while the evolution of lure-insensitivity has not been demonstrated conclusively, programs of male annihilation are most effective when applied intensely with the aim of rapid eradication.

Age-dependent response to ME has been examined in some detail for *B. dorsalis*. While all studies confirm that ME response increases with male age, they differ in their estimates of ME responsiveness among immature individuals. On one hand, several studies (Umeya et al. 1973; Itô and Iwahashi 1974; Habu et al. 1980; Tan et al. 1987) report no or very little attraction by very young males (1–5 days-old) and a close association between ME response and male sexual maturation (see also Fitt 1981b for data on *B. opiliae*). Tan et al. (1987), for example, found that less than 2 % of 5 days-old laboratory-reared males responded to ME in a wind tunnel and no wild males, emerged from naturally infested star fruits (*Averrhoa carambola* Linn.), marked, and released, at less than 7 days of age were captured in ME-baited traps. In contrast, several other studies (Steiner 1952; Steiner and Lee 1955; Wong et al. 1989; Shelly et al. 2008) reported young males (<5 days of age) showed relatively high response to ME. Wong et al. (1989), for example, found that nearly 50 % of males responded to ME before their age of first mating (13 days). Collectively, these studies involved different strains, different procedures, and different test conditions, making it impossible to draw a robust conclusion. Resolution is far from an arcane academic exercise, however, as knowledge of age-dependent response to ME is critical to predicting the success of male annihilation efforts (see Barclay and Hendrichs, Chap. 11, this volume).

In contrast to intraspecific, age-dependent variation, little attention has been given to interspecific differences in attraction to ME. In the most comprehensive study to date, Wee et al. (2002) monitored the response of males of three closely related *Bactrocera* species to serial dilutions of ME in small laboratory cages as well as consumption of ME from microcapillary pipettes. The two assays yielded the same trend, i.e., in decreasing order, the level of ME sensitivity was *B. dorsalis* > *B. papayae* > *B. carambolae*. Most notably, males of *B. carambolae* showed relatively weak attraction to the lure: the ME dose required to elicit response of 50 % of *B. carambolae* males was 9 and 17 times higher than that observed for *B. papayae* and *B. dorsalis*, respectively. Given the importance of ME in trapping programs, additional studies of this type are clearly needed to better characterize the detection sensitivity of area-wide trapping grids.

3.2.2 Fluorinated Analogs of Methyl Eugenol

One of the potential problems with the use of ME for fruit fly control is the reported carcinogenicity of this compound in mice and rats (Miller et al. 1983) and microbes (Schiestl et al. 1989; Sekizawa and Shibamoto 1982). ME has also been shown to form DNA adducts in cultured human cells and thus may contribute to human carcinogenesis (Zhou et al. 2007). However, several reviews (Smith et al. 2002; Robinson and Barr 2006) conclude that ME does not pose a significant cancer risk in humans, primarily because ME exposure in humans is as much as 1,000 times below the level utilized to produce hepatic carcinoma in rats. Human subjects fed approximately twice the daily average intake of safrole (a phenylpropene related to methyl eugenol) over a 2 year period showed no carcinogenetic symptoms (Long et al. 1963). In addition to the low exposure, ME in human blood serum is rapidly eliminated and excreted (Schecter et al. 2004). ME may, in fact, have some benefits to human health, e.g., reduction of cerebral ischemic injury (Choi et al. 2010) as well as anti-anaphylactic properties (Kim et al. 1997). ME is a regular component of the human diet (e.g., flavoring in baked goods and candy, Smith et al. 2002) and is found in most spices and some plants, particularly in the family Lamiaceae, e.g., *Ocimum basilicum* (sweet basil) and *O. sanctum* (holy basil), which have high ME contents and are regularly consumed as vegetables or used for culinary and medicinal purposes in Southeast Asian countries (Tan and Nishida 2012).

The fear of ME carcinogenicity and hepatoxicity to human health, whether legitimate or a case of overreaction, has prompted some fruit fly scientists to search for 'safer' alternative attractants for ME-responsive *Bactrocera* species and evaluate various phenylpropanoids with structural similarities to ME (Khrimian et al. 1993, 1994, 2006, 2009; Liquido et al. 1998; Metcalf et al. 1975; Mitchell et al. 1985). Two such analogs of ME are 4-[(2*E*)-3-fluoroprop-2-en-1-yl]-1,2-dimethoxybenzene (FME), an analog fluorinated at the terminal carbon of the ME side chain, and 1-fluoro-4,5-dimethoxy-2-(prop-2-en-1-yl)benzene (RFME), an analog fluorinated at the 4 position of the ME aromatic ring. In field tests, FME was as attractive to *B. dorsalis* males as ME (Khrimian et al. 1994), while RFME was only about 50 % as attractive as ME (Khrimian et al. 2009). The good performance of FME in field bioassays (Khrimian et al. 1994, 2006, 2009; Liquido et al. 1998) showed that this compound is not only equally attractive to *B. dorsalis* but has an added value as a more persistent lure. The carcinogencity of the terminal carbon fluorinated compound has not yet been determined, but, if negative, FME could serve as an excellent replacement for ME in trapping programs.

Jang et al. (2011) synthesized two additional fluorinated ME analogs, 1-(3,3-difluoroprop-2-en-1-yl)-2-fluoro-4,5-dimethoxybenzene, a ME analog trifluorinated at the 4 position of the aromatic ring and at the terminal carbon of the side chain, and 1-fluoro-2-(3-fluoroprop-2-en-1-yl)-4,5-dimethoxybenzene, a ring and side-chain difluorinated analog. Although *B. dorsalis* males were attracted strongly to and fed on the trifluoroanalog and difluoroanalog in a cage experiment, field attractiveness of male oriental fruit fly to both was markedly lower than to ME. In field bioassays, traps baited with difluoroanalog captured roughly 50 % as many flies as traps baited with ME, while the trifluoroanalog captured only about 10 % as many males. Thus, di- or tri-fluorinated ME are likely not viable replacements for ME as attractants for *B. dorsalis* and related species.

3.2.3 Plant Phenylpropanoids, Dimethoxycinnamyl Analogs as Parapheromones of ME-Responsive Species

E-3,4-dimethoxycinnamyl alcohol and *E*-3,4-dimethoxycinnamyl acetate from *Spathiphyllum cannaefolium* Schott (Araceae) (Chuah et al. 1996) and Hawaiian lei flower, *Fagraea berteriana* A. Gray ex Benth. (Loganiaceae) were characterized as attractants for *B. dorsalis* (Nishida et al. 1997). Although these compounds are less volatile than ME, the feeding stimulant activity of the former was as high as that of ME (Nishida et al. 1997). However, they will not replace ME in trapping of *B. dorsalis* because of their low volatility and attractancy.

3.2.4 Raspberry Ketone, Raspberry Ketone Formate, and Cue Lure

As noted above, among lure-responsive *Bactrocera* species, the majority is attracted to RK/CL. RK (Fig. 2.5) is found naturally as a fungal metabolite (Ayer and Singer 1980) and in many plants besides raspberries (*Rubus idaeus* L.), including other species in Rosaceae, Asteraceae, and Lamiaceae (formerly Labiatae) (Hirvi et al. 1981; Hirvi and Honkanen 1984; Lin and Chow 1984; Marco et al. 1988) as well as Orchidaceae (Nishida et al. 1993; Tan and Nishida 2005; Tan 2009). Drew (1974) reported that RK was developed as a male attractant for *B. tryoni* in Australia in 1959 but provided no additional information regarding the nature of this discovery. At approximately the same time, the United States Department of Agriculture (USDA) was engaged in a large-scale screening of thousands of chemicals as potential fruit fly baits (Beroza and Green 1963). Based on this screening process, Barthel et al. (1957) reported attraction of *B. cucurbitae* males to anisylacetone (4-(4-methoxyphenyl)-2-butanone), a synthetic aromatic ketone. In turn, Beroza et al. (1960), through continued testing of compounds related to anisylacetone, synthesized CL (4-(4-acetoxyphenyl)-2-butanone), which was a much more potent attractant for *B. curcubitae* males and is now used worldwide in detection efforts for this species and other RK/CL-responsive *Bactrocera* species (Jang et al. 2007). CL has not been isolated as a natural product but is hydrolyzed to RK (also known as rheosmin; 4-(4-hydroxyphenyl)-2-butanone), which as noted above is widespread in nature (Metcalf 1990; Metcalf and Metcalf 1992b). In field tests, CL is a more potent attractant than RK (Alexander et al. 1962; Keiser et al. 1973), likely owing to its high volatility relative to that of RK (approximately 20 times greater, Metcalf 1990).

At this juncture, it is pertinent to point out that Nishida, Howcroft and Tan (unpublished data) recently detected anisylacetone and CL (hitherto, not known as natural products as mentioned above) in certain bactrocerophillous orchid flowers (*Bulbophyllum* spp.) found in Papua New Guinea. They also showed that both the compounds have differential attraction against RK-sensitive *Bactrocera* species in field capture of wild males – with significantly more *B. cucurbitae* and *B. triangularis* (Drew) captured in RK- than anisylacetone-baited traps and vice versa for *B. atramentata* (Hering), *B. bryoniae* (Tryon) and *B. frauenfeldi* (Schiner) (unpublished data). This shows that anisylacetone and CL, along with RK and zingerone (see below), in nature may (a) play an important evolutionary role in the *Bactrocera* fruit fly-orchid interactions, and (b) affect trapping of wild flies in surveillance and quarantine detection for an areawide IPM/SIT program.

Although quantitative data are scant, it is generally accepted that CL is a weaker attractant than ME (Cunningham 1989; Jang and Light 1996). Data from a mark-release-recapture study (Shelly and Nishimoto 2011) conducted in Hawaii and California confirmed this notion. For example, among flies released 100 m from the lure source, 1–19 % of *B. dorsalis* males were captured in an ME-baited trap compared to only 0.4–1.2 % of *B. cucurbitae* males captured in a CL-baited trap. Correspondingly, with 5 ME- and 5 CL-baited traps per 2.59 km^2 (operational density in California, for example), there would be near certainty (>99.9 %) of detecting incipient *B. dorsalis* populations as small as 50–162 males, whereas the same likelihood of detection for *B. cucurbitae* would require 310–350 males in the population.

Although less potent, CL has been used in the same ways as ME, i.e., (i) detection and surveillance of invasive species (Gonzalez and Troncoso 2007; Jessup et al. 2007), quarantine surveys and delimitation (Allwood 2000), suppression and eradication (Matsui et al. 1990; Vargas et al. 2000; Sookar et al. 2008), and ecological studies, including faunal surveys (Osborne et al. 1997; Allwood 2000), population dynamics and phenology (Itô et al. 1974; Harris et al. 1986; Vargas et al. 1990), and dispersal (Fletcher 1989; Vargas et al. 1989; Kohama and Kuba 1996; Peck et al. 2005). As noted above, existing reviews address many of these topics (see, in particular, Vargas et al. 2010a), and here we briefly address two topics, namely (i) age-dependent variation in response to ME and (ii) comparative performance of CL and raspberry ketone formate.

As with the *B. dorsalis*-ME association, attraction of *B. cucurbitae* males to CL is related to sexual maturation, and findings have been inconsistent regarding the level of response displayed by very young males. In particular, whereas Beroza et al. (1960) observed attraction of newly emerged *B. cucurbitae* males to CL, other studies (Monro and Richardson 1969; Fletcher 1974; Wong et al. 1991) report no attraction until males are at least several days old. In the most comprehensive study, Wong et al. (1991) found that wild *B. cucurbitae* males did not respond to CL until 10 days of age and that the timing of CL response and mating activity were highly correlated. Based on this finding, these authors concluded that male annihilation would be less effective against *B. cucurbitae* than *B. dorsalis*, because the closer coincidence of lure response and sexual maturation in the former than the latter means that fewer *B. cucurbitae* males would be killed in lure-baited traps prior to mating than would be the case for *B. dorsalis*.

In attempting to identify a more potent lure than CL, several studies have investigated the attractancy of the formate ester of RK, formic acid 4-(3-oxobutyl) phenyl ester (RKF). In the early 1990s, Metcalf and Mitchell (1990) and Metcalf and Metcalf (1992a, b) showed that RKF was more attractive to *B. cucurbitae* males than either RK or CL. Despite this finding, no further research on RKF was undertaken for about a decade, apparently because of concern regarding the rapid hydrolytic conversion of RKF to RK (which, as noted above, is less volatile and less attractive than CL, which hydrolyzes to RK at a slower rate, Beroza et al. 1960). However, subsequent work (Casaña-Giner et al. 2003a, b) showed that rate of hydrolysis of RKF to RK was likely overestimated. Furthermore, field testing (Casaña-Giner et al. 2003a, b; Oliver et al. 2004; Jang et al. 2007) showed that RKF-baited traps generally captured more *B. cucurbitae* males than CL-baited traps. Additional field data, however, have not corroborated this result. Working with *B. cucurbitae*, Vargas et al. (2010c) reported no difference in the catch of traps baited with CL or RKF embedded in a biologically inert, waxy matrix (SPLAT), and Shelly et al. (2012a) reported that traps baited with liquid CL had significantly higher captures than traps baited with RKF presented as a liquid or in a polymeric dispenser. Reasons for these inconsistent results are unknown, though it is possible that variation in abiotic factors, which affected the conversion of CL and RKF to RK, is responsible.

RKF has also been found to attract many other RK-responsive *Bactrocera* species. Preliminary tests (Jang et al. unpublished) in Australia showed that RKF plugs recaptured 1.5 times more sterile male Queensland fruit flies compared to a CL plug. In an unpublished survey, Jang and colleagues found 19 *Bactrocera* species in traps baited with CL and RKF in the northern territories of Australia. Most of the species responded equally to either CL or RKF, but a few showed higher trap captures to RKF than CL. The results from trap evaluations in the Northern Cape York Peninsula and the Torres Strait showed that RKF had higher trap captures of *B. frauenfeldi*, *Bactrocera peninsularis* (Drew and Hancock), and *Bactrocera neohumeralis* (Hardy) compared to the CL plug.

3.2.5 Presentation of ME and RK/CL

In the early 1980s, a proprietary product of International Pheromones Ltd was marketed as 'dorsalure' (a mixture, in unknown proportions, of ME and CL) in order to capture males of both ME- and RK/CL-sensitive species. In a species survey conducted in five different ecosystems in Penang Island, Malaysia, it was shown that CL and ME traps caught five RK-responsive and two ME-responsive species, respectively, while 'dorsalure' traps caught only two RK- and one ME-responsive species of the seven total species (Tan and Lee 1982). In addition, Tan (1983) tested combinations of ME and CL (three liquid mixtures (v:v) of 2:1, 1:1 and 1:2) in the same trap, and all blends caught significantly fewer males of two ME-responsive species – *B. dorsalis* and *B. umbrosa* – when compared with ME-only baited traps. Thus, CL appeared to have caused a slight interference in the male olfactory system of the ME-responsive species. More studies (Hooper 1978; Vargas et al. 2000; Shelly et al. 2004) have corroborated a reduction in ME-responsive species with bait mixtures of ME and CL. Data regarding effects on RK/CL responding species are inconsistent, however, as ME/CL blends have been found to increase (Taiwanese data, cited by Hooper 1978), decrease (Hooper 1978), or have no effect (Vargas et al. 2000) on catch numbers of RK/CL sensitive species.

In several large-scale detection programs (e.g., California, USA), *Bactrocera* lures are applied as liquids to cotton wicks, which are then placed in Jackson traps. To minimize worker risk owing to inadvertent spillage and exposure, field tests, conducted primarily in Hawaii, have compared the efficacy of the standard liquid formulation with different solid dispensers containing ME and CL separately or in combination in the same device. In general, studies (Hiramoto et al. 2006; Suckling et al. 2008; Jang 2011; Jang et al. 2013; Vargas et al. 2009, 2010b; Shelly 2010b; Shelly et al. 2011a, b; Leblanc et al. 2011) have shown that the solid dispensers perform as well as or even better than the liquid application (but see Wee and Shelly 2013 for an exception). Interestingly, two studies (Vargas et al. 2012; Shelly et al. 2012b) conducted in Hawaii further reported that traps baited with solid wafers containing ME, RK, and TML captured as many males of *B. dorsalis*, *B. cucurbitae*, and *C. capitata* as traps baited with a single lure in liquid form. The use of such triple-lure dispensers holds promise, not only in reducing worker safety, but also in reducing costs of trapping supplies and trap monitoring and servicing.

3.2.6 Zingerone

Zingerone (4-(4-hydroxy-3-methoxy-phenyl)-2-butanone, 4-hydroxy-3-methoxybenzyl- acetone, vanillylacetone) (Fig. 2.5) is a phenylbutanoid responsible for the pungency of ginger, *Zingiber officinale* (L.) H. Karst. Field studies showed that zingerone present in flowers of *Bulbophyllum patens* King and *B. baileyi* F. Muell. attracted males of both ME- and RK-responsive *Bactrocera* species, particularly, *B. dorsalis* and *B. cucurbitae*/*B. albistrigata* (Tan and Nishida 2000, 2007). Because of the presence of a hydroxyl group and a butanone side chain (both are also found in RK) as well as a methoxy (found in ME) moiety attached to the benzene ring (Fig. 2.5), zingerone attracts males of both ME- and RK-responsive *Bactrocera* species, albeit relatively very weak attraction in comparison to ME and RK (Tan and Nishida 2007). Zingerone, when consumed by *B. dorsalis* males, is converted to zingerol, attractive to conspecific females, as a component of male sex pheromone (Tan and Nishida 2007). However, in *B. cucurbitae* males, zingerone is sequestered largely unchanged into the rectal gland (Nishida et al. 1993). Khoo and Tan (2000) and Kumaran et al. (2013) have further examined the effects of male feeding on zingerone on their success in attracting mates and obtaining matings in *B. cucurbitae* and *B. tryoni*, respectively.

After the discovery of floral zingerone attracting both ME and RK-responsive species (Tan and Nishida 2000, 2007), Fay (2010) explored the structure-activity relationships of 50 different phenylpropanoids and phenylbutanoids that might attract the non-responsive *Bactrocera* and *Dacus* to the two potent attractants. It was shown that certain non-responsive *Bactrocera* species, namely *B. aglaiae* (Hardy), *B. aurea* (May), and *B. speewahensis* Fay and Hancock (a new species), as well as a rarely trapped *Dacus secamoneae* Drew, were captured only in traps baited with zingerone and not in ME and RK/CL traps (Fay 2010). Further, a qualitative field evaluation using traps baited with zingerone, RK/CL, or ME conducted in north-eastern Australia showed that *Bactrocera jarvisi* (Tryon), previously known to be attracted to RK/CL, was strongly attracted to zingerone, with more than 97 % of flies of this species captured in traps baited with the attractant. In contrast, *B. neohumeralis* and *B. tryoni* males were caught more frequently in RK traps (Fay 2012). In north Queensland, *B. jarvisi* invariably constituted 97–99 % of the total catch, and zingerone is "now starting to be used in various places around the country (Australia) for both detection and male annihilation purposes" (Harry Fay 2012 – personal communication). These very interesting results certainly suggest that zingerone should be tested more widely throughout the Asia-Pacific region for possible attraction of other non-responsive *Bactrocera* species (which constitute approximately 50 % of the total *Bactrocera* species) to the commonly used ME and RK/CL attractants.

3.2.7 α-Ionone Analogs for *Bactrocera latifrons*

α-Ionol (latilure) and its analogs (Fig. 2.5) were found as attractants for trapping males of the solanaceous fruit fly, *B. latifrons*, which shows no affinity to either ME or CL (Flath et al. 1994a). Although the attractiveness of α-ionol is much lower than that of ME and CL for *B. dorsalis* and *B. cucurbitae*, respectively, cade oil and its ingredients (e.g., eugenol) synergistically enhanced the attraction (McQuate and Peck 2001; McQuate et al. 2004, 2008a, b). On the contrary, isophorone (3,5,5-trimethyl-2-cyclohexene-1-one) and isophorol mixed with α-ionol attracted more males than the respective individual compounds (Ishida et al. 2008). Furthermore, a series of 3-oxygenated α-ionone analogs have been found as more potent attractants/phagostimulants than α-ionone/ionol (Ishida et al. 2008; Nishida et al. 2009; Enomoto et al. 2010). These C_{13}-norterpenoid analogs, resembling raspberry ketone-type phenylbutanoid structure, are present in various fruit tissues (mostly in glycosidic-forms). Ingested 3-oxgenated α-ionones by *B. latifrons* males were selectively biotransformed to a variety of derivatives, which were eventually sequestered into the rectal gland – suggesting a possible role as sex pheromone, although the actual biological function is still unknown (Nishida et al. 2009; Enomoto et al. 2010).

3.3 *Ceratitis*

The history surrounding the development of male lures for *C. capitata* has been recounted numerous times (Chambers 1977; Cunningham 1989; Millar 1995; Jang and Light 1996), and no purpose is served in repeating it here. Instead, we focus on a few selected topics, namely (i) α-copaene and natural oils as male attractants and (ii) the chemical characterization and modification of trimedlure (TML).

3.3.1 α-Copaene and Natural Oils as Male Attractants

Ripley and Hepburn (1935) first described the attraction of male *Ceratitis*, in particular males of *C. rosa*, to angelica seed oil (*Angelica archangelica* (Linn.)). Steiner et al. (1957) later reported the attraction of *C. capitata* males to the seed oil, which was used intensively in the 1956 Florida campaign against *C. capitata* to the point of exhausting the world supply. Over a decade later, two researchers (Fornasiero et al. 1969; Guiotto et al. 1972) identified α-copaene as the main attractant in angelica seed oil and α-ylangene as a secondary attractant. In a series of field trials in Hawaii, α-copaene was found to be more attractive to medfly males than TML (Flath et al. 1994b, c). In addition, the stereochemistry of this compound was critical in determining its potency as even slight deviations from the dextrorotary form ((+)-α-copaene) led to decreased attractiveness. Although data are not provided and the enantiomer is not identified, α-copaene was reported to be 2–5 times more attractive to male medflies than TML in field tests (Cunningham 1989).

While highly attractive, (+)-α-copaene has limited practical use, because its synthesis is extremely difficult and expensive and its concentration in most natural (plant) sources is low. Methods for synthesizing (+)-α-copaene have been developed (Heathcock 1966; Heathcock et al. 1967; Corey and Watt 1973), but these are laborious and yield only small amounts. Millar (1995) noted that new synthetic pathways have been developed for copaene isomers (Kulkarni et al. 1987; Wenkert et al. 1992), thus opening the possibility that simpler, and more easily synthesized, analogs might be identified as practical alternatives. Regarding plant sources, where α-copaene occurs, the levorotatory isomer usually predominates, and in those instances where (+)-α-copaene dominates, the overall content of copaene is typically low (<1 % of the essential oil prepared from the plant – Takeoka et al. 1990). Angelica seed oil differs from most plant species sampled (Buttery et al. 1985; Elzen et al. 1985) in that (+)-α-copaene appears to be the more common stereoisomer and to be relatively abundant (0.16–0.34 % of commercial angelica seed oil) (Jacobson et al. 1987).

Another commercially available, natural product, ginger root oil, also contains relatively high amounts of (+)-α-copaene and has been investigated as a medfly attractant. A distillation procedure has been developed that increases the concentration of (+)-α-copaene from 0.4 % in commercially available oil to 8 % in the enriched oil (Shelly and Pahio 2002). However, when applied as a paste at varying doses to cotton wicks, traps baited with the enriched ginger root oil captured in significantly fewer *C. capitata* males than traps baited with liquid TML (Shelly and Pahio 2002). Moreover, the enriched oil appeared to lose its potency rather quickly: paste aged 5 days resulted in 10–20 % fewer captures than fresh paste. A second study (Shelly 2013) also conducted in Hawaii confirmed the greater attractancy of TML plugs to enriched ginger root oil applied in liquid form. In contrast, Mwatawala et al. (2013), working in Tanzania, found that trap captures with enriched ginger root oil were equal to or greater than those with TML for four *Ceratitis* species (including the medfly) and that the oil captured males of one species (*C. cosyra* not typically found in TML-baited traps). Based on these results, the authors suggest that enriched ginger root oil is a viable alternative to TML in *Ceratitis* detection programs in Africa.

The discrepancy in the results for the medfly between Hawaii and Africa could reflect differences in the composition of the oils used in the two regions. In a study of avocado varieties, Niogret et al. (2011) found that the behavioral and EAG responses of medflies were not directly related to the amount of α-copaene in the volatiles of the different varieties. For example, α-copaene comprised 31 % of the sesquiterpenes for one of the least attractive varieties but only 12 % for the most attractive variety. Thus, the presence and concentration of sesquiterpenes other than α-copaene may affect medfly response to natural oils, and variation in the chemical composition of ginger oils from different suppliers could generate different results in trapping studies.

3.3.2 Trimedlure

Since its discovery approximately 50 years ago (Beroza et al. 1961), TML (*tert*-butyl 4(and 5)-chloro-*trans* –2-methylcyclohexane-1-carboxylate) has become widely, but not universally (as noted below), adopted as the chief male attractant used in detection and surveillance programs for *C. capitata* (Jang et al. 2001). These authors recognized TML to be a mixture of isomers, with four isomers predominating (Beroza and Sarmiento 1964), but complete resolution of the isomeric constitution of trimedlure was not achieved until Leonhardt et al. (1982)

reported that the four *trans* isomers comprise 90–95 % of TML and the four *cis* isomers comprise the remaining portion (see also Sonnet et al. 1984; Warthen and McGovern 1986a, b). Nonetheless, prior to this more thorough chemical description, the structure (McGovern and Beroza 1966) and volatility and attractiveness of the *trans* isomers (McGovern et al. 1966) were described, and based on laboratory olfactometer trials, the relative attractiveness of these four isomers was C > A > B1 > B2, with the latter being essentially inactive. Concerned over variability in the relative amounts of *trans* and *cis* isomers in commercial batches of TML, and in particular, the possibility that *cis* isomers might diminish the attractiveness of the *trans* counterparts (as noted for siglure, Steiner et al. 1958). McGovern et al. (1986) conducted field tests comparing the attractancy of whole TML with pure trans and *cis* formulations, respectively, as well as the attractancy of formulations varying in the *trans-cis* ratios. Results of these tests showed conclusively that traps baited with trans-TML captured as many *C. capitata* males as whole TML and that both of these formulations were more effective than *cis*-TML. Additionally, only when mixtures contained ≥ 75 % *cis* isomers was attractancy reduced. Further field tests confirmed the above mentioned findings (McGovern et al. 1987, 1990). In addition, Jang et al. (1989a) compared the electroantennogram responses of *C. capitata* males to the four *trans* isomers of TML and found the responses were greatest to the *cis* isomer, consistent with behavioral assays. Warthen et al. (1993) later investigated the relation between the molecular structure of the different isomers and their attractiveness.

In addition, to identifying the components of TML and their relative attractiveness, several studies focused on the overall release rate of TML from trap dispensers. When originally incorporated into monitoring programs, TML was applied as a liquid (2 ml) to cotton wicks (Nakagawa et al. 1979). However, owing to its high volatility from the cotton, TML was found to be effective for only 2–4 weeks (Burditt 1975; King and Landolt 1984). Two solutions were explored to extend the effective life of TML. The first involved testing solid dispensers to control (reduce) evaporation rates. These alternatives included incorporation of TML in the adhesive, insect-catching surface of the trap (Nakagawa et al. 1975), the middle layer of laminated polymeric (plastic) sheets (Nakagawa et al. 1979; Leonhardt et al. 1989), cups covered with a semipermeable membrane (Leonhardt et al. 1984), cylindrical polymeric plugs (Leonhardt et al. 1989), or rubber septa (Leonhardt et al. 1984; Baker et al. 1988). In addition, the effectiveness of compressed discs of TML *cis* isomer has also been investigated (Heath et al. 1990). Owing to their ease of handling and their lowered release rates (thus allowing a longer interval between replacement), polymeric plugs have been adopted as standard male lure in *C. capitata* detection programs (IAEA 2003).

The second solution involves adding extenders to TML to slow volatilization (Leonhardt et al. 1984; King and Landolt 1984) as TML is costly to produce, this procedure may also reduce costs. Capilure®, which replaces a portion of TML with proprietary extenders, was developed in the early 1980s and is currently used in *Ceratitis* detection program in South Africa (T.G. Grout, Pers. Comm.). Field tests (Nakagawa et al. 1981b; Rice et al. 1984; Hill 1987; Baker et al. 1988) confirm that capilure is more persistent than TML and attracts male medflies (albeit in reduced numbers) as long as 10–36 weeks after deployment. However, these same studies have reported inconsistent results regarding the relative performance of the two lures in the initial weeks of deployment, with several studies (Hill 1987; Nakagawa et al. 1981b; Rice et al. 1984) finding equivalence between capilure and TML but one (Baker et al. 1988) finding TML outperformed capilure in the 8 weeks immediately following field deployment. Likewise, in trials in a Hawaiian coffee field, Shelly (2013) found that TML captured more *C. capitata* males than capilure during weeks 1–6 immediately following placement in the field.

3.3.3 Ceralure

In an investigation into various halogen and ester analogs of TML, ethyl 4- (and 5-) iodo-*trans*-2-methylcyclohexane-1-carboxylate (ceralure), was found to be more potent and persistent than trimedlure (McGovern et al. 1987 , DeMilo et al. 1994; Warthen et al. 1998). Ceralure, like trimedlure, is composed of 16 regio and stereoisomers, of which the B1 isomer was reported to be the most attractive (Warthen et al. 1994). This molecule was tested in the field and found to be slightly more attractive and persistent than trimedlure (Leonhardt et al. 1996).

In 2000, a novel method for synthesis of the stereoisomers of the ceralure B1 molecule was developed and tested (Raw and Jang 2000). The (−) enantiomer of ceralure B1 was shown to be more attractive and persistent to laboratory-released sterile flies than the (+) enantiomer, commercial trimedlure or commercial ceralure (Jang et al. 2001). Follow-up studies (Jang et al. 2003, 2005) reported (−)-ceralure B1 to be 4–9 times more attractive than the commercial trimedlure.

One of the problems that have prevented the adoption of (−)-ceralure B1 has been development of a commercial, cost-effective synthesis of this molecule relative to TML. Khriman et al. (2003, 2004) developed an easier synthesis of the racemic (−/+)-ceralure B1, and subsequent studies showed that > 75 % (−) optically pure ceralure B1 could be as effective as the (98 %) (−)-ceralure B1, and the racemic ceralure B1 could be almost as attractive (Jang et al. 2005). More recent research has focused on the applicability of the racemic ceralure B1 as a replacement for TML.

Recently, Jang et al. (2010) compared the persistence and attractancy of the trimedlure C isomer (racemic) with the ceralure B1 (racemic) isomer to determine which of these two were the most attractive and persistent in a field setting. This was accomplished by comparing equivocal amounts of the two racemic compounds on a standard substrate (cotton wicks) and determining attraction and the residual amounts after 0–7 days. Additionally we initiated studies on polymeric formulations of the racemic ceralure B1 to determine how much of this mixture would be needed to equal or surpass the 2 g TML standard polymer formulation. Results of this test showed that the ceralure coin captured significantly more medflies compared to TML for 6 weeks of testing (Fig. 2.6). The same treatments were tested in sterile medfly release areas in Sarasota, Florida. Although the variation was high in

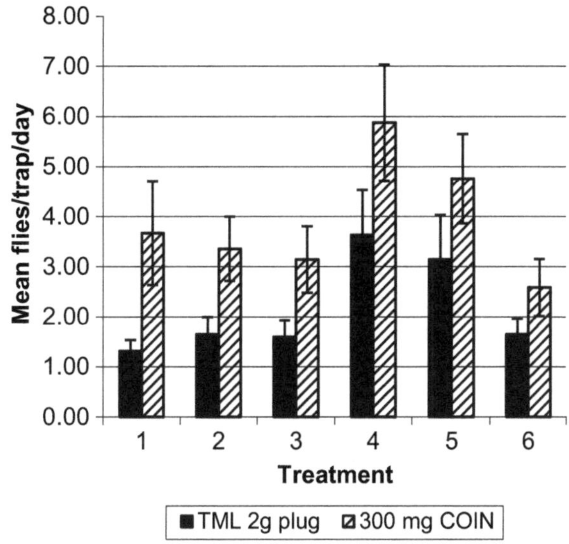

Fig. 2.6 Responses of wild Mediterranean fruit flies to ceralure B1 and trimedlure plug formulations

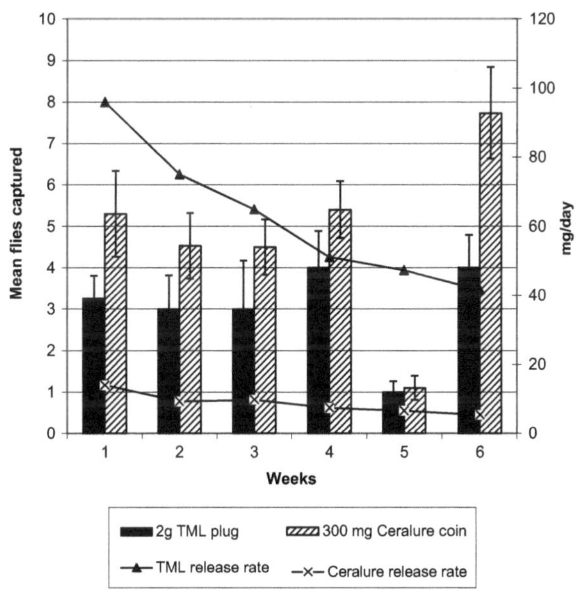

Fig. 2.7 Responses of sterile released Mediterranean fruit flies to Ceralure B1 and Trimedlure plug formulations (Adapted from Jang et al. 2010)

the weekly evaluations with released flies, results over a 6- week test period showed significantly higher trap captures with the 300 mg ceralure coin compared to the 2 g-trimedlure plug (Fig. 2.7).

On a molecule-to-molecule basis, ceralure B1 was inherently more attractive and more persistent than the C isomer of trimedlure. It also supports an earlier published results showing that when applied to cotton wicks, as little as 40 mg of the (–)

ceralure B1 (50 X less material) was as attractive as 2 g of commercial trimedlure for the first few days in the field (Jang et al. 2003).

As little as 150 mg of ceralure B1, (13 times less compound), formulated in PVC was not significantly different in trap capture compared to a 2 g also PVC-formulated commercial trimedlure plug. Further, increasing ceralure load to 300 mg per coin, which is 6.6 times less than trimedlure, captured significantly more medflies compared to a 2 g trimedlure plug in the entire 6 week test period.

Survey and detection programs are the first line of defense in keeping exotic pests such as medfly from becoming established in key agricultural states such as California, Florida and Texas. While costs of the lures used in detection programs are a consideration in overall program management, it is generally acknowledged that personnel and related costs of conducting surveys represent a much higher proportion of the total costs than the chemical lures. Further tests of ceralure B1 coins versus 2 g trimedlure plugs weathered under environmental conditions found in California, Florida and Texas are needed as to whether the increased cost of ceralure B1 synthesis and formulation are justified. Several additional uses of ceralure B1 might justify the additional costs of the product. As mentioned above labor costs for deployment of detection traps are the most costly part of a survey program. A more potent lure may reduce the number of traps required in a detection array resulting in some cost savings. We have not tested whether ceralure B1 might possibly catch younger aged flies that represent the "founder" population of an incipient introduction. Early detection is arguably more important than merely capturing the most flies, in that early detection allows for a more rapid eradication response thus reducing the overall program costs (bait sprays, fruit stripping, sterile insect releases and associated costs of quarantines). Ceralure B1, although currently expensive might also be considered for mass trapping in small outbreaks, where, when used with other control techniques would increase the likelihood of eradication.

3.4 *Dacus*

This genus consists of approximately 300 species with a handful of pest species (http://www.globalspecies.org/ntaxa/2083501). Over 40 species are known to respond to CL while only two species respond to ME (IAEA 2003). A major pest species in the Middle East region, *Dacus persicus* Hendel, but considered a beneficial insect that infests weeds in India (Kapoor 2005/2006), is attracted to ME, which has been used as bait in trapping of male flies.

Methyl paraben (methyl-4-hydroxybenzoate – detected in small quantity in the rectal gland of *B. cucurbitae* see Sect. 2.2.2.4.) was discovered to be highly attractive to the males of *Dacus vertebratus* Bezzi (Hancock 1985). It is currently marketed as "vert-lure", and used as a male attractant/lure for mass trapping of *D. vertebratus* males in control or male annihilation techniques.

For *Dacus ciliatus* Loew, pumpkin fly (a non-responder to either CL or ME), a combination of four or five acetates isolated and identified from host fruits, benzyl, hexyl, (Z)-3-hexenyl, octyl, (Z)-3-octenyl, and (Z)-3-decenyl, was most attractive, but an addition of (*E*)-β-farnesene had a deterrent effect, albeit both sexes of this species were responsive to each of the synthetic acetates in the laboratory (Alagarmalai et al. 2009). It needs to be pointed out that the host fruit acetates apparently are acting as a plant allelochemic, if they are released naturally, in the insect-plant interaction. As to whether these fruit volatiles when released act as a plant kairomone or synomone warrants further in depth chemo-ecological investigation.

3.5 Rhagoletis

No male lure has yet been identified for any *Rhagoletis* species, though male attraction to certain plant volatiles has been reported (Light and Jang 1996). Therefore, this investigation represents a potentially productive avenue for future research.

3.6 Toxotrypana

Other than the identified pheromone (see Sect. 2.2.6), there are few other attractants used routinely for detection of *T. curvicauda* in the field. Early studies on the behavior of *T. curvicauda* (Sharp and Landolt 1984) suggested that, unlike most tephritids, this species is not readily attracted to proteinaceous food baits. They further reported that both brown and white sugar had some attraction. Landolt and Reed (1990) reported oviposition attraction of females to green papaya host fruit and suggested that host odors may influence oviposition behavior. More recently, Castrejón-Gómez et al. (2004) tested brown sugar and pineapple juice as two low cost attractants for use in field trapping of *T. curvicauda*. The success of the pheromone for use in trapping of this species has limited the search for a true parapheromone or kairomone for use in applied trapping programs.

4 Conclusion

For most pestiferous species of tephritid fruit flies, aggregation and sex pheromones have limited usage in trapping and control owing to a multitude of factors, including multi-component composition of pheromone, chemo-structural complexity of each component, high cost of synthesis, low effectiveness when compared to male attractants/lures or food attractants, and other abiotic factors related to blending, chemical stability, changes in vapor pressure, and release ratio of a multicomponent bait/pheromone in the field. In contrast, male lures, particularly ME, CL, and TML, have been extensively and successfully used as bait in trapping and control of *Bactrocera* and *Ceratitis* species, respectively. Moreover, with the knowledge gained via behavioral and chemo-ecological studies, the exposure of sterile males to certain lures to enhance their mating competitiveness in the field is now gaining ground in area-wide SIT programs. There are some issues, which need to be resolved amicably, related to trapping when comparing (i) formulated and unformulated (e.g., liquid versus solid) male lures conducted in different regions/ countries, (ii) effectiveness of different colored traps, especially against clear traps, or (iii) individual against a mixture of attractants used as a trap-bait. Also, the urgency of identifying a replacement for a very potent natural male attractant (ME) deemed to be carcinogenic deserves serious consideration. Further research should also be conducted to seek new male lures from plants, like zingerone, that can attract non-responsive species to the commonly used and known male attractants. We are confident that there are a few more attractants for fruit flies, especially for the genera of *Bactrocera* and *Dacus,* may be isolated and identified through proper and in depth behavioral and chemo-ecological investigations, especially via understanding the probable co-evolution between plants and fruit flies.

Acknowledgements We are grateful to our many collaborators and colleagues, namely, A.K.W. Hee, J. Hendrichs, N.H.S. Howcroft, C.C.H. Khoo, D.O. McInnis, N. Papadopoulos, P.T. Ong, H. Ono, R. Vargas, S.L. Wee, for their kind assistance and support and/or interest in fruit flies research.

References

Alagarmalai J, Nestel D, Dragushich D, Nemny-Lavy E, Anshelevich L, Zada A, Soroker V (2009) Identification of host attractants for the Ethiopian fruit fly, *Dacus ciliatus* Loew. J Chem Ecol 35:542–551

Alexander BH, Beroza M, Oda TA, Steiner LF, Miyashita DH, Mitchell WC (1962) The development of male melon fly attractants. J Agric Food Chem 10:270–276

Allwood A (2000) Regional approach to management of fruit flies in the Pacific island countries and territories. In: Tan KH (ed) Area-wide control of fruit flies and other insect pests. Penerbit Universiti Sains Malaysia, Pulau Pinang, pp 439–448

Aluja M (1993) Unusual calling behavior of *Anastrepha robusta* flies (Diptera: Tephritidae) in nature. Fla Entomol 76:391–395

Aluja M (1994) Bionomics and management of *Anastrepha.* Annu Rev Entomol 39:155–178

Aluja M, Hendrichs J, Cabrera M (1983) Behavior and interactions between *Anastrepha ludens* (L) and *A. obliqua* (M) on a field-caged mango tree. I. Lekking behavior and male territoriality. In: Cavalloro R (ed) Fruit flies of economic importance. A.A. Balkema, Rotterdam, pp 122–133

Aluja M, Cabrera M, Guillen J, Celedonio H, Ayora F (1989) Behaviour of *Anastrepha ludens*, *A. obliqua*, and *A. serpentina* (Diptera: Tephritidae) on a wild mango tree (*Mangifera indica*) harbouring three McPhail traps. Insect Sci Appl 10:309–381

Aluja M, Jácome I, Birke A, Lozada N, Quintero G (1993) Basic patterns of behavior in wild *Anastrepha striata* (Diptera: Tephritidae) flies under field-cage conditions. Ann Entomol Soc Am 86:776–793
Aluja M, Piñero J, Jácome I, Diaz-Fleischer F, Sivinski J (2000) Behavior of flies in the genus *Anastrepha* (Trypetinae: Toxotrypanini). In: Aluja M, Norrbom AL (eds) Fruit flies (Tephritidae): phylogeny and evolution of behavior. CRC Press, Boca Raton, pp 375–406
Aluja M, Pérez-Staples D, Sivinski J, Sánchez A, Piñero J (2008) Effects of male condition on fitness in two tropical tephritid flies with contrasting life histories. Anim Behav 76:1997–2009
Arakaki N, Kuba H, Soemori H (1984) Mating behavior of the oriental fruit fly, *Dacus dorsalis* Hendel (Diptera: Tephritidae). Appl Entomol Zool 19:42–51
Ayer WA, Singer PP (1980) Phenolic metabolites of the bird's nest fungus *Nidula* nivetotomentosa. Phytochemistry 19:2717–2721
Back EA, Pemberton CE (1918) The Mediterranean fruit fly. USDA Bull 640, 43 pp
Baker R, Bacon AJ (1985) The identification of spiroacetals in the volatile secretions of two species of fruit fly *(Dacus dorsalis, Dacus cucurbitae*). Experientia 41:1484–1485
Baker JD, Heath RR (1993) NMR spectral assignment of lactone pheromone components emitted by Caribbean and Mexcian fruit flies. J Chem Ecol 19:1511–1519
Baker R, Herbert R, Howse PE, Jones OT, Franke W, Reith W (1980) Identification and synthesis of the major sex pheromone of the olive fly, *Dacus oleae*. J Chem Soc Chem Commun 1:52–53
Baker R, Herbert RH, Lomer RA (1982a) Chemical components of the rectal gland secretions of male *Dacus cucurbitae*, the melon fly. Experientia 38:232–233
Baker R, Herbert RH, Parton AH (1982b) Isolation and synthesis of 3- and 4-hydroxy-1,7dioxaspiro[5.5]undecanes from the olive fly (*Dacus oleae*). J Chem Soc Chem Commun 11:601–603
Baker R, Herbert RH, Grant GG (1985) Isolation and identification of the sex pheromone of the Mediterranean fruit fly, *Ceratitis capitata* (Wied.). J Chem Soc Chem Commun 12:824–825
Baker PS, Hendrichs J, Liedo P (1988) Improvement of attractant dispensing systems for the Mediterranean fruit fly (Diptera: Tephritidae) sterile release program in Chiapas. Mexico J Econ Entomol 81:1068–1072
Baker PS, Howse PE, Ondarza RN, Reyes J (1990) Field trials of synthetic sex pheromone components of the male Mediterranean fruit fly (Diptera: Tephritidae) in southern Mexico. J Econ Entomol 83:2235–2245
Barthel WF, Green N, Keiser I, Steiner LF (1957) Anisylacetone, synthetic attractant for male melon fly. Science 126:654
Battiste MA, Strekowski L, Vanderbilt DP, Visnick M, King RW, Nation JL (1983) Anastrephin and epianastrephin, novel lactone components from the sex pheromone blend of male Caribbbean and Mexican fruit flies. Tetrahedron Lett 24:2611–2614
Bellas TE, Fletcher BS (1979) Identification of the major components in the secretion from the rectal pheromone glands of the Queensland fruit flies *Dacus tryoni* and *Dacus neohumeralis* (Diptera: Tephritidae). J Chem Ecol 5:795–803
Benelli G, Bonsignori G, Stefanini C, Raspi A, Canale A (2013) The production of female sex pheromone in *Bactrocera oleae* (Rossi) young males does not influence their mating chances. Entomol Sci 16:47–53
Beroza M, Green N (1963) Materials tested as insect attractants. USDA-ARS, Agriculture Handbook No. 239
Beroza M, Sarmiento R (1964) Gas chromatographic analysis of insect attractants: trimedlure, medlure, siglure, butyl sorbate, cuelure, methyleugenol. J Assoc Off Agric Chem 47:838–841
Beroza M, Alexander BH, Steiner LF, Mitchell WC, Miyashita DH (1960) New synthetic lures for male melon fly. Science 131:1044–1045
Beroza M, Green N, Gertler SI (1961) New attractants for the Mediterranean fruit fly. J Agric Food Chem 9:361–365
Briceño RD, Eberhard WG, Quilici S (2005) Comparative allometry and sexual behavior of four fruit fly species in the tribe Ceratitidini (Diptera: Tephritidae). J Kansas Entomol Soc 78:20–33

Broumas T, Haniotakis GE (1994) Comparative field studies of various traps and attractants of the olive fruit fly, *Bactrocera oleae*. Entomol Exp Appl 73:145–150

Burditt AK Jr (1975) Factors affecting rate of loss of trimedlure used to bait traps for fruit flies in Florida. Fla Entomol 57:371–376

Burditt AK Jr., McGovern TP (1979) Chemicals tested as attractants for the Caribbean fruit fly, 1972–78 [*Anastrepha suspensa*]. Agric Rev Manual ARM-S, 47pp

Burk T (1983) Behavioral ecology of mating in the Caribbean fruit fly, *Anastrepha suspensa* (Loew) (Diptera: Tephritidae). Fla Entomol 66:330–344

Burk T (1984) Male-male interactions in Caribbean fruit flies, *Anastrepha suspensa* (Loew) (Diptera: Tephritidae): territorial fights and signaling stimulation. Fla Entomol 67:542–547

Buttery RG, Xu CJ, Ling LC (1985) Volatile components of wheat leaves (and stems): possible insect attractants. J Agric Food Chem 33:115–117

Cáceres C, Segura DF, Vera MT, Wornoayporn V, Cladera JL, Teal P, Sapountzis P, Bourtzis K, Zacharopoulou A, Robinson AS (2009) Incipient speciation revealed in *Anastrepha fraterculus* (Diptera; Tephritidae) by studies on mating compatibility, sex pheromones, hybridization, and cytology. Biol J Linn Soc 97:152–165

Canale A, Carpita A, Conti B, Canovai R, Raspi A (2012) Effect of age on 1,7-dioxaspiro-[5,5]-undecane production in both sexes of olive fruit fly, *Bactrocera oleae* (Rossi) (Diptera: Tephritidae). IOBC Bull 74:219–225

Carpita A, Canale A, Raffaeli A, Saba A, Benelli G, Raspi A (2012) (Z)-9-tricosene identified in rectal gland extracts of *Bactrocera oleae* males: first evidence of a male-produced female attractant in olive fruit fly. Naturwissenschaften 99:77–81

Casaña-Giner V, Primo-Millo J (1999) Identification of volatiles from *Ceratitis capitata* (Wiedemann) (Diptera: Tephritidae) by means of thermal desorption of solid samples coupled to GC-MS. IOBC/WPRS Bull 22:21–26

Casaña-Giner V, Gandía-Balaguer A, Hernández-Alamós MM, Mengod-Puerta C, Garrido-Vivas-A, Primo-Millo J, Primo-Yúfera E (2001) Attractiveness of 79 compounds and mixtures to wild *Ceratitis capitata* (Diptera: Tephritidae) in field trials. J Econ Entomol 94:898–904

Casaña-Giner V, Oliver JE, Jang EB, Carvalho L (2003a) Syntheses and behavioral evaluations of fluorinated and silylated analogs of raspberry ketone as attractants for the melon fly, *Bactrocera cucurbitae* (Coquillett). J Entomol Sci 38:111–119

Casaña-Giner V, Oliver JE, Jang EB, Carvalho L, Khrimian A, DeMilo AB, McQuate GT (2003b) Raspberry ketone formate as an attractant for the melon fly (Diptera: Tephritidae). J Entomol Sci 38:120–126

Castrejón-Gómez VR, Aluja M, Arzuffi R, Villa P (2004) Two low-cost food attractants for capturing *Toxotrypana curvicauda* (Diptera: Tephritidae) in the field. J Econ Entomol 97:310–315

Castrejón-Gómez VR, Láscares S, Malo EA, Toldeo J, Rojas JC (2007) Calling behavior of mass-reared and wild *Anastrepha serpentina* (Diptera: Tephritidae). J Econ Entomol 100:1173–1179

Chambers DL (1977) Attractants for fruit fly survey and control. In: Shorey HH, McKelvey JJ Jr (eds) Chemical control of insect behavior – theory and application. Wiley, New York, pp 327–344

Chen CC, Dong YJ, Li CT, Liu KY (2006) Movement of the oriental fruit fly, *Bactrocera dorsalis* (Hendel) (Diptera: Tephritidae), in a guava orchard with special reference to its population changes. Formosan Entomol 26:143–159

Choi YK, Cho G-S, Hwang S, Kim BW, Lim JH, Lee J-C, Kim HC, Kim W-K, Kim YS (2010) Methyleugenol reduces cerebral ischemic injury by suppression of oxidative injury and inflammation. Free Radical Res 44:925–935

Chuah CH, Ooi LE, Yong HS, Goh SH (1996) New male fruitfly attractants from spadix of *Spathiphyllum cannaefolium* (Araceae). Biochem Syst Ecol 24:189–193

Chuman T, Landolt PJ, Heath RR, Tumlinson JH (1987) Isolation, identification, and synthesis of male-produced sex pheromone of papaya fruit fly, *Toxotrypana curvicauda* Gerstaecker (Diptera: Tephritidae). J Chem Ecol 13:1979–1992

Chuman T, Sivinski J, Heath RR, Calkins CO, Tumlinson JH, Battiste MA, Wydra RL, Strkowski L, Nation JL (1988) Suspensolide, a new macrolide component of male Caribbean fruit fly [*Anastrepha suspensa* (Loew)] volatiles. Tetrahedron Lett 29:6561–6564

Corey EJ, Watt DS (1973) A total synthesis of (±)-α and (±)-β-copaenes and ylangenes. J Am Chem Soc 95:2303–2311

Cossé AA, Todd JL, Millar JG, Martínez LA, Baker TC (1995) Electroantennographic and coupled gas chromatographic-electroantennographic responses of the Mediterranean fruit fly, *Ceratitis capitata*, to male-produced volatiles and mango odor. J Chem Ecol 21:1823–1836

Cunningham RT (1989) Parapheromones. In: Robinson AS, Hooper G (eds) Fruit flies: their biology, natural enemies and control, vol 3A. Elsevier, Amsterdam, pp 221–230

Da Silva MT, Polloni YJ, Bressan S (1985) Mating behavior of some fruit flies of the genus *Anastrepha* Schiner, 1868 (Diptera: Tephritidae) in the laboratory. Rev Bras Entomol 29:155–164

De Almeida LM, Araújo A, Mendes NHD, de Souza JMGA, Menezes AAL (2011) The temporal pattern of mating behavior of the fruit fly, *Anastrepha zenildae*, in the laboratory. J Insect Sci 11(151)

De Meyer M (2000) Phylogeny of the genus *Ceratitis* (Dacinae: Ceratitidini). In: Aluja M, Norrbom AL (eds) Fruit flies (Tephritidae): phylogeny and evolution of behavior. CRC Press, Baca Raton, pp 409–428

Delrio G, Ortu S (1987) Attraction of *Ceratitis capitata* to sex pheromones, trimedlure, ammonium, and protein bait traps. Bull-SROP 11:20–25

DeMilo AB, Cunningham RT, McGovern TP (1994) Trimedlure: effects of structural modifications on its attractiveness to Mediterranean fruit fly males (Diptera: Tephritidae). J Econ Entomol 87:1494–1501

Dodson G (1982) Mating and territoriality in wild *Anastrepha suspensa* (Diptera: Tephritidae) in field cages. J Entomol Soc 17:189–200

Drew RAI (1974) The responses of fruit fly species (Diptera: Tephritidae) in the South Pacific area to male attractants. J Aust Entomol Soc 13:267–270

Drew RAI, Hooper GHS (1981) The responses of fruit fly species (Diptera: Tephritidae) in Australia to various attractants. J Aust Entomol Soc 20:201–205

Drew RAI, Tsuruta K, White IM (2005) A new species of pest fruit fly (Diptera: Tephritidae: Dacinae) from Sri Lanka and Africa. Afr Entomol 13:149–154

Elzen GW, Williams HJ, Bell AA, Stipanovic RD, Vinson SB (1985) Quantification of volatile terpenes of glanded and glandless *Gossypium hirsutum* L. cultivars and lines by gas chromatography. J Agric Food Chem 33:1079–1082

Enomoto H, Ishida T, Nishida R (2010) 3-Oxygenated α-ionone derivatives as potent male attractants for the solanaceous fruit fly, *Bactrocera latifrons* (Diptera: Tephritidae), and sequestered metabolites in the rectal gland. Appl Entomol Zool 45:551–556

Epsky ND, Heath RR (1993a) Food availability and pheromone production by males of *Anastrepha suspensa* (Diptera: Tephritidae). Environ Entomol 22:942–947

Epsky ND, Heath RR (1993b) Pheromone production by male *Anastrepha suspensa* (Diptera: Tephritidae) under natural light cycles in greenhouse studies. Environ Entomol 22:464–469

Epsky ND, Kendra PE, Peña J, Heath RR (2011) Comparison of synthetic food-based lures and liquid protein baits for capture of *Anastrepha suspensa* (Diptera: Tephritidae) adults. Fla Entomol 94:18–185

Fay HAC (2010) Exploring structure-activity relationships in the phenylpropanoids to procure new male lures for non-responsive *Bactrocera* and *Dacus*. In: Sabater-Munoz B, Navarro Llopis V, Urbaneja A (eds) Proceedings of the 8th international symposium on fruit flies of economic importance. Laimpreta CG, Valencia, pp 270–280

Fay HAC (2012) A highly effective and selective male lure for *Bactrocera jarvisi* (Tryon) (Diptera: Tephritidae). Aust J Entomol 51:189–197

Féron M (1959) Attraction chimique du male de *Ceratitis capitata* Wied. (Dipt. Trypetidae) pour la femelle. CR Acad Sci Ser D Sci Nat (Paris) 248:2403–2404

Féron M (1962) L'instinct de reproduction chez la mouche Mediterranean des fruits *Ceratitis capitata* Wied. (Dipt. Trypetidae). Comportement sexuel. Comportement de ponte. Rev Path Veg Entomol Agric Fr 41:1–129

Fitt GP (1981a) Responses by female Dacinae to "male" lures and their relationship to patterns of mating behavior and pheromone response. Entomol Exp Appl 29:87–97

Fitt GP (1981b) The influence of age, nutrition and time of day on the responsiveness of male *Dacus opiliae* to the synthetic lure, methyl eugenol. Entomol Exp Appl 30:83–90

Flath RA, Jang EB, Light DM, Mon TR, Mon TR, Carvalho L, Binder RG, John JO, John JO (1993) Volatile pheromonal emissions from the male Mediterranean fruit fly: effects of fly age and time of day. J Agric Food Chem 41:830–837

Flath RA, Cunningham RT, Liquido NJ, McGovern TP (1994a) Alpha-ionol as attractant for trapping *Bactrocera latifrons* (Diptera: Tephritidae). J Econ Entomol 87:1470–1476

Flath RA, Cunningham RT, Mon TR, John JO (1994b) Additional male Mediterranean fruit fly (*Ceratitis capitata* Wied.) attractants from angelica seed oil (*Angelica archangelica* L.). J Chem Ecol 20:1969–1984

Flath RA, Cunningham RT, Mon TR, John JO (1994c) Male lures for Mediterranean fruit fly (*Ceratitis capitata* Wied.): structural analogs of α-copaene. J Chem Ecol 20:2595–2609

Fletcher BS (1974) The ecology of a natural population of the Queensland fruit fly, *Dacus tryoni*. V. Aust J Zool 22:189–202

Fletcher BS (1989) Movements of tephritid fruit flies. In: Robinson AS, Hooper G (eds) Fruit flies: their biology, natural enemies and control. World crop pests, vol 3B. Elsevier Science Publishers, Amsterdam, pp 209–219

Fornasiero U, Guitto A, Caporale G, Baccichetti R, Musajo L (1969) Identification of the attractant of *Ceratitis capitata* males contained in the *Angelica archangelica* seed oil. Gazzetta Chimica Italiana 99:700–710

Froerer KM, Peck SL, McQuate GT, Vargas RI, Jang EB, McInnis DO (2010) Long-distance movement of *Bactrocera dorsalis* (Diptera: Tephritidae) in Puna, Hawaii: how far can they go? Am Entomol 56:88–94

Gariboldi PG, Jommi G, Rossi R, Vita G (1982) Studies on the chemical constitution and sex pheromone activity of volatile substances emitted by *Dacus oleae*. Experientia 38:441–444

Gonçalves GB, Silva CE, Dos Santos JCG, Dos Santos ES, Do Nascimento RR, da Silva EL, de Lima Mendonça A, do Rosário Tenório de Freitas M, Sant'Ana AEG (2006) Comparison of the volatile components released by calling males of *Ceratitis capitata* (Diptera: Tephritidae) with those extractable from the salivary glands. Fla Entomol 89:375–379

Gonzalez J, Troncoso P (2007) The fruit fly exclusion programme in Chile. In: Vreysen MJB, Robinson AS, Hendrichs J (eds) Area-wide control of insect pests: from research to field implementation. Springer, Dordrecht, pp 641–651

Guiotto AA, Fornasiero UU, Baccichetti FF (1972) Investigations on attractants for males of *Ceratiti capitata*. Farmaco Sci 27:663–669

Habu N, Iga M, Numazawa K (1980) Progress of the eradication program of the oriental fruit fly, *Dacus dorsalis* Hendel, on the Ogasawara (Bonin) Islands. In: Proceedings of a symposium on fruit fly problems. National Institute of Agricultural Sciences, Tsukuba, pp 123–131

Habu N, Iga M, Numazawa K (1984) An eradication program of the oriental fruit fly, *Dacus dorsalis* Hendel (Diptera: Tephritidae), in the Ogasawara (Bonin) Islands. I. Eradication field test using a sterile fly release method on small islets. Appl Entomol Zool 19:1–7

Han P, Wang X, Niu CY, Dong YC, Zhu JQ, Desneux N (2011) Population dynamics, phenology, and overwintering of *Bactrocera dorsalis* (Diptera: Tephritidae) in Hubei Province. China J Pest Sci 84:289–295

Hancock DL (1985) A specific male attractant for the melon fly, *Dacus vertebratus*. Zimbabwe Sci News 19:118–119

Hancock DL, Osborne R, Broughton S, Gleeson P (2000) Eradication of *Bactrocera papayae* (Diptera: Tephritidae) by male annihilation and protein baiting in Queensland, Australia. In: Tan KH (ed) Area-wide control of fruit flies and other insect pests. Penerbit USM, Pinang, pp 381–388

Haniotakis GE (1974) Sexual attraction in the olive fruit fly, *Dacus oleae* (Gmelin). Environ Entomol 3:82–86

Haniotakis GE, Vassiliou-Waite A (1987) Effect of combining food and sex attractants on the capture of *Dacus oleae* flies. Entomol Hellenica 5:27–33

Haniotakis G, Francke W, Mori K, Redlich H, Schurig V (1986a) Sex-specific activity of I-(−) and (S)-(+)-1,7-dioxaspiro[5.5]undecane, the major pheromone of *Dacus oleae*. J Chem Ecol 12:1559–1568

Haniotakis G, Kozyrakis E, Bonatsos C (1986b) Control of the olive fruit fly, *Dacus oleae* Gmel. (Dipt., Tephritidae) by mass trapping: pilot scale feasibility study. J Appl Entomol 101:343–352

Haniotakis G, Kozyrakis M, Fitsakis T, Antonicakj A (1991) An effective mass trapping method for the control of *Dacus oleae* (Diptera: Tephritidae). J Econ Entomol 84:564–569

Hardy DE (1979) Review of economic fruit flies of the South Pacific region. Pac Insect 20:429–432

Harris EJ, Takara JM, Nishida T (1986) Distribution of the melon fly, *Dacus cucurbitae* (Diptera: Tephritidae), and host plants on Kauai, Hawaiian Islands. Environ Entomol 15:488–493

Heath RR, Chambers DL, Tumlinson JH, Landolt PJ (1990) Controlled release of trimedlure isomer C from a compressed disk and the evaluation of its attractiveness to the Mediterranean fruit fly (Diptera: Tephritidae). J Econ Entomol 83:819–822

Heath RR, Landolt PJ, Tumlinson JH, Chambers DL, Murphy RE, Doolittle RE, Dueben BD, Sivinski J, Calkins CO (1991) Analysis, synthesis, formulation, and field testing of three major components of male Mediterranean fruit fly pheromone. J Chem Ecol 17:1925–1940

Heath RR, Manukian A, Epsky ND, Sivinski J, Calkins CO, Landolt PJ (1993) A bioassay system for collecting volatiles while simultaneously attracting tephritid fruit flies. J Chem Ecol 19:2395–2410

Heath RR, Epsky ND, Jimenez A, Dueben BD, Landolt PJ, Meyer WL, Aluja M, Rizzo J, Camino M, Jeronimo F, Baranowski RM (1996) Improved pheromone-based trapping systems to monitor *Toxotrypana curvicauda* (Diptera: Tephritidae). Fla Entomol 79:37–48

Heath RR, Landolt PJ, Robacker DC, Dueben BD, Epsky ND (2000) Sexual pheromones of tephritid flies: clues to unravel phylogeny and behavior. In: Aluja M, Norrbom AL (eds) Fruit flies (Tephritidae): phylogeny and evolution of behavior. CRC Press, Boca Raton, pp 793–809

Heathcock CH (1966) The total synthesis of (±)-α-copaene and (±)-8-isocopaene. J Am Chem Soc 88:4110–4112

Heathcock CH, Badger RA, Patterson JW Jr (1967) Total synthesis of (±)-α-copaene and (±)-ylangene. A general method for the synthesis of tricycle[4.4.0.02,7]decanes. J Am Chem Soc 89:4133–4145

Hee AKH, Tan KH (1998) Attraction of female and male *Bactrocera papayae* to conspecific males fed with methyl eugenol and attraction of females to male sex pheromone components. J Chem Ecol 24:753–764

Henning F, Matioli SR (2006) Mating time of the West Indian fruit fly *Anastrepha obliqua* (Macquart) (Diptera: Tephritidae) under laboratory conditions. Neotrop Entomol 35:145–148

Hill AR (1987) Comparison between trimedlure and capilure® – attractants for male *Ceratitis capitata* (Wiedemann) (Diptera: Tephritidae). J Aust Entomol Soc 26:35–36

Hiramoto MK, Arita-Tsutsumi L, Jang E (2006) Test of effectiveness of newly formulated plastic matrix with methyl eugenol for monitoring *Bactrocera dorsalis* (Hendel) populations. Proc Hawaii Entomol Soc 38:103–110

Hirvi T, Honkenen E (1984) The aroma of the fruit of sea buckthorn. *Hippophae rhamnoides* L. Z Lebensm-Unters Forsch 179:387–388

Hirvi T, Honkanen E, Pyysalo T (1981) The aroma of cranberries. Z Lebensm-Unters Forsch 172:365–367

Hooper GHS (1978) Effect of combining methyl eugenol and cuelure on the capture of male tephritid fruit flies. J Aust Entomol Soc 17:189–190

Howlett FM (1912) The effects of oil of citronella on two species of *Dacus*. Trans R Entomol Soc Lond 60:412–418

Howlett FM (1915) Chemical reactions of fruit flies. Bull Entomol Res 6:297–305

Howse PE, Knapp JJ (1996) Pheromones of Mediterranean fruit fly: presumed mode of action and implications for improved trapping techniques. In: McPheron BA, Steck GJ (eds) Fruit fly pests: a world assessment of their biology and management. St. Lucie Press, Delray Beach, pp 91–99

Hsu JC, Liu PF, Hertlein M, Mau RFL, Feng HT (2010) Greenhouse and field evaluation of a new male annihilation technique (MAT) product, SPLAT-MAT spinosad ME^{TM}, for the control of oriental fruit flies (Diptera: Tephritidae) in Taiwan. Formosan Entomol 30:87–101

IAEA (International Atomic Energy Agency) (2003) Trapping guidelines for area-wide fruit fly programmes. IAEA, Vienna, 47 pp

Iannotta N, Perri L, Rinaldi R (1994) Control of the olive fly by mass trapping in Calabria. Acta Hort 356:411–413

Ibañez-López A, Cruz-López L (2001) Glandulas salivales de *Anastrepha obliqua*(Macquart) (Diptera: Tephritidae): analisis quimico y morfologico, y actividad biologica de los components volatiles. Folia Entomol Mex 40:221–231

Ishida T, Enomoto H, Nishida R (2008) New attractants for males of the solanaceous fruit fly *Bactrocera latifrons*. J Chem Ecol 34:1532–1535

Itô Y, Iwahashi O (1974) Ecological problems associated with an attempt to eradicate *Dacus dorsalis* (Tephritidae: Diptera) from the southern islands of Japan with a recommendation on the use of the sterile-male technique. In: Proceedings of a panel on the sterile-insect technique and its field applications, Vienna, IAEA-PL-494/5

Itô Y, Murai M, Teruya T, Hamada R, Sugimoto A (1974) An estimation of population density of *Dacus cucurbitae* with mark-recapture methods. Res Popul Ecol 15:213–222

Iwahashi O (1972) Movement of the oriental fruit fly adults among islets of the Ogasawara Islands. Environ Entomol 1:176–179

Jacobson M, Ohinata K (1980) Unique occurrence of fenchol in the animal kingdom. Experientia 36:629–630

Jacobson M, Ohinata K, Chambers DL, Jones WA, Fujimoto MS (1973) Insect sex attractants. 13. Isolation, identification, and synthesis of sex pheromones of the male Mediterranean fruit fly. J Med Chem 16:248–251

Jacobson M, Uebel EC, Lusby WR, Waters RM (1987) Optical isomers of α-copaene derived from several plant sources. J Agric Food Chem 35:798–800

Jang EB (1995) Effects of mating and accessory gland injections on olfactory-mediated behavior in the female Mediterranean fruit fly, *Ceratitis capitata*. J Insect Physiol 41:705–710

Jang EB (2011) Effectiveness of plastic matrix lures and traps against *Bactrocera dorsalis* and *Bactrocera cucurbitae* in Hawaii. J Appl Entomol 135:456–466

Jang EB, Light DM (1996) Attraction of female Mediterranean fruit flies to identified components of the male-produced pheromone: qualitative aspects of major, intermediate, and minor components. In: McPheron BA, Steck GJ (eds) Fruit fly pests: a world assessment of their biology and management. St. Lucie Press, Delray Beach, pp 115–121

Jang EB, Light DM, Dickens JC, McGovern TP, Nagata JT (1989a) Electroantennogram responses of Mediterranean fruit fly, *Ceratitis capitata* (Diptera: Tephritidae), to trimedlure and its *trans* isomers. J Chem Ecol 15:2219–2231

Jang EB, Light DM, Flath RA, Nagata JT, Mon TR (1989b) Electroantennogram responses of the Mediterranean fruit fly, *Ceratitis capitata*, to identified volatile constituents from calling males. Entomol Exp Appl 50:7–19

Jang EB, Light DM, Binder RG, Flath RA, Carvalho LA (1994) Attraction of female Mediterranean fruit flies to the five major components of male-produced pheromone in a laboratory flight tunnel. J Chem Ecol 20:9–20

Jang EB, McInnis DO, Lance DR, Carvalho LA (1998) Mating-induced changes in olfactory-mediated behavior of laboratory-reared normal, sterile, and wild female Mediterranean fruit flies (Diptera: Tephritidae) mated to conspecific males. Ann Entomol Soc Am 91:139–144

Jang EB, Raw AS, Carvalho LA (2001) Field attraction of Mediterranean fruit fly, *Ceratitis capitata* (Wiedemann) to synthetic stereoselective enantiomers of the ceralure B1 isomer. J Chem Ecol 27:235–242

Jang EB, Holler T, Cristofaro M, Lux S, Raw AS, Moses AL, Carvalho LA (2003) Improved attractants for areawide detection and control of Mediterranean fruit fly, *Ceratitis capitata* (Weidemann): responses of sterile and wild flies to (−) enantiomer of ceralure B1. J Econ Entomol 96:1719–1723

Jang EB, Khrimian A, Holler T, Casana-Giner V, Lux S, Carvalho L (2005) Field response of Mediterranean Fruit Fly (Diptera: Tephritidae) to ceralure B1: evaluations of enantiomeric B1 ratios on fly captures. J Econ Entomol 98:1139–1143

Jang EB, Casaña-Giner V, Oliver JE (2007) Field captures of wild melon fly (Diptera: Tephritidae) with an improved male attractant, raspberry ketone formate. J Econ Entomol 100:1124–1128

Jang EB, Khriman A, Holler T (2010) Field response of Mediterranean fruit flies to ceralure B1 relative to most active isomer and commercial formulation of trimedlure. J Econ Entomol 103:1586–1593

Jang EB, Khirimian A, Siderhurst NS (2011) Di- and tri-fluorinate analogs of methyl eugenol: attraction to and metabolism in the oriental fruit fly, *Bactrocera dorsalis* (Hendel). J Chem Ecol 37:555–564

Jang EB, Ramsey A, Carvalho LA (2013) Performance of methyl eugenol + matrix + toxicant combinations under field conditions in Hawaii and California for trapping *Bactrocera dorsalis* (Diptera: Tephritidae). J Econ Entomol 106:727–734

Jessup AJ, Dominiak B, Woods B, De Lima CPF, Tomkins A, Smallridge CJ (2007) Area-wide management of fruit flies in Australia. In: Vreysen MJB, Robinson AS, Hendrichs J (eds) Area-wide control of insect pests: from research to field implementation. Springer, Dordrecht, pp 685–697

Jones OT (1989) World crop pests. In: Robinson AS, Hooper G (eds) World crop pests, vol. 3A. Fruit flies, their biology, natural enemies and control. Elsevier, Amsterdam, pp 179–183

Jones OT, Lisk SC, Longurst G, Howse PE, Ramos P, Campos M (1983) Development of a monitoring trap for the olive fly *Dacus oleae* (Gmelin) (Diptera: Tephritidae) using a component of its pheromone as lure. Bull Entomol Res 73:97–106

Kamala Jayanthi PD, Verghese A (2011) Host-plant phenology and weather based forecasting models for population prediction of the oriental fruit fly, *Bactrocera dorsalis* Hendel. Crop Prot 30:1557–1562

Kapoor VC (2005/2006) Taxonomy and biology of economically important fruit flies of India. Biotaxonomy of Tephritoidea. Isr J Entomol 35–36:459–475

Katsoyannos BI (1975) Oviposition-deterring, male-arresting, fruit-marking pheromone in *Rhagoletis cerasi*. Environ Entomol 4:801–807

Katsoyannos BI (1976) Female attraction to males in *Rhagoletis cerasi*. Environ Entomol 5:474–476

Katsoyannos BI (1982) Male sex pheromone of *Rhagoletis cerasi* L. (Diptera: Tephritidae): factors affecting release and response and its role in the mating behavior. Z Ang Entomol 94:187–198

Keiser I, Nakagawa S, Kobayashi RM, Chambers DL, Urago T, Doolittle RE (1973) Attractiveness of cue-lure and the degradation product 4-(p-hydroxyphenyl)-2-butanone to male melon flies in the field in Hawaii. J Econ Entomol 66:112–114

Khoo CCH, Tan KH (2000) Attraction of both sexes melon fruit fly, *Bactrocera cucurbitae*, to conspecific males – a comparison after pharmacophagy of cue-lure and a new attractant – zingerone. Entomol Exp Appl 97:312–320

Khoo CCH, Yuen K, Tan KH (2000) Attraction of female *Bactrocera papayae* to sex pheromone components with two different release devices. J Chem Ecol 26:2487–2496

Khrimian AP, DeMilo AB, Waters RM, Cunningham RT, Leonhardt BA (1993) Synthesis of attractants for oriental fruit fly *Dacus dorsalis* Hendel using a catalytic organcopper coupling reaction. J Chem Ecol 19:2935–2946

Khrimian AP, DeMilo AB, Waters RM, Liquido NJ, Nicholson MJ (1994) Monofluoro analogs of eugenol methyl ether as novel attractants for the oriental fruit fly. J Org Chem 59:8034–8039

Khrimian A, Margaryan AK, Schmidt WF (2003) An improved synthesis of ethyl cis-5-iodo-trans-2-methylcyclohexanecarboxylate, a potent attractant for the Mediterranean fruit fly. Tetrahedron 59:5475–5480

Khrimian A, Jang EB, Casaña-Giner V (2004) A practical synthesis of ceralure B1, a potent attractant for the Mediterranean fruit fly. In: Barnes BN (ed) Proceedings of the 6th international symposium on fruit flies of economic importance. Isteg Science Publications, Irene, pp 279–282

Khirimian A, Jang EB, Nagata J, Caravalho L (2006) Consumption and metabolism of 1,2-dimethoxy-4-(3-fluoro-2-propenyl)benzene, a fluorine analog of methyl eugenol, in the oriental fruit fly, *Bactrocera dorsalis* (Hendel). J Chem Ecol 32:1513–1526

Khrimian A, Siderhurst MS, Mcquate GT, Liquido NJ, Nagata J, Carvalho L, Guzman F, Jang EB (2009) Ring-fluorinated analog of methyl eugenol: attractiveness to and metabolism in the oriental fruit fly, *Bactrocera dorsalis* (Hendel). J Chem Ecol 35:209–218

Kim HM, Lee EH, Kim CY, Chung JG, Kim SH, Lim JP, Shin TY (1997) Antianaphylactic properties of eugenol. Pharmacol Res 36:475–480

King JR, Landolt PJ (1984) Rate of loss of trimedlure from cotton wicks under South-Florida field conditions. J Econ Entomol 77:221–224

Kobayashi RM, Ohinata K, Chambers DL, Fujimoto MS (1978) Sex pheromone of the oriental fruit fly and melon fly: mating behavior, bioassay method, and attraction of females by live males and by suspected glands of males. Environ Entomol 7:107–112

Kohama T, Kuba H (1996) Movement of sterile melon flies in Okinawa, Japan. In: McPheron BA, Steck GJ (eds) Fruit fly pests: a world assessment of their biology and management. St. Lucie Press, Delray Beach, pp 415–423

Krosch MN, Schutze MK, Armstrong KF, Boontop Y, Boykin LM, Chapman TA, Englezou A, Cameron SL, Clarke AR (2013) Piecing together an integrative taxonomic puzzle: microsatellite, wing shape and aedeagus length analyses of *Bactrocera dorsalis* s.l. (Diptera: Tephritidae) find no evidence of multiple lineages in a proposed contact zone along the Thai/Malay Peninsula. Syst Entomol 38:2–13

Kuba H, Koyama J (1985) Mating behavior of wild melon flies, *Dacus cucurbitae* Coquillett (Diptera: Tephritidae) in a field cage: courtship behavior. Appl Entomol Zool 20:365–372

Kulkarni YS, Niva M, Ron E, Snyder BB (1987) Synthesis of terpenes containing the bicycle [3.1.1]heptanes ring system by the intramolecular [2+2] cycloaddition reaction of vinylketenes with alkenes. Preparation of chrysanthenone, β-pinene, β-cis-bergamotene, β-trans-bergamotene, β-copaene, and lemnalol. J Org Chem 52:1568–1576

Kumaran N, Balagawi S, Schutze MK, Clarke AR (2013) Evolution of lure response in tephritid fruit flies: phytochemicals as drivers of sexual selection. Anim Behav 85:781–789

Landolt PJ, Averill AL (1999) Fruit flies. In: Hardie J, Minks AK (eds) Pheromones of non-lepidopteran insects associated with agricultural plants. CABI Publication, Wallingford, pp 3–25

Landolt PJ, Heath RR (1988) Effects of age, mating, and time of day on behavioral responses of female papaya fruit fly, *Toxotrypana curvicauda* Gerstaecker (Diptera: Tephritidae), to synthetic sex pheromone. Environ Entomol 17:47–51

Landolt PJ, Heath RR (1990) Effects of pheromone release rate and time of day on catches of male and female papaya fruit flies (Diptera: Tephritidae) on fruit model traps baited with pheromone. J Econ Entomol 83:2040–2043

Landolt PJ, Heath RR (1996) Development of pheromone-based trapping systems for monitoring and controlling tephritid fruit flies in Florida. In: Rosen D, Capinera J (eds) Pest management in the subtropics: integrated pest management – a Florida perspective. Intercept Ltd., Andover, pp 197–207

Landolt JJ, Hendrichs J (1983) Reproductive behavior of the papaya fruit fly, *Toxotrypana curvicauda* Gerstaecker (Diptera: Tephritidae). Ann Entomol Soc Am 76:413–417

Landolt PJ, Reed HC (1990) Behavior of the papaya fruit fly (Diptera: Tephritidae): host finding and oviposition. Environ Entomol 19:305–1310

Landolt PJ, Sivinski J (1992) Effects of time of day, adult food, and host fruit on incidence of calling by male Caribbean fruit flies (Diptera: Tephritidae). Environ Entomol 21:382–387

Landolt PJ, Heath RR, King JR (1985) Behavioral responses of female papaya fruit flies, *Toxotrypana curvicauda* (Diptera: Tephritidae), to male-produced sex pheromone. Ann Entomol Soc Am 78:751–755

Landolt PJ, Heath RR, Agee HR, Tumlinson JH, Calkins CO (1988) Sex pheromone-based trapping system for papaya fruit fly (Diptera: Tephritidae). J Econ Entomol 81:1163–1169

Landolt PJ, Gonzalez M, Chambers DL, Heath RR (1991) Comparison of field observations and trapping of papaya fruit fly in papaya plantings in Central America and Florida. Fla Entomol 74:408–414

Landolt PJ, Heath RR, Chambers DL (1992a) Oriented flight responses of female Mediterranean fruit flies to calling males, odor of calling males, and a synthetic pheromone blend. Entomol Exp Appl 65:259–266

Landolt PJ, Reed HC, Heath RR (1992b) Attraction of female papaya fruit by (Diptera: Tephritidae) to male pheromone and host fruit. Environ Entomol 21:115–1159

Leblanc L, Roger RI, Mackey B, Putoa R, Pinero JC (2011) Evaluation of cue-lure and methyl eugenol solid lure and insecticide dispensers for fruit fly (Diptera: Tephritidae) monitoring and control in Tahiti. Fla Entomol 94:510–516

Leonhardt BA, McGovern TP, Plimmer JR (1982) Capillary GC analysis of trimedlure, the attractant for the "Medfly". J High Resol Chromatogr Chromatogr Commun 5:430–431

Leonhardt BA, Rice RE, Harte EM, Cunningham RT (1984) Evaluation of dispensers containing trimedlure, the attractant for the Mediterranean fruit fly (Diptera: Tephritidae). J Econ Entomol 77:744–749

Leonhardt BA, Cunningham RT, Rice RE, Harte EM, Hendrichs J (1989) Design, effectiveness, and performance criteria of dispenser formulations of trimedlure, an attractant of the Mediterranean fruit fly (Diptera: Tephritidae). J Econ Entomol 82:860–867

Leonhardt BA, Cunninham RT, Avery JW, DeMilo AB, Warthen JD (1996) Comparison of ceralure and trimedlure attractants for the male Mediterranean fruit fly (Diptera: Tephritidae). J Entomol Sci 31:183–190

Levinson HZ, Levinson AR, Schafer K (1987) Pheromone biology of the Mediterranean fruit fly (*Ceratitis capitata* Wied.) with emphasis on the functional anatomy of the pheromone glands and antennae as well as mating behavior. J Appl Entomol 104:448–461

Light DM, Jang EB (1996) Plant volatiles evoke and modulate tephritid behavior. In: McPheron BA, Steck GJ (eds) Fruit fly pests: a world assessment of their biology and management. St. Lucie Press, Delray Beach, pp 123–133

Lima IS, Howse PE, Salles LAB (1994) Reproductive behavior of the South American fruit fly *Anastrepha fraterculus* (Diptera: Tephritidae): laboratory and field studies. Physiol Entomol 19:271–277

Lima IS, Howse PE, Stevens IDR (1996) Volatile components from the salivary glands of calling males of the South American fruit fly, *Anastrepha fraterculus*: partial identification and behavioural activity. In: McPheron BA, Steck GJ (eds) Fruit fly pests – a world assessment of their biology and management. St. Lucie Press, Delray Beach, pp 107–113

Lima IS, Howse PE, Do Nascimento RR (2001) Volatile substances from male *Anastrepha fraterculus* Wied. (Diptera: Tephritidae): identification and behavioural activity. J Braz Chem Soc 12:196–201

Lin YL, Chow CJ (1984) Studies on the constituents of aerial parts of *Scutellaria rivularis* wall. Kuo Li Chung-Kuo 1 Yao Yen Chiu So Yen Chiu Pao Kao 1984: 141–165 (in Chinese)

Liquido NJ, Cunningham RT, Nakagawa S (1990) Host plants of Mediterranean fruit fly (Diptera: Tephritidae) on the island of Hawaii (1949–1985 survey). J Econ Entomol 83:1863–1878

Liquido NJ, Khirmian AP, DeMilo AB, McQuaate GT (1998) Monofluoroanalogs of methyl eugenol: new attractants for males of *Bactrocera dorsalis* (Hendel) (Dipt., Tephritidae). J Appl Entomol 122:259–264

Long EL, Nelson AA, Fitzhugh OG, Hansen WH (1963) Liver tumors produced in rats by feeding safrole. Arch Pathol 75:595–604

López-Guillén G, Cruz-López L, Malo EA, González-Hernández H, Cazares CL, López-Collado J, Toledo J, Rojas JC (2008) Factors influencing the release of volatiles in *Anastrepha obliqua* males (Diptera: Tephritidae). Environ Entomol 37:876–882

Lu F, Teal PEA (2001) Sex pheromone components in oral secretions and crop of male Caribbean fruit flies, *Anastrepha suspensa* (Loew). Arch Insect Biochem Physiol 48:144–154

Mahmood K (1999) Intraspecific variations in two pest species of the oriental fruit fly *Bactrocera dorsalis* (Hendel) (Tephritidae: Diptera) complex. Pak J Zool 31:315–321

Mahmood K (2004) Identification of pest species in oriental fruit fly, *Bactrocera dorsalis* (Hendel) (Diptera: Tephritidae) species complex. Pak J Zool 36:219–230

Malavasi A, Morgante JS, Prokopy RJ (1983) Distribution and activities of *Anastrepha fraterculus* (Diptera: Tephritidae) flies on host and nonhost tress. Ann Entomol Soc Am 76:286–292

Marco JA, Barbera O, Rodriguez S, Domingo C, Adell J (1988) Flavonoids and other phenolics from *Artemisia hispanica*. Phytochemistry 27:3155–3159

Martelli G (1910) Alcune note intorno ai blique ed ai della mosca dele arance *Ceratitis capitata*. Boll Lab Zool Agr Portici 4:120–127

Matsui M, Nakamori H, Kohama T, Nagamine Y (1990) The effect of male annihilation on a population of wild melon flies, *Dacus cucurbitae* Coquillett (Diptera: Tephritidae) in northern Okinawa. Jap J Appl Entomol Zool 34:315–317

Mavraganis VG, Liaropoulos C, Papdopoulos NT, Kouloussis NA, Broumas T (2008) Whole body extract of Mediterranean fruit fly males elicits high attraction in virgin females. Entomol Exp Appl 127:20–29

Mavraganis VG, Papadopoulos NT, Kouloussis NA (2010) Extract of olive fruit fly males (Diptera: Tephritidae) attract virgin females. Entomol Hellenica 19:14–20

Mazomenos BE (1989) *Dacus oleae*. In: Robinson AS, Hooper G (eds) Fruit flies: their biology, natural enemies and control, vol 3A. Elsevier, Amsterdam, pp 169–178

Mazomenos BE, Haniotakis GE (1981) A multicomponent female sex pheromone of *Dacus oleae* Gmelin: isolation and bioassay. J Chem Ecol 7:437–444

Mazomenos BE, Haniotakis GE (1985) Male olive fruit fly attraction to synthetic sex pheromone components in laboratory and field tests. J Chem Ecol 11:397–405

Mazomenos BE, Pomonis GJ (1983) Male olive fruit fly pheromone: isolation, identification and lab bioassays. In: Cavalloro R (ed) Fruit flies of economic importance. A.A. Balkema, Rotterdam, pp 96–103

McGovern TP, Beroza M (1966) Structure of the four isomers of the insect attractant trimedlure. J Org Chem 31:1472–1477

McGovern TP, Beroza M, Ohinata K, Miyashita D, Steiner LF (1966) Volatility and attractiveness to the Mediterranean fruit fly of trimedlure and its isomers, and a comparison of its volatility with that of seven other insect attractants. J Econ Entomol 59:1450–1455

McGovern TP, Cunningham RT, Leonhardt BA (1986) *Cis*-trimedlure: attraction for the Mediterranean fruit fly (Diptera: Tephritidae) and isomeric structural assignments. J Econ Entomol 79:98–102

McGovern TP, Cunningham RT, Leonhardt BA (1987) Attractiveness of trans-trimedlure and its four isomers in field tests with the Mediterranean fruit fly (Diptera: Tephritidae). J Econ Entomol 80:617–620

McGovern TP, Warthen JD Jr, Cunningham RT (1990) Relative attraction of the Mediterranean fruit fly (Diptera: Tephritidae) to the eight isomers of trimedlure. J Econ Entomol 83:1350–1354

McInnis DO, Rendon P, Jang E, Sauers-Muller AV, Sugayama R, Malavasi A (1999) Interspecific mating of introduced, sterile *Bactrocera dorsalis* with wild *B. carambolae* (Diptera: Tephritdae) in Suriname: a potential case for cross-species sterile insect technique. Ann Entomol Soc Am 92:758–765

McInnis DO, Kurashima R, Shelly T, Komatsu J, Edu J, Pahio E (2011) Prerelease exposure to methyl eugenol increases the mating competitiveness of sterile males of the oriental fruit fly (Diptera: Tephritidae) in a Hawaiian orchard. J Econ Entomol 104:1969–1978

McQuate GT, Peck SL (2001) Enhancement of attraction of α-ionol to male *Bactrocera latifrons* (Diptera: Tephritidae) by addition of a synergist, cade oil. J Econ Entomol 94:39–46

McQuate GT, Keun YS, Sylva CD, Li QX, Jang EB (2004) Active ingredients in cade oil that synergize attractiveness of α-ionol to male *Bactrocera latifrons* (Diptera: Tephritidae). J Econ Entomol 97:862–870

McQuate GT, Bokonon-Ganta AH, Jang EB (2008a) Use of alpha-ionol + cade oil for detection and monitoring of *Bactrocera latifrons* (Diptera: Tephritidae) populations. In: Sugayama RL, Zucchi RA, Ovruski SM, Sivinski J (eds) Fruit flies of economic importance: from basic to applied knowledge. Proceedings of the 7th international symposium. Fruit flies of economic importance, Salvador, pp 89–95

McQuate GT, Bokonon-Ganta AH, Jang EB, Messing RH (2008b) Age-related responsiveness of male *Bactrocera latifrons* (Diptera: Tephritidae) to α-ionol + cade oil relative to age of sexual maturity. Int J Trop Insect Sci 28:12–18

Metcalf RL (1990) Chemical ecology of Dacinae fruit flies (Diptera: Tephritidae). Ann Entomol Soc Am 83:1017–1030

Metcalf RL, Metcalf ER (1992a) Fruit flies of the family Tephritidae. In: Miller TA, van Emden HF (eds) Plant kairomones in insect ecology and control. Chapman and Hall, New York, pp 139–142

Metcalf RL, Metcalf ER (1992b) Plant kairomones in insect ecology and control. Chapman and Hall, London, 184 pp

Metcalf RL, Mitchell C (1990) Development of new lures for the melon fly, *Dacus cucurbitae*. Report to California Department Food and Agriculture, Sacramento

Metcalf RL, Mitchell WC, Fukuto TR, Metcalf ER (1975) Attraction of the oriental fruit fly, *Dacus dorsalis*, by methyl eugenol and related olfactory stimulants. Proc Natl Acad Sci U S A 72:2501–2505

Metcalf RL, Metcalf ER, Mitchell WC, Lee LWY (1979) Evolution of olfactory receptor in oriental fruit fly *Dacus dorsalis*. Proc Natl Acad Sci U S A 76:1561–1565

Metcalf RL, Metcalf ER, Mitchell WC (1981) Molecular parameters and olfaction in the oriental fruit fly *Dacus dorsalis*. Proc Natl Acad Sci U S A 78:4007–4010

Metcalf RL, Mitchell WC, Metcalf ER (1983) Olfactory receptors in the melon fly *Dacus cucurbitae* and the oriental fruit fly *Dacus dorsalis*. Proc Natl Acad Sci U S A 80:3143–3147

Metcalf RL, Metcalf ER, Mitchell WC (1986) Benzyl acetates as attractants for the male oriental fruit fly, *Dacus dorsalis*, and the male melon fly, *Dacus cucurbitae*. Proc Natl Acad Sci U S A 83:1549–1553

Meza-Hernández JS, Hernández E, Salvador-Figueroa M, Cruz-López L (2002) Sexual compatibility, mating performance and sex pheromone release of mass-reared and wild *Anastrepha obliqua* (Diptera: Tephritidae) under field-cage conditions. In: Barnes BN (ed) Proceedings of the 6th international symposium on fruit flies of economic importance. Isteg Scientific Publications, Irene, pp 99–104

Millar JG (1995) An overview of attractants for the Mediterranean fruit fly. In: Morse JG, Metcalf RL, Carey JR, Dowell RV (eds) The Mediterranean fruit fly in California: defining critical research. University of California, Riverside, pp 123–143

Miller EC, Swanson AB, Phillips DH (1983) Structure-activity studies of the carcinogenicities in the mouse and rat of some naturally occurring and synthetic alkenylbenzene derivatives related to safrole and estragole. Cancer Res 43:1124–1134

Mitchell WC, Metcalf RL, Metcalf ER, Mitchell S (1985) Candidate substitutes for methy eugenol as attractants for area-wide monitoring and control of oriental fruitfly, *Dacus dorsalis* Hendel (Diptera: Tephritidae). Environ Entomol 14:176–181

Monro J, Richardson NL (1969) Traps, male lures, and a warning system for Queensland fruit fly, *Dacus tryoni* (Frogg.) (Diptera: Trypetidae). Aust J Agric Res 20:325–328

Moreno DS, Sanchez M, Robacker DC, Worley J (1991) Mating competitiveness of irradiated Mexican fruit fly (Diptera: Tephritidae). J Econ Entomol 84:1227–1234

Morgante JS, Malavasi A, Prokopy RJ (1983) Mating behavior of wild *Anastrepha fraterculus* (Diptera: Tephritidae) on a caged host tree. Fla Entomol 66:234–241

Mori K, Nakazono Y (1988) Synthesis of lactone components of the pheromone of *Anastrepha suspensa*, suspensolide, and the enantiometers of anastrephin and epianastrephin. Liebigs Ann Chem 1988:167–174

Mwatawala M, Virgilio M, Quilici S, Dominic M, De Meyer M (2013) Field evaluation of the relative attractiveness of enriched ginger root oil (EGO) lure and trimedlure for African *Ceratitis* species (Diptera: Tephritidae). J Appl Entomol 137:392–397

Nakagawa S, Farias GJ, Steiner LF (1970) Response of female Mediterranean fruit flies to male lures in the relative absence of males. J Econ Entomol 63:227–229

Nakagawa S, Farias GJ, Suda D, Chambers DL (1973) Mating behavior of the Mediterranean fruit fly following excision of the antennae. Ann Entomol Soc Am 66:583–584

Nakagawa S, Chambers DL, Bradshaw TI, Urago T, Harris EJ (1975) Performance of a sticky trap with trimedlure impregnated in the adhesive material. J Econ Entomol 68:817–818

Nakagawa S, Harris EJ, Urago T (1979) Controlled release of trimedlure from a three-layer laminated plastic dispenser. J Econ Entomol 72:625–627

Nakagawa S, Steiner LF, Farias GJ (1981a) Response of virgin female Mediterranean fruit flies to live mature normal males, sterile males, and trimedlure in plastic traps. J Econ Entomol 74:566–567

Nakagawa S, Harris EJ, Keiser I (1981b) Performance of capilure® in capturing Mediterranean fruit flies in Steiner plastic or cardboard sticky traps. J Econ Entomol 74:244–245

Nation JL (1972) Courtship behavior and evidence for a sex attractant in the male Caribbean fruit fly, *Anastrepha suspensa*. Ann Entomol Soc Am 65:1364–1367

Nation JL (1975) The sex pheromone blend of Caribbean fruit fly males: isolation, biological activity, and partial chemical characterization. Environ Entomol 4:27–30

Nation JL (1983) Sex pheromone of the Caribbean fruit fly: chemistry and fixed ecology. In: Miyamoto J, Kearney PC (eds) IUPAAC pesticide chemistry, human welfare and the economy, vol 2. Pergamon Press, New York, pp 109–110

Nation JL (1989) The role of pheromones in the mating system of *Anastrepha* fruit flies. In: Robinson AS, Hooper G (eds) Fruit flies: their biology, natural enemies and control, vol 3A. Elsevier, Amsterdam, pp 189–205

Nation JL (1990) Biology of pheromone release by male Caribbean fruit flies, *Anastrepa suspensa* (Diptera: Tephritidae). J Chem Ecol 16:553–572

Nation JL (1991) Sex pheromone components of *Anastrepha suspensa* and their role in mating behavior. In: Kawasaki K, Iwahashi O, Kaneshiro KY (eds) The international symposium of the biology and control of fruit flies. Univ. Ryukyus, Okinawa, pp 224–236

Niogret J, Montgomery WS, Kendra PE, Heath RR, Epsky ND (2011) Attraction and electroantennogram responses of male Mediterranean fruit fly to volatile chemicals from *Persea, Litchi*, and *Ficus* wood. J Chem Ecol 37:483–491

Nishida R, Tan KH, Fukami H (1988a) Cis-3.4-dimethoxycinnamyl alcohol from the rectal glands of male Oriental fruit fly, *Dacus dorsalis*. Chem Express 3(4):207–210

Nishida R, Tan KH, Serit M, Lajis NH, Sukari AM, Takahashi S, Fukami H (1988b) Accumulation of phenylpropanoids in the rectal glands of male oriental fruit fly, *Dacus dorsalis*. Experientia 44:534–536

Nishida R, Tan KH, Takahashi S, Fukami H (1990) Volatile components of male rectal glands of the melon fly, *Dacus cucurbitae* Coquillett (Diptera: Tephritidae). Appl Entomol Zool 25:105–112

Nishida R, Iwahashi O, Tan KH (1993) Accumulation of *Dendrobium* (Orchidaceae) flower fragrance in the rectal glands by males of the melon fly, *Dacus cucurbitae* (Tephritidae). J Chem Ecol 19:713–722

Nishida R, Shelly TE, Kaneshiro KY (1997) Acquisition of female-attracting fragrance by males of the oriental fruit fly from a Hawaiian lei flower, *Fagraea berteriana*. J Chem Ecol 23:2275–2285

Nishida R, Enomoto H, Shelly TE, Ishida T (2009) Sequestration of 3-oxygenated α-ionone derivatives in the male rectal gland of the solanaceous fruit fly, *Bactrocera latifrons*. Entomol Exp Appl 131:85–92

Noce ME, Belfiore T, Scalercio S, Vizzarri V, Iannotta N (2009) Efficacy of new mass-trapping devices against *Bactrocera oleae* (Diptera Tephritidae) for minimizing pesticide input in agroecosystems. J Environ Sci Health B 44:442–448

Norrbom AL, Zucchi RA, Hernández-Ortiz V (2000) Phylogeny of the genera *Anastrepha* and *Toxotrypana* (Trypetinae: Toxotrypanini) based on morphology. In: Aluja M, Norrbom AL (eds) Fruit flies (Tephritidae): phylogeny and evolution of behavior. CRC Press, Boca Raton, pp 299–342

Norrbom AL, Korytkowski CA, Zucchi RA, Uramoto K, Venable GL, McCormick J, Dallwitz MJ (2012) *Anastrepha* and *Toxotrypana*: descriptions, illustrations, and interactive keys. Version: 31st Aug 2012. http://delta-intkey.com

Obra GB, Resilva SS (2013) Influence of adult diet and exposure to methyl eugenol in the mating performance of *Bactrocera philippensis*. J Appl Entomol 137(Suppl. 1):210–216

OEPP/EPPO (2010) *Bactrocera zonata*: procedure for official control. Bull. OEPP/EPPO 40: 390–395

Ohinata K, Jacobson M, Nakagawa S, Fukimoto M, Higa H (1977) Mediterranean fruit fly: laboratory and field evaluations of synthetic sex pheromones. J Environ Sci Health A12:67–78

Ohinata K, Jacobson M, Nakagawa S, Urago T, Fujimoto M, Higa H (1979) Methyl I-6-nonenoate: a new Mediterranean fruit fly male attractant. J Econ Entomol 72:648–650

Ohinata K, Jacobson M, Kobayashi RM, Chambers DL, Fujimoto MS, Higa HH (1982) Oriental fruit fly and melon fly: biological and chemical studies of smoke produced by males. J Environ Sci Health A17:197–216

Oliver JE, Casaña-Giner V, Jang EB, McQuate GT, Carvalho L (2004) Improved attractants for the melon fly, *Bactrcoera cucurbitae*. In: Barne BN (ed) Proceedings of the 6th international symposium on fruit flies of economic importance. Isteg Science Publications, Irene, pp 283–290

Orankanok W, Chinvinijkul S, Sawatwangkhoung A, Pinkaew S, Orankanok S (2013) Methyl eugenol and pre-release diet improve mating performance of young *Bactrocera dorsalis* and *Bactrocera correcta* males. J Appl Entomol 137(Suppl. 1):200–209

Osborne R, Meats A, Frommer M, Sved JA, Drew RAI, Robson MK (1997) Australian distribution of 17 species of fruit flies (Diptera: Tephritidae) caught in cue lure traps in February 1994. Aust J Entomol 36:45–50

Payne TL, Shorey HH, Gaston LK (1973) Sex pheromones of Lepidoptera. XXXVIII. Electroantennogram responses in *Autographa californica* to *cis*-7-dodecenyl acetate and related components. Ann Entomol Soc Am 66:703–704

Peck SL, McQuate GT, Vargas RI, Seager DC, Revis HC, Jang EB, McInnis DO (2005) Movement of sterile male *Bactrocera cucurbitae* (Diptera: Tephritidae) in a Hawaiian agroecosystem. J Econ Entomol 98:1539–1550

Perdomo AJ, Baranowski RM, Nation JL (1975) Recapture of virgin female Caribbean fruit flies from traps baited with males. Fla Entomol 58:291–295

Perdomo AJ, Nation JL, Baranowski RM (1976) Attraction of female and male Caribbean fruit flies to food-baited and male-baited traps under field conditions. Environ Entomol 5:1208–1210

Perkins MV, Fletcher MT, Kitching W, Drew RAI, Moore CJ (1990a) Chemical studies of rectal gland secretions of some species of *Bactrocera dorsalis* complex of fruit flies (Diptera: Tephritidae). J Chem Ecol 16:2475–2487

Perkins MV, Kitching W, Drew RAI, Moore CJ, Konig WA (1990b) Chemistry of fruit flies: Composition of the male rectal gland secretions of some species of south-east Asian Dacinae. Re-examination of *Dacus cucurbitae* (Melon fly). J Chem Soc Perkin Trans 1:1111–1117

Petacchi R, Rizzi I, Guidotti D (2003) The 'lure and kill' technique in *Bactrocera oleae* (Gmel.) control: effectiveness indices and suitability of the technique in area-wide experimental trials. Int J Pest Manag 49:305–311

Polloni YJ, Da Silva MT (1986) Considerations on the reproductive behavior of *Anastrepha pseudoparallela* Loew 1873 (Diptera: Tephritidae). In: Economopoulos AP (ed) Fruit flies. Proceedings II International symposium, Crete, Greece. Elsevier, Amsterdam, pp 295–301

Prokopy RJ (1975) Mating behavior in *Rhagoletis pomonella* (Diptera: Tephritidae) V. Virgin female attraction to male odor. Canad Entomol 107:905–908

Prokopy RJ, Bush GL (1972) Mating behavior in *Rhagoletis pomonella* (Diptera: Tephritidae) III. Male aggregation in response to an arrestant. Canad Entomol 104:275–283

Prokopy RJ, Poramarcom R, Sutantawong M, Dokmaihom R, Hendrichs J (1996) Localization of mating behavior of released *Bactrocera dorsalis* flies on host fruit in an orchard. J Insect Behav 9:133–142

Quilici S, Franck A, Peppuy A, Dos Reis Correia E, Mouniama C, Blard F (2002) Comparative studies of courtship behavior of *Ceratitis* spp. (Diptera: Tephritidae) in Reunion Island. Fla Entomol 85:138–142

Raghu S (2004) Functional significance of phytochemical lures to dacine fruit flies (Diptera: Tephritidae): an ecological and evolutionary synthesis. Bull Entomol Res 94:385–399

Raw AS, Jang EB (2000) Enantioselective synthesis of ceralure B1, ethyl *cis*-5-iodo-*trans*-2-methylcyclohexane-1-carboxylate. Tetrahedron 56:3285–3290

Renou M, Guerrero A (2000) Insect pheromones in olfaction research and semiochemical-based pest control strategies. Annu Rev Entomol 45:605–630

Rice RE, Cunningham RT, Leonhardt BA (1984) Weathering and efficacy of trimedlure dispensers for attraction of Mediterranean fruit flies (Diptera: Tephritidae). J Econ Entomol 77:750–756

Rice RE, Phillips PA, Stewart-Leslie J, Sibbett GS (2003) Olive fruit fly populations measured in central and southern California. Calif Agric 57:122–127

Ripley LB, Hepburn GA (1935) Olfactory attractants for male fruit flies. Entomol. Memoirs, Department of Agriculture, South Africa 9: 3–17

Robacker DC (1988) Behavioral responses of female Mexican fruit flies, *Anastrepha ludens*, to components of male-produced sex pheromone. J Chem Ecol 14:1715–1726

Robacker DC, Garcia JA (1990) Responses of laboratory-strain Mexican fruit flies, *Anastrepha ludens*, to combinations of fermenting fruit odor and male-produced pheromone in laboratory bioassays. J Chem Ecol 16:2027–2038

Robacker DC, Hart WG (1984) A bioassay for investigation of sex pheromones of fruit flies. Southwest Entomol 9:134–137

Robacker DC, Hart WG (1985a) Courtship and territoriality of laboratory-reared Mexican fruit flies, *Anastrepha ludens* (Diptera: Tephritidae), in cages containing host and nonhost trees. Ann Entomol Soc Am 78:488–494

Robacker DC, Hart WG (1985b) (Z)-3-nonenol, (Z, Z)-3,6-nonadienol and (S, S)-(−) epianastrephin: male produced pheromones of the Mexican fruit fly. Entomol Exp Appl 39:103–108

Robacker DC, Hart WG (1986) Behavioral responses of male and female Mexican fruit flies, *Anastrepha ludens*, to male-produced chemicals in laboratory experiments. J Chem Ecol 12:39–477

Robacker DC, Moreno DS (1988) Responses of female Mexican fruit flies at various distances from male-produced pheromone. Southwest Entomol 13:95–100

Robacker DC, Thomas DB (2007) Comparison of two synthetic food-odor lures for captures of feral Mexican fruit flies (Diptera: Tephritidae) in Mexico and implications regarding use of irradiated flies to assess lure efficacy. J Econ Entomol 100:1147–1152

Robacker DC, Wolfenbarger DA (1988) Attraction of laboratory-reared, irradiated Mexican fruit flies to male-produced pheromone in the field. Southwest Entomol 13:75–80

Robacker DC, Garcia JA, Hart WG (1990) Attraction of a laboratory strain of *Anastrepha ludens* (Diptera: Tephritidae) to the odor of fermented chapote fruit and to pheromones in laboratory experiments. Environ Entomol 19:403–408

Robacker DC, Mangan RL, Morena DS, Tarshis-Moreno AM (1991) Mating behavior and male mating success in wild *Anastrepha ludens* (Diptera: Tephritidae) on a field-caged host tree. J Insect Behav 4:471–487

Robacker DC, Mangan RL, Moreno DS, Tarshis Moreno AM (2003) Behavior and interactions of wild *Anastrepha ludens* (Diptera: Tephritidae) on a grapefruit tree. Folia Entomol Mex 42:221–237

Robacker DC, Alija M, Bartelt RJ, Patt J (2009a) Identification of chemicals emitted by calling males of the sapote fruit fly, *Anastrepha serpentina*. J Chem Ecol 35:601–609

Robacker DC, Aluja M, Cossé AA, Sacchetti P (2009b) Sex pheromone investigation of *Anastrepha serpentina* (Diptera: Tephritidae). Ann Entomol Soc Am 102:560–566

Robinson SH, Barr DB (2006) Use of biomonitoring data to evaluate methyl eugenol exposure. Environ Health Perspect 114:1797–1801

Rocca JR, Nation JL, Strekowski L, Battiste MA (1992) Comparison of volatiles emitted by male Caribbean and Mexican fruit flies. J Chem Ecol 18:223–244

Santos Felix dos C, Uchôa-Fernandes MA, Faccenda O (2009) Capture of *Anastrepha sororcula* (Diptera: Tephritidae) in McPhail and Jackson traps with food attractant and virgin adults. Braz Arch Biol Technol 52:99–104

Schecter A, Lucier GW, Cunningham ML, Abdo KM, Blumenthal G, Silver AG, Melnick R, Portier C, Barr DB, Barr JR, Stanfill SB, Patterson DG Jr, Needham LL, Stopford W, Masten S, Mignogna J, Tung KC (2004) Human consumption of methyleugenol and its elimination from serum. Environ Health Perspect 112:678–683

Schiestl RHWS, Chan RD, Gietz RDM, Hastings PJ (1989) Safrole, eugenol and methyleugenol induce intrachromosomal recombination in yeast. Mutat Res 224:427–436

Schutze MK, Krosch MN, Armstrong KF, Chapman TA, Englezou A, Chomic A, Cameon SL, Hailstones D, Clarke AR (2012) Population structure of *Bactrocera dorsalis s.s., B. papayae* and *B. philippinensis* (Diptera:Tephritidae) in southeast Asia: evidence for a single species hypothesis using mitochondrial DNA and wing shape data. BMC Evol Biol 12:130

Schutze MK, Jessup A, Ul-Haq I, Vreysen MJB, Wornoayporn V, Vera MT, Clarke AR (2013) Mating compatibility among four pest members of the *Bactrocera dorsalis* fruit fly species complex (Diptera: Tephritidae). J Econ Entomol 106:695–707

Schutze MK, Aketarawong N, Amornsak W, Armstrong KF, Augustinos A, Barr N, Bo W, Bourtzis K, Boykin LM, Cáceres C, Cameron SL, Chapman TA, Chinvinijkul S, Chomič A, De Meyer M, Drosopoulou ED, Englezou A, Ekesi S, Gariou-Papalexiou A, Hailstones D, Haymer D, Hee AKW, Hendrichs J, Hasanuzzaman M, Jessup A, Khamis FM, Krosch MN, Leblanc L, Mahmood K, Malacrida AR, Mavragani-Tsipidou P, McInnis DO, Mwatawala M, Nishida R, Ono H, Reyes J, Rubinoff DR, San Jose M, Shelly TE, Srikachar S, Tan KH, Thanaphum S, Ul Haq I, Vijaysegaran S, Wee SL, Yesmin F, Zacharopoulou A, Clarke AR (2014) Synonymization of key pest species within the Bactrocera dorsalis complex (Diptera: Tephritidae): taxonomic changes based on 20 years of integrative morphological, genetic, behavioural, and chemoecological data. Syst Entomol (in press)

Segura D, Petit-Marty N, Sciurano R, Vera T, Calcagno G, Allinghi A, Gómez Kendra P, Cladera J, Vilardi J (2007) Lekking behavior of *Anastrepha fraterculus* (Diptera: Tephritidae). Fla Entomol 90:154–162

Sekizawa J, Shibamoto T (1982) Genotoxicity of safrole-related chemical in microbial test systems. Mutat Res 101:127–140

Sharp J, Landolt PJ (1984) Gustatory and olfactory behavior of the papaya fruit fly *Toxotrypana curvicauda* Gerst. (Diptera: Tephritidae) in the laboratory with notes on longevity. J Ga Entomol Soc 19:176–182

Shelly TE (1994) Consumption on methyl eugenol by male *Bactrocera dorsalis* (Diptera: Tephritidae): low incidence of repeat feeding. Fla Entomol 77:201–208

Shelly TE (1997) Selection for non-responsiveness to methyl eugenol in male oriental fruit flies (Diptera: Tephritidae). Fla Entomol 80:248–253

Shelly TE (2000a) Flower-feeding affects mating performance in male oriental fruit flies (Diptera: Tephritidae). Ecol Entomol 25:109–114

Shelly TE (2000b) Effects of raspberry ketone on the mating success of male melon flies (Diptera: Tephritidae). Proc Hawaiian Entomol Soc 34:163–167

Shelly TE (2000c) Male signaling and lek attractiveness in the Mediterranean fruit fly. Anim Behav 60:245–251

Shelly TE (2000d) Trapping male oriental fruit flies (Diptera: Tephritidae): does feeding on a natural source of methyl eugenol reduce capture probability? Fla Entomol 83:109–111

Shelly TE (2001a) Feeding on papaya flowers enhances mating competitiveness of male oriental fruit flies, *Bactrocera dorsalis* (Diptera: Tephritidae). Proc Hawaiian Entomol Soc 35:41–47

Shelly TE (2001b) Lek size and female visitation in two species of tephritid fruit flies. Anim Behav 62:33–40

Shelly TE (2001c) Feeding on methyl eugenol and *Fagraea berteriana* flowers increases long-range female attraction by males of the oriental fruit fly (Diptera: Tephritidae). Fla Entomol 84:634–640

Shelly TE (2010a) Effects of methyl eugenol and raspberry ketone/cue lure on the sexual behavior of *Bactrocera* species (Diptera: Tephritidae). Appl Entomol Zool 45:349–361

Shelly TE (2010b) Capture of *Bactrocera* males (Diptera: Tephritidae) in parapheromone-baited traps: A comparison of liquid versus solid formulations. Proc Hawaiian Entomol Soc 42:1–8

Shelly TE (2013) Detection of male Mediterranean fruit flies (Diptera: Tephritidae): performance of trimedlure relative to capilure and enriched ginger root oil. Proc Hawaii Entomol Soc 45:1–7

Shelly TE, Dewire A-LM (1994) Chemically mediated mating success in male Oriental fruit flies (Diptera: Tephritidae). Ann Entomol Soc Am 87:375–382

Shelly TE, Edu J (2007) Exposure to the ripe fruit of tropical almond enhances the mating success of male *Bactrocera dorsalis* (Diptera: Tephritidae). Fla Entomol 90:757–758

Shelly TE, Kaneshiro KY (1991) Lek behavior of the oriental fruit fly, *Dacus dorsalis*, in Hawaii (Diptera: Tephritidae). J Insect Behav 4:235–241

Shelly TE, Nishimoto JI (2011) Additional measurements of distance-dependent capture probabilities for released males of *Bactrocera cucurbitae* and *B. dorsalis* (Diptera: Tephritidae) in Honolulu. J Asia-Pac Entomol 14:271–276

Shelly TE, Pahio E (2002) Relative attractiveness of enriched ginger root oil and trimedlure to male Mediterranean fruit flies (Diptera: Tephritidae). Fla Entomol 85:545–551

Shelly TE, Villalobos EM (1995) Cue lure and the mating behavior of male melon flies (Diptera: Tephritidae). Fla Entomol 78:473–482

Shelly TE, Resilva S, Reyes M, Bignayan H (1996) Methyl eugenol and mating competitiveness of irradiated male *Bactrocera philippinensis* (Diptera: Tephritidae). Fla Entomol 79:481–488

Shelly TE, Pahio E, Edu J (2004) Synergistic and inhibitory interactions between methyl eugenol and cue lure influence trap catch of male fruit flies, *Bactrocera dorsalis* (Hendel) and *B. cucurbitae* (Diptera: Tephritidae). Fla Entomol 87:481–486

Shelly TE, Edu J, Pahio E, Nishimoto J (2007) Scented males and choosy females: does male odor influence female mate choice in the Mediterranean fruit fly? J Chem Ecol 33:2308–2324

Shelly TE, Edu J, Pahio E, Wee SL, Nishida R (2008) Re-examining the relationship between sexual maturation and age of response to methyl eugenol in males of the oriental fruit fly. Entomol Exp Appl 128:380–388

Shelly TE, Kurashima RS, Nishimoto J, Diaz A, Leathers J, War M, Joseph D (2011a) Capture of *Bactrocera* fruit flies (Diptera: Tephritidae) in traps baited with liquid versus solid formulations of male lures. J Asia Pac Entomol 14:463–467

Shelly TE, Kurashima RS, Nishimoto JI (2011b) Capture of *Bactrocera* males (Diptera: Tephritidae) in parapheromone-baited traps: performance of solid dispensers with different loadings of attractants and toxicant. Proc Hawaiian Entomol Soc 43:33–47

Shelly TE, Kurashima RS, Nishimoto JI (2012a) Field capture of male melon flies, *Bactrocera cucurbitae* (Coquillett), in Jackson traps baited with cue-lure versus raspberry ketone formate in Hawaii. Proc Hawaii Entomol Soc 44:63–70

Shelly TE, Nishimoto J, Kurashima R (2012b) Trap capture of three economically important fruit fly species (Diptera: Tephritidae): evaluation of a solid formulation containing multiple male lures in a Hawaiian coffee field. J Econ Entomol 105:1186–1193

Sivinski J, Calkins C (1986) Pheromones and parapheromones in the control of tephritids. Fla Entomol 69:157–168

Sivinski J, Heath RR (1988) Effects of oviposition on remating, response to pheromones, and longevity in the female Caribbean fruit fly, *Anastrepha suspensa* (Diptera: Tephritidae). Ann Entomol Soc Am 81:1021–1024

Sivinski JM, Epsky N, Heath RR (1994) Pheromone deposition on leaf territories by male Caribbean fruit flies, *Anastrepha suspensa* (Loew) (Diptera: Tephritidae). J Insect Behav 7:43–51

Smith RL, Adams TB, Doull J, Feron VJ, Goodman JI, Marnett LJ, Portoghese PS, Waddell WJ, Wagner BM, Rogers AE, Caldwell J, Sipes IG (2002) Safety assessment of allylalkoxybenzene derivatives used as flavouring substances – Methyl eugenol and estragole. Food Chem Toxicol 40:851–870

Smith PT, Kambhampati S, Armstrong KS (2003) Phylogenetic relationships among Bactrocera species (Diptera: Tephritidae) inferred from mitochondrial DNA sequences. Mol Phylogenet Evol 26:8–17

Sonnet PE, McGovern TP, Cunningham RT (1984) Enantiomers of the biologically active components of the insect attractant trimedlure. J Org Chem 49:4639–4643

Sookar P, Permalloo S, Gungha B, Alleck M, Seewooruthun SI, Soonnoo AR (2008) An area wide control of fruit flies in Mauritius. In: Sugayama RL, Zucchi RA, Ovruski SM, Sivinski J (eds) Fruit flies of economic importance: from basic to applied knowledge. Proceedings of the 7th international symposium fruit flies econ importance, Salvador, pp 261–269

Stark JD (1995) Nocturnal behavior of oriental fruit flies and melon flies (Diptera: Tephritidae) and associated parasitoids in a commercial papaya growing region on Kauai. Hawaii Proc Hawaiian Entomol Soc 32:149–151

Steiner LF (1952) Methyl eugenol as an attractant for oriental fruit fly. J Econ Entomol 45:241–248

Steiner LF, Lee RKS (1955) Large-area tests of a male-annihilation method for oriental fruit fly control. J Econ Entomol 48:311–317

Steiner LF, Miyashita DH, Christenson LD (1957) Angelica oils as Mediterranean fruit fly lures. J Econ Entomol 50:505

Steiner LF, Mitchell WC, Green N, Beroza M (1958) Effect of *cis-trans* isomerism on the potency of an insect attractant. J Econ Entomol 51:921–922

Steiner LF, Mitchell WC, Harris EJ, Kozuma TT, Fujimoto MS (1965) Oriental fruit fly eradication by male annihilation. J Econ Entomol 58:961–964

Stokes JB, Uebel EC, Warthen JD, Jacobson M, Flippen-Anderson JL, Gilardi R, Spishakoff LM, Wilzer KR (1983) Isolation and identification of novel lactones from male Mexican fruit flies. J Agric Food Chem 31:1162–1167

Suckling DM, Jang EB, Holder P, Carvalho L, Stephens AEA (2008) Evaluation of lure dispensers for fruit fly surveillance in New Zealand. Pest Manag Sci 64:848–856

Takeoka G, Flath RA, Mon TR, Buttery RG, Teranishi R, Güntert M, Lautamo R, Szejtli J (1990) Further applications of permethylated β-cyclodextrin capillary gas chromatographic columns. HRC & CC 13:202–206

Tan KH (1983) Response of *Dacus* (Diptera: Tephitidae) to *Ocimum sanctum* extracts and different synthetic attractants in Penang, Malaysia. In: Cavalloro R (ed) Fruit flies of economic importance. A.A. Balkema, Rotterdam, pp 513–515

Tan KH (1984) Description of a new attractant trap and the effect of placement height on catches of two *Dacus* species (Diptera: Tephritidae). J Plant Prot Tropics 1:117–120

Tan KH (1985) Estimation of native populations of male *Dacus* spp. by Jolly's stochastic method using a new designed attractant trap in a village ecosystem. J Plant Prot Tropics 2:87–95

Tan KH (2000) Behaviour and chemical ecology of *Bactrocera* flies. In: Tan KH (ed) Area-wide control of Fruit flies and other insect pests. Penerbit Universiti Sains Malaysia, Pulau Pinang, Malaysia, pp 647–656

Tan KH (2003) Interbreeding and DNA analysis of *Bactrocera dorsalis* sibling species. In: Recent trends on sterile insect technique and area-wide integrated pest management – economic feasibility, control projects, Farmer Organization and *Bactrocera dorsalis* Complex Control Study– Research Institute for the Subtropics, Okinawa, Japan, pp 113–122

Tan KH (2009) Fruit fly pests as pollinators of wild orchids. Orchid Digest 73:180–187

Tan KH, Jaal Z (1986) Comparison of male adult population densities of the oriental and *Artocarpus* fruit flies, *Dacus* spp. (Diptera: Tephritidae), in two nearby villages in Penang, Malaysia. Res Popul Ecol 25:85–89

Tan KH, Lee SL (1982) Species diversity and abundance of *Dacus* (Diptera: Tephritidae) in five ecosystems of Penang, Malaysia. Bull Entomol Res 72:709–716

Tan KH, Nishida R (1995) Incorporation of raspberry ketone in the male rectal glands of the Queensland fruit fly, *Bactrocera tryoni* Froggatt (Diptera: Tephritidae). Appl Entomol Zool 30:494–497

Tan KH, Nishida R (1996) Sex pheromone and mating competition after methyl eugenol consumption in *Bactrocera dorsalis* complex. In: McPheron BA, Steck GJ (eds) Fruit fly pests – a world assessment of their biology and management. St. Lucie Press, Delray Beach, pp 147–153

Tan KH, Nishida R (1998) Ecological significance of a male attractant in the defence and mating strategies of the fruit fly pest, *Bactrocera papayae*. Entomol Exp Appl 89:155–158

Tan KH, Nishida R (2000) Mutual reproductive benefits between a wild orchid, *Bulbophyllum patens*, and *Bactrocera* fruit flies via a floral synomone. J Chem Ecol 26:533–546

Tan KH, Nishida R (2005) Synomone or Kairomone? – *Bulbophyllum apertum* (Orchidaceae) flower releases raspberry ketone to attract *Bactrocera* fruit flies. J Chem Ecol 31(3):509–519

Tan KH, Nishida R (2007) Zingerone in the floral synomone of *Bulbophyllum baileyi* (Orchidaceae) attracts *Bactrocera* fruit flies during pollination. Biochem Syst Ecol 31:334–341

Tan KH, Nishida R (2012) Methyl eugenol – its occurrence, distribution, and role in nature, especially in relation to insect behaviour and pollination. J Insect Sci 20(56): 1–74. Available online: http://www.insectscience.org/12.56/i1536-2442-12-56.pdf

Tan KH, Serit M (1988) Movements and population density comparisons of native male adult *Dacus dorsalis* and *Dacus umbrosus* (Diptera: Tephritidae) between three ecosystems in Tanjong Bungah, Penang. J Plant Prot Tropics 5:17–21

Tan KH, Serit M (1994) Adult population dynamics of *Bactrocera dorsalis* (Diptera: Tephritidae) in relation to host phenology and weather in two villages of Penang Island, Malaysia. Environ Entomol 23:267–275

Tan KH, Kirton LG, Serit M (1987) Age response of *Dacus dorsailis* (Hendel) to methyl eugenol in a) a wind tunnel and b) traps set in a village, and its implication in population estimation. In: Economopoulos AP (ed) Fruit flies. G. Tsiveriotis Ltd, Athens, pp 425–432

Tan KH, Nishida R, Toong YC (2002) Floral synomone of a wild orchid, *Bulbophyllum cheiri*, lures *Bactrocera* fruit flies to perform pollination. J Chem Ecol 28:1161–1172

Tan KH, Tokushima I, Ono H, Nishida R (2011) Comparison of phenylpropanoid volatiles in male rectal pheromone gland after methyl eugenol consumption, and molecular phylogenetic relationship of four global pest fruit fly species – *Bactrocera invadens, B. dorsalis, B. correcta and B. zonata*. Chemoecology 21:25–33

Tan KH, Wee SL, Ono H, Nishida R (2013) Comparison of methyl eugenol metabolites, mitochondrial COI, and rDNA sequences of *Bactrocera philippinensis* (Diptera: Tephritidae) with those of three other major pest species within the *dorsalis* complex. Appl Entomol Zool 48: 275–282. http://link.springer.com/article/10.1007%2Fs13355-013-0183-5/lookinside/000.png

Tokushima I, Orankanok W, Tan KH, Ono H, Nishida R (2010) Accumulation of phenylpropanoid and sesquiterpenoid volatiles in male rectal pheromonal glands of the guava fruit fly, *Bactrocera correcta*. J Chem Ecol 36:1327–1334

Tumlinson JH (1988) Contemporary frontiers in insect semiochemical research. J Chem Ecol 14:2109–2130

Tychsen PH (1977) Mating behavior of the Queensland fruit fly, *Dacus tryoni* (Diptera: Tephritidae), in field cages. J Aust Entomol Soc 16:459–465

Umeya K, Sekiguchi Y, Ushio S (1973) The reproductive ability of the oriental fruit fly, *Dacus dorsalis* Hendel and the response of adults to methyl eugenol. Jap J Appl Entomol Zool 17:63–70

Vargas RI, Stark JD, Nishida T (1989) Abundance, distribution, and dispersion indices of the oriental fruit fly and melon fly (Diptera: Tephritidae) on Kauai, Hawaiian Islands. J Econ Entomol 82:1609–1615

Vargas RI, Stark JD, Nishida T (1990) Population dynamics, habitat preference, and seasonal distribution patterns of oriental fruit fly and melon fly (Diptera: Tephritidae) in an agricultural area. Environ Entomol 19:1820–1828

Vargas RI, Stark JD, Kido MH, Ketter HM, Whitehand LC (2000) Methyl eugenol and cue-lure traps for suppression of male oriental fruit flies and melon flies (Diptera: Tephritidae) in Hawaii: effects of lure mixtures and weathering. J Econ Entomol 93:81–87

Vargas RI, Stark JD, Hertlein M, Mafra Neto A, Coler R, Piñero JC (2008) Evaluation of SPLAT with spinosad and methyl eugenol or cue-lure for "attract-and-kill" of oriental and melon fruit flies (Diptera: Tephritidae) in Hawaii. J Econ Entomol 101:750–768

Vargas RI, Burns RE, Mau RFL, Stark JD, Cook P, Piñero JC (2009) Captures in methyl eugenol and cue-lure detection traps with and without insecticides and with a Farma Tech solid lure and insecticide dispenser. J Econ Entomol 102:552–557

Vargas RI, Shelly TE, Leblanc L, Piñero JC (2010a) Recent advances in methyl eugenol and cue-lure technologies for fruit fly detection, monitoring, and control in Hawaii. Vitam Horm 83:575–595

Vargas RI, Mau RFL, Stark JD, Piñero JC (2010b) Evaluation of methyl eugenol and cue-lure traps with solid lure and insecticide dispensers for fruit fly monitoring and male annihilation in the Hawaii areawide pest management program. J Econ Entomol 103:409–415

Vargas RI, Piñero JC, Jang EB, Mau RFL, Stark JD, Gomez L, Stoltman L, Mafra-Neto A (2010c) Response of melon fly (Diptera: Tephritidae) to weathered SPLAT-spinosad-cue-lure. J Econ Entomol 103:1594–1602

Vargas RI, Souder SK, Mackey B, Cook P, Morse JG, Stark JD (2012) Field trials of solid triple lure (trimedlure, methyl eugenol, raspberry ketone, and DDVP) dispensers for detection and male annihilation of *Ceratitis capitata*, *Bactrocera dorsalis*, and *Bactrocera cucurbitae* (Diptera: Tephritidae) in Hawaii. J Econ Entomol 105:1557–1565

Verghese A (1998) Methyl eugenol attracts female mango fruit fly, *Bactrocera dorsalis* Hendel. Insect Environ 4:101

Warthen JD Jr, McGovern TP (1986a) GC/FTIR analyses of trimedlure isomers and related esters. J Chromatogr Sci 24:451–457

Warthen JD Jr, McGovern TP (1986b) Purification of cis-trimedlure isomers by high-performance liquid chromatography. Chromatographia 21:651–654

Warthen JD Jr, Schmidt WF, Cunningham RT, Demilo AB, Fritz GL (1993) Quantitative structure-activity relationships (QSAR) of trimedlure isomers. J Chem Ecol 19:1323–1335

Warthen JD Jr, Cunningham RT, Demilo AB, Spencer S (1994) *Trans*-ceralure isomers: differences in attraction for Mediterranean fruit fly, *Ceratitis capitata* (Weid.) (Diptera: Tephritidae). J Chem Ecol 20:569–578

Warthen JD, Cunningham RT, Leonhardt BA, Cook JM, Avery JW, Harte EM (1998) Comparison of ceralure and trimedlure controlled-release formulations for male Mediterranean fruit flies in C & C traps. J Chem Ecol 24:1305–1314

Webb JC, Burk T, Sivinski J (1983) Attraction of female Caribbean fruit flies, *Anastrepha suspensa* (Diptera: Tephritidae), to the presence of males and male-produced stimuli in field cages. Ann Entomol Soc Am 76:996–998

Wee SL, Shelly T (2013) Capture of *Bactrocera* fruit flies in traps baited with liquid versus solid formulations of male lures in Malaysia. J Asia-Pac Entomol 16:37–42

Wee SL, Tan KH (2001) Allomonal and hepatotoxic effects following methyl eugenol consumption in *Bactrocera papayae* male against *Gekko monarchus*. J Chem Ecol 27:953–964

Wee SL, Tan KH (2005a) Female sexual response to male rectal volatile constituents in the fruit fly, *Bactrocera carambolae* (Diptera: Tephritidae). Appl Entomol Zool 40(2):365–372

Wee SL, Tan KH (2005b) Evidence of natural hybridization between two sympatric sibling species of *Bactrocera dorsalis* complex based on pheromone analysis. J Chem Ecol 31 (4):845–858

Wee SL, Hee AKW, Tan KH (2002) Comparative sensitivity to and consumption of methyl eugenol in three *Bactrocera dorsalis* (Diptera: Tephritidae) complex sibling species. Chemoecology 12:193–197

Wee SL, Tan KH, Nishida R (2007) Pharmacophagy of methyl eugenol by males enhances sexual selection of *Bactrocera carambolae* (Diptera: Tephritidae). J Chem Ecol 33:1272–1282

Wenkert E, Bookser BB, Arrhenius TA (1992) Total synthesis of (±)-α- and (±)-β-copaene and formal total syntheses of (±)-sativene, (±)-cis-sativenediol, and (±)-helminthosporal. J Am Chem Soc 114:644–645

White IM, Elson-Harris MM (1992) Fruit flies of economic significance: their identification and bionomics. CAB International, Wallingford

Wong TTY, McInnis DO, Nishimoto JI (1989) Relationship of sexual maturation rate to response of oriental fruit fly strains (Diptera: Tephritidae) to methyl eugenol. J Chem Ecol 15:1399–1405

Wong TTY, McInnis DO, Ramadan MM, Nishimoto JI (1991) Age-related response of male melon flies *Dacus cucurbitae* (Diptera: Tephritidae) to cue-lure. J Chem Ecol 17:2481–2487

Ye H, Liu JH (2005) Population dynamics of the oriental fruit fly, *Bactrocera dorsalis* (Diptera: Tephritidae) in the Kunming area, southwestern China. Insect Sci 12:387–392

Yokoyama VY, Miller GT, Stewart-Leslie J, Rice RE, Phillips PA (2006) Olive fruit fly (Diptera: Tephritidae) populations in relation to region, trap type, season, and availability of fruit. J Econ Entomol 99:2072–2079

Zhou G-D, Moorthy B, Bi J, Donnelly KC, Randerath K (2007) DNA adducts from alkoxyallylbenzene herb and spice constituents in cultured human (HepG2) cells. Environ Mol Mutagen 48:715–721

Chapter 3
History and Development of Food-Based Attractants

Nancy D. Epsky, Paul E. Kendra, and Elena Q. Schnell

Abstract Adult tephrids require sugar and protein for survival and for development of eggs, and volatile chemicals from these substances are the basis for food-based lures developed as baits for these pests. In this chapter, we discuss food-based lures that mimic food sources for adults other than host fruit. These have been primarily nitrogen sources that provide the protein needed by adult flies, although non-nitrogen-containing volatile chemicals are also included in this category. After male lures, food-based lures have been the predominant attractants used in traps for tephritid fruit flies. Although typically not as powerful as male lures, food-based lures have several advantages over male-specific attractants. They can be used for species for which there are no male lures known; they capture both females and males of target species; they tend to be female-biased, that is, they capture a higher percentage of females than males; and, at least for the Mediterranean fruit fly, traps baited with food-based lures tend to capture flies earlier than traps baited with male lure. There has been a long history of research on the development of food-based attractants for pest tephritids. Several review articles have documented the early history, which started with investigations of sugar-based food lures and lead to the development of the liquid protein baits and synthetic protein-based food lures, the standard food-lures that are currently in use. In this chapter, we discuss the development of and, as much as possible, the diversity of food-based lures that have been tested and/or are used in traps for pest tephritids. Future research directions are also discussed.

Keywords Sugar baits • Yeast baits • Protein baits • Ammonia baits • Borax • Torula yeast borax pellets • Solbait • NuLure • BioLure • Ammonium acetate • Trimethylamine • Putrescine • Preservative • Surfactant • Electroantennography

N.D. Epsky (✉) • P.E. Kendra • E.Q. Schnell
Subtropical Horticulture Research Station, USDA/ARS, 13601 Old Cutler Rd., Miami, FL 33158, USA
e-mail: nancy.epsky@ars.usda.gov; paul.kendra@ars.usda.gov; elena.schnell@ars.usda.gov

T. Shelly et al. (eds.), *Trapping and the Detection, Control, and Regulation of Tephritid Fruit Flies*, DOI 10.1007/978-94-017-9193-9_3,

1 Introduction

Adult tephritids require sugar and protein for survival and development of eggs (Christenson and Foote 1960), and volatile chemicals from these substances are the basis for food-based lures developed for these pests. Host fruit is used for both feeding and oviposition, and attractants based on host fruit are presented by Quilici et al. (Chap. 4, this volume). In this chapter, we will be discussing food-based lures that mimic adult food sources other than host fruit. These have been primarily nitrogen sources that provide the protein needed by adult flies, although non-nitrogen containing volatile chemicals are also included in this category.

After male lures, food-based lures have been the predominant attractants used in traps for tephritid fruit flies. Although typically not as powerful as male lures, food-based lures have several advantages over male lures. They can be used for a wide range of species and for species for which there are no male lures known; they capture both females and males of target species; they tend to be female-biased, that is, they capture a higher percentage of females than males (IAEA 2003); and, at least for the Mediterranean fruit fly, *Ceratitis capitata* (Wiedemann), traps baited with food-based lures tend to capture flies earlier in the season than male lure-baited traps (Papadopoulos et al. 2001). Thus, it is not surprising that there has been a long history of research on the development of food-based attractants for pest tephritids. Several review articles have documented the early history, including Crawford (1927), Baker et al. (1944), Gow (1954), Green et al. (1960), Morton and Bateman (1981), and Dominiak (2006).

Initial research focused on investigations of sugar-based food lures, which led to the development of the aqueous protein baits (also known as liquid protein baits) and synthetic protein-based food lures that are the standard food-based lures currently in use. In this chapter, we will discuss the development of and, as much as possible, the diversity of food-based lures that have been tested and/or are used in traps for pest tephritids. Summaries of these materials are presented in Table 3.1 (natural products), Table 3.2 (synthetic lures) and Table 3.3 (bacteria). We then discuss other factors to consider in deployment of food-based lures that can affect effectiveness as well as approaches used to evaluate food-based attractants. Finally, we conclude with a summary and discussion of future research needs.

2 Use of Natural Products as Bait

Natural products have long been used as bait for tephritid fruit flies, with the emphasis on low cost and on use of materials that are available locally. From the early 1900s through the 1950s, research focused on sugar sources and protein sources as fruit fly attractants due to the importance of these nutrients for survival and reproduction of adult flies. The effect of fermentation on effectiveness of sugar baits was recognized during the 1920s, and this recognition has been important to the development of food-based baits and lures used worldwide for fruit flies.

Table 3.1 Natural products that have been tested and used as food-based attractants for fruit flies

Type of bait or process	Base material for bait	Trade name (Commercial source if cited)	Species tested	1st reference
Sugar	Sugarcane	White cane sugar	*A. ludens*	Cooper (1905)
Sugar	Sugarcane	Piloncillo, Mexican brown sugar	*A. ludens*	Crawford (1927)
Sugar	Sugarcane	Molasses	*C. capitata, B. tryoni*	Gurney (1925)
Sugar	Sugarcane	Syrup	*A. ludens, A. obliqua*	McPhail (1939)
Active yeast	Brewer's yeast	Cenovis	*C. capitata, Bactrocera* spp.	Hill (1986)
Autolyzed protein	Brewer's yeast	Amber BYF (Tamogen Ltd., Tel Aviv, Israel)	*A. suspensa*	Burditt (1982)
Autolyzed Protein	Yeast Extract	Ohly STV[a] (Ohly, Hamburg, Germany)	*B. cucurbitae, B. dorsalis*	Barry et al. (2006)
Enzymatic hydrolysis	Yeast	NBC (Nutritional Biological Corp, Santa Clara, CA, USA)	*B. tryoni*	Morton and Bateman (1981)
Acid precipitated protein	Milk	Casein	*A. ludens, A. striata*	McPhail (1939)
Enzymatic hydrolysis	Milk	Lactalbumin	*B. dorsalis*	Gow (1954)
Enzymatic hydrolysis	Soy	Soy	*A. ludens*	López and Becerril (1967)
Enzymatic hydrolysis	Bovine serum albumin	Bovine serum albumin	*B. tryoni*	Morton and Bateman (1981)
Enzymatic hydrolysis	Cottonseed	Cottonseed protein	*A. ludens*	López and Becerril (1967)
Enzymatic hydrolysis	Corn	E802 Masoferm Steepwater (Corn Products, Summit Argo, IL, USA)	*B. cucurbitae*	Moreno and Mangan (1995)
Acid hydrolysis	Corn	NuLure[b](Miller Chem & Fert Corp, Hanover, PA, USA)	*B. cucurbitae*	López and Becerril (1967)
Acid hydrolyzed protein	Torula yeast	Torula yeast/borax pellets (ERA Intl., Freeport, NY, USA)	*A. suspensa*	López et al. (1971)
Hydrolyzed protein	Corn	Buminal (NABA GmbH, Germany)	*C. capitata, B. zonata*	El-Gendy (2012)
Hydrolyzed protein	Corn	Buminal (Bayer SA, Puteaux, France)	*B. cucurbitae*	Fabre et al. (2003)
Hydrolyzed protein	Corn	Buminal (Bayer A. G. W., Germany)	*B. oleae*	Zervas (1982)
Dried and purified Masoferm	Corn	Solulys (Roquette Freres, Lestrem, France)	*A. ludens*	Moreno and Mangan (2002)

(continued)

Table 3.1 (continued)

Type of bait or process	Base material for bait	Trade name (Commercial source if cited)	Species tested	1st reference
Solulys + additives, basis for GF-120	Corn	SolBait (Moreno & Mangan 2002)	*B. cucurbitae*	Fabre et al. (2003)
Hydrolyzed protein	Corn	Proprietory local product (Brasil)	*A. grandis, A. fraterculus*	Malavasi et al. (1990)
Hydrolyzed protein	Corn	Captor 300 (Paula, Promotora Agropecuaria Universal, Mexico, D.F.)	*A. ludens*	Moreno et al. (2001)
Hydrolyzed protein	Corn	Captor Plus (Agroquimica Tridenta, S.A. de C.V. Mexico, D.F.)	*A. ludens*	Piñero et al. (2003)
Hydrolyzed protein + plant extract	Proprietary	Questlure (Green Trading, Pretoria, S Africa)	*C. capitata, C. rosa, C. cosyra*	Grout et al. (2011)
Hydrolyzed protein + B-caryophyllene	Proprietary	Ceratitislure (Green Trading, Pretoria, S Africa	*C. capitata, C. rosa, C. cosyra*	Grout et al. (2011)
Hydrolyzed protein	Proprietary	Bio Nal (Bio Tec Company)	*C. capitata, B. zonata*	El-Gendy (2012)
Hydrolyzed protein	Proprietary	Cera Trap (Bioiberica, Barcelona, Spain)	*C. capitata, B. zonata*	El-Gendy (2012)
Hydrolyzed protein	Proprietary	Dacus bait (E.V.Y.P., Thessaloniki, Greece)	*B. oleae*	Zervas (1982)
Hydrolyzed protein	Proprietary	Entomozyl (Hoechst, Athens, Greece)	*B. oleae*	Haniotakis and Skyrianos (1981)
Hydrolyzed protein	Proprietary	Flavex (Halcyon Proteins Australia, Melbourne, Australia)	*C. capitata*	Broughton and de Lima (2002)
Hydrolyzed protein	Proprietary	Hym-Lure RTU (Robertson [Pty] Limited, Durban, South Africa)	*B. cucurbitae*	Fabre et al. (2003)
Hydrolyzed protein	Proprietary	Nasiman 73 (Tamogen Ltd., Tel Aviv, Israel)	*A. suspensa*	Burditt (1982)
hydrolyzed protein	proprietary	Protein Insecticide Lure - Low Salt (Mauri Flavours Pty., Homebush, New South Wales, Australia)	*C. capitata, B. cucurbitae, B. dorsalis*	Keiser and Wakabayashi (1981)
Hydrolyzed protein	Proprietary	Pinnacle Protein Insect Lure (low salt, source not cited)	*C. capitata*	Hill (1986)
Hydrolyzed protein	Proprietary	Pinnacle Protein Fruit Fly Bait (Mauri Yeast Produts, Brisbane, Australia)	*B. cucurbitae*	Fabre et al. (2003)
Hydrolyzed protein	Proprietary	Zitan 85 (Tamogen Ltd., Tel Aviv, Israel)	*A. suspensa*	Burditt (1982)

[a]Previously known as Provesta 621
[b]Previously known as PIB 7 or SIB 7

2.1 Fermenting Sugar Bait

Initially, traps for tephritids were baited with sugar solutions. In a report by J. Isaac to the California State Horticulture Commission (Cooper 1905), it was noted that aqueous solutions of sugar (79 g/L) were used as attractant bait sprays for the Mexican fruit fly, *Anastrepha ludens* (Loew), in Mexico. Crawford (1927) reported that an aqueous solution of piloncillo (36 g/L), a brown sugar available in Mexico, was used as a cheap alternative to white sugar in sweet bait sprays in studies conducted from 1913–1914 in Mexico. Tests of these solutions in traps were initiated but not completed, so no trapping results were available. Gurney (1925) reported that traps baited with fruit juice and molasses or treacle were used in Australia with varying success. At this time, there were several locally produced proprietary baits (i.e., Watson's specific and Harvey's lure) that were also in use in Australia, but ingredients for these baits were not disclosed.

During the 1920s and 1930s, there was active research on sugar-baited traps for fruit-infesting Lepidoptera (Peterson 1925; Frost 1926; Yetter and Steiner 1931; Eyer and Rhodes 1931; Eyer 1935). These baits were made as aqueous solutions and, because it was noted that microbial action occurred quickly after field deployment and seemed to increase insect attraction, there was a change in terminology from sugar bait to fermenting sugar bait. These research reports evaluated various by-products of fermentation, including CO_2, alcohol, and acetic acid. Research during this time also tested combinations of aqueous sugar solutions with various chemicals, which were known to be products of fermentation or hydrolysis of sugar, as attractants for pest moths (Peterson 1925; Frost 1937; Eyer et al. 1937). Newell (1936) noted that traps baited with a fermenting mixture of citrus juice and brown sugar were used in traps during the 1932–1933 Florida eradication efforts for the West Indian fruit fly, *Anastrepha obliqua* (Macquart), and the Caribbean fruit fly, *Anastrepha suspensa* (Loew).

2.2 Aqueous Yeast-Fermented Sugar Baits

Peterson (1924) evaluated sugar fermentation products as attractants for onion-infesting Diptera and added various types of active yeast to increase the production of attractive by-products with the goal of improving longevity of the baits. The addition of yeast to sugar baits was an active area of research for improving baits for fruit flies as well. McPhail added dry active brewer's yeast to aqueous sugar bait as an alternative to natural inoculation of wild yeast in field tests involving *A. ludens* (Baker et al. 1944). There were conflicting reports of effectiveness of natural inoculation of sugar baits, with McPhail (unpublished 1938 manuscript) finding that natural inoculation was sufficient to increase fruit fly attraction but with others reporting that wild microbes destroyed bait attractiveness (Green et al. 1960). Starr and Shaw (1944) tested capture of *A. ludens* with fermenting aqueous sugar-yeast

Table 3.2 Ammonia-based synthetic chemicals that have been tested and used as food-based attractants for fruit flies

Lure component(s)	Trade name (commercial source if cited)	Species tested	1st reference
Ammonium acetate	BioLure (Suterra LLC, Bend, OR, USA)	*R. pomonella*	Hodson (1943)
Ammonium acetate, cadaverine, trimethylamine	SEDQ (Barcelona, Spain)	*C. capitata*	Navarro-Llopis et al. (2008)
Ammonium acetate, cadaverine, trimethylamine	Trypack (Econex, Santomera, Murcia, Spain)	*C. capitata*	Navarro-Llopis et al. (2008)
Ammonium acetate, n-methyl pyrrolidine	EPAlure (EPA, Valencia, Spain)	*C. capitata*	Navarro-Llopis et al. (2008)
Ammonium acetate, putrescine	2C BioLure (Suttera LLC, Bend, OR, USA)	*C. capitata, A. ludens*	Heath et al. (1995)
Ammonium acetate, putrescine, trimethylamine	3C BioLure (Suttera LLC, Bend, OR, USA)	*C. capitata, A. ludens*	Heath et al. (1997)
Ammonium acetate, trimethylamine	TMA Susbin (Mendoza, Argentina)	*C. capitata*	Navarro-Llopis et al. (2008)
Ammonium bicarbonate	AgriSense Lure (Suterra LLC, Bend, OR, USA)	*R. pomonella*	Hodson (1943)
Ammonium bicarbonate, linolenic acid, putrescine, pyrrolidine	na	*B. cucurbitae*	Wakabayashi and Cunningham (1991)
Ammonium bicarbonate, methylamine HCl, putrescine	AFF lure (Advanced Pheromone Tech., Marylhurst, OR, USA)	*A. ludens*	Robacker and Czokajlo (2006),
Ammonium carbonate	na	*R. cingulata*	Frick (1952)
Ammonium carbonate	na	*B. tryoni*	Perkins and Hines (1934)
Ammonium carbonate	Polycon dispenser (Great Lakes IPM, Vestaburg, MI, USA)	*R. mendax*	Liburd et al. (1998)
Ammonium hydroxide	household ammonia	*R. pomonella*	Hodson (1943)
Ammonium hydroxide	household ammonia	*Z. electa*	Boucher et al. (2001)
Ammonium phosphate	na	*B. oleae*	Gow (1954)
Ammonium sulfate	na	*R. pomonella*	Hodson (1943)
Ammonium sulphate	na	*B. oleae*	Zervas (1982)

bait (8 % sucrose and 0.15 % brewer's yeast, allowed to ferment for 1–2 days at room temperature before use) presented alone or with several different additives. Of these additives, they found that pyridine provided a slightly improved lure. Steiner reported in 1936 that sassafras oil added to fermenting sugar bait also increased attractiveness (Green et al. 1960).

Table 3.3 Bacteria, spent culture media and supernatants tested as food-based attractants for fruit flies

Bacteria	Culture media	Presentation	Species tested	Test type	References
Various	Soy hydrolysate (1 %) plus dextrose (0.5 %)	Cells in culture media	*B. dorsalis*	Field cage olfactometer	Gow (1954)
Providencia rettgeri	NuLure at pH 6.5 & 9	Cells in culture media	*B. tryoni*	Field cage	Drew and Fay (1988)
Enterobacteriacea	Peptone-yeast extract broth	Cells in culture media, supernatant	*B. dorsalis*	Field cage olfactometer	Jang and Nishida (1990)
Staphylococcus spp	Tryptic soy broth	Cells in culture media	*A. ludens*	Field cage	Robacker et al. (1991)
Pantoea agglomerans[a]	Tryptic soy broth, agar	Washed cells	*R. pomonella*	Field test	MacCollom et al. (1992, 1994)
Citrobacter freundii	Tryptic soy broth	Cells in culture media	*A. ludens*	Field test	Martinez et al. (1994)
Klebsiella pneumonia	Trypticase soy broth	Supernatant	*A. ludens*	Cage top	Lee et al. (1995)
Staphylococcus aureus	Tryptic soy broth	Supernatant	*A. ludens*	Cage top	Robacker and Flath (1995)
Citrobacter freundii	Trypticase soy broth	Supernatant	*A. ludens*	Cage top	DeMilo et al. (1996)
Citrobacter freundii, Klebsiella pneumoniae	Tryptic soy broth	Solvent-extracted supernatant	*A. ludens*	Cage top	Robacker and Bartelt (1997)
Pantoea agglomerans[a]	Tryptic soy agar	Inoculated agar plates	*R. pomonella*	Field test	Lauzon et al. (1998)
Enterobacteriacea	Tryptic soy broth	Supernant	*A. ludens*	Cage top	Robacker et al. (1998)
Pantoea agglomerans[a]	Tryptic soy agar	Washed cells, inoculated agar plates	*A. suspensa*	Wind tunnel	Epsky et al. (1998)
Pantoea agglomerans[a]	tryptic soy agar	inoculated agar plates	*R. pomonella*	Field cage	Lauzon et al. (2000)
Pantoea agglomerans[a]	Uric acid agar	Inoculated agar plates	*A. ludens*	Wind tunnel	Robacker and Lauzon (2002)
Pantoea agglomerans[a]	Several liquid media	Cells in culture media, supernatant, washed cells	*A. ludens*	Cage top	Robacker et al. (2009)

[a]Previously *Enterobacter agglomerans*

2.3 Aqueous Protein Baits

McPhail (1939) began studies in Mexico on fermenting sugar baits for *A. ludens* and *Anastrepha striata* Schiner. These studies initially focused on Mexican brown sugar (piloncillo) and commercial syrup with various additives, including lye (sodium hydroxide [NaOH]) to hydrolyze the sugar. He noted the presence of protein as an impurity in the piloncillo by the smell of ammonia from the hydrolyzed solutions as well as the appearance of a protein-like foam when the piloncillo was cooked in limewater (calcium hydroxide [$Ca(OH)_2$]). The studies then shifted to tests of various proteins hydrolyzed with NaOH. Attempts were made to correlate amounts of ammonia released with attraction, but this was not confirmed. He did note, however, that ammonia was not the only attractant produced as protein baits were more attractive than aqueous ammonia bait. From this research he developed aqueous protein bait for fruit flies that contained casein (40 g/L) and NaOH (15 g/L) (McPhail 1943).

Finney (1948, 1950) and Hagen (1950) were developing methods to mass rear lacewings, *Chrysopa californica* Coquillett, and found that fecundity was increased by replacing honey with either honeydew or hydrolyzed brewer's yeast. In parallel studies, they found that hydrolyzed brewer's yeast improved fecundity of the oriental fruit fly, *Bactrocera dorsalis* (Hendel), the melon fly, *Bactrocera cucurbitae* (Coquillett), and *C. capitata* when it was used in place of dry brewer's yeast in adult diets (Hagen and Finney 1950). Steiner (1952) reported on unpublished data from Finney and Hagan that demonstrated that solutions of enzymatic yeast or soy hydrolysates were attractive to *C. capitata* and *B. dorsalis*.

2.4 Yeasts and Aqueous Protein Baits

In the context of human nutrition, Bekatorou et al. (2006) reviewed the use of yeasts for traditional fermentation processes and as alternative protein sources. *Saccharomyces cerevisiae* Meyen ex E.C. Hansen has been the most commonly cultivated yeast since ancient times. Yeast feeds on carbohydrates; the by-products of this fermentation process include carbon dioxide (CO_2) and alcohol (specifically ethanol), which are desirable to bakers and brewers, respectively. Beer brewers have selected specific strains of *S. cerevisiae* that grow slowly, produce more alcohol, and yet are able to thrive in high alcohol substrates. Similarly, bakers have selected strains that grow rapidly and produce more CO_2, which in turn gets trapped as tiny bubbles within the dough giving bread its characteristic rise. These strains are commonly known as brewer's yeast and baker's yeast, respectively. Wine makers, however, have traditionally relied on wild yeasts present in the grape skins for the fermentation, but because this produces inconsistent results, modern wine makers prefer to add a known pure yeast culture that overpowers the wild yeasts (usually strains of *S. cerevisiae*) to the grapes, thereby turning out a more consistent fermented product (González Techera et al. 2001).

Once yeast has no more available carbohydrate to feed on, it dies and undergoes autolysis, which is the process by which the yeast's own digestive enzymes break down the proteins into component peptides and amino acids. Autolyzed yeast, which is commonly sold as yeast extract, is widely used as a nutritional supplement, because it is high in protein. The manufacturing process for yeast extract products relies on the autolysis of yeast, typically accomplished by subjecting the yeast suspension to osmotic shock with the addition of NaCl (Anonymous 2009). The shriveling and dying yeast cells are then heated to complete their breakdown after which the thick cell walls are removed by centrifugation and subsequent filtration. Removing the cell walls concentrates the flavors and changes the texture.

Torula yeast, *Candida utilis* (Henneberg) Lodder & Kreger-van Rij (formerly known as *Torula utilis*), is a species of yeast widely used in its autolyzed form as a nutritional supplement or as a flavor enhancer in processed foods. It is a by-product of the paper mill industry and is propagated on wood sugars leftover after the pulp has been removed from wood for the production of paper. In a manner similar to that described above, the yeast undergoes autolysis in order to obtain peptides and amino acids. It is then spray-dried to produce a fine, light grayish-brown powder (Anonymous 1964), which is available commercially.

2.5 Role of Hydrolysis in Modifying Proteins and the Chemistry of Hydrolysis

A number of substrates tested and ultimately used as fruit fly attractants have been products of protein hydrolysis, and the type of hydrolysis can affect type and amount of chemicals released as volatiles. Hydrolysis is a process whereby chemical bonds are broken by the insertion of water between the atoms in the bond. Proteins are composed of numerous amino acids joined together with peptide bonds; hydrolysis destroys the peptide bonds resulting in a protein hydrolysate solution composed of smaller chains of amino acids (peptides), free amino acids or parts thereof, including ammonia (Univ. Waikato 2007). There are three general methods used to hydrolyze protein: acid hydrolysis, alkaline hydrolysis, and enzymatic hydrolysis. Products of acid hydrolysis and enzymatic hydrolysis have been used as fruit fly attractants. A strong acid, such as 6 M hydrochloric acid (HCl), is ordinarily used for the hydrolysis of proteins, which involves boiling the proteins in the acid for many hours (Anonymous 2011). This process attacks all peptide bonds in the protein substrate, destroying some of the individual amino acids. However, not all of them degrade to the same extent. For example, tryptophan is usually totally lost in an acid hydrolysis, while cysteine, serine, and threonine are partially broken down, and asparagine and glutamine are converted to their acidic forms. Salt may be formed during neutralization of an acid hydrolysis, resulting in a product with high salt content (Anonymous 2009). In enzymatic hydrolysis, proteins are hydrolyzed more gently than with acid hydrolysis, and the process does not require high temperatures. This type of hydrolysis, however, is target-specific depending on the enzymes used.

3 Ammonia Solutions and Salts, and Modifications to Aqueous Protein Bait

Ammonia is one of the primary products of protein hydrolysis and, as there was a redirection in use of fermenting sugar baits to use of protein-based baits for fruit fly attractants, there were investigations into the use of ammonia as fruit fly bait. Research that started in the 1940s evaluated the amino acid glycine along with proteins such as casein, which is obtained from milk. The widespread search for fruit fly attractants is cited by Hodson (1943) who describes ‘a review of extensive Italian, South African and Australian literature’ on fruit flies ‘infesting especially citrus fruits and olives, made it evident’ that ‘all of them contained ammonia and release it upon decomposition.’ Thus began an evaluation of various formulations of ammonia, including ammonium salts (e.g., ammonium carbonate, ammonium bicarbonate, ammonium acetate, ammonium sulfate, ammonium phosphate) and ammonium solutions (e.g., ammonium hydroxide which is also known as household ammonia). There were also further investigations into hydrolyzed yeast or other sources of commercially available proteins as well as modifications to aqueous protein baits to improve effectiveness. This ultimately resulted in the development of commercially available ammonia-based synthetic lures and the pelleted formulation of protein bait that facilitated field use.

3.1 Ammonia Baits

Boyce and Bartlett (1941), after discussions with McPhail and Baker, found that aqueous casein (200 g casein, 300 mL NaOH, 3,800 mL water; which they called ‘McPhail’s lure’) or aqueous glycine (2 % glycine and 3 % NaOH) were highly attractive to the walnut husk fly, *Rhagoletis completa* Cresson. Dean (1941) found that protein baits captured more apple maggot flies, *Rhagoletis pomonella* (Walsh), than sugar baits. Hodson (1943, 1948) found that more *R. pomonella* were captured in traps baited with various ammonia solutions, including household ammonia, ammonium sulfate, ammonium acetate, or ammonium carbonate, than in traps baited with solutions of glycine plus NaOH or casein plus NaOH. Baits were deployed in open pans, and he found that the addition of soap to break the surface tension improved the retention of attracted flies. Hodson also found good capture of flies in dry sticky traps baited with ammonium carbonate. Frick (1952) found the dry sticky trap with ammonium carbonate was effective for capturing the cherry fruit fly, *Rhagoletis cingulata* (Loew). Sticky traps baited with ammonium hydroxide (59 mL of 27–31 % aqueous solution placed in vial with cotton balls) have been used to capture the pepper maggot, *Zonosemata electa* (Say) (Boucher et al. 2001). Raz (1998) used traps baited with ammonium sulfate (2 %) and hexanol to monitor populations of *C. capitata* and the Mediterranean black fig fly, *Silba adipata* McAlpine (Diptera: Lonchaeidae), in Israel. Several commercial formulations of ammonia have been produced for use as lures in fruit fly traps, including ammonium acetate (BioLure, Suterra LLC, Bend, OR, USA), ammonium bicarbonate (AgriSense Lure, Suterra LLC), and ammonium carbonate (Great Lakes IPM, Vestaburg, MI, USA; ISCA technologies, Riverside, CA, USA).

Gow (1954) reviewed the use of ammonia-based lures, which he refers to as ammoniacal baits. This included the use in Australia of vanilla extract combined with ammonia for various tephritids (Jarvis 1931; 0.44 % vanilla, 1.75 % household ammonia) and ammonium carbonate for the Queensland fruit fly, *Bactrocera tryoni* (Froggatt) (Perkins and Hines 1934). He also listed Clensel, described as proprietary ammonium soap, as being attractive to the olive fruit fly, *B. oleae* (Rossi) (Bua 1933, 1938) and *C. capitata* (Newman and O'Connor 1931) and also various ammonium salts and ammonium phosphate for attraction of *B. oleae*. Membrane-based ammonium acetate lures (BioLure, Suterra LLC, Bend, OR, USA) were found to be effective for capture of *B. oleae* (Economopoulos et al. 1986). Robacker et al. (1996) provided a list of ammonium salts used as sources of ammonia in tests of a number of tephritid species. In research by Gow (1954) to develop a protein bait for *B. dorsalis*, test substances were compared to (1) van Zwaluenburg fermenting bait, which was developed at the Hawaiian Sugar Planters Association Experimental Station (raw sugar [80 g], white vinegar [13 mL], fresh yeast [1/4th cake, the equivalent of 7.4 mL dry yeast] per liter of water), and 2) Jarvis ammoniacal bait (ammonium hydroxide [0.67 %], artificial vanilla extract [0.5 %]). The fermenting bait was more attractive than the Jarvis bait. In these studies, Gow tested several yeast hydrolysates alone or in combination with the fermenting bait. He found that addition of antibiotics to inhibit mold growth resulted in improved attraction and that soy hydrolysate was more attractive than casein hydrolysate or lactalbumin hydrolysate. He also noted that attraction was due primarily to products of microbial activity and that ammonia alone was only 'mildly attractive' and could be repellent at some concentrations. Simanton (1958) noted that, at the start of the *C. capitata* eradication effort in 1957 in Florida, USA, the standard detection system was a glass McPhail trap baited with an aqueous solution of hydrolyzed yeast (5 %) and ammonium chloride (5 %).

3.2 Aqueous Protein Bait and the Role of Borax

By the 1960s, dry lures were available as male lures for *C. capitata* (trimedlure), *B. dorsalis* (methyl eugenol), and *B. cucurbitae* (cue-lure), or food-based lures (ammonium carbonate) for temperate zone tephritids (Green et al. 1960). However, aqueous protein baits continued to be the best attractants for tropical tephritid females or for capturing both sexes. Research continued in Mexico on aqueous hydrolyzed protein baits for *A. ludens*. New protein sources were evaluated, including an acid hydrolyzed corn product known first as Staley's Insecticide Bait No. 7 (SIB 7), then as Staley's Protein Insecticide Bait No. 7 (PIB 7), and subsequently (and referred to herein) as NuLure (Miller Chemical & Fertilizer Co., Hanover, PA, USA). Tests were also conducted on another protein source, enzymatic hydrolyzed cottonseed protein (CTPH; López and Becerril 1967). Early studies noted that aqueous NuLure solutions (1 %) quickly changed color and putrefied after field deployment, which increased attraction of non-target flies, and also that captured target flies disintegrated in the liquid. López and Becerril (1967) tested 76 chemical additives to NuLure solutions and found that sodium tetraborate decahydrate (borax, sodium borate) was the most promising and prevented the problems of bait discoloration and fly disintegration. Field trials showed that more *A. ludens* were captured with aqueous NuLure:borax (1:0–3 %) than with the standard fermenting aqueous sugar lure that contained light brown sugar (8 %), dry brewer's yeast (0.15 %), and pyridine (0.1 %). These authors noted, however, that the addition of 1-3 % borax to 1 % NuLure aqueous solution decreased total capture of *A. ludens* versus 1 % NuLure without borax. Addition of 2 % borax to CTPH also prevented fly disintegration but without decreasing capture. Other observations reported from this research included female-bias in protein baits, a male-bias in fermenting sugar baits, and that addition of borax to NuLure increased bait pH, which caused an immediate increase in ammonia release. Transporting and deploying bait were improved by the development of pelletized lures (López et al. 1968). Pellets included 2 parts borax by weight to 1 part by volume of either NuLure or hydrolyzed CTPH and were added as 2 pellets to 300 mL per trap. Solutions made using pelletized baits captured ~10 % fewer flies than solutions made using non-pelletized baits, likely due to the slow dissolution of the pellets, which reduced initial attractiveness.

López et al. (1971) compared capture of *A. suspensa* with solutions of NuLure: borax, CTPH:borax, or enzymatic hydrolyzed torula yeast:borax (TYB) and reported the highest capture with aqueous solution of hydrolyzed TYB (3:4 %). Burditt (1982) tested solutions made using pellets that contained hydrolyzed TYB (4:5 parts) for capture of *A. suspensa* in Florida. He found no difference in capture in traps baited with either 2 or 6 pellets and confirmed that hydrolyzed TYB solution captured equal or greater numbers of *A. suspensa* than other hydrolyzed protein solutions tested (i.e., Amber BYF, a water soluble fraction of autolyzed brewer's yeast [Amber Laboratories, Milwaukee, WI, USA]; Zitan 85; Nasiman 73 [Tel Aviv, Israel]). Malo (1992) conducted field tests in Mexico that compared TYB solutions that had aged 2, 4, 6, 8, and 10 days in the laboratory prior to field placement, and found no differences in capture of *A. ludens* and *A. obliqua*. Nakagawa et al. (1971), in tests of *C. capitata* in Hawaii, found that aqueous solutions of NuLure:borax (5:5 %) sometimes captured more but more often captured fewer flies than trimedlure-baited traps. They hypothesized that higher captures in aqueous protein-baited traps reflected a lack of nutrients in the field, which resulted in increased female response. Cunningham et al. (1978) reported that the standard lure in use in Hawaii was NuLure:borax (9:5 %), which was more effective in areas with low rainfall than in areas with high rainfall. They hypothesized that this was due to a combination of attraction to a water source as well as lack of competing adult food sources in the drier areas. Keiser and Wakabayashi (1981) found that addition of linolenic acid (0.1 %) to either NuLure:borax (9:5 %) or Protein Insecticide Lure – Low Salt:borax (9:5 %; Mauri Flavours Pty., Homebush, New South Wales, Australia) increased capture of *C. capitata, B. cucurbitae,* and *B. dorsalis* over either aqueous protein bait alone, although linolenic acid alone as a bait captured few flies. In field tests of the South American cucurbit fruit fly, *Anastrepha grandis* (Macquart), and *Anastrepha fraterculus* (Wiedemann), there was equal capture in traps baited with aqueous corn protein hydrolysate:borax (5:3 %) and aqueous TYB, and all captures were greater than in traps baited with aqueous molasses (1 %, Malavasi et al. 1990).

Although ammonia lures had been found to be the best lures for *Rhagoletis* spp. and *B. oleae* in earlier research (see 2.1 Ammonia baits), later research revisited the use of protein hydrolysates for these species. Traps baited with ammonium acetate and protein hydrolysate (Sheffield Hy Case 802, Sheffield Chem. Co., Norwich, NY, USA) alone or in combination caught more *R. pomonella* than unbaited traps early in the summer, but there was no difference in late summer (Moore 1969). In subsequent tests, various combinations of protein hydrolysates (soy, yeast, casein [Nutritional Bioch. Corp., Cleveland, OH, USA], Edamin T, NZ amine, Hy-Case Amino [Sheffield Chem., Union, NJ, USA], Seclur FF tablets [3 M Co., St. Paul, MN, USA]) alone or in combination with ammonium acetate, as well as ammonia bait alone (ammonium acetate, ammonium carbonate, ammonium phosphate) were evaluated (Reissig 1974). All lures were tested in dry traps, with aqueous protein baits placed in vials containing cotton wicks. A combination of yeast hydrolysate (5 %) and ammonium acetate solution (50 %) was found to be most attractive in tests of *R. pomonella.* Reissig (1976) later tested similar treatments of protein hydrolysates alone or in combination with ammonium acetate. Again, the combination of yeast hydrolysate (5 %) and ammonium acetate solution (50 %) was found to be most attractive for the black cherry fruit fly, *Rhagoletis fausta* (Osten Sacken), while ammonium acetate solution (50 %) alone was the most attractive for *R. cingulata.* Liburd et al. (1998) found that dry sticky traps baited with ammonium carbonate or a combination of ammonium acetate and dry protein hydrolysate could be used to capture the blueberry maggot, *Rhagoletis mendax* Curran. Barry and Polavarapu (2004) found that more *R. mendax* were attracted to aqueous solutions (vol:vol) of Solbait (50 %; Moreno and Mangan 2002) than to NuLure (9 %), with intermediate attraction to AY50% (2 %; Mauri Yeast Australia Pty. Limited). Katsoyannos et al. (2000) found that sticky traps baited with ammonium acetate (BioLure) were more effective for capture of the European cherry fruit fly, *Rhagoletis cerasi* L., than traps baited with ammonium bicarbonate or aqueous NuLure:borax (9:3 %).

Tests of *B. oleae* found that an aqueous solution of protein hydrolysate (2 %, Rodia, Rhone-Poulenc Inc., Paris, France) and borax (1.5 %) was more attractive than the standard ammonium sulfate aqueous solution (2 %) tested at 300 mL per trap (Prokopy and Economopoulos 1975). Later studies used protein hydrolysate (2 %, Entomogyl) and borax (1.5 %) for *B. oleae* capture (Fletcher and Kapatos 1981). Aqueous protein baits, such as Buminal (Bayer A.G.W., Germany), were found to be more effective than ammonium salt solutions (Economopoulos 1986). However, in tests conducted in olive orchards in Greece, Broumas and Haniotakis (1994) found no difference in capture among six bait treatments that included ammonium bicarbonate, ammonia carbonate, aqueous ammonium sulfate (2 % wgt:vol), a mixture of protein hydrolysate and molasses (Dacona, Phytophyl, Shimatari Viotias, Greece), and *Dacus* bait (Alesis S. A., Thessaloniki, Greece). In research conducted in California, USA, Yokoyama et al. (2006) used traps baited with ammonium bicarbonate (Vioryl, Athens-Lamia, Greece) or ammonium carbonate (Suterra LLC, Bend, OR, USA), and Villamil (2012) used traps baited with aqueous TYB to monitor populations of *B. oleae*.

Although hydrolyzed TYB pellets were the standard bait for *Anastrepha* spp. prior to ~1990, the hydrolyzed torula yeast was replaced by torula yeast and TYB pellets are used currently to make the standard aqueous protein bait for these species (Anonymous 2006).

3.3 Aqueous Protein Bait and the Role of pH in Fruit Fly Attraction

Matsumoto et al. (1985) and Flath et al. (1989) showed that increasing the pH of NuLure from 4.5 to 8.7 increased attraction of *C. capitata, B. dorsalis,* and *B. cucurbitae* in field tests. Laboratory bioassays of aqueous solutions of NuLure (10 %) with borax (0, 1, 5 and 10 %) showed an increase in capture of *A. suspensa* with increasing borax levels, but field tests found that traps baited with aqueous TYB solution (3 pellets per 300 mL) captured equal or higher numbers of flies than any of the NuLure solutions (Epsky et al. 1993). The same results were obtained in parallel tests of *A. ludens*, but *C. capitata* capture was highest in traps baited with aqueous solutions of NuLure (10 %) with the highest amounts of borax (5 and 10 %) (Heath et al. 1994). Subsequent studies evaluated corn steepwater (E802 Masoferm [aka Mazoferm], Corn Products, Summit Argo, IL, USA), another acid hydrolyzed corn product, for attraction of *A. suspensa*. Field tests revealed that similar numbers of flies were captured in traps baited with either TYB solutions or Masoferm (10 %) with borax (1 %), but fewer flies were captured when more borax was added (3, 5 and 10 %) (Epsky et al. 1994). When the effect of aging in the field was examined, the capture of *A. suspensa* with TYB solution decreased over the seven days in the field but remained constant or increased with Masoferm plus borax over that time period.

Duyck et al. (2004) evaluated the role of pH in attraction of *B. cucurbitae* to several aqueous protein baits in field cage tests. Borax (0, 1, 5 and 10 %) was added to aqueous NuLure (5 %) and Buminal (5 %) solutions, which increased pH from 3.5 to 9.1 and 5.6 to 9.3, respectively, but decreased fruit fly capture relative to either bait without added borax (0 %). In a follow-up study, these authors examined the effect of pH modification on NuLure, Buminal, and torula yeast alkalized with the addition of NaOH, and Buminal and torula yeast acidified with nitric acid (HNO_3). The alkalization using NaOH had no effect on the response of *B. cucurbitae* to NuLure for solutions with pH 4 to pH 10, but again response to Buminal decreased for solutions with pH 6 to pH 9. Torula yeast solution increased in attractiveness as pH increased from pH 9 to pH 10.5 and remained high at pH 12. Acidification of Buminal increased attraction as pH decreased from pH 6 to pH 3, and acidification of torula yeast solution decreased attraction as pH decreased from pH 9 to pH 7 and stayed low for pH 6 and pH 3 solutions. A number of factors may affect the final pH of protein bait solutions and hence efficacy for various fruit flies species. The most important factors likely include the pH of water used to make the solutions (Epsky et al. 1993), the initial pH of protein bait, which may vary among source of the bait and/or storage conditions prior to use (Epsky, unpublished data), and the substance used to modify bait pH (Duyck et al. 2004).

4 Multiple Component Synthetic Lures

As noted above, one of the problems with aqueous protein baits is the high variability in the source material, which increases the difficulty in using information from traps for management decisions. Availability of synthetic lures with controlled release of attractive chemicals would overcome this problem and would allow more direct comparisons among results of tests conducted in different areas or in different host plants. Although single component ammonia lures were found to be equal to or more effective than aqueous protein baits for most *Rhagoletis* spp. (see Sect. 3.1), research with *Anastrepha* spp., *C. capitata* and some *Bactrocera* spp. typically found that traps baited with aqueous protein baits captured more flies than traps baited with ammonia alone. Research on identification of volatile chemicals from aqueous protein baits, in addition to ammonia, led to the development of multiple component synthetic food-based lures.

4.1 Identification of Volatile Chemicals from Aqueous Protein Baits

Baker et al. (1944) described the early research efforts primarily as empirical tests of materials that were likely attractants or known attractants for other types of flies. For the most part, the materials tested were various products of microbial action (e.g., alcohol, acetic acid, etc.), protein degradation, (e.g., amino acids), or were based on odors perceived from the test material (e.g., ammonia). Once attractiveness was confirmed, analyses were undertaken to relate it to chemical structure

(Green et al. 1960). With improved analytical chemistry techniques available by the 1980s, there was a shift to identification and quantification of volatile chemicals emitted from aqueous protein baits in addition to the continuation of empirical tests of promising compounds. Morton and Bateman (1981) analyzed various aqueous protein baits (including NuLure, enzymatic yeast hydrolysate [NBC], and bovine serum albumin) using gas chromatography-mass spectrometry of methylene chloride extracts and head-space volatiles to identify volatile chemical constituents. A total of 39 chemicals were identified in the aqueous protein baits, and the role of these chemicals was hypothesized to be primarily as feeding stimulants, which increased capture of flies that had been attracted to the bait by the ammonia. Additional analyses of NuLure and autolyzed brewer's yeast resulted in the identification of 43 volatile components obtained from headspace volatiles and from vacuum steam distillation (Buttery et al. 1983). Matsumoto et al. (1985) identified major chemicals from analysis of vacuum steam distillation extracts of NuLure. Lee et al. (1997) identified 19 compounds from headspace collections from Masoferm concentrate (pH 3.9) or Masoferm adjusted to pH 8 by addition of NaOH.

The importance of ammonia as the primary attractant released from aqueous protein baits was documented in studies of the *B. tryoni* (Bateman and Morton 1981). Mazor et al. (1987) showed the importance of ammonia for *C. capitata* attraction and, while showing a direct correlation between ammonia release and fruit fly attraction, confirmed that additional chemicals added to the attractiveness of aqueous protein baits. Keiser et al. (1976) found that acetic acid and acetic anhydride, identified as contaminants of the male attractant cue-lure (Jacobson et al. 1976), were attractive to *C. capitata*, *B. dorsalis,* and *B. cucurbitae* in laboratory bioassays as 0.1 % solutions but repellent as 1 % solutions. Subsequently, Buttery et al. (1983) identified acetic acid as one of the major components released from NuLure.

Casaña-Giner et al. (2001) evaluated the attractiveness of protein baits and 79 chemicals identified from protein baits, host fruit, and *C. capitata* male emissions in field tests conducted in Spain. Chemical groups tested included (1) heterocyclic nitrogen compounds, (2) male compounds, and (3) proteinaceous and ammonia compounds, which included corn steep liquor, Buminal, ammonia, methylamine-HCL, putrescine and cadaverine. They found that the highest capture was in traps baited with mixtures of corn steep liquor (source not given), ammonia compounds and amines followed by traps baited with fruit volatiles. Low capture was obtained with traps baited with chemicals emitted by males.

Mazor (2009) confirmed the role of ammonia in *C. capitata* attraction and documented potential competition from ammonia released from manure or other agricultural supplements applied as fertilizer that may interfere with fly response to traps baited with food-based lures. Laboratory bioassays using a six-choice olfactometer found the highest response to pelletized poultry manure and aqueous ammonium nitrite. The next highest response was to ammonium acetate, guano and poultry litter, followed by Entomela (Vioryl, Athens, Greece), Buminal, and cattle manure. Poor attractants included NuLure, Corn Steepwater Liquor (Roquette, Lestrem, France), and Nasiman. The commercial baits were tested at the original concentration or as aqueous solutions (10 %). More flies were attracted to the original concentrations of Entomela, Buminal, Corn Steepwater Liquor and Nasiman than diluted baits, but they preferred diluted NuLure over original concentration. Mazor (2009) speculated that this was due to reduction in repellant chemicals from the concentrated NuLure when it was diluted.

4.2 Development of Synthetic Chemical Blends with Ammonia and Putrescine

In a series of field tests of synthetic chemicals that had been identified from chemical analysis of aqueous protein baits or that were from known degradation products of amino acids and fats, Wakabayashi and Cunningham (1991) found that traps baited with an aqueous blend of ammonium bicarbonate, linolenic acid, putrescine, and pyrrolidine were as effective in capturing sterile *B. cucurbitae* as traps baited with NuLure:borax (9:5 %). Research by Robacker and Warfield (1993) and Robacker (1995) showed that traps baited with a blend of ammonium bicarbonate, methylamine HCl, and putrescine as an aqueous solution (10:10:1 ratio) or ammonium carbonate, methylamine HCl, and putrescine (AMPu) mixed into agar (6:10:1) were equal to or better than traps baited with TYB aqueous solutions in laboratory and field tests of sterile *A. ludens*. Heath et al. (1995) found that the combination of ammonium acetate (BioLure) and a vial-formulation of putrescine could be used in traps to capture *C. capitata* and *A. ludens*. Subsequently, a membrane-based putrescine lure was also commercially available for use with the membrane-based ammonium acetate lure as a two component attractant (2C BioLure, Suterra LLC) (Epsky et al. 1995). Traps baited with 2C BioLure captured similar or greater numbers of flies than traps baited with TYB solution in tests of *A. suspensa* (Florida), *A. ludens* (Texas, Mexico; Thomas et al. 2001), and in tests that included 19 *Anastrepha* spp. and *C. capitata* (Guatemala; Martinez et al. 2007). Comparisons of *A. ludens* capture in traps baited with 2C BioLure and AFF lure, a commercial formulation of AMPu (Advanced Pheromone Tech., Marylhurst, OR, USA), found equal or greater capture with 2C BioLure (Robacker and Czokajlo 2006; Robacker and Thomas 2007). Traps baited with TYB solution, however, tended to capture more sterile *A. ludens* than traps baited with 2C BioLure (Conway and Forrester 2007).

4.3 Addition of Trimethylamine to Ammonia and Putrescine Blend

Robacker and Flath (1995) identified methylamine, dimethylamine, and trimethylamine from a microbial supernatant that was attractive to *A. ludens* in laboratory bioassays. Although they had demonstrated previously that methylamine was attractive to this fly, dimethylamine HCl and trimethylamine HCl were not attractive. Heath et al. (1997), in field trials of *C. capitata* and *A. ludens* conducted in Guatemala, tested these chemicals in combination with 2C BioLure. They found that trimethylamine HCl was synergistic as traps with trimethylamine HCl alone were not attractive, but traps baited with 2C BioLure plus trimethylamine HCl were more attractive to *C. capitata* than traps baited with either 2C BioLure or TYB solution. Trimethylamine HCl in combination with 2C BioLure was less attractive to *A. ludens* than aqueous TYB, with intermediate capture with 2C BioLure alone. Trimethylamine HCl, formulated in a membrane-based lure, is commercially available in combination with ammonium acetate and putrescine lures (3C BioLure, Suterra LLC). In tests conducted in several continents, McPhail-type traps baited with 3C BioLure and used with aqueous retention fluid containing triton as a surfactant captured equal or more *C. capitata* than traps baited with aqueous NuLure:borax (9:5 %) (Epsky et al. 1999; Miranda et al. 2001) or traps baited with other aqueous protein baits (Broughton and de Lima 2002). There was also equal capture of *A. suspensa* in traps baited with either 3C BioLure or aqueous TYB, although both of those baits captured fewer flies than in traps baited with 2C BioLure (Holler et al. 2006; Epsky et al. 2011). Leblanc et al. (2010a), however, found that 3C BioLure-baited traps captured fewer *B. cucurbitae* and *B. dorsalis* than aqueous TYB-baited traps and equal numbers of *C. capitata* in Hawaii. Additional comparisons of 2C BioLure, 3C BioLure (called FA-2 and FA-3, respectively), NuLure:borax (9:3 %), and trimedlure tested in various traps and environments and against various target species were conducted as part of an International Atomic Energy Agency Cooperative Research Programme (IAEA CRP; IAEA 1999). Economopoulos (2002) and Robacker and Landolt (2002) presented overviews of the role of 3C BioLure and other attractants for *C. capitata* detection and monitoring. In field tests conducted in South Africa, 3C BioLure was more effective for capturing males and females of the Natal fruit fly, *Ceratitis rosa* Karsch, and females of the mango fruit fly (also known as the marula fly), *Ceratitis cosyra* (Walker), than aqueous protein baits Questlure (Green Trading, Pretoria, S. Africa) and Ceratitislure (Green Trading, Pretoria, S Africa) (Grout et al. 2011). However, in that study, Ceratitislure captured more male *C. cosyra* than 3C BioLure.

2C and 3C BioLures were originally formulated as separate lures, however, users requested that a single formulation containing ammonium acetate, putrescine, and trimethylamine be developed to replace the separate components. Jang et al. (2007) found that 3C BioLure and a "cone" solid matrix (3C cone, Scentry Biologicals, Billings, MT, USA) were equal for capture of wild and sterile *C. capitata* and wild *A. suspensa*. Holler et al. (2009) found no difference in capture between individual lure and unipak formulations (Suterra LLC, Bend, OR, USD) of either 2C BioLure or 3C BioLure for capture of sterile *C. capitata* and wild *A. suspensa*. Similar results were found by Epsky et al. (2011) in field tests of *A. suspensa*. In field tests conducted in Spain, Navarro-Llopis et al. (2008) evaluated a single formulation of ammonium acetate, putrescine, and trimethylamine

(Trypak; Econex, Santomera, Murcia, Spain) and found that it captured as many *C. capitata* as 3C BioLure although not as many as 3C BioLure Medfly 100, a higher release rate formulation of 3C BioLure.

4.4 Role of Putrescine in Synthetic Food-Based Lures

The combination of putrescine and ammonia formed the basis for synthetic lures that were found to be as attractive as aqueous protein baits for *B. cucurbitae*, *A. ludens*, and *C. capitata* (see Sect. 4.2). Alternatives to putrescine that have been found to be as effective include pyrrolidine for *B. cucurbitae* (Wakabayashi and Cunningham 1991), cadaverine for *A. suspensa* (Kendra et al. 2008), and cadaverine and n-methyl pyrrolidine for *C. capitata* (Navarro-Llopis et al. 2008). Robacker (2001) noted that 1-pyrroline can occur as a contaminant in technical putrescine and can contribute to *A. ludens* attraction to putrescine lures. Initial studies found that putrescine was a synergist when added to ammonium acetate for *C. capitata*, and subsequent studies evaluated putrescine when added to ammonium acetate and trimethlyamine. Heath et al. (2004) and Leblanc et al. (2010b) found no difference in capture of *C. capitata* with ammonium acetate and trimethylamine alone or in combination with putrescine unless population levels were very low (<1.0 and 0.3 females per trap per day, respectively) in field tests conducted in Guatemala and Hawaii, respectively. Typically, there is greater discrimination found for both *C. capitata* and *A. suspensa* among baits when population levels are low (Epsky, unpublished data), but the basis for this is unknown. It could be due to changes in physiological state of flies at the start of the growing season or at the end of the growing season or when tests are conducted in less suitable hosts (i.e., conditions that result in low population levels) versus during the middle of the growing season or in preferred hosts (i.e., conditions that result in high population levels). Navarro-Llopis et al. (2008) found no differences between the two lure blends in field trials in Spain that had population levels >2 females per trap per day. Similarly, Grout et al. (2011) found no differences between the two blends in field tests conducted in South Africa even though populations were very low and there were <0.1 females per trap per day.

4.5 Role of Acetic Acid in Synthetic Food-Based Lures

Acetic acid is emitted with ammonia from ammonium acetate, which distinguishes ammonium acetate from other ammonium salts tested as fruit fly attractants. As noted in the previous paragraphs, acetic acid is a by-product of microbial fermentation of sugar, is a contaminant in cue-lure, is emitted from NuLure, and was found to be attractive to *C. capitata*. Robacker et al. (1996) found that sticky traps baited with acetic acid (17 mg/lure in agar [1 %]) were attractive to sterile *A. ludens* and that flies deprived of both protein and sugar were more attracted to acetic acid than non-deprived flies in laboratory bioassays. However, acetic acid tested at a range of concentrations (4.3–68 mg/lure) was as attractive to sterile *A. ludens* as AMPu alone or combined with acetic acid. Hall et al. (2005) found that a greater number of wild *A. suspensa* were captured in traps baited with 2C BioLure than with either ammonium bicarbonate and putrescine or AMPu but that there was no difference in capture of sterile flies. In tests of wild flies, Thomas et al. (2008) found more *A. ludens, A. suspensa,* and *A. obliqua* were captured in traps baited with 2C BioLure than with ammonium bicarbonate and putrescine, although the differences were not significant for *A. ludens,* indicating that acetic acid is not as attractive to *A. ludens* as the other species. There were also differences in ammonia release from the lures tested that may have affected the responses in the above tests. The role of acetic acid along with ammonia concentration was evaluated as part of an IAEA CRP (IAEA 2007), which confirmed that ammonium acetate versus ammonium bicarbonate alone or in combination with other components (e.g., putrescine and/or trimethlyamine), improved capture of *C. capitata*, *C. rosa*, *C. cosyra*, *Bactrocera zonata* (Saunders) (the peach fruit fly), *B. cucurbitae*, *Bactrocera invadens* Drew, Tsuruta, and White, and *Dacus ciliatus* (Loew) (the Ethiopian fruit fly). These studies also indicated that acetic acid did not increase capture of *B. oleae*, however, all synthetic lures worked poorly for this fly in comparison with NuLure:borax (9:3 %).

5 Additional Aspects of Food-Based Lure Types and Use

Other aspects of food-based lure use will be addressed in the following sections. These include preservatives that are used in traps with aqueous bait or retention fluid but may provide additional attractant volatile chemicals, proprietary and/or low cost baits that have been tested or are in use, and bacteria and/or bacterial by-products. Although the research on these materials is more limited, information from these studies may provide additional avenues of research that could be pursued. The last two aspects to be discussed include results of tests that combine food-based lures with other types of attractants, and the non-target capture that has been documented for traps baited with food-based lures.

5.1 Role of Preservatives/Surfactants in Traps with Food-Based Lures

With the development of synthetic food-based lures for tropical tephritids, it was hoped that highly effective dry traps for females of these species would be available. This has been true for temperate tephritids for which ammonia-baited sticky traps can be used. However, studies of tropical tephritids have found that McPhail-type traps with some type of aqueous retention fluid are more effective than McPhail traps used with internal sticky panels or fumigant or in other types of dry traps, including Jackson or delta traps (Heath et al. 1997; IAEA 1999; Thomas et al. 2001). Surfactants, such as triton or liquid soap, can be added to water to reduce escape of attracted flies. Some types of liquid soap contain ammonia and thus may contribute to fruit fly attraction. As discussed above (Sect. 3.2), borax has been added to aqueous protein baits to preserve both the bait and the captured flies. Subsequently, propylene glycol (PG, environmentally-friendly antifreeze and food additive) was added to the retention fluid. An aqueous solution of PG (10 %) was found to both reduce evaporation of water from the trap and preserve trapped flies. Thomas et al. (2001) found that the PG solutions increased capture of *A. suspensa* and *A. ludens* in traps baited with 2C BioLure versus traps with 2C BioLure and water alone, indicating that the PG added to fruit fly attraction. Robacker and Czokajlo (2006) confirmed synergism for *A. ludens* of PG solution and 2C BioLure in tests that compared retention fluid with PG versus retention fluid with triton. Thomas and Robacker (2006) tested the use of PG with TYB solutions and found improvement in capture of wild but not sterile *A. ludens*.

5.2 Proprietary Aqueous Protein Baits

Protein sources tested as food-based baits include proteins that are commercially available as products for use as feeding supplements. These include inactive and/or hydrolyzed yeasts, such as nutritional yeast, brewer's yeast and baker's yeast, and proteins that are often used in insect artificial diets, such as casein or soy. Other materials are by-products of manufacturing processes, such as corn (e.g., NuLure, Masoferm) and wood/paper processing (e.g., torula yeast). However, there are a number of other aqueous protein baits that have been tested and found to be attractive to fruit flies. Because they are available locally they can provide a low-cost alternative to the more expensive synthetic lures or protein baits that may need to be imported. Often these products are only listed as hydrolyzed protein with little information provided about original source or type of hydrolysis used in the process, which makes it harder to compare results among baits. Some of the proprietary baits that have been tested for fruit fly attraction are summarized in Table 3.1.

Zervas (1982) found that more *B. oleae* were captured in traps baited with Entomosyl (Höchst Hellas) than with ammonium sulfate, Buminal (Bayer SA, Puteaux, France), or *Dacus* bait (E.V.Y.P., Thessaloniki, Greece) tested at a ratio of 3 % bait:2 % borax. Fabre et al. (2003) conducted field cage tests of capture of laboratory-reared *B. cucurbitae* in traps baited with six commercially available aqueous protein baits, including NuLure, Masoferm, SolBait (modified Masoferm; Moreno and Mangan 2002), Buminal, Hym-Lure RTU (Robertsons [Pty] Limited, Durban, South Africa), and Pinnacle Protein Fruit Fly Bait (Mauri Yeast Products, Brisbane, Australia). The highest capture was obtained in traps baited with SolBait

(10 % aqueous solution). Vargas and Prokopy (2006) found that more female *B. dorsalis* were attracted to Provesta 621 autolyzed yeast extract (Integrated Ingredients, Bartlesville, OK, USA; product is now known as Ohly STV, Hamburg, Germany), GF-120 Fruit Fly Bait (Dow AgroSciences, Indianapolis, IN, USA), and Masoferm than to water, with intermediate attraction to NuLure, while male attraction to all four protein baits was greater than attraction to water in field cage tests conducted with laboratory-reared flies. In parallel tests with *B. cucurbitae*, all protein baits captured more females and males than water. Additionally, more females were attracted to Provesta and GF-120 than to the other two baits, and more males were attracted to GF-120 than NuLure, with intermediate attraction to Provesta and Masoferm. Barry et al. (2006) found no difference in attraction of *B. cucurbitae* or *B. dorsalis* to GF-120, Provesta 621, and Masoferm in bioassays of F1 generation laboratory flies, and response to all baits was greater than to water. They also found higher numbers of *B. cucurbitae* responded to any protein bait than *B. dorsalis*.

El-Gendy (2012) found the highest capture of *C. capitata* in traps baited with Buminal (NABA GmbH, Germany) and *B. zonata* in traps baited with Cera Trap bait (Bioiberica, Barcelona, Spain) when these baits were compared with Bio Nal bait (Bio Tec Company) for capture of *C. capitata* and *B. zonata* in field tests in Egypt. Moustafa (2009) found that Glan, Pro-lure, Agrisense, and Bioprox captured more flies of both species than Amadene, Buminal, Norlan, and Agrinal (commercial sources not given). Manrakhan and Kotze (2012) conducted field cage tests of HymLure (Savoury Food Industries [Pty] Limited, Industria, South Africa), which is a protein hydrolysate, (ii) GF-120, and (iii) M3 bait (also known as Questlure) (River Bioscience [Pty] Ltd., Port Elizabeth, South Africa), which is used in the M3 bait station for capture of *C. capitata, C. rosa,* and *C. cosyra*. All baits were equally attractive to *C. capitata* and *C. rosa*, but there was lower attraction of *C. cosyra* to HymLure than the other two baits.

5.3 Low Cost Fruit Fly Baits

Choices of protein material used for tests for fruit fly attraction were often dictated not only by what was readily or commercially available, but also by what was the lowest in cost. Thus, many baits are the end-product of some type of processing, which explains the wide variety of materials tested and also the variation inherent in batches produced over time, from different substrates, or from different processing methods. Efforts have also been directed toward identifying other readily available, low cost materials that could be used locally by growers. Hendrichs and Hendrichs (1990) observed *C. capitata* adults feeding on avian fecal material, and Prokopy et al. (1993) found that bird and lizard droppings (diluted as 3 parts droppings to 1 part water) were as attractive as aqueous NuLure (80 %) to *C. capitata* in field cage bioassays. *A. suspensa* adults were attracted to aqueous avian fecal material preparations in laboratory bioassays (Epsky et al. 1997). Most of the response was directly correlated with amount of ammonia emitted from the preparation, although additional unidentified chemicals were thought to be responsible for attraction to preparations that had aged for three days and were low in ammonia release. Robacker et al. (2000) quantified response of *A. ludens* to volatile chemicals from avian fecal material, and chemical analysis identified ethanol, propanol, phenol, ammonia, low-molecular weight amines, and pyrazines. A blend of ammonia, methylamine, dimethylamine, trimethylamine, 1-pyrroline, phenol, and 2-ethylhexanol was as attractive as the original material.

Piñero et al. (2003) found that traps baited with aqueous solutions of avian fecal material (25 %) or human urine (HU, 50 %) could be used to capture *A. obliqua* and *A. serpentina* (Wiedemann), although they were not as effective as aqueous protein baits (i.e., Captor Plus [Agroquimica Tridenta, S.A. de C.V. Mexico, D.F.] and TYB) in field tests conducted in Mexico. Subsequent research found that HU-baited traps also captured *A. ludens* and *A. fraterculus* and that, under some orchard conditions, captured equal or greater numbers of flies than Captor Plus-baited traps (Aluja and Piñero 2004).

Grape products have also been evaluated for fruit fly attraction as low cost alternative baits. Mangan and Thomas (2014) conducted field tests in Mexico that compared three types of grape products, including juice, mixed concentrate, and mixed powder (all available locally in Mexico), and aqueous TYB. Traps baited with grape products captured *A. ludens, A. striata*, *Anastrepha serpentina* (Wiedemann), and the papaya fruit fly, *Toxotrypana curvicauda* Gerstaeker. They found that the grape products often captured equal or greater numbers of *A. ludens* than aqueous TYB depending on the specific comparison or time of year. Mangan and Thomas (2014) also reviewed results of various field tests conducted in Brazil that showed that grape products can also be used for capture of *A. fraterculus*. Although not always as effective as aqueous protein-baited traps, Castrejón-Gómez et al. (2004) found that traps baited with aqueous solutions of brown sugar (1 kg/L) could be used to capture *T. curvicauda*, and that the highest capture was obtained after bait solutions had aged 3–4 days in the field. Such low cost, readily available materials may provide alternatives for growers for population suppression and improved crop protection.

5.4 Bacteria and Bacterial Fermentation as Fruit Fly Attractants

Most of the studies evaluating the role of microorganisms in attraction of fruit flies to aqueous protein baits have focused on yeasts, either through natural inoculation or by introduction of active yeast cultures. However, there have also been studies on the role of bacteria, either added to protein baits or tested alone, as a source of volatile attractants. The bacterial species, the substrates used, and the target species are summarized in Table 3.3. Microorganisms on fruit or leaves are used by adults as a protein source (Drew et al. 1983), and mutualistic or symbiotic roles for bacteria associated with tephritid fruit flies have been proposed (Drew and Lloyd 1987a). Drew and Lloyd (1987b) identified several species of Enterobacteriaceae from *B. tryoni* and *Bactrocera cacuminatus* (Hering) and hypothesized that chemicals emitted from bacteria on leaves attract fruit flies to host trees.

Drew and Fay (1988) revisited the role of ammonia in *B. tryoni* attraction to aqueous NuLure (5 %). They adjusted the pH of the aqueous solution by first adding NaOH to increase the pH from the original pH 3.94 (as obtained from the manufacturer) to pH 9. They then added HCl to reduce the pH to 6.5. NuLure solutions at both pHs were then inoculated with bacteria cultured from wild flies and were tested in comparison to uninoculated solutions of NuLure as well as aqueous ammonium bicarbonate at the same pH levels but without bacteria. Results of field cage tests with colony flies indicated that inoculated NuLure at pH 6.5, which had the greatest amount of bacterial growth and the lowest amount of ammonia released, captured the highest number of flies and that the capture was male-biased. They speculated that bacterial-produced metabolites other than ammonia had a sex-specific role, possibly attracting males to female feeding and oviposition sites to increase their mating success.

Jang and Nishida (1990) observed attraction of *B. dorsalis* in olfactometer bioassays to Enterobacteriaceae isolated from lab-reared and wild flies. They also found greater responses to both cultures with and without washed cells (i.e., broth and broth-free cultures) than to water blanks but less than to aqueous NuLure (5 %). Robacker et al. (1991) found that bacteria isolated from laboratory *A. ludens* and presented as unwashed cells were equally attractive as aqueous TYB in flight chamber and simulated field tests with sterile flies. In field tests using bacterium isolated from *R. pomonella*, MacCollom et al. (1992) reported that traps baited with washed cells were more attractive than unbaited traps or traps baited with apple volatiles when the washed cells were presented alone or in combination with apple volatiles. Subsequent tests revealed that traps baited with washed cells in combination with apple volatiles were more attractive than traps baited with apple volatiles alone or with ammonium acetate in combination apple volatiles (MacCollom et al. 1994). Martinez et al. (1994), in field trials of wild *A. ludens*, found that traps baited with autoclaved supernatants from three bacterial species were as attractive as traps baited with aqueous TYB or aqueous NuLure (10 %).

A series of tests evaluated attraction of *A. ludens* to supernatants obtained from cultures of several bacterial species, and these were all found to be equal to aqueous protein baits. In addition, volatile chemicals were identified from headspace collections of supernatant from broth used to culture bacteria. Lee et al. (1995) identified 21 volatile chemicals and found that the five most abundant were 3-methyl-1-butanol, phenethyl alcohol, 2,3-dimethylpyrazine, 2-methyl-1-propanol, and 3-(methlthio)-1-propanol. Robacker and Flath (1995) identified ammonia, trimethylamine, isoamylamine, 2-methyl-butylamine, 2,5-dimethylpyrazine, and acetic acid and found all attracted *A. ludens* when tested as single component synthetics in laboratory bioassays. Their research also identified dimethylamine, obtained from an altered preparation of supernatant, as the most effective attractant. DeMilo et al. (1996) identified 22 volatile chemicals and observed that 3-methly-1-butanol was the most abundant. Robacker and Bartelt (1997) found that a synthetic chemical blend (ammonia, trimethylamine, 1-pyrroline, 3-methylbutanamine, pyrazine, 2,3,4,5-tetrahydropyridine, 2,5-dimethylpyrazine, and trimethylpyrazine) was 73–87 % as attractive as the original bacterial supernatants. Epsky et al. (1998) reported that culture plates with bacterial cultures attracted *A. suspensa* in laboratory bioassays and that the attractants were ammonia and 3-methyl-1-butanol. Subsequent research has revealed that a variety of bacteria produce fruit fly attractant chemicals (Robacker et al. 1998) and that there may be within species variation in production of attractant chemicals (Lauzon et al. 1998). For example, within species differences in enzymatic capability, specifically the ability to metabolize uric acid (Lauzon et al. 2000), were found to be related to attraction of *A. ludens* (Robacker and Lauzon 2002) and *A. suspensa* (Epsky and Lauzon, unpublished data) to bacteria originally isolated from *R. pomonella*. Type of culture media as well as preparation of test materials can also affect volatile chemical production (Robacker et al. 2009).

5.5 Combination of Food-Based Lures with Other Types of Attractants

One of the advantages of food-based lures is that they capture both females and males, although they are female-biased, that is, they tend to capture more females than males. In contrast, male lures capture males almost exclusively. For example, in comparative tests conducted in seven countries, percentage female *C. capitata* of total capture in trimedlure-baited traps was 0–4.4 % but was 43–90 % in traps baited with food-based lures (Epsky et al. 1999). There have been studies conducted to evaluate combining food-based lures with male lures. Nadel traps baited with the combination of trimedlure and NuLure:borax (5:5 %) captured fewer total flies than Nadel traps baited with trimedlure alone in tests of *C. capitata* in Hawaii (Nakagawa et al. 1971). Hill (1986), in tests conducted in Australia, found that combining aqueous protein bait with male lures increased capture of male *C. capitata, B. tryoni, Bactrocera neohumeralis* (Hardy), and *B. cacuminatus* but decreased capture of females versus traps baited with only one type of attractant. Liquido et al. (1993) conducted field trials of wild and sterile *C. capitata* that compared capture of flies in Jackson traps baited with trimedlure alone or in combination with a vial containing aqueous ammonium carbonate (2 mL saturated solution). They observed that the addition of ammonia increased capture of both wild and sterile males and tended to capture more sterile females, although the difference was not statistically significant. No wild females were captured. The authors hypothesized that the increase in male capture was due to males remaining near the lure longer and increasing probability of retention. They also cited unpublished data by Chambers et al. showing that suspending a trimedlure plug 3.8–5.1 cm below a McPhail trap baited with NuLure increased capture of male

C. capitata over traps baited with either lure alone. But, they cited other studies showing that the combination of aqueous protein bait and trimedlure in Jackson traps increased female but not male capture (Zervas 1987; Hendrichs et al. 1989). Katsoyannos (1994) found that trimedlure combined with NuLure:borax (9:3 %) captured mostly males, with decreased female and non-target capture. Broughton and De Lima (2002) observed that the combination of trimedlure and 3C BioLure captured the same number of flies as 3C BioLure alone, but the percentage of females per trap decreased. Tóth et al. (2004) confirmed, in field tests conducted in Italy, that there were decreases in both female and male *C. capitata* capture when trimedlure and synthetic food-based lures were used in a single trap versus each attractant deployed in a separate trap. Similarly, Yee et al. (2005) found that sticky traps baited with combinations of host fruit lures and ammonium carbonate caught fewer *R. pomonella* than sticky traps baited with ammonium carbonate alone.

The combination of food-based lures and sex pheromones has been tested for *B. oleae*. Traps baited with the combination of aqueous protein bait (Entomozyl [Hoechst, Athens, Greece]:borax, 3 %:1.5 % wgt:vol) and solvent (diethyl ether) extracts of virgin flies increased male capture over traps baited with either lure alone in field tests of released laboratory flies (Haniotakis and Skyrianos 1981). There was no effect on capture of females. In subsequent research, Haniotakis and Vassiliou-Waite (1987) found that the combination of ammonium bicarbonate and synthetic female-produced pheromone lures increased female capture over traps baited with ammonium bicarbonate alone but decreased male capture over traps baited with pheromone alone in field tests of wild flies. Burrack et al. (2008), however, found that traps baited with aqueous TYB captured more *B. oleae* than traps baited with ammonium bicarbonate and synthetic pheromone lures. For *B. tryoni*, the combination of orange juice solution plus ammonium was no more effective than protein hydrolysate in McPhail traps (Dominiak et al. 2003).

5.6 Nontarget Capture

One of the disadvantages of food-based baits, especially aqueous protein baits, has been the high capture of nontarget insects. Ammonia attracts muscid dipterans that are associated with animal excrement (Richardson 1916), and a number of dipteran families as well as hymenopterans and other insect orders have been collected in aqueous protein-baited traps (e.g., Steyskal 1977). Katsoyannos et al. (1999) found that 3C BioLure captured fewer nontarget insects than 2C BioLure, with the lowest capture in NuLure:borax (9:3 %). In comparison with TYB solutions, fewer total nontarget insects were captured in traps baited with 2C BioLure, although the synthetic lure captured more chrysopids and halictid bees (Thomas et al. 2001; Thomas 2003; Conway and Forrester 2007). Leblanc et al. (2010a, b, c) conducted a series of experiments that evaluated capture of nontarget insects in traps baited with food-based lures in Hawaii. They reviewed reports of nontarget capture and noted that nontarget capture increased the time needed to sort through the samples for target flies and also increased capture of beneficial and endemic insects. Initial studies evaluated the components of 3C BioLures alone and in combination in field trials conducted in a variety of habitats, including native and non-native forests, residential areas, and farmlands (Leblanc et al. 2010b). They found that ammonium acetate was primarily responsible for nontarget capture, with putrescine contributing to a lesser effect. Most of the nontarget captures were saprophagous flies, with few beneficials attracted. 3C BioLure was found to capture more nontarget insects than either aqueous solulys (20 %) or aqueous TYB mixed with aqueous PG (20 %), although they noted different responses among different insect families (Leblanc et al. 2010c). Subsequent studies, which still recorded higher nontarget capture with 3C BioLure than with aqueous TYB, confirmed that use of PG (20 %) further suppressed nontarget capture (Leblanc et al. 2010a).

6 Approaches for Evaluation of Food-Based Attractants

Field tests have been widely used to evaluate food-based attractants and to determine preferences of wild fruit flies. Widespread use historically of field tests has contributed greatly to the development of the highly effective food-based attractants currently in use. Standard trapping procedures are used for trap placement within a site (IAEA 2003). Treatments to be compared are placed typically in replicated blocks within a planting, although spacing both among traps within a block and among blocks within a study site may be variable based on type of attractant as well as spacing among host plants and size of the field. Tests may be conducted as choice tests, with all treatments placed less than 3 m apart around the periphery of a tree, bush or planting (e.g., Epsky et al. 1993) or as no-choice tests with traps placed greater than 5 m apart within a row (e.g., Heath et al. 1994; IAEA 1999; 2007). Spacing should be dictated by effective sampling range for an attractant and additional tests may be needed to make this determination (Epsky et al. 2010; Kendra et al. 2010). However, because variation in age structure and density among wild populations that may affect response (e.g. Heath et al. 2004, see Díaz-Fleischer et al., Chap. 5, this volume) and because wild populations are not always available for tests, laboratory and simulated field tests are widely used to evaluate lures and attractants. Use of behavioral bioassays and electrophysiological analysis are two approaches that are used to fill this research gap.

6.1 Behavioral Bioassays

Behavioral bioassays are tests designed to quantify attraction to a specific bait or determine preference among several baits. The advantages of these bioassays include ability to manipulate factors, such as fly source (laboratory strain versus wild adults obtained from field-infested fruit), population level (number per unit area), sex, sex ratio and physiological state. It is not unusual, however, for flies to be attracted to materials in laboratory bioassays (especially if compared to response to a blank or un-baited control) that elicit no response in the field. Therefore, it is important to conduct parallel or subsequent field tests to confirm response observed in laboratory tests and to compare response to standard baits. In addition to numbers trapped, it is important to evaluate sex ratio of target flies and to document capture of non-target and/or beneficial insects. Trapped females can be dissected to evaluate differences in capture of immature versus mature females among the test baits.

Release/recapture studies have been widely used as simulated field tests. Flies may be marked to distinguish released flies from wild flies, which is a standard procedure for sterile flies released as part of the Sterile Insect Technique (SIT). Flies can be marked to test simultaneously different physiological states within the same field cage, further improving the usefulness of comparative tests. Typically release/recapture field tests use sterile flies to avoid release of crop-damaging fertile flies. Recapture rates may be very low and sterile flies are less responsive to food-based lure due to reduced need for protein (e.g., Midgarden et al. 2004). For tests of fertile flies, field cage tests are often employed. Field cages are screened or mesh cages of various sizes that enclose potted plants or are placed over field-planted material. Fertile flies can be used in these tests, which increase the usefulness of the results while preventing the escape of fertile flies. Additional information can be obtained from fruit fly behavior easily observed in field cage tests (e.g., Prokopy et al. 1993). Also, the larger the field cage, the better the assessment of long range attraction to bait. Field cage tests can be run as choice tests, with multiple treatments tested within the field cage, or as no choice tests, with one treatment deployed per field cage.

Wind tunnel or flight tunnel bioassays have also been used to assess attraction to test baits. Bait can be tested in traps hung within a tunnel or volatile chemicals can be introduced from chambers placed outside of the tunnel (Heath et al. 1993). Typically, these are conducted as choice tests of preference between two test substrates. This limits the ability to determine preference among more than two test substrates at a time since not all can be presented simultaneously. Flies tend to respond to any substrate over a clean air blank, so if there is no attraction of flies in a wind tunnel bioassay there will probably be no response in field tests. However, response to substrates in a wind tunnel bioassay does not guarantee response in the field. Concentration of the bait can also affect response in wind tunnel bioassays. Materials presented in too high of a concentration may be repellant and typically when this occurs, the flies may be observed to move to the downwind end of the wind tunnel. This was observed in tests of ammonia, although it did confirm the difference in antennal sensitivity between immature and mature *A. suspensa* (Kendra et al. 2005b).

Y-tube olfactometer bioassays are rarely used for tests of tephritids. These bioassays assess walking responses primarily, and the small diameter of a typical y-tube olfactometer increases potential problems of volatile chemicals that would be attractive at an appropriate concentration becoming repellant due to being presented at a concentration that is too high. Small cage bioassays such as the cage top bioassay (Robacker et al. 1991) have also been used to quantify response.

6.2 Electroantennography

The insect antenna is the primary organ responsible for chemoreception and transduction of olfactory stimuli. Functionally, antennae serve as the interface between environmental odors and insect behavior. The first electrophysiological investigations of insect olfaction were conducted in 1954, using tungsten electrodes to record peripheral olfactory responses from the cockroach antenna (Roys 1954). That pioneering work documented both the fast nerve potentials ('spikes' generated by individual olfactory receptors) and the slow potentials, now known as the electroantennogram (EAG), which represents the summation of multiple receptor potentials over the length of the antenna. The study also showed that an increase in concentration of odor source resulted in an increase in amplitude of the EAG response (i.e., EAG is a graded response due to recruitment of additional receptors). Thus, comparative EAG recordings provide a useful method for ranking the relative potencies of volatile stimuli, which in turn can provide insight into the potential behavioral significance of those compounds (Mayer 2001). In current applications, EAG is typically coupled with gas chromatography (GC), referred to as electroantennal detection (EAD), which facilitates initial screening and identification of potential attractants from a complex mixture of chemicals (e.g., host plant volatiles). With this technique, a sample is first separated by GC and then split for simultaneous delivery to the GC detector and the insect antenna. The GC trace shows all of the chemical constituents present in the sample, and the antennal response identifies those peaks of biological relevance (Ryan 2002).

The antennae of higher dipterans, including the Tephritidae, are particularly conducive to EAG analysis. The majority of olfactory sensilla are located on the enlarged third antennal segment (Shanbhag et al. 1999), but good quality EAG recordings can be obtained by using simple whole head mounts, requiring minimal dissection (Fig. 3.1a–d). EAG has been used to evaluate tephritid olfactory response to a variety of behavior-mediating chemicals (semiochemicals), including pheromones, male lures, host volatiles, and food-based attractants, and studies have been conducted on numerous pest species, including *B. dorsalis* (Light and Jang 1987), *B. tryoni* (Hull and Cribb 2001), *C. capitata* (Light et al. 1988; Jang et al. 1989a, b; Niogret et al. 2011), *A. ludens* (Robacker et al. 1986), *A. obliqua* (López-Guillén et al. 2011; Jenkins et al. 2012), and *A. suspensa* (Kendra et al. 2005a, b, 2008, 2009).

Although the relationship between behavioral response and amplitude of EAG response is not always clear, (e.g., Cha et al. 2008), studies with *A. suspensa* have shown good correlation between EAG and tephritid behavior with food-based attractants. As part of an ongoing effort to develop better lures for pest *Anastrepha*, USDA-ARS (Miami, FL) initiated a research program to address tephritid olfactory ecology by integrating electroantennography and developmental physiology with behavioral response to olfactory attractants. The goal is to identify the principal factors influencing attraction to chemical cues and ultimately use that information to develop improved female-targeted trapping systems compatible with sterile male

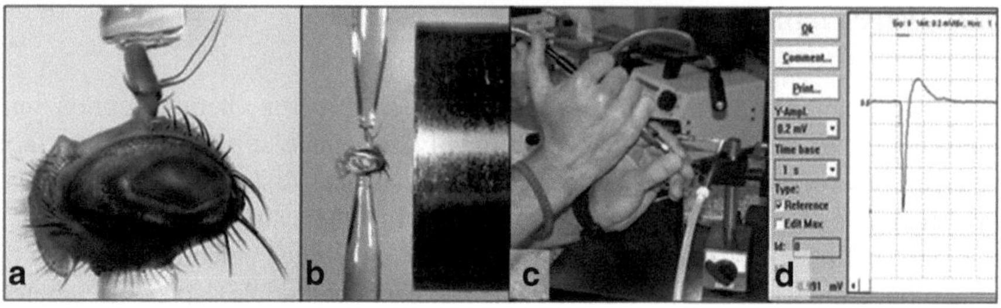

Fig. 3.1 Electroantennography technique. A freshly dissected fly head is mounted between micropipette electrodes with conductive gel (**a**) and placed under a stream of purified air (**b**). Using gas-tight syringes, test samples are injected into the airstream and delivered to the antennae (**c**). Upon binding specific olfactory receptors, test chemicals evoke an electrical response, the electroantennogram or EAG (**d**)

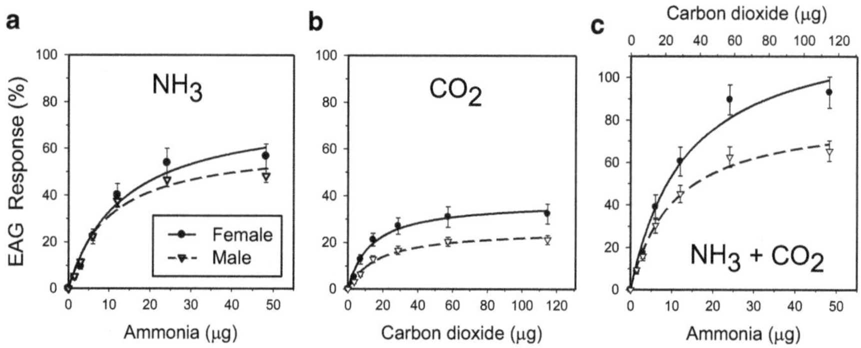

Fig. 3.2 Mean EAG response of female and male *Anastrepha suspensa* to quantified vapor samples of ammonia (**a**), carbon dioxide (**b**), and an equimolar mixture of the two gases (**c**). Responses were normalized and expressed as a percentage of the standard reference response (20 µL 2-butanone saturated vapor) (Adapted from Kendra et al. 2005a)

release programs. Factors to be evaluated include the sex, age, nutritional requirements, sexual maturity, and mating status of the adult fly as well as the dose, formulation, and potential interaction of the chemical components that comprise the attractant lure. EAG technology (using quantified vapor samples) was utilized in several ways, and three examples of those applications are presented here.

- Example 1. In an initial study (Kendra et al. 2005a), EAG was used to construct dose-response curves for pure ammonia and carbon dioxide, the two volatiles released from ammonium bicarbonate field lures. There was no difference in female versus male response to ammonia alone (Fig. 3.2a), but female response was significantly greater than male response to carbon dioxide (Fig. 3.2b) and to a mixture of ammonia + carbon dioxide (Fig. 3.2c). For both sexes, response to ammonia was greater than response to carbon dioxide, and EAG responses were additive when the two gases were combined and presently concurrently.

These results suggest that there are separate antennal receptors for the two chemicals, that both sexes have more receptors for ammonia than for carbon dioxide, and that female antennae have more receptors for carbon dioxide than male antennae. The results also suggest that the carbon dioxide component is responsible for the female-biased attraction obtained with ammonium bicarbonate lures.

- Example 2. Comparative analyses using known-aged females and fixed doses of ammonia and carbon dioxide revealed that EAG response was not static but varied depending upon age and maturity status (Kendra et al. 2005b). Dissections at 1-day intervals indicated that fully developed ovaries were not present in the laboratory strain of *A. suspensa* until 8 days post-eclosion (Fig. 3.3a). Maximum EAG response to ammonia was observed in immature females, 4–6 days old (Fig. 3.3b), just prior to synthesis and deposition of yolk proteins (vitellogenesis) and rapid ovary development (Kendra et al. 2006). Conversely, peak EAG response to carbon dioxide occurred in sexually mature females, 10–12 days old, at the onset of oviposition (Fig. 3.3c). The antennal responses correlated well with results obtained in behavioral bioassays. In two-choice tests conducted in flight tunnels (Fig. 3.3d), more mature (gravid) females were attracted to a mixture of ammonia + carbon dioxide than to the same dose of ammonia alone. This difference was not observed with immature females. These combined results support the functional roles of ammonia as a tephritid protein cue (Bateman and Morton 1981) and carbon dioxide as a short-range oviposition cue (Stange 1999). Another finding from this study was that immature females, which displayed the stronger EAG response to ammonia, were also more sensitive to ammonia dose in flight tunnel assays. In a series of two-choice test evaluating a range of ammonia release rates, mature and immature females were captured in equal numbers when low doses of ammonia were presented. However, at higher doses, ammonia became repellent to the immature females and significantly fewer were captured relative to mature females. Therefore, a strong EAG response must be interpreted with caution, as this is not necessarily an indicator of attraction. EAG screening should always be complemented with appropriate bioassays to determine behavioral response.
- Example 3. EAG with a series of related diamine compounds identified a new attractant for *A. suspensa* (Kendra et al. 2008). EAG analyses were used to quantify antennal response to a known synergistic attractant, putrescine (1,4-diaminobutane, C4), and to four homologous diamines that differed only in the length of the carbon chain (C5–C8). This comparative approach indicated that cadaverine (1,5-diaminopentane, C5) elicited an antennal response comparable to that of putrescine (Fig. 3.4a). When evaluated under field conditions (Fig. 3.4b), cadaverine was found to be just as efficacious as putrescine for capture of female *A. suspensa* when deployed in combination with ammonium acetate (AA) lures. The 1,6-diaminohexane (C6) also conferred synergistic attraction when combined with AA, but captures were less than those obtained with putrescine or cadaverine.

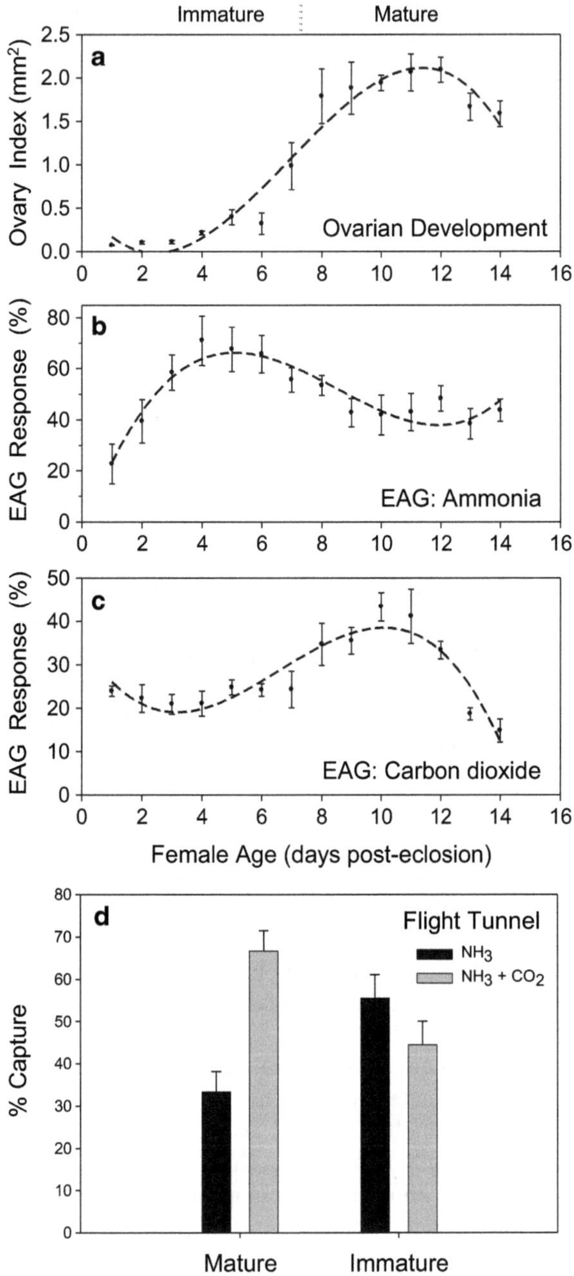

Fig. 3.3 Relationships among stage of ovary development (**a**), EAG response to a fixed 24 μg dose of ammonia (**b**), and EAG response to a fixed 57 μg dose of carbon dioxide (**c**) measured from female *Anastrepha suspensa* 1–14 days after emergence; and mean captures of mature (10–13 days) and immature (3–6 days) females in two-choice bioassay presenting ammonia alone or ammonia + carbon dioxide (**d**). Ovarian development assessed by ovary index (ovary length × width). EAG responses normalized and expressed as a percentage of the standard reference response (20 μL 2-butanone saturated vapor) (Adapted from Kendra et al. 2005b)

Through comparative EAG analyses using a synchronous population of *A. suspensa*, it has been shown that antennal responses to specific olfactory stimuli are not constant throughout the life of an adult fly but vary according to the

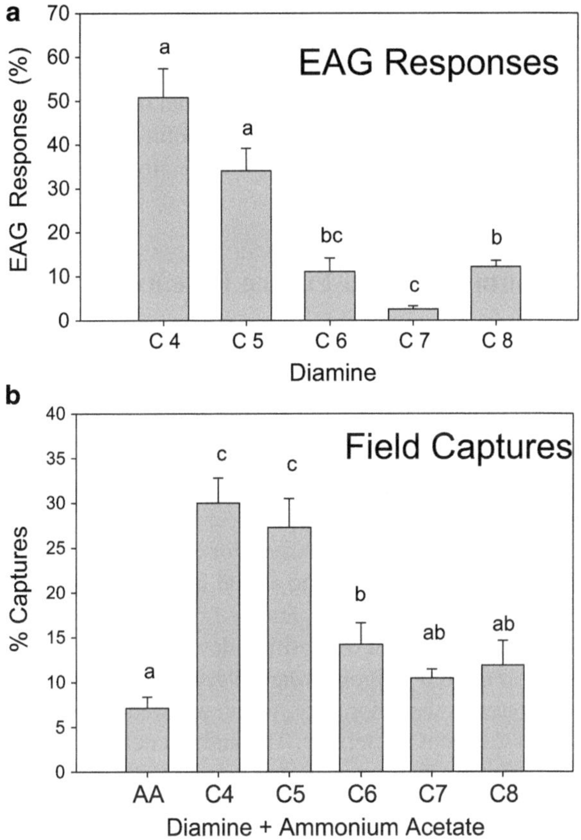

Fig. 3.4 Mean EAG response (**a**) and field captures (**b**) of female *Anastrepha suspensa* obtained with a series of homologous terminal diamines. Diamines consisted of 1,4-diaminobutane (putrescine, C4), 1,5-diaminopentane (cadaverine, C5), 1,6-diaminohexane (C6), 1,7-diaminoheptane (C7), and 1,8-diaminooctane (C8). EAG responses were normalized and expressed as a percentage of the standard reference response (20 μL 2-butanone saturated vapor). Field test used MultiLure traps baited with ammonium acetate plus diamine, or ammonium acetate alone (AA) (Adapted from Kendra et al. 2008)

physiological state and nutritional needs of the insect. Thus far, pure ammonia vapor, pure carbon dioxide vapor, and emissions from commercial lures of ammonium bicarbonate and putrescine have been evaluated, and age-related changes in EAG response have been observed with each attractive substrate. Plasticity in the olfactory system based on changing ecological needs is intuitively adaptive for tephritids and can be correlated with developmental events in the life of a fly. However, the underlying cellular and physiological processes have not yet been studied in the Tephritidae. Possible mechanisms include hormone-mediated temporal regulation of the protein components (and/or the corresponding encoding genes) that comprise the peripheral olfactory system. Those proteins include the transmembrane olfactory receptors themselves as well as a variety of soluble protein constituents of the sensillum lymph, including odorant-binding proteins, chemosensory proteins, and enzymes that remove active odorants from the dendritic membrane (de Bruyne and Baker 2008).

Despite the complexity of the system, quantitative EAG research with *A. suspensa* suggests that development of improved female-target lures may be realized by combining olfactory attractants that (1) elicit higher EAG responses in females, (2) convey the same functional message, since multiple feeding cues may synergize attraction (e.g., ammonia plus putrescine), (3) have additive EAG responses (recruit additional receptor types to send a stronger signal to the central nervous system processing centers), and (4) elicit peak EAG responses at different stages (physiological ages) of the adult female's life, thereby achieving broad attraction of flies regardless of the age structure of the population.

7 Summary and Future Research Needs

Availability of highly effective food-based lures has increased the opportunities beyond use of these attractants for population monitoring and detection. Some of these applications are discussed in other chapters in this volume, including bait sprays (Mangan, Chap. 12), bait stations (Piñero et al., Chap. 13), and mass trapping (Navarro-Llopis and Vaca, Chap.15). Additionally, traps baited with food-based synthetic attractants, as opposed to aqueous protein baits, can be used to capture live females for other purposes. For example, traps can be used to obtain and assess fertility of wild flies during sterile insect technique (SIT) programs (Katsoyannos et al. 1999). In that study, female *C. capitata* captured in 3C BioLure-baited traps failed to lay eggs in oviposition devices placed inside the traps, but females from those traps did oviposit when placed individually in chambers with oviposition substrates in the laboratory, which allowed quantification of percentage egg hatch and assessment of sterility. The authors noted that this technique would be further improved by development of female-specific lures that would reduce the number of sterile males that could enter the trap and mate with wild females. Live trapping of fruit flies could also be used to determine the age structure and reproductive potential of a pest population (Kouloussis et al. 2009, 2011). Traps baited with 3C BioLure were also used to obtain wild *C. capitata* for use in release/recapture studies to determine effective sampling range (Epsky et al. 2010).

Response to food-based attractants is variable among different species and habitat (e.g., Epsky et al. 2004; IAEA 2007), and additional research is needed to understand this variation and to identify new or additional substrates and chemicals that may improve capture of target fruit flies. This continues to be an active area of research. Flies responding to food-based lures are seeking protein primarily, and the need for protein varies with factors such as species, gender, physiological state, and availability of alternative protein sources in the habitat among other parameters (Díaz-Fleischer et al., Chap. 5, this volume). For example, larvae of *T. curvicauda* feed on seeds, and so adult females do not need protein for egg development (Drew and Yuval 1999) and thus do not respond as strongly to food-based lures, although they are occasionally captured in these traps (Heath et al. 1996). Higher capture in aqueous protein-baited traps over synthetic lure-baited traps, which has been observed in some tests of *Anastrepha* spp., for example, indicates that the identification of additional chemicals from the protein bait may provide an improved lure for these species. However, overall poor response to food-based attractants may indicate that other cues are needed to obtain an optimal trapping system. These cues may be chemicals from pheromones (Tan et al., Chap. 2, this volume); host fruit (Quilici et al., Chap. 4, this volume) or visual cues that can be incorporated into a trap design (Díaz-Fleischer et al., Chap. 5, this volume).

Use of a food-based attractant as bait in a trap for multiple species may not be possible or even desirable (Díaz-Fleischer et al. 2009). A single multi-species trap for detection of new invasions of fruit flies in areas currently fly-free may be preferred to deploying multiple single species-targeted traps as this would decrease overall costs and number of personnel needed to maintain the traps. However, variations in bait efficacy among the different species and habitats may require optimization for different conditions. Capture of unmated females before they have the opportunity to develop eggs and oviposit would increase effectiveness of attractants for fruit fly population suppression and control. Food-based attractants tend to attract mated females with mature eggs, although changes in release rate of ammonium acetate were found to affect the ratio of unmated, immature female to mated, mature female *C. capitata* (Heath et al. 1995). Thus, it may be possible to target flies with different physiological states by modifying the release rate/formulation of food-based attractants or by combining these baits with other semiochemicals. This would increase effectiveness of food-based attractants for both fruit fly detection and control.

Acknowledgements We would like to thank Robert Mangan (USDA/ARS, Weslaco, TX and Miami, FL) for valuable discussions on this topic and for access to documents from the USDA Bureau of Entomology laboratory that was located in Mexico City, Mexico, that chronicled the research on food-based baits for *A. ludens*. This included original reports and an unpublished manuscript by M. McPhail. We also thank Robert Heath (USDA/ARS, Miami, FL, deceased) for his mentorship and for introducing the authors to research on food-based fruit fly attractants.

References

Aluja M, Piñero J (2004) Testing human urine as a low-tech bait for *Anastrepha* spp. (Diptera: Tephritidae) in small guava, mango, sapodilla and grapefruit orchards. Fla Entomol 87:41–50

Anonymous (1964) Forest Products Laboratory, Forest Service, U.S. Department of Agriculture. U.S. Forest Service Research Note FPL-065. http://www.fpl.fs.fed.us/documnts/fplrn/fplrn065.pdf

Anonymous (2006) International panel for review of fruit fly surveillance programs. February 2006. Review of Fruit Fly Surveillance Programs in the United States. USDA/APHIS/PPQ/Fruit Fly Program, Riverdale, MD, USA.

Anonymous (2009) Hydrolysis to hydrolysate. Becton, Dickinson and Co., Franklin Lakes. http://www.bdbiosciences.com/documents/Hydrolysis_to_Hydrolysate.pdf

Anonymous (2011) Protein hydrolysis: acid and alkaline method. http://biotechcrunch.blogspot.com/2011/06/protein-hydrolysis-acid-and-alkaline.html

Baker AC, Stone WE, Plummer CC, McPhail M (1944) A review of studies on the Mexican fruitfly and related Mexican species. USDA Misc. Publ. No. 531. 151 pp

Barry JD, Polavarapu S (2004) Feeding and attraction of blueberry maggot flies to protein baits, ammonium acetate, and sucrose. J Econ Entomol 97:1269–1277

Barry JD, Miller NW, Piñero JC, Tuttle A, Mau RFL, Vargas RI (2006) Effectiveness of protein baits on melon fly and oriental fruit fly: attraction and feeding. J Econ Entomol 99:1161–1167
Bateman MA, Morton TC (1981) The importance of ammonia in proteinaceous attractants for fruit flies (Family: Tephritidae). Aust J Agric Res 32:883–903
Bekatorou A, Psarianos C, Koutinas AA (2006) Production of food grade yeasts. Food Technol Biotechnol 44:407–415
Boucher TJ, Ashley RA, Adams RG Jr, Morris TF (2001) Effect of trap position, habitat, and height on the capture of pepper maggot flies (Diptera: Tephritidae). J Econ Entomol 94:455–461
Boyce AM, Bartlett BR (1941) Lures for the walnut husk fly. J Econ Entomol 34:318
Broughton S, de Lima CPF (2002) Field evaluation of female attractants for monitoring *Ceratitis capitata* (Diptera: Tephritidae) under a range of climatic conditions and population levels in Western Australia. J Econ Entomol 95:507–512
Broumas T, Haniotakis GE (1994) Comparative field studies of various traps and attractants of the olive fruit fly. Entomol Exp Appl 73:145–150
Bua G (1933) Experimenti del 1933 con sostanze attrattive per la mosca delle olive. Partic R I Super Agric Ann 6:125–145
Bua G (1938) Serie di esperimenti con sostanze attractive per la masca delle olive. L'Olivicolt 15:1–19
Burditt AK Jr (1982) *Anastrepha suspensa* (Loew) (Diptera: Tephritidae) McPhail traps for survey and detection. Fla Entomol 65:367–373
Burrack HJ, Connell JH, Zalom FG (2008) Comparison of olive fruit fly (*Bactrocera oleae* (Gmelin)) (Diptera: Tephritidae) captures in several commercial traps in California. Intl J Pest Manag 54:227–234. doi:10.1080/09670870801975174
Buttery RG, Ling LC, Teraishi R, Mon TR (1983) Insect attractants: volatiles of hydrolyzed protein insect baits. J Agric Food Chem 31:689–692
Casaña-Giner V, Gandía-Balaguer A, Hernández-Alamós MM, Mengod-Puerta C, Garrido-Vivas A, Primo-Millo J, Primo-Yúfera E (2001) Attractiveness of 79 compounds and mixtures to wild *Ceratitis capitata* (Diptera: Tephritidae) in field trials. J Econ Entomol 94:898–904
Castrejón-Gómez VR, Aluja M, Arzuffi R, Villa P (2004) Two low-cost food attractants for capturing *Toxotrypana curvicauda* (Diptera: Tephritidae) in the field. J Econ Entomol 97:310–315
Cha D, Nojuma S, Hesler S, Zhang A, Linn C, Roelofs W, Loeb G (2008) Identification and field evaluation of grape shoot volatiles attractive to female grape berry moth (*Paralobesia viteana*). J Chem Ecol 34:1180–1189
Christenson LE, Foote RE (1960) Biology of fruit flies. Annu Rev Entomol 5:171–192
Conway HE, Forrester OT (2007) Comparison of Mexican fruit fly (Diptera: Tephritidae) capture between McPhail traps with torula and MultiLure traps with BioLures in south Texas. Fla Entomol 90:579–580
Cooper E (1905) Report of the Commissioner appointed to investigate the prevalence of *Trypeta ludens* in Mexico. California State Horticultural Commission, Sacramento, 48 pp
Crawford DL (1927) Investigation of Mexican fruit fly (*Anastrepha ludens* Loew) in Mexico. Mon Bull Dep Agric State Calif, August:422–445.
Cunningham RT, Nakagawa S, Suda DY, Urago T (1978) Tephritid fruit fly trapping: liquid food baits in high and low rainfall climates. J Econ Entomol 71:762–763
de Bruyne M, Baker TC (2008) Odor detection in insects: volatile codes. J Chem Ecol 34:882–897
Dean RW (1941) Attraction of *Rhagoletis pomonella* adults to protein baits. J Econ Entomol 34:123
DeMilo AB, Lee C, Moreno DS, Martinez AJ (1996) Identification of volatiles from *Citrobacter* fermentation of trypticase soy broth. J Agric Food Chem 44:607–612
Díaz-Fleischer F, Arredondo J, Flores S, Montoya P, Aluja M (2009) There is no magic fruit fly trap: multiple biological factors influence the response of adult *Anastrepha ludens* and

Anastrepha obliqua (Diptera: Tephritidae) individuals to MultiLure traps baited with BioLure or NuLure. J Econ Entomol 102:86–94

Dominiak B (2006) Review of the use of protein food based lures in McPhail traps for monitoring Queensland fruit fly *Bactrocera tryoni* (Froggatt) (Diptera: Tephritidae). Gen Appl Entomol 35:7–12

Dominiak BC, Gilmour AR, Kerruish B, Whitehead D (2003) Detecting low populations of Queensland fruit fly, *Bactrocera tryoni* (Froggatt) with McPhail and Lynfield traps. Gen Appl Entomol 32:49–53

Drew RAI, Fay HAC (1988) Comparison of the roles of ammonia and bacterial in the attraction of *Dacus tryoni* (Froggatt) (Queensland fruit fly) to proteinaceous suspensions. J Plant Prot Trop 5:127–130

Drew RAI, Lloyd AC (1987a) Bacteria associated with fruit flies and their host plants. In: Robinson AS, Hooper GHS (eds) Fruit flies, their biology, natural enemies and control, vol 3A. Elsevier, Amsterdam, pp 131–140

Drew RAI, Lloyd AC (1987b) Relationship of fruit flies (Diptera: Tephritidae) and their bacteria to host plants. Ann Entomol Soc Am 80:629–636

Drew RAI, Yuval B (1999) The evolution of fruit fly feeding behavior. In: Aluja M, Norrbom AL (eds) Fruit Flies (Tephritidae): pylogeny and evolution of behavior. CRC Press, Boca Raton, pp 731–749

Drew RAI, Courtice AC, Teakle DS (1983) Bacteria as a natural resource of food for adult fruit flies (Diptera: Tephritidae). Oecologia (Berl) 60:279–284

Duyck PF, Rousse P, Ryckewaert P, Fabre F, Quilici S (2004) Influence of adding borax and modifying pH on effectiveness of food attractants for melon fly (Diptera: Tephritidae). J Econ Entomol 97:1137–1141

Economopoulos AP (1986) Evaluation of color and food-odor trapping methods in the olive fruit fly. In: Mengel M, Carey JR, Plant RE (eds) Pest control: operations and systems analysis in fruit fly management, vol G11, NATO ASI series. Springer, Berlin, pp 111–115

Economopoulos AP (2002) Mediterranean fruit fly: attraction/trapping for detection, monitoring and control. Phytoparasitica 30:115–118

Economopoulos AP, Raptis A, Stavropoulou-Delivoria A, Papadopoulos A (1986) Control of *Dacus oleae* by yellow sticky traps combined with ammonium acetate slow-release dispensers. Entomol Exp Appl 41:11–16

El-Gendy IR (2012) Evaluating attractency of some protein derivatives for the Mediterranean fruit fly, *Ceratitis capitata* (Wiedmann) and the peach fruit fly, *Bactrocera zonata* (Saunders). Int J Agric Res 7:185–194

Epsky ND, Heath RR, Sivinski JM, Calkins CO, Baranowski RM, Fritz AH (1993) Evaluation of protein bait formulations for the Caribbean fruit fly (Diptera: Tephritidae). Fla Entomol 76:626–635

Epsky ND, Heath RR, Holler TC, Harris DL, Mullins T (1994) Corn steepwater as protein bait for *Anastrepha suspensa* (Diptera: Tephritidae). Environ Entomol 23:827–831

Epsky ND, Heath RR, Guzman A, Meyer WL (1995) Visual cue and chemical cue interactions in a dry trap with food-based synthetic attractant for *Ceratitis capitata* and *Anastrepha ludens* (Diptera: Tephritidae). Environ Entomol 24:1387–1395

Epsky ND, Dueben BD, Heath RR, Lauzon CR, Prokopy RJ (1997) Attraction of *Anastrepha suspensa* (Diptera: Tephritidae) to volatiles from avian fecal material. Fla Entomol 80:270–277

Epsky ND, Heath RR, Dueben BD, Lauzon CR, Proveaux AT, MacCollom GB (1998) Attraction of 3-methyl-1-butanol and ammonia identified from *Enterobacter agglomerans* to *Anastrepha suspensa*. J Chem Ecol 24:1867–1880

Epsky ND, Hendrichs J, Katsoyannos BI, Vásquez LA, Ros JP, Zümreoglu A, Pereira R, Bakri A, Seewooruthun SI, Heath RR (1999) Field evaluation of female-targeted trapping systems for *Ceratitis capitata* (Diptera: Tephritidae) in seven countries. J Econ Entomol 92:156–164

Epsky ND, Kendra PE, Heath RR (2004) Development of lures for detection and delimitation of invasive *Anastrepha* fruit flies. Proc Caribb Food Crops Soc 39:84–89

Epsky ND, Espinoza HR, Kendra PE, Abernathy R, Midgarden D, Heath RR (2010) Effective sampling range of a synthetic food-based attractant for *Ceratitis capitata* (Diptera: Tephritidae). J Econ Entomol 103:1886–1895

Epsky ND, Kendra PE, Peña JE, Heath RR (2011) Comparison of synthetic food-based lures and liquid protein baits for capture of *Anastrepha suspensa* (Diptera: Tephritidae) adults. Fla Entomol 94:180–185

Eyer JR (1935) Further observations on the attractiveness of esters of the ethyl acetate series to the codling moth. J Econ Entomol 28:940–942

Eyer JR, Rhodes H (1931) Preliminary notes on the chemistry of codling moth baits. J Econ Entomol 24:702–711

Eyer JR, Medler JT, Linton HL (1937) Analysis of attractant factors in fermenting baits used for codling moth. J Econ Entomol 30:750–756

Fabre F, Ryckewaert P, Duyck PF, Chiroleu F, Quilici S (2003) Comparison of the efficacy of different food attractants and their concentration for melon fly (Diptera: Tephritidae). J Econ Entomol 96:231–2003

Finney GL (1948) Culturing *Chrysopa californica* and obtaining eggs for field distribution. J Econ Entomol 41:719–721

Finney GL (1950) Mass-culturing *Chrysopa californica* to obtain eggs for field distribution. J Econ Entomol 43:97–100

Flath RA, Matsumoto KE, Binder RG, Cunningham RT, Mon RT (1989) Effect of pH on the volatiles of hydrolyzed protein insect baits. J Agric Food Chem 37:814–819

Fletcher BS, Kapatos E (1981) Dispersal of the olive fly, *Dacus oleae*, during the summer period on Corfu. Entomol Exp Appl 29:1–8

Frick KE (1952) Determining emergence of the cherry fruit fly with ammonium carbonate bait traps. J Econ Entomol 45:262–263

Frost SW (1926) Bait-pails as a possible control for the oriental fruit moth. J Econ Entomol 19:441–450

Frost SW (1937) Tests on baits for oriental fruit moths, 1936. J Econ Entomol 30:693–695

González Techera A, Jubany S, Carrau FM, Gaggero C (2001) Differentiation of industrial wine yeast strains using microsatellite markers. Lett Appl Microbiol 33:71–75. doi:10.1046/j.1472-765X.2001.00946.x

Gow PL (1954) Proteinaceous bait for the Oriental fruit fly. J Econ Entomol 47:153–160

Green N, Beroza M, Hall SA (1960) Recent developments in chemical attractants for insects. Adv Pest Control Res 3:129–179

Grout TG, Daneel JH, Ware AB, Beck RR (2011) A comparison of monitoring systems used for *Ceratitis* species (Diptera: Tephritidae) in South Africa. Crop Prot 30:617–622

Gurney WB (1925) The control of fruit fly. In Agric. Gazette of New South Wales, December, 1995, Sydney, pp. 879–887

Hagen KS (1950) Fecundity of *Chrysopa californica* as affected by synthetic foods. J Econ Entomol 43:101–104

Hagen KS, Finney GL (1950) A food supplement for effectively increasing the fecundity of certain tephritid species. J Econ Entomol 43:735

Hall DG, Burns RE, Jenkins CC, Hibbard KL, Harris DL, Sivinski JM, Nigg HN (2005) Field comparison of chemical attractants and traps for Caribbean fruit fly (Diptera: Tephritidae) in Florida citrus. J Econ Entomol 98:1641–1647

Haniotakis GE, Skyrianos G (1981) Attraction of the olive fruit fly to pheromone, McPhail, and color traps. J Econ Entomol 74:58–60

Haniotakis GE, Vassiliou-Waite A (1987) Effect of combining food and sex attractants on the capture of *Dacus oleae* flies. Entomol Hellenica 5:27–33

Heath RR, Manukian A, Epsky ND, Sivinski J, Calkins CO, Landolt PJ (1993) A bioassay system for collecting volatiles while simultaneously attracting tephritid fruit flies. J Chem Ecol 19:2395–2410

Heath RR, Epsky ND, Bloem S, Bloem K, Acajabon F, Guzman A, Chambers D (1994) pH effect on the attractiveness of a corn hydrolysate to the Mediterranean fruit fly, and several *Anastrepha* species (Diptera: Tephritidae). J Econ Entomol 87:1008–1013

Heath RR, Epsky ND, Guzman A, Dueben BD, Manukian A, Meyer WL (1995) Development of a dry plastic insect trap with food-based synthetic attractant for the Mediterranean and the Mexican fruit fly (Diptera: Tephritidae). J Econ Entomol 88:1307–1315

Heath RR, Epsky ND, Jimenez A, Landolt PJ, Dueben BD, Meyer WL, Aluja M, Rizzo J, Camin M, Jeronimo F, Baranowski R (1996) Development of an improved pheromone based system to monitor papaya fruit flies *Toxotrypana curvicauda* Gerstaecker. Fla Entomol 79:37–48

Heath RR, Epsky ND, Dueben BD, Rizzo J, Felipe J (1997) Adding methyl-substituted ammonia derivatives to food-based synthetic attractant on capture of the Mediterranean and Mexican fruit flies (Diptera: Tephritidae). J Econ Entomol 90:1584–1589

Heath RR, Epsky ND, Midgarden D, Katsoyannos B (2004) Efficacy of 1,4- diaminobutane (putrescine) in a food-based synthetic attractant for capture of Mediterranean and Mexican fruit flies (Diptera: Tephritidae). J Econ Entomol 97:1126–1131

Hendrichs J, Hendrichs MA (1990) Mediterranean fruit fly (Diptera: Tephritidae) in nature: location and diel pattern of feeding and other activities on fruiting and nonfruiting hosts and nonhosts. Ann Entomol Soc Am 83:632–641

Hendrichs J, Reyes J, Aluja M (1989) Behaviour of female and male Mediterranean fruit flies, *Ceratitis capitata*, in and around Jackson traps placed on fruiting host trees. Insect Sci Appl 10:285–294

Hill AR (1986) Reduction in trap captures of female fruit flies (Diptera: Tephritidae) when synthetic male lures are added. J Aust Entomol Soc 25:211–214

Hodson AC (1943) Lures attractive to the apple maggot. J Econ Entomol 36:545–548

Hodson AC (1948) Further studies on lures attractive to the apple maggot. J Econ Entomol 41:61–68

Holler T, Sivinski J, Jenkins C, Fraser S (2006) A comparison of yeast hydrolysate and synthetic food attractants for capture of *Anastrepha suspensa* (Diptera: Tephritidae). Fla Entomol 89:419–420

Holler T, Peebles M, Young A, Whiteman L, Olson S, Sivinski J (2009) Efficacy of the Suterra BioLure individual female fruit fly attractant packages vs. the unipak version. Fla Entomol 92:667–669

Hull CD, Cribb BW (2001) Olfaction in the Queensland fruit fly. *Bactrocera tryoni*. I: Identification of olfactory receptor neuron types responding to environmental odors. J Chem Ecol 27:871–887

IAEA (International Atomic Energy Agency) (1999) Development of female medfly attractant systems for trapping and sterility assessment, IAEA-TECDOC-1099. International Atomic Energy Agency, Vienna, 228 pp

IAEA (International Atomic Energy Agency) (2003) Trapping guidelines for area-wide fruit fly programmes. International Atomic Energy Agency, Vienna, 47 pp

IAEA (International Atomic Energy Agency) (2007) Development of improved attractants and their integration into fruit fly SIT, IAEA-TECDOC-1574. International Atomic Energy Agency, Vienna, 230 pp

Jacobson M, Keiser I, Harris EJ, Miyashita DH (1976) Impurities in cue-lure attractive to female Tephritidae. J Agric Food Chem 24:782–783

Jang EB, Nishida KA (1990) Identification and attractancy of bacteria associated with *Dacus dorsalis* (Diptera: Tephritidae). Environ Entomol 19:1726–1731

Jang EB, Light DM, Dickens JC, McGovern TP, Nagata JT (1989a) Electroantennogram responses of Mediterranean fruit fly, *Ceratitis capitata* (Diptera: Tephritidae) to trimedlure and its *trans* isomers. J Chem Ecol 15:2219–2231

Jang EB, Light DM, Flath RA, Nagata JT, Mon TR (1989b) Electroantennogram responses of the Mediterranean fruit fly, *Ceratitis capitata* to identified volatile constituents from calling males. Entomol Exp Appl 50:7–19
Jang EB, Holler TC, Moses AL, Salvato MH, Frazer S (2007) Evaluation of a single-matrix food attractant tephritid fruit fly bait dispenser for use in federal trap detection programs. Proc Hawaiian Entomol Soc 39:1–8
Jarvis H (1931) Experiments with a new fruit fly lure. Qld Agric J 36:758–761
Jenkins DA, Kendra PE, Epsky ND, Montgomery WS, Heath RR, Jenkins DM, Goenaga RJ (2012) Antennal responses of West Indian and Caribbean fruit flies (Diptera: Tephritidae) to ammonium bicarbonate and putrescine lures. Fla Entomol 95:28–34
Katsoyannos BI (1994) Evaluation of Mediterranean fruit fly traps for use in sterile-insect-technique programmes. J Appl Entomol 118:442–452
Katsoyannos BI, Papadopoulos NT, Kouloussis NA, Heath RR, Hendrichs J (1999) Method of assessing the fertility of wild *C. capitata* (Diptera: Tephritidae) females for use in sterile insect technique programs. J Econ Entomol 92:590–597
Katsoyannos BI, Papadopoulos NT, Stavridis D (2000) Evaluation of trap types and food attractants for *Rhagoletis cerasi* (Diptera: Tephritidae). J Econ Entomol 93:1005–1010
Keiser I, Wakabayashi N (1981) Fermentation per se as a biological slow delivery mechanism for releasing cofactors of fruit fly attractants. In: Lee PI, Lewis DH (eds) Controlled release of pesticide and pharmaceuticals. Plenum Press, New York, pp 295–301
Keiser I, Jacobson MJ, Nakagawa S, Miyashita DH, Harris EJ (1976) Mediterranean fruit fly: attraction of females to acetic acid and acetic anyhydride, to two chemical intermediates in the manufacture of cue-lure, and to decaying Hawaiian tephritids. J Econ Entomol 69:517–520
Kendra PE, Vázquez A, Epsky ND, Heath RR (2005a) Ammonia and carbon dioxide: quantitation and electroantennogram responses of Caribbean fruit fly, *Anastrepha suspensa* (Diptera: Tephritidae). Environ Entomol 34:569–575
Kendra PE, Montgomery WS, Mateo DM, Puche H, Epsky ND, Heath RR (2005b) Effect of age on EAG response and attraction of female *Anastrepha suspensa* (Diptera: Tephritidae) to ammonia and carbon dioxide. Environ Entomol 34:584–590
Kendra PE, Montgomery WS, Epsky ND, Heath RR (2006) Assessment of female reproductive status in *Anastrepha suspensa* (Diptera: Tephritidae). Fla Entomol 89:144–151
Kendra PE, Epsky ND, Heath RR (2008) Response of *Anastrepha suspensa* (Diptera: Tephritidae) to terminal diamines in a food-based synthetic attractant. Environ Entomol 37:1119–1125
Kendra PE, Montgomery WS, Epsky ND, Heath RR (2009) Electroantennogram and behavioral responses of *Anastrepha suspensa* (Diptera: Tephritidae) to putrescine and ammonium bicarbonate lures. Environ Entomol 38:1259–1266
Kendra PE, Epsky ND, Heath RR (2010) Effective sampling range of food-based attractants for female *Anastrepha suspensa* (Diptera: Tephritidae). J Econ Entomol 103:533–540
Kouloussis NA, Papadopoulos NT, Müller H, Wang J, Mao M, Katsoyannos BI, Duyck P, Carey JR (2009) Life table assay of field-caught Mediterranean fruit flies, *Ceratitis capitata*, reveals age bias. Entomol Exp Appl 132:172–181
Kouloussis NA, Papadopoulos NT, Katsoyannos BI, Müller H, Wang J, Su Y, Molleman F, Carey JR (2011) Seasonal trends in *Ceratitis capitata* reproductive potential derived from live-caught females in Greece. Entomol Exp Appl 140:181–188
Lauzon CR, Sjogren RE, Wright SE, Prokopy RJ (1998) Attraction of *Rhagoletis pomonella* (Diptera: Tephritidae) flies to odor of bacteria: apparent confinement to specialized members of Enterobacteriaceae. Environ Entomol 27:853–857
Lauzon CR, Sjogren RE, Prokopy RJ (2000) Enzymatic capabilities of bacteria associated with apple maggot flies: a postulated role in attraction. J Chem Ecol 26:953–967
Leblanc L, Vargas RI, Rubinoff D (2010a) Captures of pest fruit flies (Diptera: Tephritidae) and nontarget insects in BioLure and torula yeast traps in Hawaii. Environ Entomol 39:1626–1630

Leblanc L, Vargas RI, Rubinoff D (2010b) Attraction of *Ceratitis capitata* (Diptera: Tephritidae) and endemic and introduced nontarget insects to BioLure bait and its individual components in Hawaii. Environ Entomol 39:989–998

Leblanc L, Vargas RI, Rubinoff D (2010c) A comparison of nontarget captures in BioLure and liquid protein bait lures in Hawaii. Proc Hawaiian Entomol Soc 42:15–22

Lee C, DeMilo AB, Moreno DS, Martinez AJ (1995) Analysis of the volatile components of a bacterial fermentation that is attractive to the Mexican fruit fly, *Anastrepha ludens*. J Agric Food Chem 43:1348–1351

Lee C, DeMilo AB, Moreno DS, Mangan RL (1997) Identification of the volatile components of E802 Mazoferm steepwater, a condensed fermented corn extractive highly attractive to the Mexican fruit fly (Diptera: Tephritidae). J Agric Food Chem 45:2327–2331

Liburd OE, Alm SR, Casagrande RA, Polavarapu S (1998) Effect of trap color, bait, shape, and orientation in attraction of blueberry maggot (Diptera: Tephritidae) flies. J Econ Entomol 91:243–249

Light DM, Jang EB (1987) Electroantennogram responses of the oriental fruit fly, *Dacus dorsalis*, to a spectrum of alcohol and aldehyde plant volatiles. Entomol Exp Appl 45:55–64

Light DM, Jang EB, Dickens JC (1988) Electrontennogram responses of the Mediterranean fruit fly, *Ceratitis capitata*, to a spectrum of plant volatiles. J Chem Ecol 14:159–180

Liquido NJ, Teranishi R, Kint S (1993) Increasing the efficiency of catching Mediterranean fruit fly (Diptera: Tephritidae) males in trimedlure-baited traps traps with ammonia. J Econ Entomol 86:1700–1705

López DF, Becerril OH (1967) Sodium borate inhibits decomposition of two protein hydrolysates attractive to the Mexican fruit fly. J Econ Entomol 60:137–140

López DF, Spishakoff LM, Hernández O (1968) Pelletized lures for trapping the Mexican fruit fly. J Econ Entomol 61:316–317

López-D F, Steiner LF, Holbrook FR (1971) A new yeast hydrolysate-borax bait for trapping the Caribbean fruit fly. J Econ Entomol 64:1541–1543

López-Guillén G, Cruz López L, Malo EA, Rojas JC (2011) Olfactory responses of *Anastrepha obliqua* (Diptera: Tephritidae) to volatiles emitted by calling males. Fla Entomol 94:874–881

MacCollom GB, Lauzon CR, Weires RW Jr, Rutkowski AA (1992) Attraction of adult apple maggot (Diptera: Tephritidae) to microbial isolates. J Econ Entomol 85:83–87

MacCollom GB, Lauzon CR, Payne EB, Currier WW (1994) Apple maggot (Diptera: Tephrididae) trap enhancement with washed bacterial cells. Environ Entomol 23:354–359

Malavasi A, Duarte AL, Cabrini G, Engelstein M (1990) Field evaluation of three baits for South American cucurbit fruit fly (Diptera: Tephritidae) using McPhail traps. Fla Entomol 73:510–512

Malo EA (1992) Effect of bait decomposition time on capture of *Anastrepha* fruit flies. Fla Entomol 75:272–274

Mangan RL, Thomas DB (2014) Comparison of torula yeast and various grape juice products as attractants for Mexican fruit fly (*Anastrepha ludens* (Loew)). J Econ Entomol 107:591–600

Manrakhan A, Kotze C (2012) Attraction of *Ceratitis capitata, C. rosa* and *C. cosyra* (Diptera: Tephritidae) to proteinaceous baits. J. Appl Entomol 135:98–105

Martinez AJ, Robacker DC, Garcia JA, Esau KL (1994) Laboratory and field olfactory attraction of the Mexican fruit fly (Diptera: Tephritidae) to metabolites of bacterial species. Fla Entomol 77:117–126

Martinez AJ, Salinas EJ, Rendon P (2007) Capture of *Anastrepha* species (Diptera: Tephritidae) with MultiLure traps and BioLure attractants in Guatemala. Fla Entomol 90:258–263

Matsumoto KE, Buttery RG, Flath RA, Mon TR, Teranishi R (1985) Protein hydrolysate volatiles as insect attractants. In: Hedin PA (ed) Bioregulators of pest control, ACS symposium series. American Chemical Society, Washington, DC, pp 353–366

Mayer MS (2001) Nomographic correspondence between specialist cell responses, electroantennograms and the probability of behavioral response: the importance of comparable stimuli and a benchmark. J Insect Physiol 47:19–29

Mazor M (2009) Competitiveness of fertilizers with proteinaceous baits applied in Mediterranean fruit fly, *Ceratitis capitata* Wied. (Diptera: Tephritidae) control. Crop Prot 28:314–318
Mazor M, Gothilf S, Galun R (1987) The role of ammonia in the attraction of females of the Mediterranean fruit fly to protein hydrolyzate baits. Entomol Exp Appl 43:25–29
McPhail M (1939) Protein lures for fruit flies. J Econ Entomol 32:758–761
McPhail M (1943) Linseed oil soap – a new lure for the melon fly. J Econ Entomol 36:426–429
Midgarden D, Ovalle O, Epsky ND, Puche H, Kendra PE, Rendon P, Heath RR (2004) Capture of Mediterranean fruit flies (Diptera: Tephritidae) in dry traps baited with a food-based attractant and Jackson traps baited with trimedlure during sterile male release in Guatemala. J Econ Entomol 97:2137–2143
Miranda MA, Alonso R, Alemany A (2001) Field evaluation of medfly (Dipt., Tephritidae) female attractants in a Mediterranean agrosystem (Balearic Islands, Spain). J Appl Entomol 125:333–339
Moore RC (1969) Attractiveness of baited and unbaited lures to apple maggot and beneficial flies. J Econ Entomol 62:1076–1078
Moreno DS, Mangan RL (1995) Responses of the Mexican fruit fly (Diptera: Tephritidae) to two hydrolyzed proteins and incorporation of Phloxine B to kill adults. In: ACS Symposium Series Number 616, pp 257–279
Moreno DS, Mangan RL (2002) A bait matrix for novel toxicants for use in control of fruit flies (Diptera: Tephritidae). In: Hallman G, Schwalbe CP (eds) Invasive arthropods in agriculture. Science Publishers, Enfield, pp 333–362
Moreno DS, Celedonio H, Mangan RL, Zavala JL, Montoya P (2001) Field evaluation of a phototoxic dye, phloxine b, against three species of fruit flies (Diptera: Tephritidae) J Econ Entomol 94:1419–1427
Morton TC, Bateman MA (1981) Chemical studies on proteinaceous attractants for fruit flies, including the identification of volatile constituents. Aust J Agric Res 32:905–916
Moustafa SA (2009) Response of the Mediterranean fruit fly, *Ceratitis capitata* (Wied.) and peach fruit fly, *Bacrocera zonata* (Saund.) to some food attractants. Egypt Acad J Biol Sci 2:111–118. http://entomology.eajbs.eg.net/pdf/vol2-num2/10.pdf
Nakagawa S, Chambers DL, Urago T, Cunningham RT (1971) Trap-lure combinations for surveys of Mediterranean fruit flies in Hawaii. J Econ Entomol 64:1211–1213
Navarro-Llopis V, Alfaro F, Domínguez J, Sanchis J, Primo J (2008) Evaluation of traps and lures for mass trapping of Mediterranean fruit fly in citrus groves. J Econ Entomol 101:126–131
Newell W (1936) Progress report on the Key West (Florida) fruit fly eradication project. J Econ Entomol 29:116–120
Newman LJ, O'Connor BA (1931) Fruit fly (*Ceratitis capitata*). A further series of trapping or luring experiments. J Dep Agric West Aust 8(2):316–318
Niogret J, Montgomery WS, Kendra PE, Heath RR, Epsky ND (2011) Attraction and electroantennogram responses of male Mediterranean fruit fly to volatile chemicals from *Persea*, *Litchi*, and *Ficus* wood. J Chem Ecol 37:483–491
Papadopoulos NT, Katsoyannos BI, Kouloussis NA, Hendrichs J, Carey JR, Heath RR (2001) Early detection and population monitoring of *Ceratitis capitata* (Diptera: Tephritidae) in a mixed-fruit orchard in northern Greece. J Econ Entomol 94:971–978
Perkins FA, Hines HJ (1934) A note on some preliminary experiments with ammonia as a lure for the Queensland fruit fly (*Chaetodacus tryoni* Frogg). R Soc Qld Proc 45:29
Peterson A (1924) Some chemicals attractive to adults of the onion maggot, (*Hylemyia antique* Meig.) and the seed corn maggot (*Hylemyia cilicrura* Rond.). J Econ Entomol 17:87–94
Peterson A (1925) A bait which attracts the oriental peach moth (*Laspeyresia molesta* Busck). J Econ Entomol 18:181–190
Piñero J, Aluja M, Vázquez A, Equihua M, Varón J (2003) Human urine and chicken feces as fruit fly (Dipera: Tephritidae) attractants for resource-poor fruit growers. J Econ Entomol 96:334–340

Prokopy RJ, Economopoulos AP (1975) Attraction of laboratory-cultured and wild *Dacus oleae* flies to sticky-coated McPhail traps of different colors and odors. Environ Entomol 4:187–192

Prokopy RJ, Hsu CL, Vargas RI (1993) Effect of source and condition of animal excrement on attractiveness to adults of *Ceratitis capitata* (Diptera: Tephritidae). Environ Entomol 22:453–458

Raz, D. 1998. The phenology of the fig fly and its control. International Society for Horticultural Science (ISHS) Acta Hort. 480:207–208 http://www.actahort.org/books/480/480_35.htm

Reissig WH (1974) Field tests of traps and lures for the apple maggot. J Econ Entomol 67:484–486

Reissig WH (1976) Comparison of traps and lures for *Rhagoletis fausta* and *R. cinculata*. J Econ Entomol 69:639–643

Richardson CH (1916) The attraction of Diptera to ammonia. Ann Entomol Soc Am 9:408–413

Robacker DC (1995) Attractiveness of a mixture of ammonia, methylamine and putrescine to Mexican fruit flies (Diptera: Tephritidae) in a citrus orchard. Fla Entomol 78:571–578

Robacker DC (2001) Roles of putrescine and 1-pyrroline in attractiveness of technical-grade putrescine to the Mexican fruit fly (Diptera: Tephritidae). Fla Entomol 84:679–685

Robacker DC, Bartelt RJ (1997) Chemicals attractive to Mexican fruit fly from *Klebsiella pneumonia* and *Citrobacter freundii* cultures sampled by solid-phase microextraction. J Chem Ecol 23:2897–2915

Robacker DC, Czokajlo D (2006) Effect of propylene glycol antifreeze on captures of Mexican fruit flies (Diptera: Tephritidae) in traps baited with BioLures and AFF lures. Fla Entomol 89:286–287

Robacker DC, Flath RA (1995) Attractants from *Staphylococcus aureus* cultures for Mexican fruit fly, *Anastrepha ludens*. J Chem Ecol 21:1861–1874

Robacker DC, Landolt PJ (2002) Importance and use of attractants. In: Hallman GJ, Schwalbe CP (eds) Invasive arthropods in agriculture. Science Publishers, Enfield, pp 169–199

Robacker DC, Lauzon CR (2002) Purine metabolizing capability of *Enterobacter agglomerans* affects volatiles production and attractiveness to Mexican fruit fly. J Chem Ecol 28:1549–1563

Robacker DC, Thomas DB (2007) Comparison of two synthetic food-odor lures for captures of feral Mexican fruit flies (Diptera: Tephritidae) in Mexico and implications regarding use of irradiated flies to assess lure efficacy. J Econ Entomol 100:1147–1152

Robacker DC, Warfield WC (1993) Attraction of both sexes of the Mexican fruit fly, *Anastrepha ludens*, to a mixture of ammonia, methylamine, and putrescine. J Chem Ecol 19:2999–3016

Robacker DC, Chapa BE, Hart WG (1986) Electroantennograms of Mexican fruit flies to chemicals produced by males. Entomol Exp Appl 40:123–127

Robacker DC, Garcia JA, Martinez AJ, Kaufman MG (1991) Strain of *Staphylococcus* attractive to laboratory strain *Anastrepha ludens* (Diptera: Tephritidae). Ann Entomol Soc Am 84:555–559

Robacker DC, Moreno DS, DeMilo AB (1996) Attractiveness to Mexican fruit flies of combinations of acetic acid with ammonium/amino attractants with emphasis on effects of hunger. J Chem Ecol 22:499–511

Robacker DC, Martinez AJ, Garcia JA, Bartelt RJ (1998) Volatiles attractive to the Mexican fruit fly (Diptera: Tephritidae) from eleven bacteria taxa. Fla Entomol 81:497–508

Robacker DC, Garcia JA, Bartelt RJ (2000) Volatiles from duck feces attractive to Mexican fruit fly. J Chem Ecol 26:1849–1867

Robacker DC, Lauzon CR, Patt J, Margara F, Sacchetti P (2009) Attraction of Mexican fruit flies (Diptera: Tephritidae) to bacteria: effects of culturing medium on odour volatiles. J Appl Entomol 133:155–163

Roys C (1954) Olfactory nerve potentials as a direct measure of chemoreception in insects. Ann NY Acad Sci 58:250–255

Ryan MF (2002) Insect Chemoreception: Fundamental and Applied. Kluwer Academic Publishers, Boston, MA, USA

Shanbhag S, Muller B, Steinbrecht A (1999) Atlas of olfactory organs of *Drosophila melanogaster*. 1. Types, external organization, innervations and distribution of olfactory sensilla. Int J Insect Morphol Embryol 28:377–397

Simanton WA (1958) Studies of Mediterranean fruit fly lures in Florida. J Econ Entomol 51:679–682
Stange G (1999) Carbon dioxide is a close-range oviposition attractant in the Queensland fruit fly *Bactrocera tryoni*. Naturwissenschaften 86:190–192
Starr DF, Shaw JG (1944) Pyridine as an attractant for the Mexican fruitfly. J Econ Entomol 37:760–763
Steiner LF (1952) Fruit fly control in Hawaii with poisoned-bait sprays containing protein hydrolysates. J Econ Entomol 45:838–843
Steyskal GC (1977) History and use of the McPhail trap. Fla Entomol 60:11–16
Thomas DB (2003) Nontarget insects captured in fruit fly (Diptera: Tephritidae) surveillance traps. J Econ Entomol 96:1732–1737
Thomas DB, Robacker DC (2006) Trapping for Mexican fruit fly (Diptera: Tephritidae) with torula yeast and propylene glycol. Subtrop Plant Sci 58:24–26
Thomas DB, Holler TC, Heath RR, Salina EJ, Moses AL (2001) Trap-lure combinations for surveillance of *Anastrepha* fruit flies (Diptera: Tephritidae). Fla Entomol 84:344–351
Thomas DB, Epsky ND, Serra CA, Hall DG, Kendra PE, Heath RR (2008) Ammonia formulations and capture of *Anastrepha* fruit flies (Diptera: Tephritidae). J Entomol Sci 43:76–85
Tóth M, Nobili P, Tabilio R, Ujáry I (2004) Interference between male-targeted and female-targeted lures of the Mediterranean fruit fly *Ceratitis capitata* (Dipt., Tephritidae) in Italy. J Appl Entomol 128:64–69
Univ. Waikato (New Zealand Biotechnology Learning Hub – the University of Waikato) (2007) http://www.biotechlearn.org.nz/focus_stories/mining_milk/hydrolysis_of_protein
Vargas RI, Prokopy R (2006) Attraction and feeding response of melon flies and oriental fruit flies (Diptera: Tephritidae) to various protein baits with and without toxicants. Proc Hawaiian Entomol Soc 38:49–60
Villamil SC (2012) Behaviors related to reproduction of the olive fruit fly, *Bactrocera oleae* (Rossi) (Diptera: Tephritidae), in Northern California. Dissertation, Univ. of California-Davis, 66 pp
Wakabayashi N, Cunningham RT (1991) Four-component synthetic food bait for attracting both sexes of the melon fly (Diptera: Tephritidae). J Econ Entomol 84:1672–1676
Yee WL, Landolt PJ, Darnell TJ (2005) Attraction of *Rhagoletis pomonella* (Diptera: Tephritidae) and nontarget flies to traps baited with ammonium carbonate and fruit volatile lures in Washington and Oregon. J Agric Urban Entomol 22:133–149
Yetter WP, Steiner LF (1931) A preliminary report on large-scale bait trapping of the oriental fruit moth in Indiana and Georgia. J Econ Entomol 24:1181–1197
Yokoyama VY, Miller GT, Stewart-Leslie J, Rice RE, Phillips PA (2006) Olive fruit fly (Diptera: Tephritidae) populations in relation to region, trap type, season, and availability of fruit. J Econ Entomol 99:2072–2079
Zervas GA (1982) A new long-life trap for olive fruit fly, *Dacus oleae* (Gmelin) (Diptera, Tephritidae) and other Diptera. J Appl Entomol 94:522–529
Zervas GA (1987) Trapping Mediterranean fruit flies in Delta and McPhail traps in the field. In: Economopoulos AP (ed) Fruit Flies: Proc. 2nd Intl. Symp., 16–21 Sept. 1986, Crete, Greece. Elsevier, New York, pp 475–481

Chapter 4
Plant Odors as Fruit Fly Attractants

Serge Quilici, Toulassi Atiama-Nurbel, and Thierry Brévault

Abstract Plant odors consist of a mixture of volatile compounds that are conveyed by diffusion through air and may disperse over a long distance. They play a major role in mediating insect-plant relationships, particularly food location and selection of suitable sites for mating or oviposition. This chapter presents state-of-the-art research on the response of fruit flies (Diptera, Tephritidae) to plant odors and their potential for the development of trapping systems. Main research results from Tephritids of economic importance (i.e., *Rhagoletis*, *Ceratitis*, *Bactrocera/Dacus*, and *Anastrepha*) show evidence of response to (i) general plant volatiles from host or non-host plants, the so-called 'green leaf volatiles', (ii) essential oils from host or non-host plants, and (iii) fruit odors (whole fruit, wounded or crushed fruit, extracts, etc.). Synergies between plant odors and food odors or sex pheromones are also addressed. Factors including insect physiology (age, mating status, egg load, etc.), experience (learning), and genetic background can substantially modify the response pattern to plant odors.

One of the main challenges of using plant odors as fruit fly attractant is to improve the technology for identification (analysis), synthesis and emission (dispensers) of key compounds that may compete with natural volatile blends in the field. Further research should include the role of microorganisms in host location and recognition by fruit flies. Synthetic plant odors could be used either as kairomones for trapping systems, as allomones to push flies away from the crop or to disrupt host location, or as synomones to attract natural enemies to the crop.

Keywords Semiochemical • Kairomone • Plant odor • Blend • Volatile • Attractant • Olfactory stimuli • Host location

---

S. Quilici (✉) • T. Atiama-Nurbel
CIRAD, UMR PVBMT, 7 Chemin de l'IRAT, Saint-Pierre 97410, La Réunion, France
e-mail: serge.quilici@cirad.fr; toulassi.atiama@wanadoo.fr

T. Brévault
CIRAD, UR AIDA, Avenue Agropolis, 34398, Montpellier, France
e-mail: thierry.brevault@cirad.fr

T. Shelly et al. (eds.), *Trapping and the Detection, Control, and Regulation of Tephritid Fruit Flies*, DOI 10.1007/978-94-017-9193-9_4,

1 Introduction

Semiochemicals are chemical signals that mediate interactions between living organisms of the same or different species (Price et al. 2011). They are naturally occurring and can be used by insects for intra- or interspecific communication and for resource location. At the intraspecific level, pheromones are a group of semiochemicals that play a major role in mediating interactions between conspecifics, e.g., the location of a sexual partner (Wyatt 2003). By contrast, allelochemicals play a role in the chemical communication between species. Reflecting the co-evolutionary history between plants and insects, they are classified as allomones (advantage to the producer), kairomones (advantage to the receiver), or synomones (advantage to both) (Kogan 1982; Metcalf and Metcalf 1992). In this review, we will focus mainly on plant volatile compounds that play a major role in tephritid-plant relationships, primarily in food location or selection of suitable sites for mating or oviposition. Generally, plant odors consist of a mixture of volatile compounds (Metcalf and Metcalf 1992), mostly terpenoids, phenylpropanoids, alcohols, aldehydes, esters, acid, and sulphur compounds (Metcalf and Metcalf 1992; Birkett et al. 2004). They are conveyed by diffusion through air and may disperse relatively long distances (Metcalf and Metcalf 1992). From an applied perspective, these plant odors can be used for monitoring or controlling tephritids of economic importance.

A comprehensive understanding of the resource-foraging behavior of an insect pest as well as the identification of chemical and/or visual stimuli eliciting this behavior is central for the development of effective trapping systems to monitor and/or control its populations (Foster and Harris 1997). Visser (1986) proposed two hypotheses regarding the attractiveness of volatile, plant-derived semiochemical cues to foraging insects: (i) plant odors are highly specific due to specific compounds and/or (ii) plant odors are highly specific due to the particular ratio between ubiquitous constituents. Identification of plant volatiles involved in host plant location by phytophagous insects can be achieved through different techniques. A first step is to assess the behavioral response of insects to plant odors using bioassays (olfactometer, wind tunnel, etc.) (Haynes and Millar 1998). Concurrent odor collection and chemical analysis of plant odors can be achieved to identify volatile compounds (Millar and Haynes 1998; Goodner and Rousseff 2011). For example, coupled gas chromatography-electroantennogram detection (GC-EAD) analysis (Bjostad 1998) is a widely used technique to identify specific compounds from plants. Further steps include formulation and test of attractive blends of volatile compounds under laboratory and field conditions. The effectiveness of trapping systems depends, not solely on the quality of chemical stimuli, but also on visual characteristics and placement of the trap (Epsky et al. 2004).

The main objective of this chapter is to present state-of-the-art research on tephritid attraction to plant odors, focusing on results that might be relevant for the development or improvement of trapping systems. We successively examine the main genera of tephritids of economic importance, i.e., *Rhagoletis*, *Ceratitis*,

Bactrocera/Dacus, and *Anastrepha*. We have compiled scientific literature showing evidence of fruit fly response to i) general plant volatiles from host or non-host plants, such as the so-called 'green leaf volatiles'; (ii) essential oils from host or non-host plants, and (iii) fruit odors (i.e., whole fruit, wounded or crushed fruit, extracts, varying ripeness). Research showing synergy between plant odors and food odors or sex pheromone is also addressed. Lastly, the potential use of plant odors as bait for trapping systems is discussed. Various parameters, including a fly's physiology (e.g., age, mating status, etc.), experience (learning), and genetic background, can substantially modify response pattern to host stimuli (Papaj 2000; Schoonhoven et al. 2005), and relevant examples are noted (Díaz-Fleisher et al., Chap. 5, this volume).

2 Plant Odors for Trapping Fruit Flies

Frugivorous fruit flies have evolved mechanisms to use plant volatiles and visual stimuli from the plant during the host plant location process (Roitberg 1985). Plant volatiles are used at long or medium distance, whereas visual cues mediate host location at close range (Aluja and Prokopy 1992; Zhang et al. 1999; Brévault and Quilici 2010b). Research efforts on plant volatiles have been directed primarily at species for which no male-specific attractant is available (i.e., the male lures methyl-eugenol, cue-lure, trimedlure, and terpinyl acetate). Most studies on fruit fly attraction to host plant odor concern stenophagous species (e.g., *Rhagoletis* spp.) for which specific plant volatile attractants are much likely to be found than for polyphagous species. Here, we review the current knowledge on plant odors as fruit fly attractants, focusing on fruit flies of economic importance.

2.1 Genus Rhagoletis

Plant kairomones have been shown to play a significant role in the host plant selection process of various *Rhagoletis* species. Aluja and Prokopy (1992) used synthetic apple fruit volatiles to characterize the host searching behavior of Apple Maggot Fly (AMF), *Rhagoletis pomonella* (Walsh), in the field. They showed that flies released in the center of a patch containing host trees (*Crataegus mollis* Scheele var. toba) permeated with synthetic apple fruit volatiles moved faster and in more linear paths than flies released in the same plot with clean air. This suggests that flies are able to use host odor to optimize their foraging efficiency. Orchard studies clearly demonstrated that sexually mature females and males are strongly attracted to the odor of ripe apple (*Malus domestica* Borkh.) relative to unripe and non-host fruit (Prokopy and Bush 1973; Reissig 1974) and that this attraction may occur over a considerable distance. In wind tunnel bioassays, a blend of volatile compounds from whole Red Delicious and Red Astrachan apples was shown to be attractive to sexually mature AMF of both sexes (Fein et al. 1982). This blend consisted of a mixture of 7 esters from ripening fruit: hexyl acetate, (E)-2-hexen-1-yl acetate, butyl 2-methylbutanoate, propyl hexanoate, hexyl propanoate, butyl hexanoate and hexyl-butanoate (35:2:8:12:5:28:10 ratio). Synthetics of the identified compounds and the natural extract elicited directed upwind movement towards the source and EAG responses, but none of the components elicited full activity when presented alone. In contrast, foliar volatiles were supposed to play a secondary role in the host location process (Fein et al. 1982). Carle et al. (1987) conducted a comparative study of volatiles produced by whole hawthorn fruit (*Crataegus coccinea* L.) and four cultivars of apple from early- to late-ripening cultivars. They reported a total of 52 esters (31 esters in the hawthorn extract and 48 in the apple extracts) and some similarity between the volatile profiles of apple cultivars and hawthorn, but significant quantitative and qualitative changes of volatiles associated with fruit ripening. Response of AMF to these compounds was not tested. Further chemical studies using solid phase microextraction (SPME) and GC-EAD identified a new five component blend, including two of the previously identified volatiles (hexyl butanoate, propyl hexanoate) and three compounds not previously recorded (butyl butanoate, butyl hexanoate and pentyl hexanoate) in the respective ratios of 44:4:10:37:5 (Zhang et al. 1999).

In two other *Rhagoletis* species, *Rhagoletis mendax* Curran and *Rhagoletis cingulata* (Loew), Pelz-Stelinsky et al. (2005) showed that feral adults are attracted to the volatiles of their host fruit, blueberry (*Vaccinium myrtillus* L.), and cherry (*Prunus avium* (L.)), respectively, which could be exploited to improve the monitoring methods for these pests.

2.2 Genus Ceratitis

In the Mediterranean fruit fly, *Ceratitis capitata* (Wiedemann), the perception of plant volatiles has been investigated in both electroantennogram (EAG) and behavioral studies. Light et al. (1988) reported positive EAG responses from unmated, laboratory-reared males and females in response to a range of C1 and C2 to C12 carbon chain-length aliphatic alcohols, aldehydes, acetates, acids, and lactones, some of which are known volatiles from leaves and fruits. The greatest EAG responses of all compounds tested were elicited by C6 alcohols and aldehydes that are constituents of the general 'green-leaf odor'.

EAGs or electrophysiological recordings from olfactory sensilla on the antennal funiculi of *C. capitata* showed that both males and females detect blends of citrus peel essential oils (Levinson 1990; Hernández et al. 1996) as well as most individual compounds (Light et al. 1992; Hernández et al. 1996). Total airborne volatiles from fresh oranges elicited greater response of females than males (Levinson 1990; Hernández et al. 1996). Additional tests conducted in large field cages housing naturally planted orange trees showed that both sexes of *C. capitata* respond to chemicals released from artificial cuts made in the pulp of peeled oranges (*Citrus sinensis* L.) and also to natural or commercial orange juice applied to the surface of yellow 7.0 cm diameter spheres (Katsoyannos et al. 1997). However, volatiles released from artificial cuts made in the oily region (flavedo) of the orange peel were found to be much more attractive to male than to female medflies (Katsoyannos et al. 1997; Papadopoulos et al. 2001). Papaj et al. (1989) showed that medfly females were more likely to land on and attempt to oviposit into oranges that were artificially pricked than into unpricked control oranges. Deep wounds that pierced the fruit pulp and caused the release of juices elicited more female landings than shallow ones that only pierced the flavedo. These results suggest that the active compounds of orange juice, if isolated, identified and synthesized, may prove useful as a monitoring or control tool for *C. capitata* females (Heath et al. 1996).

Limonene, the most abundant chemical in citrus essential oils, stimulates oviposition in *C. capitata* females, whereas linalool, a compound representative of immature citrus fruit associated with high toxicity against immature stages of fruit flies and considered as an important compound conferring resistance against fruit fly larval development, has a significant deterrent effect (Ioannou et al. 2012). As citrus fruits mature from an immature green to an orange/yellow mature stage, the linalool content of the peel oil declines progressively, and the less toxic limonene becomes the major component. Limonene content acts as a potential stimulus for the ovipositional responses observed with sweet orange oil, whereas high linalool contents could mask or disrupt those effects in citrus oils (Ioannou et al. 2012).

Volatiles from coffee (*Coffea arabica* L.), the presumed ancestral host of the medfly, are also attractive. The odor of ripe intact or crushed coffee fruit was significantly more attractive to *C. capitata* than the odor of ripe intact or crushed fruit of five lower-ranking hosts and three nonhosts (Prokopy and Vargas 1996). Odor of crushed coffee fruit was significantly more attractive than odor of intact coffee fruit, and odor of ripe or near-ripe coffee fruit was significantly more attractive than odor of unripe coffee fruit. The odor of ripe (red) *C. arabica* fruit was found to be more attractive than the odor of ripe fruit of several other *Coffea* spp. (Prokopy et al. 1997). The odor of ripe *C. arabica* fruit was also more attractive than the odor of less mature fruit or the odor of foliage or twigs of this species (Prokopy et al. 1997). In addition, Prokopy et al. (1997) showed that the odor of a 24-h-old water extract of ripe *C. arabica* fruit was more attractive than the odor of 24-h-old extracts of such fruit with methanol, methylene chloride, or hexane and that the odor of ripe *C. arabica* fruit that had been frozen, thawed, and crushed was just as attractive as the odor of crushed unfrozen fruit. Based on the above information, Warthen et al. (1997) used headspace analysis techniques to identify 28 volatile compounds emitted by crushed ripe *C. arabica* fruit that had been frozen after picking and thawed just prior to volatile collection. They used a wind tunnel to assess the attractiveness of each compound to mature, protein-fed laboratory-cultured female medflies under dual choice conditions, wherein response to the odor of each compound was compared to response to clean air. In these assays, medflies responded positively to nine of the compounds: 3-methyl-1-butanal, decanal, 3-methyl-1-butanol, 2-(Z)-pentenol, 2-(E)-hexenol, 2-heptanone, 2-(Z)-hexanol, 2-heptanol, and 3-octanol. Follow-up assays in which an indoor olfactometer was used showed that among these nine compounds, the first six were the most attractive (Jang et al., unpublished data). Among the six coffee fruit volatile compounds tested in field cages, 2-heptanone always elicited greater attraction of protein-fed females than did water, conferring to 2-heptanone a possible role of oviposition-site signal (Prokopy et al. 1998).

Aside from citrus and coffee, few other fruits have been studied in relation to fruit fly attraction. Jang (1995) observed female *C. capitata* to be strongly attracted to the odor of ripe guavas (*Psidium guajava* L.) and to oviposit into spherical dummies emitting guava odor. Cossé et al. (1995) identified three compounds from mango (*Mangifera indica* L.) volatiles consistently triggering significant responses in combined GC-EAG analyses with the antennae of *C. capitata* females. Marked differences in fruit susceptibility to the medfly among peach cultivars (*Prunus persica* L.) suggest that the composition of volatile molecules may have an influence. Repeated field observations confirmed a clear preference of the medfly for nearly ripe fruit. A lower relative content of methyl esters, such as methyl hexanoate and methyl octanoate, known to act as medfly pheromone and attractant, respectively, was found in the least susceptible peach cultivars (Tabilio et al. 2013).

Natural substances have been found to attract male *C. capitata*, including angelica seed oil, which was used to bait traps during the early eradication program conducted in Florida in the 1950s (Steiner et al. 1957). The sesquiterpene α-copaene is the chemical found to be primarily responsible for this response, although other co-occurring chemicals may contribute to attraction (Jacobson et al. 1987; Flath et al. 1994a, b). α-Copaene is a complex, highly-volatile, widely-distributed plant compound found as a minor component in the essential oils of various plant species, including medfly hosts such as orange, guava, and mango (Nishida et al. 2000). The compound could play a role in the mating behavior of *C. capitata* as a signal for potential rendezvous sites for courtship and mating and is involved in mating success (Shelly 2001; Shelly and Villalobos 2004). Male *C. capitata* respond to material from both hosts (e.g., *Litchi chinensis* Sonn.) and non-hosts (e.g., *Ficus benjamina* L.) that contain α-copaene (Warthen and McInnis 1989; Niogret et al. 2011). While α-copaene is reported to be 2–5 times more attractive than trimedlure, difficulties in obtaining this compound in quantities sufficient for large scale trap deployment have prevented its use as a field lure (Cunningham 1989). Enriched ginger root oil (EGROlure), which contains the male attractant α-copaene may be a suitable alternative for monitoring and control of African *Ceratitis* species, including *C. capitata*, *Ceratitis rosa* Karsch and *Ceratitis cosyra* (Walker) males (Mwatawala et al. 2013; Shelly and McInnis 2001). However, further experiments in Hawaii showed that trimedlure-baited traps captured more *C. capitata* males than EGRO-baited traps when lures were aged 3–6 weeks (Shelly 2013).

2.3 Genera Bactrocera/Dacus

The attractiveness of various materials of plant origin (from host and non-host plants) has been investigated for many *Bactrocera* and *Dacus* species, both polyphagous and monophagous (Fletcher and Kitching 1995). Early investigations include the pioneer work of Howlett (1912), who observed that citronella oil was attractive to males of *Dacus* (*Bactrocera*) spp. and subsequently identified methyl eugenol as the active attractant (Howlett 1915). Methyl eugenol is now widely used in detection and/or control of the oriental fruit fly (*Bactrocera dorsalis* (Hendel)) (Jang and Light 1996), and its strong attractancy prompted its use in several successful eradication programs based on the Male Annihilation Technique (MAT; Koyama et al. 1984; Nakamori et al. 1988). However, its attractiveness is limited to males. The value of methyl eugenol as a male lure has encouraged efforts to develop female lures of comparable attractiveness, with significant efforts focused on plant materials that are associated with flies in their environment.

A number of studies have demonstrated the attractiveness of non-host plants to *Bactrocera* species. For *Bactrocera cucurbitae* (Coquillett), many non-host plants have been shown to be attractive for both males and females, such as corn, *Zea mays* L. (Nishida and Bess 1957; McQuate et al. 2003; Atiama-Nurbel et al. 2012), guava, *P. guajava*, and some citrus varieties (Kazi 1976). Other non-host plants include border (windbreak) plants, such as tiger's claw, *Erythrina tahitensis* Nadeaud (Stark 1995) and even weeds, such as castor bean, *Ricinus communis* L., spiny amaranth, *Amaranthus spinosus* L., and fuzzy rattlepod, *Crotalaria incana* L. (Nishida and Bess 1957; Kazi 1976). However, no research has been conducted on the potential causes of attraction to non-hosts, with the exception of McQuate et al. (2003), who presented evidence supporting the potential importance of corn pollen as food for *B. cucurbitae* and *B. dorsalis*. The general assumption is that these non-host plants release volatiles attractive to the flies. Indeed, many plant volatiles are more or less ubiquitous in various host as well as non-host plants and provide insects with a plethora of semiochemicals in their environment. Plant odors, such as common "green leaf volatiles" (GLVs) present in leaves of many plant species, have been shown to modulate or enhance tephritid behavior, while other semiochemicals, such as the odor of ripening fruit, serve as primary kairomones (Jang and Light 1996). A study on electroantennogram (EAG) responses of the oriental fruit fly to a spectrum of alcohols and aldehydes of plant origin proved that GLVs are among the compounds that elicited the highest EAG responses for both males and females (Light and Jang 1987). Jang et al. (1997) investigated the attractiveness of volatile semiochemicals from leaves and extracts of a non-host plant, Panax (*Polyscias guilfoylei* (W. Bull) L.H. Bailey) for the females of *B. dorsalis*. An extract of Panax was attractive to mated females but less attractive to males or unmated females. In an ambitious project, Keiser et al. (1975) studied the attractiveness of ethyl ether extracts of 232 botanicals for the oriental fruit fly, the melon fly, and the Mediterranean fruit fly. They recorded extracts eliciting the greatest response for females of the three species: 61 extracts for the

Mediterranean fruit fly, 31 for the melon fly, and only 7 for the oriental fruit fly. Some extracts were attractive for both sexes of a given species, while some were more attractive for females than for males. Only extracts of two plants (*Coffea robusta* Pierre ex Froehne and *Iva axillaries* Pursh) were attractive for the three species.

Most studies on female attractants have focused primarily on host plants as a source of such attractants (Chiu 1990; Jang and Light 1991; Jang and Light 1996). Host fruits are of particular interest since some tephritid females utilize host fruit volatiles when searching for oviposition sites (Drew 1989; Fletcher and Prokopy 1991; Landolt et al. 1992; Jang and Light 1996; Jang et al. 1998, 1999; Jang 2002). However, the isolation and identification of attractants from fruits is often difficult as fruit odors are a complex blend of volatiles, changing in composition during ripening (Chyau et al. 1992) and also differing among variety evaluated (Kamala Jayanthi et al. 2012; Atiama-Nurbel et al. in press). The degree of host fruit ripening influences its physical and chemical traits, such as color, tissue firmness, aroma, proportion of starch to free sugars, and quantities of other organic compounds (Bidwell 1979; Medlicott and Thompson 1985; Lalel et al. 2003; Yashoda et al. 2007). Not surprisingly, such physiological changes during ripening influence fruit fly oviposition behavior (Messina and Jones 1990; Messina et al. 1991) but in different ways depending on the tephritid species considered. For instance, ripeness stage of mango fruit was significant for oviposition decisions of *C. cosyra*, in particular, ripe and fully ripe fruits had more probability of oviposition than unripe ones, while ripeness stage appeared to have no significant effect on oviposition decisions in *Bactrocera invadens* Drew, Tsuruta & White (Migani et al. 2013).

Syed (1969) found that *B. dorsalis* adults remained in orchards as long as ripe fruits were present on the trees. Stark et al. (1991) also reported that the ratio of females to males foraging in guava trees increased as the season progressed and guava ripened. This is in agreement with the observations of Alyokhin et al. (2000b), who suggested that areas with plentiful ripe guava fruit attract females searching for oviposition sites. When offered papaya (*Carica papaya* L.) fruits of different maturation stages, the females of *B. dorsalis* preferred spheres with ripe papaya odor over blank air controls and, when presented with a choice of three ripeness stages of fruit, females landed equally on the three stages but spent more time and laid more eggs on spheres with the odor of the ripest stage (Jang and Light 1991). With mango and common guava, females of *B. dorsalis* were most attracted to odors of soft, ripe fruit over those of unripe fruit (Cornelius et al. 2000a; Rattanapun et al. 2009). Therefore, gravid females of *B. dorsalis* searching for suitable oviposition sites may use the odors of overripe fruits as their long-distance orienteering cues. Comparing the attractiveness of different fruit odors for *B. dorsalis* females in field cages, Cornelius et al. (2000a) showed that the odor of common guava was more attractive than papaya and starfruit odors and equally attractive as strawberry (*Fragaria vesca* L.), guava, orange, and mango odors. In additional field tests, McPhail traps baited with mango, guava, and orange captured equal numbers of *B. dorsalis* females.

Coupled GC-EAD studies (Siderhurst and Jang 2006) were conducted on *B. dorsalis* females with volatiles from tropical almond (*Terminalia catappa* L.) fruit, which showed that 22 compounds are detected by the antennae of the female. A nine-component subset of compounds showed relatively small EAD responses but attracted mainly females. In field cage experiments with McPhail traps, this blend was as attractive to females as torula yeast. A recent coupled GC-EAD study on the response of gravid females of *B. dorsalis* using two varieties of mango also revealed that 7 compounds from the variety 'Alphonso' and 15 from the variety 'Chausa' elicited an EAD; these compounds were subsequently identified using GC-MS (Kamala Jayanthi et al. 2012). The attractiveness of individual compounds was confirmed in olfactometer tests, but field tests are still needed to evaluate the potential practical use of these kairomones.

For *B. cucurbitae*, the odor of cucumber, *Cucumis sativus* L. (crushed skin and flesh) and cantaloupe, *Cucumis melo* L., (crushed flesh) was more attractive than the odor of tomato, *Lycopersicon esculentum* L. (crushed skin and flesh). Also, the odor of kabocha, *Cucurbita maxima* Duchesne (crushed flesh) was more attractive than the odor of bittermelon, *Momordica charantia* L. (crushed skin and flesh) but not more attractive than zucchini squash, *Cucurbita pepo* L. (crushed skin and flesh) (Miller et al. 2004). The reason why females respond more strongly to some fruit odors over others is not known. A correlation between female preference for host fruits and larval performance still has to be demonstrated.

Coupled GC-EAD studies using fresh and aged puréed cucumbers enabled (Siderhurst and Jang 2010) to identify 31 compounds detected by the females of *B. cucurbitae*. Among various blends tested in an outdoor olfactometer in McPhail traps, a nine-component blend was found to be the most promising. In subsequent field tests, this blend showed a female-biased attraction and was twice as attractive as 'Solulys' protein bait (Siderhurst and Jang 2010). Working with another cucurbit pest, *Dacus ciliatus* Loew, Alagarmalai et al. (2009) conducted GC-EAD studies with ripe melon volatiles and showed that 14 compounds elicited similar antennal responses of both sexes. Twelve of them were identified by GC-MS and in bioassays the most attractive blend was a mixture of four or five acetates.

For polyphagous *Bactrocera* spp., such as *B. dorsalis*, trapping studies targeting females have generally focused more on combining an attractive color with protein odor than on the use of kairomones (Alyokhin et al. 2000a). A few studies, however, considered the use of kairomones for trapping both sexes. In field-cage tests, Cornelius et al. (2000b) showed that mature protein-fed females of *B. dorsalis* were more attracted to orange odors than to protein odors, whereas protein-deprived females were equally attracted to both. However, in field tests, traps baited with 'Nu-Lure' were more effective for capturing females than traps baited with orange puree. In further field tests, Cornelius et al. (2000a) showed that McPhail traps baited with fruit purees (mango, common guava, and orange) were equally attractive to wild females of *B. dorsalis*. McPhail traps baited with mango captured more females than visual fruit-mimicking sticky traps ('Ladd traps') and equal numbers of females as McPhail traps baited with protein odors. These authors conclude that an effective strategy may require a combination of food odors to attract young females and host fruit odors to attract gravid females. Unfortunately, no long-lasting synthetic fruit volatiles attractive to *B. dorsalis* females and capable of competing with the odors of naturally occurring ripening fruit are yet available (Alyokhin et al. 2000a). Clarke and Dominiak (2010) used orange-ammonia traps (pulped orange based associated with ammonium carbonate, liquid lure used in McPhail traps) to catch both sexes of *Bactrocera tryoni* (Froggatt).

2.4 Genus Anastrepha

Robacker et al. (1990a) showed that odor from fermenting fruit of yellow chapote, *Casimiroa greggii* S. Wats (Rutaceae), acted as a food attractant for both sexes of the Mexican fruit fly, *Anastrepha ludens* (Loew). These authors developed a 3-component mixture of chemicals (1,8-cineole, ethylhexanoate, and hexanol at a 10:1:1 ratio) that proved to be more attractive than torula yeast in greenhouse experiments (Robacker et al. 1990b). Further studies showed that the addition of a fourth component, ethyl octanoate, further increased attractiveness (Robacker et al. 1992). More recently, Gonzalez et al. (2006) studied the response of both sexes of *A. ludens* to volatiles of white sapote, *Casimiroa edulis* Oerst. GC-EAD analysis of white sapote extracts revealed that antennae of both sexes respond to eight compounds among which GC-MS allowed identification of styrene, myrcene, 1,2,4-trimethylbenzene, 1,8-cineole, linalool, and β-trans-ocimene. However, in field-cage tests, the number of flies captured by traps baited with the white sapote blend was not different from the catches with hydrolyzed protein. Studying the response of *A. ludens* to volatiles of the fruit of the bitter orange *Citrus aurantium* L. (Rutaceae), Rasgado et al. (2009) found that both sexes were more attracted to mature green bitter orange fruit extracts than to controls (unbaited spheres or traps) in both flight tunnel and field cage assays. Among the ten compounds identified by GC-MS, limonene was the most abundant, while linalool, β-pinene, and methyl salicylate were found in lesser proportions. In field cage tests, MultiLure traps (Better World Manufacturing, Fresno, CA, USA) baited with this four-component blend captured significantly more flies of both sexes than traps baited with hydrolyzed protein.

Malo et al. (2005) studied the response of both sexes of *A. ludens* to guava (*P. guajava*) volatiles. GC-EAD analysis of guava extracts showed that eight and seven single compounds elicited antennal response from males and females, respectively. These compounds included ethyl butyrate, (E)-3-hexenol, (Z)-3-hexenol, hexanol, ethyl hexanoate, hexyl acetate, (Z)-3-hexenyl butyrate, and ethyl octanoate. Though both sexes showed a positive response to a blend of these eight components in wind tunnel experiments, field tests are still needed to better evaluate the potential of this mixture. In field tests, Loera-Gallardo et al. (2006) found that both sexes of *A. ludens* were attracted to commercial grape juice. Massa et al. (2008) sampled the volatile compounds of another commercial grape juice with SPME and developed a nine-component synthetic grape essence mixture that appeared to be 70 % as attractive as the juice in field tests. More recently, Robacker et al. (2011), using another grape juice, concluded that propylene glycol, acetic acid, methyl anthranilate, water, and at least one of three of methyl-branched esters are essential to induce the observed attraction.

Nigg et al. (1994) tested extracts of 22 fruits for their attractiveness to both sexes of the Caribbean fruit fly, *Anastrepha suspensa* (Loew) and showed that Box-orange (*Severinia buxifolia* (Poiret) Ten.), calamondin (*Citrofortunella microcarpa* (Bunge) Wijnands), carambola (*Averrhoa carambola* L.), Cattley guava (*Psidium littorale* Raddi), loquat (*Eriobotrya japonica* (Thunb.) Lindley), and Surinam cherry (*Eugenia uniflora* L.) were equally attractive to males and females. Chemical analysis revealed that farnesol, α-phellandrene, and 3-carene were highest in attractiveness to both males and females. These data suggest that host chemicals serve as attractants and that female and male specific attractants and traps could possibly be developed for this species from host kairomone data.

Additional studies have been conducted with the West Indian fruit fly, *Anastrepha obliqua* (Macquart). GC-EAD studies on volatile compounds from ripe fruits of *Spondias mombin* L. showed that nine chemicals (ethyl butyrate, isopropyl butyrate, hexan-1-ol, propyl butyrate, isobutyl butyrate, ethyl hexanoate, isopentyl butyrate, ethyl benzoate, and ethyl octanoate) individually elicited antennal response in both sexes. Further field cage bioassays with MultiLure traps showed that those baited with a blend of the nine compounds were more attractive than those baited with hydrolyzed protein (Cruz-Lopez et al. 2006). Recently, Malo et al. (2012) studied the response of *A. obliqua* to mature green fruits of three cultivars of mango and identified various compounds from each cultivar. In field-cage tests, traps baited with a blend of three synthetic components identified from Amate mango (myrcene, α-pinene, and trans-β-ocimene) were as attractive to *A. obliqua* as traps baited with the Amate mango volatiles.

3 Interactions Among Plant Odors, Other Attractants and Traps

In addition to olfactory chemical cues, visual cues play an important role in the host-finding behavior of fruit flies (Roitberg 1985; Jang and Light 1991; Vargas et al. 1991; Cornelius et al. 1999; Alyokhin et al. 2000b). Once they arrive at the host plant, visual cues are the main or sole stimuli guiding fruit detection. However, Aluja et al. (1993) showed that olfactory cues may be important in short range searching in the tree canopy when fruit are less apparent or scarce. Their results indicate that upon arrival on host trees, AMF females find host fruit of high density solely by visual stimuli. However, at low host fruit density, the females locate host fruit both by chemical and visual stimuli. In field tests in commercial orchards, sticky-red spheres baited with a blend of synthetic apple volatiles (Reissig et al. 1982, 1985) or only with butyl hexanoate (Duan and Prokopy 1992) captured significantly more male and female AMF than unbaited spheres, indicating the relevance of fruit odor. The red sticky sphere probably provides visual stimuli mimicking a potential mating or oviposition site (Prokopy 1968). In flight-tunnel choice tests and field trial captures involving red sticky spheres with odor sources, the new five-component blend of apple volatiles attracted more flies than the previous seven-component blend or the single compound (butyl hexanoate) used with commercial apple maggot monitoring spheres (Zhang et al. 1999).

In a wind tunnel, mature females of the tomato fruit fly, *Neoceratitis cyanescens* Bezzi, move upwind in response to specific olfactory cues from blends of host flowers or host fruits (Brévault and Quilici 2010a). In the absence of wind, mature females mainly use visual information to locate the host fruit. In wind, host fruit odor significantly increases the probability and speed of locating the host fruit (Brévault and Quilici 2010b). When odor source and fruit model were spatially decoupled (90 or 180°), > 50 % flies that landed on the fruit model initially performed an orienting flight toward the odor source and then turned back to the fruit model while in flight or after one landing on the floor, suggesting visual information to be the ultimate indicator of host fruit. Visual stimuli are sufficient to elicit orientation response to host fruit; the integration of more specific olfactory cues can improve the host finding efficiency in terms of speed and accuracy, especially when visual cues are inadequate or poor (obstructed or weakly attractive), or when visual cues are similar to those of non-host plants. Although *N. cyanescens* females do not lay their eggs in flowers or ripe fruit, the fragrance emanating from these structures may constitute a good indicator of the presence of a suitable host at a long distance. In another wind tunnel assay, Brévault and Quilici (2010b) observed that leaf odor of host plants elicited a significant response of *N. cyanescens* females (but lower than fruit odor), including orientation and upwind towards source. Combining sets of host and non-host plants in the same field cage or spraying leaf extract of host plants on non-host plants, Brévault and Quilici (2007) showed that leaf volatiles of host plants assisted females of the tomato fruit fly, *N. cyanescens*, in finding host fruit. Moreover, the response was specific to mature females with a high oviposition drive, while starved mature females, immature females, and males did not show preference for fruit-mimicking spheres hung in plant foliage. Olfactory signals emitted by the host foliage could be an indicator of an appropriate habitat, leading flies to engage in searching behavior.

Several studies have been conducted on host-associated visual stimuli for *Bactrocera* spp. (Prokopy and Haniotakis 1975; Hill and Hooper 1984; Vargas et al. 1991; Drew et al. 2003). Indeed, traps that combine visual and olfactory cues may prove to be the most effective for capturing tephritid fruit fly pests (Prokopy and Economopoulos 1975; Epsky and Heath 1998; Piñero et al. 2006). Cornelius et al. (2000a) used a mango bait in two commercially available fruit fly traps. McPhail traps baited with mango puree captured more females than visual fruit-mimicking sticky traps ('Ladd traps') and equal numbers of females as McPhail traps baited with protein odors. Piñero et al. (2006) showed that the addition of cucumber odor strongly enhanced the attractiveness of yellow-colored hemispheres, which indicates that both visual and olfactory stimuli are synergistic in eliciting responses of sexually mature melon fly females.

Some studies suggest that combinations of attractants are more attractive for fruit flies than the individual attractants presented alone (Zervas 1989; Landolt et al. 1992; MacCollom et al. 1994; Robacker and Heath 1996). Additive or synergistic effects from combining host or plant odors with pheromones or bacteria odors have, for instance, been reported for medfly (Dickens et al. 1990), papaya fruit fly (Landolt et al. 1992), and AMF (MacCollom et al. 1994). In contrast, for *A. ludens*, the combination of fermenting chapote odor and male produced pheromone was never more attractive than chapote odor alone. For immature females, the presence of pheromone inhibited the response to chapote, while for virgin females chapote odor inhibited the response to pheromone (Robacker and Garcia 1990). A decrease of attraction of *A. ludens* to combinations of two synthetic lures, AMPu (food attractant) (Robacker et al. 1992) and CEHO (fruit odor) (Robacker 1992), was also demonstrated using different traps (McPhail and sticky traps) in outdoor conditions (Robacker and Heath 1997). For this species, as for *Bactrocera oleae* (Rossi) (Haniotakis and Vassiliou-Waite 1987), adding food odor to pheromone did not improve the response at least for the sex most responsive to the pheromone (males in *B. oleae* and females in *A. ludens*). In wind tunnel bioassays with the Mexican fruit fly, Robacker and Rios (2005) showed that female attraction to grapefruit oil was not enhanced if they had prior experience with grapefruit oil, but in field experiments captures of females were higher in traps baited with grapefruit oil than in unbaited ones. Furthermore, a combination of a nitrogenous food odor (*Anastrepha* fruit fly lures) and grapefruit oil in field traps enhanced captures of females but not those of males. For *A. obliqua*, males and females (virgin and mated) were more attracted to *Spondias monbin* volatiles than to putrescine and ammonium acetate, whereas sugar-fed virgin flies preferred putrescine and ammonium acetate over fruit odor (López-Guillén 2008).

In *A. ludens*, Robacker (1991) evaluated the effects of specific hunger on attractiveness of proteinaceous and fruit-derived lures. Sugar hunger led flies to be more responsive to fruit odor than to food odors, while protein hunger led to a higher response to bacteria (presumed protein source) than to fruit odor. The combination of Torula yeast and fruit odor was never more attractive than fruit odor alone for any age-feeding history groups of flies, while the combination of bacterial odor and fruit odor was never more attractive to protein-hungry flies than bacteria alone. Similarly, the combination of bacterial odor and fruit odor was less attractive than fruit odor alone to sugar-hungry flies. This study suggests that fruit odor bait (favored by sugar-hungry flies) is more effective in orchards where few fruits are present and trees are rich in fly-type bacteria, while torula yeast (attractive to protein-hungry flies) is more effective than fruit odor in orchards with few fly type bacteria but with trees laden with fruit.

4 Parameters Influencing Fruit Fly Response to Plant Stimuli

The decision to engage in searching a plant is based not only on the perception of suitable olfactory information, but also on the insect's physiology ('internal state'), experience, and genetic background (Schoonhoven et al. 2005).

Physiological state, such as ovarian maturation, egg production, and mating status, can greatly influence the response of females to host cues (Browne 1993). In *R. pomonella*, age and sexual maturity of females significantly affected the probability and time to discover fruit (Duan and Prokopy 1994). Rull and Prokopy (2000) released marked AMF of different physiological states in blocks of apple trees ringed by sticky red spheres. Spheres were unbaited, baited with butyl hexanoate, or baited with both butyl hexanoate and ammonium carbonate. Large proportions (25–40 %) of released mature male and female AMF were recovered in blocks having traps baited with butyl hexanoate. The presence of ammonium carbonate had no significant effect on the response to synthetic fruit odors by mature AMF. Immature flies of each sex responded weakly to traps and to both types of synthetic lures.

Egg load (e.g., the number of mature oocytes available in the ovaries) also influences female response to host fruits, such as the time invested in host searching, the probability that the host is accepted once found, and the size of the female clutch (Minkenberg et al. 1992; Bjorksten and Hoffmann 1998; Papaj 2000). In studies conducted on potted host trees in field cages and in the laboratory, Prokopy et al. (1994a) examined the influence of egg load on the finding and acceptance of high-ranking (kumquat) and lower-ranking (grapefruit) hosts for oviposition by wild-origin Mediterranean fruit flies. Egg load had no discernible effect on behavior associated with finding either type of fruit. In another experiment with potted nonfruiting host trees in outdoor field cages, immature females (without eggs) were significantly more attracted to odor of a proteinaceous food lure than to odor of ripe coffee fruit, whereas the reverse was true for mature females carrying a high egg load (Prokopy and Vargas 1996). By contrast, for *B. invadens* and *C. cosyra*, female egg load was the most important factor influencing host acceptance in both species (Migani et al. 2013).

Nutritional status can also influence female response to host odors. In field-cage bioassays, protein-fed females always responded to a greater extent to the odor of coffee fruit extract than to the odor of NuLure (Miller Chemical & Fertilizer Corporation, Hanover, PA, USA), whereas the reverse was true for protein-deprived females, which did not exhibit greater attraction to odor of any of the six coffee fruit volatile compounds tested than to water. All types of mature, protein fed-females tested (laboratory-cultured virgin, laboratory-cultured mated, wild mated) in field-cage assays responded similarly to 2-heptanone, whereas protein-deprived females of the same age (9- to 11-day-old) did not respond significantly to 2-heptanone (Prokopy et al. 1998). Mated females of *C. capitata* were found to be attracted to the same extent by fragrant orange fruits and odorless sham oranges, while unmated females were notably less attracted than mated females to oranges and odorless orange dummies (Levinson et al. 2003). Immature, protein-hungry females of *B. tryoni* and *B. dorsalis* were more responsive to odors of bacteria than to odors of host fruit (Prokopy et al. 1991; Cornelius et al. 2000b). By contrast, protein-fed females showed a greater level of attraction to host fruit odor over protein odor (Cornelius et al. 2000b; Miller et al. 2004).

Females of the Mediterranean fruit fly exhibit a preferential switch in certain olfactory-mediated behaviors as a result of mating. Unmated, laboratory-reared, virgin females chose the odor of male-produced pheromone over host fruit odor (guava) in a dual-choice flight tunnel bioassay (Jang 1995). Females continued to respond preferentially to the male pheromone for several weeks if not allowed to mate. Mated females chose the host fruit odor over the male-produced pheromone. Females of *N. cyanescens* respond to host fruit odor regardless of their age, egg load, or mating status, and do so more consistently in the afternoon, which is the peak time of day for egg-laying (Brévault and Quilici 2010a). In another study, Brévault and Quilici (1999) showed that females become responsive to a fruit-mimicking sphere only when they have completed their reproductive maturity.

Learning can influence the visual ability of AMF to detect host fruit. For example, the ability of AMF females to find fruit of unfamiliar color was significantly affected by prior experience with host fruit color (Prokopy et al. 1994b). Specifically, females exposed to red hawthorns or red apples were less able to find green hawthorns or green apples than were females experienced with either of the latter fruit types. Papaj and Prokopy (1989) demonstrated that females exposed to a particular host fruit species tended to remain longer in test trees harboring fruits than did inexperienced females or females exposed to another fruit. This could have negative effects on the efficacy of traps depending upon the prior experience of females, particularly the type(s) of fruits previously used for oviposition. Robacker and Fraser (2005) found no evidence that *A. ludens* females learn fruit color or size after experience with host fruit, including oviposition. However, females with grapefruit experience were more attracted to fruit models with extract of either grapefruit peel or pulp than to models without extract. Females with no experience with grapefruit were not attracted to models treated with grapefruit extract. These results suggest that the flies learn fruit odors after encountering host fruit during general host foraging, then may increase their searching efficiency by responding to the learned host odor.

Response to host fruit odor may also change according to genetic background as shown by AMF populations associated with different hosts (Jones and Davis 1989). Nojima et al. 2003b showed that hawthorn-infesting AMF have a preference for a blend of four volatiles identified from hawthorn fruit that differs from the blend previously identified from apple fruits. Further studies showed that *R. pomonella* originating from flowering dogwood (*Cornus florida* L.) preferentially respond to the dogwood volatile blend than to volatile blends from apple or hawthorn (Nojima et al. 2003a). The recent shift of *R. pomonella* from its native host downy hawthorn, *C. mollis*, to introduced domesticated apple in the eastern United States is a model for sympatric host race formation (Linn et al. 2012). Apple- and hawthorn-native

R. pomonella flies use fruit odor to distinguish between apples and hawthorns (Linn et al. 2003; Forbes et al. 2005; Forbes and Feder 2006), leading to pre-mating reproductive isolation (Dambroski et al. 2005). In addition to apple, *R. pomonella* also infest two hawthorn spp. in the western United States, one the native black hawthorn, *Crataegus douglasii* Lindl, and the other the introduced English ornamental hawthorn, *Crataegus monogyna* Jacq. Linn et al. (2012) reported that western apple, black hawthorn, and ornamental hawthorn flies show significantly increased levels of upwind-directed flight to their respective natal compared to non-natal fruit volatile blends, consistent with host race status. Recent studies on the response of AMF from different *Crataegus* spp. also confirm that AMF respond maximally to their natal fruit volatile blends and are relatively unresponsive to the alternative non-natal blends (Cha et al. 2011a, b, 2012). This sharp behavioral distinction underscores the diversity of odor response phenotypes to host fruit odor in *R. pomonella*.

5 Deployment Strategies for Traps Baited with Plant Odors

Up to now, traps baited with plant odors for use in tephritid control have been deployed only for the apple maggot fly, *R. pomonella*. An IPM method achieving good control of AMF was developed based on the use of red spheres baited with butyl hexanoate placed on perimeter trees ("perimeter trapping"), while the other trees were only protected by sticky-coated red spheres or sugar/flour pesticide-treated red spheres (Prokopy et al. 2000). Field experiments in orchards with varieties of different susceptibility showed that traps and lures should be deployed on preferred rather than on less preferred cultivar trees (Rull and Prokopy 2005). Comparing various trap/lure combinations, AliNiazee et al. (1987) showed that a yellow rectangle with a red hemisphere in the center and apple volatile attractant (consisting of a mixture of hexyl acetate, butyl 2-methyl butyrate, propyl hexanoate, hexyl propionate, butyl hexanoate, and hexyl butanoate in a 36:7:12:5:29:11 ratio) captured the largest number of AMF. Yellow board traps sandwiched between the two halves of red spheres sprayed with pyrethroid insecticide loaded with butyl hexanoate in semi-permeable sachets and hung on branches 1.2–1.7 m above the ground at the orchard perimeter were an effective "attract and kill" technique to control AMF in Quebec apple orchards (Bostanian et al. 1999; Bostanian and Racette 2001). Agnello et al. (1990) reported that the addition of apple odor increased catches for both yellow panel traps and red sphere traps. In a test using these baited traps to time control sprays in commercial orchards, 70 % fewer sprays (2.8 fewer applications) were applied than in a calendar-based program. A possible factor, however, that may limit the effectiveness of trapping systems using such synthetic blends is the increasing competition with natural odors emitted by ripening fruits during the maturation process (Carle et al. 1987).

Fruit volatiles have replaced ammonia as the AMF attractant for use with traps in the northeastern United States (Prokopy et al. 2005). Release-recapture studies showed that larger proportions of released mature adults of both sexes were recovered with sticky spheres baited with butyl hexanoate, whereas adding ammonium carbonate to butyl hexanoate did not enhance trap captures (Rull and Prokopy 2000). The five-component blend identified by Zhang et al. (1999) also attracted more AMF than ammonium acetate or ammonium carbonate mixed in adhesive on spheres (Stelinski and Liburd 2002). In contrast, ammonium carbonate with sticky yellow panels is preferred to detect *R. pomonella* in the northwestern United States (Klaus 2003). Field trials in Oregon and Washington showed that *R. pomonella* was more attracted to spheres and yellow panels baited with various doses of ammonium carbonate than to those baited with apple volatile blends (Yee et al. 2005). A change in the attractiveness of different traps during the season has been frequently reported, yellow panels being more effective at the beginning of the season, while red spheres are more effective later in the season (Prokopy 1972; Reissig 1974; Neilson et al. 1981).

6 Conclusion and Perspectives

One of the main challenges of using kairomones for trapping systems in the field is to identify optimal concentrations of key compounds so that an artificial mixture may act as a "super-stimulus" that may compete with natural volatile blends in the field. For both polyphagous and stenophagous species, comparisons of the relative attractiveness of blends of volatile compounds from different host fruit species, or from different ripeness stages, should enable identification of key components. In addition, GC-EAD-MS studies should also be conducted to identify or confirm the optimal concentrations. Further research is also needed on dispensers that ensure the emission of the different compounds at a suitable rate on a sufficiently long period of time. Optimal plant odor blends should ideally be associated with appropriate visual stimuli (e.g., shape, size, color, contrast with the background) to provide a synergistic combination of stimuli. Though yellow is the main color used for fruit fly traps, it is known that in some species, such *N. cyanescens* (Brévault and Quilici 2007), sex- and maturity-specific response of females may be used for improving trap specificity and efficacy.

Further development of semiochemicals for insect control should involve closer attention on the potential role of microorganisms in host location and recognition by fruit flies. For example, the codling moth *Cydia pomonella* (L.) uses yeast volatiles in addition to plant volatiles to find a suitable host plant (Witzgall et al. 2012). *Drosophila melanogaster* Meigen flies mainly use microbial, and not plant cues, to locate feeding and oviposition sites (Becher et al. 2012). Several of the compounds that mediate attraction of *Rhagoletis* flies to host fruits (e.g., apple, hawthorn, and flowering dogwood), such as isoamylacetate, ethylacetate, 3-methylbutan-1-ol, and 1-octen-3-ol are known to be produced by fungi and yeasts (Witzgall et al. 2012).

Novel technologies designed at identifying, reproducing, and dispensing plant volatile blends open up a large and promising avenue of research for fruit fly management. Synthetic plant odors could be used either as kairomones (attractant or aggregation stimulant) for trapping systems or attract-and-kill devices (bait sprays), as allomones (repellents) to push flies away from the crop or to disrupt host location (push-pull systems), as synomones to attract natural enemies to the crop, or even as pheromones (plant communication) to induce plant defense against herbivores in the crop by production of toxic secondary metabolites, deterrents or repellents.

Acknowledgments The authors wish to present all their thanks to Todd Shelly and Nancy Epsky for their major input in improving an earlier version of the manuscript.

References

Agnello AM, Spangler SM, Reissig WH (1990) Development and evaluation of a more efficient monitoring system for apple maggot (Diptera: Tephritidae). J Econ Entomol 83:539–546

Alagarmalai J, Nestel D, Dragushich D, Nemny-Lavy E, Anshelevich L, Zada A, Soroker V (2009) Identification of host attractants for the Ethiopian Fruit Fly, *Dacus ciliatus* Loew. J Chem Ecol 35:542–551

AliNiazee M, Mohammad A, Booth S (1987) Apple maggot (Diptera: Tephritidae) response to traps in an unsprayed orchard in Oregon. J Econ Entomol 80:1143–1148

Aluja M, Prokopy RJ (1992) Host Search behaviour by *Rhagoletis pomonella* flies: inter-tree movement patterns in response to wind-borne fruit volatiles under field conditions. Physiol Entomol 17:1–8

Aluja M, Jacome I, Birke A, Lozada N, Quintero G (1993) Basic patterns of behaviour in wild *Anastrepha striata* flies under field-cage conditions. Ann Entomol Soc Am 86:776–793

Alyokhin AV, Messing RH, Duan JJ (2000a) Visual and olfactory stimuli and fruit maturity affect trap captures of oriental fruit flies (Diptera: Tephritidae). J Econ Entomol 93:644–649

Alyokhin AV, Messing RH, Duan JJ (2000b) Visual and olfactory stimuli and fruit maturity affect trap captures of oriental fruit flies (Diptera: Tephritidae). J Econ Entomol 93:644–649

Atiama-Nurbel T, Deguine J-P, Quilici S (2012) Maize more attractive than Napier grass as non-host plants for *Bactrocera cucurbitae* and *Dacus demmerezi*. Arthropod Plant Interact 6:395–403

Atiama-Nurbel T, Quilici S, Boyer E, Deguine JP, Glenac S, Bialecki A (in press) Volatile constituents of cucumber: differences between five tropical cultivars. Chem Nat Comp (in press)

Becher B, Flick G, Rozpedowska E, Schmidt A, Hagman A, Lebreton S, Larsson MC, Hansson BS, Piskur J, Witzgall P, Bengtsson M (2012) Yeast, not fruit volatiles mediate attraction, oviposition and development of the fruit fly *Drosophila melanogaster*. Funct Ecol 26:822–828

Bidwell RGS (1979) Plant physiology, 2nd edn. Macmillan Publishing, New York

Birkett M, Bruce TA, Martin J, Smart L, Oakley J, Wadhams L (2004) Responses of female orange wheat blossom midge, *Sitodiplosis mosellana*, to wheat panicle volatiles. J Chem Ecol 30:1319–1328

Bjorksten TA, Hoffmann AA (1998) Separating the effects of experience, size, egg load, and genotype on host response in *Trichogramma* (Hymenoptera: Trichogrammatidae). J Insect Behav 11:129–148

Bjostad LB (1998) Electrophysiological methods. In: Millar JG, Haynes KF (eds) Chemical ecology – methodology, vol 1. Kluwer, Boston pp 339–369

Bostanian NJ, Racette G (2001) Attract and kill, an effective technique to manage apple maggot, *Rhagoletis pomonella* [Diptera: Tephritidae] in high density Quebec apple orchards. Phytoprotection 82:25–34

Bostanian NJ, Vincent C, Chouinard G, Racette G (1999) Managing apple maggot, *Rhagoletis pomonella* [Diptera: Tephritidae], by perimeter trapping. Phytoprotection 80:21–33

Brévault T, Quilici S (1999) Factors affecting behavioural responses to visual stimuli in the tomato fruit fly, *Neoceratitis cyanescens*. Physiol Entomol 24:333–338

Brévault T, Quilici S (2007) Visual response of the tomato fruit fly, *Neoceratitis cyanescens*, to colored fruit models. Entomol Exp Appl 125:45–54

Brévault T, Quilici S (2010a) Flower and fruit volatiles assist host-plant location in the Tomato fruit fly *Neoceratitis cyanescens*. Physiol Entomol 35:9–18

Brévault T, Quilici S (2010b) Interaction between visual and olfactory cues during host finding in the tomato fruit fly *Neoceratitis cyanescens*. J Chem Ecol 36:249–259

Browne LB (1993) Physiologically induced changes in resource-oriented behavior. Annu Rev Entomol 38:1–23

Carle SA, Averill AL, Rule GS, Reissig WH, Roelofs WL (1987) Variation in host fruit volatiles attractive to apple maggot fly, *Rhagoletis pomonella*. J Chem Ecol 13:795–805

Cha DH, Powell THQ, Feder JL, Linn CE Jr (2011a) Identification of fruit volatiles from green hawthorn (*Crataegus viridis*) and blueberry hawthorn (*Crataegus brachyacantha*) host plants attractive to different phenotypes of *Rhagoletis pomonella* flies in the Southern United States. J Chem Ecol 37:974–983

Cha DH, Powell THQ, Feder JL, Linn CE Jr (2011b) Identification of host fruit volatiles from three mayhaw species (*Crataegus Series* Aestivales) attractive to mayhaw-origin *Rhagoletis pomonella* flies in the Southern United States. J Chem Ecol 37:961–973

Cha DH, Powell THQ, Feder JL, Linn CE (2012) Geographic variation in fruit volatiles emitted by the hawthorn *Crataegus mollis* and its consequences for host race formation in the apple maggot fly, *Rhagoletis pomonella*. Entomol Exp Appl 143:254–268

Chiu H (1990) Bioassay of ovipositional attraction for *Dacus dorsalis* Hendel. Chin J Entomol 10:343–352

Chyau CC, Chen SY, Wu CM (1992) Differences of volatile and nonvolatile constituents between mature and ripe guava (*Psidium guajava* Linn.) fruits. J Agric Food Chem 40:846–849

Clarke AR, Dominiak BC (2010) Positive correlation of male and female *Bactrocera tryoni* (Froggatt)(Diptera: Tephritidae) catches in orange-ammonia traps. Gen Appl Entomol 39:9–13

Cornelius ML, Duan JJ, Messing RH (1999) Visual stimuli and the response of female oriental fruit flies (Diptera: Tephritidae) to fruit-mimicking traps. J Econ Entomol 92:121–129

Cornelius ML, Duan JJ, Messing RH (2000a) Volatile host fruit odors as attractants for the oriental fruit fly (Diptera: Tephritidae). J Econ Entomol 93:93–100

Cornelius ML, Nergel L, Duan JJ, Messing RH (2000b) Responses of female oriental fruit flies (Diptera: Tephritidae) to protein and host fruit odors in field cage and open field tests. Environ Entomol 29:14–19

Cossé AA, Todd JL, Millar JG, Martínez LA, Baker TC (1995) Electroantennographic and coupled gas chromatographic-electroantennographic responses of the Mediterranean fruit fly, *Ceratitis capitata*, to male-produced volatiles and mango odor. J Chem Ecol 21:1823–1836

Cruz-Lopez L, Malo EA, Toledo J, Virgen A, del Mazo A, Rojas JC (2006) A new potential attractant for *Anastrepha obliqua* from Spondias mombin fruits. J Chem Ecol 32:351–365

Cunningham RT (1989) Parapheromones. In: Robinson AS, Hooper G (eds) World crop pests, fruit flies: their biology, natural enemies and control, vol 3A. Elsevier, Amsterdam

Dambroski HR, Linn C, Berlocher SH, Forbes AA, Roelofs W, Feder JL (2005) The genetic basis for fruit odor discrimination in *Rhagoletis* flies and its significance for sympatric host shifts. Evolution 59:1953–1964
Dickens J, Jang E, Light D, Alford A (1990) Enhancement of insect pheromone responses by green leaf volatiles. Naturwissenschaften 77:29–31
Drew RA (1989) The tropical fruit flies (Diptera: Tephritidae: Dacinae) of the Australasian and Oceanian regions. Mem Qld Mus 26, 521 pp
Drew RAI, Prokopy RJ, Romig MC (2003) Attraction of fruit flies of the genus *Bactrocera* to colored mimics of host fruit. Entomol Exp Appl 107:39–45
Duan JJ, Prokopy RJ (1992) Visual and odor stimuli influencing effectiveness of sticky spheres for trapping apple maggot flies *Rhagoletis pomonella* (Walsh)(Dipt., Tephritidae). J Appl Entomol 113:271–279
Duan JJ, Prokopy RJ (1994) Apple maggot fly response to red sphere traps in relation to fly age and experience. Entomol Exp Appl 73(3):279–287
Epsky ND, Heath RR (1998) Exploiting the interactions of chemical and visual cues in behavioral control measures for pest tephritid fruit flies. Fla Entomol 81:273–282
Epsky ND, Morrill WL, Mankin R (2004) Traps for capturing insects. In: Capinera JL (ed) Encyclopedia of entomology. Kluwer Academic Publishers, Dordrecht
Fein BL, Reissig WH, Roelofs WL (1982) Identification of apple volatiles attractive to the apple maggot, *Rhagoletis pomonella*. J Chem Ecol 8:1473–1487
Flath RA, Cunningham RT, Mon TR, John JO (1994a) Additional male Mediterranean fruitfly (*Ceratitis capitata* Wied.) attractants from angelica seed oil (*Angelica archangelica* L.). J Chem Ecol 20:1969–1984
Flath RA, Cunningham RT, Mon TR, John JO (1994b) Male lures for Mediterranean fruitfly (*Ceratitis capitata* Wied.): structural analogs of α-copaene. J Chem Ecol 20:2595–2609
Fletcher MT, Kitching W (1995) Chemistry of fruit flies. Chemi Rev 95:789–828
Fletcher BS, Prokopy R (1991) Host location and oviposition in Tephritid fruit flies. In: Bailey WJ, Ridsdill-Smith J (eds) Reproductive behaviour of insects: individuals and populations. Chapman & Hall, Melbourne, pp 139–171
Forbes AA, Feder JL (2006) Divergent preferences of *Rhagoletis pomonella* host races for olfactory and visual fruit cues. Entomol Exp Appl 119:121–127
Forbes AA, Fisher J, Feder JL (2005) Habitat avoidance: overlooking an important aspect of host-specific mating and sympatric speciation? Evolution 59:1552–1559
Foster SP, Harris MO (1997) Behavioral manipulation methods for insect pest-management. Annu Rev Entomol 42:123–146
Gonzalez R, Toledo J, Cruz-Lopez L, Virgen A, Santiesteban A, Malo EA (2006) A new blend of white sapote fruit volatiles as potential attractant to *Anastrepha ludens* (Diptera: Tephritidae). J Econ Entomol 99:1994–2001
Goodner K, Rousseff R (2011) Practical analysis of flavor and fragrance materials. Blackwell Publishing Ltd, Hoboken
Haniotakis G, Vassiliou-Waite A (1987) Effect of combining food and sex attractants on the capture of *Dacus oleae* flies. Entomol Hellen 5:27–33
Haynes KF, Millar JG (1998) Bioassay methods, vol 2. Kluwer Academic, Norwell, MA, USA
Heath RR, Epsky ND, Dueben BD, Meyer WL (1996) Systems to monitor and suppress *Ceratitis capitata* (Diptera: Tephritidae) populations. Fla Entomol 79:144–153
Hernández MM, Sanz I, Adelantado M, Ballach S, Primo E (1996) Electroantennogram activity from antennae of *Ceratitis capitata* (Wied.) to fresh orange airborne volatiles. J Chem Ecol 22:1607–1619
Hill AR, Hooper AM (1984) Attractiveness of various colours to Australian tephritid fruit flies in the field. Entomol Exp Appl 35:119–128
Howlett F (1912) The effect of oil of citronella on two species of *Dacus*. Trans Royal Entomol Soc Lond 60:412–418
Howlett F (1915) Chemical reactions of fruit-flies. Bull Entomol Res 6:297–305

Ioannou CS, Papadopoulos NT, Kouloussis NA, Kouloussis NA, Tananaki CI, Katsoyannos BI (2012) Essential oils of citrus fruit stimulate oviposition in the Mediterranean fruit fly *Ceratitis capitata* (Diptera: Tephritidae). Physiol Entomol 37:330–339

Jacobson M, Uebel EC, Lusby WR, Waters RM (1987) Optical isomers of alpha-copaene derived from several plant sources. J Agric Food Chem 35:798–800

Jang EB (1995) Effects of mating and accessory gland injections on olfactory-mediated behavior in the female Mediterranean fruit fly, *Ceratitis capitata*. J Insect Physiol 41:705–710

Jang E (2002) Physiology of mating behavior in Mediterranean fruit fly (Diptera: Tephritidae): chemoreception and male accessory gland fluids in female post-mating behavior. Fla Entomol 85:89–93

Jang EB, Light DM (1991) Behavioral responses of female oriental fruit flies to the odor of papayas at three ripeness stages in a laboratory flight tunnel (Diptera: Tephritidae). J Insect Behav 4:751–762

Jang EB, Light DM (1996) Olfactory semiochemicals of tephritids. In: McPheron BA, Steck GJ (eds) Fruit fly pests: a world assessment of their biology and management. St. Lucie Press, Delray Beach, pp 73–90

Jang EB, Carvalho LA, Stark JD (1997) Attraction of female oriental fruit fly, *bactrocera dorsalis* to volatile semiochemicals from leaves and extracts of a nonhost plant, Panax (*Polyscias guilfoylei*) in laboratory and olfactometer assays. J Chem Ecol 23:1389–1401

Jang EB, Jang EB, McINNIS DO, Lance DR, Carvalho LA (1998) Mating-induced changes in olfactory-mediated behavior of laboratory-reared normal, sterile, and wild female Mediterranean fruit flies (Diptera: Tephritidae) mated to conspecific males. Ann Entomol Soc Am 91:139–144

Jang EB, McInnis DO, Kurashima R, Carvalho LA (1999) Behavioural switch of female Mediterranean fruit fly. *Ceratitis capitata*: mating and oviposition activity in outdoor field cages in Hawaii. Agric For Entomol 1:179–184

Jones VP, Davis DW (1989) Evaluation of traps for apple maggot (Diptera: Tephritidae) populations associated with cherry and hawthorn in Utah. Environ Entomol 18:521–525

Kamala Jayanthi PD, Woodcock CM, Caulfield J, Birkett MA, Bruce TJA (2012) Isolation and identification of host cues from mango, *Mangifera indica,* that attract gravid female oriental fruit fly, *Bactrocera dorsalis*. J Chem Ecol 38:361–369

Katsoyannos BI, Kouloussis NA, Papadopoulos NT (1997) Response of *Ceratitis capitata* to citrus chemicals under semi-natural conditions. Entomol Exp Appl 82:181–188

Kazi A (1976) Studies on the field habits of adult melon fruit fly *Dacus* (*Strumeta*) *cucurbitae*, Coquillet. Pak J Sci Ind Res 19:71–76

Keiser I, Harris E, Miyashita K, Jacobson M, Perdue R Jr (1975) Attraction of ethyl ether extracts of 232 botanicals to oriental fruit flies, melon flies, and Mediterranean fruit flies. Lloydia 38:141

Klaus MW (2003) WSDA apple maggot survey summary report. Department of Agriculture, Washington State, 24 pp

Kogan M (1982) Plant resistance in pest management. In: Metcalf L, Luckman HW (eds) Introduction to insect pest management. Wiley, New York

Koyama J, Teruya T, Tanaka K (1984) Eradication of the oriental fruit fly (Diptera: Tephritidae) from the Okinawa Islands by a male annihilation method. J Econ Entomol 77:468–472

Lalel HJ, Singh Z, Tan SC (2003) Aroma volatiles production during fruit ripening of 'Kensington Pride'mango. Postharv Biol Technol 27:323–336

Landolt PJ, Reed HC, Heath RR (1992) Attraction of female papaya fruit fly (Diptera: Tephritidae) to male pheromone and host fruit. Environ Entomol 21:1154–1159

Levinson H (1990) Influence of some olfactory and optical properties of fruits on host location by the Mediterranean fruit fly (*Ceratitis capita* Wied.). J Appl Entomol 109:44–54

Levinson H, Levinson A, Osterried E (2003) Orange-derived stimuli regulating oviposition in the Mediterranean fruit fly. J Appl Entomol 127:269–275
Light DM, Jang EB, Flath RB (1992) Electroantennogram responses of the Mediterranean fruit fly, *Ceratitis capitata*, to the volatile constituents of nectarines. Entomol Exp Appl 63:13–26
Light DM, Jang EB (1987) Electroantennogram responses of the oriental fruit fly, *Dacus dorsalis*, to a spectrum of alcohol and aldehyde plant volatiles. Entomol Exp Appl 45:55–64
Light DM, Jang EB, Dickens JC (1988) Electroantennogram responses of the Mediterranean fruit fly, *Ceratitis capitata*, to a spectrum of plant volatiles. J Chem Ecol 14:159–180
Linn CJ, Feder JL, Nojima S, Dambroski H, Berlocher SH, Roelofs W (2003) Fruit odor discrimination and sympatric race formation in *Rhagoletis*. Proc Natl Acad Sci U S A 100:11490–11493
Linn CE, Yee WL, Sim SB, Cha DH, Powell THQ, Goughnour RB, Feder JL (2012) Behavioral evidence for fruit odor discrimination and sympatric host races of *Rhagoletis pomonella* flies in the western united states. Evolution 66:3632–3641
Loera-Gallardo J, Lopez-Arroyo, JI, Miranda-Salcedo MA, Canales-Cruz R, Thomas DB, Mangan RL (2006). Attraction of Mexican fruit flies to colored traps and lures. Subtrop Plant Sci 58:23–25
López-Guillén G (2008) Estõ´mulos quõ´micos y visuales como potenciales atrayentes de *Anastrepha obliqua* (Macquart) (Diptera: Tephritidae). PhD dissertation, Mexico Texcoco
MacCollom G, Lauzon C, Payne E, Currier W (1994) Apple maggot (Diptera: Tephritidae) trap enhancement with washed bacterial cells. Environ Entomol 23:354–359
Malo EA, Cruz-Lopez L, Toledo J, Del Mazo A, Virgen A, Rojas JC (2005) Behavioral and electrophysiological responses of the Mexican fruit fly (Diptera: Tephritidae) to guava volatiles. Fla Entomol 88:364–371
Malo EA, Gallegos-Torres I, Toledo J, Valle-Mora J, Rojas JC (2012) Attraction of the West Indian fruit fly to mango fruit volatiles. Entomol Exp Appl 142:45–52
Massa MJ, Robacker DC, Patt J (2008) Identification of grape juice aroma volatiles and attractiveness to the Mexican fruit fly (Diptera: Tephritidae). Fla Entomol 91:266–276
McQuate GT, Jones GD, Sylva CD (2003) Assessment of corn pollen as a food source for two tephritid fruit fly species. Environ Entomol 32:141–150
Medlicott AP, Thompson AK (1985) Analysis of sugars and organic acids in ripening mango fruits (*Mangifera indica* L. var Keitt) by high performance liquid chromatography. J Sci Food Agric 36:561–566
Messina FJ, Jones VP (1990) Relationship between fruit phenology and infestation by the apple maggot (Diptera: Tephritidae) in Utah. Ann Entomol Soc Am 83:742–752
Messina FJ, Alston DG, Jones VP (1991) Oviposition by the Western cherry fruit fly (Diptera: Tephritidae) in relation to host development. J Kans Entomol Soc 64:197–208
Metcalf RL, Metcalf ER (1992) Fruit flies of the family Tephritidae. In: Metcalf RL, Metcalf ER (eds) Plant kairomones in insect ecology and control. Chapman and Hall, New York, pp 109–152
Migani V, Ekesi S, Hoffmeister T (2013) Physiology vs. environment: what drives oviposition decisions in mango fruit flies (*Bactrocera invadens* and *Ceratitis cosyra*)? J Appl Entomol 138(6):395–402. doi:10.1111/jen.12038
Millar JG, Haynes KF (1998) Chemical methods, vol 1. Kluwer, New York
Miller NW, Vargas RI, Prokopy RJ, Mackey BE (2004) State-dependent attractiveness of protein bait and host fruit odor to *Bactrocera cucurbitae* (Diptera: Tephritidae) females. Ann Entomol Soc Am 97:1063–1068
Minkenberg OP, Tatar M, Rosenheim JA (1992) Egg load as a major source of variability in insect foraging and oviposition behavior. Oikos 65:134–142
Mwatawala M, Virgilio M, Quilici S, Dominic M, De Meyer M (2013) Field evaluation of the relative attractiveness of enriched ginger root oil (EGO)lure and trimedlure for African *Ceratitis* species (Diptera: Tephritidae). J Appl Entomol 137:392–397
Nakamori H, Nishimura M, Kakinohana H (1988) Eradication of the oriental fruit fly, *Dacus dorsalis* Hendel (Diptera: Tephiritidae), from Miyako and Yaeyama Islands by the male annihilation method. In: Proceedings of the first international symposium on fruit flies in

tropics. Malaysian Agricultural Research and Development Institute, Kuala Lumpur, pp 220–231

Neilson W, Knowlton A, Whitman R (1981) Capture of apple maggot adults on pherocon, rebell, and sticky sphere traps. J Econ Entomol 74:203–206

Nigg HN, Mallory LL, Simpson SE, Callaham SB, Toth JP, Fraser S, Klim M, Nagy S, Nation JL, Attaway JA (1994) Caribbean fruit fly, *Anastrepha suspensa* (Loew), attraction to host fruit and host kairomones. J Chem Ecol 20:727–743

Niogret J, Montgomery WS, Kendra PE, Heath RR, Epsky ND (2011) Attraction and electroantennogram responses of male Mediterranean fruit fly to volatile chemicals from *Persea*, *Litchi* and *Ficus* wood. J Chem Ecol 37:483–491

Nishida T, Bess HA (1957) Studies on the ecology and control of the melon fly *Dacus* (*Strumeta*) *cucurbitae* Coquillet (Diptera: Tephritidae). Tech Bull Hawaii Agr Exp Station 34:1–44

Nishida R, Shelly TE, Whittier TS, Kaneshiro KY (2000) α-Copaene, a potential rendezvous cue for the Mediterranean fruit fly, *Ceratitis capitata*? J Chem Ecol 26:87–100

Nojima S, Linn C, Roelofs W (2003a) Identification of host fruit volatiles from flowering dogwood (*Cornus florida*) attractive to dogwood-origin *Rhagoletis pomonella* flies. J Chem Ecol 29: 2347–2357

Nojima S, Linn C Jr, Morris B, Zhang A, Roelofs W (2003b) Identification of host fruit volatiles from hawthorn (*Crataegus* spp.) attractive to hawthorn-origin *Rhagoletis pomonella* flies. J Chem Ecol 29:321–336

Papadopoulos NT, Katsoyannos BI, Kouloussis NA, Hendrichs J (2001) Effect of orange peel substances on mating competitiveness of male *Ceratitis capitata*. Entomol Exp Appl 99: 253–261

Papaj DR (2000) Ovarian dynamics and host use. Annu Rev Entomol 45:423–448

Papaj DR, Prokopy RJ (1989) Ecological and evolutionary aspects of learning in phytophagous insects. Annu Rev Entomol 34:315–350

Pelz-Stelinski KS, Gut LJ, Stelinski LL, Liburd OE, Isaacs R (2005) Captures of *Rhagoletis mendax* and *R. cingulata* (Diptera: Tephritidae) on sticky traps are influenced by adjacent host fruit and fruit juice concentrates. Environ Entomol 34:1013–1018

Piñero JC, Jacome I, Vargas R, Prokopy RJ (2006) Response of female melon fly, *Bactrocera cucurbitae*, to host-associated visual and olfactory stimuli. Entomol Exp Appl 121:261–269

Price PW, Denno RF, Eubanks MD, Finke DL, Kaplan I (2011) Insect ecology: behavior, populations and communities. Cambridge University Press, New York

Prokopy RJ (1968) Visual responses of apple maggot flies, *Rhagoletis pomonella* (Diptera: Tephritidae): orchard studies. Entomol Exp Appl 11:403–422

Prokopy RJ (1972) Evidence for a marking pheromone deterring repeated oviposition in apple maggot flies. Environ Entomol 1:326–332

Prokopy RJ, Bush GL (1973) Ovipositional responses to different sizes of artificial fruit by flies of *Rhagoletis pomonella* species group. Ann Entomol Soc Am 66:927–929

Prokopy RJ, Economopoulos AP (1975) Attraction of laboratory-cultured and wild *Dacus oleae* flies to sticky-coated McPhail traps of different colors and odors. Environ Entomol 4:187–192

Prokopy RJ, Haniotakis GE (1975) Responses of wild and laboratory-cultured *Dacus oleae* to host plant color. Ann Entomol Soc Am 68:73–77

Prokopy RJ, Vargas RI (1996) Attraction of *Ceratitis capitata* (Diptera: Tephritidae) flies to odor of coffee fruit. J Chem Ecol 22:807–820

Prokopy RJ, Drew RAI, Sabine BNE, Lloyd AC, Hamacek E (1991) Effect of physiological and experiential state of *Bactrocera tryoni* flies on intra-tree foraging behavior for food (bacteria) and host fruit. Oecologia 87:394–400

Prokopy R, Roitberg B, Vargas R (1994a) Effects of egg load on finding and acceptance of host fruit in *Ceratitis capitata* flies. Physiol Entomol 19:124–132

Prokopy R, Bergweiler C, Galarza L, Schwerin J (1994b) Prior experience affects the visual ability of *Rhagoletis pomonella* flies (Diptera: Tephritidae) to find host fruit. J Insect Behav 7: 663–677

Prokopy RJ, Phillips TW, Vargas RI, Jang EB (1997) Defining sources of coffee plant odor attractive to *Ceratitis capitata* flies. J Chem Ecol 23:1577–1587
Prokopy RJ, Hu XP, Jang EB, Vargas RI, Warthen JD (1998) Attraction of mature *Ceratitis capitata* females to 2-heptanone, a component of coffee fruit odor. J Chem Ecol 24:1293–1304
Prokopy RJ, Wright SE, Black JL, Hu XP, McGuire MR (2000) Attracticidal spheres for controlling apple maggot flies: commercial-orchard trials. Entomol Exp Appl 97:293–299
Prokopy RJ, Jácome I, Bigurra E (2005) An index for assigning distances between odor-baited spheres on perimeter trees of orchards for control of apple maggot flies. Entomol Exp Appl 115:371–377
Rasgado MA, Malo EA, Cruz-Lopez L, Rojas JC, Toledo J (2009) Olfactory response of the Mexican fruit fly (Diptera: Tephritidae) to *Citrus aurantium* volatiles. J Econ Entomol 102: 585–594
Rattanapun W, Amornsak W, Clarke AR (2009) *Bactrocera dorsalis* preference for and performance on two mango varieties at three stages of ripeness. Entomol Exp Appl 131:243–253
Reissig W (1974) Field tests of traps and lures for the apple maggot. J Econ Entomol 67:484–486
Reissig WH, Fein BL, Roelofs WL (1982) Field tests of synthetic apple volatiles as apple maggot (Diptera: Tephritidae) attractants. Environ Entomol 11:1294–1298
Reissig W, Stanley B, Roelofs W, Schwarz M (1985) Tests of synthetic apple volatiles in traps as attractants for apple maggot flies (Diptera: Tephritidae) in commercial apple orchards. Environ Entomol 14:55–59
Robacker DC (1991) Specific hunger in *Anastrepha ludens* (Diptera: Tephritidae): effects on attractiveness of proteinaceous and fruit-derived lures. Environ Entomol 20:1680–1686
Robacker DC (1992) Effects of shape and size of colored traps on attractiveness to irradiated, laboratory-strain Mexican fruit flies (Diptera: Tephritidae). Fla Entomol 75:230–241
Robacker DC, Fraser I (2005) What do Mexican fruit flies learn when they experience fruit? J Insect Behav 18:529–542
Robacker D, Garcia J (1990) Responses of laboratory-strain Mexican fruit flies, *Anastrepha ludens*, to combinations of fermenting fruit odor and male-produced pheromone in laboratory bioassays. J Chem Ecol 16:2027–2038
Robacker DC, Heath RR (1996) Attraction of Mexican fruit flies (Diptera: Tephritidae) to lures emitting host-fruit volatiles in a citrus orchard. Fla Entomol 79:600–602
Robacker D, Heath R (1997) Decreased attraction of *Anastrepha ludens* to combinations of two types of synthetic lures in a citrus orchard. J Chem Ecol 23:1253–1262
Robacker DC, Rios C (2005) Grapefruit oil enhances attraction of Mexican fruit flies to a synthetic food-odor lure. J Chem Ecol 31:1039–1049
Robacker DC, Garcia JA, Hart WG (1990a) Attraction of a laboratory strain of *Anastrepha ludens* (Diptera, Tephritidae) to the odor of fermented chapote fruit and to pheromones in laboratory experiments. Environ Entomol 19:403–408
Robacker DC, Moreno AMT, Garcia JA, Flath RA (1990b) A novel attractant for Mexican fruit fly, *Anastrepha ludens* (Diptera, Tephritidae) from fermented host fruit. J Chem Ecol 16: 2799–2815
Robacker D, Warfield W, Flath R (1992) A four-component attractant for the Mexican fruit fly, *Anastrepha ludens* (Diptera: Tephritidae), from host fruit. J Chem Ecol 18:1239–1254
Robacker DC, Massa MJ, Sacchetti P, Bartelt RJ (2011) A novel attractant for *Anastrepha ludens* (Diptera: Tephritidae) from a concord grape product. J Econ Entomol 104:1195–1203
Roitberg BD (1985) Search dynamics in fruit-parasitic insects. J Insect Physiol 31:865–872
Rull J, Prokopy R (2000) Attraction of apple maggot flies, *Rhagoletis pomonella* (Diptera: Tephritidae) of different physiological states to odour-baited traps in the presence and absence of food. Bull Entomol Res 90:77–88

Rull J, Prokopy RJ (2005) Interaction between natural and synthetic fruit odor influences response of apple maggot flies to visual traps. Entomol Exp Appl 114:79–86

Schoonhoven LM, van Loon JJA, Dicke M (2005) Insect-plant biology. Oxford University Press, Oxford

Shelly TE (2001) Exposure to alpha-copaene and alpha-copaene-containing oils enhances mating success of male Mediterranean fruit flies (Diptera: Tephritidae). Ann Entomol Soc Am 94:497–502

Shelly T (2013) Detection of male Mediterranean fruit flies (Diptera: Tephritidae): performance of trimedlure relative to capilure and enriched ginger root oil. Proc Haw Entomol Soc 45:1–7

Shelly TE, McInnis DO (2001) Exposure to ginger root oil enhances mating success of irradiated, mass-reared males of Mediterranean fruit fly (Diptera: Tephritidae). J Econ Entomol 94: 1413–1418

Shelly TE, Villalobos EM (2004) Host plant influence on the mating success of male Mediterranean fruit flies: variable effects within and between individual plants. Anim Behav 68:417–426

Siderhurst MS, Jang EB (2006) Female-biased attraction of oriental fruit fly, *Bactrocera dorsalis* (Hendel), to a blend of host fruit volatiles from *Terminalia catappa* L. J Chem Ecol 32: 2513–2524

Siderhurst MS, Jang EB (2010) Cucumber volatile blend attractive to female melon fly, *Bactrocera cucurbitae* (Coquillett). J Chem Ecol 36:699–708

Stark JD (1995) Nocturnal behavior of oriental fruit flies and melon flies (Diptera: Tephritidae) and associated parasitoids in a commercial papaya growing region on Kaua'i, Hawai'i. Proc Haw Entomol Soc 32:149–151

Stark JD, Vargas RI, Thalman RK (1991) Diversity and abundance of oriental fruit fly parasitoids (Hymenoptera: Braconidae) in guava orchards in Kauai, Hawaii. J Econ Entomol 84: 1460–1467

Steiner LF, Miyashita DH, Christenson LD (1957) Angelica oils as Mediterranean fruit fly lures. J Econ Entomol 50:505

Stelinski LL, Liburd OE (2002) Attraction of apple maggot flies, *Rhagoletis pomonella* (Walsh) (Diptera: Tephritidae), to synthetic fruit volatile compounds and food attractants in Michigan apple orchards. Great Lakes Entomol 35:37–46

Syed RA (1969) Studies on the ecology of some important species of fruit flies and their natural enemies in West Pakistan. Pakistan Commonwealth Institute – Biological Control Station, pp 12

Tabilio MR, Fiorini D, Marcantoni E, Materazzi S, Delfini M, Salvador RD, Musmeci S (2013) Impact of the Mediterranean fruit fly (Medfly) *Ceratitis capitata* on different peach cultivars: the possible role of peach volatile compounds. Food Chem 140:375–381

Vargas RI, Stark JD, Prokopy RJ, Green TA (1991) Response of oriental fruit fly (Diptera: Tephritidae) and associated parasitoids (Hymenoptera: Braconidae) to different-color spheres. J Econ Entomol 84:1503–1507

Visser JH (1986) Host odor perception in phytophagous insects. Annu Rev Entomol 31:121–144

Warthen JD, McInnis DO (1989) Isolation and identification of male medfly attractive components in *Litchi chinensis* stems and *Ficus* spp. stem exudates. J Chem Ecol 15:1931–1946

Warthen JD, Lee CJ, Jang EB, Lance DR, McInnis DO (1997) Volatile, potential attractants from ripe coffee fruit for female Mediterranean fruit fly. J Chem Ecol 23:1891–1900

Witzgall P, Proffit M, Rozpedowska E, Becher PG, Andreadis S, Coracini M, Lindblom TUT, Ream LJ, Hagman A, Bengtsson M (2012) "This is not an apple" – yeast mutualism in codling moth. J Chem Ecol 38:949–957

Wyatt TD (2003) Pheromones and animal behaviour: communication by smell and taste. Cambridge University Press, Cambridge

Yashoda H, Prabha T, Tharanathan R (2007) Mango ripening – role of carbohydrases in tissue softening. Food Chem 102:691–698

Yee WL, Landolt PJ, Darnell TJ (2005) Attraction of *Rhagoletis pomonella* (Diptera: Tephritidae) and non-target flies to traps baited with ammonium carbonate and fruit volatile lures in Washington and Oregon. J Agric Urb Entomol 23

Zervas G (1989) Significant increase of *Dacus oleae* trapping using sex and food attractants separately in the same tree. Fruit flies of economic importance. Balkema, Rotterdam, pp 433–442

Zhang AJ, Linn C, Wright S, Prokopy R, Reissig W, Roelofs W (1999) Identification of a new blend of apple volatiles attractive to the apple maggot, *Rhagoletis pomonella*. J Chem Ecol 25: 1221–1232

Chapter 5
Interactions Between Tephritid Fruit Fly Physiological State and Stimuli from Baits and Traps: Looking for the Pied Piper of Hamelin to Lure Pestiferous Fruit Flies

Francisco Díaz-Fleischer, Jaime C. Piñero, and Todd E. Shelly

Abstract The development of effective fruit fly trapping methods depends on knowing the factors that affect the temporal and spatial activity of the target species. Several endogenous factors, such as nutritional and mating status, sexual development, age, and gender, influence fly physiological condition and directly impact the effectiveness of baits and trapping systems, since only a small portion of the population is usually attracted to particular stimuli. Therefore, the identification of signals and cues used by fruit flies to locate the resources that satisfy their needs is the basis for developing effective lures and traps. Exogenous factors known to impact fruit fly captures include abiotic (e.g., temperature, precipitation, and relative humidity) and biotic conditions (e.g., resource availability). In this chapter, we first discuss ways in which these factors affect the behavioral response of fruit flies to traps and lures. Then, we analyze the specific response of fruit flies to natural and synthetic attractants used for trapping them and also discuss aspects of life history among fruit fly species in an attempt to explain variations in responses to visual and olfactory cues associated with traps. Finally, directions of future research are discussed.

F. Díaz-Fleischer (✉)
Instituto de Biotecnología y Ecología Aplicada (INBIOTECA), Universidad Veracruzana, CP 91090, Xalapa, Veracruz, Mexico
e-mail: fradiaz@uv.mx

J.C. Piñero
Lincoln University, Cooperative Research and Extension, Allen Hall 212, 900 Chestnut Street, Jefferson City, MO 65101, USA
e-mail: PineroJ@lincolnu.edu

T.E. Shelly
USDA-APHIS, 41-650 Ahiki Street, Waimanalo, HI 96795, USA
e-mail: todd.e.shelly@aphis.usda.gov

T. Shelly et al. (eds.), *Trapping and the Detection, Control, and Regulation of Tephritid Fruit Flies*, DOI 10.1007/978-94-017-9193-9_5,

Keywords Physiological state • Chemical stimuli • Physical stimuli • Color spectrum • Behavioral response • Homeostatic shifts • Sexual development • Semiochemicals • Nutrition • Dietary history • Food lures • Fly age • Mating status • Kairomones • Fruit volatiles • Olfactory cues • Foraging behavior • Hormones • Odors

1 Introduction

Trapping systems for pestiferous fruit flies provide useful information about the presence, seasonal abundance, and spatial distribution of the adults in order to make predictions of host fruit infestation levels (IAEA 2007). The development of effective trapping methods depends on knowing the factors that affect the temporal and spatial activity of the target species. These factors can be divided into two types according to their origin: exogenous and endogenous. Those considered exogenous include various abiotic and biotic environmental conditions, such as temperature, precipitation, humidity, risk of predation, and resource availability (e. g., host phenology and seasonality). Endogenous factors are those related to the insect's physiological state that can modify the motivational level to perform different tasks or forage for different resources. Physiological state is probably the most important endogenous factor influencing resource-oriented behavior (Tschinkel 1985; Barton-Browne 1993). For example, deprivation of a specific resource will increase the probability of responding to some resource-related cues, possibly leading to acceptance of a previously unacceptable resource (Bell 1990). Thus, specific motivation for resources, such as mates, oviposition site, or food, can result in the expression of a variety of behaviors. This behavioral plasticity allows individuals to satisfy their needs. A clear example of behavioral changes in response to physiological condition has been observed in the Mediterranean fruit fly, *Ceratitis capitata* (Wiedemann), where females prefer the male sex pheromone to fruit odors when virgin but show the opposite preference after mating (Jang 1995). Another example of a physiologically-mediated behavioral change has been observed in the apple maggot fly, *Rhagoletis pomonella* (Walsh). In this species, mating takes place on host tree foliage ('lek-like' mating strategy) at the beginning of the fruiting season, whereas mating occurs on fruits (male resource defense strategy) later in the season (Prokopy and Bush 1973; Smith and Prokopy 1980). This strategic switch occurs in response to changes in female physiological condition, since females, once mated, primarily search for host fruits for oviposition. Also, it has been observed that sexually mature, presumably mated, protein-fed females of the melon fly, *Bactrocera cucurbitae* (Coquillett), tend to respond primarily to cucumber, *Cucumis sativus* L., odor over protein odor (Miller et al. 2004). In these and other instances, physiological state influences behavioral decisions in a manner that maximizes reproductive fitness through the expression of specific behaviors at the appropriate time and place.

Endogenous processes that alter the physiological state of animals, and consequently their behavior, can be divided arbitrarily into those triggered by hormones or by homeostatic shifts (physiological adjustments in response to stress in order to maintain its internal equilibrium). The latter reflects factors, such as development, nutrition, and change in physical condition because of daily activities (Tschinkel 1985). In the case of adults, the behaviors activated by hormones are those related with sexual maturation. As they mature sexually, flies respond to specific stimuli emanating from those resources necessary for producing eggs, sperm, and pheromones (reviewed by Wheeler 1996; Drew and Yuval 2000; Hee and Tan 2004). It has been observed that behaviors, such as oviposition, pheromone release (calling), and copulation, show discrete circadian rhythms in most fruit flies (Hendrichs and Hendrichs 1990; Warburg and Yuval 1997a, b; Aluja et al. 2000). Although these behaviors are under hormonal control, exogenous factors, such as light intensity and temperature, entrain their expression presumably as an adaptation to increase reproductive success (Kaspi and Yuval 1999; Tychsen and Fletcher 1971; Tschinkel 1985; Díaz-Fleischer and Arredondo 2011a, b). In the case of homeostatic shifts, changes in energy budget are manifest both long-term as a function of fly age and short-term as a function of homeostatic circadian variation in daily nutrient requirements and energy expenditures. In other words, hormonal and homeostatic effects on behavior are all endogenous, though behavioral expression may be triggered by exogenous cues. Interactions between the two processes (hormonal and homeostatic) affect the physiological state of flies and consequently their response to traps. Changes in the need for specific resources of an adult insect thus occur along two axes (Fig. 5.1): those related to sexual maturation and reproduction in general with clear pre- and post-maturation periods and those concerning homeostatic conditions, e.g., the need to replenish the energy spent searching for resources, fighting, mating, which in many cases follows a circadian pattern. Thus, individual fly response to a specific stimulus varies within a day as well as over the complete life (Brieze-Stegeman et al. 1978; Malo and Zapien 1994; Robacker 1998; Thomas et al. 2001). For example, insects need to replenish the energy consumed performing activities undertaken at specific periods of the day (Warburg and Yuval 1996, 1997a). However, immature flies exhibit a different demand of nutrients in comparison with mature ones. In the Mexican fruit fly, *Anastrepha ludens* (Loew), for example, flies less than 4 days old prefer sugar over protein, while 5–9 day-old flies respond to protein and sugar equally (Robacker 1991). Thus, trapping programs encounter a variable mosaic of fly physiological states, with certain lures and traps likely biased to capture only a subset of the population (Thomas et al. 2001; Kouloussis et al. 2009).

In this chapter, we discuss the different ways some endogenous factors affect the physiological state of tephritid fruit flies and the concomitant behavioral response to cues used to locate their resources. We analyze the specific response of the flies to natural and synthetic attractants used in trapping. We also consider aspects of life history among fruit fly species in an attempt to explain interspecific differences in responses to visual and olfactory cues associated with traps. Finally, directions of future research are examined.

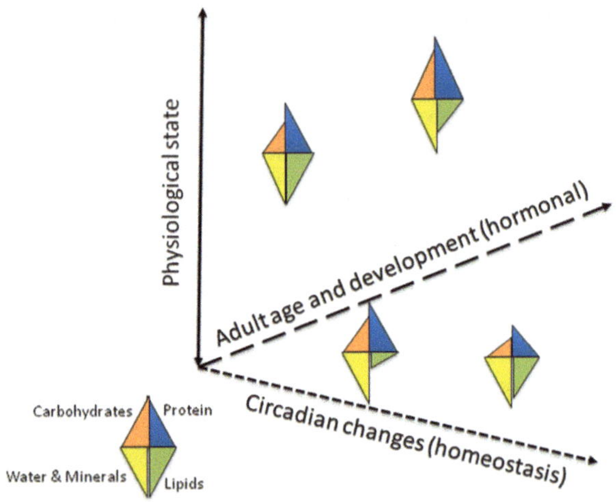

Fig. 5.1 The axes that modulate the physiological state of fruit flies. The hormonal mechanisms depend of factors as mating and sexual development, and normally moves in longer periods than the homeostatic mechanisms that depends of fly activity

2 Fly Response to Lures According to Physiological State

Several variables associated with the physiological state of pest tephritid fruit flies can influence the probability of a fly responding to visual and olfactory stimuli associated with trapping devices. Nutritional status, mating status, sexual development, age, and gender generate natural sources of physiological variation that directly impact the effectiveness of baits and trapping. Additionally, sterilized insects and those treated with juvenile hormone or aromatherapy represent non-natural variants in the catalogue of physiological conditions that may generate variable response to the common trapping lures.

2.1 *Effect of Dietary History and Nutritional Status*

According to their life history strategies, tephritid flies of economic importance can be considered, in parallel, as "income breeders" (refers to the use of concurrent intake to pay for a reproductive attempt) and synovigenic (females continuously produce and develop eggs throughout their adult period) (Papaj 2000; Houston et al. 2007). Therefore, economically important tephritid species flies have a continuous demand of resources for egg development as well as for somatic maintenance and locomotion and, consequently, they must search continuously for resources that satisfy their nutritional demands (Warburg and Yuval 1997b; Aluja et al. 2011). Typically, insects with nutritional deficiencies will behave in ways that increase the likelihood of encountering foods that satisfy their requirements, i.e., they will exhibit a higher responsiveness to food cues. Among nutrients, flies need carbohydrates, especially sucrose, as a primary source of energy vital for survival (Bateman 1972; Hagen et al. 1984; Hendrichs et al. 1993; Binder 1996;

Jacome et al. 1995). However, flies can metabolize carbohydrates so rapidly that they need to replenish their reserves at least every other day simply to remain active, and they cannot survive more than 2–4 days when deprived of carbohydrates (Yee 2003; Teal et al. 2004). Nevertheless, carbohydrates are not a limiting resource in nature, since sources, such as fruit juices, honeydews, and extra-floral nectaries, are available most of the year in subtropical and tropical regions. Although less important for simple survival, protein sources provide the essential amino acids that females require for developing eggs and are apparently less abundant resources in the field (Drew and Yuval 2000). For example, body protein level and oviposition of female *A. ludens* oscillate dramatically according to the availability of protein (hydrolyzed yeast) (Aluja et al. 2011). In the case of the West Indian fruit fly, *Anastrepha obliqua* (Macquart), protein deprivation severely alters the discrimination threshold for protein (i.e., the smallest quantity of a given nutrient that can be perceived by an insect in a given volume of food) leading flies to search intensely for this resource (Cresoni-Pereira and Zucoloto 2001). Many studies have demonstrated that protein-deprived flies offset this deficiency by preferring meals that include proteins (reviewed in Drew and Yuval 2000 and in Cresoni-Pereira and Zucoloto 2012).

Given that nearly all fruit fly species need protein for reproduction, it is puzzling that species do not respond with the same intensity to a common source of proteins. This variation is evident when testing the response of fruit flies to proteinaceous lures. For example, *A. obliqua* and the guava fruit fly, *Anastrepha striata* Schiner, respond more consistently to natural and synthetic protein lures than *A. ludens* and the sapote fruit fly, *Anastrepha serpentina* (Wiedemann) (Piñero et al. 2002; Díaz-Fleischer et al. 2009; Utgés et al. 2011). Differences in their life histories could explain such variable responses. For example, *A. obliqua* females exhibit a shorter lifespan and a shorter reproductive period, with a peak of egg production during the first 6 weeks of adult life. Conversely, *A. ludens* and *A. serpentina* have a longer lifespan, a longer reproductive period, and produce fewer eggs per week (Liedo et al. 1992). Thus, it is conceivable that temporally concentrated reproduction in *A. obliqua* females results in a lower threshold for protein than in the other two species. In *A. striata* proteins are also important for longevity and sexual selection, since those males that have proteins in their diets are more successful in procuring mates than protein-deprived males (Pérez-Staples and Aluja 2004).

Responses to a proteinaceous lure may also depend upon the type of diet a fly ingested previously. In *C. capitata*, Prokopy et al. (1996) observed that 85 % of protein-deprived females entered a food-baited McPhail trap (bell shaped glass traps with a water reservoir [Newell 1936]), while only about 50 % of protein–fed females did so. In general, fly responses to lures are highly influenced by their nutritional state. As in *C. capitata*, individuals of *Anastrepha* spp. deprived protein likewise show a stronger response to proteinaceous lures than previously protein fed flies (Robacker 1998; Díaz-Fleischer et al. 2009). This pattern of response to protein sources was also observed in females of *Bactrocera tryoni* (Froggatt) whose levels of body carbon and body nitrogen were analyzed. It was observed that females with high body nitrogen reduced protein foraging and increased oviposition activity, however, high total body carbon levels also reduced protein hunger but do not increased oviposition response (Balagawi et al. 2014). Also, a reduction in the amount of available carbohydrates may also predispose flies to respond to proteinaceous lures as observed in the western cherry fruit fly, *Rhagoletis indifferens* Curran, with spinosad bait (GF-120® Naturalyte® Fruit Fly Bait; Dow AgroSciences, Indianapolis, IN, USA) (Yee and Chapman 2009). These authors observed that *R. indifferens* must feed on carbohydrates several times during the day to maintain their energy levels and that their responses to spinosad bait will increase if they do not.

Nutritional status also alters fly response to non-proteinaceous lures. For example, if males of the Queensland fruit fly, *B. tryoni* are deprived of protein for 24 h, their response to the male lure cue-lure, [CL, 4-(p-hydroxyphenyl)-2-butanone acetate], is significantly lower compared to males with continuous access to protein (Weldon et al. 2008). Similarly, Fitt (1981a) reported that, for *Bactrocera opiliae* (Drew and Hardy), male responsiveness to the male lure methyl eugenol (ME) is higher if they were protein fed. In contrast, physiological state of *C. capitata* (age or nutritional status [protein-fed vs. protein-deprived]) had little or no effect on the propensity of flies to enter Jackson or Nadel-Harris traps baited with the male lure trimedlure (TML) (Prokopy et al. 1996).

2.2 Effect of Age and Mating Condition

Hormonal effects on behavior are clearly observed during aging. Physiological age (often characterized by chronological age) may influence an animal's behavior either through differing levels of maturation or senescence (Carey and Molleman 2010). Aging and reproduction act in combination to influence nutritional requirements and may be responsible for dietary shifts once a fly reaches sexual maturity. Commonly, post-mating physiological and behavioral changes in insects include egg-production, oviposition, increased feeding behavior, changes in food preferences, and reduced mating readiness (Tschinkel 1985). In female fruit flies, mating does not appear to increase egg production (Sivinski 1993; Chapman et al. 1998) but may induce sexual refraction after their first mating. Mating-induced sexual refraction has been found in *C. capitata* (Chapman et al. 1998; Miyatake et al 1999; Mossinson and Yuval 2003; Kraaijeveld and Chapman 2004), the olive fruit fly, *Bactrocera oleae* (Rossi), (Cavalloro and Delrio 1970; Tsiropoulos and Tzanakakis 1970), various *Anastrepha* species (Aluja et al. 2000) and *B. tryoni* (Barton Browne 1957; Harmer et al. 2006; Pérez-Staples et al. 2008; Radhakrishnan et al. 2008) but not in *R. pomonella* (Opp and Prokopy 2000). Searching behavior is modified after mating in *C. capitata*, since virgin females are preferentially attracted to the male pheromone over host fruit odors, while the opposite characterizes mated females (Jang 1995). Substances within the seminal fluid that are transferred together with the sperm during copulation often induce behavioral changes in females (Jang 1995, 2002). In *C. capitata*, the effects of mating are expressed transcriptionally after copulation in both males and females with very different patterns. Broad transcriptional changes were detected during female maturation, while post-mating transcriptional changes in females were small in contrast. In male *C. capitata*, transcriptional changes were consistent both during maturation and as a consequence of mating (Gomulski et al. 2012). In spite of the small transcriptional changes observed in females after copulation, dramatic behavioral changes are observed (Jang 1995). It could be assumed that changes happened at the level of expression of odorant-binding proteins (extracellular proteins found in insect chemosensilla, where they participate in the sensing of odors, tastes, and pheromones). This would involve a change from ones specialized in perceiving sexual pheromones to those specialized in fruit volatiles (Pelosi and Maida 1995; Christophides et al. 2000).

Studies on the response of flies to food lures that compare mature vs. immature flies or virgin vs. mated flies are somewhat conflicting. For example, in a laboratory study with four *Anastrepha* species it was observed that sexually mature and presumably mated individuals (primarily females) responded more strongly to protein baits than did sexually immature individuals (Piñero et al. 2002). However, in a field cage study with *A. ludens* and *A. obliqua*, no differences were observed between mature (also presumably mated) and immature female flies in their response to commercial proteinaceous lures (Díaz-Fleischer et al. 2009). Using electroantennography and behavioral bioassays with Caribbean fruit fly, *Anastrepha suspensa* (Loew), Kendra et al. (2005) studied the effects of age on fly response to ammonia (food related attractant) and carbon dioxide (host related attractant), two volatile chemicals released from commercial ammonium bicarbonate lures. Females were found to be more responsive to ammonia when sexually immature (3-6 days old) and more responsive to carbon dioxide when mature and mated (10–13 days old) (Kendra et al. 2005). In release-recapture studies with sterile insects, there was higher recapture of immature than mature *A.* suspensa in traps baited with either two-component synthetic food based lure or with aqueous torula yeast/borax (Kendra et al. 2010). Also, in *A. suspensa*, female response to the liquid protein bait NuLure (Miller Chemical & Fertilizer Corp., Hanover, PA, USA) + sugar lures was higher at 8–9 days of age (when egg maturation normally begins) compared to presumed unmated immature females and males (Nigg et al. 1995).

Heath et al. (1995) found that female *C. capitata* captured in traps baited with a low dose of ammonium acetate (AmAc) and putrescene were predominantly unmated (55–69 %), while few were captured when a high dose of the same lure was tested (4–13 %). In a study with the mango fruit fly, *Ceratitis cosyra* (Walker), *Ceratitis fasciventris* (Bezzi), and *C. capitata*, Manrakhan and Lux (2008) demonstrated that, for both males and females of all three species, nutritional state was more important than mating status in influencing response to food odors. These authors also found that protein deprivation had variable effects among the three species. Mature, protein-deprived virgin *C. capitata* females and males had a higher response to food odors than comparable *C. fasciventris* and *C. cosyra* flies, possibly indicating a higher need for protein for *C. capitata* compared to the two other species (Manrakhan and Lux 2008). For *B. dorsalis*, in field and field-cage studies, it was found that mated, protein-deprived females (10–12 days old) and unmated, protein-fed females (2–3 days old) were equally attracted to fruit and protein odors (Cornelius et al. 2000).

Of particular interest are the interactions between age (related to egg load) and nutritional status (specifically protein hunger) and the response to particular compounds present in baits, since these are two key components of insect physiological state known to influence the foraging behavior (Prokopy et al. 1995). However, fly responses can vary among fly species independently of its age. For example, in a comparative field cage study conducted in Hawaii, Piñero et al. (2011) reported that the response to different amounts of AmAc present in GF-120 (0 %, 1 %, or 2 %, respectively) can be modulated by factors, such as age and level of protein starvation, and that these effects can differ among fly species. These authors observed that young females *B. cucurbitae* that were protein deprived since emergence responded equally to GF-120 regardless of the presence and amount of AmAc. In contrast, for young *B. dorsalis* females that were protein deprived since emergence, the highest response recorded was to GF-120 with 2 % AmAc, whereas for young females that were protein deprived for three days, there was a significant effect of the presence, regardless of amount, of AmAc. The commercial (1 % AmAc) formulation of GF-120 was found to be unattractive to old (35–38 days old) female *B. dorsalis* that were protein deprived for only 15 h. In contrast, for *B. cucurbitae*, all GF-120 formulations were highly attractive to females regardless of female age and level of protein hunger. Ammonia is a key olfactory component involved in fruit fly attraction to sources of protein (Mazor et al. 1987). In general, the response of female *B. cucurbitae* to GF-120 was consistently greater than that of *B. dorsalis* over the various ages and levels of protein starvation regimes evaluated. The stronger overall response of female *B. cucurbitae* to GF-120 compared to *B. dorsalis*, for the various ages and levels of protein starvation regimes tested, was hypothesized to result in more effective control of the former species.

While differences among species could reflect differences in life history as observed in other tephritids (Aluja et al. 2001), differences between old and young females may depend on the energy demands associated with egg production. One of the main tasks for a newly eclosed female is to forage for the nutrients required during oogenesis. A surge in protein feeding was observed in *B. tryoni* at this stage (Meats and Leighton 2004). Behavioral studies have confirmed that female tephritids with developing ovaries have a stronger response to proteinaceous odors compared to mature females, which respond more intensely to host-fruit odors (Prokopy et al. 1991; Nigg et al. 1995; Cornelius et al. 2000; Rull and Prokopy 2000).

Differential age-related response to non-proteinaceous lures has also been observed in fruit flies. In the case of host-related lures, for example, it was shown that the proportion of female *R. pomonella* responding to red spheres baited with the synthetic fruit odor lure butyl hexanoate was age-related, with 25, 68, 61, 71, and 64 % of 3, 7, 11, 15 and 19 day-old flies responding, respectively (Duan and

Prokopy 1994). Evidently, the relationship of age and fly egg load was related to this skewed response.

In the case of male lures, the response of *B. dorsalis* and *B. opiliae* males to ME and *B. cucurbitae* males to CL increased with age and corresponded with sexual maturity (Fitt 1981a; Wong et al. 1989, 1991; Shelly et al. 2008). In another example, females of three species of *Bactrocera, Bactrocera* sp. A*, *Bactrocera aquilonis* (May), and *Bactrocera tenuifascia* (May) respond to male lures (CL or ME) when sexually immature but do not respond at all after maturing and mating (Fitt 1981b). In the case of *C. capitata*, virgin females responded to the male lures TML, medlure, and angelica seed oil but stopped responding after mating (Nakagawa et al. 1970). The response of male *C. capitata* to TML appears independent of age as immature (1 day old) and mature (9–13 days old) males were captured in similar numbers in TML-baited traps (Shelly and Pahio 2002). Age also affects male response to female pheromone in *B. oleae*. In this fly, pheromone response began on day 3 after adult emergence; increased gradually up to the day 6, and then dropped gradually thereafter up a minimum after 35 days (no more measurements were done after this period) (Haniotakis and Pittara 1994).

2.3 *Effect of Gender*

Behavioral priorities may differ between the two sexes even within the context of a single common behavior. Thus, preference for specific food items or food-related odors may differ significantly between the sexes. For instance, *B. tryoni* males do not require dietary protein to attain complete sexual maturity, however, in females there is a threshold ration of approximately 0.7 mg per fly per day for a normal oocyte development (Drew 1987). Male demand for protein is lower, presumably because the cost of sperm production is lower. Since females of frugivorous tephritid fruit flies are synovigenics and income-breeders, they require constant consumption of carbohydrates, protein, and other nutrients, such as minerals, vitamins, and sterols, for egg maturation and daily oviposition (Teran 1977; Webster and Stoffolano 1978; Tsitsipis 1989; Aluja et al. 2001, 2011). Specifically, protein consumption by *C. capitata, A. ludens, A. suspensa,* and *A. serpentina* females is associated with ovarian development (Sharp and Chambers 1983; Robacker 1991; Landolt and Davis-Hernández 1993; Aluja et al. 2001). Given this sexual difference, it is not surprising that females show greater attraction to protein baits than do males (Houston 1981; Aluja et al 1989; Hendrichs et al. 1991; Robacker 1991, 1999, Robacker and Warfield 1993; Landolt and Davis-Hernández 1993). Nonetheless, post-teneral diet can play an important role in the sexual competitiveness of males (Blay and Yuval 1997; Taylor and Yuval 1999; Kaspi and Yuval 2000; Kaspi et al. 2000; Yuval and Hendrichs 2000; Maor et al. 2004; Niyazi et al. 2004). These gender-characteristic needs are expressed in fly foraging behavioral patterns. For example, *C. capitata* females feed more than males and also forage for considerable periods off the primary host in search of a more diverse diet, while males feed for a shorter period, mostly during the afternoon after courtship and mating, generally do not forage away from the primary host, and also feed on a comparatively narrow diet (Hendrichs et al. 1991). In many lekking tephritids, males display courtship behavior at dusk, thus feeding behavior takes place earlier than for females, who, once inseminated, spend midday searching for hosts and then later replenish their energy and nutrients (Hendrichs et al. 1991). These different schedules of behaviors are reflected in the response to lures according to the time of day. Schedules of response to feeding lures seems to be a generalized pattern in fruit flies in spite of their different mating times. For example, in *A. ludens*, which mate at dusk, and *A. obliqua*, which exhibit mating peaks at dawn and at dusk, the greatest captures in protein baited McPhail traps of females occur at 1600 h, whereas for males captures peak at 1400 h (Malo and Zapien 1994).

In the case of male lures CL and ME, immature males and mature virgin females show little attraction, mature males show strong attraction and feed on the lure, and mated females exhibit no response at all (Fitt 1981b; Wong et al. 1989; Shelly et al. 2008). For *C. capitata* flies, virgin females as well as one-month-old virgin fertile females responded to TML, medlure, and angelica seed oil when males were scarce in release/recapture studies (Nakagawa et al. 1970). Response of virgin females to TML was equivalent to response to a trap baited with 1 sterile male (Nakagawa et al. 1981). Interestingly, there is also a diurnal fluctuation in the response of *B. cucurbitae* to CL, since males responded mainly in the morning (Nakamori and Soemori 1985; Manoukis and Jang 2013).

2.4 *Effect of Sterilization, Exogenous Juvenile Hormone, and Aromatherapy on Mass Reared Flies*

Fly sterilization, aromatherapy, and the use of topically applied juvenile hormone (JH) may introduce variants into the constellation of potential physiological states that result in altered foraging activity and resource demands. These procedures are generally confined to males, and while their effect on male maturation and courtship behavior has received considerable attention, their effect on male foraging behavior, and hence trap capture, has not. Mass rearing and sterilization via gamma radiation may alter the response of male *C. capitata* to TML. Reduced response to TML was observed in mass-reared (unirradiated) males of *C. capitata* from a *tsl* genetic sexing strain, which were less likely to be captured in TML-baited traps than males from a recently established (wild-like) strain (Shelly and Edu 2009). In general, irradiation reduces male response to the lure. For example, in a field test using Steiner traps (horizontal clear cylinder with openings on both ends of the cylinder) baited with TML, it was observed that irradiated males were captured in lower numbers than either irradiated ones or wild males (Wong et al. 1982). Further, those males irradiated at a higher dose were trapped in lower numbers than those exposed to a lower dose (Wong et al. 1982). A similar pattern was observed in *C. capitata tsl* all-male Vienna-7 strain where unirradiated flies showed a greater response to TML than irradiated flies (Barry et al. 2003). This result implies that the use of TML-baited traps to compare the relative abundance of sterile and wild males may underestimate population levels of sterile males, potentially leading to misguided actions in the control process. Additionally, it also could be an indication of a reduction in flight ability of sterile flies (Barry et al. 2003).

Sterilization has also been found to reduce fly response to food baits. Studies on the feeding behavior on *C. capitata* and *A. suspensa* demonstrated that irradiation affects response to and consumption of protein. In *C. capitata*, irradiation resulted in reduced olfactory response, reduced total food intake by flies of both sexes, and a significant reduction in aggregation on and intake of protein by females and of sugar by males (Galun et al. 1985). The ratio of *C. capitata* captures in TML-baited traps versus 3C BioLure-baited (Suttera LLC, Bend, OR, USA) traps was 6.5:1 for sterile males and 1.7:1 for wild males in tests conducted in Guatemala (Midgarden et al. 2004), again indicating lower response of sterile males to food-based lures. In irradiated *A. suspensa*, there was a significant reduction in olfactory response of females to protein (Galun et al. 1985), and lower recapture of sterile mature females than wild mature females in traps baited with AmAc and putrescine (Kendra et al. 2010). Gamma radiation was also found to greatly reduce the response of *A. ludens* and *A. obliqua* to synthetic food lures (Robacker 1998; Díaz-Fleischer et al. 2009).

The term aromatherapy was used originally for human alternative medicine and was introduced in tephritid control jargon by Shelly et al. (2004b). It refers to the use of volatile plant materials, essential oils or other aromatic compounds, for the purpose of improving male sexual performance. Topical application of the JH analog methoprene on the dorsal surface of adults accelerates sexual maturation by several days (Teal and Gomez-Simuta 2002) and also increases male sexual success apparently because it stimulates the production of male sex pheromone (Pereira et al. 2009). Few studies have investigated the response of sterile insects treated with aromatherapy or topical applications of JH to traps. Two studies on the effect of aromatherapy on fly response to lures yielded differing results: pre-release exposure of male *C. capitata* to ginger root oil (GRO) reduced the recapture probability of mass-reared, sterile males in TML-baited traps in Hawaii but not in Florida (Shelly et al. 2006, 2007). In Spain, GRO-treated sterile males exhibited a higher response to proteinaceous baits than non-treated flies (San Andrés et al. 2009), a tendency that might inhibit the effectiveness of sterile insect release programs, since protein-baited traps are run concurrently with sterile releases. Thus, even though GRO exposure may benefit SIT through enhanced male mating ability, the effect of male response to lures must be carefully analyzed to obtain the best results of combining two or more strategies of control. *Bactrocera dorsalis* males have a greatly reduced tendency to visit an ME source if they fed previously on this compound (Shelly 1994), and consequently concurrent use of pre-release feeding on ME by sterile males and male annihilation appears feasible against this species (McInnis et al. 2011, see Shelly and Villalobos 1995 for similar results with *B. cucurbitae*). In *Anastrepha*, immature (2–4 day old) male *A. ludens* and

A. obliqua treated with JH were captured less frequently than non-treated immature males but more frequently than non-treated mature males (15–18 day old) (Arredondo et al. 2014).

3 Fly Responses to Trap and Bait Stimuli

Capturing the target insect is the objective of trapping activities in any insect control program. Attracting insects to the trap is the *sine qua non* for capturing them. Thus, the effectiveness of a trap will depend upon how well traps can artificially mimic those cues or signals used by the insect to locate food, potential mates, and oviposition sites (Katsoyannos 1994; Epsky and Heath 1998; Epsky et al. 2008). Cues used to lure fruit flies may be divided in physical and chemical. Physical cues include the visual cues of shape, size, and color of the trapping device as well as the correct placement of traps inside the tree canopy where flies congregate when foraging. Chemical cues involve pheromones, kairomones, proteinaceous food, and host-based volatiles.

3.1 Physical Cues

Studies on a variety of tephritid species have demonstrated that visual cues associated with hosts are important in host location and are important components of trapping devices. Size and shape are two of the physical characteristics that have received special attention. In general, flies exhibit a common preference for spheres over any other shape. They also tend to accept larger spheres over smaller ones (reviewed by Prokopy 1977b; Prokopy and Owens 1983; Katsoyannos 1989; Epsky and Heath 1998; Cornelius et al. 1999). This preference is apparently independent of natural host shape as seen in *B. cucurbitae* whose females showed a preference for spherical objects compared to cylindrical objects even though the latter closely mimic the shape of cucumber fruits, the preferred host fruit of this fly species (Piñero et al. 2006).

The role of insect vision in host plant detection and selection has been studied intensively in some fruit fly species, but especially in *R. pomonella* (Prokopy and Owens 1983). The relative influence of the components of color cues, such as hue and surface reflectance, of fruit mimics has been demonstrated. With respect to spectral range, tephritid species respond to a bandwidth of the color spectrum from near-UV to red (360–650 nm) and respond most intensely to colors reflecting most of their energy between 500 and 600 nm (Prokopy 1968b). *Rhagoletis pomonella* flies respond to monochromatic light stimuli from ultraviolet (350 nm) to red (675 nm) wavelengths, with the peak of response occurring from 380 to 570 nm (blue-green to yellow), with a non-response plateau from 600 to 625 nm (orange-red) (Agee 1985). In early studies, Prokopy (1968a, b) hypothesized that flies reacted to yellow on the basis of true color discrimination and suggested that large yellow surfaces represented a super-normal foliage-type stimulus eliciting food-seeking behavior in the cherry fruit fly, *Rhagoletis cerasi* L., and *R. pomonella*. Research by Duan and Prokopy (1992) on *R. pomonella*, Mayer et al. (2000) on *R. indifferens*, Katsoyannos et al. (2000) on *R. cerasi*, and Barry et al. (2004) on *Rhagoletis mendax* Curran demonstrated that each species has a preference for spheres of a particular diameter and color. Studies under field conditions on the spectral sensitivities of *C. capitata, B. oleae*, and *R. cerasi*, showed that the spectral sensitivities of all three species are basically similar, with a broad major peak at 485–500 nm (yellow-green) and a secondary peak at 365 nm (ultraviolet) (Agee et al. 1982). In tests involving the Rebell trap (a yellow sticky trap that consists of two intersecting rectangular panels), only *R. cerasi* displayed a distinct color preference, with the yellow polypropylene (Rebell 78) trap being the most attractive of all colors tested (Boller 1969). Neither *C. capitata* nor *B. oleae* exhibited a specific preference for any color (Agee et al. 1982). In the case of *B. dorsalis*, yellow or green attracted flies equally (Cornelius et al. 1999; Wu et al. 2007). However, when the wavelength was measured, fly preference was high in both UV and green spectrum (300–380 nm and 500–570 nm, respectively), while the intermediate blue reflection (380–500 nm) diminished the attractiveness (Wu et al. 2007). Additionally, inter-specific differences in the visual responses expressed by fruit flies have been documented. For example, laboratory and field studies have shown that *A. ludens* females are attracted to yellow and green (Robacker et al. 1990; Robacker 1992), whereas *A. suspensa* females showed a preference for orange (580–590 nm), an indication of specific fruit-seeking behavior since the former species exhibits a wider range of hosts than the later species (Greany et al. 1977). In the case of *B. oleae*, among sticky traps of different colors, the yellow ones were found to be the most attractive, and flies were particularly attracted to hues reflecting maximally between 520 and 580 nm and minimally below 520 nm (Haniotakis 1986; Prokopy et al. 1975).

Specific gender responses were observed in *B. oleae*, since more males were captures by spheres with lower wavelength, 580 and 600 nm, with peak response at 590 nm (yellow to orange) than those that attracted females, 610 and 650 nm, with peak response at 650 nm (orange to red) (Katsoyannos and Koulousis 2001). However, no differences were detected between males and females of *B. dorsalis* with respect to their response to color and spectrum (Wu et al. 2007). *Bactrocera cucurbitae* females are particularly attracted to pigments that offer high reflectance values (white, yellow, orange) regardless of hue, and conversely, they respond less to objects associated with low-reflecting pigments (e.g., black, blue, and sap green) (Piñero et al. 2006). Thus, the visual response of *B. cucurbitae* females seems to be more related to the amounts of light reflected by objects rather than the specific hue associated with those objects. Sexual and age differences in fly response to shapes and colors have also been documented for the Ethiopian fruit fly, *Dacus ciliatus* Loew (Vayssières and Dal 2004). Both sexes preferred spherical shapes to ovoid or rectangular shapes, but sexually mature males responded more frequently than females of the same age. Sexually mature females preferred orange colors, whereas immature females preferred yellow colors. In males, yellow colors were the most attractive to both mature and immature individuals. In the case of the tomato fruit fly, *Neoceratitis cyanescens* (Bezzi), females responded strongly to a visual stimulus (bright orange spheres) after day six of emergence, when egg load increased, apparently as a behavioral response related to host searching (Brevault and Quilici 1999). In *A. obliqua*, both sexes were similarly attracted to wavelengths ranging from 340 to 670 nm (ultraviolet and visible spectrum light), although the broad major peak of attraction occurred between 380 and 570 nm, corresponding to violet, blue, green and yellow (López-Guillén et al. 2009).

While the physical features of host fruit may trigger attraction as a simple behavioral response, they may also stimulate egg development, which in turn influences fly foraging behavior. For example, the color and shape of host fruit stimuli apparently stimulate oogenesis in the first maturation cycle of *Rhagoletis juglandis* Cresson and consequently have a large effect on the physiological state of females (Alonso-Pimentel et al. 1998). This phenomenon could underlie the temporal shift noted in this species from a preference for extremely large host fruit models early in the season to smaller, more natural-size models later in the fruiting season after considerable oviposition had occurred (Prokopy 1977a).

Although trap position within the tree canopy is not an intrinsic physical trap characteristic, more flies will be captured in traps placed in sites where flies forage for food or mates in higher frequency. Basically, trap sites must offer resting and feeding areas in trees that provide, food, shelter from strong winds and predators and, favours mating encounters (Drew 1987; IAEA 2003). Generally, fruit host trees offer such characteristics, however, when host fruit trees are not available, traps should be placed in trees that can provide shaded areas, protection, and food to adult fruit flies (IAEA 2003). Disregarding absolute tree height, studies generally report that traps placed in the inside and in the upper half of the tree canopy capture more flies than any other sector (Drummond et al. 1984; Robacker et al. 1990; Boucher et al. 2001; Pelz-Stelinski et al. 2006; Ragab El-Gendy 2012, but see Haniotakis 1986 for no effect of trap height). Also, to increase *R. pomonella* captures it has been recommended that all fruit and leaves must be removed 30 cm around the trap to enhance the contrast against background (Rull and Prokopy 2004). Additionally, it has been demonstrated that placing traps in the north side of the tree attract more *A. ludens* flies than those placed in the south (Robacker et al. 1990). Thus, when deciding the site to place a trap, factors such as orchard design, tree architecture, season and latitude must be considered (Katsoyannos et al. 1999; Thomas et al. 2001; Dimou et al. 2003).

3.2 Chemical Cues

Chemical lures used to attract tephritids can be divided into those related to feeding behavior, those focused on host searching behavior (kairomones), and those related to mating behavior (pheromones and parapheromones).

3.2.1 Food-Based Attractants

Historically, lures related to feeding behavior have been based on protein volatiles to attract females of different fruit fly species (López and Hernández-Becerril 1967; Houston 1981; Malo 1992; Heath et al. 1993; Thomas et al. 2001; Kouloussis et al. 2009; Díaz-Fleischer et al. 2009, Epsky et al., Chap. 3, this volume). Traditionally, liquid proteins have been the standard bait for tropical fruit flies (McPhail 1939). Lately, some protein based synthetic attractants have been developed using ammonia and its derivatives as a foundation. Thus, a two-component attractant composed of AmAc and putrescine (e.g., 2C BioLure) is used in traps that target *Anastrepha* spp. (Heath et al. 1995; Epsky et al. 1995, 2011, Thomas et al. 2001; Holler et al. 2006). The addition of trimethylamine to these two components improves capture of *C. capitata* (Heath et al. 1997), and McPhail-type traps baited with the three-component attractant (3C BioLure) are equal to or even better than liquid protein-baited traps for capture of *C. capitata* females (Epsky et al. 1999; IAEA 2003, 2007). Ammonium carbonate and/or AmAc lures are used to lure several *Rhagoletis* species (Katsoyannos et al. 2000; Mayer et al. 2000).

3.2.2 Host-Based Lures (Kairomones)

Host-based lures are especially effective for monitoring and even suppressing monophagous and oligophagous species. For example, apple esters are very attractive to both male and female *R. pomonella* (Averill et al. 1988; Zhang et al. 1999). Traps with a complex five-component blend containing butyl butanoate (10 %), propyl hexanoate (4 %), butyl hexanoate (37 %), hexyl butanoate (44 %), and pentyl hexanoate (5 %) captured more flies than traps baited with butyl hexanoate alone, which has been used with commercially available apple maggot monitoring spheres (Zhang et al. 1999). The level of response to butyl hexanoate depends on sexual development, since more mature than immature flies (of both sexes) are attracted to this lure (Rull and Prokopy 2000). Interestingly, ammonium carbonate, a food-based odor, was not attractive to mature or immature flies (Rull and Prokopy 2000).

Host-based lures that attract host-seeking females seem to have a different effect in tephritids with a lekking mating system (e.g., *Anastrepha*, *Bactrocera* and *Ceratitis*) from those that use a resource defense mating strategy (*Rhagoletis*). In the first strategy, hosts are valuable primarily for females as an oviposition resource, while in flies with resource defense strategy host fruits are important for both males and females, since fruits represent oviposition resources and potential sites for mating (Emlen and Oring 1977). Thus, in *R. pomonella*, males and females are similarly attracted to traps that use plant volatiles cues as baits (Bostanian et al. 1993; Rull and Prokopy 2000). Alternatively, traps baited with host fruit odors favor female attraction in *B. tryoni* and *A. obliqua* (Dalby-Ball and Meats 2000; Toledo et al. 2009). Also, in *B. dorsalis* and *C. capitata*, more females than males were captured in traps baited with juice from coffee berries (Vargas et al. 1997). Recently, Siderhust and Jang (2006, 2010) demonstrated that blends of host plant volatiles of *Terminalia cattapa* L. and cucumber fruit were highly attractive to *B. dorsalis* and *B. cucurbitae* females and slightly to males, respectively. Non-host fruit volatiles can also attract females and males as documented in *A. ludens* species for which positive responses to volatiles emitted from commercial grape juice were recorded (Loera-Gallardo et al. 2006). Investigation on non-host attractants may lead to the development of more specific lures. For example, mated females of *B. dorsalis* were more attracted to the volatile semiochemicals from leaves and extracts of a nonhost plant, panax (*Polyscias guilfoylei* (Bull)) than flies of any other physiological state (Jang et al. 1997). Sometimes, lures based on host volatiles blends resulted more attractive to flies than single compounds as observed in *R. pomonella* and *A. obliqua* (Zhang et al. 1999; Cruz-López et al. 2006; Quilici et al., Chap. 4, this volume).

3.2.3 Mating-Related Attractants

The role of sex pheromones and male lures in trapping tephritid fruit flies is discussed at length by Tan et al. (Chap. 2, this volume), and here we briefly describe efforts to combine (i) sex-related attractants with food lures and (ii) different male lures in individual traps.

Tests on the effectiveness of combining food baits and sex pheromones have been restricted to the olive fly, with mixed results. Working at an olive orchard in Greece, Haniotakis and Vassiliou-Waite (1987) compared male and female captures in traps baited with ammonia (A) alone, sex pheromone (P) alone (which in *B. oleae* is produced by the females), or a combination of food and pheromone (A + P) odors. Based on data from the peak mating period, the sexes responded differently to the combination baits. The greatest number of males was found in P traps, followed by P + A traps and then A traps. In contrast, the greatest number of females was found in P + A traps, followed by A traps, and then P traps. Thus, P + A traps caught fewer males than P traps but more females than A traps. For males, the authors proposed that the reduced catch in P + A traps could have resulted from the interference of ammonia with pheromone or to increased female abundance in the vicinity of the trap (owing to its effect as an arrestant on females), which in turn decreased male entry into the traps. For females, the increased catch in the P + A traps was considered to reflect the combined response to food odors plus the arrestant effect of the pheromone.

Subsequent studies have produced mixed results. Although the only comparison drawn was between P + A and P traps (i.e., traps with food baits alone were not included), Yokoyama et al. (2006) presented data from California that were consistent with Haniotakis and Vassiliou-Waite (1987): female captures were greater in P + A traps than P traps, while the opposite was true for males. In contrast, however, Broumas and Haniotakis (1994) reported that P + A traps captured greater numbers of both sexes than did traps baited with P or A only. Additional data from California were also at odds with the original findings. Rice et al. (2003) compared captures in P + A versus A traps (i.e., traps baited with pheromone alone were not included in the test) and reported higher male captures in the P + A traps but essentially no difference in female captures between P + A and A traps.

Seeking cost-cutting measures, various studies have tested the effect of mixing the *Bactrocera* male lures ME and CL, which, if effective, could reduce the number of traps deployed by half. Results from several studies (Hooper 1978; Vargas et al. 2000; Shelly et al. 2004a), however, indicate that the mixture reduces the response of ME-responding species. In contrast, data on CL-responding species are inconsistent. Results on *B. cucurbitae* from Taiwan (cited by Hooper 1978) showed that adding ME to CL nearly doubled the number of males captured compared with traps baited with CL alone. On the contrary, Hooper (1978) found the ME-CL mixture reduced trap capture of CL-responding species in Queensland, Australia, and more recent studies (Vargas et al. 2000; Shelly et al. 2004a) found captures of *B. cucurbitae* in ME-CL-baited traps did not differ from traps baited with CL alone. Similarly, traps baited with solid dispensers containing both ME and raspberry ketone (a plant-borne equivalent of CL) captured equal or greater captures of ME- and CL-responding *Bactrocera* males as traps baited with liquid ME or CL (Vargas et al. 2000; Leblanc et al. 2011; Shelly et al. 2012).

TML has been used routinely as the best practical attractant for survey and detection traps for *C. capitata* (IAEA 2007). It functions as a male *C. capitata* attractant and possibly an arrestant, and confers some short lasting mating advantages, but its sexual function is unclear (Shelly and Pahio 2002). It has been reported that the hydrocarbon sesquiterpene α-copaene is highly attractive to wild male medflies (Flath 1994a, b) and that pure samples of the compound confer a mating advantage to the males that are exposed to it, even 6 days after exposure (Shelly 2001). However, α-copaene, presented in enriched ginger root oil and used as a lure, is less attractive than TML (Shelly and Pahio 2002).

4 Influence of Environmental Conditions

The role that abiotic factors, specifically rainfall and temperature, play in tephritid capture in traps is poorly understood (Aluja et al. 2011). Cunningham et al. (1978) documented an effect of rainfall on liquid protein-baited trap captures of three fruit fly species in Hawaii. These authors reported that captures of *B. dorsalis* were about 20 times greater in comparatively dry areas than in comparatively wet areas and about seven times greater than in areas with intermediate rainfall. Similar results were noted for *C. capitata* and *B. cucurbitae*, species that were captured in comparatively greater numbers in dry areas compared to intermediate rainfall areas (18 times and four times more for *C. capitata* and *B. cucurbitae*, respectively). In contrast, Aluja et al. (1996) found no relationship between rainfall and seasonal variation in population size of *Anastrepha* spp. as indicated by capture in liquid protein-baited traps. Rather, the population fluctuations of the dominant *Anastrepha* species seemed to be related to the availability of host fruits. For *A. ludens*, *A. serpentina*, and *A. obliqua* under subtropical conditions during the dry season (lower relative humidity and higher temperature) in Colombia, Mexico, and Honduras, a synthetic food-based lure (AmAc and putrescine) was equally or more effective than the conventional protein baits torula yeast/borax and NuLure/borax (IAEA 2007). However, in these same locations during the rainy season (higher relative humidity and lower temperature), the conventional protein baits were most effective. Considering that the synthetic food-based lures are easier to handle, are more likely to catch target species, and tend to be more consistent than liquid protein baits, these attractants are considered to be a better choice under dry and hot conditions.

5 Conclusions and Future Research Needs

Any insect population presents a mosaic of individual physiological states. Thus, identifying a single lure to attract most of the individuals in the population is problematic, since that lure likely attracts only a fraction of the population. In the case of food lures, even individuals of the same sex and age vary in their immediate nutritional needs depending on the local availability of food sources and an individual's ability to find these resources. These factors alone will result in the variable effectiveness of any lure as a trapping tool. In the case of parapheromones, the number of potential trapping targets is even smaller, since only males are strongly attracted.

In lekking flies, sex pheromones offer a very specific lure that would target females primarily and possibly sexually active males. However, the complexity of these pheromones renders their use in trapping problematic. As Heath et al. (2000) wrote, " "*Not only are numerous chemical compounds released from calling males, but the amounts and ratios may vary over time or among different populations of flies of the same species. In addition, the range in volatilities of compounds produced by male fruit flies has increased the difficulties in formulating synthetic blends that mimic the release rates and ratios of pheromones from live males. Thus, it is not known if lack of field efficacy of synthetic compounds that have been tested is due to absence of biological activity or to inadequate formulation*". Perhaps studies on odorant binding proteins, insect olfactory receptors genes, and the transcriptomic response would help to determine with precision those volatiles that elicit high fly response (Jang 1995; Bohbot and Dickens 2012; Gomulsky et al. 2012; Siciliano et al. 2014).

Exploitation of the multiple sensory modalities used to detect essential resources can have important implications for fruit fly monitoring (e.g., using traps) and control (e.g., mass trapping). Integrated pest management (IPM) tools that do not rely on a single cue are likely to work more reliably under changing environmental conditions (Dorn and Piñero 2009). For example, improving visual attraction by adding some host fruit stimuli, like UV surfaces, may stimulate flies to alight on traps. This has been demonstrated in *Bactrocera cacuminata* (Hering), *B. dorsalis* and *B. tryoni* (Drew et al. 2003; Wu et al. 2007). Studies on fly and host fruit phenologies, which include surveys aimed at assessing the physiological condition of the female flies, would help to decide the most effective bait/trap combination as well as deployment strategies, including trap deployment patterns and trap density.

In summary, there are clearly interactions between the environment of an orchard and the physiological state and behavior of flies. Research on fly responses to a specific signal in a complex sensory environment must be considered to enhance trap and lure attractiveness. In order to accomplish this goal, it is necessary to improve the understanding of the mechanisms involved in the recognition of an olfactory signal and its decodification leading to a behavioral response.

Acknowledgments This chapter is dedicated to the memory of Robert R. Heath, pioneer in the study and development of baits and traps used in the control of agricultural insect pests. We thank Nancy Epsky and the anonymous referees who helped to improve chapter quality.

References

Agee HR (1985) Spectral response of the compound eye of the wild and laboratory-reared apple maggot fly, *Rhagoletis pomonella*. J Agric Ent 2:147–154

Agee HR, Boller E, Remund U, Davis JC, Chambers DL (1982) Spectral sensitivities and visual attractant studies on the Mediterranean fruit fly, *Ceratitis capitata* (Wiedemann), olive fly, *Dacus oleae* (Gmelin), and the European cherry fruit fly, *Rhagoletis cerasi* (L.) (Diptera, Tephritidae). Z Angew Entomol 93:403–412

Alonso-Pimentel H, Korer JB, Nufio C, Papaj DR (1998) Role of color and shape stimuli in host-enhanced oogenesis in the walnut fly, *Rhagoletis juglandis*. Physiol Ent 23:97–104

Aluja M, Cabrera M, Guillén J, Celedonio H, Ayora F (1989) Behaviour of *Anastrepha ludens*, *A. obliqua* and *A. serpentina* (Diptera: Tephritidae) on a wild mango tree (*Mangifera indica*) harbouring three McPhail traps. Insect Sci Appl 10:309–318

Aluja M, Celedonio-Hurtado H, Liedo P, Cabrera M, Castillo F, Guillen J, Rios E (1996) Seasonal population fluctuations and ecological implications for management of *Anastrepha* fruit flies (Diptera: Tephritidae) in commercial mango orchards in southern Mexico. J Econ Ent 89:654–667

Aluja M, Piñero J, Jácome I, Díaz-Fleischer F, Sivinski J (2000) Behavior of flies in the genus *Anastrepha* (Trypetinae: Toxotrypanini). In: Aluja M, Norrbom A (eds) Fuit flies (Tephritidae): phylogeny and evolution of behavior. CRC Press, Boca Raton, pp 375–406

Aluja M, Díaz-Fleischer F, Lagunes G, Papaj DR, Sivinski J (2001) Effects of age, diet, female density and the host resource on egg load in *Anastrepha ludens* and *A. obliqua* (Diptera: Tephritidae). J Insect Physiol 47:975–988

Aluja M, Birke A, Guillen L, Díaz-Fleischer F, Nestel D (2011) Coping with an unpredictable and stressful environment: the life history and metabolic response to variable food and host availability in a polyphagous tephritid fly. J Insect Physiol 57:1592–1601

Arredondo J, Flores MP, Díaz-Fleischer F (2014) Effect of multiple endogenous biological factors on the response of the tephritids *Anastrepha ludens* and *Anastrepha obliqua* to MultiLure traps baited with BioLure or NuLure in mango orchards. J Econ Ent 107:1022–1031

Averill AL, Reissing WH, Roelofs WL (1988) Specificity of olfactory responses in the tephritid fruit fly *Rhagoletis pomonella*. Ent Exp Appl 47:211–222

Balagawi S, Jackson K, Haq IU, Hood-Nowotny R, Resch C, Clarke AR (2014) Nutritional status and the foraging behavior of *Bactrocera tryoni* with particular reference to protein bait spray. Physiol Ent 39:33–43
Barry JD, McInnis DO, Gates D, Morse JG (2003) Effects of irradiation on Mediterranean fruit flies (Diptera: Tephritidae): emergence, survivorship, lure attraction, and mating competition. J Econ Ent 96:615–622
Barry JD, Polavarapu S, Teixeira LAF (2004) Evaluation of traps and toxicants in an attract-and-kill system for *Rhagoletis mendax* (Diptera: Tephritidae). J Econ Ent 97:2006–2014
Barton Browne L (1957) The effect of light on the mating behaviour of the Queensland fruit fly, *Strumeta tryoni* (Frogg.). Australian J Zool 5:145–158
Barton Browne L (1993) Physiologically induced changes in host oriented behavior. Annu Rev Ent 38:1–25
Bateman MA (1972) The ecology of fruit flies. Annu Rev Ent 17:493–518
Bell WJ (1990) Searching behavior patterns in insects. Annu Rev Ent 35:447–467
Binder BF (1996) Effect of carbohydrate on age-related feeding behaviors and longevity in adult black cutworm, *Agrotis ipsilon* (Lepidoptera: Noctuidae). J Insect Behav 9:215–222
Blay S, Yuval B (1997) Nutritional correlates to reproductive success of male Mediterranean fruit flies. Anim Behav 54:59–66
Bohbot JD, Dickens JC (2012) Selectivity of odorant receptors in insects. Front Cell Neurosci 6:1–4
Boller E (1969) Neues uber die Kirschenfliege: Freilandversuche im Jahre 1969 (Testing new traps and dispersion studies). Schweiz Z Obst-Weinbau 105:566–572
Bostanian NJ, Vincent C, Roy M (1993) Comparative effectiveness of three trap models to monitor the apple maggot, *Rhagoletis pomonella* (Walsh), (Diptera: Tephritidae), in Quebec. J Agric Ent 10:73–82
Boucher TJ, Ashley RA, Adams RG, Morris TF (2001) Effect of trap position, habitat, and height on the capture of pepper maggot flies (Diptera: Tephritidae). J Econ Ent 94:455–461
Brevault T, Quilici S (1999) Factors affecting behavioural responses to visual stimuli in the tomato fruit fly, *Neoceratitis cyanescens*. Physiol Ent 24:333–338
Brieze-Stegman R, Rice MA, Hooper GHS (1978) Daily periodicity in attraction of male tephritid fruit flies to synthetic chemical lures. J Aust Ent Soc 17:341–346
Broumas T, Haniotakis GE (1994) Comparative field studies of various traps and attractants of the olive fruit fly, *Bactrocera oleae*. Entomol Exp Appl 73:145–150
Carey JR, Molleman F (2010) Reproductive aging in tephritid fruit flies. Ann New York Acad Sci 1204:139–148
Cavalloro R, Delrio G (1970) Studi sulla radiosterilizzazione di *Ceratitis capitata* Wiedemann e sul comportamento dell'insetto normale e sterile. Redia 52:511–547
Chapman T, Takahisa M, Smith HK, Partridge L (1998) Interactions of mating, egg production and death rates in females of the Mediterranean fruit fly, *Ceratitis capitata*. Proc R Soc Lond B 265:1879–1894
Christophides GK, Mintzas AC, Komitopoulou K (2000) Organization, evolution and expression of a multigene family encoding putative members of the odourant-binding protein family in the medfly *Ceratitis capitata*. Insect Mol Biol 9:185–195
Cornelius ML, Duan JJ, Messing RH (1999) Visual stimuli and the response of female oriental fruit flies (Diptera: Tephritidae) to fruit-mimicking traps. J Econ Ent 92:121–129
Cornelius ML, Nergel L, Duan JJ, Messing RI (2000) Responses of female oriental fruit flies (Diptera: Tephritidae) to protein and host fruit odors in field cage and open field tests. Environ Ent 29:14–19
Cresoni-Pereira C, Zucoloto FS (2001) Dietary self-selection and discrimination threshold in *Anastrepha obliqua* females (Diptera, Tephritidae). J Insect Physiol 47:1127–1132
Cresoni-Pereira C, Zucoloto FS (2012) Fruit flies (Diptera). In: Parra JRP (ed) Insect bioecology and nutrition for integrated pest management. CRC Press, Boca Raton, pp 451–472

Cruz-López L, Malo EA, Toledo J, Virgen A, Del Mazo A, Rojas JC (2006) A new potential attractant for *Anastrepha obliqua* from *Spondias mombin* fruit. J Chem Ecol 32:351–365
Cunningham RT, Nakagawa S, Suda DY, Urago T (1978) Tephritid fruit-fly trapping: liquid food baits in high and low rainfall climates. J Econ Ent 71:762–763
Dalby-Ball G, Meats A (2000) Influence of the odour of fruit, yeast and cuelure on the flight activity of the Queensland fruit fly, *Bactrocera tryoni* (Froggatt) (Diptera: Tephritidae). Aust J Ent 39:195–200
Díaz-Fleischer F, Arredondo J (2011a) Light conditions affect sexual performance in a lekking tephritid fruit fly. J Exp Biol 214:2595–2602
Díaz-Fleischer F, Arredondo J (2011b) Effects of post-eclosion light–dark regimes on mating performance of the mass reared tephritid fly *Anastrepha ludens*. Ent Exp Appl 141:52–58
Díaz-Fleischer F, Arredondo J, Flores S, Montoya P, Aluja M (2009) There is no magic fruit fly trap: multiple biological factors influence the response of adult *Anastrepha ludens* and *Anastrepha obliqua* (Diptera: Tephritidae) individuals to Multilure traps baited with BioLure or NuLure. J Econ Ent 102:86–94
Dimou I, Koutsikopoulos C, Economopoulos A, Lykakis J (2003) The distribution of olive fruit fly captures with McPhail traps within an olive orchard. Phytoparasitica 31:124–131
Dorn S, Piñero JC (2009) How do key tree-fruit pests detect and colonize their hosts: mechanisms and application for IPM. In: Aluja M, Leskey TC, Vincent C (eds) Biorational tree-fruit pest management. CABI Publishers, Wallingford, pp 85–109, 295 p
Drew RAI (1987) Behavioural strategies of fruit flies of the genus *Dacus* (Diptera: Tephritidae) significant in mating and host plant relationships. Bull Ent Res 77:73–81
Drew RAI, Yuval B (2000) The evolution of fruit fly feeding behavior. In: Aluja M, Norrbom AL (eds) Fruit flies (Tephritidae): phylogeny and evolution of behavior. CRC Press, Boca Raton, pp 731–749
Drew RAI, Prokopy RJ, Romig MC (2003) Attraction of fruit flies of the genus *Bactrocera* to colored mimics of host fruit. Ent Exp Appl 107:39–45
Drummond F, Groden E, Prokopy RJ (1984) Comparative efficacy and optimal positioning of traps for monitoring apple maggot flies (Diptera: Tephritidae). Environ Ent 13:232–235
Duan JJ, Prokopy RJ (1992) Visual and odor stimuli influencing effectiveness of sticky spheres for trapping apple maggot flies *Rhagoletis pomonella* (Walsh) (Diptera, Tephritidae). J Appl Ent 113:271–279
Duan JJ, Prokopy RJ (1994) Apple maggot fly response to red sphere traps in relation to fly age and experience. Ent Exp Appl 73:279–287
Emlen ST, Oring LW (1977) Ecology, sexual selection and the evolution of mating systems. Science 197:215–223
Epsky ND, Heath RR (1998) Exploiting the interactions of chemical and visual cues in behavioral control measures for pest tephritid fruit flies. Fla Ent 81:273–282
Epsky ND, Heath RR, Guzman A, Meyer WL (1995) Visual cue and chemical cue interactions in a dry trap with food-based synthetic attractant for *Ceratitis capitata* and *Anastrepha ludens* (Diptera: Tephritidae). Environ Ent 24:1387–1395
Epsky ND, Hendrichs J, Katsoyannos BI, Vásquez LA, Ros JP, Zümreoglu A, Pereira R, Bakri A, Seewooruthun SI, Heath RR (1999) Field evaluation of female-targeted trapping systems for *Ceratitis capitata* (Diptera: Tephritidae) in seven countries. J Econ Ent 92:156–164
Epsky ND, Kendra PE, Peña JE, Heath RR (2011) Comparison of synthetic food-based lures and liquid protein baits for capture of *Anastrepha suspensa* (Diptera: Tephritidae) adults. Fla Entomol 94:180–185
Epsky ND, Morrill WL, Mankin RW (2008) Traps for capturing insects. In: Capinera JL (ed) Encyclopedia of entomology. Springer Science and Business Media B.V, Dordrecht
Fitt GP (1981a) The influence of age, nutrition and time of day on the responsiveness of male *Dacus opiliae* to the synthetic lure methyl eugenol. Ent Exp Appl 30:83–90
Fitt GP (1981b) Responses by female dacinae (Diptera:Tephritidae) to male lures and their relationship to patterns of mating behaviour and pheromone response. Ent Exp Appl 29:87–97

Flath RA, Cunningham RT, Mon TR, John JO (1994a) Additional male Mediterranean fruitfly (*Ceratitis capitata* Wied.) attractants from angelica seed oil (*Angelica archangelica* L.). J Chem Ecol 20:1969–1984
Flath RA, Cunningham RT, Mon TR, John JO (1994b) Male lures for Mediterranean fruit fly (*Ceratitis capitata* Wied.): structural analogues of α-copaene. J Chem Ecol 20:2595–2609
Galun R, Gothilf S, Blondheim S, Sharp JL, Mazor M, Lachman A (1985) Comparison of aggregation and feeding responses by normal and irradiated fruit flies, *Ceratitis capitata* and *Anastrepha suspensa* (Diptera: Tephritidae). Environ Ent 14:726–732
Gomulski LM, Dimopoulos G, Xi Z, Scolari F, Gabrieli P, Siciliano P, Clarke AR, Malacrida AR, Gasperi G (2012) Transcriptome profiling of sexual maturation and mating in the Mediterranean fruit fly, *Ceratitis capitata*. PLoS ONE 7:e30857
Greany PD, Agee HR, Burditt AK, Chambers DL (1977) Field studies on color preferences of the Caribbean fruit fly, *Anastrepha suspensa* (Diptera: Tephritidae). Exp Appl 21:63–70
Hagen KS, Dadd RH, Reese I (1984) The food of insects. In: Huffaker CB, Rabb RL (eds) Ecological entomology. Wiley, New York, pp 79–112
Haniotakis G (1986) Effect of size, color and height of sticky traps on captures of *Dacus oleae*. Ent Hellenica 4:55–61
Haniotakis GE, Pittara IS (1994) Response of *Bactrocera* (*Dacus*) *oleae* males (Diptera: Tephritidae) to pheromones as affected by concentration, insect age, time of day, and previous exposure. Environ Ent 23:726–731
Haniotakis GE, Vassiliou-Waite A (1987) Effect of combining food and sex attractants on the capture of *Dacus oleae* flies. Ent Hellenica 5:27–33
Harmer AMT, Radhakrishnan P, Taylor PW (2006) Remating inhibition in female Queensland fruit flies: effects and correlates of sperm storage. J Insect Physiol 52:179–186
Heath RR, Epsky ND, Landolt PJ, Sivinski J (1993) Development of attractants for monitoring Caribbean fruit flies (Diptera: Tephritidae). Fla Ent 76:233–244
Heath RR, Epsky ND, Guzman A, Dueben BD, Manukian A, Meyer WL (1995) Development of a dry plastic insect trap with food-based synthetic attractant for the Mediterranean and Mexican fruit flies (Diptera: Tephritidae). J Econ Ent 88:1307–1315
Heath RR, Epsky ND, Deuben BD, Rizzo J, Jeronimo F (1997) Adding methyl-substituted ammonia derivatives to a food-based synthetic attractant on capture of the Mediterranean and Mexican fruit flies (Diptera: Tephritidae). J Econ Ent 90:1584–1589
Heath RR, Landolt PJ, Robacker DC, Dueben BD, Epsky ND (2000) Sexual pheromones of tephritid flies: clues to unravel phylogeny and behavior. In: Aluja M, Norrbom AL (eds) Fruit flies (Tephritidae): phylogeny and evolution of behavior. CRC, Boca Raton, pp 731–749
Hee AKW, Tan KH (2004) Male sex pheromonal components derived from methyl eugenol in the hemolymph of the fruit fly *Bactrocera papayae*. J Chem Ecol 30:2127–2138
Hendrichs J, Hendrichs MA (1990) Mediterranean fruit fly (Diptera: Tephritidae) in nature: location and diel pattern of feeding and other activities on fruiting and non-fruiting hosts and non hosts. Ann Ent Soc Am 83:632–641
Hendrichs J, Katsoyannos BI, Papaj DR, Prokopy RJ (1991) Sex differences in movement between natural feeding and mating sites and tradeoffs between food consumption, mating success and predator evasion in Mediterranean fruit flies (Diptera: Tephritidae). Oecologia 86:223–231
Hendrichs J, Lauzon CR, Cooley SS, Prokopy RJ (1993) Contribution of natural food sources to adult longevity and fecundity of *Rhagoletis pomonella* (Diptera, Tephritidae). Ann Entomol Soc Am 15:250–264
Holler T, Sivinski J, Jenkins C, Fraser S (2006) A comparison of yeast hydrolysate and synthetic food attractants for capture of *Anastrepha suspensa* (Diptera: Tephritidae). Fla Ent 89:419–420
Hooper GHS (1978) Effect of combining methyl eugenol and cuelure on the capture of male tephritid fruit flies. J Aust Entomol Soc 17:189–190
Houston W (1981) Fluctuations in numbers and the significance of the sex ratio of the Mexican fruit fly *Anastrepha ludens* caught in McPhail traps. Ent Exp Appl 30:140–150

Houston AI, Stephens PA, Boyd IL, Harding KC, McNamara JM (2007) Capital or income breeding? A theoretical model of female reproductive strategies. Behav Ecol 18:241–250

IAEA (International Atomic Energy Agency) (2003) Trapping guidelines for area-wide fruit fly programmes. IAEA, Vienna, 2003

IAEA (International Atomic Energy Agency) (2007) Development of improved attractants and their integration into fruit fly sit management programmes. In: Proceedings of a final Research Coordination Meeting organized by the Joint FAO/IAEA Programme of Nuclear Techniques in Food and Agriculture, Vienna, 5–7 May 2005 IAEA-TECDOC-1574

Jacome I, Aluja M, Liedo P, Nestel D (1995) The influence of adult diet and age on lipid reserves in the tropical fruit fly *Anastrepha serpentina* (Diptera:Tephritidae). J Insect Physiol 41:1079–1086

Jang EB (1995) Effects of mating and accessory gland injections on olfactory- mediated behavior in the female Mediterranean fruit fly, *Ceratitis capitata*. J Insect Physiol 41:705–710

Jang EB (2002) Physiology of mating behavior in Mediterranean fruit fly (Diptera: Tephritidae): chemoreception and male accessory gland fluids in female post-mating behavior. Fla Ent 85:89–93

Jang EB, Carvalho LA, Stark JD (1997) Attraction of female oriental fruit fly, *Bactrocera dorsalis*, to volatile semiochemicals from leaves and extracts of a nonhost plant, panax (*Polyscias guilfoylei*) in laboratory and olfactometer assays. J Chem Ecol 23:1389–1401

Kaspi R, Yuval B (1999) Mediterranean fruit fly leks: factors affecting male location. Funct Ecol 13:539–545

Kaspi R, Yuval B (2000) Post-teneral protein feeding improves sexual competitiveness but reduces longevity of mass-reared sterile male Mediterranean fruit flies (Diptera: Tephritidae). Ann Ent Soc Am 93:949–955

Kaspi R, Taylor PW, Yuval B (2000) Diet and size influence sexual advertisement and copulatory success of males in Mediterranean fruit fly leks. Ecol Ent 25:279–284

Katsoyannos BI (1989) Response to shape, size and color. In: Robinson AS, Hooper GHS (eds) Fruit flies: their biology, natural enemies and control, vol 3A. Elsevier, Amsterdam, pp 307–324

Katsoyannos BI (1994) Evaluation of Mediterranean fruit fly traps for use in sterile-insect-technique programmes. J Appl Ent 118:442–452

Katsoyannos BI, Kouloussis NA (2001) Captures of the olive fruit fly *Bactrocera oleae* on spheres of different colours. Entomol Exp Appl 100:165–172

Katsoyannos BI, Heath RR, Papadopoulos NT, Epsky ND, Hendrichs J (1999) Field evaluation of Mediterranean fruit fly (Diptera: Tephritidae) female selective attractants for use in monitoring programs. J Econ Ent 92:583–589

Katsoyannos BI, Papadopoulos NT, Stavridis D (2000) Evaluation of trap types and food attractants for *Rhagoletis cerasi* (Diptera: Tephritidae). J Econ Ent 93:1005–1010

Kendra PE, Montgomery WS, Mateo DM, Puche H, Epsky ND, Heath RR (2005) Effect of age on EAG response and attraction of female *Anastrepha suspensa* (Diptera: Tephritidae) to ammonia and carbon dioxide. Environ Ent 34:584–590

Kendra PE, Epsky ND, Heath RR (2010) Effective sampling range of food-based attractants for female *Anastrepha suspensa* (Diptera: Tephritidae). J. Econ Entomol 103:533–540

Kouloussis NA, Papadopoulos NT, Müller HG, Wang JL, Mao M, Katsoyannos BI, Duyck PF, Carey JR (2009) Life table assay of field-caught Mediterranean fruit flies, *Ceratitis capitata*, reveals age bias. Ent Exp Appl 123:172–181

Kraaijeveld K, Chapman T (2004) Effects of male sterility on female remating in the Mediterranean fruit fly, *Ceratitis capitata*. Proc Royal Soc Lond B (Suppl Biol Lett) 271:209–211

Landolt PJ, Davis-Hernández KM (1993) Temporal patterns of feeding by Caribbean fruit flies (Diptera: Tephritidae) on sucrose and hydrolyzed yeast. Ann Ent Soc Am 86:749–755

Leblanc L, Vargas RI, MacKey B, Putoa R, Piñero JC (2011) Evaluation of cue-lure and methyl eugenol solid lure and insecticide dispensers for fruit fly (Diptera: Tephritidae) monitoring and control in Tahiti. Fla Ent 94:510–516

Liedo P, Carey JR, Celedonio H, Guillen J (1992) Size specific demography in three species of *Anastrepha* fruit flies. Ent Exp Appl 63:135–142
Loera-Gallardo J, López-Arroyo JI, Miranda-Salcedo MA, Canales-Cruz R, Thomas DB, Mangan RL (2006) Attraction of Mexican fruit flies to colored traps and lures (abstract). Subtrop Plant Sci 58
López DF, Hernández-Becerril O (1967) Sodium borate inhibits decomposition of two protein hydrolysates attractive to Mexican fruit fly. J Econ Ent 60:136–140
López-Guillén G, Virgen A, Rojas JC (2009) Color preference of *Anastrepha obliqua* (Diptera, Tephritidae). Rev Brasileira Ent 53:157–159
Malo EA (1992) The effect of bait decomposition on the capture of *Anastrepha* fruit flies. Fla Ent 75:272–274
Malo EA, Zapien GI (1994) McPhail trap captures of *Anastrepha obliqua* and *Anastrepha ludens* (Diptera: Tephritidae) in relation to time of day. Fla Ent 77:290–294
Manoukis NC, Jang EB (2013) The diurnal rhythmicity of *Bactrocera cucurbitae* (Diptera: Tephritidae) attraction to cuelure: insights from an interruptable lure and computer vision. Ann Ent Soc Am 106:136–142
Manrakhan A, Lux SA (2008) Effect of food deprivation on attractiveness of food sources, containing natural and artificial sugar and protein, to three African fruit flies: *Ceratitis cosyra*, *Ceratitis fasciventris*, and *Ceratitis capitata*. Ent Exp Appl 127:133–143
Maor M, Kamensky B, Shloush S, Yuval B (2004) Effects of post-teneral diet on foraging success of sterile male Mediterrenean fruit flies. Ent Exp Appl 110:225–230
Mayer DF, Long LE, Smith TJ, Olsen J, Riedl H, Heath RR, Leskey TC, Prokopy RJ (2000) Attraction of adult *Rhagoletis indifferens* (Diptera: Tephritidae) to unbaited and odor-baited red spheres and yellow rectangles. J Econ Ent 93:347–351
Mazor M, Gothilf S, Galun R (1987) The role of ammonia in the attraction of females of the Mediterranean fruit fly to protein hydrolyzate baits. Entomol Exp Appl 43:25–29
McInnis D, Kurashima R, Shelly T, Komatsu J, Edu J, Pahio E (2011) Prerelease exposure to methyl eugenol increases the mating competitiveness of sterile males of the oriental fruit fly (Diptera: Tephritidae) in a Hawaiian orchard. J Econ Ent 104:1969–1978
McPhail M (1939) Protein lures for fruit flies. J Econ Ent 32:758–761
Meats A, Leighton SM (2004) Protein consumption by mated, unmated, sterile and fertile adults of the Queensland fruit fly, *Bactrocera tryoni* and its relation to egg production. Physiol Ent 29:176–182
Midgarden D, Ovalle O, Epsky ND, Puche H, Kendra PE, Rendon P, Heath RR (2004) Capture of Mediterranean fruit flies (Diptera: Tephritidae) in dry traps baited with a food-based attractant and Jackson traps baited with trimedlure during sterile male release in Guatemala. J Econ Entomol 97:2137–2143
Miller NW, Vargas RI, Prokopy RJ, Mackey BE (2004) State-dependent attractiveness of protein bait and host fruit odor to *Bactrocera cucurbitae* (Diptera: Tephritidae) females. Ann Ent Soc Am 97:1063–1068
Miyatake T, Chapman T, Partridge L (1999) Mating-induced inhibition of remating in female Mediterranean fruit flies *Ceratitis capitata*. J Insect Physiol 45:1021–1028
Mossinson S, Yuval B (2003) Regulation of sexual receptivity of female Mediterranean fruit flies: old hypotheses revisited and a new synthesis proposed. J Insect Physiol 49:561–567
Nakagawa S, Farias GJ, Steiner LF (1970) Response of female Mediterranean fruit flies to male lures in the relative absence of males. J Econ Ent 63:227–229
Nakagawa S, Steiner LF, Farias GJ (1981) Response of virgin female Mediterranean fruit flies to live mature normal males, sterile males, and trimedlure in plastic traps. J Econ Ent 74:566–567
Nakamori H, Soemori H (1985) Diurnal changes in the attraction of the melon fly, *Dacus cucurbitae* Coquillett (Diptera: Tephritidae), in different habitats. Jpn J Appl Ent Zool 29:216–222
Newell W (1936) Progress report on the Key West (Florida) fruit fly eradication project. J Econ Ent 29:116–120

Nigg HN, Simpson SE, Attaway JA, Fraser S, Burns E, Littell RC (1995) Age-related response of *Anastrepha suspensa* (Diptera: Tephritidae) to protein hydrolysate and sucrose. J Econ Ent 88:669–677

Niyazi N, Lauzon CR, Shelly TE (2004) Effect of probiotic adult diets on fitness components of sterile male Mediterranean fruit flies (Diptera: Tephritidae) under laboratory and field cage conditions. J Econ Ent 97:1570–1580

Opp SB, Prokopy RJ (2000) Multiple mating and reproductive success of male and female apple maggot flies, *Rhagoletis pomonella* (Diptera: Tephritidae). J Insect Behav 13:901–914

Papaj DR (2000) Ovarian dynamics and host use. Annu Rev Ent 45:423–448

Pelosi P, Maida R (1995) Odorant-binding proteins in insects. Comp Biochem Physiol 111 (3):503–514

Pelz-Stelinski KS, Gut LJ, Isaacs R (2006) Vertical position of traps influences captures of eastern cherry fruit fly (Diptera: Tephritidae). Fla Ent 89:80–82

Pérez-Staples D, Aluja M (2004) *Anastrepha striata* females mated to virgin males live longer. Ann Ent Soc Am 97:1336–1341

Pereira R, Sivinski J, Teal PEA (2009) Influence of methoprene and dietary protein on male *Anastrepha suspensa* (Diptera: Tephritidae) mating aggregations. J Insect Physiol 55:328–335

Pérez-Staples D, Harmer AMT, Collins SR, Taylor PW (2008) Potential of pre-release diet supplements to increase the sexual performance and longevity of male Queensland fruit flies. Agric Forest Ent 10:255–262

Piñero J, Aluja M, Equihua M, Ojeda MM (2002) Feeding history and age influence the response of four *Anastrepha* species (Diptera: Tephritidae) to human urine and hydrolyzed protein. Folia Ent Mex 41:283–298

Piñero JC, Jácome I, Vargas RI, Prokopy RJ (2006) Response of female melon fly, *Bactrocera cucurbitae*, to host-associated visual and olfactory stimuli. Ent Exp Appl 121:261–269

Piñero JC, Mau RFL, Vargas RI (2011) A comparative assessment of the response of three fruit fly species (Diptera: Tephritidae) to a spinosad-based bait: effect of ammonium acetate, female age, and protein hunger. Bull Ent Res 101:373–381

Prokopy RJ (1968a) Orientation of the apple maggot flies *Rhagoletis pomonella* (Walsh) and European cherry fruit flies *R. cerasi* L. (Diptera: Tephritidae) to visual stimuli. In: Proceedings of the 13 international congress of entomology, Moscow, 2–9 Aug 1968; pp 34–35

Prokopy RJ (1968b) Visual responses of apple maggot flies, *Rhagoletis pomonella*: orchard studies. Ent Exp Appl 11:403–422

Prokopy RJ (1977a) Attraction of *Rhagoletis* flies (Diptera: Tephritidae) to red spheres of different sizes. Can Ent 109:593–596

Prokopy RJ (1977b) Stimuli influencing trophic relations in Tephritidae. In: Labeyrie V (ed) Comporement des insects et millieu trophique. Coll. Int. CNRS 265: 305–336

Prokopy RJ, Bush GL (1973) Mating behavior of *Rhagoletis pomonella.* IV Courtship. Can Ent 105:879–891

Prokopy RJ, Owens ED (1983) Visual detection of plants by herbivorous insects. Annu Rev Ent 28:337–364

Prokopy RJ, Economopoulos AP, McFadden MW (1975) Attraction of wild and laboratory-cultured *Dacus oleae* flies to small rectangles of different hues, shades, and tints. Ent Exp Appl 18:141–152

Prokopy RJ, Drew RAI, Sabine BNE, Lloyd AC, Hamacek E (1991) Effect of physiological and experiential state of *Bactrocera tryoni* flies on intra-tree foraging behavior for food (bacteria) and host fruit. Oecologia 87:394–400

Prokopy RJ, Cooley SS, Luna I, Duan JJ (1995) Combined influence of protein hunger and egg load on the resource foraging behavior of *Rhagoletis pomonella* flies (Diptera: Tephritidae). Eur J Ent 92:655–666

Prokopy RJ, Resilva SS, Vargas RI (1996) Post-alighting behavior of *Ceratitis capitata* flies on odor baited traps. Fla Ent 79:422–428

Radhakrishnan P, Nair S, Raftos D, Taylor PW (2008) Transfer and fate of male ejaculate in female Queensland fruit flies. Physiol Ent 33:302–309

Ragab El-Gendy I (2012) Elevation of attraction efficiency of Jackson trap on peach fruit fly, *Bactrocera zonata* (Saunders). Int J Agric Res 7:223–230

Rice RE, Phillips PA, Stewart-Leslie J, Sibbett GS (2003) Olive fruit fly populations measured in central and southern California. Calif Agric 57:122–127

Robacker DC (1991) Specific hunger in *Anastrepha ludens* (Diptera: Tephritidae): effects on attractiveness of proteinaceous and fruit-derived lures. Environ Ent 20:1680–1686

Robacker DC (1992) Effects of shape and size of colored traps on attractiveness to irradiated, laboratory-strain Mexican fruit flies (Diptera: Tephritidae). Fla Ent 75:230–240

Robacker DC (1998) Effects of food deprivation, age, time of day, and gamma irradiation on attraction of Mexican fruit flies (Diptera: Tephritidae) to two synthetic lures in a wind tunnel. Environ Ent 27:1303–1309

Robacker DC (1999) Attraction of wild and laboratory strain of Mexican fruit flies (Diptera: Tephritidae) to two synthetic lures in a wind tunnel. Fla Ent 82:87–96

Robacker DC, Warfield WC (1993) Attraction of both sexes of Mexican fruit fly, *Anastrepha ludens*, to a mixture of ammonia, methylamine, and putrescine. J Chem Ecol 19:2999–3016

Robacker DC, Moreno DS, Wolfenbarger DA (1990) Effects of trap color, height, and placement around trees on capture of Mexican fruit flies (Diptera: Tephritidae). J Econ Ent 83:412–419

Rull J, Prokopy RJ (2000) Attraction of apple maggot flies, *Rhagoletis pomonella* (Diptera: Tephritidae) of different physiological states to odour-baited traps in the presence and absence of food. Bull Ent Res 90:77–88

Rull J, Prokopy RJ (2004) Revisiting within-tree trap positioning for apple maggot fly (Dipt., Tephritidae) behavioural control. J Appl Ent 128:195–199

San Andrés V, Pérez-Panadés J, Carbonell EA, Castañera P, Urbaneja A (2009) Effects of post-teneral nutrition and ginger root oil exposure on longevity and mortality in bait treatments of sterile male *Ceratitis capitata*. Ent Exp Appl 132:256–263

Sharp JL, Chambers DE (1983) Aggregation response of *Anastrepha suspensa* (Diptera: Tephritidae) to proteins and amino acids. Environ Ent 12:923–928

Shelly TE (1994) Consumption of methyl eugenol by male *Bactrocera dorsalis* (Diptera: Tephritidae): low incidence of repeat feeding. Fla Ent 77:201–208

Shelly TE (2001) Exposure to α-copaene and α-copaene-containing oils enhances mating success of male Mediterranean fruit flies (Diptera: Tephritidae). Ann Ent Soc Am 94:497–502

Shelly TE, Edu J (2009) Capture of mass-reared vs. wild-like males of *Ceratitis capitata* (Dipt., Tephritidae) in trimedlure-baited traps. J Appl Ent 133:640–646

Shelly TE, Pahio E (2002) Relative attractiveness of enriched ginger root oil and trimedlure to male Mediterranean Fruit Flies (Diptera: Tephritidae). Fla Ent 85:545–551

Shelly TE, Villalobos EM (1995) Cue lure and the mating behavior of male melon flies (Diptera: Tephritdae). Fla Ent 78:473–482

Shelly TE, Pahio E, Edu J (2004a) Synergistic and inhibitory interactions between methyl eugenol and cue lure influence trap catch of male fruit flies, *Bactrocera dorsalis* (Hendel) and *B. cucurbitae* (Diptera: Tephritidae). Fla Ent 87:481–486

Shelly TE, McInnis DO, Pahio E, Edu J (2004b) Aromatherapy in the Mediterranean fruit fly (Diptera: Tephritidae): sterile males exposed to ginger root oil in prerelease storage boxes display increased mating competitiveness in field-cage trials. J Econ Ent 97:846–853

Shelly TE, Holler TC, Stewart JL (2006) Mating competitiveness of mass-reared males of the Mediterranean fruit fly (Diptera: Tephritidae) from eclosion towers. Fla Ent 89:380–387

Shelly TE, Edu J, Pahio E (2007) Exposure to ginger root oil decreases capture of male Mediterranean fruit flies (Diptera: Tephritidae) in trimedlure-baited traps Proc. Hawaiian Ent Soc 39:27–32

Shelly TE, Edu J, Pahio E, Wee SL, Nishida R (2008) Re-examining the relationship between sexual maturation and age of response to methyl eugenol in males of the oriental fruit fly. Ent Exp App 128:380–388

Shelly T, Nashimoto J, Kurashima R (2012) Trap capture of three economically important fruit fly species (Diptera: Tephritidae): evaluation of a solid formulation containing multiple male lures in a Hawaiian coffee field. J Econ Entomol 105:1186–1193

Siciliano P, Scolari F, Gomulski LM, Falchetto M, Manni M, Gabrieli P, Field LM, Zhou J-J, Gasperi G, Malacrida AR (2014) Sniffing out chemosensory genes from the Mediterranean fruit fly, *Ceratitis capitata*. PLoS ONE 9(1):e85523. doi:10.1371/journal.pone.0085523

Siderhurst MS, Jang EB (2006) Female-biased attraction of oriental fruit fly, *Bactrocera dorsalis* (Hendel), to a blend of host fruit volatiles from *Terminalia catappa* L. J Chem Ecol 32:2513–2524

Siderhurst MS, Jang EB (2010) Cucumber volatile blend attractive to female melon fly, *Bactrocera cucurbitae* (Coquillett). J Chem Ecol 36:699–708

Sivinski JM (1993) Longevity and fecundity in the Caribbean fruit fly (Diptera: Tephritidae) effects of mating, strain and body-size. Fla Ent 76:635–644

Smith DC, Prokopy RJ (1980) Mating behavior of *Rhagoletis pomonella* (Diptera: Tephritidae). VI Site of early-season encounters. Can Ent 112:585–590

Taylor PW, Yuval B (1999) Postcopulatory sexual selection in Mediterranean fruit flies: advantages for large and protein-fed males. Anim Behav 58:247–254

Teal PEA, Gomez-Simuta Y (2002) Juvenile hormone: action in regulation of sexual maturity in Caribbean fruit flies and potential use in improving efficacy of sterile insect control technique for Tephritid fruit flies. IOBC-WPRS Bull 25:167–180

Teal PEA, Gavilanez-Slone JM, Dueben BD (2004) Effects of sucrose in adult diet on mortality of males of *Anastrepha suspensa* (Diptera: Tephritidae). Fla Ent 87:487–492

Teran HR (1977) Comportamiento alimentario y su correlación a la reproducción en hembras de *Ceratitis capitata* (Wied.) (Diptera, Trypetidae). Rev Agron NO Argent 14:17–34

Thomas DB, Holler TC, Heath RR, Salinas EJ, Moses AL (2001) Trap-lure combinations for surveillance of *Anastrepha* fruit flies (Diptera: Tephritidae). Fla Ent 84:344–351

Toledo J, Malo EA, Cruz-Lopez L, Rojas JC (2009) Field evaluation of potential fruit-derived lures for *Anastrepha obliqua* (Diptera: Tephritidae). J Econ Ent 102:2072–2077

Tschinkel WR (1985) Behavior and physiology. In: Blum MS (ed) Fundamentals of insect physiology. Wiley, New York, pp 391–435

Tsiropoulos GJ, Tzanakakis ME (1970) Mating frequency and inseminating capacity of radiation-sterilized and normal males of the olive fruit fly. Ann Ent Soc Am 63:1007–1010

Tsitsipis JA (1989) Nutrition: requirements. In: Robinson AS, Hooper G (eds) Fruit flies: their biology, natural enemies and control, vol 3 A, World Crop Pests. Elsevier Science, Amsterdam, pp 103–120

Tychsen PH, Fletcher BS (1971) Studies on the rhythm of mating in the Queensland fruit fly, *Dacus tryoni*. J Insect Physiol 17:2139–2156

Utgés ME, Vilardi JC, Oropeza A, Toledo J, Liedo P (2011) Pre-release diet effect on field survival and dispersal of *Anastrepha ludens* and *Anastrepha obliqua* (Diptera: Tephritidae). J Appl Ent 135:1–15

Vargas RI, Prokopy RJ, Duan JJ, Albrecht C, Qing XL (1997) Captures of wild Mediterranean and oriental fruit flies (Diptera: Tephritidae) in Jackson and McPhail traps baited with coffee juice. J Econ Ent 90:165–169

Vargas RI, Stark JD, Kido MH, Ketter HM, Whitehand LC (2000) Methyl eugenol and cuelure traps for suppression of male oriental fruit flies and melon flies (Diptera: Tephritidae) in Hawaii: effects of lure mixtures and weathering. J Econ Ent 93:81–87

Vayssières JF, Dal F (2004) Responses of the Ethiopian fruit fly, *Dacus ciliatus* Loew (Diptera: Tephritidae), to coloured rectangles, spheres and ovoids. In: Barnes BN (ed) Proceedings of the 6th international symposium on fruit flies of economic importance, Stellenbosch, 6–10 May 2004 pp 111–116

Warburg MS, Yuval B (1996) Effects of diet and activity on lipid levels of adult Mediterranean fruit flies. Physiol Ent 21:151–158

Warburg MS, Yuval B (1997a) Circadian patterns of feeding and reproductive activities of Mediterranean fruit flies (Diptera: Tephritidae) on various hosts in Israel. Ann Ent Soc Am 86:305–308

Warburg MS, Yuval B (1997b) Effects of energetic reserves on behavioral patterns of Mediterranean fruit flies (Diptera, Tephritidae). Oecologia 112:314–319

Webster RP, Stoffolano JG Jr (1978) The influence of diet on the maturation of the reproductive system of the apple maggot, *Rhagoletis pomonella* (Walsh). Ann Ent Soc Am 71:844–849

Weldon CW, Pérez-Staples D, Taylor PW (2008) Feeding on yeast hydrolysate enhances attraction to cuelure in Queensland fruit flies, *Bactrocera tryoni*. Ent Exp Appl 129:200–209

Wheeler D (1996) The role of nourishment in oogenesis. Annu Rev Ent 41:407–431

Wong TTY, Whitehand LC, Kobayashi RM, Ohinata K, Tanaka N, Harris EJ (1982) Mediterranean fruit fly: dispersal of wild and irradiated and untreated laboratory-reared males. Environ Ent 11:339–343

Wong TTY, McInnis DO, Nishimoto JI (1989) Relationship of sexual maturation rate to response of oriental fruit fly strains (Diptera: Tephritidae) to methyl eugenol. J Chem Ecol 15:1399–1405

Wong TTY, McInnis DO, Ramadan MM, Nishimoto JI (1991) Age-related response of male melon flies *Dacus cucurbitae* (Diptera: Tephritidae) to cuelure. J Chem Ecol 17:2481–2487

Wu WY, Chen YP, Yang EC (2007) Chromatic cues to trap the oriental fruit fly, *Bactrocera dorsalis*. J Insect Physiol 53:509–516

Yee WL (2003) Effects of sucrose concentrations and fly age on feeding responses and survival of female and male western cherry fruit flies, *Rhagoletis indifferens*. Physiol Ent 28:122–131

Yee WL, Chapman PS (2009) Food deprivation effects on carbohydrate levels and their relation to feeding responses of western cherry fruit fly *Rhagoletis indifferens* to spinosad bait. Physiol Ent 34:163–173

Yokoyama VY, Miller GT, Stewart-Leslie J, Rice RE, Phillips PA (2006) Olive fruit fly (Diptera: Tephritidae) populations in relation to region, trap type, season, and availability of fruit. J Econ Entomol 99:2072–2079

Yuval B, Hendrichs J (2000) Behavior of flies in the genus *Ceratitis*. In: Aluja M, Norrbom A (eds) Fruit flies (tephritidae): phylogeny and evolution of behavior. CRC Press, Boca Raton, pp 429–456

Zhang A, Linn Jr C, Wright S, Prokopy R, Reissig W, Roelofs W (1999) Identification of a new blend of apple volatiles attractive to the apple maggot, *Rhagoletis pomonella*. J Chem Ecol 25:1221–1232

Part III
Ecology and Detection

Chapter 6
Trapping to Monitor Tephritid Movement: Results, Best Practice, and Assessment of Alternatives

Christopher W. Weldon, Mark K. Schutze, and Minette Karsten

Abstract Movement of tephritid flies underpins their survival, reproduction, and ability to establish in new areas and is thus of importance when designing effective management strategies. Much of the knowledge currently available on tephritid movement throughout landscapes comes from the use of direct or indirect methods that rely on the trapping of individuals. Here, we review published experimental designs and methods from mark-release-recapture (MRR) studies, as well as other methods, that have been used to estimate movement of the four major tephritid pest genera (*Bactrocera*, *Ceratitis*, *Anastrepha*, and *Rhagoletis*). In doing so, we aim to illustrate the theoretical and practical considerations needed to study tephritid movement. MRR studies make use of traps to directly estimate the distance that tephritid species can move within a generation and to evaluate the ecological and physiological factors that influence dispersal patterns. MRR studies, however, require careful planning to ensure that the results obtained are not biased by the methods employed, including marking methods, trap properties, trap spacing, and spatial extent of the trapping array. Despite these obstacles, MRR remains a powerful tool for determining tephritid movement, with data particularly required for understudied species that affect developing countries. To ensure that future MRR studies are successful, we suggest that site selection be carefully considered

C.W. Weldon (✉)
Department of Zoology and Entomology, University of Pretoria, Private Bag X20, Hatfield 0028, South Africa
e-mail: cwweldon@zoology.up.ac.za

M.K. Schutze
School of Earth, Environmental and Biological Sciences, Queensland University of Technology, GPO Box 2434, Brisbane, QLD 4001, Australia
e-mail: m.schutze@qut.edu.au

M. Karsten
Department of Conservation Ecology and Entomology, University of Stellenbosch, Private Bag X1, Matieland 7602, South Africa
e-mail: minettek@sun.ac.za

T. Shelly et al. (eds.), *Trapping and the Detection, Control, and Regulation of Tephritid Fruit Flies*, DOI 10.1007/978-94-017-9193-9_6,

and sufficient resources be allocated to achieve optimal spacing and placement of traps in line with the stated aims of each study. An alternative to MRR is to make use of indirect methods for determining movement, or more correctly, gene flow, which have become widely available with the development of molecular tools. Key to these methods is the trapping and sequencing of a suitable number of individuals to represent the genetic diversity of the sampled population and investigate population structuring using nuclear genomic markers or non-recombinant mitochondrial DNA markers. Microsatellites are currently the preferred marker for detecting recent population displacement and provide genetic information that may be used in assignment tests for the direct determination of contemporary movement. Neither MRR nor molecular methods, however, are able to monitor fine-scale movements of individual flies. Recent developments in the miniaturization of electronics offer the tantalising possibility to track individual movements of insects using harmonic radar. Computer vision and radio frequency identification tags may also permit the tracking of fine-scale movements by tephritid flies by automated resampling, although these methods come with the same problems as traditional traps used in MRR studies. Although all methods described in this chapter have limitations, a better understanding of tephritid movement far outweighs the drawbacks of the individual methods because of the need for this information to manage tephritid populations.

Keywords Area-wide management • Assignment tests • Dispersal • Gene flow • Insect movement • Mark-release-recapture • Marking methods • Molecular markers • Monitoring • Remote sensing • Sterile insect technique • Surveillance

The movement of pest species in the family Tephritidae (Diptera) has important consequences for their establishment, survival, and reproduction, and understanding their movement is crucial for the development and implementation of effective management strategies. At the most fundamental level, the dynamics of tephritid populations, like those of all other populations, are a function of fecundity (births), mortality (deaths), and movement (either into or out of a defined area). Consequently, the size of a tephritid population under control is influenced, not only by control strategies that aim to reduce fecundity (e.g., sterile insect technique) or increase mortality (e.g., bait sprays and male annihilation technique), but also the tendency of individuals to move into and recolonize the treated area. Tephritid survival and population growth also rely on the movement of individuals to forage for food and water, thermoregulate, avoid predators, locate mates, and search for suitable hosts for oviposition. These movements at the individual level, when scaled up to include the entire population, have important consequences for population size in a given area and consequently how the population should be managed.

This chapter will address the role of traps in monitoring tephritid movement. Traps are most useful for addressing dispersive movements of insects, so dispersal and the methodological variables that influence its measurement using mark-release-recapture studies will form the focus of much of this discussion. Further, trapped individuals have been used in population genetic studies using nuclear and mitochondrial genes to determine the genetic structure of populations of a small number of pest tephritids. The use of these data from trapped individuals to infer gene flow and the minimum distance required for populations to remain isolated under particular sets of environmental conditions will be addressed. Finally, recent developments in electronics that offer the tantalizing possibility to track individual movements of insects will be discussed as an alternative to trapping for the study of tephritid movement.

1 Definitions

There is a large literature addressing insect movement, including that of tephritid flies. These studies, however, rarely explicitly define the types of movement that they address, which has resulted in the inappropriate use of some terms. Turchin (1998) proposed the following set of refined, interrelated definitions for the concepts of movement, population redistribution, and dispersal in ecology: *movement* is the process by which individual organisms are displaced in space over time; *population redistribution* is the population-level consequence of movement by individual organisms; and *dispersal* is population redistribution that leads to spatial spread. This definition of dispersal is more specific than that given by Southwood and Henderson (2000), who define dispersal as "any movement away from the initial locality", but differentiates dispersal from other forms of spatial redistribution, such as *aggregation* (movement that results in non-uniform spatial distribution at some locality) and *congregation* (aggregation as a result of behavioral processes of organisms to conspecifics) (Turchin 1998). Further, using the lexicon of Turchin (1998), dispersal can also be distinguished from *migration*, which can be defined as population redistribution in response to environmental stimuli resulting from individual movements with directional bias that leads to a net displacement of the population. Dispersal should not be confused with *dispersion*, which is a property of static spatial patterns (Turchin 1998).

Dispersing populations generally exhibit a characteristic pattern of density over time. The majority of insects within a cohort will initially be found around a central point (e.g., site of eclosion, release point of sterile insects), and density will drop to immeasurably low levels within a short distance. The density of insects around the origin drops over time, but individuals are detected at greater and greater distances from the origin. The *dispersal tail* represents individuals that exhibit long-distance movement and, as later described, may often represent the segment of the population of most concern in the management of fruit fly populations. This spatio-temporal pattern can be described mathematically using a diffusion framework (Okubo 1980; Turchin and Thoeny 1993), which models population density based on a normal distribution but with variance increasing linearly over time. Such a distribution can also be regarded as a probability function that indicates the likelihood of an individual within a population moving a particular distance within a certain time frame. These models, however, need to be parameterized appropriately to be accurate given that dispersal of flying insects from a single point can vary with time, species, phenotype, and environmental variables (Baker and Chan 1991a; Banks et al. 1985; Gilchrist and Meats 2012; Turchin 1998; Weldon and Meats 2010).

2 Ecological and Applied Consequences of Dispersal

Dispersal plays a fundamental role in the population dynamics of all organisms, including tephritid flies. In addition to births and deaths, movement into and out of an area are the key factors that determine population size (Turchin 1998). Immigration can, in turn, affect population size by altering density-dependent, intra- and interspecific interactions (Bowler and Benton 2005). Dispersal also has important implications for population connectivity, which enables the persistence of populations in habitat regions and patches (Dempster et al. 1995; Eber and Brandl 1996; Halley and Dempster 1996). In addition, dispersal permits gene flow, leading to the maintenance of genetic diversity or limitation of local adaptation (Eber and Brandl 1994). The role of dispersal in these fundamental ecological processes positions it as a key parameter in population forecasting, conservation, and management. In fragmented habitats, knowledge of dispersal and how it is affected by habitat heterogeneity plays a key role in defining protected areas (Svensson et al. 2011) and planning for movements driven by forecasted climate change (Le Galliard et al. 2012).

Beyond its importance for population dynamics and genetic diversity, dispersal has a range of implications for the management of insect populations and their invasions.

2.1 Delimitation of Quarantine and Treatment Zones in Pest-Free Areas

Quantifying the mean and maximum distance for dispersal establishes the boundaries for management activities that attempt to limit the spread and impact of an incursion of pests in newly invaded areas. The absence of particular pest species reduces the cost of agricultural production and guarantees preferential access to the markets of other pest-free countries. Countries importing biological material from pest-affected countries understandably take measures to limit the risk of pest incursions (Dominiak 2012). Trade restrictions may be imposed when the threshold number of a pest deemed to be indicative of a breeding population is detected. The duration of these trade restrictions and associated control measures is usually a function of the life cycle of the pest, with the condition that an area must remain pest free for the equivalent of several generations before pest-free trade can recommence. International Standards for Phytosanitary Measures (ISPM No. 26) on establishment of pest-free areas for fruit flies recommend that a pest-free area can be reinstated when no further detection of the species occurs for at least three life cycles based on the prevailing temperature in the area (FAO 2006). Similarly, the size of the area affected by trade restrictions, or quarantine distance, is determined by the dispersal capacity of the pest (Clarke et al. 2011). For example, a quarantine area with radius of 5 km around a trapping point was recommended for the invasive fruit fly, *Bactrocera invadens* Drew, Tsuruta & White, in South Africa (Manrakhan et al. 2009), and in the USA, a quarantine area with radius 8 km is applied when *Anastrepha* species are detected (Papadopoulos et al. 2013). For the Queensland fruit fly, *Bactrocera tryoni* (Froggatt), it has been noted (Dominiak 2012) that the quarantine distance imposed by importing countries is dependent on their level of risk aversion and is often unsupported by empirical data on the dispersal of this species. This has led to a situation where different countries accept different quarantine distances for *B. tryoni* (between 15, 50, and 80 km accepted by 10, 1, and 10 countries, respectively; Dominiak 2012), which is untenable from an administrative perspective. Further, the imposition of unnecessarily large quarantine radii can lead to excessive levels of pesticide application and places a financial burden for pre- and post-harvest treatment on producers that are unlikely to be affected by *B. tryoni*. Dominiak (2012) consequently presented a case for quarantine distances for all tephritid flies to be based on their mean dispersal distance in relation to the size of the outbreak as indicated by trap captures.

2.2 Location of the Epicentre of Incursions

Detailed knowledge of the pattern of dispersal enables accurate calculation of the origin and size of an incipient population based on trap captures. Based on an expected pattern of dispersal from a point source where density is initially highest close to the point of origin, it can be assumed that a trap at or near the point of introduction will catch the highest number of insects (Meats 1998a). Density, and therefore trap captures, will then decline with distance from the point of origin according to the shape of the dispersal curve. Using the known pattern of dispersal and geographic location of traps, Meats (1998a) defined a Cartesian method for locating the origin of single infestations based on data obtained from a trapping array targeting the papaya fruit fly, *Bactrocera papayae* Drew & Hancock. Briefly, the mean coordinates for fly captures were determined for a cluster of traps within a 2 km radius of traps catching more than 20 flies, and then the expected density at the epicentre was estimated according to the change in trap catch with distance from the epicentre, assuming that this relationship declined exponentially (Meats 1998a). It was noted that more confident estimates of the epicentre of incursions could be made with more precise knowledge of the decline in density with distance from the point of origin (Meats 1998a).

2.3 Optimal Spacing of Monitoring Traps

Dispersal influences the ability of an array of surveillance traps to detect an infestation, which should determine the choice of effective trap density. This is because the probability of trapping is related to insect density, trap density, and trap efficiency and therefore to the probability of a trap being close enough to the centre of an incipient population (Cunningham and Couey 1986; Meats 1998b; Meats and Clift 2005). For example, Lance and Gates (1994) determined optimal density of traps in a surveillance array for detection of the Mediterranean fruit fly, *Ceratitis capitata* (Wiedemann), by empirically determining distance-dependent capture rates and then using models based on binomial probabilities to define the trap density required to detect a local population. Using this approach, they concluded that ten rotating traps per 2.6 km^2, as used in the fruit fly detection grid in California, USA, was sufficient to trap at least one fly from a population of several hundred adults. For *B. tryoni*, by assuming that the decline in density with distance from the origin of an incursion follows the 'inverse square rule' (as proposed by Fletcher 1974a), it has been determined that estimating the size of an incursion is not possible with an array of cue-lure baited traps with spacing greater than 1 km (Meats 1998b). At this trap spacing, the likelihood that a fly dispersing from its origin would be intercepted is so low and unpredictable that the exercise is meaningless, especially if the purpose of trapping is to determine the presence of a *B. tryoni* population large enough to breed.

2.4 Optimal Spacing of Sterile Insect Releases

Dispersal is also relevant to the control of pest fruit fly populations through the mass release of sexually sterilized conspecific individuals (the 'Sterile Insect Technique', SIT). Too little dispersal by released sterilized individuals may result in uneven coverage or no coverage of patches of the target area (Meats 2007; Meats et al. 2006), whereas too much may result in sterile individuals rapidly leaving the target area. An equally important consideration is that dispersal of wild adults from untreated areas into a treated one can have a dramatic influence of the success of sterile insect technique programmes (Knipling 1959; Meats et al. 2003). Based on a combination of field dispersal experiments and a mathematical model that included parameters for diffusion, convection, "settling," and mortality, Plant and Cunningham (1991) suggested that release points or lines of sterile *C. capitata* be spaced no more than 200–250 m apart, because approximately 70 % of individuals surviving more than 3 days after release remain within 150 m of the release point. A spacing of release lines of 200–250 m for sterile *C. capitata* was verified using aerial releases and recommended for such an application (Vargas et al. 1995). More recently, aerial release trials from fixed-wing aircraft have found that release lines spaced 402 m apart provide sufficient coverage of sterile males in California, USA (Andress et al. 2013); this is permitted by the tendency for flies released from fixed-wing aircraft to disperse further than ground or even helicopter releases (Vargas et al. 1995). Field releases of sterile Mexican fruit flies, *Anastrepha ludens* (Loew), found that 240 m was the typical distance that flies would move from a release point within their lifespan (Thomas and Loera-Gallardo 1998). While Thomas and Loera-Gallardo (1998) did not recommend an optimal spacing for release of sterile *A. ludens*, their results aligned with the standardized flight lane spacing of 320 m used to release these flies in Mexico.

3 Measuring Dispersal

Theoretical and empirical studies on the dispersal of organisms can take one or a combination of two approaches termed *Lagrangian* and *Eulerian* (Turchin 1998). Each term originates from, and is widely used in, fluid mechanics to describe different approaches to investigate motion and the redistribution of particles (Nathan et al. 2003). The Lagrangian and Eulerian approaches to studying and modelling movement are related, but each has its own merits and limitations. The Lagrangian approach is centered on movements made by the individual that can be characterized by velocity, acceleration and turning, and the effect of habitat structure and interactions with competitors and predators on these parameters (Turchin 1998). Taken together, the movements of many individuals modelled or monitored in this way can provide a mechanistic approach to understanding population spread with time. Experiments using a Lagrangian approach are generally more difficult to accomplish and have not usually been feasible for insects with small body size (Nathan et al. 2003), although as discussed later in this chapter, technological advances are changing this situation. Experiments of this kind are also limited to a small number of individuals at a time (Nathan et al. 2003) so are not likely to detect long distance dispersal events.

By contrast, the Eulerian approach is centered on a point in space that is characterized by densities and fluxes of moving organisms with time (Turchin 1998). Eulerian methods (such as mark-release-recapture, 'MRR', discussed below) are generally much more feasible for estimating the pattern of dispersal but require large source strength to increase the likelihood of detecting individuals that travel long distances. Additionally, Eulerian methods do not provide information about events between the source and the recovery site (Nathan et al. 2003).

3.1 Mark-Release-Recapture Studies

Mark-release-recapture methods represent the most practical means for studying movement of organisms, particularly over long distances (Southwood and Henderson 2000). In the case of dispersal studies, this method entails the marking of large numbers of individuals to distinguish them from those already present in the area of the release and the release of these marked individuals from a single point or single small area within an array of traps that extend away from the release point. After a period of time, based on the speed of movement of the study organism, traps are inspected and the number of recaptures determined with distance from the release point. The relative numbers of recaptures in different traps are assumed to reflect the density of released individuals in the vicinity of each trap. Traps are then inspected at several later time points to monitor the spread of the released population over time.

3.1.1 Considerations of MRR

Marking of Individuals

Marking of animals destined to be released is the single most important methodological consideration of MRR studies. Depending on the aim of the study, marking is required to discriminate between released individuals or to differentiate between a released cohort and conspecifics already in the field. A successful marking technique should be easily applied and cost-effective, persist for the full duration of a study, and not affect the competitiveness, survival, longevity, or behavior of the marked individuals (Hagler and Jackson 2001; Southwood 1978). There are a wide range of techniques available to mark insects that are thoroughly reviewed elsewhere (Hagler and Jackson 2001). The aim of this discussion is to summarize the methods that have been, or show promise to be, applied to MRR studies measuring tephritid dispersal.

Where the aim of a MRR study is to track movement of individual insects, a number of variants on individual marking have been developed. These include mutilation (Severin and Hartung 1912), hand painting (Aluja and Prokopy 1992; Fletcher 1973, 1974a; Senger et al. 2009), labelled and/or colored tags (Robacker et al. 1991), and microdots (Whitehead and Peakall 2012). Individual marking is laborious and time-consuming, which limits the number of individuals that can be released. Furthermore, available techniques are not amenable to very small insects. Individual marking does, however, have the distinct advantage of permitting re-release to generate data on longevity and persistence in an area and can enable a Lagrangian approach, albeit crudely, to establish how the movement of individuals contributes to dispersal. Because of these benefits, tephritid dispersal has been successfully monitored using individual marking techniques, but because of their drawbacks they are not commonly used. Robacker et al. (1991), while not measuring dispersal, attached 2 mm-diameter plastic tags bearing a single black symbol to the thorax of flies for individual identification of male *A. ludens* released into field cages (see also McInnis et al. 2002). Small spots of enamel paint of different colors and locations on the thorax were used by Fletcher (1974a) to identify cohorts of released, recaptured, and re-released flies in a study of the dispersal of *B. tryoni* within and from an orchard. Differentiation of cohorts of the cherry maggot fly,

Rhagoletis indifferens Curran, was also achieved by the application of up to two different colors on the thorax (Senger et al. 2009).

By far the most common method for marking tephritids and other small insects for MRR studies is mass-marking. Mass-marking of a released cohort can be achieved using a range of techniques that, in general, can be grouped into the use of recognisable phenotypes (genetic mutants), the application of colored marks, isotopic markers, and molecular markers. The release of flies with a recognisable phenotype has been reported most often in studies on the dispersal of *B. tryoni*, where *white marks* and *bent wings* strains have been bred (Meats and Edgerton 2008; Meats et al. 2002; Weldon and Meats 2007, 2010). *White marks* is a strain exhibiting a natural color mutation in adults caused by a homozygous recessive allele on chromosome 2 (Zhao et al. 2003). This strain possesses white markings rather than yellow markings typical of wild-type *B. tryoni*, which is a feature that has been used to differentiate it from flies already in the field and other cohorts of released flies (Meats and Edgerton 2008; Weldon and Meats 2007, 2010). The *bent wings* strain also results from a recessive mutation on chromosome 2 (Zhao et al. 2003), but the deformity from which its name derives renders it a poor choice for dispersal studies. Both recapture rate and maximum recapture distance of released *bent wings* were far lower than that of wild-type flies (Meats et al. 2002).

Colored paints, enamels, dyes and powders are used extensively to mass-mark insects in MRR studies. Colored paints and enamels can be applied by hand after subduing the animals by chilling or anaesthesia with carbon dioxide or ether (Hamada 1980; Phipps and Dirks 1932, 1933). Like individual marking, this technique is laborious and time-consuming, and the means by which individuals are subdued can have adverse side effects on behavior and mortality (Barron 2000; Champion de Crespigny and Wedell 2008; Phipps and Dirks 1932). However, Froerer et al. (2011) suggest hand application of paints and enamels is one way in which wild-caught insects can be marked and their movements tracked. Paints or dyes can also be applied as an aerosol over large numbers of insects in a container. Gilchrist and Meats (2012) applied fast-drying fluorescent acrylic paint onto unsubdued adult *B. tryoni* in a fly wire cage using a spray can. With experience, they could apply the paint in a way that left small spots of paint on the wings (Gilchrist and Meats 2012). The dispersal of flies marked in this way did not differ from that of conspecifics marked using fluorescent pigment powders (discussed below) but did lead to a large proportion of 'non-fliers' because of the stickiness of the paint, and for this reason, spray marking was not used for large-scale release (Gilchrist and Meats 2012). A more promising example of aerosol application involves the use of readmission ink, which dries rapidly and is invisible under white light but fluoresces yellow under ultra-violet light (Froerer et al. 2010; Hagler et al. 1992). In a topical application, Froerer et al. (2011) used 90 mL to mark 1,000 adult oriental fruit flies, *Bactrocera dorsalis* (Hendel), but they suggest that the required volume can be reduced substantially with aerosol application and have used it in a large-scale MRR study (Froerer et al. 2010).

The use of fluorescent pigment powder remains the dominant means for marking large numbers of tephritid flies prior to release in MRR studies and for identifying flies used in SIT programmes. The pertinent methods were first developed for the marking of the Australian sheep blowfly, *Lucilia cuprina* (Wiedemann), by Norris (1957) but can potentially be used to mark any Diptera in the section Schizophora due to the fact that flies from this group possess a *ptilinum*, an eversible sac on the head used to break out from the puparium. As originally described, the method involves covering the puparia with a layer of dry sand through which fluorescent pigment powder is well mixed. Flies emerging from this treatment can be readily identified under an ultra-violet light even when other traces of the dust have been groomed away from the body surface, because they are 'self-marked' with a quantity of fluorescent pigment powder that is retained in the ptilinal suture and frontalia (Norris 1957). A similar self-marking method was used by Steiner (1965) for melon flies, *Bactrocera cucurbitae* (Coquillett), *B. dorsalis*, and *C. capitata* with Calco blue oil-soluble dye, with the added step of crushing the head and thorax of recaptured flies on filter paper with an acetone rinse. Subsequent applications of self-marking have used a range of different media in which the fluorescent pigment powder is mixed, including sawdust (e.g., Gilchrist and Meats 2012; Macfarlane et al. 1987), as well as direct application of the powder to puparia (e.g., Campbell et al. 2009; Paranhos et al. 2010; Peck et al. 2005). Fly heads may be crushed to detect powder even if fluorescence is visible on the body surface (e.g., Baker et al. 1986; Bloem et al. 1994; Shelly and Edu 2010). The main benefits of self-marking with fluorescent pigment powders are that it minimizes handling effort, removes the need to hold and subdue adults prior to marking with the methods described earlier, and makes identification relatively easy and rapid. However, there are a number of drawbacks that should be considered prior to the use of self-marking with fluorescent pigment powders, primary of which is that they can dramatically reduce adult emergence rates and flight ability (Campbell et al. 2009; Dominiak et al. 2000, 2010; Weldon 2005).

The amount of fluorescent pigment powder used per volume of pupae is one factor that can affect adult emergence and flight ability (Dominiak et al. 2010). As an example, adult eclosion rates of *B. tryoni* declined from 85.7 to 77.4 % and were significantly different from a control with pigment concentrations of 1.5–4.5 g/L. Over the same range of pigment concentrations, flight ability indices ranged from 92.1 to 83.3 %. As a generic standard for SIT operations, it has been recommended that 1.5 g of fluorescent pigment powder be applied per litre of pupae (FAO/IAEA/USDA 2003). However, there has been wide variation in the concentrations used even after this recommendation was proposed. For example, Peck et al. (2005) marked *B. cucurbitae* with 5 g of pigment powder per liter, Meats (2007) and Meats and Edgerton (2008) used 50 g of pigment powder per 100,000 pupae (approx. 1 kg) to mark *B. tryoni*, Shelly and Edu (2010) marked *B. cucurbitae* and *B. dorsalis* with 3 g per liter, and Rempoulakis and Nestel (2012) marked olive fruit flies, *Bactrocera oleae* (Rossi), with 2 g per liter. Other reports do not indicate the concentration at all (e.g., Dominiak et al. 2011; Hernández et al. 2007; Kendra et al. 2010; Paranhos et al. 2010; Peck and McQuate 2004). Further, the concentration that negatively influences adult emergence and flight ability may vary with particle size of the selected fluorescent pigment powder: Weldon (2005) reported poor emergence of *B. tryoni* from puparia associated with powders with a particle size of 4–5 μm. Another issue that must be considered when using fluorescent pigment powders is the visibility of different colors, the ability to discriminate between different colors under ultra-violet light when multiple cohorts are released, and their persistence (Dominiak et al. 2000; Schroeder and Mitchell 1981; Weldon 2005). Needless to say, attempts should be made to optimize the appropriate dose, particle size, and color for the species to be marked using fluorescent pigment powders.

Internal marking with a range of vital dyes is possible. Like all other marking techniques, the key to choosing a successful vital dye is to ensure that it does not negatively affect survival, modify behavior, and is persistent for the duration of the experiment (Schroeder and Mitchell 1981; Sharp and Ashley 1984). Internal marking of trapped individuals is often difficult and time consuming to evaluate because of the need for dissection and internal examination (Schroeder and Mitchell 1981). Sudan Deep Black BB (1 g dissolved in vegetable oil) when added to 1 L of larval diet colors adult *B. cucurbitae* deep black, because it becomes incorporated into the hemolymph of the larva and adult (Schroeder and Mitchell 1981). Over a period of 2 weeks, the black color is gradually eliminated from the hemolymph, but the rectal papillae are permanently dyed deep blue. Unfortunately, the behavior of adults dyed internally with Sudan Deep Black BB differs from that of undyed adults; there was evidence of assortative mating of undyed and dyed *B. cucurbitae*, and flight propensity was reduced in dyed flies. Interestingly, however, dyed male *B. cucurbitae* exhibited improved flight performance on a flight mill (Schroeder et al. 1974). Another vital dye, fat red 7B, was found to internally mark adults of *A. suspensa* (Loew) for 2 days after eclosion when incorporated into the larval diet (125 mg dye in 125 g diet; Sharp and Ashley 1984). Fluorescent dyes fed to adults have also been used to mark tephritid flies for dispersal studies. Arévalo et al. (2009) fed adult blueberry maggot fly, *Rhagoletis mendax* Curran, a 1-mMol solution mixture of Fluorescent Brightener 28 in honey. Trapped adults were then homogenized, and the fluorescence of each homogenate was determined using a microplate fluorometer with a 355-nm excitation filter and a 460-nm emission filter. Marks were persistent for at least 7 days (Arévalo et al. 2009).

Isotopes of a range of elements have been used to mark insects in MRR studies. An isotope of an element has the same atomic number as the element but a different number of neutrons and thus a different atomic mass. Both radioisotopes (e.g., phosphorus 32, Barnes 1959; Jones and Wallace 1955; strontium 89, Neilson 1971) and stable isotopes (e.g., carbon 13, nitrogen 15, Hagler and Miller 2002) have been used to mark tephritid flies, but the majority of recent studies have relied on stable isotopes to discriminate released insects from their wild counterparts. Stable isotopes are preferred, because they are safe, non-radioactive, and hence do not decay (Hagler 1997). In comparison with other forms of marking, the use of isotopes is very non-invasive, because stable isotopes are easily incorporated into feeding regimens and, depending on the stable-isotope enriched compounds used, remain in the tissues of the animal (Hagler 1997). The cost and handling required for isotope analysis is very competitive compared with endogenous molecular markers

(Hagler 1997), but processing of samples is more time consuming than sorting insects with a visual mark.

A range of molecular markers have become available to distinguish released insects from their wild counterparts. These are discussed in more detail later in this chapter (Sect. 3.2.2), but in the context of marking insects for MRR studies, here we introduce the use of immunomarking. This approach involves the marking of insects with mammalian (Hagler 1997; Hagler et al. 1992; Hagler and Miller 2002) or plant proteins (Jones et al. 2006) that are then detected on trapped individuals using an enzyme-linked immunosorbent assay (ELISA). A plate reader generates ELISA optical density values for each insect sample, and the key result is the presence or absence of a positive reaction to the presence of the marking protein (Hagler et al. 1992). Application of the protein mark can be achieved by topical application or incorporation into the insect diet (Hagler et al. 1992) or even by walking across treated plant surfaces (Jones et al. 2006). Marks can be retained for over 20 days (Hagler 1997; Jones et al. 2006). Immunomarking of insects has been most successfully achieved using rabbit immunoglobulin G (IgG; Hagler and Miller 2002) and chicken egg albumin, and individuals marked in this way are best detected using sandwich ELISA rather than direct ELISA (Hagler and Miller 2002). It has been argued that the method is inexpensive, especially when using chicken egg albumin or plant proteins as the immunomarker (between US$0.12 and $0.26 per litre; Jones et al. 2006), but it does require considerable processing of trapped insects and a capital outlay for a plate reader, ELISA plates, and reagents (Jones et al. 2006). Immunomarking has been used in a study on the dispersal of solanum fruit flies, *Bactrocera latifrons* (Hendel), where rabbit IgG was applied to adults both in drinking water and topically (Peck and McQuate 2004). Recaptures of *B. latifrons* marked in this way were much higher than those with ptilinal fluorescent pigment marks: 3.1 % of immunomarked flies were recaptured, whereas only 0.92 % of those with fluorescent pigment marks were recaptured (Peck and McQuate 2004).

Design of Trap Arrays

Trap array design has important consequences for the conclusions drawn from a dispersal experiment. The spatial arrangement of traps needs to be carefully considered to ensure that it meets experimental aims while recognizing practical limitations, such as the range of options available to track individual movement and the costs associated with increased sampling in space and time (Skarpaas et al. 2005). A number of alternative sampling designs of equivalent total trap area have been assessed with Monte Carlo simulation. For a known release rate, transects (linear or cross-shaped trap arrays radiating outwards from the release point) and sectors (wedge-shaped trap arrays radiating outwards from the release point) provided better data for estimating the distribution of dispersal distances (the 'dispersal kernel') than random placement, grid arrays, and annuli (Skarpaas et al. 2005). In this scenario, 'better data' relates to the precision of model estimates of the dispersal kernel relative to the true model. If dispersal was directional but unknown, annuli (i.e., concentric circles) or grid arrays performed better than transects or sectors, but the performance of transects and sectors was once again superior to annuli and grids if they were aligned with known directional movement tendencies (Skarpaas et al. 2005). Considering these results, scientists conducting dispersal studies that aim to determine the mean, median, and maximum limit of dispersal should deploy their traps as transects or sectors to maximize the potential for accurate sampling of the dispersal tail. Studies that aim to determine population displacement as a consequence of environmental variables (e.g., prevailing wind, habitat and resource heterogeneity) will be better served by a grid array. Baker and colleagues advocated this latter approach for quantification of sterile tephritid fly dispersal (based on replicated releases of *A. ludens* and *C. capitata*; Baker and Chan 1991a; Baker et al. 1986). However, data obtained using this approach are not adequate if the ultimate goal is to define the dispersal distance of a species for the purpose of setting quarantine radii or quantifying population connectivity. In this circumstance, the methods of Baker and colleagues (Baker and Chan 1991a; Baker et al. 1986) represent the preliminary stage of a rigorous program that aims to ascertain factors causing directional displacement of marked individuals, which could then be followed by releases on a transect or sector trap array aligned with identified movement tendencies.

Trap array design also needs to take into consideration the attractiveness of the traps being used. The majority of traps used in studies of tephritid movement rely on an attractant, whether visual or chemical, to increase the likelihood of recaptures, but the properties of attractive traps led Baker and Chan (1991a) to question their use in the quantification of dispersal and explanation of processes that shape observed patterns for four reasons. First, flies do not enter traps by random movement, therefore random movement can not be assumed or investigated when using them. In relation to the regression-based null dispersal models employed by Baker and Chan (1991a), this assertion is warranted. Most empirical dispersal studies, however, have demonstrated that tephritid dispersal is not random, but associated with ecological and physiological variables (e.g., habitat suitability, oviposition sites, and wind direction) and that they likely override the localized attraction of flies to traps. Second, trap efficiency may be density-dependent; therefore, they can not be used to study density-dependent dispersal. It is certainly the case that studies on trap performance recapture far fewer flies than are released (e.g., Lance and Gates 1994; Shelly and Nishimoto 2011), so it stands to reason that only traps in areas with high population density will detect flies. Third, the active space of traps is continually changing, so that it is impossible to ascertain whether variation in catch is real or apparent. Finally, food traps catch hungry flies, not dispersing flies (Baker and Chan 1991a). This has certainly been proven to be important in relation to the effectiveness of food-based traps when females have been fed protein (e.g., in *Anastrepha* species, Díaz-Fleischer et al. 2009a, b; in *C. capitata*, Rousse et al. 2005).

It has also been suggested that the use of strongly attractive traps in the vicinity of the release point may have a dramatic effect on the density-distance curve produced from a MRR study. Strong traps located close to the release point will capture a high proportion of marked individuals and reduce the number of insects that would otherwise be caught in more distant traps (Turchin 1998). Additionally, strong traps may arrest movement away from the release point by most insects that are primed to respond to the attractant (Turchin 1998). The attractiveness of the various lures for tephritid flies is discussed by Tan et al. (Chap. 2, this volume). At this point, however, it is important to note that early work explicitly determined the effectiveness of available traps and lures for some species: for *B. dorsalis*, 50 - non-competitive traps (baited with methyl eugenol) are required per mile^2 (2.6 km^2) to exhaust the male population, whereas 80 traps (baited with cue-lure) are required for *B. cucurbitae* and 500 traps (baited with trimedlure) are required for *C. capitata* (Steiner 1969). It is likely that these values are influenced by dispersal capacity of the species involved, but it is still apparent that traps baited with methyl eugenol are considerably more attractive to *B. dorsalis* than those baited with trimedlure are to *C. capitata*, and this should be considered when designing trap arrays. More recently, it has been demonstrated that McPhail traps baited with ammonium acetate and putrescine do not have a long-distance attractive action for the West Indian fruit fly, *Anastrepha obliqua* (Macquart), or the Caribbean fruit fly, *Anastrepha suspensa* (Loew) (Jenkins et al 2013). As a result, these traps must be placed in host trees where flies are already present.

Maximum Distance

Intimately associated with the spatial arrangement of traps for the purpose of dispersal studies is the maximum distance sampled. Long distance dispersal events are rare, and the probability of their detection decreases with distance from the release point, but such events are important for the empirical assessment of population connectivity and gene flow. In many cases, the maximum distance sampled by a trap array is determined by practical constraints, including the costs associated with maintaining a large trapping array and access to property on which to place the traps. However, if a MRR study aims to estimate the dispersal capacity of an insect, considerable effort should be made to adequately sample the dispersal tail. Counter to this theoretical ideal, many studies (e.g., Fletcher 1974a, b; Hamada 1980; Neilson 1971; Paranhos et al. 2010; Rempoulakis and Nestel 2012; Weldon and Meats 2010; Wong et al. 1982) on tephritid dispersal are characterized by declining sampling effort (i.e., fewer traps per unit area) with increasing distance from the release point. One exception is a study reported by Barry et al. (2002) on *C. capitata* that involved a concentric circular array of trimedlure-baited Jackson traps arranged so that there was an equal distance between traps in each circle of traps (although radius of the outer ring of traps was only 366 m). Another is a MRR study on the dispersal capacity of *B. dorsalis* designed with the intent of recording rare, long-distance, movements. Sampling effort in the immediate vicinity of the release point was almost entirely absent with flies often released at least 2 km from the closest trap of a haphazard grid array (trap spacing approximately 1 km) in Puna, Hawaii (Froerer et al. 2010). The result was the detection of 23 flies moving greater than 5 km in less than 4 days (Froerer et al. 2010).

If few data are available on the dispersal capacity of a species, it is difficult to ascertain the distance over which dispersal should be sampled to detect long-distance events. It is clear from published data from MRR studies that the mean dispersal distance of tephritid flies is usually well below 1 km (Table 6.1, Fig. 6.1a). But, it is also apparent that the maximum dispersal distance reported in many studies was limited to the distance of the trap placed furthest from the release point (Fig. 6.1b). One way to select the appropriate spatial scale for dispersal studies is to use the relationship between spatial scale and recapture distance reported in earlier studies (Jones et al. 2006). Plotting the relationship between sampling area and mean recapture distance in reported MRR studies for a range of tephritid species (although biased towards *Bactrocera* species, see Table 6.1) indicates that mean recapture distance increases with sampling area up to a point where further increases in spatial scale yield little benefit (Fig. 6.1a). Solving the equation for the approximate asymptote of this relationship (mean recapture distance = 1.4 km) yields an optimal sampling area of approximately 107 km^2. Assuming that this sampling area is a circle surrounding a single release point, the maximum distance that should be sampled is 5.8 km. This approaches the maximum sampling distance required to detect the mean dispersal distance from most published MRR studies (Fig. 6.1c). Outliers above the logarithmic fit in Fig. 6.1a and c are from a single study on the movement of *B. dorsalis* between islets (Iwahashi 1972).

Another approach that objectively optimizes the spatial extent of a trapping array to estimate dispersal involves subsampling MRR data generated by a pilot study. This approach has been used successfully by Franzén and Nilsson (2007) and Hassall and Thompson (2012) to verify the minimum landscape scale used in dispersal studies of burnet moths (Lepidoptera: Zygaenidae) and a damselfly (Odonata: Coenagrionidae), respectively. In summary, cohorts of insects marked with distinctive colors are released into a study area at different locations and later recaptured. The study area is then divided into smaller compartments, the mean recapture distance is calculated within each compartment, and then after the serial addition of adjoining compartments, the optimal spatial scale for sampling is set at the distance or area where mean recapture distance no longer increases with the addition of further sampling effort.

Time Scale

Dispersal is a process involving changes in abundance over both space and time. It is not surprising, therefore, that the time scale of sampling in MRR studies can influence estimates of the shape of the density-distance relationship (e.g., Hassall and Thompson 2012). Studies on the dispersal of species representing *Anastrepha* (Baker and Chan 1991a; Kovaleski et al. 1999), *Bactrocera* (Gilchrist and Meats 2012; Weldon and Meats 2010), and *Ceratitis* (Baker et al. 1986; Paranhos et al. 2010; Plant and Cunningham 1991) that incorporate repeated sampling through time indicate that mean dispersal distance increases with time after release.

Table 6.1 Summary of mark-release-recapture study designs and results for tephritid flies

	Trap											Recapture distance (km)			
Taxon	Type	Attractant	Toxin	Spacing (km)	Pattern	Area covered (km^2)	Maximum sampled (km)	Study duration	Strain/treatment	Adults released (N)	Recapture rate (%)	Mean	0.95	Maximum	Source
Anastrepha															
fraterculus	MPh	Grape juice	None	?	Haphazard	0.7	0.96	40 d	Wild, bisex adult	2,154	7.1	?	<0.2	0.8	Kovaleski et al. (1999)
	MPh	Grape juice	None	?	long. grid		1.7	60 d		3,284	37.1	?	<0.2	0.8–1	
ludens	MPh	Protein + water	None	0.02	Grid	0.0324	0.13	4 d	Bisex adult	15,000	2	?	?	?	Baker et al. (1986)
ludens	MPh	prot. hyd., borax + water	None	0.008–0.012	Radial	?	?	4 d	Ster., bisex adult	15–20,000	?	?	<0.1	?	Baker and Chan (1991a)
ludens	MPh	prot. hyd., borax + water	None	0.06	Grid	0.176	0.59	15 d	Ster., bisex adult	6,000, 9 replicates	2.88	0.111	?	>0.15	Hernández et al. (2007)
									Wild, bisex adult		7.62	0.112	?	>0.15	
ludens	MPh	Torula yeast + water	None	0.1–4	Transects	N/A	10	10 wk	Ster, bisex adult	250,000 in 10 weekly releases	0.7	?	<0.1	9	Thomas and Loera-Gallardo (1998)
	MPh	Torula yeast + water	None	0.1–1	Haphazard	0.7	1	11 wk	Ster, bisex adult	275,000 in 11 weeks	0.7	?	<0.1	0.3	
	MPh	Torula yeast + water	None	?	Haphazard	1.8	1.7	9 wk	Ster, bisex adult	275,000 in 11 weeks	1	0.236	<0.4	?	
obliqua	MPh	prot. hyd., borax + water	None	0.06	Grid	0.176	0.59	15 d	Ster, bisex adult	6,000, 9 replicates	1.78	0.139	?	>0.2	Hernández et al. (2007)
									Wild, bisex adult		6.7	0.152	?	>0.2	
Bactrocera															
cucurbitae	Bkt	CL	Naled	0.05–0.5	Radial	?	1	7 d	Wild, bisex adult	600	9.07	0.088	0.25	1	Hamada (1980)
							1	6 d	Wild, bisex adult	869	4.35	0.05	0.15	1	
				0.05	Concentric circles	0.785	0.25	10 d	Ster., bisex adult	?	~1	?	?	0.25	
cucurbitae	Ste	CL	Naled	?	Haphazard	5,913	110	6 mth	Ster., bisex adult	300,000,000	n/a	?	?	56	Kawai et al. (1978)

cucurbitae	Ste	CL	Naled	>0.05	Haphazard	1.17	0.92	45 d	Wild, bisex adult	1,582	15.17	0.172	?	?	Nakamori and Soemori (1981)
									M-r, bisex adult	830	9.16	0.149	?	?	
								20 d	Wild, bisex adult	1,008	2.38	0.09	?	?	
									M-r, bisex adult	1,076	1.02	0.058	?	?	
								24 d	Wild, bisex adult	1,716	13.33	0.07	?	?	
									M-r, bisex adult	1,515	5.06	0.051	?	?	
dorsalis	Ste	ME	DDVP	1	Grid	51	12.5	2 wk	Ster., bisex adult	43,259	0.0005	?	?	11.39	Froerer et al. (2010)
	MPh	Torula yeast	None				15.9		Ster., bisex adult	26,507	0	?	?	?	
							13		Ster., bisex adult	57,716	0.1	?	?	?	
							9.2		Ster., bisex adult	90,078	0.98	?	?	?	
dorsalis	Ste	ME	?	0.5–8	Haphazard	~63	8	?	Wild, bisex adult	2,275	4.5	4.5	?	8	Iwahashi (1972)
						~63	7.5	?		1,801	2.4	3.3	?	7.5	
						~63	15	?		1,572	1.4	4.5	?	15	
						~63	16.3	?		1,458	5.5	6.3	?	16.3	
						?	50	3 mth		3,000	0.3	?	?	50	
latifrons	MPh	Autolysed yeast extract	None	?	Haphazard	?	?	6 wk	Bisex adult	1,405	3.27	?	?	0.5	Peck and McQuate (2004)
	Jac	Alpha-ionol +cade oil	None									?	?		
oleae	ST-y	None	None		Concentric circles	~0.028	~0.17	3, 15 d	Ster. bisex adult	5,000, 8 replicates	0.3–6.6	0.019–0.068	?	?	Rempoulakis and Nestel (2012)
tryoni	Lyn	CL	Malathion	0.4–5	Haphazard	707	17	5 wk	Bisex adult	1,147,000	0.314	?	0.4	0.4	Dominiak et al. (2003)
tryoni	Lyn	CL	Malathion	0.1–1	Haphazard	7	?	13 wk	Ster., bisex adult	810,000	0.198	0.1	0.3	1.5	Dominiak et al. (2011)
									Ster., bisex pupal	600,000	0.023	0.2	0.4	1.2	
									Ster., chilled bisex adult	300,000	0.01	0.2	0.7	1.4	
tryoni	Ste	CL	Malathion	0.4	Transects	1,810	24	7 wk	Adults from fruit	66,856	0.026	0.6	?	22.7	Fletcher (1974a)
									Wild-caught adults	4,673	0.335	0.6	?	22.7	

(continued)

Table 6.1 (continued)

	Trap											Recapture distance (km)			
Taxon	Type	Attractant	Toxin	Spacing (km)	Pattern	Area covered (km^2)	Maximum sampled (km)	Study duration	Strain/ treatment	Adults released (N)	Recapture rate (%)	Mean	0.95	Maximum	Source
tryoni	Lyn	CL	Malathion	0.02	long. grid	~0.18	0.5	25 d	M-r, bisex adult	90,000	26	0.24	?	0.5	Gilchrist and Meats (2012)
									M-r, bisex adult	114,000	16	0.264	?	0.5	
								29 d	Out-bred, bisex pupal	136,000	5.5	0.368	?	0.5	
									M-r, bisex pupal	136,000	7.5	0.365	?	0.5	
								29 d	Out-bred, bisex pupal	56,000	8.5	0.306	?	0.5	
									M-r, bisex pupal	62,000	6	0.284	?	0.5	
tryoni	Jac	CL	None	0.5–1	Transects	14,000	6	9 wk	Ster. bisex adult	540,000	0.3	1.1	?		MacFarlane et al. (1987)
	Jac	CL	None	≥5	Haphazard	14,000	112	9 wk		400,000	0.2	1.1	?	94	
										420,000	0.03	1.4	?		
										500,000	?	?	?		
tryoni	Lyn	CL	Malathion	0.02	long. grid	~0.13	0.48	5 wk	Bisex pupal	50,000	14.8	?	?	?	Meats and Edgerton (2008)
									Bisex pupal	7,300	12.8	?	?	?	
									Bisex pupal	12,000	17.5	?	?	?	
									Bisex adult	12,000	14	?	?	?	
tryoni	Lyn	CL	Malathion	0.02	long. grid		1	?	Bisex adult	8,000	23.6	?	?	0.22	Meats et al. (2002)
	ST	None	None	0.02	long. grid		1	?			1.6	?	?	0.11	
	Lyn	CL	malathion	0.02	long. grid		1	?	*bent wings*, bisex adult	2000	4	?	?	0.05	
	ST	None	None	0.02	long. grid		1	?			1.7	?	?	0.04	
tryoni	Lyn	CL	Malathion	0.02	Grid	~0.009	0.088	7 d	*wm*, bisex adult	21000	6.3	?	?	0.088	Weldon and Meats (2007)
	ST-y	prot. auto.	None								0.5	?	?	0.057	
	ST-b	prot. auto.	None								0.3	?	?	0.088	
	Lyn	CL	Malathion	0.02	Grid	~0.009	0.088	7 d	*wm*, bisex adult	12,000	6.7	?	?	0.088	
	ST-y	prot. auto.	None								0.1	?	?	0.049	
	ST-b	prot. auto.	None								0.2	?	?	0.057	
	Lyn	CL	Malathion	0.02	Grid	~0.009	0.088	7 d	Ster., bisex adult	12,000	17.1	?	?	0.088	
	ST-y	prot. auto.	None								0.3	?	?	0.041	
	ST-b	prot. auto.	None								0.2	?	?	0.039	

tryoni	Lyn	CL	Malathion	0.02	long. grid	~0.55	1.1	2 wk	Wild, bisex adult	1,500–4,500, 5 replicates	3	0.1	<0.5	1.1	Weldon and Meats (2010)
									wm, bisex adult	1,992–10,500, 3 replicates	33.2	0.1	<0.5	1.1	
									Ster., bisex adult	6,000–24,000, 4 replicates	13.4	0.1	<0.5	1.1	
Ceratitis															
capitata	Jac	TML	None	0.268	Linear, perpendicular to aerial release axis	932	4.824	11 d	Ster, male only	645,249–922,424, 4 replicates	0.0119–0.0316	?	0.8–1.9	>3.216	Andress et al. (2013)
				?	Haphazard	618.24	?	?			0.0796–0.1558	?	<3.216	>3.216	
capitata	Jac	TML	None	0.025	Grid	0.12	?	10 d	Ster., bisex adult	45–50,000	12.4	?	?	?	Baker et al. (1986)
	Jac	TML	None	0.05	Grid	0.12	?				7.3-11.1	?	?	?	
	MPh	Protein + water	None	0.02	Grid	0.2	?	4 d		20,000	2.7	?	?	?	
capitata	MPh	prot. hyd., borax + water	None	0.008–0.012	Radial	?	?	4 d	Ster., bisex adult	15–20,000	?	?	?	?	Baker and Chan (1991a)
capitata	Jac	TML	None	0.092–0.144	Concentric circles	0.421	0.366	10 d	Ster. bisex adult (Maui-93)	12,000, 5 replicates	3.1	~0.13	<0.366	0.366	Barry et al. (2002)
									Ster. male only (Vienna 4)	12,000, 5 replicates	2.0	~0.13	<0.366	0.366	
capitata	Jac	TML	None	N/A	Single trap, releases at different distances	10.4	0.32	5 wk	Ster., bisex adult	6,500	?	?	?	0.32	Lance and Gates (1994)
capitata	Lyn	Capilure	Dichlorvos	N/A	Single trap, releases at different distances	N/A	0.16	?	Ster. male only (Vienna 7 Mix 99)	~84,000, 3 replicates	?	?	<0.8	?	Meats and Smallridge (2007)
	Lyn	Capilure	Dichlorvos	0.4	Grid		10	?	Ster. male only (Vienna 7 Mix 99)	~38,800,000	0.02	?	?	10	
capitata	Jac	TML	None	0.025–0.05	Radial	0.2	0.25	11 d	Ster. male only (Vienna 8)	10,000	8.34	0.06–0.09	<0.15	0.25	Paranhos et al. (2010)

(continued)

Table 6.1 (continued)

Taxon	Trap Type	Attractant	Toxin	Spacing (km)	Pattern	Area covered (km^2)	Maximum sampled (km)	Study duration	Strain/ treatment	Adults released (N)	Recapture rate (%)	Recapture distance (km) Mean	0.95	Maximum	Source
capitata	Jac	Capilure	None	0.1295	Grid	1.07	0.73	3–10 d	Ster., bisex adult	20,000, 9 replicates	?	0.11–0.26	?	?	Plant and Cunningham (1991)
capitata	Bkt	TML	Naled	0.05	Linear, perpendicular to aerial release axis	0.96	0.3	2 d	Ster., bisex adult	3,000,000, 2 replicates	0.06–0.52	?	<0.3	0.3	Vargas et al. (1995)
capitata	Glass traps	Diammonium phosphate or TML	None	0.125	?	0.035	0.35	3 wk	Ster., bisex adult	5,000	0.7	0.1	?	0.35	Wakid and Shoukry (1976)
							0.7		Ster., bisex adult	2,500	1.9	0.26	?	0.7	
							0.7		Ster., bisex adult	2,500	0.9	0.24	?	0.525	
							0.7		Ster., bisex adult	1,500	1.5	0.12	?	0.525	
capitata	Ste	TML	None	0.03	Concentric circles	0.1	0.24	5 wk	Wild, maie adult	21,770	8	?	?	0.24	Wong et al. (1982)
									L-r, 0 Gy, male adult	26,600	12.7	?	?	0.24	
									L-r, 50 Gy, male adult	26,800	11.9	?	?	0.24	
									L-r, 100 Gy, male adult	26,550	11	?	?	0.24	
									L-r, 150 Gy, male adult	26,490	10	?	?	0.24	
									L-r, 200 Gy, male adult	26,380	9.2	?	?	0.24	
Rhagoletis															
cingulata	MPh + fly paper	Ammonium carbonate	None	0.015	Haphazard	?	0.287	24 d	Bisex adult	2,010	1.9	0.058	0.17	0.287	Jones and Wallace (1955)
completa	bait pans	Glycine sodium hydroxide	None	Not reported	Haphazard	0.26	1.45	3 wk	Wild, bisex adult	Unknown: wild flies fed radio-isotopes	n/a	?	?	1.45	Barnes (1959)
mendax	ST-y	None	None	0.003–0.030	Radial	?	0.091	7 d		100	72.0	0.027	<0.091	0.091	Arévalo et al. (2009)

									Wild, bisex adult						
pomonella	ST	None	None	?	long. grid	1.61	1.57	Not reported	Wild, bisex adult	38,614	1.39	0.27	<1.38	1.57	Maxwell and Parsons (1968)
pomonella	ST	None	None	0.009	Grid	0.2	0.19	10 wk	Wild, bisex adult	Unknown: wild flies fed radio-isotopes	n/a	0.04	0.06	0.08	Neilson (1971)
							0.17	11 wk			?	0.04	0.07	0.09	
							0.2	12 wk			?	0.03	0.06	0.08	
pomonella	hand collect.	None	None	N/A	N/A	0.06	N/A	2 wk	Wild, bisex adult	1,035	11.9	0.07	0.14	0.14	Phipps and Dirks (1932)
pomonella	hand collect.	None	None	N/A	N/A	0.14	N/A	1 mth	Wild, bisex adult	3,152	4.1	0.14	0.17	0.21	Phipps and Dirks (1933)

Where not explicitly stated, values for trap spacing, area covered by the trap array and maximum distance sampled by the array were estimated using maps of the array (where published) that were copied into ImageJ image analysis software (version 1.43r). Similarly, recapture rate, mean recapture distance and the distance encompassing at least 95 % of recaptures were calculated from tabulated values or by using ImageJ to estimate recaptures at each distance from published figures. Blank cells within the same source indicate the same conditions as initially described

Key to codes

Trap type: *Bkt* bucket trap, *hand collect* hand collection, *Jac* Jackson trap, *Lyn* Lynfield trap, *MPh* McPhail trap, *ST* sticky trap, *ST-b* black sticky trap, *ST-y* yellow sticky trap, *Ste* Steiner trap

Trap attractant: *CL* cue-lure, *TML* trimedlure, *ME* methy eugenol, *prot. aut.* protein autolysate, *prot. hyd.* protein hydrolysate

Trap pattern: *long. grid* longitudinal grid

Study duration: day (d); week (wk); month (mth)

Strain/treatment: *L-r* lab-reared, *M-r* mass-reared, *ster.* sterile

N/A: not applicable to study design

?: not reported or difficult to ascertain from report

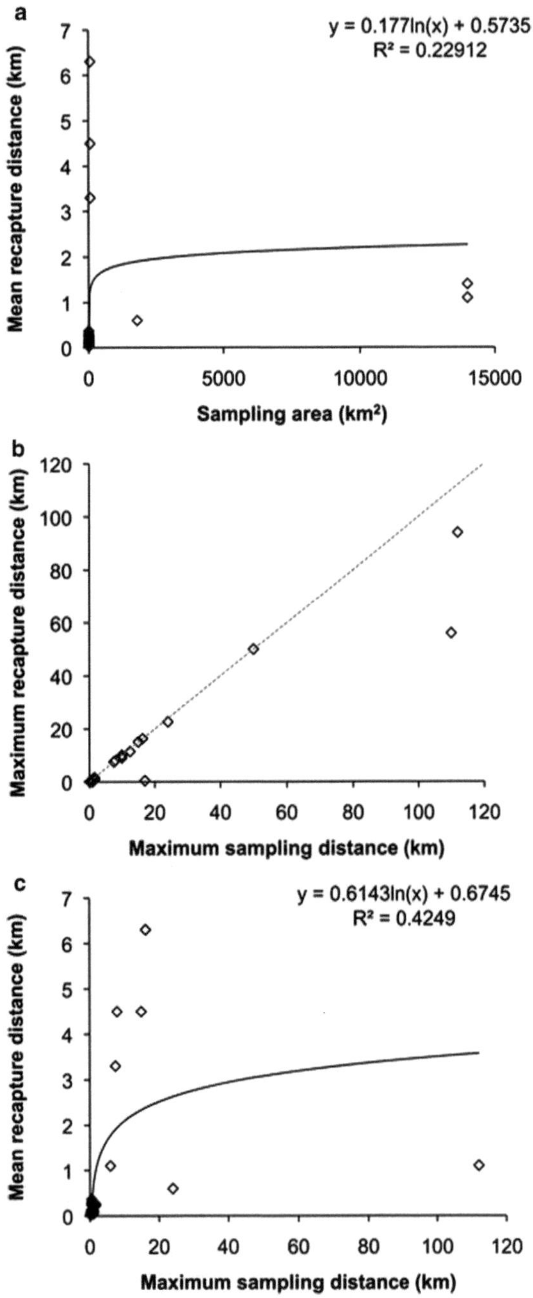

Fig. 6.1 Effects of sampling area and maximum sampling distance of trapping arrays on mean and maximum dispersal distances of tephritid flies determined from mark-release-recapture studies. (**a**) Relationship between sampling area and mean recapture distance. The *solid black line* indicates the logarithmic fit for the data. (**b**) Relationship between maximum sampling distance and maximum recapture distance. Points lying on the *gray dotted line* indicate recaptures in the trap furthest from the release point. (**c**) Relationship between maximum sampling distance and mean recapture distance. The *black line* indicates the logarithmic fit for the data

However, it is also equally important to point out that tephritid flies can disperse over relatively large distances very quickly following release. Shelly and Edu (2010) reported that male *B. cucurbitae* and *B. dorsalis* released 500 m from a trap baited with cue-lure and methyl eugenol, respectively, were recaptured within 1–3 days. Even more remarkable is that only 1 day after release, *B. dorsalis* have been recaptured in traps more that 10 km from the release point (Hagler et al. 1992).

The interval at which traps are emptied or replaced has important implications for the type of models that can be used to explain changes in abundance over space and time, because traps integrate density over time (Turchin 1998). If the change in density of a dispersing population is small in the time period between trap collections, the density data can be considered 'instantaneous' and fit to a Gaussian curve (normal distribution). The Gaussian distribution is the null model for movement assuming that the movement pattern of the study organism is approximated by simple diffusion. The emptying of traps at daily intervals to track dispersive movement of *A. ludens* by Baker and Chan (1991a) is a good example of instantaneous density data from the tephritid dispersal literature, although the range of phenomenological empirical models that were used to fit the data do little to aid in understanding movement processes (Turchin 1998). It is more common to encounter time-integrated density data in dispersal studies. The key difference between instantaneous and time-integrated analyses is that the latter include loss of organisms in the diffusion model that arises from mortality, long distance dispersal, and loss of marks (Turchin 1998). Of the numerous MRR studies to quantify dispersal of tephritid flies, only Plant and Cunningham (1991) have paired empirical data (for sterile *C. capitata*) with a diffusion model with loss terms to assess the movement patterns of a release cohort. The diffusion-convection-settling-mortality model predicted recapture values qualitatively similar to actual average trap captures over time (Plant and Cunningham 1991).

Release and Recapture Rates

To determine population redistribution, MRR studies rely on the release of large numbers of individuals. In general, Eulerian approaches require high source strength to increase the number of recaptures and thereby increase the probability of detecting insects that travel long distances (Nathan et al. 2003; Turchin 1998). This is also evident, although weakly, in the data from tephritid dispersal studies (Fig. 6.2). With the exception of the study by Iwahashi (1972) on *B. dorsalis*, which are the outliers on Fig. 6.2, as source strength increases, so too does mean recapture distance. However, the release of large numbers of insects from a point or small area may bias measured dispersal distances: a higher incidence of long-distance dispersal events may not be due to increased probability of detection but rather an artefact of over-crowding. It has been reported that insect dispersal can be driven by high density as a result of resource depletion or interference by conspecifics (reviewed by Bowler and Benton 2005). The potential for overcrowding at the release point to influence the results of studies on the dispersal of tephritid flies is considerable. It has been suggested that dispersal of some species is related to the availability of food and shelter (Fletcher 1973, 1979). Further, high density is known to lead to reductions in mating performance and survival of males in some

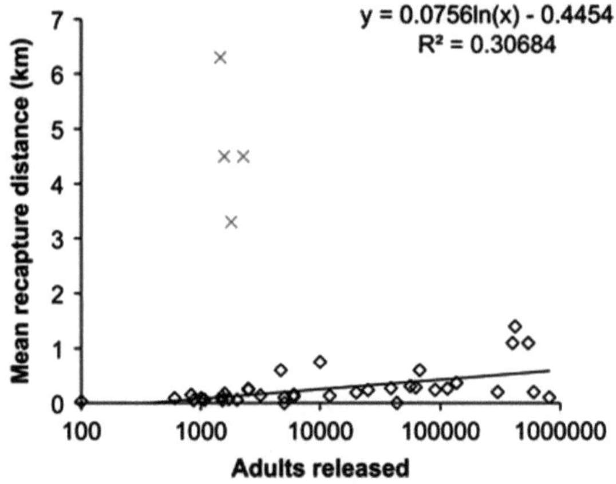

Fig. 6.2 Relationship between the number of adults released in mark-release-recapture studies and mean recapture distance of tephritid flies. The *black line* indicates the logarithmic fit for the data marked with *black diamonds*. *Grey crosses* represent data taken from a study on the inter-island movement of the oriental fruit fly, *Bactrocera dorsalis* (Iwahashi 1972), and are not included in the logarithmic fit

species (Díaz-Fleischer et al. 2009a, b; Gaskin et al. 2002), which may act as a selective pressure to avoid such conditions as would prevail at a release point during dispersal studies or when sterile insects are set loose.

3.1.2 Key Results from MRR Studies

Dispersal Distance

A review of the results of MRR studies is a frustrating exercise. While many studies have been performed (Table 6.1), it is surprising how many of them do not clearly state mean dispersal distance, or its variance, at specific time points. It could be argued that measures of mean and maximum dispersal distance are irrelevant, because the key aspect of dispersal is the shape of the density-distance relationship. However, measures of central tendency and their variability can play an important and easily understood role in defining quarantine distances for the purposes of trade restrictions (Dominiak 2012). From those papers that do report mean dispersal distances, it is evident that *Rhagoletis* are the most sedentary of the genera (mean dispersal distance = 0.03–0.14 km; Table 6.1), although this observation is based exclusively on the results of studies on the movement of the apple maggot, *Rhagoletis pomonella* (Walsh) (Neilson 1971; Phipps and Dirks 1932, 1933). Several studies (Table 6.1) have shown that species of *Ceratitis* do not disperse far from their point of origin. The same can be said for *Anastrepha* spp., because mean and maximum dispersal distances for species in this genus overlap considerably with that of *Ceratitis* spp. (Table 6.1). Species of the genus *Bactrocera* are by far the most mobile of the economically important fruit flies that have been studied (Table 6.1), although as discussed earlier, it is important to interpret these values in relation to the size of trap arrays that have been used, attractancy of traps, release rates, and the time scale of the study.

Influence of Environment and Fly Condition

Factors that have been implicated in variation in dispersal distance of fruit flies include sexual maturity (Fletcher 1973), the availability of fruiting host plants (Drew and Hooper 1983; Drew et al. 1984; Sonleitner and Bateman 1963), which interacts with seasonal changes in temperature and rainfall (Fletcher 1973), and wind (Baker and Chan 1991b; Baker et al. 1986). It has been hypothesized that there is a post-teneral dispersive phase in the life history of female and male *B. tryoni* that probably includes periods of undistracted flight (Fletcher 1973). Fletcher (1973) found that around 75 % of newly emerged *B. tryoni* disappeared from an orchard within 1 week of release and that this rapid decline in abundance could not be attributed to mortality (which was only ~20 % during the same stage). Drew et al. (1984) suggested that arrival of *B. tryoni* and lesser Queensland fruit flies, *Bactrocera neohumeralis* (Hardy), at a rainforest patch could be attributed to post-teneral dispersal of adults that completed their larval development in hosts located approximately 60 km away. Recapture of *B. dorsalis* greater than 10 km from the release point after only one day of release may also be attributed to high activity in post-teneral flies (Froerer et al. 2010).

Facultative dispersal in the absence of oviposition or other resources has been proposed to explain some instances of long-distance dispersal. Steiner et al. (1961) reported that the absence of ripening fruit stimulates dispersal of *C. capitata*. Dispersal of *B. cucurbitae* appears to be related to habitat heterogeneity and suitability of resources, with recaptures linearly related to the capture of resident wild conspecifics and higher mean recapture distance when released in an area unsuitable for this species (Hamada 1980). Fletcher (1973, 1974a) suggested that long distances travelled by marked *B. tryoni* may be related to the absence of fruiting trees near the release point, such that this species continues to disperse even after sexual maturation. Laboratory flight mill studies have verified field observations, showing that tethered flight increased in the absence of fruit prior to testing (Chapman 1982). Conversely, there is evidence that tephritids tend to move into, and remain in, areas containing trees bearing ripe fruit (*B. tryoni*, Bateman and Sonleitner 1967; *B. dorsalis*, Iwahashi 1972; *A obliqua* and *A. suspensa*, Jenkins et al. 2013). Iwaizumi and Shiga (1989) reported that released sterile *B. cucurbitae* moved into areas over time where wild conspecifics were abundant, which presumably indicated areas of high habitat suitability. The tendency for flies to remain in areas with oviposition resources, and higher dispersal than anticipated in the absence of these resources, has also been observed in *B. latifrons* (Peck and

McQuate 2004) and *B. oleae* (Fletcher and Economopoulos 1976; Fletcher and Kapatos 1981; Rempoulakis and Nestel 2012).

The role of wind in dispersal of tephritid flies is still a subject of considerable debate. Very early observations were made of wind direction influencing the initial direction of flight and a tendency for recaptures downwind of release of *C. capitata* (Severin and Hartung 1912). Baker et al. (1986) reported a directional bias in dispersal. The direction of this drift was aligned with the prevailing wind direction, leading Baker et al. (1986) to suggest that, in *C. capitata*, both wind direction and wind strength interact to affect population redistribution. In the same study, however, it was noted that drift was not as apparent in *A. ludens*, which may have been due to these flies being larger than *C. capitata* and exhibiting different activity patterns (Baker et al. 1986). Prevailing wind direction has been associated with remarkable movement distances in *B. cucurbitae* (Kawai et al. 1978) and *B. dorsalis* (Iwahashi 1972), with ordinary wind speeds leading to the recapture of marked flies on islands over 50 km from where they were released. The longest recorded movement of *B. cucurbitae* involved the recapture of a single marked sterile fly on an island 200 km from the release that was presumably transported by cyclonic winds (Miyahara and Kawai 1979). It is important to note for the long-distance movement events reported for *B. cucurbitae* that they followed the release of approximately 300 million sterile flies during an eradication program on Kume Island, Japan (Iwahashi 1977). There is no consensus on the influence of wind direction or speed on the distribution patterns of released *B. tryoni*. Fletcher (1974a, b) found no relation between the direction of prevailing winds and trap recaptures. MacFarlane et al. (1987) found no consistent correlation between strong winds and trap captures; strong south-westerly winds preceded long-distance recoveries in areas north-east of the release point, yet long distance travel was also detected in the absence of strong winds, which indicated multiple means of such dispersal. Conversely, prevailing south-westerly winds with speeds of more than 4 km/h tended to be associated with the recapture of sterile flies in traps to the north and east of their point of release (Dominiak et al. 2003). However, no studies on the dispersal of *B. tryoni* have employed a trap array that would definitively establish the role of wind as a determinant of directional bias in dispersal.

The number of studies that have sought to document the dispersal of sterile tephritid flies underscores the importance of this information for the success of SIT programs. Of the 38 reports summarized in Table 6.1, 21 have involved the release of sterile flies. Sterilization with gamma radiation may have deleterious effects on locomotion, because it can result in mutations that lead to changes in the structure of enzymes and proteins, including those involved with energy metabolism (Allen and Sohal 1982) and neural signal transduction (Haddad et al. 1997). Despite this, sterilized tephritid flies have been recorded moving very large distances (Fletcher 1974a, b; Froerer et al. 2010; MacFarlane et al. 1987), especially when dispersal is aided by wind (Iwahashi 1972; Kawai et al. 1978; Miyahara and Kawai 1979). It is surprising, however, that relatively few studies have directly compared dispersal of wild and sterile tephritid flies. Further, some of these do not control for laboratory-adaptation by simultaneously releasing a mass-reared fertile cohort. In a comparison of wild and mass-reared *B. cucurbitae*, Nakamori and Soemori (1981) found that wild flies were consistently recaptured further away from the release point than their mass-reared counterparts. Wong et al. (1982) reported that wild *C. capitata* were recaptured less and did not move as far as laboratory-reared conspecifics. In the same study, they demonstrated that dispersal declined with increasing radiation dose. In *A. ludens* and *A. obliqua*, however, the dispersal of mass-reared, sterile flies was not significantly different from that of wild flies (Hernández et al. 2007). Weldon and Meats (2010) found no evidence from direct comparison of recaptures within 2 weeks of release that the dispersal distance of sterile *B. tryoni* differs from that of their wild counterparts or a laboratory-reared *white marks* strain. A later study involving the release of much larger numbers, while not using sterile flies, did show that an out-bred laboratory strain of *B. tryoni* dispersed further than the strain mass-reared for a sterile insect technique program (Gilchrist and Meats 2012). Taken together, no overall trends are evident for the effects of mass-rearing or sterilization on tephritid dispersal.

3.1.3 Limitations of Existing Studies and Suggestions for the Future

It is evident from a review of the MRR studies used to measure dispersal of tephritid flies with traps that the design of trap arrays has been influenced more by convenience than the stated aims of these studies. There is a clear relationship between maximum distance sampled by trap arrays and maximum recapture distance (Fig. 6.1b), which suggests that the dispersal capacity of some species may be underestimated. This situation is acceptable if the aim of the study is to determine whether environmental conditions, abiotic or biotic, shape patterns of spatial distribution over time (Baker and Chan 1991a). However, the data required for this purpose are best acquired using a regular grid of traps, which is often not the case (Table 6.1). It is therefore important to stress that the design of trap arrays should be planned carefully to match the aims of future studies on tephritid dispersal that utilize MRR methods. To this end, site selection and appropriate allocation of resources to achieve optimal spacing and placement of traps when resources are limited are essential considerations.

Trap spacing used in MRR studies on the dispersal of tephritid flies is highly variable within and between species (Table 6.1). This is a concern, because little attention has been paid to the effective radius of the traps used. In addition, there is often a lack of consideration given to the potential influence of traps baited with an attractant and located close to the release point on the pattern of dispersal. As discussed earlier, strong traps (e.g., those baited with methyl eugenol) when located close to the release point will likely capture a high proportion of marked individuals, arrest movement of responsive males away from the release point, and consequently reduce the number of insects that would otherwise be caught in more distant traps (Turchin 1998). In many cases, there are few published data on the effective radius of traps, which contributes to differences and potential confusion when designing trap arrays. For example, while a regular grid of traps baited with cue-lure and spaced 400 m apart will recapture 8–9 % of sexually mature male *B. tryoni* (Fletcher 1974b; Monro and Richardson 1969), the average distance over which a single trap is attractive for this species has not been determined. This type of information has been determined for *B. cucurbitae* responding to cue-lure and *B. dorsalis* responding to methyl eugenol in Hawaii (Shelly and Edu 2010; Shelly et al. 2010), and similar data would benefit the design of trap arrays for future studies on tephritid dispersal.

Not surprisingly, most attention has been paid to the dispersal of tephritid species of economic importance. This is particularly the case where a species is regarded a severe quarantine concern by importing countries or regions where the species is absent. Good examples of these are *B. tryoni* and *C. capitata*, which have been more extensively studied than any other species (Table 6.1). The level of interest in dispersal of *B. tryoni* is particularly notable considering that it is present only in Australia and several Pacific islands. This research has been driven by the desire of Australian producers and administrative bodies to simplify quarantine conditions imposed by importing nations and to ensure that these conditions are evidence-based (Dominiak 2012). It is disconcerting, however, that very little research on dispersal has been undertaken (or at least reported in the English literature) on tephritids of economic significance from developing economies in Africa, Asia, and Central and South America. These regions are home to a very large contingent of species in the genera *Anastrepha*, *Bactrocera*, *Ceratitis*, *Dacus*, and *Toxotrypana* that represent considerable biosecurity risks for importing countries. However, these regions also have much to gain in terms of economic and social development from the introduction of area-wide integrated pest management plans for these pests. Dispersal of tephritid species in developing countries represents an important first step in devising strategies to protect pest-free areas and to limit the spread of invasive fruit fly species, which can then increase production and capacity to export to lucrative international markets.

3.2 Molecular Techniques

The limitations of directly measuring movement (such as MRR), coupled with the development of molecular technologies, have resulted in the increased use of molecular methods to infer dispersal of individuals in natural populations (Raybould et al. 2001; Slatkin 1985; Whitlock and McCauley 1999). Molecular marker data are often used to resolve population structure, which describes interrelatedness among groups of interbreeding individuals (populations) by revealing the extent of effective gene flow, while accounting for processes such as genetic drift, natural selection, and mutation (Bohonak 1999). Consequently, likely movement patterns within the metapopulation can be inferred (e.g., sources and sinks or range expansions and local extinctions) as gene flow is predicted, and often demonstrated, to be positively correlated with dispersal rate (i.e., greater gene flow means more dispersal) (Bohonak 1999; Peterson and Denno 1998). As a result, these techniques are called 'indirect' measures of dispersal as they are inferred from molecular data rather than having been directly measured *via* tracking of individuals in the environment (Osborne et al. 2002; Slatkin 1985).

Elucidating changes in gene flow dynamics across different temporal scales is often difficult and represents one of the major challenges (and criticisms) in using molecular data to infer dispersal (Berry et al. 2004; Bossart and Prowell 1998; Pavlacky et al. 2009; Vandewoestijne and Baguette 2004). Indeed, whereas direct methods of assessing dispersal track individuals over one or two generations, indirect molecular techniques resolve population genetic structure that can be averaged over thousands of generations and may therefore be of limited contemporary ecological value (Bohonak 1999). Although it is unclear whether studies using estimates of indirect measurements of dispersal *via* gene flow can be directly compared with those from direct measurements using MRR (Bohonak 1999), it is often shown that those from indirect estimates are higher (Koenig et al. 1996). For example, Karsten et al. (2013) showed that *C. capitata* disperse far greater distances (possibly human-mediated) based on estimates from molecular markers than what have previously been shown using MRR (Meats and Smallridge 2007). Possible reasons for the discrepancy between MRR and molecular marker estimates include dispersal in life-stages not measured by MRR studies (larvae or eggs) because most MRR studies involve the release of marked adult flies from a central point that are then recaptured. Also, estimates of effective dispersal (migration and successful reproduction) using molecular markers disregard the mode of dispersal (Broquet and Petit 2009), whereas in MRR studies, human-mediated dispersal is less likely to play a role in estimates of dispersal as trap arrays do not typically cover vast distances.

Here, we will briefly discuss some of the most important considerations when designing a population genetic study that incorporates estimates of dispersal *via* gene flow. We will present information on the numbers of markers or individuals to include and protocols for sample collection and preservation, describe available markers, and outline emerging statistics based on assignment tests that can be used to analyse the data. Finally, we touch on some new developments in genetics that have implications for the study of insect movement.

3.2.1 Sampling Individuals

The design of molecular studies can have a marked influence on detection of gene flow in a study system. Therefore, careful consideration should be given to the sampling regime used in data collection (for sampling regimes see Storfer et al. 2007; Broquet and Petit 2009). It is often easy to sample high numbers of pest tephritids as they can occur in high numbers. The available sample size for analysis would be considerably lower, however, if species have a limited geographical distribution or host range and are difficult to obtain. Despite the number of individuals sampled, however, it is quite often the case that a research budget allows sequencing or genotyping of only a set number of individuals. Under these circumstances, to gain as much knowledge from a study system as possible a trade-off must be made between the number of individuals used and the number of markers used. Studies have shown that most of the time it is better to include more markers rather than more individuals to increase the power of analyses (Landguth et al. 2012), and this guideline is also likely to apply where sample size is low.

Tephritid flies are often collected for molecular studies by the use of traps or from the rearing of infested fruit. The use of traps, however, may raise challenges for molecular studies, because the preservation of sampled individuals can affect the quality of genetic data. There have been a number of studies investigating long-term storage of samples for genetic work, and the proper preservation of samples substantially increases the chances of amplification (Dawson et al. 1998; Murphy et al. 2002). One of the easiest ways to store samples for future genetic work is to place flies in 95 % (or higher) alcohol in airtight vials. Vials should be checked regularly as alcohol does evaporate quite easily. These vials can also be frozen in a −80 °C freezer although it is not essential. If substantial numbers of samples were not properly stored, it is still possible in some instances to extract DNA from these samples. For example, samples stored dry for 5 years can still be amplified, but the rate of amplification failure is much higher than for fresh samples or samples stored in alcohol.

3.2.2 Molecular Markers

A range of molecular markers (Table 6.2) are available for resolving gene flow (and inferring movement), and they can be broadly categorized into one of two groups: nuclear genomic markers (nDNA, including allozymes) or non-recombinant mitochondrial DNA markers (mtDNA) (Osborne et al. 2002). Not all markers provide the same information, however, as mutation rates vary from relatively slow (e.g., allozymes) to relatively fast (e.g., microsatellites). Therefore, different markers are better suited to resolving historical versus contemporary patterns of dispersal. Markers are discussed below in order of increasing mutation rate.

Allozymes

Allozymes (proteins encoded by genes and hence useful as Mendelian markers) have long been used to measure gene flow and represent one of the most commonly used molecular markers for assessing population structure of insects (Loxdale and Lushai 2001). However, having a relatively slow rate of evolution (10^{-6}–10^{-9}/gene/generation), they are best suited to resolving historical population events (e.g., vicariance) rather than contemporary dispersal patterns (Loxdale and Lushai 2001; Whitlock and McCauley 1999). Consequently, allozyme data have often been used for assessing broader questions relating to population displacement rather than recent dispersal *per se*. In the gall-forming tephritid, *Urophora cardui* (L.), for example, allozyme data revealed high levels of gene flow in an assessment of biogeographic population displacement since Pleistocene glaciation in Europe

Table 6.2 Summary of molecular markers characteristically used in studies of population structure of tephritids

Marker type	Acronym	Variability	Application
Organellar			
Mitochondrial DNA	mtDNA	Low	Only maternal lineage, phylogenies, phylogeography and population genetic
Nuclear			
Amplified fragment length polymorphism	AFLP	High	Linkage mapping, population genetic
Allozymes		Low	Linkage mapping, population genetic
Microsatellites	SSR	High	Linkage mapping, population genetic, parentage analysis

Adapted from Parker et al. (1998), Le Roux and Wieczorek (2009)

(Eber and Brandl 1994, 1997). These historical data corroborated direct dispersal measures (presence/absence of galls in the field) and supported the hypothesis that *U. cardui* is a highly mobile species with at least 1 % of individuals dispersing up to 8 km with rapid range expansion up to almost 7 km/year (Jannson 1992). Yet, more recent dispersal estimates have been made among populations of *C. capitata*, where a high proportion of alleles were shared among individuals from different geographic regions, indicating more active dispersal than estimated by earlier MRR studies (Kourti 2004).

Mitochondrial DNA (mtDNA)

Continued advances in genetic technologies have stimulated the use of direct DNA sequence analysis over allozyme studies (Bossart and Prowell 1998). One such marker, mtDNA, often evolves at faster rates than allozymes and may be better suited to the detection of more contemporary dispersal (Bossart and Prowell 1998; Loxdale and Lushai 2001). Further, mtDNA is maternally inherited, non-recombinant, and abundant in the flight muscles of winged insects, such as fruit flies (Osborne et al. 2002). These characteristics have led to its wide use in measuring gene flow (but see Galtier et al. 2009). While faster evolving than allozymes, patterns of genetic connectivity also reflect historical processes rather than present-day dispersal activities, and as different regions of the mtDNA genome evolve at different rates, specific genomic data will be more (or less) useful depending on the question being asked (Zhang and Hewitt 1997). In the case of *B. oleae*, for example, mtDNA data (NADH dehydrogenase subunit I) resolved historical movement patterns, information which was coupled with microsatellite data to describe recent dispersal patterns, including identification of likely sources of range expansions (Nardi et al. 2005).

Microsatellites

Microsatellites (tandem repeat sequences in the nuclear genome) have been the preferred marker for detecting recent population displacement and consequently dispersal, despite the drawback of high specificity (Vandewoestijne and Baguette 2004, but see Baliraine et al. 2003 for application of *C. capitata* primers across other *Ceratitis* species). This is because microsatellites evolve two to three orders of magnitude faster than allozymes (Balloux and Lugon-Moulin 2002; Loxdale and Lushai 2001) and hence can resolve recent changes in population structure that may reflect contemporary dispersal patterns. Furthermore, microsatellites are abundant in the genome, easily scored, and have high levels of polymorphism (Bruford and Wayne 1993). In a study of *A. suspensa* (Boykin et al. 2010), microsatellite variation revealed gene flow between Florida and Caribbean populations from which it was inferred that frequent movement (presumably human mediated) occurred between these two regions. This contradicted earlier suggestions that *A. suspensa* was distributed in the region as the result of only a few introduction events. Much broader scale movement patterns were similarly examined using microsatellites for the widely distributed pest, *B. dorsalis*, and migration estimates indicated a predominantly unidirectional dispersal of flies from mainland China to Taiwan and southeast Asia (Aketarawong et al. 2007). This provided evidence that southern China was the original source of *B. dorsalis* (Aketarawong et al. 2007), a hypothesis further supported by mtDNA studies (Schutze et al. 2012).

3.2.3 Assignment Tests

Unlike the frequency-based methods outlined above which are often better suited to measuring historical dispersal, assignment methods are a promising approach for revealing contemporary movement of individuals to within a few generations. These approaches use genetic information, such as microsatellite data, to identify the likely original source population of an individual based on expected probabilities of that individual's multilocus genotype occurring in a range of potential sources (Manel et al. 2005). This approach, therefore, allows for a more *direct* measure of dispersal unlike indirect estimates based on gene flow. One of the drawbacks of this approach, however, is that all potential source populations from which an individual may have dispersed must be identified in advance – a situation difficult to achieve and resulting in assignment tests often being conducted on species that rarely disperse under natural conditions, such as identifying the origin of smuggled animals or tracing the translocations of endangered species (Berry et al. 2004; Manel et al. 2005). Further, demarcation of geographical populations is challenging for species that occur over continuous environments. Bayesian clustering methods (e.g., Pritchard et al. 2000) provide a way around this problem, yet such approaches may encounter their own problems when genetic differentiation among subpopulations is very low, as the performance of these approaches, specifically their ability to pinpoint a source location, relies on structure within the metapopulation (Latch et al. 2006). Two examples involving such analyses include that of *B. dorsalis*, in which no population structure was evident in Thailand following examination of 10 microsatellite loci and hence active dispersal (probably human mediated) is presumed ongoing (Krosch et al. 2013); and *B. invadens*, where multiple distinct populations were resolved in Africa following Bayesian cluster analysis on 11 microsatellite loci, and Sri Lanka was identified as part of the native range and the possible origin of this species (albeit there being few Sri Lankan genotypes found in Africa, Khamis et al. 2009). The ability to pinpoint the source population of a new invasive pest is important, for example, to identify natural enemies present in the native range of the species for use in integrated pest management programs (Kirk et al. 2013). One such example is *B. oleae*, which is now widespread in the Mediterranean region and has a native range believed to be in Africa or Western Asia. Parasitoid diversity of this species in the Mediterranean area has been shown to be unspecialized and low in contrast with those found in the proposed native range (Hoelmer et al. 2011). The parasitoids found in the native range can therefore be screened for possible successful biological control agents.

Clearly, molecular techniques have come a long way in the past 70 years since the early days of allozyme electrophoresis, allowing resolution of population structure and hence inference regarding modes of dispersal in organisms, like tephritids. The future of molecular markers will most definitely be influenced by the wave of Next Generation Sequencing (NGS) technologies that are being developed. Understandably, the costs incurred are substantial and may prove unaffordable to some labs, although companies that offer these services have packages that have been scaled down to attempt to address this (Taylor and Harris 2012). The use of NGS allows the user to characterize the transcriptome, perform gene expression profiling, find candidate genes, or sequence the entire genome. This, in turn, helps to develop large numbers of molecular markers, including SNPs (single nucleotide polymorphisms) and microsatellites (Ekblom and Galindo 2010). Despite these advances, however, it is worth remembering the limitations of these approaches, particularly the assumptions made with respect to population structure (often overly simplistic), the temporal scales over which markers are sensitive, and the confounding effects of other evolutionary factors, such as genetic drift and selection. Furthermore, although lure-based trapping (e.g., methyl eugenol and cue lure) allows easy collection of males (Tan et al., Chap. 2, this volume) and subsequent focus on a single sex, the importance of sex-biased dispersal must also be appreciated under such circumstances (Prugnolle and de Meeus 2002). Despite these limitations, however, molecular techniques represent an ongoing and valuable tool towards resolving movement patterns in tephritids and complement traditional direct approaches.

3.3 Remote Sensing and Computer-Based Methods

Existing methods for monitoring the movement of tephritid flies, whether direct (i.e., MRR) or indirect (i.e., molecular techniques), rely on trapping, and thus usually involve permanent removal of part of the population. By doing so, it is not possible to measure the potential future movement, resource use, and survival of those individuals, which prevents the generation of data that can be used to definitively ascribe a mechanism for observed patterns of spatial redistribution (Nathan et al. 2003). This shortcoming of field studies of insect movement is beginning to be addressed as a consequence of research and development of a range of methods to recognize or track free-living individuals in the environment. In the context of studying the movement of tephritid flies, which represent a considerable challenge to track individually due to their relatively small size and mobility, three technologies have to date received some attention: harmonic radar, machine vision, and radio-frequency identification tags.

Harmonic radar utilizes transponders fitted to an object (e.g., an animal) that respond to a high frequency microwave emission by immediately emitting a pulse on another (harmonic) frequency. The radar carries a narrow-band receiver tuned to the shifted frequency and so detects the target. Harmonic radar can be used at low altitudes, because it overcomes the problem of strong radar echoes reflected off ground features, such as buildings and vegetation (Reynolds and Riley 2002), and successfully detects tagged individuals in low row crops, tall row crops, and tall but well-separated orchard trees (Boiteau et al. 2011). The reply pulse can also carry a code identifying the target, even giving its altitude (Riley and Smith 2002). This technique monitors uninterrupted movement of tagged individuals throughout a landscape without the need to trap them after release, meaning that all released individuals will be tracked unless tags fail or are lost after release or the insect moves beyond the range of the transmitter (Boiteau et al. 2011).

Developments in microelectronics enable the use of harmonic radar to track the movement of small insects in simplified landscapes by fitting them with passive, lightweight transponder tags. The weight of some transponder tags used to study insect movements ranges between 0.6 and 12 mg, which represents between 2.2 and 14.7 % of the mean body weight of the study species (Boiteau et al. 2010; Cant et al. 2005; Capaldi et al. 2000; Gui et al. 2011; Riley and Smith 2002). The use of harmonic radar to track movement of small insects is still in its infancy and requires further development and optimization to minimize the effects of tags on flight and other behavior (e.g., Boiteau et al. 2011; Gui et al. 2011). However, harmonic radar offers the opportunity to adopt a Lagrangian approach to the study of tephritid movement and begin to understand how individual movement behavior and resource utilization influence population-level patterns of dispersal. Harmonic radar has been used successfully to monitor movement of walking Colorado potato beetles, *Leptinotarsa decemlineata* (Say) (Gui et al. 2012) and flight of honeybees, *Apis mellifera* L. (Capaldi et al. 2000; Reynolds et al. 2007), and some butterflies (Cant et al. 2005; Ovaskainen et al. 2008). Transponder tags weighing 3.5–3.8 mg have been developed for the Chinese fruit fly, *Bactrocera minax* (Enderlein), which represents 8 % of average fly weight (Gui et al. 2011). Tags with a weight of 3.8 mg had no significant effect on flight propensity, flight duration, feeding or longevity, suggesting that use of harmonic radar may be a valid method for monitoring flight of this tephritid (Gui et al. 2011). At the time of writing, however, no field trials to track flight of *B. minax* using harmonic radar had been conducted.

Computer vision is the automated processing and recognition of subjects from digital images. This technology offers the possibility of replacing traps: insects crossing the field of view of an image sensor could be photographed, transferred to a computer, identified by automatic image processing techniques, and the results transmitted back to the laboratory (Reynolds and Riley 2002). This method has already been developed to count the number of *B. cucurbitae* visiting a cue-lure dispenser (Manoukis and Jang 2013). It is also possible to program software to recognize *B. tryoni* in images from a sensor within a trap baited with cue-lure based on shape, pattern, and color (Liu et al 2009). Given the ability to recognize pattern and color using machine vision techniques, it may be possible in the future to monitor the movements of fruit flies individually marked with, for example, microdots (Whitehead and Peakall 2012).

Radio frequency identification (RFID) technology, like harmonic radar, relies on the attachment of a transponder tag to individuals being studied. RFID tags reply with a coded signal when stimulated by a radio signal received from a 'scanner', and this allows insects to be uniquely identified (Reynolds and Riley 2002). A major limitation of this system is that scanners have a reading distance of approximately 3–4 mm (Schneider et al. 2012; Streit et al. 2003), which necessitates the movement of tagged individuals through narrow channels at a correct orientation with regard to the scanner. RFID is providing new insight into social insect behavior (e.g., Schneider et al. 2012), but it remains to be seen whether the logistical challenges associated with its use will permit its use with tephritid flies.

Both computer vision and RFID technology offer interesting opportunities for studying the movement of tephritid flies in place of traps. It is important, however, to recognize their inherent limitations. Earlier in this chapter we highlighted that the use of traps to study tephritid movement is limited by several issues, such as marking method, the design of arrays with regard to the properties of traps, and maximum sampling distance. The use of computer vision and RFID as currently envisaged does not overcome any of these concerns: it will be necessary to maximize interception of tagged flies by carefully considering the spatial arrangement of image sensors or scanners and with the use of chemical and/or visual cues.

4 Conclusions

Traps form an essential role in monitoring movement of tephritid flies. They enable MRR studies to determine long distance dispersal as well as more localized patterns of population redistribution following release of marked flies from a point. These studies need to be designed carefully to address a number of potential limitations, but they are still required for a large number of species of economic concern, particularly from the developing world. Traps are also necessary for the collection of specimens used in molecular methods to determine 'effective dispersal', which leads to population connectivity and gene flow. It is, however, necessary to recognize the limitations of traps as a means of monitoring tephritid movement and be open to the possibility of using new technologies that are being developed to track individual insect movement.

References

Allen RG, Sohal RS (1982) Life-lengthening effects of gamma-radiation on the adult housefly, Musca domestica. Mech Ageing Dev 20:369–375

Aketarawong N, Bonizzoni M, Thanaphum S, Gomulski LM, Gasperi G, Malacrida AR, Gugliemino CR (2007) Inferences on the population structure and colonization process of the invasive oriental fruit fly, *Bactrocera dorsalis* (Hendel). Mol Ecol 16:3522–3532

Aluja M, Prokopy RJ (1992) Host search behavior by *Rhagoletis pomonella* flies – inter-tree movement patterns in response to wind-borne fruit volatiles under field conditions. Physiol Entomol 17:1–8

Andress E, Walters I, del Toro M, Shelly T (2013) Release-recapture of sterile male Mediterranean fruit flies (Diptera: Tephritidae) in southern California. Proc Hawaii Entomol Soc 45:11–29

Arévalo HA, Collins J, Groden E, Drummond F, Simon K (2009) Marking bluberry maggot flies (Diptera: Tephritidae) with fluorescent diet for recapture studies. Fla Entomol 92:379–381

Baker PS, Chan AST (1991a) Appetitive dispersal of sterile fruit-flies – aspects of the methodology and analysis of trapping studies. Zeitschrift für Angewandte Entomologie 112:263–273

Baker PS, Chan AST (1991b) Quantification of tephritid fruit-fly dispersal – guidelines for a sterile release program. Zeitschrift für Angewandte Entomologie 112:410–421

Baker PS, Chan AST, Jimeno Zavala MA (1986) Dispersal and orientation of sterile *Ceratitis capitata* and *Anastrepha ludens* (Tephritidae) in Chiapas, Mexico. J Appl Ecol 23:27–38

Baliraine FN, Bonizzoni M, Osir EO, Lux SA, Mulaa FJ, Zheng L, Gomulski LM, Gasperi G, Malacrida AR (2003) Comparative analysis of microsatellite loci in four fruit fly species of the genus *Ceratitis* (Diptera: Tephritidae). Bull Entomol Res 93:1–10

Balloux F, Lugon-Moulin N (2002) The estimation of population differentiation with microsatellite markers. Mol Ecol 11:155–165

Banks HT, Kareiva PM, Lamm PK (1985) Modelling insect dispersal and estimating parameters when mark-release techniques may cause initial disturbances. J Math Biol 22:259–277

Barnes MM (1959) Radiotracer labelling of a natural tephritid population and flight range of the walnut husk fly. Ann Entomol Soc Am 52:90–92

Barron AB (2000) Anaesthetising *Drosophila* for behavioural studies. J Insect Physiol 46:439–442

Barry JD, Dowell RV, Morse JG (2002) Comparison of two sterile Mediterranean fruit fly (Diptera: Tephritidae) strains released in California's preventative release program. J Econ Entomol 95: 936–944

Bateman MA, Sonleitner FJ (1967) The ecology of a natural population of the Queensland fruit fly, *Dacus tryoni*. I. The parameters of the pupal and adult populations during a single season. Aust J Zool 15:303–335

Berry O, Tocher MD, Sarre SD (2004) Can assignment tests measure dispersal? Mol Ecol 13:551–561

Bloem KA, Bloem S, Chambers DL (1994) Field assessment of quality – release recapture of mass-reared Mediterranean fruit-flies (Diptera, Tephritidae) of different sizes. Environ Entomol 23: 629–633

Bohonak AJ (1999) Gene flow and population structure. Q Rev Biol 74:21–45

Boiteau G, Vincent C, Meloche F, Leskey TC (2010) Harmonic radar: assessing the impact of tag weight on walking activity of Colorado potato beetle, plum curculio, and western corn rootworm. J Econ Entomol 103:63–69

Boiteau G, Vincent C, Meloche F, Leskey T, Colpitts BG (2011) Harmonic radar: efficacy at detecting and recovering insects on agricultural host plants. Pest Manag Sci 67:213–219

Bossart JL, Prowell DP (1998) Genetic estimates of population structure and gene flow: limitations, lessons and new directions. Trends Ecol Evol 13:202–206

Bowler DE, Benton TG (2005) Causes and consequences of animal dispersal strategies: relating individual behaviour to spatial dynamics. Biol Rev 80:205–225

Boykin LM, Shatters RG Jr, Hall DG, Dean D, Beerli P (2010) Genetic variation of *Anastrepha suspensa* (Diptera: Tephritidae) in Florida and the Caribbean using microsatellite DNA markers. J Econ Entomol 103:2214–2222

Broquet T, Petit EJ (2009) Molecular estimation of dispersal for ecology and population genetics. Annu Rev Ecol Evol Syst 40:193–216

Bruford MW, Wayne RK (1993) Microsatellites and their application to population genetic studies. Curr Opin Genet Dev 3:939–943

Campbell AJ, Lynch AJ, Dominiak B, Nicol H (2009) Effects of radiation, dye, day of larval hopping and vibration on eclosion of Queensland fruit fly, *Bactrocera tryoni* (Froggatt) (Diptera: Tephritidae). Gen Appl Entomol 38:49–53

Cant ET, Smith AD, Reynolds DR, Osborne JL (2005) Tracking butterfly flight paths across the landscape with harmonic radar. Proc R Soc B 272:785–790

Capaldi EA, Smith AD, Osborne JL, Fahrbach SE, Farris SM, Reynolds DR, Edwards AS, Martin A, Robinson GE, Poppy GM, Riley JR (2000) Ontogeny of orientation flight in the honeybee revealed by harmonic radar. Nature 403:537–540

Champion de Crespigny F, Wedell N (2008) The impact of anaesthetic technique on survival and fertility in *Drosophila*. Physiol Entomol 33:310–315

Chapman MG (1982) Experimental analysis of the pattern of tethered flight in the Queensland fruit fly, *Dacus tryoni*. Physiol Entomol 7:143–150

Clarke AR, Powell KS, Weldon CW, Taylor PW (2011) The ecology of *Bactrocera tryoni* (Diptera: Tephritidae): what do we know to assist pest management? Ann Appl Biol 158:26–54

Cunningham RT, Couey HM (1986) Mediterranean fruit fly (Diptera: Tephritidae): distance/ response curves to trimedlure to measure trapping efficiency. Environ Entomol 15:71–74

Dawson MN, Raskoff KA, Jacobs DK (1998) Field preservation of marine invertebrate tissue for DNA analyses. Mol Mar Biol Biotechnol 7:145–152

Dempster JP, Atkinson DA, French MC (1995) The spatial population-dynamics of insects exploiting a patchy food resource. 2. Movements between patches. Oecologia 104:354–362

Díaz-Fleischer F, Arredondo J, Aluja M (2009a) Enriching early adult environment affects the copulation behaviour of a tephritid fly. J Exp Biol 212:2120–2127

Díaz-Fleischer F, Arredondo J, Flores S, Montoya P, Aluja M (2009b) There is no magic fruit fly trap: multiple biological factors influence the response of adult *Anastrepha ludens* and *Anastrepha obliqua* (Diptera: Tephritidae) individuals to MultiLure traps baited with BioLure or NuLure. J Econ Entomol 102:86–94

Dominiak B (2012) Review of dispersal, survival, and establishment of *Bactrocera tryoni* (Diptera: Tephritidae) for quarantine purposes. Ann Entomol Soc Am 105:434–446

Dominiak BC, Schinagl L, Nicol H (2000) Impact of fluorescent marker dyes on emergence of sterile Queensland fruit fly, *Bactrocera tryoni* (Froggatt) (Diptera: Tephritidae). Gen Appl Entomol 29:45–47

Dominiak BC, Westcott AE, Barchia IM (2003) Release of sterile Queensland fruit fly, *Bactrocera tryoni* (Froggatt) (Diptera: Tephritidae), at Sydney, Australia. Aust J Exp Agric 43:519–528

Dominiak BC, Sundaralingam S, Jiang L, Jessup AJ, Barchia IM (2010) Impact of marker dye on adult eclosion and flight ability of mass produced Queensland fruit fly *Bactrocera tryoni* (Froggatt) (Diptera: Tephritidae). Aust J Entomol 49:166–169

Dominiak BC, Campbell AJ, Worsley P, Nicol HI (2011) Evaluation of three ground release methods for sterile Queensland fruit fly *Bactrocera tryoni* (Froggatt) (Diptera: Tephritidae). Crop Prot 30:1541–1545

Drew RAI, Hooper GHS (1983) Population studies of fruit flies (Diptera: Tephritidae) in southeast Queensland (Australia). Oecologia 56:153–159

Drew RAI, Zalucki MP, Hooper GHS (1984) Ecological studies of Eastern Australian fruit flies (Diptera: Tephritidae) in their endemic habitat. I. Temporal variation in abundance. Oecologia 64:267–272
Eber S, Brandl R (1994) Ecological and genetic spatial patterns of *Urophora cardui* (Diptera: Tephritidae) as evidence for population structure and biogeographical processes. J Anim Ecol 63:187–199
Eber S, Brandl R (1996) Metapopulation dynamics of the tephritid fly *Urophora cardui*: an evaluation of incidence-function model assumptions with field data. J Anim Ecol 65
Eber S, Brandl R (1997) Genetic differentiation of the tephritid fly *Urophora cardui* in Europe as evidence for its biogeographical history. Mol Ecol 6:651–660
Ekblom R, Galindo J (2010) Applications of next generation sequencing in molecular ecology of non-model organisms. Heredity 107:1–15
FAO (Food and Agriculture Organization) (2006) Establishment of pest free areas for fruit flies (Tephritidae). ISPM No. 26: 15. FAO, Rome
FAO/IAEA/USDA (Food and Agriculture Organization/International Atomic Energy Agency/ United States Department of Agriculture) (2003) Manual for product quality control and shipping procedures for sterile mass-reared tephritid fruit flies, Version 5.0. International Atomic Energy Agency, Vienna
Fletcher BS (1973) The ecology of a natural population of the Queensland fruit fly, *Dacus tryoni*. IV. The immigration and emigration of adults. Aust J Zool 21:541–565
Fletcher BS (1974a) The ecology of a natural population of the Queensland fruit fly, *Dacus tryoni*. V. The dispersal of adults. Aust J Zool 21:541–565
Fletcher BS (1974b) The ecology of a natural population of the Queensland fruit fly, *Dacus tryoni*. VI. Seasonal changes in fruit fly numbers in the areas surrounding the orchard. Aust J Zool 22:353–363
Fletcher BS (1979) The overwintering survival of adults of the Queensland fruit fly, *Dacus tryoni*, under natural conditions. Aust J Zool 27:403–412
Fletcher BS, Economopoulos AP (1976) Dispersal of normal and irradiated laboratory strains and wild strains of the olive fly *Dacus oleae* in an olive grove. Entomol Exp Appl 20:183–194
Fletcher BS, Kapatos E (1981) Dispersal of the olive fly, *Dacus oleae*, during the summer period on Corfu. Entomol Exp Appl 29:1–8
Franzén M, Nilsson SG (2007) What is the required minimum landscape size for dispersal studies? J Anim Ecol 76:1224–1230
Froerer KM, Peck SL, McQuate GT, Vargas RI, Jang EB, McInnis DO (2010) Long-distance movement of *Bactrocera dorsalis* (Diptera: Tephritidae) in Puna, Hawaii: how far can they go? Am Entomol 56:88–95
Froerer KM, Peck SL, McQuate GT (2011) Evaluation of readmission ink as a marker for dispersal studies with the Oriental fruit fly, *Bactrocera dorsalis*. J Insect Sci 11:125
Galtier N, Nabholz B, Glémin S, Hurst GDD (2009) Mitochondrial DNA as a marker of molecular diversity: a reappraisal. Mol Ecol 18:4541–4550
Gaskin T, Futerman P, Chapman T (2002) Increased density and male-male interactions reduce male longevity in the medfly, *Ceratitis capitata*. Anim Behav 63:121–129
Gilchrist AS, Meats AW (2012) Factors affecting the dispersal of large-scale releases of the Queensland fruit fly, *Bactrocera tryoni*. J Appl Entomol 136:252–262
Gui L-Y, Xiu-Qin H, Chuan-Ren L, Boiteau G (2011) Validation of harmonic radar tags to study movement of Chinese citrus fly. Can Entomol 143:415–422
Gui L-Y, Boiteau G, Colpitts BG, MacKinley P, McCarthy PC (2012) Random movement pattern of fed and unfed adult Colorado potato beetles in bare-ground habitat. Agric Forest Entomol 14:59–68
Haddad GG, Sun Y, Wyman RJ, Xu T (1997) Genetic basis of tolerance to O2 deprivation in Drosophila melanogaster. Proc Natl Acad Sci 94:10809–10812
Hagler JR (1997) Field retention of a novel mark-release-recapture method. Environ Entomol 26: 1079–1086
Hagler JR, Jackson CG (2001) Methods for marking insects: current techniques and future prospects. Annu Rev Entomol 46:511–543

Hagler JR, Miller E (2002) An alternative to conventional insect marking procedures: detection of a protein mark on pink bollworm by ELISA. Entomol Exp Appl 103:1–9

Hagler JR, Cohen AC, Bradley-Dunlop D, Enriques FJ (1992) New approach to mark insects for feeding and dispersal studies. Environ Entomol 21:20–25

Halley JM, Dempster JP (1996) The spatial population dynamics of insects exploiting a patchy food resource: a model study of local persistence. J Appl Ecol 33:439–454

Hamada R (1980) Studies on the dispersal behavior of melon flies, *Dacus cucurbitae* Coquillett (Diptera: Tephritidae), and the influence of gamma-irradiation on dispersal. Appl Entomol Zool 15:363–371

Hassall C, Thompson DJ (2012) Study design and mark-recapture estimates of dispersal: a case study with the endangered damselfly *Coenagrion mercuriale*. J Insect Conserv 16:111–120

Hernández E, Orozco D, Breceda SF, Dominguez J (2007) Dispersal and longevity of wild and mass-reared *Anastrepha ludens* and *Anastrepha obliqua* (Diptera : Tephritidae). Fla Entomol 90:123–135

Hoelmer KA, Kirk AA, Pickett CH, Daane KM, Johnson MW (2011) Prospects for improving biological control of olive fruit fly, *Bactrocera oleae* (Diptera: Tephritidae), with introduced parasitoids (Hymenoptera). Biocontrol Sci Technol 21:1005–1025

Iwahashi O (1972) Movement of the Oriental fruit fly adults among islets of the Ogasawara Islands. Environ Entomol 1:176–179

Iwahashi O (1977) Eradication of the melon fly, *Dacus cucurbitae*, from Kume Is., Okinawa with the sterile insect release method. Res Popul Ecol 19:87–98

Iwaizumi R, Shiga M (1989) Spatial relationship between wild and released sterile melon flies, *Dacus cucurbitae* Coquillett (Diptera: Tephritidae). Appl Entomol Zool 24:147–149

Jannson A (1992) Distribution and dispersal of *Urophora cardui* (Diptera: Tephritidae) in Finland in 1985–1991. Entomologia Fennica 2:211–216

Jenkins DA, Kendra PE, van Bloem S, Whitmire S, Mizell R III, Goenaga R (2013) Forest fragments as barriers to fruit fly dispersal: *Anastrepha* (Diptera: Tephritidae) populations in orchards and adjacent forest fragments in Puerto Rico. Environ Entomol 42:283–292

Jones SC, Wallace L (1955) Cherry fruit fly dispersion studies. J Econ Entomol 48:616

Jones VP, Hagler JR, Brunner JF, Baker CC, Wilburn TD (2006) An inexpensive immunomarking technique for studying movement patterns of naturally occurring insect populations. Environ Entomol 35:827–836

Karsten M, van Vuuren BJ, Barnaud A, Terblanche JS (2013) Population genetics of *Ceratitis capitata* in South Africa: implications for dispersal and pest management. PLoS One 8:e54281

Kawai A, Iwahashi O, Itô Y (1978) Movement of the sterilized melon fly from Kume Is. to adjacent islets. Appl Entomol Zool 13:314–315

Kendra PE, Epsky ND, Heath RR (2010) Effective sampling range of food-based attractants for female *Anastrepha suspensa* (Diptera: Tephritidae). J Econ Entomol 103:533–540

Khamis FM, Karam N, Ekesi S, De Meyer M, Bonomi A, Gomulski LM, Scolari F, Gabrieli P, Siciliano P, Masiga D, Kenya EU, Gasperi G, Malacrida AR, Guglielmino CR (2009) Uncovering the tracks of a recent and rapid invasion: the case of the fruit fly pest *Bactrocera invadens* (Diptera: Tephritidae) in Africa. Mol Ecol 18:4798–4810

Kirk H, Dorn S, Mazzi D (2013) Molecular genetics and genomics generate new insights into invertebrate pest invasions. Evol Appl 6:842–856

Knipling EF (1959) Sterile-male method of population control. Science 130:902–904

Koenig WD, Van Vuren D, Hooge PN (1996) Detectability, philopatry, and the distribution of dispersal distances in vertebrates. Trends Ecol Evol 11:514–517

Kourti A (2004) Estimates of gene flow from rare alleles in natural populations of medfly *Ceratitis capitata* (Diptera: Tephritidae). Bull Entomol Res 94:449–456

Kovaleski A, Sugayama RL, Malavasi A (1999) Movement of *Anastrepha fraterculus* from native breeding sites into apple orchards in Southern Brazil. Entomol Exp Appl 91:457–463

Krosch MN, Schutze MK, Armstrong KF, Boontop Y, Boykin LM, Chapman TA, Englezou A, Cameron SL, Clarke AR (2013) Piecing together an integrative taxonomic puzzle: microsatellite, wing shape and aedeagus length analyses of *Bactrocera dorsalis* s.l. (Diptera:

Tephritidae) find no evidence of multiple lineages in a proposed contact zone along the Thai/ Malay Peninsula. Syst Entomol 38:2–13

Lance DR, Gates DB (1994) Sensitivity of detection trapping systems for Mediterranean fruit flies (Diptera: Tephritidae) in southern California. J Econ Entomol 87:1377–1383

Landguth EL, Fedy BC, Oyler-McCance SJ, Garey AL, Emel SL, Mumma M, Wagner HH, Fortin M-J, Cushman SA (2012) Effects of sample size, number of markers, and allelic richness on the detection of spatial genetic pattern. Mol Ecol Resour 12:276–284

Latch EK, Dharmarajan G, Glaubitz JC, Rhodes OE (2006) Relative performance of Bayesian clustering software for inferring population substructure and individual assignment at low levels of population differentiation. Conserv Genet 7:295–302

Le Galliard JF, Massot M, Clobert J (2012) Dispersal and range dynamics in changing climates: a review. In: Clobert J, Baguette M, Benton TG, Bullock JM, Ducatez S (eds) Dispersal ecology and evolution. Oxford University Press, Oxford, pp 317–336

Le Roux J, Wieczorek AM (2009) Molecular systematics and population genetics of biological invasions: towards a better understanding of invasive species management. Ann Appl Biol 154: 1–17

Liu Y, Zhang J, Richards MA, Pham BL, Roe P, Clarke AR (2009) Towards continuous surveillance of fruit flies using sensor networks and machine vision. In: Proceedings of the 5th international conference on wireless communications, networking and mobile computing, Beijing, 24–26 September 2009

Loxdale HD, Lushai G (2001) Use of genetic diversity in movement studies of flying insects. In: Woiwood IP, Reynolds DR, Thomas CD (eds) Insect movement: mechanisms and consequences. CABI, Wallingford, pp 361–386

Macfarlane JR, East RW, Drew RAI, Betlinski GA (1987) Dispersal of irradiated Queensland fruit flies, *Dacus tryoni* (Froggatt) (Diptera: Tephritidae), in south-eastern Australia. Aust J Zool 35:275–281

Manel S, Gaggiotti OE, Waples RS (2005) Assignment methods: matching biological questions with appropriate techniques. Trends Ecol Evol 20:136–142

Manoukis NC, Jang EB (2013) The diurnal rhythmicity of *Bactrocera cucurbitae* (Diptera: Tephritidae) attraction to cuelure: insights from an interruptable lure and computer vision. Ann Entomol Soc Am 106:136–142

Manrakhan A, Venter J-H, Hattingh V (2009) Combating the African invader fly *Bactrocera invadens* Drew, Tsuruta & White, The African Invader Action Plan. Online at: http://www.citrusres.com/

Maxwell CW, Parsons EC (1968) The recapture of marked apple maggot adults in several orchards from one release point. J Econ Entomol 61:1157–1159

McInnis DO, Rendon P, Komatsu J (2002) Mating and remating of medflies (Diptera: Tephritidae) in Guatemala: individual fly marking in field cages. Fla Entomol 85:126–137

Meats A (1998a) Cartesian methods of locating spot infestations of the papaya fruit fly *Bactrocera papayae* Drew and Hancock within the trapping grid at Mareeba, Queensland, Australia. Gen Appl Entomol 28:57–60

Meats A (1998b) The power of trapping grids for detecting and estimating the size of invading propagules of the Queensland fruit fly and risks of subsequent infestations. Gen Appl Entomol 28:47–55

Meats A (2007) Dispersion of fruit flies (Diptera: Tephritidae) at high and low densities and consequences of mismatching dispersions of wild and sterile flies. Fla Entomol 90:136–146

Meats A, Clift AD (2005) Zero catch criteria for declaring eradication of tephritid fruit flies: the probabilities. Aust J Exp Agric 45:1335–1340

Meats A, Edgerton JE (2008) Short- and long-range dispersal of the Queensland fruit fly, *Bactrocera tryoni* and its relevance to invasive potential, sterile insect technique and surveillance trapping. Aust J Exp Agric 48:1237–1245

Meats A, Smallridge CJ (2007) Short-and long-range dispersal of medfly, *Ceratitis capitata* (Dipt., Tephritidae), and its invasive potential. J Appl Entomol 131:518–523

Meats A, Maheswaran P, Frommer M, Sved J (2002) Towards a male-only release system for SIT with the Queensland fruit fly, *Bactrocera tryoni*, using a genetic sexing strain with a temperature-sensitive lethal mutation. Genetica 116:97–106

Meats AW, Duthie R, Clift AD, Dominiak BC (2003) Trials on variants of the sterile insect technique (SIT) for suppression of populations of the Queensland fruit fly in small towns neighbouring a quarantine (exclusion) zone. Aust J Exp Agric 43:389–395

Meats A, Smallridge CJ, Dominiak BC (2006) Dispersion theory and the sterile insect technique: application to two species of fruit fly. Entomol Exp Appl 119:247–254

Miyahara Y, Kawai A (1979) Movement of sterilized melon fly from Kume Is. to the Amami Islands. Appl Entomol Zool 14:496–497

Monro J, Richardson NL (1969) Traps, male lures, and a warning system for Queensland fruit fly, *Dacus tryoni* (Frogg.) (Diptera: Trypetidae). Aust J Agr Res 20:325–338

Murphy MA, Waits LP, Kendall KC, Wasser SK, Higbee JA, Bogden R (2002) An evaluation of long-term preservation methods for brown bear (*Ursus arctos*) faecal DNA samples. Conserv Genet 3:435–440

Nakamori H, Soemori H (1981) Comparison of dispersal ability and longevity for wild and mass-reared melon flies, *Dacus cucurbitae* Coquillett (Diptera: Tephritidae), under field conditions. Appl Entomol Zool 16:321–327

Nardi F, Carapelli A, Dallai R, Roderick GK, Frati F (2005) Population structure and colonization history of the olive fly, *Bactrocera oleae* (Diptera, Tephritidae). Mol Ecol 14:2729–2738

Nathan R, Perry G, Cronin JT, Strand AE, Cain ML (2003) Methods for estimating long-distance dispersal. Oikos 103:261–273

Neilson WT (1971) Dispersal studies of a natural population of apple maggot adults. J Econ Entomol 64:648–653

Norris KR (1957) A method of marking Calliphoridae (Diptera) during emergence from the puparium. Nature 180:1002

Okubo A (1980) Diffusion and ecological problems: mathematical models. Springer, Heidelberg

Osborne JL, Loxdale HD, Woiwod IP (2002) Monitoring insect dispersal: methods and approaches. In: Bullock JM, Kenward RE, Hails RS (eds) Dispersal ecology: the 42nd symposium of the British Ecological Society. Blackwell Publishing, Oxford, pp 24–49

Ovaskainen O, Smith AD, Osborne JL, Reynolds DR, Carreck NL, Martin AP, Niitepold K, Hanski I (2008) Tracking butterfly movements with harmonic radar reveals an effect of population age on movement distance. Proc Natl Acad Sci 105:19090–19095

Papadopoulos NT, Plant RE, Carey JR (2013) From trickle to flood: the large-scale, cryptic invasion of California by tropical fruit flies. Proc R Soc B 280:20131466

Paranhos BJ, Papadopoulos NT, McInnis D, Gava C, Lopes FSC, Morelli R, Malavasi A (2010) Field dispersal and survival of sterile medfly males aromatically treated with ginger root oil. Environ Entomol 39:570–575

Parker PG, Snow AA, Schug MD, Booton GC, Fuerst PA (1998) What molecules can tell us about populations: choosing and using a molecular marker. Ecology 79:361–382

Pavlacky DC, Goldizen AW, Prentis PJ, Nicholls JA, Lowe AJ (2009) A landscape genetics approach for quantifying the relative influence of historic and contemporary habitat heterogeneity on the genetic connectivity of a rainforest bird. Mol Ecol 18:2945–2960

Peck SL, McQuate GT (2004) Ecological aspects of *Bactrocera latifrons* (Diptera: Tephritidae) on Maui, Hawaii: movement and host preference. Environ Entomol 33:1722–1731

Peck SL, McQuate GT, Vargas RI, Seager DC, Revis HC, Jang EB, McInnis DO (2005) Movement of sterile male *Bactrocera cucurbitae* (Diptera: Tephritidae) in a Hawaiian agro-ecosystem. J Econ Entomol 98:1539–1550

Peterson MA, Denno RF (1998) The influence of dispersal and diet breadth on patterns of genetic isolation by distance in phytophagous insects. Am Nat 152:428–446

Phipps CR, Dirks CO (1932) Dispersal of the apple maggot. J Econ Entomol 25:576–582

Phipps CR, Dirks CO (1933) Dispersal of the apple maggot – 1932 studies. J Econ Entomol 26: 344–349

Plant RE, Cunningham RT (1991) Analyses of the dispersal of sterile Mediterranean fruit-flies (Diptera, Tephritidae) released from a point-source. Environ Entomol 20:1493–1503

Pritchard JK, Stephens M, Donnelly P (2000) Inference of population structure using multilocus genotype data. Genetics 155:945–959

Prugnolle F, de Meeus T (2002) Inferring sex-biased dispersal from population genetic tools: a review. Heredity 88:161–165

Raybould AF, Clarke RT, Bond JM, Welters RE, Gliddon CJ (2001) Inferring patterns of dispersal from allele frequency data. In: Bullock JM, Kenward RE, Hails RS (eds) Dispersal ecology: the 42nd symposium of the British Ecological Society. Blackwell Publishing, Oxford, pp 89–110

Rempoulakis P, Nestel D (2012) Dispersal ability of marked, irradiated olive fruit flies [*Bactrocera oleae* (Rossi) (Diptera: Tephritidae) in arid regions. J Appl Entomol 136:171–180

Reynolds DR, Riley JR (2002) Remote-sensing, telemetric and computer-based technologies for investigating insect movement: a survey of existing and potential techniques. Comput Electron Agric 35:271–307

Reynolds AM, Smith AD, Reynolds DR, Carreck NL, Osborne JL (2007) Honeybees perform optimal scale-free searching flights when attempting to locate a food source. J Exp Biol 210: 3763–3770

Riley JR, Smith AD (2002) Design considerations for an harmonic radar to investigate the flight of insects at low altitude. Comput Electron Agric 35:151–169

Robacker DC, Mangan RL, Moreno DS, Tarshis Moreno AM (1991) Mating behavior and male mating success in wild *Anastrepha ludens* (Diptera: Tephritidae) on a field-caged host tree. J Insect Behav 4:471–487

Rousse P, Duyck PF, Quilici S, Ryckewaert P (2005) Adjustment of field cage methodology for testing food attractants for fruit flies (Diptera: Tephritidae). Ann Entomol Soc Am 98:402–408

Schneider CW, Tautz J, Grünewald B, Fuchs S (2012) Tracking of sublethal effects of two neonicotinoid insecticides on the foraging behaviour of *Apis mellifera*. PLoS One 7:e30023

Schroeder WJ, Mitchell WC (1981) Marking Tephritidae fruit fly adults in Hawaii for release-recovery studies. Proc Hawaii Entomol Soc 23:437–440

Schroeder WJ, Mitchell WC, Miyabara RY (1974) Dye-induced changes in melon fly behavior. Environ Entomol 3:571

Schutze MK, Krosch MN, Armstrong KF, Chapman TA, Englezou A, Chomic A, Cameron SL, Hailstones D, Clarke AR (2012) Population structure of *Bactrocera dorsalis* s.s., *B. papayae* and *B. philippinensis* (Diptera: Tephritidae) in southeast Asia: evidence for a single species hypothesis using mitochondrial DNA and wing-shape data. BMC Evol Biol 12:130. doi:10.1186/1471-2148-1112-1130

Senger SE, Tyson R, Roitberg BD, Thistlewood HMA, Harestad AS, Chandler MT (2009) Influence of habitat structure and resource availability on the movements of *Rhagoletis indifferens* (Diptera: Tephritidae). Environ Entomol 38:823–835

Severin HHP, Hartung WJ (1912) The flight of two thousand marked male Mediterranean fruit flies (*Ceratitis capitata* Wied.). Ann Entomol Soc Am 5:400–407

Sharp JL, Ashley TR (1984) Chemicals tested as internal dye markers for the Caribbean fruit fly, *Anastrepha suspensa* (Loew) (Diptera: Tephritidae). Fla Entomol 67:575–577

Shelly T, Edu J (2010) Mark-release-recapture of males of *Bactrocera cucurbitae* and *B. dorsalis* (Diptera: Tephritidae) in two residential areas of Honolulu. J Asia-Pacific Entomol 13:131–137

Shelly TE, Nishimoto J (2011) Additional measurements of distance-dependent capture probabilities for released males of *Bactrocera cucurbitae* and *B. dorsalis* (Diptera: Tephritidae) in Honolulu. J Asia-Pacific Entomol 14:271–276

Shelly T, Nishimoto JI, Diaz A, Leathers J, War M, Shoemaker R, Al-Zubaidy M, Joseph D (2010) Capture probability of release males of two *Bactrocera* species (Diptera: Tephritidae) in detection traps in California. J Econ Entomol 103:2042–2051

Skarpaas O, Shea K, Bullock JM (2005) Optimizing dispersal study design by Monte Carlo simulation. J Appl Ecol 42:731–739

Slatkin M (1985) Gene flow in natural populations. Annu Rev Ecol Syst 16:393–430

Sonleitner FJ, Bateman MA (1963) Mark-recapture analysis of a population of Queensland fruit-fly, *Dacus tryoni* (Frogg.) in an orchard. J Anim Ecol 32:259–269

Southwood TRE (1978) Ecological methods: with particular reference to the study of insect populations, 2nd edn. Chapman and Hall, London

Southwood TRE, Henderson PA (2000) Ecological methods, 3rd edn. Blackwell Publishing Ltd., Oxford

Steiner LF (1965) A rapid method for identifying dye-marked fruit flies. J Econ Entomol 58: 374–375

Steiner LF (1969) A method of estimating the size of native populations of Oriental, melon, and Mediterranean fruit flies, to establish the overflooding ratios required for sterile-male releases. J Econ Entomol 62:4–7

Steiner LF, Rowher GG, Ayers EL, Christenson LD (1961) The role of attractants in the recent Mediterranean fruit fly eradication program in Florida. J Econ Entomol 54:30–35

Storfer A, Murphy MA, Evans JS, Goldberg CS, Robinson S, Spear SF, Dezzani R, Demelle E, Vierling L, Waits LP (2007) Putting the "landscape" in landscape genetics. Heredity 98:128–142

Streit S, Bock F, Pirk CWW, Tautz J (2003) Automatic life-long monitoring of individual insect behaviour now possible. Zoology 106:169–171

Svensson GP, Sahlin U, Brage B, Larsson MC (2011) Should I stay of should I go? Modelling dispersal strategies in saproxylic insects based on pheromone capture and radio telemetry: a case study on the threatened hermit beetle *Osmoderma eremita*. Biodivers Conserv 20:2883–2902

Taylor HR, Harris WE (2012) An emergent science on the brink of irrelevance: a review of the past 8 years of DNA barcoding. Mol Ecol Resour 12:377–388

Thomas DB, Loera-Gallardo J (1998) Dispersal and longevity of mass-released, sterilized Mexican fruit flies (Diptera: Tephritidae). Environ Entomol 27:1045–1052

Turchin P (1998) Quantitative analysis of movement: measuring and modelling population redistribution in animals and plants. Sinauer Associates, Sunderland

Turchin P, Thoeny WT (1993) Quantifying dispersal of southern pine beetles with mark-recapture experiments and a diffusion model. Ecol Appl 3:187–198

Vandewoestijne S, Baguette M (2004) Demographic versus genetic dispersal measures. Popul Ecol 46:281–285

Vargas RI, Whitehand L, Walsh WA, Spencer JP, Hsu C-L (1995) Aerial releases of sterile Mediterranean fruit fly (Diptera: Tephritidae) by helicopter: dispersal, recovery, and population suppression. J Econ Entomol 88:1279–1287

Wakid AM, Shoukry A (1976) Dispersal and flight range of the Mediterranean fruit fly, *Ceratitis capitata* Wied. in Egypt. Zeitschrift für Angewandte Entomologie 81:214–218

Weldon CW (2005) Marking Queensland fruit fly, *Bactrocera tryoni* (Froggatt) (Diptera: Tephritidae) with fluorescent pigments: pupal emergence, adult mortality, and visibility and persistence of marks. Gen Appl Entomol 34:7–13

Weldon C, Meats A (2007) Short-range dispersal of recently emerged males and females of *Bactrocera tryoni* (Froggatt) (Diptera: Tephritidae) monitored by sticky sphere traps baited with protein and Lynfield traps baited with cue-lure. Aust J Entomol 46:160–166

Weldon C, Meats A (2010) Dispersal of mass-reared sterile, laboratory-domesticated and wild male Queensland fruit flies. J Appl Entomol 134:16–25

Whitehead MR, Peakall R (2012) Microdot technology for individual marking of small arthropods. Agric Forest Entomol 14:171–175

Whitlock MC, McCauley DE (1999) Indirect measures of gene flow and migration: $FST \neq 1/(4\ Nm + 1)$. Heredity 82:117–125

Wong TTY, Whitehand LC, Kobayashi RM, Ohinata K, Tanaka N, Harris EJ (1982) Mediterranean fruit fly: dispersal of wild and irradiated and untreated laboratory-reared males. Environ Entomol 11:339–343

Zhang D-X, Hewitt GM (1997) Insect mitochondrial control region: a review of its structure, evolution and usefulness in evolutionary studies. Biochem Syst Ecol 25:99–120

Zhao JT, Frommer M, Sved JA, Gillies CB (2003) Genetic and molecular markers of the Queensland fruit fly, *Bactrocera tryoni*. J Hered 94:416–420

Chapter 7
Fruit Fly Invasion: Historical, Biological, Economic Aspects and Management

Nikos T. Papadopoulos

Abstract Enhanced by global warming as well as by intense human mobility and trading of agricultural goods, pest invasions have profound effects on national and regional economies, entire ecosystems, agricultural cropping patterns, sustainable production of agricultural goods, pesticide use, and conservation. Fruit flies (Diptera: Tephritidae) comprise a major group of pests including several invasive species, such as the Mediterranean fruit fly and the oriental fruit fly, that threaten sustainable fruit and vegetable production worldwide. The current paper covers several aspects of fruit fly invasion biology, including (a) historical perspectives for major genera and species of fruit flies, (b) the enormous impact on the economies at state, national and regional levels, (c) effects of global warming on invasion dynamics and range expansion, (d) detection and monitoring of invasion events, and (e) methods and strategies to confront invasive fruit flies. Last, but not least, a discussion is provided regarding prospects for research and policy regarding fruit fly invasions.

Keywords Propagule pressure • Eradication • Early detection • Containment • Quarantine regulations • Range expansion • Naturalization • Lag-phase • Low prevalence areas • Pest free zones • Sterile Insect Technique • Area-Wide Pest Management • Allee effect

N.T. Papadopoulos (✉)
Department of Agriculture, Laboratory of Entomology and Agricultural Zoology, Crop Production and Rural Environment, University of Thessaly, Phytokou St. 38446 N. Ionia, Magnisias, Greece
e-mail: nikopap@uth.gr

T. Shelly et al. (eds.), *Trapping and the Detection, Control, and Regulation of Tephritid Fruit Flies*, DOI 10.1007/978-94-017-9193-9_7,

1 Introduction

Accelerated by global warming as well as increased human travelling and product trading worldwide, biological invasions of insect pests, broadly defined as "...the geographical expansion of a species into an area not previously occupied by that species" (Vermeij 1996), precipitate changes occurring on a global scale. Migration of invasive species outside their natural range constitutes a major threat to biodiversity, ecosystem function, sustainable agricultural production, agricultural cropping patterns, pesticide use, both national and regional economies, and public health. The rapid increase in world trading activities combined with recent advances in agriculture and transportation have resulted in an increasing number of accidental invasion events (Perrings et al. 2005). Climate change may also affect biological invasions in a dramatic way. A wealth of studies attempting to predict the identity, impact, and distribution of possible invaders is available in the literature (Parker et al. 1999; Kolar and Lodge 2001; Mooney and Cleland 2001). Reflecting interest in the geographic spread of numerous insect species, many recent studies have used bioclimatic and other models to (a) determine suitable areas and predict the future distribution of invasive species (Vera et al. 2002a; De Meyer et al. 2008; Li et al. 2009; De Meyer et al. 2010) and (b) assess possible impacts of climate warming on the expansion of the geographic range of invasive species (Gutierrez et al. 2009a; Ponti et al. 2009; Trumble and Butler 2009; Ladanyi and Horvath 2010). On the other hand, invasive species provide excellent models to study the adaptation process and the evolution of life history, behavioral, and physiological traits in novel environments (Diamantidis et al. 2008a, 2009). For example, Diamantidis and co-workers (2008a, 2009, 2011b) showed that large differences in major fitness traits, such as pre-adult survival and developmental rates, adult life span and fecundity, and the intrinsic rate of population increase, may even occur among biotypes of the same invasive species. Variation in life history traits among populations of an invasive species may also indicate a differential invasion potential for the respective populations. However, intra- and interspecific competition as well as geographical, climatic, and environmental barriers may also determine the fate of an invasion event (Diamantidis et al. 2011b) (Box 7.1).

There are several hurdles that alien species must overcome in order to become established in a new environment. Successful invasion requires four distinct population processes: (a) arrival, (b) establishment, (c) naturalization, and (d) spread (Liebhold and Tobin 2008; Carey 2010). Arrival involves the dispersal of individuals to a previously unoccupied region and for most invasive insects is directly related to the dispersal ability of the species, while human movement and trading of goods account for longer distance arrival events. Establishment means that the immigrant population is able to sustain itself in a newly colonized area through local reproduction. Establishment is dependent upon habitat suitability, size of the founder population, frequency of founder events, demographic and other life

Box 7.1: Glossary Terms

Aggressive invader. Those species that have either colonized multiple areas at a global scale or/and exhibit rapid spread after being established in the invaded area.

Allee Effect. Positive relationship between individual fitness and population size or density. Component Allee Effect, defined as a decrease in one or more fitness components as a result of reduced population density, can lead to demographic Allee Effect, which is the per capita decline in population growth resulting from low population densities.

Containment. Targets established populations that exist at low densities and restricted areas towards preventing future spread and establishment and therefore keeping population densities at very low levels.

Eradication. Permanent elimination-removal of all individuals of established populations from a broader area by means of relatively time-limited campaign

Extirpation. Removal of all individuals of a local population from a specific site without affecting neighbouring populations from a broader area.

Lag phase. A rather long period that involves the time from establishment to completion of the naturalization process, which results in the incorporation of the invasive species into invaded environment.

Propagule pressure. Frequency and number of individuals of a species that arrive in a new site.

history traits of the invading species, and interactions with the biotic environment. There is a minimum density of a population that assures its persistence in a specific area. A positive relationship between individual fitness and population size is known as the "Allee Effect", and manifestation of this process in invading populations is greatly affected by the rate at which propagules arrive to a specific area (Liebhold and Tobin 2008). Naturalization, which is often overlooked (or covered under establishment), is a relatively long process that involves establishment of self-perpetuating populations and adaptation (including genetic changes driven by differential selection pressures in the invaded habitat compared to the area of origin) within the local ecosystem and is related to genetic changes that enable a species to overcome geographic, environmental, and reproductive barriers and hence adapt to the new environment (Richardson et al. 2000). Spread, on the other hand, is the process by which an invasive species expands its range from the current habitat into unoccupied ones (Liebhold and Tobin 2008). During this last phase of invasion, newcomers interact with native species ecologically, while climatic and other biological factors, such as host fruit quality and availability, may drive adaptation and evolutionary changes (Aluja et al. 2014). The dispersion of an established population can follow a simple diffusion model or more complex ones that incorporate environmental influences, such as topography or human social factors and movement. Topography of mountains, valleys, rivers, and shorelines, as well as antagonistic interactions with other species, channel dispersion of an established population through paths of least resistance, resulting in stream-like movements (Carey 1996a, b; Duyck et al. 2006a, b). Therefore, successful invasion is a complex biological process that is constrained by climatic, geographical, biological, and community barriers (ecological resistance, ecosystems resilience) (Harmon et al. 2009) (Box 7.2).

Box 7.2: Phases of Invasion

Introduction. Arrival of invasive species in a new area beyond its current geographic range. Transportation of propagules over long distance is usually human-or weather-mediated, while range expansion includes dispersion on a smaller spatial scale by natural means

Establishment. Self-perpetuating breeding populations that can potentially grow and disperse or remain at low densities.

Naturalization. A process that involves establishment of self-perpetuating populations, incorporation within the resident fauna and genetic changes that enables a species to overcome geographic, environmental, and reproductive barriers and hence become adapted to the new environment (Richardson et al. 2000). In many papers the characteristics of naturalization have been included into establishment.

Spread. The process by which an invasive species expands its range by natural means from the current habitat into unoccupied ones (Liebhold and Tobin 2008)

Climate change may affect both abiotic and biotic parameters, rendering the prediction of invasion patterns, under a climate change scenario, a rather complex procedure. Several modeling techniques have been employed to predict the potential distribution of invasive species (Gallien et al. 2010; Gutierrez et al. 2009a). However, there are still problems associated with accurately predicting the expansion of the geographic distribution of insect species. Phenomenological habitat suitability models are applied to predict the potential distribution of non-indigenous species, while mechanistic ones are used to understand invasion dynamics after the invaders' introduction. Habitat suitability models use characteristics of the native distribution to identify suitable areas for establishment, assuming that underlying processes are indirectly captured by analyzing patterns at large spatial scales. Mechanistic models focus on questions related with demographic response and spread of the invader and the effect of invasive species on the native animal and plant communities. Hybrid models are believed to provide a more comprehensive account for predicting and understanding patterns of invasion (Gallien et al. 2010).

As far as notorious invasive insect species are concerned, certain management activities correspond to each of the four invasion phases. For example, to prohibit arrival, international quarantines and intensive inspection protocols must be established (Liebhold and Tobin 2008). Sensitive detection networks followed by eradication programs are implemented to deal with incipient and established populations, while domestic quarantines and barrier zones are applied to deal with an invasive population in the last phase of spread. As mentioned above, naturalization is generally overlooked, and there are no specific measures connected with this component of invasion. The complex nature of invasion events makes decision-making difficult and population management projects risky and costly. Therefore, understanding the invasion biology of notorious insect pests that are expanding their geographic range is of enormous importance for pest management policy makers at national, regional, and international levels.

Fruit flies of the family Tephritidae represent one of the most economically important groups of insects. Approximately one third of the 4,000 species (500 genera) of Tephritidae oviposit in soft fruits where larval development takes place (White and Elson-Harris 1992). Fruit fly pests exert a huge economic impact on fruit and vegetable production because of direct damage on fruit and vegetable commodities and quarantine regulations that restrict fruit trading from infested areas to fruit fly free countries or areas. For example, the Mediterranean fruit fly (medfly), *Ceratitis capitata* (Wiedemann), alone threatens a fruit producing industry with a gross value of products totaling approximately $12 billion in the United States of America (USA) (Anonymous 2012). There are many invasive fruit fly species, most of which are tropical, that appear to be expanding their current geographic distribution mainly within tropics but also to the temperate zone (Table 7.1). Among the most pestiferous are the Mediterranean fruit fly, the oriental fruit fly, *Bactrocera dorsalis* (Hendel), the peach fruit fly, *Bactrocera zonata* (Saunders), *Bactrocera invadens* (*B. dorsalis* complex) Drew, Tsuruta & White, and the Mexican fruit fly, *Anastrepha ludens* (Loew). A recent study demonstrates that at least five tropical fruit flies (including *C. capitata*, *B. dorsalis*, *Bactrocera correcta* (Bezzi), *B. zonata* and *A. ludens*) have become self – sustaining populations and therefore established populations in California (Papadopoulos et al. 2013). On the other hand, the list of invasive tephritids also includes species of the temperate zone, such as *Rhagoletis completa* (Cresson) and *Rhagoletis cingulata* (Loew), that have relatively recently invaded several countries in central Europe and the northern Mediterranean coast. In addition to the huge economic impact, and its apparent social effects (collapse of previously stable commodities trading, market restrictions for individual growers), invasive fruit flies may have major effects on the ecology of native species by displacing and outcompeting indigenous tephritids (Vargas et al. 1995; Duyck et al. 2004; Ekesi et al. 2009). In addition, intensive eradication campaigns, which may involve wide range of pesticide use, contribute to emergence of secondary pests, because natural enemies dwindle, and entire ecosystems become altered (Aluja et al. 2011).

Table 7.1 Most important invasive Tephritidae

Common name	Scientific name	Origin	Invasive areas
Mexican fruit fly	*Anastrepha ludens*	Central America, Mexico	Southern USA, California, Florida
West Indian fruit fly	*A. obliqua*	Caribbean	Southern and Central America, Southern USA (Texas), California, Florida
Sapote fruit fly	*A. striata*	North America	Southern and Central America, California, Southern Texas
South American fruit fly	*A. fraterculus*	Central – South America	Northern Argentina, Trinidad, Galapagos Islands, Southern Texas
Carambola fruit fly	*Bactrocera carambolae*	South East Asia	South America (Surinam, Brazil),
Melon fly	*B. cucurbitae*	India	Southeastern Asia, Oceania, Africa, Hawaii
Guava fruit fly	*B. correcta*	South East Asia	North America (USA, California)
Oriental fruit fly	*B. dorsalis*	South East Asia	Japan, USA (California, Florida), islands of the Pacific Ocean
	B. invadens	India – Shri Lanka	Sub Saharan Africa, Islands of the Indian Ocean
Solanum fruit fly	*B. latifrons*	South East Asia	Africa, Hawaii
Olive fruit fly	*B. oleae*	Africa	North America (USA, Mexico)
Queensland fruit fly	*B. tryoni*	North East Australia	South and West Australia
Peach fruit fly	*B. zonata*	Southeastern Asia	Africa (Egypt, Mauritius, Reunion), Asia (Middle East, India, Indochina), U.S.A. (California)
Mediterranean fruit fly	*Ceratitis capitata*	Central East Africa	North and Central America, Islands of the Pacific Ocean, South Australia, Central Europe
	C. falsiventris	Sub Saharan Africa	Within Africa
Natal fruit fly	*C. rosa*	Sub Saharan Africa	Within Africa, Islands of the Indian Ocean
Ethiopian fruit fly	*Dacus ciliatus*	Central East Africa	Africa, Asia (Middle East, India)
Walnut husk fly	*Rhagoletis completa*	North America	Central Europe, northern Mediterranean countries
Eastern Cherry fruit fly	*Rhagoletis cingulata*	North East America	Central Europe, northern Mediterranean countries

The literature on fruit fly invasion biology is quite large and scattered and includes several disciplines, different genera, and a long list of major agricultural pests. The current paper does not provide an exhaustive review of the field but presents a comprehensive synopsis of the major invasive fruit fly species, their economic importance, possible effects of climate change on their invasion dynamics, expansion of their current geographic range, and some measures that are widely adopted in order to confront fruit fly invasions. Last, but not least, implications and suggestions for policy regarding invasive fruit fly species are also reported.

2 Historic Perspectives of Fruit Fly Invasions

2.1 Ceratitis *spp.*

Medfly is considered one of the most destructive pests for fresh fruit and vegetable production (White and Elson-Harris 1992; Papadopoulos 1999; Papadopoulos et al. 2001a), since it is widespread, multivoltine, and highly polyphagous, infesting more than 300 plant species (Liquido et al. 1991). *Ceratitis capitata* today exhibits an almost cosmopolitan geographical distribution due to its high invasive potential (Malacrida et al. 2007; De Meyer et al. 2008; Diamantidis et al. 2009; Papadopoulos et al. 2013). Because of its economic importance and cosmopolitan distribution, there are several records concerning its global invasion history and extensive knowledge of its population genetics (Gasperi et al. 1995; Malacrida et al. 1998; De Meyer et al. 2008 and references therein). It appears that medfly has dispersed from the ancestral areas of the eastern part of sub-Saharan Africa to almost all parts of Africa (De Meyer et al. 2004). Historical records elucidate the expansion of medfly distribution from Africa to the Mediterranean either through the Nile valley or the west coast of Africa (De Breme 1842; Malacrida et al. 2007) and from there to South America, with subsequent northern expansion into Central America. Genetic studies utilizing extensive sampling throughout the medfly's range of distribution suggest that the region of sub-Sahara East Africa (Kenya) represents the source area of the species, since Kenyan populations carry the highest levels of genetic variability (Bonizzoni et al. 2000). An alternative, though less plausible analysis, suggests the western Africa as the ancestral area for the Mediterranean fruit fly (Gasparich et al. 1997; De Meyer et al. 2004). Independent and repeated colonization events from both the Mediterranean region and Africa, due to increased human mobility and trading activities, probably account for the more recent invasion of medfly to Latin America and the Pacific. Within continent dispersion may also account for the wide distribution of medfly in Central and South America. Medfly was detected in Western Australia in 1895, in Argentina in the early 1900s, in Hawaii, USA, in 1910, and in Central America and California in 1975 (Carey 1991; Vera et al. 2002a; Bonizzoni et al. 2004; Papadopoulos et al. 2013). Since its first detection in 1975, medfly is frequently detected in California. The probability of a repeat outbreak the first, fifth and 10th year following a detection at county's level is around 0.65 and 0.75 and 0/91, respectively (Papadopoulos et al. 2013). In recent years, medfly has frequently been detected along the northern coast of the Mediterranean Sea, such as northern Italy and Slovenia, and is considered a major pest of citrus, pome, and stone fruits in Croatia (Bjeliš 2008a). Frequent detections have been reported in southern Germany where significant, though sporadic, infestations have been reported in late ripening fruits (Vogt and Koeppler, personal communication). Over the last 4 years repeated detections have been reported in Austria (in a lower elevation area close to Vienna) as well (Lethmayer, personal communication). Medfly was detected in 1930 in Austria and from 1952 to 1957 was continuously reported in the fruit producing area near Vienna, where recent detections have been reported (Bohm 1958; Lethmayer 2011). Sporadic invasion events in other countries of central Europe have also been reported (Anonymous 2013). Medfly has been detected in 2007 in Romania and in 2013 has been reported to develop substantial populations in the continental area near Bucharest (Chireceanu, personal communication).

There are several studies that adopt bioclimatic models to identify areas suitable for medfly establishment (Baker et al. 2000; Vera et al. 2002a; De Meyer et al. 2008). Climatic mapping, the most common output of these analyses, examines the climate in the home range of the invasive pest and compares it with climatic variables in an area at risk for invasion (Baker et al. 2000). Similar approaches can be employed to predict habitat suitability under a climate change scenario. Moreover, the phenotypic plasticity regarding thermal tolerance of the Mediterranean fruit fly has been recently studied in an effort to understand its invasive potential (Nyamukondiwa and Terblanche 2009; Nyamukondiwa et al. 2010; Terblanche et al. 2010). Medfly develops faster rapid cold hardening that lasts longer compared with *Ceratitis rosa* Karsch, a less aggressive invasive species. This variation in rapid cold hardening may enable survival of *C. capitata* in novel cooler habitats and therefore promote invasion success (Nyamukondiwa et al. 2010). Demographic components of six medfly biotypes and their differential invasive potential have also been determined (Diamantidis et al. 2008a, b, 2009, 2011a). Differences among biotypes were evident in adult longevity and age-specific egg laying patterns as well as in the development rate of the immature stages. Overall, biotypes from Kenya, Guatemala, and Hawaii exhibited higher intrinsic rates of population increase at optimal laboratory conditions than those from Greece, Portugal, and Brazil. However, population increase parameters of the above biotypes may change under field conditions in a novel "stressful" environment, including interspecific interaction with other frugivorous species and natural enemies. Interspecific interactions of the Mediterranean fruit fly with other invasive and native Tephritids have been studied in great detail on La Reunion island (Duyck et al. 2004, 2006a). Considering competitive interspecific interactions among larvae of different species within fruit and adult females for egg laying sites, Duyck and colleagues concluded that recently arrived species systematically outcompete earlier ones. Although *C. capitata* expressed higher intrinsic rate of population increase, it was the least competitive compared to *B. zonata* and *C. rosa* (Duyck et al. 2007). Hence, the invasion potential of the Mediterranean fruit fly may be related to intrinsic biological properties, interspecific competitive interactions, and physiological adjustments. Additional traits that define a successful invader may include rapid plastic and long-term genetic responses to the newly invaded habitat.

All in all, medfly represents the most important invasive species of all tephritids and has been the target of intensive eradication (California, South Mexico,

Australia and elsewhere) and suppression (Middle East, Spain, Brazil, and Western Australia) efforts over several decades. In addition, detailed records of medfly invasion history, its economic and social impact on several areas all over the globe, and a wealth of fundamental studies on its biological traits render medfly an important model species for addressing several questions in invasion biology.

Other notorious species in Africa include the natal fruit fly, *C. rosa* and *Ceratitis fasciventris* Bezzi (Baliraine et al. 2004). Of the two, *C. rosa* seems to be a more aggressive invader (based, at least, on the size of the area colonized and/or the rate of spread after becoming established) as it is widely dispersed in continental sub-Sahara Africa and has colonized Mauritius and Reunion islands in the Indian Ocean. By comparison, *C. fasciventris* is distributed in many sub-Saharan countries but not outside of the continent. Molecular data suggest significant clustering and geographic differentiation within both *C. rosa* and *C. fasciventris* and rather complex genetic relationships among these two species and *Ceratitis anonae* Graham (Virgilio et al. 2013). Although there are two clusters for both *C. rosa* and *C. fasciventris*, the genetic divergence between conspecific groups is higher or comparable with that between heterospecific groups. Recent studies have explored the invasion dynamics of *C. rosa* trying to define the bioclimatic factors as well as the biological traits that determine the lower invasion potential of *C. rosa* compared with *C. capitata* (De Meyer et al. 2008; Nyamukondiwa and Terblanche 2009; Nyamukondiwa et al. 2010; Terblanche et al. 2010). Although both species showed similar tolerance to low temperatures, medfly was able to withstand more extremes of high temperatures than *C. rosa*. Besides a better response to static conditions (constant temperature regimes), plasticity in acute thermal tolerance was higher in medfly. However, the two *C. rosa* genetic entities reported by Virgilio et al. (2013) may exhibit different environmental thresholds and therefore differential invasion potential. Using the CLIMEX model, de Villiers et al. (2013) suggested that, besides the Sub-Saharan Africa, large parts of South America, Central America, Mexico, and the southern USA may be suitable for *C. rosa* establishment. Likewise, major areas in South and South East Asia and southeastern Australia may be suitable as well. On the other hand, the model predicts that prevailing cold temperatures restrict suitable areas for *C. rosa* distribution in Europe in coastal areas of the Iberian peninsula, Italy, and Greece (but see De Meyer et al. 2008).

2.2 Bactrocera *spp.*

In addition to the Mediterranean fruit fly, which is rather the exception within the *Ceratitis* genus in terms of invasiveness, fruit flies of the genus *Bactrocera* represent a highly invasive taxon, which has dispersed to many countries worldwide. The genus, which originated in southeastern Asia (Nardi et al. 2010; Wan et al. 2011), includes agricultural pests of huge importance, such as the olive fly, *B. oleae* (Rossi), the oriental fruit fly, *B. dorsalis,* the peach fruit fly, *B. zonata*, the guava fruit fly, *B. correcta*, the melon fly, *B. cucurbitae* (Coquillett), *B. carambolae* (Drew and Hancock), *B. latifrons* (Hendel) and the recently described *B. invadens* that has dispersed rapidly over much of Africa. Recent evidence from both molecular and behavioral (mating compatibility) studies as well as apparent niche similarity and overlap suggest that *B. invadens* is not distinct from *B. dorsalis* (Khamis et al. 2012a; San Jose et al. 2013; Hill and Terblanche 2014). Nevertheless, I treat them as separate entities in the current paper until all taxonomic issues have been resolved.

The oriental fruit fly is the most cosmopolitan and highly aggressive, invasive species of the genus. Originated in South East Asia, *B. dorsalis* has dispersed into almost all countries of the area, posing a great threat to the economy of the whole region. Several recent studies, drawing largely on genetic data, suggest a westward dispersion from China to neighboring countries (Yang et al. 1994; Wan et al. 2011; Li et al. 2012). The oriental fruit fly was reported in Taiwan in 1912 and has colonized almost all of southeastern Asia, westward to Pakistan (Wan et al. 2011). It was first detected in 1934 in the Hainan island of China, and, since then, the detections continued sporadically in the southern part of the country until the 1970s. However, since the 1980s, populations have exploded and spread long distances to most fruit growing areas south of the 32°N parallel (Wan et al. 2011). In addition to within-area dispersion, the fly has spread to several island complexes of the Pacific Ocean, including the northern Mariana (1935), Hawaii (1945), Guam (1945), Nauru (1980), Tahiti (1996) (Leblanc and Putoa 2000), and the Okinawa islands of Japan (Ohno et al. 2009). In Hawaii the oriental fruit fly has been reported to competitively displace medfly from lowland areas; however, medfly is still the most abundant species in newly planted coffee fields (Vargas et al. 1995). Successful eradication was declared in Okinawa in 1986; however, there have been several new detections since then attributed to reinvasion events (Ohno et al. 2009). *Bactrocera dorsalis* has also been detected in the mainland USA, especially in California and less frequently in Florida. In fact, it was first detected in southern California in 1960, and since 1970 it has been detected in low numbers every year in the greater Los Angeles area, despite eradication campaigns launched after every detection event (Papadopoulos et al. 2013).

Similar invasion dynamics have been reported for *B. invadens*. The African populations of *B. invadens* show a strong genetic affinity to the Sri Lankan populations, indicating this area as the origin of the African invasion (Khamis et al. 2009). From the east coast of Kenya, where it was first detected in 2003 (Lux et al. 2003), *B. invadens* has dispersed rapidly to almost all sub-Saharan countries, including South Africa (Manrakhan et al. 2012; Hill and Terblanche 2014), southern parts of the Nile river (Abdelmagid et al. 2012), northern part of Sudan close to Libya borders (Mohamed et al. 2012), and islands of the Indian Ocean (Comoro archipelago) (Khamis et al. 2012b). The pest causes enormous damage in fruit production in these areas and is described as a "devastating pest" by the Inter-African Phytosanitary Council (French 2005).

Dispersion of the monophagous olive fly closely follows the distribution of olive tree (*Olea europea* L.). The species has dispersed from the eastern and southern parts of Africa, including South Africa (Nardi et al. 2010), to all Mediterranean countries and Portugal, countries of the Middle East, east to Pakistan (White and

Elson-Harris 1992), and recently California and Mexico (Yokoyama et al. 2006; Zygouridis et al. 2009; Burrack et al. 2011). Dispersion within California was very rapid, with populations spreading within few years from southern parts of the state to practically every locality where olive trees are cultivated for commercial or ornamental purposes. The rapid spread of the olive fly in California may reflect the undetected presence (over multiple years) of small and geographically widespread populations or the species' capability for long-distance flight that enabled it to disperse rapidly. Recent studies on the genetics of the olive fly identify the eastern Mediterranean countries as the source of the North American invasion (Nardi et al. 2005; Zygouridis et al. 2009; Schrader et al. 2006). Also, demographic bioclimatic modeling has attempted to estimate future population densities in a recently invaded area, such as California, and in an endemic one, such as Italy, under climate warming scenarios (Gutierrez et al. 2009a). The study predicts that the abundance of olive fly will decrease in the central valley of California because of increased summer temperatures and will increase in the coastal areas. In Italy and other Mediterranean countries, the olive fly is expected to follow the expansion of the range of olive to higher elevations and northern areas following a global warming scenario (Ponti et al. 2014).

Regarding other *Bactrocera* species, the guava fruit fly originates from south-eastern Asia and has been detected in California almost every year since its first occurrence in 1987 (Papadopoulos et al. 2013). The peach fruit fly has dispersed from the original habitats of Pakistan and India to Mauritius and Reunion islands in the Indian Ocean and countries of north-eastern Africa, such as Somalia and Egypt (Ni et al. 2012). It has been detected in the southern coast of the Mediterranean basin, in extremely high numbers in Egypt, and more recently in Libya (Mohamed et al. 2012). *Bactrocera zonata* has also been detected in North America, with highest frequency in California (Papadopoulos et al. 2013). *Bactrocera cucurbitae*, of Indian origin, was reported from Tanzania in 1936 and has been recently reported in West Africa as well (Virgilio et al. 2010), although it displays a much slower rate of dispersion than *B. invadens*. It has also invaded several islands of the Indian Ocean, such as Mauritius and Reunion many decades ago and more recently the Seychelles (White et al. 2001; Virgilio et al. 2010), and was detected in Hawaii in 1897 (Nishida and Bess 1950). *Bactrocera cucurbitae* has recently been detected several times in the central valley of California, though its first occurrence was in 1956 in the southern part of the state (Papadopoulos et al. 2013). The only species of the genus that has colonized South America is *B. carambolae*, which occurs in Surinam, Guiana, French Guiana, and some isolated areas of north-east Brazil (Amapà state) (Vayssières et al. 2007). The Carambola fruit fly was first detected in Surinam in 1975, with some additional detections in 1981 until an extensive survey (1986–1990) revealed established populations at high densities (Sauers-Muller 1991). Remarkable, though within continent, invasion dynamics are exhibited by the Queensland fruit fly, *B. tryoni* (Froggatt). *Bactrocera tryoni*, a tropical – subtropical species has spread from the northeast tropical areas throughout the eastern coast of the continent and is frequently detected in South East

Australia (last outbreak was reported in Riverland, SA, on the 15th of January 2014: http://www.pir.sa.gov.au/biosecuritysa/planthealth/fruit_fly/fruit_fly_outbreak). It has also been detected with increasing incidence in more temperate areas of Western Australia (Yonow et al. 2004; Gilchrist and Meats 2010). Historical perspectives and details of the Queensland fruit fly detection and occurrence in different parts of Australia are given in a recent review paper (Dominiak and Daniels 2012).

2.3 Anastrepha *spp.*

All species of this genus are native to the Americas, and they have never been detected in other continents. Even within the New World, the dispersion of *Anastrepha* species is rather restricted (in tropical and subtropical areas) and *Anastrepha* species are considered less aggressive compared to other tephritids. The Mexican fruit fly, *A. ludens*, the South American fruit fly, *A. fraterculus* (Wiedemann), the West Indian fruit fly, *A. obliqua* (Macquart), the sapote fruit fly, *A. serpentina* (Wiedemann), and the new world guava fruit fly, *A. striata* Schiner are considered among the important species threatening almost all fruit growing regions of the world situated in tropics and subtropics and especially for the USA (Aluja 1994). Since its first detection in 1954 in California, the Mexican fruit fly has been frequently detected (almost every year since the 1980s) in southern California. On the other hand, southern Texas has recently been declared free from *A. ludens* following intensive eradication efforts (although there was a recent detection in 2012). A detailed pest risk analysis revealed that all southern states of the USA are suitable for the establishment of *Anastrepha* species (Sequeira et al. 2001). Nevertheless, detections have only been reported in California, Florida, and a restricted area of south Texas (Rio Grande Valley and Willacy County in 2012). In fact, in this specific part of Texas, detections of the sapote fruit fly are observed almost every year (Sequeira et al. 2001). *Anastrepha fraterculus,* has dispersed to almost all countries of South and Central America, from the southern USA and Mexico to Buenos Aires, Argentina (White and Elson-Harris 1992; Alberti et al. 1999). Detection of this fly in other parts of North America, such as California, is rare.

2.4 Dacus *spp.*

Dacus species are indigenous to Africa and Asia and are of local importance. However, the Ethiopian fruit fly, *D. ciliatus* (Loew), a cucurbit-attacking species native to the sub-Saharan Africa, has dispersed to many countries of southeast Asia, Pakistan, Iran, Saudi Arabia, and Yemen and has more recently spread to Israel and Jordan (White 2006; Drosopoulou et al. 2011). This species is also present in some islands of the Indian Ocean, where it has been found to compete with other cucurbit-attacking flies (Vayssières et al. 2008).

2.5 Rhagoletis *spp.*

The genus *Rhagoletis* contains more than 60 species widely distributed from Eurasia to the New World (both Nearctics and Neotropics regions) (Bush 1966). In general, this genus is not considered among the most aggressive invaders in the family and, contrary to all other genera analyzed above, it includes mainly univoltine, monophagous or stenophagous species of temperate – cooler areas that show a great affinity to their host plants, where all activities take place (Smith and Bush 1997). Nevertheless, there are two North American species, the eastern American cherry fruit fly, *R. cingulata,* and the walnut husk fly, *R. completa,* that have relatively recently invaded central Europe, and, together with the native *R. cerasi* (L.), comprise an interesting group of fruit flies that is slowly dispersing throughout Central and South Europe.

Rhagoletis cingulata was reported for the first time in Switzerland in 1983, and 10, 15, and 18 years later in Germany, Italy, and the Netherlands, respectively (Lampe et al. 2005). In recent years, it has dispersed to Belgium, France, Austria, Hungary, and the northern countries of the Balkan Peninsula (Egartner et al. 2010; Anonymous 2014). It is considered an important pest of sour and sweet cherries and also attacks other *Prunus* species (secondary hosts), such as *Prunus serotina* Ehrh. (Black Cherry), *Prunus mahaleb* L. (St. Lucie Cherry), and *Prunus virginiana* L. (Choke Cherry). Earlier reports of the western American cherry fruit fly, *Rhagoletis indifferens* Curran, in Europe are now attributed to misidentification of *R. cingulata* samples (Johannesen et al. 2013).

Another indigenous species of North America, the walnut husk fly, *R. completa,* was first detected in Europe near Venice, Italy (Duso 1991) and has since probably spread to other areas of Italy, Switzerland, southern Germany, Slovenia, Croatia, and Austria (Bjeliš 2008b; Aluja et al. 2011). This species is considered a pest of several species of walnuts (*Juglans* spp.) and causes significant damage to mid- and late-maturing varieties (Duso and Dal Lago 2006). Recently, Aluja et al. (2011) studied the details of *R. completa* distribution in Switzerland and invoked the phenomenon of global warming to explain the relaxing of climatic barriers and the expansion of its distribution in Europe. Despite the recent evidence of expansion in Europe, no serious efforts have been made to delimit the spread of the two *Rhagoletis* species (*R. cerasi* and *R. completa*), probably due to their relatively minor economic significance.

In contrast to global dispersion, several *Rhagoletis* species display regional or within-country invasions. For example, *R. indifferens* has invaded California from neighboring states and has dispersed to major cherry producing areas (Dowell and Penrose 2012). Likewise, the presence of *R. pomonella* Walsh in the northwestern

USA is also attributed to invasive, established populations arriving in infested apples from eastern states (Hood et al. 2013).

3 Economic Aspects of Fruit Fly Invasions

Many invasive insect species cause enormous damage on their host crops during the first years of their establishment in an area, but their impact soon levels off, mainly because of the activity of natural enemies, both introduced and native species, as well as other aspects of the "resistance" of the biotic and abiotic environment. *Aleurothrixus flocossus* Maskell (Homoptera: Aleurodidae) and *Phyllocnistis citrella* Stainton (Lepidoptera: Gracillariidae) are prominent examples among citrus crops, at least in countries of the Mediterranean basin (Tzanakakis and Katsoyannos 2003). However, invasive fruit flies follow a completely different pattern and become key pests of the fruit or vegetable crop after being established in a new area. Establishment of a tephritid species causes continuous devastation in invaded areas that may even lead to abandonment of cultivation of some sensitive crops and requires regular, intensive management efforts to sustain fruit and vegetable production. For example, soon after establishment the medfly became the key pest of the local fruit production in Iran (Mirsardoo et al. 2010). The aggressive and rapid invasion and establishment of *B. invadens* in the sub-Saharan Africa and *B. zonata* in Egypt, Sudan, and currently Libya are additional examples regarding the economic importance of recently invading fruit flies (Mohamed et al. 2012; Ali et al. 2012).

Fruit flies threaten the fresh fruit industry in many tropical and temperate areas of the globe. As noted by White and Elson Harris (1992), 20 years ago, fruit flies threatened the approximately one billion (AUS$) fruit industry of Australia, and the cost associated with an uncontrolled infestation of fruit flies was estimated at AUS $100 million. Incursion of species of the *B. dorsalis* complex into Australia, America, and Oceania is estimated to result in losses of billions of dollars because of direct and indirect damage. It has been estimated that the spread of *Bactrocera papayae* Drew and Hancock into North Queensland in mid 1990s (first detection in 1995) caused losses of AUS$100 million (Clarke et al. 2005) (Anonymous 1986 from White and Elson-Harris 1992), and its eradication a few years later (by 2000) cost an additional AUS$34 million (De Meyer et al. 2008). Annual losses in the Middle East due to the activity of only one species (*C. capitata*) of fruit fly were estimated at $192 million (Enkerlin and Mumford 1997). It is widely believed that fruit flies would have a devastating effect on the $43 billion California agricultural industry. The cost caused by fruit flies infestation if remained uncontrolled was estimated to reach $910 million in California plus $290 million was spent for control interventions (Dowell and Wange 1986). However, recent assessments estimate annual losses because of failure to eradicate the oriental fruit fly from California up to US$176 million (http://www.cdfa.ca.gov). These numbers increase dramatically in the case of the Mediterranean fruit fly and may reach the amount of US$1.8 billion per year if it becomes established and spread in agricultural areas in California (http://www.cdfa.ca.gov). An economic analysis has estimated a potential impact of $927.75 million of *A. ludens* establishment in the USA (Erikson 2000 cited in Sequeira et al. 2001). Remarkably, USDA-APHIS is estimated to have spent approximately US$63 million (from Congressional Appropriations) for fruit fly exclusion and detection in 2010 (Anonymous 2011b). On the other hand, eradication campaigns are extremely laborious and costly as are fruit fly exclusion and detection efforts. On average, a single eradication campaign is estimated to cost approximately US$32 million, and there have been more than 250 eradication campaigns against fruit flies in California since 1982 (Papadopoulos et al. 2013). However, the eradication effort in 1980–81 in the San Francisco Bay Area reached US$100 million (Carey 2010). Interestingly, California has launched more than 60 emergency projects since 1982, and at least 17 of them over the past 10 years against a single species, the medfly (Carey 2010). Another point that highlights the importance of tropical fruit flies as invasive species in California is the fact that 90 % of the eradication projects initiated in the state between 1982 and 2007 were directed against them (Papadopoulos et al. 2013). In addition, four eradication campaigns against the melon fly, involving the Sterile Insect Techniche (SIT), conducted in the Okinawa prefecture of Japan from 1973 to 1992 reached a total cost of 177.2 million US$ (Ito et al. 2003).

Despite direct losses on fruits and vegetables as well as management efforts, establishment of invasive fruit flies is expected to have huge impact on fresh fruit and vegetable trading because of embargos, loss of markets and quarantine regulations, and subsequent job losses. For example, Siebert and Cooper (1995) have estimated a revenue loss following an embargo on Californian fruits by Asian countries at 564 million US$ (approximately 50 % of the gross state product loss) and loss of more than 14,000 jobs because of medfly establishment in the state.

4 Global Climate Change and Invasion Dynamics

Global climate change is unequivocal. The Fourth Assessment Report of the United Nations Intergovernmental Panel on Climate Change (IPCC) noted that 11 of the 12 years within the period 1995–2006 were among the 12 warmest years recorded since 1850 and that an increase of about 0.74 °C in global average temperature has been observed as a result of warming in the last 100 years. The same report predicted that a rise in global average temperature of about 0.2 °C per decade should be expected for the next two decades. Global warming is mainly attributed to changes in atmospheric concentrations of greenhouse gases (GHGs) and aerosols that alter the energy balance of the climate system. Among anthropogenic GHGs, carbon dioxide (CO_2) is the most important with its annual emissions having increased by about 80 % between 1970 and 2004 (IPCC 2007). Fossil fuel use combined with land-use changes are largely responsible for the increase in CO_2, whose amount in the atmosphere during the year 2005 (379 ppm) is far beyond the natural range of the last 650,000 years (180–300 ppm) (IPCC 2007). Understanding how these changing climate patterns affect the life history, phenology, and distribution of species is a prerequisite to better predict the overall impact of invasive species to ecosystems. In insect species, for example, climate changes may affect several important biological traits, such as developmental rate of immature stages, adult life span, and reproduction as well as population size and density, host exploitation, and geographical distribution, which is directly linked to colonization and extinction events (Bale et al. 2002).

In a constantly warming world, biological invasions hold a central position. Climate change may result in (a) the expansion of the geographic range of insect pests, making previously unsuitable areas more susceptible for their establishment, (b) shift of species' geographic ranges towards the poles (Estay et al. 2009), (c) an increasing number of arrivals of new pests to a region, and (d) dramatic increases of some insect pest populations that may impose huge economic impact on crops and possibly force other species into extinction. Therefore, a period of climate change would likely trigger a series of concurrent biological phenomena (Altermatt 2010). Elevated temperatures may affect all four stages of invasion. For example, an increase in temperature may influence the dispersal ability of an invasive insect population, since flight thresholds will be reached earlier in the season (Bale et al. 2002). Temperature increases may also affect important life history traits, such as developmental rate, life span, and fecundity, and therefore the persistence of an invasive population during its establishment and naturalization process in a newly occupied area. Finally, climate change may affect several important aspects of the last stage of an invasion (spread), such as ecological displacement of antagonistic species (Duyck et al. 2004, 2006b). Therefore, climate change may affect both abiotic and biotic constraints, making prediction of invasion patterns, under a global warming scenario, rather complex. Consequently, decision-making regarding control efforts and population management projects may be more difficult and the likelihood of costly missteps increased. Several modeling techniques have been employed to predict the potential distribution of invasive species (Coviella and Trumble 1999; Gutierrez et al. 2009a; Ponti et al. 2009; Ladanyi and Horvath 2010). However, accurately predicting the expansion of the geographic distribution of insect species is not an easy task, since it encompasses a complex interaction of social, biological, ecological, and climatic factors. Nevertheless, almost all models (Ni et al. 2012; Stephens et al. 2007) that explore a global warming scenario conclude that temperate habitats will become suitable for tropical fruit flies and, therefore, the risk of invasion events by tropical fruit flies in these areas will increase dramatically. On the other hand, the current geographic distribution of the univoltine *Rhagoletis* species that undergo obligatory pupae diapause are expected to shrink into higher altitudes and northern cooler areas (Moraiti et al. 2014). Therefore, negative effects of the global warming on fruit fly invasion and range expansion cannot be excluded.

There are several bioclimatic studies predicting expansion of fruit fly distribution as a result of climate warming (Stephens et al. 2007; Gutierrez et al. 2009a; Ponti et al. 2009; Gutierrez et al. 2010; Gutierrez and Ponti 2011; Gutierrez et al. 2009b). However, species are treated as having a homogeneous and constant invasive potential, and in many cases predictions are based on biological data obtained from laboratory-adapted populations (Gutierrez and Ponti 2011). This may result in unlikely predictions, such as most of California is unsuitable for medfly establishment (Gutierrez and Ponti 2011). In Mediterranean areas that are climatically similar to California, medfly sustains thriving populations (Papadopoulos et al. 2001a; Katsoyannos et al. 1998; Mavrikakis et al. 2000; Penarrubia-Maria et al. 2012), and detections of medfly re-occur in both southern and northern California (Papadopoulos et al. 2013). Future studies that consider the within species differential invasive potential (e.g., biotypes differentiated in major life history traits), as well as components of plasticity in life history traits and biological data from wild population, might provide more accurate predictions.

Although dominated by temperature increases, climate change includes humidity and precipitation shifts that may affect the potential geographic distribution of invasive fruit flies (Vera et al. 2002b; de Villiers et al. 2013). For example, medfly experiences both wet-dry and hot-cold stresses for prolonged periods in several areas in Argentina, and the combination of these stresses may explain low abundance of this pest in some areas (Vera et al. 2002a). Irrigation seems to locally relax effects resulting from otherwise dry conditions.

Gradual expansion of a species' geographic range may involve a relatively small spatial scale and occupation of new marginal habitats that were formerly unfavorable for the permanent residency of the specific fruit fly species. Global warming may relax some of the barriers that prohibit dispersion and establishment of fruit flies in cooler, more temperate areas (Aluja et al. 2011). For example, the expansion of *R. completa* distribution in Switzerland is related to the average spring temperatures, which affect the length of growing season and not to average winter temperatures (Aluja et al. 2011). Changes in absolute climatic values (e.g., minima and maxima of temperatures) and/or seasonal shifts may be equally or even more critical than average values for winter survival and successful establishment (Lynch et al. 2014), and small temperature changes may have large impact. Although expansion of the geographic distribution of invasive species is not necessarily related to suitable climatic conditions (Hill et al. 2011), climate-driven range expansion may be more successful and frequent for aggressive invasive species. The capability of adapting to a multitude of newly invaded environments, as well as high levels of plasticity of an invasive fruit fly, will both contribute towards expanding its geographic range. For example, medfly biotypes obtained from geographically isolated areas on a global scale show high levels of differentiation in major life history traits, such as longevity and reproduction (Diamantidis et al. 2008a, b, 2009, 2011a). Longer adult life span of temperate biotypes relative to tropical ones seems to be one of the key traits enabling survival and establishment in cooler areas. Increased cold tolerance (adaptation to cooler environments) of the same biotypes combined with high levels of thermal plasticity (adjusting thermal tolerance within a single generation) may facilitate occupation of marginal areas and contribute further to expansion of the range of global invaders into marginal areas (Nyamukondiwa and Terblanche 2009; Terblanche et al. 2010).

Invasive potential may differ a great deal among biotypes of the same species (Diamantidis et al. 2011a, b), and colonization of new available habitats will be more successful for populations that are adapted to marginal environments and share a suite of specific life history traits associated with colonization ability (Hill et al. 2011).

The Mediterranean fruit fly in Europe provides a good example of gradual range expansion to marginal areas. The Mediterranean fruit fly has expanded its geographic distribution in Europe along the Adriatic Sea to Slovenia (Bjelis, personal communication). Established populations have been reported in Southern France (Cayol and Causse 1993) and northern Spain (Penarrubia-Maria et al. 2012). Over the last few years, the medfly has been detected frequently in climatically milder areas of central Europe, such as South West Germany (Voght and Koeppler, personal communication) and Austria (Lethmayer, personal communication). Interestingly, over the last four years medfly has been detected in the same area in Austria (close to Vienna), suggesting a persistent population for this time period (Lethmayer 2011). Likewise, Rigamonti et al. (2002) and Rigamonti (2004) reported on the extensive occurrence of established medfly populations in the continental area of Lombardia, northern Italy, and winter survival in this same area. As shown by Papadopoulos et al. (1996, 1998, 2001a), medfly can successfully overwinter in cold temperate areas, where subfreezing temperatures are frequently observed over the long winter period accompanied by a long absence of suitable host trees. Expansion of medfly range into cooler temperate areas of the northern Mediterranean coasts, as well as in continental areas of central Europe, may follow a "trial and error" pattern that, along with a climate-warming scenario, may eventually lead to permanent residency in these areas. Direct fruit damage of medfly and similarly aggressive invasive species in these marginal areas might prove to be of limited importance. Nevertheless, this case may still impose an economic burden on the trading of fresh fruits and vegetables.

Climatic, social, and economic factors interact with biological traits of invasive species and contribute to range expansion into new areas. Global warming, as stated above, is expected to relax several barriers for establishment, such as winter survival and host fruit availability over longer periods of time. A small increase in average temperature might reflect increases in the minimum temperature and shortened duration of cooling periods and may have a major impact on survival rates of tropical flies in a temperate setting. Moreover, global warming promotes the cultivation of tropical and subtropical fruit and vegetable crops in currently cooler areas and extends the fruiting season; therefore, it contributes to survival and development of fruit flies. Increased human mobility and fresh fruit trading is expected to increase the "propagule pressure" and the frequency of arrival of new tropical species into more temperate settings. Interactions of the above factors may shape the biological traits of invasive species and lead to a faster adaptation to cooler environments. Adaptation and plasticity in key life history traits are fundamental features characterizing invasive species. Therefore, as shown in Fig. 7.1, there is a complex physical, social, and biological setting that contributes to the geographic expansion of fruit flies.

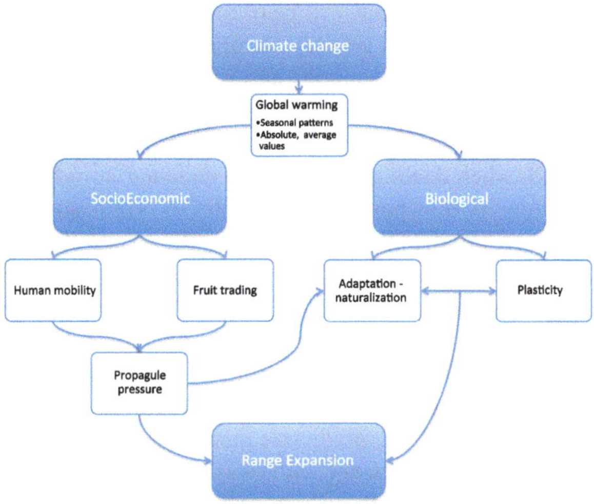

Fig. 7.1 Socioeconomic, climatic and biological factors contributing to fruit fly range expansion

5 Monitoring and Detecting Invasion Events

Long distance invasion may be either human or weather mediated (movement of insect propagules with strong winds and storms), with the second factor being far from proven at least for tephritids. Understanding how human activities mediate transport of exotic fruit flies, especially through international transport networks and hubs, is of paramount importance for the management of biological invasions (Tatem 2009). International transport networks involve both commodities shipment and human travel. Despite strict quarantine regulations and surveillance efforts that are conducted in fruit producing areas to assure shipment of fruit fly-free fruits and vegetables, there are still reports of infested cargo fruits. For example, EUROPHYT database, the European notification system for plant health interception, revealed that one third of the total number of interceptions of harmful organisms in plants and plant products imported into the European Union in 2011 (534 out of 1,600) were tephritids (Papadopoulos et al. 2013). In addition, Work and co-authors (2005) analyzing data from 1997 to 2001 regarding interceptions of alien insect in cargo shipments to USA report that a significant number of nonindigenous insect species remain undetected under current inspection regimes, and the number of interceptions increases with inspection effort. Even under very low infestation rates, large shipments may result in a huge propagule pressure that is highly correlated with successful invasion events (Liebhold et al. 2006). However, a recent study suggests that propagule cannot alone explain invasion success and "becomes increasingly important the better the climate matches the arthropod's requirements and the larger the food source" (host availability) (Bacon et al. 2014).

Human travel via ground or air has been considered the most important path of "propagule" delivery (Liebhold et al. 2006). Survival of exotic fruit flies in spatially distant but climatically similar regions is more likely if transported via the international airlines networks (Tatem 2009). Ground transportation may also contribute to invasion events but will be of major importance for range expansion of invasive species, i.e., within country or region dispersion. Ground transportation can be of massive scale. For example, almost 350 million persons crossed 45 borders stations and 330 ports of entry along the USA-Mexico borderline during 2011 (http://en.wikipedia.org/wiki/Mexico-US_border). An important point that should be considered, when dealing with human mediated introduction of exotic fruit flies, is the direction of transport and the country of origin, since there is a negative correlation between gross national income of the exporting country and interception rates (Liebhold et al. 2006), owing to limited surveillance and control of fruit flies in poorer countries.

The threat of fruit fly invasions is likely to escalate given increasing human and agricultural commodities movement over long distances. Restricting and minimizing invasion events is one of the most challenging issues in pest management and relies on optimizing predictive models as well as surveillance and control methods at international passage entryways. In fact, predictive models determining the spatial and temporal risks of introduction events as well as the mode of introductions are increasingly used by plant protection authorities to allocate resources and optimize interception (Tatem et al. 2006).

In attempting to minimize the probability of invasion and establishment of pestiferous tephritids, interception efforts alone are insufficient, and investment in a rigorous detection program is imperative. Again, modeling can be useful in directing limited resources for detection at both spatial and temporal levels. Despite their limitations, bioclimatic predictive models provide an important tool for pest management authorities and for decision-making regarding detection of small fruit fly populations. Detection of invasive fruit flies relies largely on adult trapping and in some cases on extensive surveys for infested fruits. Trapping tools for fruit flies, especially invasive species, have been advanced a great deal over the last decades; nonetheless, the perfect trapping tool is an illusion for fruit flies (Diaz-Fleischer et al. 2009). Strong adult attractants and sophisticated dispensers for most invasive *Bactrocera* spp. and the medfly are currently available; however, there are no strong species-specific attractants for *Anastrepha* spp., several invasive *Dacus* spp., and *Rhagoletis* spp. Besides, attractiveness ranges from 30 to 50 m, at least for trapping systems employing food baits, and a high density of traps (with concomitant cost) is essential to provide high probability of detecting small populations of invasive species (Epsky et al. 2010; Kendra et al. 2010). It seems that both high trap density as well as strong species-specific attractants contribute to early detection of fruit fly populations (Papadopoulos et al. 2001b). On the other hand, fruit fly populations may remain below detection levels for long periods extending several years or even decades (Carey 2010 but see also Liebhold et al. 2006). Under a stable age distribution assumption, the proportion of adults (main target of detection efforts) in a tropical fruit fly population is estimated to be less than 10 % (Papadopoulos et al. 2002). This means that a single adult detection indicates a 9-fold higher number of undetected individuals in the area. Efficacy rates of even the most successful fruit fly trapping systems for feral individuals are largely undetermined. There are inherent difficulties in determining the proportion of feral flies that are captured – detected, since it is almost impossible to estimate the number of existing wild flies in a given area, especially under low population densities, which is the case for invasive species at least during the first phases of invasion. Release-recapture studies provide some insights (Wong et al. 1982); however, they rather estimate the distance of dispersion of mass reared flies and not the detection efficacy of feral flies (Lance and Gates 1994; Paranhos et al. 2010; Gavriel et al. 2012; Shelly and Nishimoto 2011; Shelly et al. 2010). Responsiveness of mass reared males to synthetic male-specific lures is usually inferior to wild or wild-like males (wild flies reared in laboratory for few generations), and different wild strains may also express differential response (Shelly and Edu 2009; Wong et al. 1982).

Small populations of invasive species may also remain localized and exhibit extremely low mobility, despite their ability to fly long distances in release-recapture studies, rendering detection an even more difficult task. The concept of a "sleeper pest", which is widely recognized for invasive plants, that can apparently remain innocuous for years and below detection levels before exhibiting explosive population growth might be of importance in fruit flies as well (Gewin 2005). Sleeper pests remain in a lag phase for years or decades before becoming fully naturalized and subsequently proliferating and spreading. Invasive tropical fruit flies exhibit high population growth rates under benign laboratory environments (Vargas et al. 1984; Vargas and Nishida 1985; Krainacker et al. 1987; Vargas and Carey 1990; Vargas et al. 1997). Nevertheless, population growth rates may vary greatly when breeding on fruits (Papadopoulos et al. 2002; Papachristos and Papadopoulos 2009) and may be extremely low in the wild especially in a "hostile" invaded area. Control measures and eradication campaigns, following a detection of a single individual, may further decrease population growth parameters and render invasive fruit flies into a "sleeper mode". Understanding the ecology of small, localized fruit fly populations may greatly contribute to advancing detection technology and strategy and should be considered in developing new policy for invasive fruit flies.

Overall, monitoring and detecting an invasion event is a complicated, multitask project that involves cooperation and coordination among many government bodies and stakeholder groups. This task is even more difficult when real or potential invasions involve countries with differing response capabilities and/or response strategies. Any response to fruit fly invasion risks should be based on an integrated system that involves domestic and offshore surveillance activities, port of entry mitigation, and detection efforts.

6 Confronting Invasive Species

There are three broad actions that should be considered when dealing with biological invasions: prevention, early detection, and management (Venette and Koch 2009; Venette et al. 2010; Simberloff et al. 2013). Prevention, all those actions that aim to reduce or eliminate "propagule pressure", includes pest risk assessment and bioclimatic modeling to determine vulnerable areas, as well as legislative and quarantine measures. Pest risk encompasses both the likelihood that a species will successfully invade an area and the magnitude of resulting harm (Venette and Koch 2009). Despite prevention measures, invasive fruit flies circumvent them and arrive in new areas. Early detection, as mentioned above, targets mainly adults of breeding populations and provides critical information at both temporal and spatial scales that triggers emergency actions for prompt removal and therefore reduced establishment risk. Besides important improvements in fruit fly detection technologies, there are inherent issues related to detection of very small populations (see discussion on sleeper pest concept above) that should be considered within a general strategy for countering fruit fly invasions. Historical detection data, such as those collected in California and Florida USA, should also be considered to assess efficacy of detection strategies and understand long-term patterns of invasion. Management, an extremely costly and difficult task, is always the final response and entails, in succession, eradication, containment, and long-term control when previous efforts have failed.

Eradication, defined as "the removal of every potentially reproducing individual of a species or the reduction of their population density below sustainable levels" (Myers et al. 2000) has been applied with success against several fruit fly species, including the oriental fruit fly on islands in Japan and the Mediterranean fruit fly in the USA and Mexico (Hendrichs et al. 2002 but see Carey 1991, 2010). On the other hand, eradication efforts failed to remove *B. carambolae* from South America and *B. tryoni* from southeast Australia (http://www.freshplaza.com/news_detail.asp?id=100633). Eradication campaigns are extremely costly (albeit less than controlling established populations) and are usually efficient when they target small localized populations that are detected early prior to local adaptation and dispersion over large areas (Myers et al. 2000). To consider an eradication campaign as successful, there are certain criteria that should be met. For example, a region is declared, or certified, fruit fly-free when no flies are detected for three generations according to International Phytosanitary Commission and the United States Department of Agriculture (USDA) (Floyd et al. 2002). However, later detection of fruit flies in previously eradicated areas several years later may indicate lack of complete eradication (Myers et al. 2000).

Long-term management efforts are applied when eradication is not attempted or has failed and includes both area-wide suppression and farm-by-farm control activities. Preventive measures to reduce the risk of fruit fly outbreaks, such as those performed in southern California against the Mediterranean fruit fly, may be well regarded as part of a long term management strategy (http://www.cdfa.ca.gov). There are several recent advances in pest management and fruit fly management, in general, that have been adopted to assist both eradication and suppression efforts. In fact, most eradication and suppression programs rely on the SIT and Male Annihilation (Mau et al. 2007). Numerous technological and biological advances have improved the efficacy of SIT for fruit flies, especially the medfly and *Anastrepha* spp. (Dyck et al. 2005). These include, among others, the development of stable, competent genetic sexing strains, efficient releasing systems, and pre-release treatments to enhance mating competitiveness. On the other hand, a new generation of dispensers has come to advance male annihilation, which is considered as the main eradication and control tool for several *Bactrocera* spp. (Vargas et al. 2012; Vargas et al. 2010). Advances in Geographical Information Systems (GIS) and their adoption by almost all international fruit fly control programs is a big asset towards fighting invasive fruit flies (Cox and Vreysen 2005).

7 Prospects – New Policy for Invasive Fruit Flies

Fruit flies are a unique group of insect pests, encompassing more than 16 invasive species (see Table 7.1) and several genera of tropical and temperate origin distributed in almost all continents and threatening the global fruit and vegetable production. Invasive fruit flies follow various life history strategies and feeding habits. As noted earlier, fruit fly pests include monophagous, stenophagous, and polyphagous species. The majority includes tropical species that are multivoltine, polyphagous or oligophagous species, while there are few temperate, univoltine species whose life cycle includes an obligatory – long dormancy. Because of their economic importance, there is a wealth of data regarding interceptions and detections for many areas all over the globe, such as the USA and especially California (Carey 1991, 2010; Liebhold et al. 2006). Additionally, many studies have been conducted on tephritid genetics as well as basic biological, ecological, and ethological attributes of many invasive fruit fly species. Therefore, fruit flies may represent one of the most important groups of insects – even organisms – to address basic questions regarding the biology, ecology, and economics of biological invasions. Comparative studies across species with similar life histories within single or different genera would provide insights regarding the invasive properties of the species and possible phylogenetic constraints. Conversely, biological traits can be compared between species that share different life histories.

Fruit fly invasion events are expected to intensify in years to come as a result of human mobility, fresh fruit and vegetables trading, climate change, and probably other factors. Therefore, there is an urgent need for a long lasting extensive and intensive strategy against invasive fruit fly species that should be established at regional, continental, or even the global level as part of a general policy against invasive pests. Regardless of the obvious economic interests at the state or country level, transparency and data sharing should be promoted, and detection and interception data should become freely available to the public through web-based platforms. There are currently several websites providing maps of pest distribution. Nevertheless, it is hard to retrieve original data regarding detections and interceptions. Data sharing on a "real time" basis (or the soonest possible) would enable a fast reaction to invasion events in a regional level and allow an in-depth analysis from the scientific community.

Full advantage of technological developments should also be used. In recent years, almost all detection and eradication programs have adopted GIS. However, further developments are available in the field, including automatic trapping, data transfer and management, and real time availability on web platforms that should be considered and incorporated into programs dealing with invasive fruit flies. Additionally, certain advances in the field of molecular biology and genetics may facilitate a fast and thorough analysis of the genetics of detected individuals, providing precious information regarding the origin and the pathways of detected invasive species. All detection events (even individuals when possible) should be analyzed using the most up-to-date molecular and other tools. So far, a wealth of knowledge has been acquired on the genetics of invasion of the Mediterranean fruit fly and the olive fly in California and Australia for medfly (Villablanca et al. 1998; Davies et al. 1999; Bohonak et al. 2001; Bonizzoni et al. 2001, 2004; Meixner et al. 2002; Zygouridis et al. 2009), for *B. cucurbitae* in Africa and the Reunion Islands (Jacquard et al. 2013; Virgilio et al. 2010), and the oriental fruit fly in southeast Asia (Wan et al. 2011; Shi et al. 2010; Aketarawong et al. 2007). However, data for other invasive fruit fly species are rather scarce. The recent release of the medfly genome is an asset for respective analyses, not only for this species, but others as well (http://listserv-public.bcm.tmc.edu/cgi-bin/wa?A0=MEDFLY-GENOME).

Early detection and interception will definitely continue to be an important part of the strategy against invasive fruit flies. Nevertheless, regardless of the expected technological advances, the difficulty of discovering exceedingly rare, scattered, and ultra-small populations of tephritids that are mostly in pre-adult stages should be recognized and acknowledged. This is analogous to "rare-event detection problem", which is well established in cancer diagnostics (Willyard 2012) and merits attention in modeling and analyzing existing data. Other concepts that should be considered in fruit flies regard the "mysterious lag phase" (Simberloff 2009) in which new populations experience an unexplained delayed growth, naturalization, and cryptic population persistence. Understanding the ecology of very small invasive populations, which admittedly represents a difficult problem, will provide basic elements for developing new policy against invasive fruit flies.

Acknowledgements Thanks are expressed to Alex Diamantidis and Stella Papanastasiou (University of Thessaly, Greece) for various inputs throughout the writing process, and to James R. Carey (University of California Davis, USA) for long discussions on several ideas regarding fruit fly invasion. I would also like to thank Heidrun Vogt (Julius Kühn-Institut, Germany) Kirsten Köppler (Landwirtschaftliches Technologiezentrum Augustenberg, Germany), Christa Lethmayer (Austrian Agency for Health and Food Safety, Austria) and Mario Bjelis (Croatian Centre for Agriculture Food and Rural Affairs) for sharing unpublished information regarding detection of fruit flies in central Europe and the Balkan Peninsula. Thanks are also extended to Marc De Meyer (Royal Museum of Central Africa, Belgium) and Todd Shelly (APHIS USDA) for providing extensive constructive comments on earlier drafts of the current chapter. This study was partially supported by OECD (Organization for Economic Cooperation and Development) grant awarded to Nikos T. Papadopoulos.

References

Abdelmagid FM, Abass AM, Ali AA (2012) Seasonal abundance of *Bactrocera invadens*, *Ceratitis cosyra* and *Ceratitis capitata* (Diptera: Tephritidae) in River Nile State (Shendi Area), Sudan TEAM – 2nd International Meeting "Biological Invasions of Tephritidae: ecological and economic impacts", 2012 Kolymbari, Crete, Greece. Thessaloniki: Ziti, 144

Aketarawong N, Bonizzoni M, Thanaphum S, Gomulski LM, Gasperi G, Malacrida AR, Gugliemino CR (2007) Inferences on the population structure and colonization process of the invasive oriental fruit fly, *Bactrocera dorsalis* (Mendel). Mol Ecol 16:3522–3532

Alberti AC, Calcagno G, Saidman BO, Vilardi JC (1999) Analysis of the genetic structure of a natural population of *Anastrepha fraterculus* (Diptera: Tephritidae). Ann Entomol Soc Am 92: 731–736

Ali AN, Awad AA, Mohamed HO (2012) Population fluctuation of the peach fruit fly, *Bactrocera zonata* (Saunders) (Diptera: Tephritidae) in relation to prevailing weather factors in Assuit Governorate. Regional symposium on the management of fruit flies in Near East countries, 2012 Hammamet, Tunisia. 80

Altermatt F (2010) Climatic warming increases voltinism in European butterflies and moths. Proceedings of the Royal Society B-Biological Sciences, 277:1281–1287

Aluja M (1994) Bionomics and management of *Anastrepha*. Ann Rev Entomol 39:155–178

Aluja M, Guillen L, Rull J, Hohn H, Frey J, Graf B, Samietz J (2011) Is the alpine divide becoming more permeable to biological invasions? – Insights on the invasion and establishment of the Walnut Husk Fly, *Rhagoletis completa* (Diptera: Tephritidae) in Switzerland. Bull Entomol Res 101:451–465

Aluja M, Birke A, Ceymann M, Guillen L, Arrigoni E, Baumgartner D, Pascacio-Villafan C, Samietz JR (2014) Agroecosystem resilience to an invasive insect species that could expand its geographical range in response to global climate change. Agric Ecosyst Environ 186:54–63

Anonymous (2011b) Exotic fruit fly strategic plan FY 2011–2015. Fruit fly exclusion and detection programs. APHIS-USDA, Riverdale

Anonymous (2012) Crop values 2011 Summary. In: United States Department of Agriculture N. A. S. S. (ed) http://usda.mannlib.cornell.edu/usda/nass/CropValuSu//2010s/2012/CropValuSu-02-16-2012.pdf

Anonymous (2013) EPPO quarantine pest. http://www.eppo.int/QUARANTINE/insects/Ceratitis_capitata/CERTCA_ds.pdf

Anonymous (2014) Invasive species compendium, Cabi http://www.cabi.org/isc/datasheet/47051

Bacon SJ, Aebi A, Calanca P, Bacher S (2014) Quarantine arthropod invasions in Europe: the role of climate, hosts and propagule pressure. Divers Distrib 20:84–94

Baker RHA, Sansford CE, Jarvis CH, Cannon RJC, Macleod A, Walters KFA (2000) The role of climatic mapping in predicting the potential geographical distribution of non-indigenous pests under current and future climates. Agric Ecosyst Environ 82:57–71

Bale JS, Masters GJ, Hodkinson ID, Awmack C, Bezemer TM, Brown VK, Butterfield J, Buse A, Coulson JC, Farrar J, Good JEG, Harrington R, Hartley S, Jones TH, Lindroth RL, Press MC, Symrnioudis I, Watt AD, Whittaker JB (2002) Herbivory in global climate change research: direct effects of rising temperature on insect herbivores. Glob Change Biol 8:1–16

Baliraine FN, Bonizzoni M, Guglielmino CR, Osir EO, Lux SA, Mulaa FJ, Gomulski LM, Zheng L, Quilici S, Gasperi G, Malacrida AR (2004) Population genetics of the potentially invasive African fruit fly species, *Ceratitis rosa* and *Ceratitis fasciventris* (Diptera: Tephritidae). Mol Ecol 13:683–695

Bjeliš M (2008a) Landscape ecology of species from subfamily Tphritinae (Tephritidae, Diptera) along a north-east Adriatic lowlands. First meeting of Tephritid workers of Europe, Africa and the Middle East, 2008 Palma de Mallorca. 50

Bjeliš M (2008b) Fruit flies from the family *Rhagoletis* (Tephritidae) in Croatia. Glasilo Biljne Zaštite 8:25–28

Bohm H (1958) Zum vorkommen der Mittelmeerfruchtfliege, *Ceratitis capitata* Wied., im Wiener ohsthaugebiet. Pflanzenschutzberichte 21(9/10):129–158

Bohonak AJ, Davies N, Villablanca FX, Roderick GK (2001) Invasion genetics of New World medflies: testing alternative colonization scenarios. Biol Invasions 3:103–111

Bonizzoni M, Malacrida AR, Guglielmino CR, Gomulski LM, Gasperi G, Zheng L (2000) Microsatellite polymorphism in the Mediterranean fruit fly, *Ceratitis capitata*. Insect Mol Biol 9: 251–261

Bonizzoni M, Zheng L, Guglielmino CR, Haymer DS, Gasperi G, Gomulski LM, Malacrida AR (2001) Microsatellite analysis of medfly bioinfestations in California. Mol Ecol 10:2515–2524

Bonizzoni M, Guglielmino CR, Smallridge CJ, Gomulski M, Malacrida AR, Gasperi G (2004) On the origins of medfly invasion and expansion in Australia. Mol Ecol 13:3845–3855

Burrack HJ, Bingham R, Price R, Connell JH, Phillips PA, Wunderlich L, Vossen PM, O'Connell NV, Ferguson L, Zalom FG (2011) Understanding the seasonal and reproductive biology of olive fruit fly is critical to its management. Calif Agric 65:14–20

Bush GL (1966) The taxonomy, cytology, and evolution of the Genus *Rhagoletis* in Northern America (Diptera, Tephritidae). Bull Museum Comp Zool 134:562

Carey JR (1991) Establishment of the Mediterranean fruit fly in California. Science 253: 1369–1373

Carey JR (1996a) The future of the Mediterranean fruit fly *Ceratitis capitata* invasion of California: a predictive framework. Biol Conserv 78:35–50

Carey JR (1996b) The incipient Mediterranean fruit fly population in California implications for invasion biology? Ecology 77:1690–1697

Carey JR (2010) The Mediterranean fruit fly (*Ceratitis capitata*). Am Entomol 56:116–121

Cayol JP, Causse R (1993) Mediterranean fruit fly *Ceratitis capitata* Wiedemann (Dipt, Trypetidae) back in southern France. J Appl Entomol 116:94–100

Clarke AR, Armstrong KF, Carmichael AE, Milne JR, Raghu S, Roderick GK, Yeates DK (2005) Invasive phytophagous pests arising through a recent tropical evolutionary radiation: the *Bactrocera dorsalis* complex of fruit flies. Ann Rev Entomol 50:293–319

Coviella C, Trumble J (1999) Elevated atmospheric CO_2 and insect-plant interactions: implications for insect conservation. Conserv Biol 13:12

Cox JSH, Vreysen MJB (2005) Use of geographic information systems and spatial analysis in area-wide integrated pest management programmes that integrate the Sterile Insect Technique. In: Dyck VA, Hendrichs J, Robinson AS (eds) Sterile Insect Technique: Principles and Practice in Area- Wide Integrated Pest Management. Springer, Vienna

Davies N, Villablanca FX, Roderick GK (1999) Bioinvasions of the medfly *Ceratitis capitata*: source estimation using DNA sequences at multiple intron loci. Genetics 153:351–360

De Breme F (1842) Note sur le gentre *Ceratitis* de M. Mc Leavy (Diptera). Ann Soc Entomol Fr 11:183–190

De Meyer M, Copeland RS, Wharton RA, McPheron BA (2004) On the geographic origin of the Medfly *Ceratitis capitata* (Wiedemann) (Diptera: Tephritidae). In: Barnes B (ed) Proceedings of the 6th international fruit fly symposium, 6–10 May 2002, Stellenbosch, South Africa. Isteg Scientific Publications, Irene, pp 45–53

De Meyer M, Robertson MP, Peterson AT, Mansell MW (2008) Ecological niches and potential geographical distributions of Mediterranean fruit fly (*Ceratitis capitata*) and Natal fruit fly (*Ceratitis rosa*). J Biogeogr 35:270–281

De Meyer M, Robertson MP, Mansell MW, Ekesi S, Tsuruta K, Mwaiko W, Vayssieres JF, Peterson AT (2010) Ecological niche and potential geographic distribution of the invasive fruit fly *Bactrocera invadens* (Diptera, Tephritidae). Bull Entomol Res 100:35–48

De Villiers M, Hattingh V, Kriticos DJ (2013) Combining field phenological observations with distribution data to model the potential distribution of the fruit fly *Ceratitis rosa* Karsch (Diptera: Tephritidae). Bull Entomol Res 103:60–73

Diamantidis AD, Carey JR, Papadopoulos NT (2008a) Life-history evolution of an invasive tephritid. J Appl Entomol 132:695–705

Diamantidis AD, Papadopoulos NT, Carey JR (2008b) Medfly populations differ in diel and age patterns of sexual signalling. Entomol Exp Appl 128:389–397

Diamantidis AD, Papadopoulos NT, Nakas CT, Wu S, Muller HG, Carey JR (2009) Life history evolution in a globally invading tephritid: patterns of survival and reproduction in medflies from six world regions. Biol J Linn Soc 97:106–117

Diamantidis A, Carey J, Nakas C, Papadopoulos N (2011a) Ancestral populations perform better in a novel environment: domestication of Mediterranean fruit fly populations from five global regions. Biol J Linn Soc 102:334–345

Diamantidis AD, Carey JR, Nakas CT, Papadopoulos NT (2011b) Population-specific demography and invasion potential in medfly. Ecol Evol 1:479–488

Diaz-Fleischer F, Arredondo J, Flores S, Montoya P, Aluja M (2009) There is no magic fruit fly trap: multiple biological factors influence the response of adult *Anastrepha ludens* and *Anastrepha obliqua* (Diptera: Tephritidae) individuals to Multi Lure traps baited with BioLure or NuLure. J Econ Entomol 102:86–94

Dominiak BC, Daniels D (2012) Review of the past and present distribution of Mediterranean fruit fly (*Ceratitis capitata* Wiedemann) and Queensland fruit fly (*Bactrocera tryoni* Froggatt) in Australia. Aust J Entomol 51:104–115

Dowell RV, Penrose RL (2012) Distribution and phenology of *Rhagoletis fausta* (Osten Sacken 1877) and *Rhagoletis indifferens* Curren 1932 (Diptera: Tephritidae) in California. Pan-Pac Entomol 88:130–150

Dowell RV, Wange LK (1986) Process analysis and failure avoidance in fruit fly programs. In: Mangel M, Carey JR, Plant RE (eds) Pest control: operations and systems analysis in fruit fly management. Springer, New York

Drosopoulou E, Nestel D, Nakou I, Kounatidis I, Papadopoulos NT, Bourtzis K, Mavragani-Tsipidou P (2011) Cytogenetic analysis of the Ethiopian fruit fly *Dacus ciliatus* (Diptera: Tephritidae). Genetica 139:723–732

Duso C (1991) Sulla comparsa in Italia di un Tefritide neartico del noce: *Rhagoletis completa* Cresson (Diptera: Tephritidae). Boll Zool Agr Bachicoltura 23:203–209

Duso C, Dal Lago G (2006) Life cycle, phenology and economic importance of the walnut husk fly *Rhagoletis completa* Cresson (Diptera: Tephritidae) in Northern Italy. Ann Soc Entomol Fr 42: 245–254

Duyck PF, David P, Quilici S (2004) A review of relationships between interspecific competition and invasions in fruit flies (Diptera: Tephritidae). Ecol Entomol 29:511–520

Duyck PF, David P, Junod G, Brunel C, Dupont R, Quilici S (2006a) Importance of competition mechanisms in successive invasions by polyphagous tephritids in La Reunion. Ecology 87: 1770–1780

Duyck PF, David P, Quilici S (2006b) Climatic niche partitioning following successive invasions by fruit flies in La Reunion. J Anim Ecol 75:518–526

Duyck PF, David P, Quilici S (2007) Can more K-selected species be better invaders? A case study of fruit flies in La Reunion. Divers Distrib 13:535–543

Dyck VA, Hendrichs J, Robinson AS (2005) Sterile insect technique. Springer, Dordrecht

Egartner A, Zeisner N, Hausdorf H, Blümel S (2010) First record of *Rhagoletis cingulata* (Loew) (Dipt., Tephritidae) in Austria. EPPO Bull 40:158–162

Ekesi S, Billah MK, Nderitu PW, Lux SA, Rwomushana I (2009) Evidence for competitive displacement of *Ceratitis cosyra* by the invasive fruit fly *Bactrocera invadens* (Diptera: Tephritidae) on mango and mechanisms contributing to the displacement. J Econ Entomol 102:981–991

Enkerlin W, Mumford J (1997) Economic evaluation of three alternative methods for control of the Mediterranean fruit fly (Diptera: Tephritidae) in Israel, Palestinian territories, and Jordan. J Econ Entomol 90:1066–1072

Epsky ND, Espinoza HR, Kendra PE, Abernathy R, Midgarden D, Heath RR (2010) Effective sampling range of a synthetic protein-based attractant for *Ceratitis capitata* (Diptera: Tephritidae). J Econ Entomol 103:1886–1895

Estay SA, Lima M, Labra FA (2009) Predicting insect pest status under climate change scenarios: combining experimental data and population dynamics modelling. J Appl Entomol 133: 491–499

Floyd J, Lee BG, Schall R, Wade D (2002) Emergency programs manual. In: S. A. E. P., APHIS, PPQ (ed) Department of Agriculture. Washington, DC

French C (2005) The new invasive *Bactrocera* species. Insect pest control newsletter. International Atomic Energy Agency, Vienna

Gallien L, Munkemuller T, Albert CH, Boulangeat I, Thuiller W (2010) Predicting potential distributions of invasive species: where to go from here? Divers Distrib 16:331–342

Gasparich GE, Silva JG, Han HY, Mcpheron BA, Steck GJ, Sheppard WS (1997) Population genetic structure of Mediterranean fruit fly (Diptera: Tephritidae) and implications for worldwide colonization patterns. Ann Entomol Soc Am 90:790–797

Gasperi G, Baruffi L, Gomulski L, Guglielmino CR, Bandi C, Damiani G, Malacrida AR (1995) Studies on the genome organization and population-genetics of *Ceratitis capitata* using molecular and biochemical markers. J Cell Biochem 21A:198

Gavriel S, Gazit Y, Leach A, Mumford J, Yuval B (2012) Spatial patterns of sterile Mediterranean fruit fly dispersal. Entomol Exp Appl 142:17–26

Gewin V (2005) Eco-Defense against invasions. PLoS Biol 3:e429

Gilchrist AS, Meats AW (2010) The genetic structure of populations of an invading pest fruit fly, *Bactrocera tryoni*, at the species climatic range limit. Heredity 105:165–172

Gutierrez AP, Ponti L (2011) Assessing the invasive potential of the Mediterranean fruit fly in California and Italy. Biol Invasions. doi:10.1007/s10530-011-9937-6

Gutierrez A, Ponti L, Cossu Q (2009a) Effects of climate warming on olive and olive fly (*Bactrocera oleae* (Gmelin)) in California and Italy. Climatic Change 95:195–217

Gutierrez AP, Ponti L, Cossu QA (2009b) Effects of climate warming on olive and olive fly (*Bactrocera oleae* (Gmelin)) in California and Italy. Climatic Change 95:195–217

Gutierrez AP, Ponti L, Gilioli G (2010) Climate change effects on plant-pest-natural enemy interactions. In: Hillel D, Rosenzweig C (eds) Handbook of climate change and agroecosystems. Impact, adaptation, and migration. Imperial College Press, London, pp 209–237

Harmon JP, Moran NA, Ives AR (2009) Species response to environmental change: impacts of food web interactions and evolution. Science 323:1347–1350

Hendrichs J, Robinson AS, Cayol JP, Enkerlin W (2002) Medfly area wide sterile insect technique programmes for prevention, suppression or eradication: the importance of mating behavior studies. Fla Entomol 85:1–13

Hill MP, Terblanche JS (2014) Niche overlap of congeneric invaders supports a single-species hypothesis and provides insight into future invasion risk: implications for global management of the *Bactrocera dorsalis* complex. PLoS One 9:e90121. doi:10.1371/journal.pone.0090121

Hill JK, Griffiths HM, Thomas CD (2011) Climate change and evolutionary adaptations at species' range margins. Ann Rev Entomol 56:143–159

Hood GR, Yee W, Goughnour RB, Sim SB, Egan SP, Arcella T, Saint-Jean G, Powell THQ, Xu CCY, Feder JL (2013) The geographic distribution of *Rhagoletis pomonella* (Diptera: Tephritidae) in the Western United States: introduced species or native population? Ann Entomol Soc Am 106:59–65

IPCC (Intergovernmental Panel on Climate Change) (2007) Climate change 2007: synthesis report. [Online]. Available: http://www.ipcc.ch. Accessed 8 Aug 2011

Ito Y, Kakinohana H, Yarnagishf M, Koham T (2003) Eradication of the Melon Fly, *Bactrocera cucurbitae*, from Okinawa, Japan, by means of the Sterile Insect Technique, with special emphasis on the role of basic studies. J Asia Pac Entomol 6:119–129

Jacquard C, Virgilio M, David P, Quilici S, De Meyer M, Delatte H (2013) Population structure of the melon fly, *Bactrocera cucurbitae*, in Reunion Island. Biol Invasions 15:759–773

Johannesen J, Keyghobadi N, Schuler H, Stauffer C, Vogt H (2013) Invasion genetics of American cherry fruit fly in Europe and signals of hybridization with the European cherry fruit fly. Entomol Exp Appl 147:61–72

Katsoyannos BI, Kouloussis NA, Carey JR (1998) Seasonal and annual occurrence of Mediterranean fruit flies (Diptera: Tephritidae) on Chios Island, Greece: differences between two neighboring citrus orchards. Ann Entomol Soc Am 91:43–51

Kendra PE, Epsky ND, Heath RR (2010) Effective sampling range of food-based attractants for female *Anastrepha suspensa* (Diptera: Tephritidae). J Econ Entomol 103:533–540

Khamis FM, Karam N, Ekesi S, De Meyer M, Bonomi A, Gomulski LM, Scolari F, Gabrieli P, Siciliano P, Masiga D, Kenya EU, Gasperi G, Malacrida AR, Guglielmino CR (2009) Uncovering the tracks of a recent and rapid invasion: the case of the fruit fly pest *Bactrocera invadens* (Diptera: Tephritidae) in Africa. Mol Ecol 18:4798–4810

Khamis FM, Masiga DK, Mohamed SA, Salifu D, De Meyer M, Ekesi S (2012a) Taxonomic identity of the invasive fruit fly pest, *Bactrocera invadens*: concordance in morphometry and DNA barcoding. PLoS One 7:e44862. doi:10.1371/journal.pone.0044862

Khamis FM, Masiga DK, Mohamed SA, Salifu D, De Meyer M, Ekesi S (2012b) Taxonomic identity of the invasive fruit fly pest, *Bactrocera invadens*: concordance in morphometry and DNA barcoding. Plos One 7:1443–1448

Kolar CS, Lodge DM (2001) Progress in invasion biology: predicting invaders. Trends Ecol Evol 16:199–204

Krainacker DA, Carey JR, Vargas RI (1987) Effect of larval host on life history traits of the Mediterranean fruit fly, *Ceratitis capitata*. Oecologia 73:583–590

Ladanyi M, Horvath L (2010) A review of the potential climate change impact on insect populations – general and agricultural aspects. Appl Ecol Environ Res 8:143–152

Lampe I, Burghause F, Krauthausen HJ (2005) Introduction and distribution of the American Eastern Cherry Fruit Fly, *Rhagoletis cingulata*, in the Rhine Valley, Germany. In: Alford DV, Backhaus GF (eds) Plant protection and plant health in Europe: introduction and spread of invasive species. British Crop Protection Council, Berlin, pp 135–140

Lance DR, Gates DB (1994) Sensitivity of detection trapping systems for Mediterranean fruit-flies (Diptera, Tephritidae) in Southern California. J Econ Entomol 87:1377–1383

Leblanc L, Putoa R (2000) Fruit flies in French Polynesia and Pitcairn Islands. Secretariat of the Pacific Community Pest Advisory Leaflet. Suva, Fiji

Lethmayer C (2011) Drei invasive lnsekten gefahrden den heimischen Obstbau. Pflanzenschutz 12:4–5

Li BN, Ma J, Hu XN, Liu HJ, Zhang RJ (2009) Potential geographical distributions of the fruit flies *Ceratitis capitata*, *Ceratitis cosyra*, and *Ceratitis rosa* in China. J Econ Entomol 102: 1781–1790

Li Y, Wu Y, Chen H, Wu J, Li Z (2012) Population structure and colonization of *Bactrocera dorsalis* (Diptera: Tephritidae) in China, inferred from mtDNA COI sequences. J Appl Entomol 136:241–251

Liebhold AM, Tobin PC (2008) Population ecology of insect invasions and their management. Ann Rev Entomol 53:387–408

Liebhold AW, Work TT, Mcgullough DG, Cavey JF (2006) Airline baggage as a pathway for alien insect species invading the United States. Am Entomol Spring 2006:48–54

Liquido NJ, Shinoda LA, Cunningham RT (1991) Host plants of the Mediterranean fruit fly (Diptera, Tephritidae) an annotated world review, Miscellaneous publications 77. Entomological Society of America. Entomological Society of America, Lanham, MD

Lux SA, Copeland RS, White IM, Manrakhan A, Billah MK (2003) A new invasive fruit fly species from the *Bactrocera dorsalis* (Hendel) group detected in East Africa. Int J Trop Insect Sci 23:355–361

Lynch HJ, Rhainds M, Calabrese JM, Cantrell S, Cosner C, Fagan WF (2014) How climate extremes-not means-define a species' geographic range boundary via a demographic tipping point. Ecol Monogr 84:131–149

Malacrida AR, Marinoni F, Torti C, Gomulski LM, Sebastiani F, Bonvicini C, Gasperi G, Guglielmino CR (1998) Genetic aspects of the worldwide colonization process of *Ceratitis capitata*. J Hered 89:501–507

Malacrida AR, Gomulski LM, Bonizzoni M, Bertin S, Gasperi G, Gugliclmino CR (2007) Globalization and fruitfly invasion and expansion: the medfly paradigm. Genetica 131:1–9

Manrakhan A, Hattingh V, Grout TG, De Villiers M, Venter, J-H (2012) Combating *Bactrocera invadens* in South Africa. TEAM – 2nd international meeting "Biological Invasions of Tephritidae: ecological and economic impacts", 2012 Kolymbari, Crete, Greece. Thessaloniki: Ziti, 144

Mau RFL, Jang EB, Vargas RI (2007) The Hawaii area-wide fruit fly pest management programme: influence of partnerships and a good education programme. In: Vreysen MJB, Robinson AS, Hendrichs J (eds) Area-wide control of insect pests. Springer, Vienna

Mavrikakis PG, Economopoulos AP, Carey JR (2000) Continuous winter reproduction and growth of the Mediterranean fruit fly (Diptera: Tephritidae) in Heraklion, Crete, southern Greece. Environ Entomol 29:1180–1187

Meixner MD, Mcpheron BA, Silva JG, Gasparich GE, Sheppard WS (2002) The Mediterranean fruit fly in California: evidence for multiple introductions and persistent populations based on microsatellite and mitochondrial DNA variability. Mol Ecol 11:891–899

Mirsardoo S, Mafi-Pashakolaei SA, Barari H (2010) Preliminary investigation on the geographical distribution of Mediterranean fruit fly, *Ceratitis capitata* (Wiedemann) (Dip., Tephritidae), in Mazandaran province, Iran. J Entomol Res 2:143–154

Mohamed S, Ekesi S, Khamis F (2012) Biology and management of fruit flies in Africa and their potential impact in Near East countries. Regional symposium on the management of fruit flies in Near East countries, Hammamet, Tunisia. 80

Mooney HA, Cleland EE (2001) The evolutionary impact of invasive species. Proceedings of the National Academy of Sciences of the United States of America 98:5446–5451

Moraiti CA, Nakas CT, Papadopoulos NT (2014) Diapause termination of *Rhagoletis cerasi* pupae is regulated by local adaptation and phenotypic plasticity: escape in time through bet-hedging strategies. J Evol Biol 27:43–54

Myers JH, Simberloff D, Kuris AM, Carey JR (2000) Eradication revisited: dealing with exotic species. Trends Ecol Evol 15:316–320

Nardi F, Carapelli A, Dallai R, Roderick GK, Frati F (2005) Population structure and colonization history of the olive fly, *Bactrocera oleae* (Diptera, Tephritidae). Mol Ecol 14:2729–2738

Nardi F, Carapelli A, Boore JL, Roderick GK, Dallai R, Frati F (2010) Domestication of olive fly through a multi-regional host shift to cultivated olives: comparative dating using complete mitochondrial genomes. Mol Phylogenet Evol 57:678–686

Ni WL, Li ZH, Chen HJ, Wan FH, Qu WW, Zhang Z, Kriticos DJ (2012) Including climate change in pest risk assessment: the peach fruit fly, *Bactrocera zonata* (Diptera: Tephritidae). Bull Entomol Res 102:173–183

Nishida T, Bess HA (1950) Applied ecology in melon fly control. J Econ Entomol 43:877–883

Nyamukondiwa C, Terblanche JS (2009) Thermal tolerance in adult Mediterranean and Natal fruit flies (*Ceratitis capitata* and *Ceratitis rosa*): effects of age, gender and feeding status. J Therm Biol 34:406–414

Nyamukondiwa C, Kleynhans E, Terblanche JS (2010) Phenotypic plasticity of thermal tolerance contributes to the invasion potential of Mediterranean fruit flies (*Ceratitis capitata*). Ecol Entomol 35:565–575

Ohno S, Tamura Y, Haraguchi D, Matsuyama T, Kohama T (2009) Re-invasions by *Bactrocera dorsalis* complex (Diptera: Tephritidae) occurred after its eradication in Okinawa, Japan, and local differences found in the frequency and temporal patterns of invasions. Appl Entomol Zool 44:643–654

Papachristos DP, Papadopoulos NT (2009) Are citrus species favorable hosts for the Mediterranean fruit fly? A demographic perspective. Entomol Exp Appl 132:1–12

Papadopoulos NT (1999) Study on the biology and ecology of the Mediterranean fruit fly *Ceratitis capitata* (Diptera: Tephritidae) in northern Greece. PhD dissertation, (in Greek with English summary). Aristotle University of Thessaloniki

Papadopoulos NT, Carey JR, Katsoyannos BI, Kouloussis NA (1996) Overwintering of the Mediterranean fruit fly (Diptera: Tephritidae) in northern Greece. Ann Entomol Soc Am 89: 526–534

Papadopoulos NT, Katsoyannos BI, Carey JR (1998) Temporal changes in the composition of the overwintering larval population of the Mediterranean fruit fly (Diptera: Tephritidae) in northern Greece. Ann Entomol Soc Am 91:430–434

Papadopoulos NT, Katsoyannos BI, Carey JR, Kouloussis NA (2001a) Seasonal and annual occurrence of the Mediterranean fruit fly (Diptera: Tephritidae) in northern Greece. Ann Entomol Soc Am 94:41–50

Papadopoulos NT, Katsoyannos BI, Kouloussis NA, Hendrichs J, Carey JR, Heath RR (2001b) Early detection and population monitoring of *Ceratitis capitata* (Diptera: Tephritidae) in a mixed-fruit orchard in northern Greece. J Econ Entomol 94:971–978

Papadopoulos NT, Katsoyannos BI, Carey JR (2002) Demographic parameters of the Mediterranean fruit fly (Diptera: Tephritidae) reared in apples. Ann Entomol Soc Am 95:564–569

Papadopoulos NT, Plant RE, Carey JR (2013) From trickle to flood: the large-scale, cryptic invasion of California by tropical fruit flies. Proc R Soc B Biol Sci 280:20131466

Paranhos BJ, Papadopoulos NT, Mcinnis D, Gava C, Lopes FSC, Morelli R, Malavasi A (2010) Field dispersal and survival of sterile medfly males aromatically treated with ginger root oil. Environ Entomol 39:570–575

Parker IM, Simberloff D, Lonsdale WM, Goodell K, Wonham M, Kareiva PM, Williamson MH, Von Holle B, Moyle PB, Byers JE, Goldwasser L (1999) Impact: toward a framework for understanding the ecological effects of invaders. Biol Invasions 1:3–19

Penarrubia-Maria E, Avilla J, Escudero-Colomar LA (2012) Survival of wild adults of *Ceratitis capitata* (Wiedemann) under natural winter conditions in North East Spain. Psyche, 6 pages

Perrings C, Dehnen-Schmutz K, Touza J, Williamson M (2005) How to manage biological invasions under globalization. Trends Ecol Evol 20:212–215

Ponti L, Cossu QA, Gutierrez AP (2009) Climate warming effects on the *Olea europaea-Bactrocera oleae* system in Mediterranean islands: Sardinia as an example. Glob Change Biol 15:2874–2884

Ponti L, Gutierrez AP, Ruti PM, Dell Aquila A (2014) Fine-scale ecological and economic assessment of climate change on olive in the Mediterranean Basin reveals winners and losers (in press). doi:10.1073/pnas.1314437111)

Richardson DM, Pyšek P, Rejmánek M, Barbour MG, Panetta FD, West CJ (2000) Naturalization and invasion of alien plants: concepts and definitions. Divers Distrib 6:93–107

Rigamonti IE (2004) Contributions to the knowledge of *Ceratitis capitata* Wied. (Diptera, Tephritidae) in Northern Italy. Observations on the biology. Boll Zool Agr Bachicoltura 36: 89–100

Rigamonti IE, Agosti M, Malacrida AR (2002) Distribuzione e danni della Mosca Mediterranea della Frutta *Ceratitis capitata* Wied. (Diptera Tephritidae) in Lombardia (Italia settentrionale). XIX Congresso nazionale italiano di Entomologia, Catania, Italy, pp 581–587

San Jose M, Leblanc L, Geib SM, Rubinoff D (2013) An evaluation of the species status of *Bactrocera invadens* and the systematics of the *Bactrocera dorsalis* (Diptera: Tephritidae) Complex. Ann Entomol Soc Am 106:684–694

Sauers-Muller A (1991) An overview of the carambola fruit fly *Bactrocera* species (Diptera: Tephritidae), found recently in Suriname. Fla Entomol 74:432–440

Schrader J, Luders S, Kulschewski A, Hammersen F, Plate K, Berger J, Zidek W, Dominiak P, Diener HC, MOSES Study Group (2006) MOSES raises questions – Response. Stroke 37: 339–340

Sequeira R, Millar L, Bartels D (2001) Identification of susceptible areas for the establishment of Anastrepha spp. Fruit flies in the United States and analysis of selected pathways. [Online]. Available: http://www.aphis.usda.gov/plant_health/plant_pest_info/fruit_flies/.../isa.pdf
Shelly TE, Edu J (2009) Capture of mass-reared vs. wild-like males of *Ceratitis capitata* (Dipt., Tephritidae) in trimedlure-baited traps. J Appl Entomol 133:640–646
Shelly TE, Nishimoto J (2011) Additional measurements of distance-dependent capture probabilities for released males of *Bactrocera cucurbitae* and *B. dorsalis* (Diptera: Tephritidae) in Honolulu. J Asia Pac Entomol 14:271–276
Shelly T, Nishimoto J, Diaz A, Leathers J, War M, Shoemaker R, Al-Zubaidy M, Joseph D (2010) Capture probability of released males of two *Bactrocera* species (Diptera: Tephritidae) in detection traps in California. J Econ Entomol 103:2042–2051
Shi W, Kerdelhue C, Ye H (2010) Population genetic structure of the oriental fruit fly, *Bactrocera dorsalis* (Hendel) (Diptera: Tephritidae) from Yunnan province (China) and nearby sites across the border. Genetica 138:377–385
Siebert JB, Cooper T (1995) Embargo on California produce would cause revenue, job loss. Calif Agric 49:7–12
Simberloff D (2009) The role of propagule pressure in biological invasions. Ann Rev Ecol Evol Systemat 40:81–102
Simberloff D, Martin J-L, Genovesi P, Maris V, Wardle DA, Aronson J, Courchamp F, Galil B, García-Berthou E, Pascal M, Pyšek P, Sousa R, Tabacchi E, Vilà M (2013) Impacts of biological invasions: what's what and the way forward. Trends Ecol Evol 28:58–66
Smith JJ, Bush GL (1997) Phylogeny of the genus *Rhagoletis* (Diptera: Tephritidae) inferred from DNA sequences of mitochondrial cytochrome oxidase II. Mol Phylogenet Evol 7:33–43
Stephens AEA, Kriticos DJ, Leriche A (2007) The current and future potential geographical distribution of the oriental fruit fly, *Bactrocera dorsalis* (Diptera: Tephritidae). Bull Entomol Res 97:369–378
Tatem AJ (2009) The worldwide airline network and the dispersal of exotic species: 2007–2010. (vol 32, pg 94, 2009). Ecography 32:703–704
Tatem AJ, Hay SI, Rogers DJ (2006) Global traffic and disease vector dispersal. Proc Natl Acad Sci 103:6242–6247
Terblanche JS, Nyamukondiwa C, Kleynhans E (2010) Thermal variability alters climatic stress resistance and plastic responses in a globally invasive pest, the Mediterranean fruit fly (*Ceratitis capitata*). Entomol Exp Appl 137:304–315
Trumble JT, Butler CD (2009) Climate change will exacerbate California's insect pest problems. Calif Agric 63:73–78
Tzanakakis ME, Katsoyannos BI (eds) (2003) Insects of fruit trees and grapevine. Agrotypos, Marousi
Vargas RI, Carey JR (1990) Comparative survival and demographic-statistics for wild Oriental fruit-fly, Mediterranean fruit-fly, and Melon fly (Diptera, Tephritidae) on Papaya. J Econ Entomol 83:1344–1349
Vargas RI, Nishida T (1985) Life history and demographic parameters of *Dacus latifrons* (Diptera: Tephritidae). J Econ Entomol 78:1242–1244
Vargas RI, Miyashita D, Nishida T (1984) Life-history and demographic parameters of three laboratory-reared tephritids (Diptera, Tephritidae). Ann Entomol Soc Am 77:651–656
Vargas RI, Walsh WA, Nishida T (1995) Colonization of newly planted coffee fields – dominance of Mediterranean fruit fly over oriental fruit fly (Diptera, Tephritidae). J Econ Entomol 88: 620–627
Vargas RI, Walsh WA, Kanehisa D, Jang EB, Armstrong JW (1997) Demography of four Hawaiian fruit flies (Diptera: Tephritidae) reared at five constant temperatures. Ann Entomol Soc Am 90:162–168
Vargas RI, Pinero JC, Jang EB, Mau RFL, Stark JD, Gomez L, Stoltman L, Mafra-Neto A (2010) Response of melon fly (Diptera: Tephritidae) to weathered SPLAT-Spinosad-Cue-Lure. J Econ Entomol 103:1594–1602

Vargas RI, Souder SK, Mackey B, Cook P, Morse JG, Stark JD (2012) Field trials of solid triple lure (trimedlure, methyl eugenol, raspberry ketone, and DDVP) dispensers for detection and male annihilation of *Ceratitis capitata*, *Bactrocera dorsalis*, and *Bactrocera cucurbitae* (Diptera: Tephritidae) in Hawaii. J Econ Entomol 105:1557–65

Vayssières JF, Cayol JP, Perrier X, Midgarden D (2007) Impact of methyl eugenol and malathion bait stations on non-target insect populations in French Guiana during an eradication program for *Bactrocera carambolae*. Entomol Exp Appl 125:55–62

Vayssières JF, Carel Y, Coubes M, Duyck PF (2008) Development of immature stages and comparative demography of two cucurbit-attacking fruit flies in Reunion Island: *Bactrocera cucurbitae* and *Dacus ciliatus* (Diptera Tephritidae). Environ Entomol 37:307–314

Venette R, Koch R (2009) IPM for invasive species. In: Radcliffe E, Hutchison W, Cancelado R (eds) Integrated pest management: concepts tactics strategies, and case studies. Cambridge University Press, New York

Venette RC, Kriticos DJ, Magarey RD, Koch FH, Baker RHA, Worner SP, Raboteaux NNG, Mckenney DW, Dobesberger EJ, Yemshanov D, De Barro PJ, Hutchison WD, Fowler G, Kalaris TM, Pedlar J (2010) Pest risk maps for invasive alien species: a roadmap for improvement. Bioscience 60:349–362

Vera MT, Rodriguez R, Segura DF, Cladera JL, Sutherst RW (2002a) Potential geographical distribution of the Mediterranean fruit fly, *Ceratitis capitata* (Diptera: Tephritidae), with emphasis on Argentina and Australia. Environ Entomol 31:1009–1022

Vera MT, Wood RJ, Cladera JL, Gilburn AS (2002b) Factors affecting female remating frequency in the Mediterranean fruit fly (Diptera: Tephritidae). Fla Entomol 85:156–164

Vermeij GJ (1996) An agenda for invasion biology. Biol Conserv 78:3–9

Villablanca FX, Roderick GK, Palumbi SR (1998) Invasion genetics of the Mediterranean fruit fly: variation in multiple nuclear introns. Mol Ecol 7:547–560

Virgilio M, Delatte H, Backeljau T, De Meyer M (2010) Macrogeographic population structuring in the cosmopolitan agricultural pest *Bactrocera cucurbitae* (Diptera: Tephritidae). Mol Ecol 19:2713–2724

Virgilio M, Delatte H, Quilici S, Backeljau T, De Meyer M (2013) Cryptic diversity and gene flow among three African agricultural pests: *Ceratitis rosa*, *Ceratitis fasciventris* and *Ceratitis anonae* (Diptera, Tephritidae). Mol Ecol 22:2526–2539

Wan XW, Nardi F, Zhang B, Liu YH (2011) The oriental fruit fly, *Bactrocera dorsalis*, in China: origin and gradual inland range expansion associated with population growth. PLoS One 6: e25238

White IM (2006) Taxonomy of the Dacinae (Diptera: Tephritidae) of Africa and the Middle East. Afr Entomol Mem 2:1–156

White IM, Elson-Harris MM (1992) Fruit flies of economic significance their identification and bionomics. CAB International, Wallingford

White IM, De Meyer M, Stonehouse J (2001) A review of the native and introduced fruit flies (Diptera, Tephritidae) in the Indian Ocean Islands of Mauritius, Réunion, Rodrigues and Seychelles. In: Price NS, Seewooruthun I (eds) Proceedings of the Indian Ocean Commission Regional Fruit Fly Symposium, Mauritius, 5–9th June 2000, pp 15–21. Indian Ocean Commission, Mauritius

Willyard C (2012) Playing detective. Nature 491:S64–S65

Wong TTY, Whitehand LC, Kobayashi RM, Ohinata K, Tanaka N, Harris EJ (1982) Mediterranean fruit-fly (Diptera, Tephritidae) – dispersal of wild and irradiated and untreated laboratory-reared males. Environ Entomol 11:339–343

Work TT, Mccullough DG, Cavey JF, Komsa R (2005) Arrival rate of nonindigenous insect species into the United States through foreign trade. Biol Invasions 7:323–332

Yang PJ, Carey JR, Dowell RV (1994) Tephritid fruit-flies in China – historical background and current status. Pan Pac Entomol 70:159–167

Yokoyama VY, Miller GT, Stewart-Leslie J, Rice RE, Phillips PA (2006) Olive fruit fly (Diptera: Tephritidae) populations in relation to region, trap type, season, and availability of fruit. J Econ Entomol 99:2072–2079

Yonow T, Zalucki MP, Sutherst RW, Dominiak BC, Maywald GF, Maelzer DA, Kriticos DJ (2004) Modelling the population dynamics of the Queensland fruit fly, *Bactrocera* (*Dacus*) *tryoni*: a cohort-based approach incorporating the effects of weather. Ecol Model 173:9–30

Zygouridis NE, Augustinos AA, Zalom FG, Mathiopoulos KD (2009) Analysis of olive fly invasion in California based on microsatellite markers. Heredity 102:402–412

Chapter 8
Fruit Fly Detection Programs: The Potentials and Limitations of Trap Arrays

A. Meats

Abstract Detection programs are a specialized aspect of sampling and include quarantine inspections, surveys for rare organisms of conservation value, and surveillance trapping for exotic or locally quarantined pests. The aim in each case is to establish whether a given species is there or not with a reasonable degree of certainty. To detect incursions of exotic tephritid fruit fly species into a country, arrays of widely spaced sentinel traps are deployed around points of entry for people and goods, main centres of population, and commercial fruit production areas. Response to detection is according to a protocol (code of practice). This includes the installation of a higher density trap array that is used to (a) discover the spatial limits of the infestation, (b) to monitor the effectiveness of the eradication process, and (c) to confirm that eradication has, in fact, occurred when zero flies are caught in the trap array. It takes a very large number of trapping weeks (well over a year) to achieve confidence limits on zero of useful size. However, a much shorter period of zero trapping is needed to calculate a useful probability for some density (or index of density) that we know to be non-viable or would find acceptable for other reasons. This chapter deals with such problems in terms of rationally argued risk levels using examples of management of the Mediterranean fruit fly in South Australia and California. Because risk arguments involve the length of time when trapping arrays catch no flies (fly-free periods), attention is also given to the techniques of temperature and development summation and how daily temperature records, calendar time, and generation time are related. Finally, the impacts of any improvements in trap efficiency, trap placement and data management are considered, especially with respect to telemetry, delimitation, and extinction modelling.

Keywords Area freedom • By-catch • *Bactrocera tryoni* • *Ceratitis capitata* • Day degrees • Eradication • Extinction • Medfly • Probit 9 • Telemetry • Trap shifting • Quarantine • Viable density • Zero catch period

A. Meats (✉)
School of Biological Sciences A12, University of Sydney, Sydney, NSW 2006, Australia
e-mail: alan.meats@sydney.edu.au

T. Shelly et al. (eds.), *Trapping and the Detection, Control, and Regulation of Tephritid Fruit Flies*, DOI 10.1007/978-94-017-9193-9_8,

1 Introduction

Invasion of a country by an exotic species of fruit fly (Diptera: Tephritidae) can cause expense through loss of production and restrictions on the export of fruit due to international trade agreements. Infestations of most fruit flies of economic importance can be dealt with at the incipient stage if they are detected early by arrays of sentinel traps. These should consist of at least three kinds of trap, baited to attract the main 'lure groups' of tephritid species, those attracted to methyl eugenol, cue-lure, and trimedlure, respectively (Tan et al., Chap. 2, this volume). Also, where relevant, the array can be augmented with traps containing proteinaceous baits (Epsky et al., Chap. 3, this volume) for species that do not respond to any of the main lure groups. To detect incursions of exotic species, these arrays are deployed around points of entry for people and goods, main centres of population, and commercial fruit production areas. Response to detection is according to a protocol (code of practice). This protocol includes restriction of the movement of fruit, eradication measures, and the installation of a higher density trap array that is used to (a) discover the spatial limits of the infestation, (b) to monitor the effectiveness of the eradication process, and (c) to confirm that eradication has, in fact, occurred when zero flies are caught on the trap array. Protocols vary by species and also can differ between and within countries (e.g., see Table 8.1).

Species that are endemic within a country can also affect trade. Trade barriers may not apply, however, if the fruit is subjected to post-harvest treatment if it is grown within a production area that is not free from infestation but post-harvest treatment may be waived for fruit grown in production areas that are be deemed to be 'fly-free zones'. The 'area freedom' status of each zone is audited, in part, by a sentinel trap array, and each zone must comply with a management protocol (as above) when an incursion is detected.

2 Surveillance for Incipient Incursions

The detection of an incipient infestation would seem to be quite straightforward, because the flies announce their arrival by getting trapped. However, a population could have been present some time before the first trapping (especially if it started between widely spaced traps or the lure was weak). The rate of trapping may indicate that a population is not large enough to be viable, and the regulatory authority may require a sufficiently high rate of trapping (the trigger level or action threshold) before any action is required. The inference here is that, if the trapping rate was lower than this trigger level, the infestation would die out – presumably due to some aspect of the Allee Effect, such as a lowered chance of finding mates (Bateman 1977; Meats 1998b; Meats et al. 2003a; Meats and Edgerton 2008). If this were not the case, the numbers would rise to the action threshold, and extinction would be imposed by human intervention rather than occurring naturally. Such

Table 8.1 Sentinel and monitoring traps for Medfly in two jurisdictions

Specification	California	South Australia
Normal sentinel density (male lure traps only)		
Per sq. mile	5	16.19
Per sq. km	1.93	6.25
Spacing (m)	720	400
Action threshold (flies trapped)	2 within 3 mile Radius (141 traps)	3 within 1 km Radius[a] (20 traps)
	Within 1 life cycle (~6–8 weeks)	Within 2 weeks
Delimitation/monitoring (male lure traps only)		
Traps per sq. mile	[b]100	99.1
Traps per sq. km	[b]39	38.25
Spacing (m)	[b]161	162

[a]1 km radius = 3.142 km^2 = 1.21 mi^2
[b]Core square mile (2.59 km^2). The latter is surrounded by 4 buffer zones, each with fewer traps per square mile than the zone more central to it
For further detail, see relevant codes of practice

assumptions are only possible with experience and thus can only apply to species familiar to the authorities. In the case of a rarely encountered exotic species, there may be no information on what the action threshold should be, so the default level could well be one fly.

3 The Efficiency of a Detection System

To obtain the lowest possible risk of missing the earliest stage of an incipient infestation, it would be ideal to have a surveillance trap array that was a sensitive as possible with trap spacing reduced to (say) 20 m or less. However, the areas to be monitored are typically very large, and the potential expense and logistic problems force a compromise with trap spacing between 400–1,000 m (Tables 8.1 and 8.2). The probability of a fruit fly being caught by a given trap in an array depends upon how near it is to that trap (Meats 1998a, b). If flies emerge from pupae at a point within a trap array, or adults are experimentally released at a point, the catch declines with distance. The rate of decline is slow within ~100 m of the origin (Weldon and Meats 2007; Meats and Edgerton 2008) and beyond that according to a power model (Cunningham and Couey 1986; Plant and Cunningham 1991; Lance and Gates 1994; Meats 1998a, b; Meats and Smallridge 2007; Meats and Edgerton 2008; Shelly et al. 2010; Shelly and Nishimoto 2011). The distribution over the whole range of distance is best described by the Cauchy distribution (Mayer and Atzeni 1993; Clift et al. 1998; Meats and Smallridge 2007; Meats and Edgerton

Table 8.2 Grid calibration and surveillance sensitivity

Species	Test flies[a] flown (males)	Trap spacing (m)	Percentage[b] recaptured	Prediction limit[c]	Reference[d]
B. dorsalis	11,142	720	22.9	37	1
B. cucurbitae	1,904	720	4.25	220	2
B. tryoni	367 (wild)	400	4.1	229	3
B. tryoni	Mass release	400	0.47	2,027	4
B. tryoni	Mass release	400	1.69	561	5
B. tryoni	Mass release	400	0.34	2,804	6
C. capitata	9,600	400	0.6	1,587	7
C. capitata	Mass release	400	0.9	1,057	8

[a]Sterile flies unless otherwise indicated

[b]Recaptures as % males flown in all cases

[c]Prediction from recapture rate (using Eq. 8.2) of number of mature wild males required to give 99.9 % chance of detection

[d]References: 1, 2, Shelly et al. (2010); 3, Monro and Richardson (1969); 4, 5 Meats et al. (2003b); 5, 6 Reynolds et al. 2012; 7, Lance and Gates (1994); 8, Smallridge and Hopkins (2004)

2008). Experiments with released tephritid fruit flies have shown that almost all of the recaptures are likely to be made within 1 km of the point of release (Plant and Cunningham 1991; Meats and Smallridge 2007; Meats and Edgerton 2008).

The importance of trap spacing and the distance/response relationship has been illustrated by an heuristic model using what amounts to an approximation of the Cauchy distribution (Meats 1998b). This model shows how the detection of a wild male fruit fly within a grid array could signal the existence there of a number of mature flies and that the error of the estimate depends upon how far the origin of the infestation was from the nearest trap. In particular, the model shows that the upper limit of estimated range increases with increasing inter-trap distance.

The question then arises as to how big would an infestation have to be for there to be a given chance (say 99.9 %) of catching at least one fly. Distance-response data (from releases of flies at different distances from given traps) have been used to estimate the number of wild mature males within a grid array that would have a given chance of detection (Cunningham and Couey 1986; Lance and Gates 1994; Shelly et al. 2010; Shelly and Nishimoto 2011). Unfortunately, few investigators have used this method so for the purposes of this review, another method is given for comparing the detectability of various species. If releases of flies (usually sterile) are made within an array of surveillance traps, there is an easier (if more approximate) computation using the overall recapture rate (p) and the binomial theorem that is justified as follows.

The confidence limit (CL) of a proportion p can be found by

$$\mathrm{CL} = t \cdot (p\, q/n)^{0.5} \tag{8.1}$$

Where $q = (1 - p)$, $n =$ sample size, and the value of t depends upon $n - 1$ and the desired significance level.

The value of n that would be significantly different from zero (say a 99.9 % chance of detection or a significance level of $= 0.001$) for a given proportion can be estimated because p –lower CL = 0, hence p = lower CL and re-arranging (8.1) gives

$$n = p\,q/(p/t)^2 \qquad (8.2)$$

Equation (8.2) is solved for n using the recapture rate as a proportion (p) and a value of $t = 3$ that corresponds approximately to the required level of significance. A more accurate value of t (one-tailed) can then be found with a t calculator (several are available on the internet) using the estimate of n and a significance level of $a = 0.001$. The new value of t can then be used in Eq. (8.2) to find a more accurate estimate of n. This process can be repeated one or more times but it is hardly worth it, because it makes little difference, and there are probably unknown factors with greater effect on the recapture rate used in the equation. Note also that the use of one-tailed' t as opposed to two-tailed t makes very little difference to the result. The results of such calculations are given in Table 8.2, where a value of $n > 300$ indicates that an infestation probably would be more than a generation old before detection is 99.9 % certain.

Results of release-recapture trials indicate that the chance of recapture in a lure trap is typically highest when the flies become mature enough to respond to the lure, and then it diminishes rapidly to zero with time over 2–3 weeks (e.g., see Fletcher 1974a; Meats and Edgerton 2008; Gilchrist and Meats 2012). Thus, any method using recapture rates that would apply to a single cohort would have to use recapture rates based on accumulated trappings over a period determined by trials. Real populations have more complex dynamics. They can be affected by immigration as well as emigration (Fletcher 1973), and they can generate co-existing cohorts of different ages and sizes with these becoming mature (trappable) at different times (Meats 1983). Thus, wild flies from some of the overlapping cohorts will be exposed to trapping over a given period during one stage of their trappable lives, and flies from other cohorts will be exposed at other stages. So it could be said that, collectively, they experience the same distribution of age-related trapping rates over that period as a single cohort of sterile flies that is exposed to trapping from the time they start to be responsive to traps. Thus, in theory, the chances of a wild fly being trapped during a given period are the same as a released one in a recapture trial that extends for the same period.

The conclusion above depends on the assumption that wild and sterile flies have identical abilities to survive and disperse and respond to lures. Indeed, any difference could be used as a measure of sterile fly quality (Meats et al. 2003b). Tests in laboratory cages and field enclosures have revealed various physiological differences between irradiated and unirradiated flies that could be due to domestication, artificial conditions of culture or the stress of irradiation, packing, or transport before release (Barry et al. 2003; Dominiak et al. 2007a; Worsley et al. 2008; Collins et al. 2009; Weldon et al. 2010; Gilchrist et al. 2012; Rull et al. 2012).

Ideally, the recapture rate of wild and sterile flies should be compared in open field conditions on the same trap array, so that a correction factor could be obtained for the use of 'sterile only' recapture rates on a given grid to estimate the potential detectability of wild flies on the same grid. Shelly and Edu (2009) came close to this by simultaneously comparing recapture rates of 'mass-reared' and 'wild-like' (recently domesticated) *Ceratitis capitata* (Wiedemann) or medfly, finding that the recapture rates of 'wild-like' flies were higher than those obtained for mass-reared ones. Similar trials with *Bactrocera tryoni* (Froggatt) found no difference between recapture rates of irradiated (mass-reared) and unirradiated 'wild-like' flies (Weldon and Meats 2010), and Gilchrist and Meats (2012) found no difference in terms of recapture rates and distances flown between mass-reared and cultured (but outbred) flies. Notwithstanding this, it is apparent that the recapture rate of sterile flies can vary from site to site even when flies from the same source are irradiated and released simultaneously (Reynolds et al. 2012).

It is also important to note that for species for which there is no strong male lure, the next best thing is a natural or synthetic 'food- type' lure ((Epsky et al., Chap. 3, this volume). When closely spaced (say 10 m) food-baited traps can achieve recapture rates of up to ~30–40 % in trials (Kendra et al. 2010). This means that, whereas food-baited traps would be suitable for monitoring eradication in small plots, the required spacing makes them unsuitable for use as sentinel traps over large areas and that the best tactic in such circumstances would be to place sentinel traps in certificated crops within production zones (Simpson 1993).

4 Surveillance for Resurgence After Eradication Procedures

The end of an infestation is hard to determine because the absence of flies in the traps is not proof that they are completely absent in the area being monitored. This is a problem that has its counterparts in other fields from wildlife conservation to quarantine inspections. It is possible to determine the confidence limits on zero, but for feasible sample sizes, the limits are too wide to be useful (Venette et al. 2002; Sauro 2005; Lewis and Sauro 2006). Even for quarantine inspections, when sampling is of a small (finite) amount of goods, the only way to be 100 % certain that there are no pests is to sample the whole consignment (Venette et al. 2002) and when sampling is destructive or without replacement, this defeats the object of the exercise. When trapping a fruit fly population of unknown size, it takes a very large number of trapping weeks (or rather months) to achieve confidence limits of useful size and an infinite number of samples for complete certainty (Fig. 8.1). So as with any sampling to determine the presence of a rare species, we can abandon the pursuit of confidence limits on zero but instead calculate limits for some density (or index of density) that we would find acceptable.

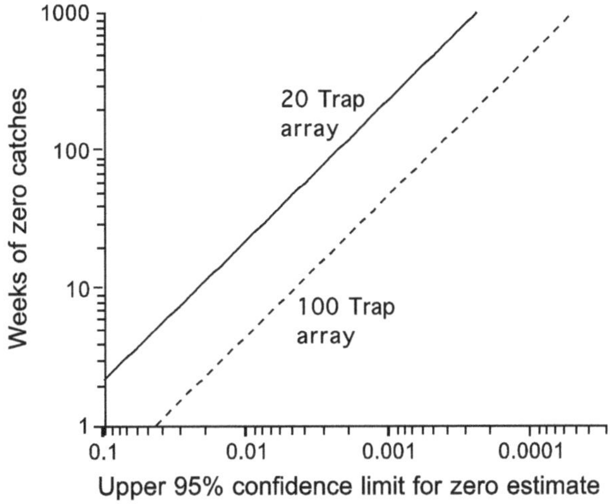

Fig. 8.1 The number of weeks of zero catches by a given trap array that are required to attain a given confidence level for an estimate of zero flies. Note the law of diminishing returns for each extra week of trap surveillance and the impossibility of obtaining a confidence limit equal to zero (This figure is based on the results of the 'exact method' option of the calculator of Sauro (2005) using $t \times$ weeks for sample size where $t =$ number of traps in the array)

For rare and endangered species, McArdle (1990) used simple probability equations for the estimation of the number of sampling units required for sufficient precision given any expected frequency of occurrence. The rationale has analogies with that used for trapping data in this chapter. McArdle (1990) proposed that a decision must be made as to the degree of rarity worth detecting. If rarity is defined as the chance (p) of detecting a species in a sampling unit, the probability (a) of detecting it in the whole sample (all the sampling units used) is

$$a = 1 - (1 - p)^N \tag{8.3}$$

The sample size (number of sampling units, N) needed for a desired value of a is found by rearranging (8.3).

$$N = \left(\log(1 - a)/(\log(1 - p)\right) \tag{8.4}$$

The acceptability of the indicated sampling effort would depend on the perceived conservation value (of either the species or the site) or the difficulty of rescuing the population from such a low level. In the case of sampling to confirm the eradication of an exotic weed, the optimal stopping time would be a trade-off between the cost of continued sampling and the cost of eradication if resurgence of the weed occurs from an undetected low density (Regan et al. 2006). For quarantine inspections, the tolerable level of a pest in cargo could be related to the size of inoculum required to establish in the importing country. However the tolerable level appears to be arbitrary and adjustable according to trade volume but is probably based on what seems to have worked in the past (Venette et al. 2002).

For establishing 'area freedom' status from fruit flies an 'acceptable density' would be indicated by the trapping rate that is considered too low to warrant any regulatory action (see above).

5 Establishing Effective Extinction Within a Trap Array

The mean expectation of the number of flies (m) caught with time on a trap array of given size rises with time and is found by

$$m = c_r t w \tag{8.5}$$

where c_r = catch per trap per week, t = number of traps and w = number of weeks.

At first sight, Eq. (8.5) has redundant terms because c_r is calculated using t and w. However, c_r plays a role that is equivalent to that of p in Eqs. (8.3) and (8.4). When trap catches are reduced to zero, c_r can be set at a level signifying a population density that does not require regulatory action. It can then be used to determine how many weeks of trapping with zero captures would signify that population density is below that level, and area freedom could be declared again. The rationale is as follows.

The probability (P_0) of a zero catch with time on a given array falls with time and is found by the zero term of the Poisson equation

$$P_0 = \exp(-c_r t\, w) \tag{8.6}$$

The number of weeks of zero catches required to achieve a desired value of P_0 (such as 0.01, 0.001 or 0.00003) with a given value of c_r is found by

$$w = (-\ln P_{0\ crit})/(c_r t) \tag{8.7}$$

6 Effective Size of Trap Arrays

Arrays of sentinel traps can extend over large areas (Smith 2000; USDA 2006; Shelly et al. 2010), but only those close to the origin of a new infestation are augmented with extra traps during and after the eradication procedure. As trapping protocols were established before much research on fruit fly dispersal, some of the intensification of trapping effort would appear to cover a wider area than is necessary. However, a large area of trap intensification may be necessary if sentinel traps are so widely spaced that an infestation has time to spread and establish extra foci at a distance from the origin before any detection happens. However, this chapter is only concerned with single foci and hence with traps within a 1 km radius of the origin or within a square mile with the origin at the centre. Thus, Eqs. 8.5, 8.6, and 8.7 only apply to such arrays. When there are multiple foci, each must be dealt with as a separate infestation with treatments and trapping arrays organised with respect to every apparent point of origin. This is not to say that such separate foci are never part of the same introduction event. The use of molecular markers and sib analysis has shown that outbreak flies with a high probability of being from the same source can occur in foci separated by distances that appear large enough to implicate spread by human agency rather than natural dispersal (Gilchrist et al. 2004).

7 Effect of Changing Trap Density

Any change from sentinel trap density to that used for delimiting the infestation and monitoring its extinction (e.g., see Table 8.1) does not mathematically alter the outcomes of Eqs. (8.6) and (8.7), because the effect of a change in the number of traps (*t*) in Eqs. (8.6) and (8.7) is cancelled out by a countervailing change in c_r (which stands for the tolerable density in terms of catch per trap per week). Thus, in theory, data from a sparse sentinel array could be used for the whole exercise. However, the equations assume that the traps are equally accessible to all flies, and this is certainly not true for most sentinel arrays (see earlier). So, a denser array is more accurate for monitoring extinction. In practice, the equations should use settings based on the density of male-lure traps in the monitoring array within the 1 km radius or the central square mile (Table 8.1) but can ignore the density of other supplementary traps, which are basically there to help the field operators locate the actual sources of the infestation.

8 Example Using Medfly Monitoring Arrays in South Australia

A search of any comprehensive electronic database of science literature reveals that *C. capitata* is by far the most widespread and researched tephritid fruit fly. The USA and Australia have long experience with extensive detection systems, particularly in the states of California and South Australia. The essential differences between the two systems are given in Table 8.1.

In South Australia, the sentinel array for medfly covers the whole of the capital city (Adelaide), the production areas beyond, and their associated towns (BSA 2012). It has a density of 6.25 traps per km^2 (trap spacing 400 m) and a trigger level of 3 or more medfly trapped within a 1 km radius (i.e., a cluster of 20 traps) within 2 weeks. The trap density is not increased apart from the placement of supplementary traps for delimiting purposes (i.e., to identify the origin and perimeter of the infestation more precisely). Thus, the density of traps for surveillance purposes is the same as it is for confirming extinction. The trigger level indicates that a level of only 2 flies trapped within a 1 km radius within 2 weeks is the highest tolerable trapping rate (i.e., $c_r = 2/20/2 = 0.05$). If c_r was $1/20/2 = 0.025$, it would make for a large difference when it comes to calculation the required fly- free period as follows.

The fly-free period currently required to re-instate area freedom in South Australia is 12 weeks, which is about the time taken for 1 generation plus 28 days in mid-summer. Equation (8.7) calculates the number of weeks for P_0 to drop to the probit 9 criterion of $P_0 = 0.00003$ when $c_r = 0.05$ as 10.5 weeks (very close to the current requirement). But, if $c_r = 0.025$ (half the current acceptable level), then Eq. (8.7) calculates the probit 9 criterion as 21 weeks.

The example above emphasizes the importance of selecting an appropriate level of $c_{r,}$ bearing in mind that it is an index of the highest tolerable density, one that should indicate that the population should die out of its own accord. Presumably, the choice of c_r and hence the required period of fly-free trapping should be based on evidence or at least on the principle that what has worked up till now should probably work in future. However, if c_r were set unnecessarily low (to be extra cautious), then Eq. (8.7) would require an unnecessarily long 'zero catch' period. However, if the highest tolerable value of c_r is set too high as a result of negotiations between industry and the national government, it may be acceptable for local markets but may not be acceptable for international trade and especially for the purposes of declaring extinction of an exotic pest. The risks of being wrong are much greater in such circumstances, because they would apply to exports of the whole country (or a large geographical area of it) rather than a small local radius of a few km.

9 Example Using Medfly Monitoring Arrays in California

California has opted for a very low-density sentinel array of 5 traps per square mile (Table 8.1) and a very low trigger level of 2 or more medfly trapped within a 3 mile radius (28.3 sq. miles, 141 traps) within one generation time (~6–8 weeks), indicating that a level of only one fly trapped within the same radius and period is tolerable (USDA 2003). This gives c_r values of 0.00118 and 0.000887 for 6 and 8 weeks, respectively. However, if c_r values are calculated only from traps in the core square mile (for reasons explained earlier under 'effective size of trap arrays'), they become 1/5/6 = 0.033 flies per trap per week for a generation time of 6 weeks and 1/5/8 = 0.025 flies per trap per week for 8 weeks.

This has great significance to the restoration of fly-free status after an infestation. When the trigger level is reached, an infestation is declared, and the trap density is increased to 100 traps in a square mile around the site and is also boosted in surrounding areas (Table 8.1). This density is also maintained after the last wild fly is trapped until eradication is declared. Thus c_r values based on the core square mile are 1/100/6 = 0.00167 and 1/100/8 = 0.00125 for generation times of 6 and 8 weeks, respectively. The usual period of zero catches required for restoration of area freedom status also relies on generation time (about 6-8 weeks) and is, in fact, the equivalent of three generations (~18–24 weeks). From Eq. (8.6) we can estimate that the probability (P_0) of zero flies being caught on a 100-trap array with the two tolerable trapping rates above is 0.05 in both cases. That level is not very low, considering the risk to trade and the standards usually applied to quarantine inspection (Venette et al. 2002). Using a P_0 value of 0.00003 (the 'probit 9 criterion'), Eq. (8.7) indicates that the required zero catch periods would be 62 and 83 weeks, respectively, if generations were 6 and 8 weeks, respectively. Those times appear to be far too long for restoration of area freedom status when compared to the current standards given above (which are presumably effective because they have been retained over decades). Also, they are most likely too long for a fruit grower to be out of business due to quarantine restrictions. The same consideration applies if we drop our standard to a P_0 value of 0.001, because we would need zero catch periods of 42 and 55 weeks, respectively, if generations were 6 or 8 weeks long respectively.

Why is the risk analysis in such discord with practice? It is not likely to be due to the accepted 'fly-free period' criterion being unrealistic, because the time equivalent to three generations makes biological sense and is only twice as long as the period used in South Australia. That leaves the criterion c_r used as an index of highest tolerable population level, which is defined by the highest rate of trapping below action threshold (trigger level).

As explained earlier, the method used in this chapter is confined to one square mile at a time, because the likelihood of a new incursion dispersing over a larger area has been shown to be very low, and any occurrences in a wider area should be treated as separate foci. If the infestation were confined to a square mile, the highest tolerable trapping rate would be only one fly per 5 traps in one generation (~6–8 weeks). That would mean a c_r value between $1/5/6 = 0.033$ and $1/5/8 = 0.025$. However, using Eqs. (8.6) and (8.7) with the sentinel trap density (8.5) and c_r values of 0.033 or 0.025 gives exactly the same answers as the previous ones based on 100 traps per square mile. This because (as mentioned earlier) when using Eqs. (8.6) and (8.7) any changes in trap density are balanced by countervailing changes in c_r.

We can only conclude that the c_r value for California has been set too low for a rational prediction of the length of the necessary fly-free period. However, it has to be low, because with trap spacing so large, the infestation could also be large (see earlier). It therefore seems that a price has to be paid for saving money by using a low density for sentinel traps. However, a cost-benefit analysis may indicate that a low c_r is more economical if outbreaks are historically at low frequency. Figure 8.2 gives an overall picture of the relation between c_r, risk levels, and the required fly-free period.

10 Generation Time or Calendar Time?

Most codes of practice apply the concept of generation time for the fly-free period required after an eradication treatment, and some also use it for the period over which the required number of flies are trapped to trigger treatment in the first place (e.g., see Table 8.1). Other uses in agricultural and medical entomology include the

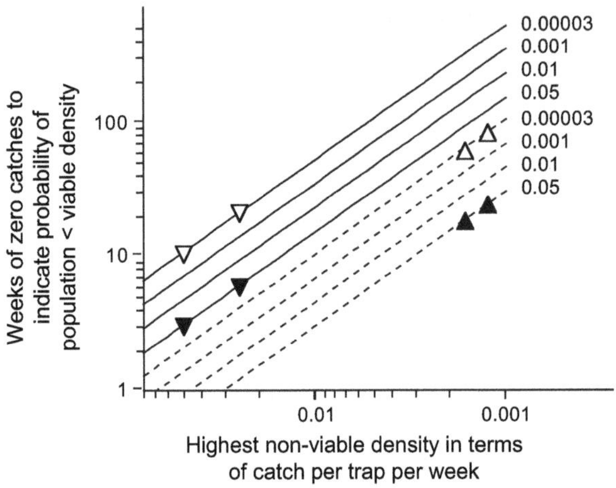

Fig. 8.2 How long does it take to establish with a given probability that a medfly population is at or below the highest non-viable density in terms of catch per trap per week (c_r)? The relation of the number of weeks of zero catches required is related to c_r at a given set of probabilities (P_0) represented by the lines labelled (0.00003–0.05). The *solid lines* pertain to a 20-trap array at the core of an infestation in South Australia. When $c_r = 0.05$ (the accepted highest non-viable rate in South Australia) the time pertaining to a probability of 0.00003 or probit 9 is 10.5 weeks (lower inverted white triangle) and is very close to the current requirement for Medfly in that state. But if $c_r = 0.025$ (half the current highest acceptable level) then the probit 9 criterion is 21 weeks (upper inverted *white triangle*). The inverted *black triangles* represent the analogous case for $P_0 = 0.05$. The *dashed lines* pertain to a 100-trap array in the core square mile of an infestation in California. California has opted for a very low-density sentinel array (5 traps per square mile) and a trigger level of 2 or more Medfly trapped within a radius of 3 miles within ~ 6–8 weeks. After the trigger level is reached, the number of traps in the core square mile is increased to 100. The period of trapping yielding zero flies pertaining to a probability of 0.00003 or probit 9 are 62 and 83weeks for generation times of 6 and 8 weeks respectively (*white upright triangles*). The times actually used for required fly-free periods are respectively around 18 and 24 weeks corresponding a probability of 0.05 (*black upright triangles*) and which are only about twice as long as the times actually used in South Australia

timing of development-based (as opposed to calendar based) insecticide applications, flowering and harvesting times for crops, and (in forensic entomology) the back-calculation of the age of maggots in corpses. With fruit-fly regulations (as for the other uses) the intention is to achieve realism and accuracy at an acceptable cost.

11 Temperature Summation

Development time can be related to many factors, such as temperature, photoperiod, soil moisture, and fruit condition (Meats 1974a, b; Fletcher 1975; Fletcher et al. 1978; Fletcher and Kapatos 1983; Filchak et al. 2001; Raspi et al. 2002;

Gonçalves and Monteiro-Torres 2011; Baumgartner et al. 2012). The original method of calculating development time was restricted to cases where development rate was thought to be affected only by temperature (T) above an estimated threshold (T_0) and that the development rate in the temperature range of interest was linearly related to temperature. Bodenheimer (1925) predicted the generation time of medfly (from egg to egg) under given climatic conditions in different parts of the world, using 'Blunck's rule', which states: 'the product of duration of development and the difference between the surrounding temperature and the critical cold-point is constant.' However, using just the average temperature for a location is rather crude, and so the practice of summating daily average temperatures was used (Watzl 1927; Headlee 1928). The relationship between temperature T and development rate ($1/d$) can be predicted by a linear regression of $1/d$ on T, where $1/d = a + bT$, and where d = the total development time in days, b is the slope of the regression line, and a = the intercept (where development rate is predicted to be zero). The lower developmental threshold (T_0) is estimated by $T_0 = -$ (a/b). Thus, the relationship $1/\mathrm{d} = b\,(T - T_0)$ can be re-arranged to $d \cdot (T - T_0) = 1/b$ = the total number of degree days (DD) over threshold that are required for development.

One can predict the development time from any particular date in any particular district so long as historical daily average temperatures are available. This will be only a mean expectation, because future temperatures can not be predicted with any more accuracy. Thus, starting at the required date, one can calculate and successively sum the DD until the point at which the total indicates the date of the next generation.

Many modifications can be made to this procedure. Temperature data for various districts are usually available only as daily maximum and minimum temperatures, thus mean daily temperature can be estimated as the average of the two. This is still the method used in the fruit fly codes of practice (USDA 2003; DPIPWE 2011; BSA 2012). Worner (1988) reviewed various algorithms for improving the estimate of effective daily mean temperature but noted that none of them are satisfactory in certain climates when the mean temperature is close to the developmental threshold. She suggested that the use of hour degrees rather than day degrees would be better if sufficient temperature data were available (which is impractical except for forensic cases that estimate the timing of a past generation using actual historical records rather than the timing of a future generation using mean expectations of temperature trends).

Another improvement (also seen in the fruit fly codes) is the prediction of the length of each life history stage separately if those stages have different developmental thresholds. However, threshold estimates tend to differ with author or place, and only rarely is a range of error reported. For instance, there are many estimates in the case of *C. capitata*. On Réunion Island, the lower development thresholds for egg, larval, pupal stages, and the ovarian maturation period were estimated as 11.6, 10.2, 11.2 and 8.9 °C, respectively (Duyck and Quilici 2002). In Western Australia the equivalent values were estimated as 9.3 °C, 11.1 °C, 8.4 °C, and 12.8 °C, respectively (De Lima 2008), whereas at three different sites in Brazil development thresholds were, respectively, 7.26 °C, 5.10 °C, and 5.9 °C for eggs, 8.7 °C, 9.6 °C, and 8.1 °C for larvae and 10.4 °C, 10.5 °C, and 10.3 °C for pupae (Ricalde et al. 2012). One reason for this may be the use of different methods (Fletcher 1989). The lower development threshold is estimated by extrapolating the linear regression (of development rate on temperature) to the intercept. The error of the estimate would thus depend upon the error of the regression slope and the distance from the data points to the intercept. Some variation between the results of different authors is likely if they use different numbers of replicates for each data point, use different numbers of data points, or have to extrapolate across different distances to the intercept. However, even after using the error of the intercepts for statistical comparison, differences between sites can be found when the methods used are identical (Ricalde et al. 2012).

More seriously, a fundamental failing of the temperature summation method would occur if the regression was non-linear close to the lower temperature threshold. Moreover, there is no provision for an upper threshold or the slowing of development at high temperatures (Garcia-Ruiz et al. 2011). These failings can be avoided by using development summation.

12 Development Summation

The relationship between $1/d$ and T can be non-linear (approximately sigmoid and sometimes with a downward trend at high temperatures). Instead of summing day degrees or hour degrees as above, increments of d should be summed until the total of 1 is reached. For this, an empirical model for 1/d on t should be established experimentally over a range of constant temperatures (Meats 1974a, b; 1981; O'Loughlin et al. 1984; Garcia-Ruiz et al. 2011). However, such a model may have some failings, because under constant conditions the more extreme temperatures (high or low) may slow development more when experienced on a continuous basis than when experienced for brief periods in the field when temperature fluctuates on a daily basis (Meats and Khoo 1976; Mironidis and Savopoulou-Soultani 2008).

The relative merits of temperature summation and development summation thus depend upon the problem and the species involved. The DD model requires minimal data and is easy to calculate and apply; it has often been used successfully with the desired accuracy in many IPM and research programs (Stark and Aliniazee 1982; Fan et al. 1992). It has also been noted that an attempt to attain more realism by adding extra features or processes to a model can lead to less rather than more accuracy when the model is tested with real data (Mertz 1972; Tassan et al. 1983).

13 Choosing a Method and Conversion of Units for Fly-Free Period

First, one must choose whether or not to use generation time. Presumably, multiples of generation time (Table 8.1) are used to prescribe the fly-free period required for the lifting of quarantine, because any resurgence after eradication treatment would be most likely to occur at generational intervals, although at very low fly densities this periodicity may not be noticeable, especially if flies can survive for some weeks. If it is decided to use generation time, then we must be able to convert the calendar times given by the risk analysis model (Eq. 8.7) to generation times and *vice-versa*. One suggestion would be to express calendar time as a multiple of generation time at an optimum or otherwise defined constant temperature. If the multiple were, say 3.2, then it would be a matter of estimating how many days that 3.2 generations would require in the field.

As far as frequency of trapping goes, trigger levels (and therefore c_r) are set to be tolerant of temperature fluctuations of the kind usually found throughout for the peak flight season. If it is established that ambient temperature is important to weekly catch rate, then trigger levels and c_r can be expressed in terms of catch rate per generation (as for medfly in California – see Table 8.1). The rationale for this is not stated in the relevant Code of Practice, but if flight activity and tendency to enter traps are both related to temperature, then the accumulated total degree-days could be experimentally related to accumulated trapping (recapture) rate of marked flies, just as it is to development time (generation length) of flies in laboratory cultures. More research is needed on this point, because it is important to establish the degree of similarity between the relationships of accumulated trapping rate to degree-days and generation time to degree days. If they were not similar, then it would be appropriate to find another method of calculating the required fly-free trapping period.

14 Consequences of Improving Traps and Their Placement

Despite the contention that the old saying about building a better mousetrap was apparently not said, new traps and lures for fruit flies are always being devised or sought after (e.g., see Vargas et al. 2010; Shelly et al. 2011). For each new trap/lure combination used for detection surveillance, a new value of c_r may have to be estimated. This requirement remains, whether the methods of risk assessment given here are used or not because c_r is derived from action threshold (trigger level). It is possible that action threshold and c_r would not change if the rate of trapping flies at low densities were limited by the rate at which they arrived within trapping range. The same considerations would apply if the placement of traps were improved to increase chances of detection.

A trap array tends to have a grid-like layout to make servicing easier (often following the pattern of roads and streets). Despite the fact that population level can only be estimated indirectly (from trapping rate), the array's purpose may be the (a) monitoring of population trends and dispersion of an existing infestation, (b) monitoring the survival and dispersion of released sterile insects, (c) detection of a new infestation in a presently fly-free area, and (d) confirmation of extinction (successful eradication) with reference to the statutory fly-free period required as proof. The non-random (but arbitrary and unbiased) placement of traps should be no hindrance to analysis of the catch data if either or both (a) and (b) are the objectives (Milne 1959; Cole et al. 2001). These considerations also apply to (c) and (d) if one wants an unbiased estimate of trigger levels and c_r. However, if one wanted to have the very best chance of detection at the earliest possible stage, then surveillance could be done with traps placed in the most likely sites for capture if these can be identified.

The most logical place for a fruit fly trap would be a tree or crop bearing host fruit. Medfly traps in Hawaii caught more flies in fruit-bearing host trees than traps in non-host trees (Wong et al. 1985), whereas *Rhagoletis pomonella* (Walsh) is more readily trapped on ripe rather that unripe apples (Murphy et al. 1991). Most medfly individuals do not disperse very far (Plant and Cunningham 1991; Meats and Smallridge 2007), and it is perhaps inevitable that most will be found near to a breeding site, especially if they can be produced over a time equivalent to two or more successive generations at such a site. Release-recapture trials with sterile *B. tryoni* have shown that more flies can be recaptured in non-host trees than in host trees at a given distance from the release site (Weldon and Meats 2007; Gilchrist and Meats 2012).

Some heterogeneous distributions appear to have no obvious cause. Plant and Cunningham (1991) released sterile medfly in an essentially homogeneous host-free plantation of macadamia (*Macadamia integrifolia* Maiden and Betche), and although mean recapture rate declined with the distance from the release point, there was also wide unexplained spatial variation in recapture rates in traps at any given distance. Similar observations have been made with respect to releases of *Bactrocera cucurbitae* (Coquillett) and *B. tryoni* (Teruya 1986; Horwood and Keenan 1994).

There may be more chance of trapping wild flies at some heights above the ground than others, depending upon the type of vegetation (Hooper and Drew 1979; Robacker et al. 1990; Liburd et al. 2000; Boucher et al. 2001; Sarada et al. 2001; Pelz-Stelinski et al. 2006). However, the probability of trapping *Bactrocera oleae* (Rossi) may not depend upon height (Haniotakis 1986). Some species apparently migrate to or through non-host areas and can be trapped there on a seasonal basis (Fletcher 1974b; Hooper and Drew 1979; Zalucki et al. 1984). In the case of *B. tryoni*, up to 3–10 flies per trap per week were caught in sclerophyll bushland and non-host trees adjacent to open pasture (Fletcher 1974b). Release-recapture trials with sterile *B. tryoni* have revealed that long distance dispersal (>5 km) may follow watercourses in dry areas (MacFarlane et al. 1987). Other patterns of distribution can also be related to features of the habitat (Zalucki et al. 1984; Vargas et al. 1990; Gaul et al. 1995; Dimou et al. 2003; Papadopoulos et al. 2003). Thus, it appears that although there may be a single dominant factor (such as fruiting host tree) related to the distribution of a fruit fly species, it is best to obtain more detail with a general trapping survey spread in space and time.

Spatial-temporal analyses of natural populations at a particular site can identify the places (or tree species) and times where there is the greatest chance of trapping the target species at that site or may help to target sites for treatment (Papadopoulos et al. 2003; Jang et al. 2008; Sciarretta et al. 2009; Castrignano et al. 2012). For presence/absence monitoring (with sentinel traps) this could well entail shifting traps throughout the year (Sciarretta et al. 2009; De Lima et al. 2011). However, to date there is no instance of a cost-benefit study or regular implementation of a trap-shifting strategy especially with regard to extinction monitoring and the confirmation of fly-free periods required to the declaration of eradication.

In Western Australia, trap-shifting ('dynamic trapping') to target seasonally fruiting trees for medfly detection was more successful than the traditional 'static' trapping grid except in places with very low abundance resembling recently invaded areas (De Lima et al. 2011). This exception is due to the fact that comparison statistics lack the power to yield unambiguous results if only one or two flies are trapped in the whole trial area. Thus, it appears that comparisons of the two strategies relevant to sentinel traps in rarely infested areas may have to be continued for many years in order to obtain a significant result either way. Similar dynamic trapping in fruiting trees in New South Wales revealed that *B. tryoni* was not usually caught in greater numbers than it was by static trapping even in towns where numbers were relatively high (De Lima et al. 2011). In this case, we can conclude that some spatial analysis of traditional trapping data should have been done first, or at least some tests directly comparing traps in fruiting and non-fruiting trees.

15 Telemetry and Data Management

With the advent of telemetry (remote sensing) based on images or sound (Jiang et al. 2008; Liu et al. 2009; Mankin et al. 2011; Okuyama et al. 2011; Mankin 2012; Philimis et al. 2013), the development of a reliable and affordable trap that is based upon such techniques could mean that placement may not necessarily be determined by the requirement of frequent access but instead be based mainly on the chances of early detection. Delimitation trapping and extinction monitoring may also use a telemetry array but if unbiased placement is required (e.g., for map-making), a conventional grid should be used. However, there are many obstacles to overcome apart from cost. At present, telemetry may be feasible where there is only one responding tephritid species in the area. It has yet to be demonstrated how it would work in places, such as Australia, where the by-catch in cue lure and methyl eugenol traps can be both species-rich and voluminous (Osborne et al. 1997; Hancock et al. 2000). This would be particularly onerous when the target species is very rare as expected in sentinel traps in a surveillance zone.

Every surveillance program should have a data management system (Dominiak et al. 2007b), so if telemetry is feasible it could feed directly into that and there should be no reason why delimitation maps should not be available in real time enabling immediate and correctly targeted treatment.

16 General Conclusions

There are no risk-free methods of surveillance and extinction monitoring, just as are there are no risk-free methods of quarantine inspection and post-harvest treatment. There are many ways in which surveillance with traps and the analysis of results can be improved, but they may not lead to significant improvements in the timely and efficient treatment of new infestations. This should be the perspective when consideration is given to investing in new methods or improvements to existing methods of risk management. Thus, improvements should be introduced with caution, preferably after some comparison with existing protocols.

References

Barry JD, McInnis DO, Gates D, Morse JG (2003) Effects of irradiation on Mediterranean fruit flies (Diptera: Tephritidae): emergence, survivorship, lure attraction, and mating competition. J Econ Entomol 96:615–622

Bateman MA (1977) Dispersal and species interaction as factors in the establishment and success of tropical fruit flies in new areas. Proc Ecol Soc Aust 10:106–112

Baumgartner J, Gutierrez AP, Pesolillo S, Severini M (2012) A model for the overwintering process of European grapevine moth *Lobesia botrana* (Denis and Schiffermuller) (Lepidoptera, Tortricidae) populations. J Entomol Acarol Res 44:8–16

Bodenheimer FS (1925) On predicting the development cycles of insects. I: *Ceratitis capitata*, wied. Bull Soc Entomol Egypte 2–4:149–157

Boucher TJ, Ashley RA, Adams RG, Morris TF (2001) Effect of trap position, habitat, and height on the capture of pepper maggot flies (Diptera: Tephritidae). J Econ Entomol 94:455–461

BSA (Biosecurity South Australia, Plant Health Operations) (2012) Fruit fly detection and eradication manual (Draft). South Australian Government, Adelaide.

Castrignano A, Boccaccio L, Cohen Y, Nestel D, Kounatidis I, Papadopoulos NT, De Benedetto D, Mavragani-Tsipidou P (2012) Spatio-temporal population dynamics and area-wide delineation of Bactrocera oleae monitoring zones using multi-variate geostatistics. Prec Agric 13:421–441

Clift AD, Meats A, Gleeson PJ (1998) A dispersal model for papaya fruit fly *Bactrocera papayae* Drew and Hancock and its application to treatment priorities in an eradication protocol. In: Zalucki MP, Drew RAI, White GG (eds) Pest management – future challenges, vol 2. Proceedings of 6th Australian applied entomology research conference, Brisbane, Australia', pp 27–31. (Australian Entomological Society)

Cole RG, Healy TR, Wood ML, Foster DM (2001) Statistical analysis of spatial pattern: a comparison of grid and hierarchical sampling approaches. Environ Monit Assess 69:85–99

Collins SR, Weldon CW, Banos C, Taylor PW (2009) Optimizing irradiation dose for sterility induction and quality of *Bactrocera tryoni*. J Econ Entomol 102:1791–1800

Cunningham RT, Couey HM (1986) Mediterranean fruit fly (Diptera: Tephritidae): distance/response curves to trimedlure to measure trapping efficiency. Environ Entomol 15:71–74

De Lima CPF (2008) Area wide management of Mediterranean fruit fly. In: Samietz J (ed) VIII international symposium on modelling in fruit research and Orchard Management. Acta Horticulturae, International Society for Horticultural Science, vol 803, pp 51–59. www.actahort.org

De Lima CPF, Poogoda S, Smallridge C, Caldwell A, Reynolds O, van der Rijt V, Clark S, Orchard B (2011) Trapping to prove area freedom: final report CRC30039. Cooperative Research Centre for National Plant Biosecurity, Canberra, ACT, Australia

Dimou I, Koutsikopoulos C, Economopoulos A, Lykakis J (2003) The distribution of olive fruit fly captures with McPhail traps within an olive orchard. Phytoparasitica 31:124–131

Dominiak BC, Campbell AJ, Barchia IM (2007a) Changes in emergence parameters as a result of transporting sterile Queensland fruit fly, *Bactrocera tryoni* (Froggatt) (Diptera: Tephritidae) pupae. Gen Appl Entomol 36:7–10

Dominiak BC, Gillespie PS, Toma G (2007b) Changes in data management and the impediments to change – a case study for fruit fly surveillance. Plant Prot Quart 22:95–99

DPIPWE (2011) Review of import requirements for fruit fly host produce from mainland Australia. Department of Primary Industries Parks and Water, Tasmania, Devonport, TAS, Australia, p 118. ISBN: 978-0-7246-6610-2

Duyck PF, Quilici S (2002) Survival and development of different life stages of three *Ceratitis* spp. (Diptera: Tephritidae) reared at five constant temperatures. Bull Entomol Res 92:461–469

Fan Y, Groden E, Drummond FA (1992) Temperature-dependent development of Mexican bean beetle (Coleoptera: Coccinellidae) under constant and variable temperatures. J Econ Entomol 85:1762–1770

Filchak KE, Roethele JB, Feder JL (2001) Effects of photoperiod and light intensity on the genetics of diapause in the apple maggot (Diptera: Tephritidae). Ann Entomol Soc Am 94:902–908

Fletcher BS (1973) The ecology of a natural population of the Queensland fruit fly, *Dacus tryoni* IV. The immigration and emigration of adults. Aust J Zool 21:541–565

Fletcher BS (1974a) The ecology of a natural population of the Queensland fruit fly, *Dacus tryoni*, V. The dispersal of adults. Aust J Zool 22:189–202

Fletcher BS (1974b) The ecology of a natural population of the Queensland fruit fly, *Dacus tryoni* VI. Seasonal changes in fruit fly numbers in the areas surrounding the orchard. Aust J Zool 22:353–363

Fletcher BS (1975) Temperature-regulated changes in the ovaries of overwintering females of the Queensland fruit fly, *Dacus tryoni*. Aust J Zool 23:91–102

Fletcher BS (1989) Temperature-development rate relationships of the immature stages and adults of tephritid fruit flies. In: Robinson AS, Hooper G (eds) Fruit flies, their biology, natural enemies and control, vol 3B, World cup pests. Elsevier, Amsterdam, pp 273–289

Fletcher BS, Kapatos E (1983) The influence of temperature, diet and olive fruits on the maturation rates of female olive flies at different times of the year. Entomol Exp Appl 33:244–252

Fletcher BS, Pappas S, Kapatos E (1978) Changes in ovaries of olive flies (*Dacus oleae* (Gmelin)) during summer, and their relationship to temperature, humidity and fruit availability. Ecol Entomol 3:99–107

Garcia-Ruiz E, Marco V, Perez-Moreno I (2011) Effects of variable and constant temperatures on the embryonic development and survival of a new grape pest, *Xylotrechus arvicola* (Coleoptera: Cerambycidae). Environ Entomol 40:939–947

Gaul SO, Neilson WTA, Estabrooks EN, Crozier LM, Fuller M (1995) Deployment and utility of traps for management of *Rhagoletis mendax* (Diptera, Tephritidae). J Econ Entomol 88:134–139

Gilchrist AS, Meats A (2012) Factors affecting the dispersal of large-scale releases of the Queensland fruit fly, *Bactrocera tryoni*. J Appl Entomol 136:252–262

Gilchrist AS, Sved JA, Meats A (2004) Genetic relations between outbreaks of the Queensland fruit fly, *Bactrocera tryoni* (Froggatt) (Diptera: Tephritidae), in Adelaide in 2000 and 2002. Aust J Entomol 43:157–163

Gilchrist AS, Cameron EC, Sved JA, Meats A (2012) Genetic consequences of domestication and mass rearing of pest fruit fly *Bactrocera tryoni* (Diptera: Tephritidae). J Econ Entomol 105:1051–1056

Gonçalves MF, Monteiro-Torres L (2011) The use of the cumulative degree-days to predict olive fly, *Bactrocera oleae* (Rossi), activity in traditional olive groves from the northeast of Portugal. J Pest Sci 84:187–197

Hancock DL, Osborne R, Broughton S, Gleeson PJ (2000) Eradication of *Bactrocera papayae* (Diptera: Tephritidae) by male annihilation and protein baiting in Queensland, Australia. In: Tan KH (ed) Area-wide control of fruit flies and other insect pests. USM Press, Penang, pp 381–388

Haniotakis GE (1986) Effect of size color and height of pheromone baited sticky traps on captures of *Dacus oleae* flies. Entomol Hellenica 4:55–61

Headlee TJ (1928) Some data relative to the relationship of temperature to codling moth activity. J N Y Entomol Soc 36:147–163

Hooper GHS, Drew RAI (1979) Effect of height of trap on capture of tephritid fruit flies with cuelure and methyl eugenol in different environments. Environ Entomol 8:786–788

Horwood MA, Keenan PJ (1994) Eradication of the Queensland fruit fly. Report CT 336 to Horticulture Research & Development Corporation, NSW Agriculture, Orange, NSW, Australia

Jang EB, McQuate GT, McInnis DO, Harris EJ, Vargas RI, Bautista RC, Mau RF (2008) Targeted trapping, bait-spray, sanitation, sterile-male, and parasitoid releases in an area wide integrated melon fly (Diptera: Tephritidae) control program in Hawaii. Am Entomol 54:240–250

Jiang JA, Tsengb CL, Lua FM, Yangc EC, Wua ZS, Chena CP, Lina SH, Linb KC, Liaob CS (2008) A GSM-based remote wireless automatic monitoring system for field information: a case study for ecological monitoring of the oriental fruit fly, *Bactrocera dorsalis* (Hendel). Comput Electron Agric 62:243–259

Kendra PE, Epsky ND, Heath RR (2010) Effective sampling range of food-based attractants for female *Anastrepha suspensa* (Diptera: Tephritidae). J Econ Entomol 103:533–540

Lance DR, Gates DB (1994) Sensitivity of detection trapping systems for Mediterranean fruit flies (Diptera: Tephritidae) in southern California. J Econ Entomol 87:1377–1383

Lewis JR, Sauro J (2006) When 100 % really isn't 100 %: improving the accuracy of small-sample estimates of completion rates. J Usability Stud 1:136–150

Liburd OE, Polavarapu S, Alm SR, Casagrande RA (2000) Effect of trap size, placement, and age on captures of blueberry maggot flies (Diptera: Tephritidae). J Econ Entomol 93:1452–1458

Liu Y, Zhang J, Richards M, Pham B, Roe P, Clarke A (2009) Towards continuous surveillance of fruit flies using sensor networks and machine vision. In: WiCom 09. 5th international conference on wireless communications, networking and mobile computing. Institute of Electrical and Electronics Engineers (IEEE), vol 1–8, pp 4705–4709. http://www.ieee.org/conferences

MacFarlane JR, East RW, Drew RAI, Betlinski GA (1987) Dispersal of irradiated Queensland fruit flies, Dacus tryoni (Froggatt) (Diptera: Tephritidae), in south-eastern Australia. Aust J Zool 35:275–281

Mankin RW (2012) Applications of acoustics in insect pest management. CAB Rev 7:1–7

Mankin RW, Hagstrum DW, Smith MT, Roda AL, Kairo MTK (2011) Perspective and promise: a century of insect acoustic detection and monitoring. Am Entomol 57:30–44

Mayer DG, Atzeni MG (1993) Estimation of dispersal distances for Cochliomyia hominivorax (Diptera: Calliphoridae). Environ Entomol 22:368–374

McArdle BH (1990) When are rare species not there? Oikos 57:276–277

Meats A (1974a) Population model for 2 species of *Tipula* (Diptera; Nematocera) derived from data on their physiological relations with their environment. Oecologia 16:119–138

Meats A (1974b) Simulation of population trends of *Tipula paludosa* using a model fed with climatological data. Oecologia 16:139–147

Meats A (1981) The bioclimatic potential of the Queensland fruit fly, *Dacus tryoni*, in Australia. Proc Ecol Soc Aust 11:151–161

Meats A (1983) Strategies for maximising the potential of the sterile insect release method: experiments with *Dacus tryoni*. In: Cavolloro R (ed) Fruit flies of economic importance. Balkema, Rotterdam, pp 371–377

Meats A (1998a) Predicting or interpreting trap catches resulting from natural propagules or releases of sterile fruit flies. An actuarial and dispersal model tested with data on *Bactrocera tryoni*. Gen Appl Entomol 28:29–38

Meats A (1998b) The power of trapping grids for detecting and estimating the size of invading propagules of the Queensland fruit fly and risks of subsequent infestation. Gen Appl Entomol 28:47–55

Meats A, Edgerton JE (2008) Short- and long-range dispersal of the Queensland fruit fly, *Bactrocera tryoni* and its relevance to invasive potential, sterile insect technique and surveillance trapping. Aust J Exp Agric 48:1237–1245

Meats A, Khoo KC (1976) The dynamics of ovarian maturation and oocyte resorption in the Queensland fruit fly (*Dacus tryoni*) in constant and rhythmic temperature regimes. Physiol Entomol 1:213–221

Meats A, Smallridge CJ (2007) Short and long range dispersal of Medfly, *Ceratitis capitata* (Diptera: Tephritidae) and its invasive potential. J Appl Entomol 8:518–523

Meats AW, Clift AD, Robson MK (2003a) Incipient founder populations of Mediterranean and Queensland fruit flies in Australia: the relation of trap catch to infestation radius and models for quarantine radius. Aust J Exp Agric 43:407–417

Meats A, Duthie R, Clift AD, Dominiak BC (2003b) Trials with variants of the Sterile Insect Technique (SIT) for suppression of populations of the Queensland fruit fly in small towns neighbouring a quarantine (exclusion) zone. Aust J Exp Agric 43:389–395

Mertz DB (1972) The *Tribolium* model and the mathematics of population growth. Annu Rev Ecol Systemat 3:51–78

Milne A (1959) The centric systematic area-sample treated as a random sample. Biometrics 15:270–297

Mironidis GK, Savopoulou-Soultani M (2008) Development, survivorship, and reproduction of *Helicoverpa armigera* (Lepidoptera: Noctuidae) under constant and alternating temperatures. Environ Entomol 37:16–28

Monro J, Richardson NL (1969) Traps, male lures, and a warning system for Queensland fruit fly, *Dacus tryoni* (Frogg.) (Diptera, Trypetidae). Aust J Agr Res 20:325–338

Murphy BC, Wilson LT, Dowell RV (1991) Quantifying apple maggot (Diptera, Tephritidae) preference for apples to optimize the distribution of traps among trees. Environ Entomol 20:981–987

O'Loughlin GTO, East RA, Meats A (1984) Survival, development rates and generation times of the Queensland fruit fly, Dacus tryoni, in a marginally favourable climate: experiments in Victoria. Aust J Zool 32:353–361

Okuyama T, Yang EC, Chen CP, Lin TS, Chuang CL, Jiang JA (2011) Using automated monitoring systems to uncover pest population dynamics in agricultural fields. Agr Syst 104:666–670

Osborne R, Meats A, Frommer M, Sved JA, Drew RAI, Robson MK (1997) Australian distribution of 17 species of fruit flies (Diptera: Tephritidae) caught in cue lure traps in February 1994. Aust J Entomol 36:45–50

Papadopoulos NT, Katsoyannos B, Nestel D (2003) Spatial autocorrelation analysis of a *Ceratitis capitata* (Diptera: Tephritidae) adult population in a mixed deciduous fruit orchard in northern Greece. Environ Entomol 32:319–326

Pelz-Stelinski KS, Gut LJ, Isaacs R (2006) Vertical position of traps influences captures of eastern cherry fruit fly (Diptera: Tephritidae). Fla Entomol 89:80–82

Philimis P, Psimolophitis E, Hadjiyiannis S, Giusti A, Perrello J, Serrat A, Avila P (2013) A centralised remote data collection system using automated traps for managing and controlling the population of the Mediterranean (*Ceratitis capitata*) and Olive (*Dacus oleae*) fruit flies. In: Hadjimitsis DG, Themistocleous K, Michaelides S, Papadavid G (eds) Proceedings of SPIE (Society of Photo-Optical Instrumentation Engineers) Article Number: UNSP 87950X. doi:10.1117/12.2028244

Plant RE, Cunningham RT (1991) Analyses of the dispersal of sterile Mediterranean fruit flies (Diptera: Tephritidae) released from a point source. Environ Entomol 20:1493–1503

Raspi A, Iacono E, Canale A (2002) Variable photoperiod and presence of mature eggs in olive fruit fly, *Bactrocera oleae* (Rossi) (Diptera Tephritidae). Redia 85:111–119

Regan TJ, McCarthy MA, Baxter PWJ, Panetta FD, Potteringham HP (2006) Optimal eradication: when to stop looking for an invasive plant. Ecol Lett 9:759–766

Reynolds OL, Smallridge CJ, Cockington VG, Penrose LD (2012) The effect of release method and trial site on recapture rates of adult sterile Queensland fruit fly, *Bactrocera tryoni* (Froggatt) (Diptera: Tephritidae). Aust J Entomol 51:116–126

Ricalde MP, Nava DE, Loeck AE, Donatti MG (2012) Temperature-dependent development and survival of Brazilian populations of the Mediterranean fruit fly, *Ceratitis capitata*, from tropical, subtropical and temperate regions. J Insect Sci 12:33

Robacker DC, Moreno DS, Wolfenbarger DA (1990) Effects of trap color, height, and placement around trees on capture of Mexican fruit flies (Diptera: Tephritidae). J Econ Entomol 83:412–419

Rull J, Birke A, Ortega R, Montoya P, Lopez L (2012) Quantity and safety vs. quality and performance: conflicting interests during mass rearing and transport affect the efficiency of sterile insect technique programs. Entomol Exp Appl 142:78–86

Sarada G, Maheswari TU, Purushotham K (2001) Effect of trap colour, height and placement around trees in capture of mango fruit flies. J Appl Zool Res 12:108–110

Sauro J (2005) Confidence interval calculator for a completion rate. http://www.measuringusability.com/wald.htm

Sciarretta A, Cesare D, Salvador R, Tabilio MR, Trematerra P (2009) Spatio-temporal distribution of *Ceratitis capitata* trap catches in an agricultural landscape. IOBC/WPRS Bull 41:123–129

Shelly TE, Edu J (2009) Capture of mass-reared vs. wild-like males of *Ceratitis capitata* (Dipt., Tephritidae) in trimedlure-baited traps. J Appl Entomol 133:640–646

Shelly TE, Nishimoto J (2011) Additional measurements of distance-dependent capture probabilities for released males of *Bactrocera cucurbitae* and *B. dorsalis* (Diptera: Tephritidae) in Honolulu. J Asia Pac Entomol 14:271–276

Shelly T, Nishimoto J, Diaz A, Leathers J, War M, Shoemaker R, Al-Zubaidy M, Joseph D (2010) Capture probability of released males of two *Bactrocera* species (Diptera: Tephritidae) in detection traps in California. J Econ Entomol 103:2042–2051

Shelly TE, Kurashima R, Nishimoto J, Diaz A, Leathers J, War M, Joseph D (2011) Capture of *Bactrocera* fruit flies (Diptera: Tephritidae) in traps baited with liquid versus solid formulations of male lures. J Asia Pac Entomol 14:463–467

Simpson SE (1993) Development of the Caribbean fruit-fly free zone certification protocol in Florida. Fla Entomol 76:228–233

Smallridge CJ, Hopkins DC (2004) Preventative sterile fly release for the management of a Mediterranean fruit fly outbreak in South Australia. In: Barnes BN (ed) Proceedings of the 6th international symposium on fruit flies of economic importance, Stellenbosch, South Africa, 6–10 May 2002. Isteg Scientific Pub. Irene, South Africa, pp 213–215

Smith ESC (2000) Internal and external fruit fly quarantine in Australia: In: Price NS, Seewooruthun I (eds) Proceedings of the Indian Ocean commission, regional fruit fly symposium, Flic en Flac, Mauritius, 5th-9th June 2000. Indian Ocean Commission, Quatre Bornes, Mauritius, pp 171–177

Stark SB, Aliniazee MT (1982) Evaluation of modifications to a basic thermal summation model for predicting the time of emergence of the adult western cherry fruit fly, Rhagoletis indifferens Curran (Dipt., Tephritidae). Z Angew Entomol 94:401–407

Tassan RL, Hagen KS, Cheng A, Palmer TK, Feliciano G, Blough TL (1983) Mediterranean fruit fly life cycle estimations for the California eradication program. In: Cavalloro R (ed) Fruit flies of economic importance. Balkema, Rotterdam, pp 564–570

Teruya T (1986) Trials to establish quality control method of mass-reared and irradiated flies in the field. In: Mangel M, Carey JR, Plant R (eds) Pest control: operations and systems analysis in fruit fly management. Springer, Berlin, pp 211–224

USDA (United States Department of Agriculture) (2003) Mediterranean fruit fly action plan. www.aphis.usda.gov/import_export/plants/manuals/domestic/downloads/medfly_action_plan.pdf

USDA (United States Department of Agriculture) (2006) International Panel for Review of Fruit Fly Surveillance Programs: review of fruit fly surveillance programs in the United States. USDA/APHIS/PPQ/Fruit Fly Program, Riverdale

Vargas RI, Stark JD, Nishida T (1990) Abundance, distribution, and dispersion indices of the Oriental fruit fly and melon fly (Diptera: Tephritidae) on Kauai, Hawaiian Islands. J Econ Entomol 82:1609–1615

Vargas RI, Mau RFL, Stark JD, Piñero JC, LeBlanc L, Souder SK (2010) Evaluation of methyl eugenol and cue-lure traps with solid lure and insecticide dispensers for fruit fly monitoring and male annihilation in the Hawaii area-wide pest management program. J Econ Entomol 103:409–415

Venette RC, Moon RD, Hutchison WD (2002) Strategies and statistics of sampling for rare individuals. Annu Rev Entomol 47:143–174

Watzl O (1927) Thermal summation rule and its relation to agricultural pests. Fortschr Landwirt 2:350–354

Weldon CW, Meats A (2007) Short-range dispersal of recently emerged males and females of *Bactrocera tryoni* (Froggatt) (Diptera: Tephritidae) monitored by sticky sphere traps with protein odour and pot traps with male lure. Aust J Entomol 46:160–166

Weldon C, Meats A (2010) Dispersal of mass-reared sterile, laboratory-domesticated and wild male Queensland fruit flies. J Appl Entomol 134:16–25

Weldon C, Prenter J, Taylor PW (2010) Activity patterns of Queensland fruit flies (*Bactrocera tryoni*) are affected by both mass-rearing and sterilization. Physiol Entomol 35:148–153

Wong TTY, McInnis DO, Kobayashi RM, Nishimoto JI (1985) Distribution and seasonal abundance of adult male Mediterranean fruit flies (Diptera: Tephritidae) in Kula, Maui, Hawaii. J Econ Entomol 78:552–557

Worner SP (1988) Evaluation on diurnal temperature models and thermal summation in New Zealand. J Econ Entomol 81:9–13

Worsley PM, Dominiak BC, Arthur R, Gilmour AR, James DG (2008) Pilot study at Cowra of intra-town dynamics of Queensland Fruit Fly (*Bactrocera tryoni*) (Froggatt) populations based on trap catch data. Plant Prot Quart 23:86–89

Zalucki MP, Drew RAI, Hooper GHS (1984) Ecological studies of eastern Australian fruit flies (Diptera: Tephritidae) in their endemic habitat. II. The spatial pattern of abundance. Oecologia 64:273–279

Chapter 9
Spatial Analysis of Tephritid Fruit Fly Traps

David Midgarden, Estuardo Lira, and Micha Silver

Abstract In tephritid fruit fly programs, spatial analysis attempts to determine the spatial distribution of target fly populations, recognize and understand geographic patterns, and identify relationships with ecological factors and control activities. Traps provide the majority of information about the presence and size of tephritid fruit fly populations, and spatial analysis of the trap data elucidates tephritid fruit fly population patterns and trends. Geographic Information Systems (GIS) serve as a bridge between the traps and the spatial analysis methods; although GIS is widely available and in use in many tephritid management programs, spatial analysis *per se* is used less frequently. The Medfly (*Ceratitis capitata* (Wiedemann)) Program in Guatemala, Mexico, and Belize has been using spatial analysis since 2004, and data from this program was used here to illustrate the basic steps to conduct this kind of analysis. These steps include standardization, representation, and exploration of trap data from a spatial and temporal perspective. Geographic database formats (including vectors and rasters), geo-referencing, projections and symbology are key concepts for GIS implementation in tephritid fruit fly programs. Analyses can include combining geographic data to reveal patterns, trends, or cycles that might otherwise be difficult to discern. Results of analyses should be presented in a manner that is accessible and easy-to-interpret. If analyses include a temporal component, animations can show changes over time. One of the applications of spatial analysis is to evaluate trap captures of flies released under sterile insect technique (SIT) programs. The use of spatial analysis in the Medfly Program in Guatemala resulted in improved reports that summarize information in a more meaningful way, identify previously unrecognized patterns of population growth and spread, and promote the development of improved integrated pest management

D. Midgarden (✉) • E. Lira
USDA APHIS IS Medfly Program, Guatemala City, Guatemala
e-mail: david.g.midgarden@aphis.usda.gov; estuardo.lira@aphis.usda.gov

M. Silver
Arava Development Company, Sapir, Israel
e-mail: micha@arava.co.il

T. Shelly et al. (eds.), *Trapping and the Detection, Control, and Regulation of Tephritid Fruit Flies*, DOI 10.1007/978-94-017-9193-9_9,

strategies. These new strategies allowed the Medfly Program to advance 150 km in less than 4 years despite severe budget reductions. Spatial analysis has been used in other countries and tephritid species to support management decisions, e.g., the medfly in Spain, *Bactrocera carambolae* Drew & Hancock in the guiana shield of South America, and *Anastrepha ludens* (Loew) in the USA. As internet-based database technologies improve and become more widely available, the possibility exists for expanding local, national, and regional information into a world database.

Keywords Geographic Information Systems (GIS) • GIS Data formats • Geo-reference(ing) • Map symbology • Standard distance/ellipse • Density analysis • Spatial dependence/autocorrelation • Inverse Distance Weighted (IDW) • Kriging • Interpolation • Raster • Vector

1 Introduction

Tephritid fruit fly programs, researchers and fruit producers use traps to determine the presence and size of populations of these pests. Trapping is particularly suited to tephritid sampling due to the high motility of the flies and their preference for specific baits that aid in attracting them to the traps. Among sampling methods, trapping is categorized as a relative technique, because traps do not directly measure the number of individuals per area or volume (e.g., per hectare or cubic meter or per unit of habitat) as do absolute measures. Relative methods produce a measure of abundance based on the number caught per unit of effort (e.g., sweep-netting or trapping) and are used when either resources are limited, when it is necessary to obtain a sample of the species present quickly, or when the organism is difficult to sample (Pedigo 1996). Given constant sampling protocols and efficiency, relative methods can be used to describe population trends or compare abundance between habitats (Southwood and Henderson 2000). In brief, traps do not directly measure the population as do absolute methods, but they are frequently used to estimate, monitor, and compare population levels. In this chapter, we will use trap capture as a comparative measure of tephritid populations and assume that trap capture reflects the tephritid adult population density.

Traps are the sensory system of tephritid fruit fly programs and provide the majority of information about the target population as well as the distribution of sterile flies when the Sterile Insect Technique (SIT) is used (Knipling 1960). Despite the resources and importance placed on trapping, the data collected are often not used to full advantage. The advent of Geographic Information Systems (GIS) facilitates the activity of organizing the data and making reports, but, like many users of GIS, tephritid fruit fly program managers may be unfamiliar with the more recently integrated tools available to conduct spatial analyses that could improve decision-making and promote more effective control measures (Mitchell

2009). Trap data are rich in information that could be used more thoroughly to give a better representation of reality and guide the management of a program.

The purpose of this chapter is to introduce some of the analyses available for fruit fly trap data that can add to understanding the behavior and ecology of the target populations and thus improve decisions related to all areas of fruit fly management. The chapter will introduce the reader to the main concepts of maps and spatio-temporal analyses, including a description of GIS and its implementation. We will discuss such important considerations as the data types and coordinate systems and then describe some basic spatial analyses using examples from the Medfly Program in Guatemala, Mexico, and Belize, which targets the Mediterranean fruit fly, *Ceratitis capitata* (Weidemann). Finally, we will present some ways spatial analyses have been used in different parts of the world and look at possible trends for the future.

2 Main Concepts

2.1 Maps

A map is a visual representation of an area or a model of reality at a given scale and location. Maps are the oldest and most generally used representation of spatial data. Mapping dates to 6000 BC if not earlier (Utrilla et al. 2009; Schmitt et al. 2014), and maps are constructed for a wide variety of purposes. They are used in tephritid pest management to show detections, report actions taken, and estimate infested areas and legal action required. The information used for tephritid fruit fly management/control programs includes the insect's biology and distribution as well as the relative effectiveness of control activities used in the past. The various control strategies work best when they are integrated to support one another (Pedigo 1996). The use of maps can improve the integration of information for effective decision-making.

2.2 Spatial Analysis

Spatial analysis is the description of data relating to a process operating in space, the exploration of patterns and relationships in that data, and the search for explanations of the identified patterns and relationships (Bailey and Gatrell 1995). In general, the objective of spatial analysis is to examine properties and relationships of the available data while taking into account their spatial location in a direct way (Camara et al. 2004). Spatial analysis of tephritid fruit fly trap counts attempts to answer questions about the presence of the flies by identifying geographic patterns in trap data and looking for relationships with ecological and human factors present in the area. Spatial analysis might increase understanding of a process (e.g., invasion and establishment of a population) or assess hypotheses to describe tephritid presence or behavior in relationship to something else (e.g., hosts, climate, modes of transportation) (Bailey and Gatrell 1995). The methods can be simple, such as plotting the data on a map, or more complex using multiple sources or groupings of information as well as other techniques, such as spatial statistics and interpolation.

2.3 Spatio-temporal Analysis

Detecting change in the location and size of tephritid fruit fly populations over time is one key objective of trapping programs. Analyzing tephritid fruit fly data in only the spatial dimension or only in time is often incomplete. As population density increases, there may be a tendency for the geographic distribution to increase over time. The flies can spread both to nearby and distant locations either by natural means (e.g., direct or wind assisted flight) or by transport of fruit that serve as hosts that may follow routes of human movement (Gould and Wallace 1994). Likewise, as populations decrease in density, they may have a tendency to reduce their spatial distribution. One objective of spatio-temporal analysis is to monitor and perhaps predict changes in spatial distribution over time. In general, there are three types of temporal spatial analysis: trends, pre and post, and cyclical (Mitchell 2009). Trends indicate whether the population is increasing or decreasing or the direction and pattern of insect movement. Pre and post patterns show conditions before and after an event or action and attempt to evaluate the impact. Cycles show recurring events or patterns and how they may be related to other geographic features.

Spatio-temporal analysis takes advantage of ‘when’ as well as ‘where’ information is collected (Cressie and Wilke 2011). Fortunately, temporal data are inherently collected in trapping programs, and these analyses are well suited to trap data. However, the trapping information itself is not the only data useful in understanding tephritid fruit fly population patterns. Space and time act differently on the factors important to tephritid fruit fly populations. Soil, elevation, and host species, for example, may be spatially and temporally stable, while other factors, such as seasonal changes in temperature, soil moisture, and host phenology, may be more variable. Unfortunately, trapping is rarely conducted at a spatial or temporal scale optimal for analysis. Management programs may not have the luxury of determining and implementing optimal sampling regimes for spatio-temporal analysis but can make every attempt to make the best use of the data that have been collected.

2.4 A Trap as a Geographic Unit

There are intrinsic properties of fruit fly traps that make them useful as the basis for spatial analysis. Trapping inherently averages the sampling over time (the amount of time a trap is operating) as well as space (the range of attraction of the trap combined with the movement of the tephritid fruit fly adults) and can result in less variable observations than an absolute measurement (e.g., adults per leaf). This is not to say that the trap results are uniform, and as we will see there is a great deal of variability in fruit fly trap catch both spatially and temporally.

Traps are installed at a given location, and thus they can most easily be seen as points in space. However, as traps are geographically separated – regularly or irregularly – the resulting captures represent the tephritid population from a larger area surrounding the traps. There are several spatial data formats that can be used to display trap data, each of which has a different set of methods available to conduct analyses. These methods, when applied to the trap data saved in spatial formats comprise a GIS, which is the best option to organize, display, and analyze trap data.

3 Geographic Information Systems

The storage, display, and analysis of spatial data are traditionally done in a GIS (Bivand et al. 2008), which is often viewed as a toolbox for collecting, storing, retrieving, transforming, and displaying spatial data to support decisions (Burrough 1996). GIS provides a framework for gathering and organizing spatial data and related information so that it can be displayed and analyzed. One clear use for GIS in entomology is to assist in the management of area-wide Integrated Pest Management (AW-IPM) or eradication programs. Commercial fruit producers can use GIS to support their management strategies and researchers in their analysis of experimental studies (Dminić et al. 2010). While it is possible to conduct spatial analysis without a GIS – using only statistical software, such as S-plus, SAS, or R, this technology is used in most tephritid fruit fly management programs, and spatial analysis can be conducted using the tools available in the GIS software or in collaboration with it. GIS adds a compelling visual component to the data analysis.

3.1 Planning Before Deployment of a GIS

A GIS has all the basic elements of any information system, with the additional capability to hold and manipulate spatial data. The deployment of a data management system generally goes through four stages:

3.1.1 Characterization

Careful thought and planning before the implementation of an information system enhances its usefulness. Time spent considering the following questions help ensure that the resulting information system fulfills expectations.

- What will be the format for reports?
- How often will reports be produced? Who receives them? How will they be delivered?
- How much data will the information system contain? What size computer storage will be needed?
- What size of backup storage will be needed?
- How often will data be added? Who enters new data? How will they access the database?
- What is the spatial extent of the program? What time span is to be covered?
- What queries will be required? How often?
- What precision is required for the raw data? For the queries?

3.1.2 Design

An outline of the dataflow showing links between tables, entry forms, and other critical features should be designed using the data structure options together with the information system characterization. A flow chart of the movement of raw data through to final reports should also be presented to the program managers for review. After this critical step, the database management personnel can build entry forms and formats for reports and establish communication channels as well as determine schedules for data transfer and entry and report production. Sources of spatial data can be located and suitable software chosen.

3.1.3 Implementation

Once these tools, design prototypes, and characterizations are complete, the managers can acquire software and hardware, create the required data tables, obtain the spatial layers, and formulate database links and queries. Preliminary trial data should be entered into the tables, and queries tested to verify that results are as expected. After determining the backup strategy, a trial run of the backup routine should be conducted before it is established. Data entry forms need to be tested by staff to insure that they are usable and error free. After a trial period, the system can be put into production.

3.1.4 Operation

Once a GIS is in use, an ongoing feedback loop should be put in place to follow the process from new data to queries to reports. Timely distribution of reports should be verified and the consistency of backups checked.

3.2 Data Types

In a GIS, the basic unit of data for display is known as a layer. A layer is a slice or stratum of a particular feature in an area; on a road map, for example, roads, national parks, political boundaries, and rivers might be considered different layers as can shading or isolines to indicate elevation, rainfall, or temperature. The core characteristic of a GIS is that layers can be superimposed on one another to look for spatial coincidence and potential relationships. There are two groups of spatial data types: *vector* and *raster*. Vector data layers show discrete locations together with an associated table of alphanumeric data known as *attributes*. In a GIS, attributes are the characteristics of the feature at that particular location, i.e., if the features represent traps, the attributes can be the name of the place, trap type, number of flies captured, or any other information of interest. In most implementations, vector data layers are either point features (i.e., traps, fruit sampling sites, quarantine stations, and detections), line features, (e.g., roads, trap routes, and boundaries), or polygon features (SIT release or insecticide application blocks, land use, water bodies, or phytosanitary designations, such as free or low prevalence areas). A point is a pair of coordinates. Each feature in a line vector is composed of an ordered array of two or more points. Each polygon in an area vector is also an ordered array of points – at least three – where the first and the last points are the same. Each vector layer can contain many individual features, e.g., many road segments in a line layer or many trap locations in a point layer. All individual features have associated attributes that describe them.

Raster based data, on the other hand, consist of a regular matrix of values throughout a region similar to pixels in an image. A raster data layer is defined by its extent and resolution. The extent sets the north, south, east, and west bounds of the region covered by the raster layer, and the resolution is the size of each grid or matrix cell. Thus, a raster layer is always composed of a set number of rows and columns (the extent divided by the resolution). Each raster cell has a single value associated with it. This value can represent elevation, fruit fly capture, land use, host presence, and any other phenomenon of interest. The single cell value may also be interpreted as a Red-Green-Blue color combination. In this case, known as a three-band raster, the data layer becomes an image, such as used in aerial photography or satellite imagery.

A raster layer can be a continuous surface where each pixel can have any value, either integer or real number, or areal data (i.e., of or related to an area) where pixels are categorized as one of a limited number of preset integer values. Examples of continuous rasters include elevation, rainfall, and temperature data. A continuous raster can also be estimated or interpolated from a set of field samples (e.g., traps) that are distributed regularly or irregularly. A typical example of an aerial raster is landcover classification in which each pixel holds one of only 20 or so classification categories (e.g., pine forest, urban, or citrus). Many tools are available in GIS software to conduct analyses using raster layers, such as combining several raster layers by a mathematical or logical formula or reclassifying by some criteria (e.g., a raster of elevations in meters could be reclassified into five different ranges) as well as neighborhood analyses.

3.2.1 Available Data Formats

Both vector and raster data can be stored in a number of different ways. It is vital to the success of any project to adopt a standardized format or set of formats to ensure data compatibility between GIS data users. GIS data formats can be categorized into two groups. The first contains formats that support single features and allow access to only a single user. The second includes full multi-user databases with a collection of spatial functions built into the database itself.

Single Feature – Single User

Single feature – single user is a data format that stores only one type of geometry in a local file in the computer. The ESRI[1] shapefile is a good example of a simple, single user GIS vector format. All contemporary GIS applications and mapping software support ESRI shapefiles, and they can manage layers with hundreds of thousands of features easily. The main limitation of this format is that each file contains only one kind of feature: Point, Line, or Polygon. Another issue is that the database management characteristics of the shapefile, as well as most single user formats, suffer from limitations of older database formats: column headers cannot exceed 10 characters, only "simple" data types are accepted ("Character", "Numeric", and "Date"), and there is very limited support for queries with no option to link to other tables. Therefore, in order to perform sophisticated linking or querying with data stored in shapefiles, the GIS software must supply this functionality. A program with modest GIS requirements and **no need to share data** among a group of GIS technicians can use a single user format. Beyond the shapefile, there are many other single-user vector formats used by various GIS software, such as the Mapinfo *.mif and *.tab files (http://www.pbinsight.com), *.

[1] ESRI: Environmental Systems Research Institute (http://www.esri.com/) is an international supplier of GIS software, database management applications, and training.

gpx files used by the global position system, and the *.kml format made popular by GoogleEarth (www.google.com).

Multi-user Database

Multi-user databases are able to store all geometric types in the same file. This is an advantage compared to single-user formats, which must be defined as point, line or polygon. In addition, they offer all the flexibility of a modern full relational database, such as joining spatial layers with alpha-numeric tables, creating database "views" to display filtered sets of data, and running a variety of spatial operations directly. Some popular examples of full featured, multi-user databases include Spatialite (https://www.gaia-gis.it/fossil/libspatialite/index), PostGIS (https://postGIS.net), ESRI's GeoDatabase, and Oracle Spatial (Oracle Corporation, Redwood Shores, CA, USA). A project that shares spatial data among many users and locations should choose one of the multi-user database solutions. Storing all data in a single shared spatial database ensures that everyone is using the same updated layers and allows centralized backups.

3.3 Geo-referencing Spatial Data

The main difference between an IS and a GIS is that the latter uses geographical information and because of that requires geo-referencing. In our daily life, we use places or noticeable objects to give references (e.g., rivers, roads, buildings, etc.). For example, we can say, "drive north two kilometers from the lake" or "turn left at the red building and walk three blocks." A geo-reference works in a similar way, using the center of the earth as a reference to find any location on earth's surface. Coordinate systems have been developed using this center to place spatial information. There are two kinds of coordinate systems: geographic and projected.

Geographic Coordinates Systems (GCS) are based on a three-dimensional model (sphere or spheroid) to reference locations on the earth's surface. A GCS is made by making angular measurements from the center of the earth (ESRI 2004). One of the measurements uses this center and the equator to measure the angles to the north or to the south, and these angles are called Latitudes. The other measure uses the center of the earth and a main meridian (often the Greenwich Meridian) to measure angles to the west or the east, and these angles are called longitudes. The latitudes and longitudes create a network that allows any position on the earth to be uniquely identified. A very important component of the GCS is the datum, which defines where the center of the earth is located, and therefore the origin of the angular measurements. In summary, the GCS is composed of (i) angular measurements (latitudes and longitudes), (ii) a sphere or spheroid, (iii) the equator and a primary meridian, and (iv) a datum.

Projected Coordinate Systems (PCS) are based on a GCS and are then "flattened" using mathematical formulae to create a two-dimension representation of a portion of the earth (ESRI 2004). The coordinates in a PCS, instead of angular measures, are distance measurements from an arbitrary point of reference (the origin of the PCS) on an X and Y axis. Representing the earth's three-dimensional surface in two dimensions causes distortion in the shape, area, orientation, and distance, and each PCS tries to minimize those distortions. The PCS are derived from a GCS, and they have a projection type, one or two standard parallels (the points or lines of tangency between the sphere or spheroid and the projection), and a central meridian. The advantage of PCS is that accurate distance and area can be calculated within a small region, but estimated distances outside the coverage area are distorted and have more error.

Defining the coordinate system to be used for the spatial information of a tephritid fruit fly program is an important step to ensure the standardization and accuracy of the data and to avoid errors. Most countries have a government mapping agency that chooses a local PCS that most suits the needs of that country. However, tephritid fruit fly programs should take into account the following considerations when determining which coordinate system to use. First, if the program's main sources of data layers (e.g., elevation, land use, terrain, and transportation) are already in a certain coordinate system, then that coordinate system should be considered for all data layers. Second is the size of the working area. If the area is not larger than a few hundred kilometers, then the data should be stored in a projected coordinate system. Keeping data in Lon/Lat is convenient, but using un-projected (GCS) data for calculating area and distance may lead to errors. If the area is very large (over 600–700 km east to west), then it will likely span more than one projected coordinate system. In this case, base data should be kept in a geographic coordinate system and be re-projected as needed when making local maps and calculations using measurements. Third, most GIS software today can display layers in different coordinate systems and overlay them properly by performing "on the fly" re-projection. Even though this capability exists, it should not be relied upon, because it slows down rendering of layers and can introduce errors in analysis that are difficult to trace. Thus, it is best to keep all data in the same coordinate system if possible.

3.4 Trapping and GIS

The data collected and stored for each trap (the attributes in GIS terminology) vary from one program to another and can include administrative information (e.g., name of property owner, trapper, trap route, etc.). We have found the following attributes most useful for spatial analysis and recommend that these be part of any fruit fly database:

- Number of flies captured. This information should be recorded in as much detail as possible to differentiate sex, fertility and marking color (if SIT is being used), and any other factor that is possible to collect defining the characteristics of the specimen (e.g., age, presence of eggs in the ovarian, mating status, genetic markers, possible origin, etc.). The more information that is kept in separate fields (or columns) the more analysis options available.
- Time the trap is active in the field. Programs may have a variety of schedules for checking traps, e.g., all traps serviced weekly, bi-weekly, or monthly, or may have several different schedules for different regions within the total coverage area. It is important to be able to calculate the amount of time traps are operating as accurately as possible and retain the ability to be flexible. There are always some traps that cannot be serviced on schedule, and it is important to adjust active collection times for individual traps to account for these discrepancies. Using a pre-filled database that reports a default value even if a trap was not serviced can cause serious data quality issues.
- The host or location where the trap is placed. A complete record includes the actual tree, shrub, or substrate where the trap is placed, a categorization of the hosts in the area, or both (pecan tree near peach orchard) using more than one field (column). The host should be recorded as unambiguously as possible, using species name or more detailed if necessary (i.e., variety or cultivar).
- The location where the trap is placed in the standardized coordinate system. Altitude may also be useful in some geographic areas of the world and can be estimated using the GPS. Elevation data can be added or validated later using a digital elevation model.

Any information stored in the database should follow strict standardization rules to ensure that retrieving, summarizing, or querying can be done with integrity and efficiency. Catalogs with the standardized names and/or codes for the different values of the attributes should be constructed and used exclusively, for example, a list of host or trap types with no spelling or letter (e.g., accents) variations (café, cafe, coffee will each be recognized as different hosts, resulting in errors).

While the trap information is the foundation of the spatial analyses of most fruit fly programs, it is equally important to have key ecological or geographic layers that may explain the presence, population dynamics, and movement of tephritid fruit flies. The specific layers that are important for a given program depend on the geography and climate of the area, the tephritid species, and the objectives of the program. Layers potentially important for tephritid fruit flies include host presence, land use, organic waste sites, altitude, transportation, soil type, temperature, and rainfall as well as areas that might presumably preclude fly presence (e.g., deserts, bodies of water, large areas of field crops, or pastureland). These layers may be available in a variety of formats and coordinate systems on internet sites (see Addendum), government agencies, universities, the private sector, and other organizations or may have to be generated by the program itself.

4 Spatial Data Analysis

One advantage of using a GIS to manage data is the ability to conduct spatial analyses once the data are organized into a series of compatible layers. Despite the widespread use of GIS in the scientific, business, and agriculture community throughout the world, spatial analysis is generally underused (Mitchell 2009). This is partly because GIS itself is only about 35 years old, is in constant development, and technicians and managers may not be familiar with newly integrated methods and capabilities other than maps and reports. This is not to imply that GIS is not useful without spatial analysis; in fact, the maps and reports from GIS provide the basic questions that spatial analysis may be able to answer. One of the main benefits of presenting a map of tephritid data is to provoke thoughtful inquiries from the program decision makers. Why is the population so high here? Why didn't we catch sterile flies there? What is present at this location where there have been detections the last 3 years in the same month? These are the questions that direct spatial analysis, and they come from well-presented maps that show the trap information clearly.

As briefly discussed in the main concepts, spatial analysis is the description of data relating to a process operating in space, the exploration of patterns and relationships in that data, and the search for explanations of the identified patterns and relationships (Bailey and Gatrell 1995). More simply, it is the process of looking for geographic patterns in data and relationships among geographic features (Mitchell 2009). Spatial analysis includes a collection of methods and relies on a good knowledge of the tephritid trapping system and the environment in which the program operates as well as tephritid fruit fly biology, behavior, and ecology. The analyses should include a tephritid specialist's input to develop the questions, define the parameters to use in the analyses, and evaluate the results.

4.1 Medfly Program Background

We will use trap data from the Medfly Program in Southern Mexico, Guatemala, and Belize to introduce some basic steps and concepts of spatial analysis. First, we will outline the main characteristics of the program. Its primary objective is to prevent the spread of this pest into Mexico and Belize, reduce the risk of introduction to the USA, and create an increasing number of fly-free areas in Guatemala, eventually leading to eradication. These goals are accomplished by reducing *C. capitata* population densities at the leading edge followed by consistent use of SIT at densities sufficient for eradication, moving the front of infestation in Guatemala away from the border of Mexico and freed areas in Guatemala through an area-wide rolling carpet approach as described by Barclay et al. (2011). The program is located in an area with a wide diversity of topography, climate, and host presence. Lower elevations, primarily coastal, support tropical hosts such as mango

(*Mangifera indica* L.), guava (*Psidium guajava* L.), citrus (*Citrus* spp.), tropical almond (*Terminalia catappa* L.), sapote (Sapotaceae, Ebenaceae, Rutaceae; species of various genera), and hog plum (*Spondias purpurea* L.). High elevations are cool, and the primary hosts are temperate, such as peaches (*Prunus persica* (L.) Batsch), pears (*Pyrus* spp.), apples (*Malus domestica* Borkh.) and plums (*Prunus* spp.). The intermediate elevations are home to some of the richest and most concentrated coffee (*Coffea arabica* L.) production areas in the world. Trapping is the main detection method, and between 2004 and 2011 an average of 30,000 dry traps were in service on a weekly basis (Lira 2010). Approximately 45 % are sticky traps baited with male lures (i.e., trimedlure), and the remainder are Phase 4 traps baited with synthetic food-based lures (i.e., three component BioLure, Suterra LLC, Bend, OR, USA).

The eradication of *C. capitata* is based on an area-wide Integrated Pest Management (AW-IPM) SIT approach. Male flies are reared in large numbers and sterilized. After rearing, the pupae are shipped to emergence centers, where the adult male flies are held for an average of 3.5 days, chilled to dormancy at 3–5 °C, packed into release boxes that are kept at a constant temperature, and loaded onto airplanes where release machines are installed. These machines release the flies into the environment at a metered rate. Once in the environment, they mate with wild females and thereby reduce the overall reproductive potential of the medfly population within the target areas (Knipling 1960; IAEA/FAO 2004). The trapping information is used to assess the spatial distribution and relative density of the wild populations. Based on this distribution, the location and densities for sterile releases are defined. Traps also provide a means to evaluate the distribution of sterile flies in the field after their release.

Figure 9.1 shows the Working Area of the program, and the portion within this that we have selected as the area of interest, which coincides with the location of the program's historically most persistent infestation of *C. capitata*. The area of interest is representative of the region as a whole, encompassing a similar range of conditions and consistent distribution of traps over the 8-year period. The altitude varies from sea level to 4,200 m. The average distance between traps was 450 m, which meets the program protocol of 2 traps per square km, however, the trap density varied from 0.5 to 10 traps per square km based on host presence, accessibility, and risk of establishment.

The initial steps in conducting spatial analysis are to define the objective of the analysis or the question to answer, the exploration of the data to understand it better, and the selection of an analysis method. Mitchell (2009) summarizes the steps in spatial analysis succinctly:

- Frame the question and define the objective
- Select and characterize the data
- Choose a method and parameters
- Run the analysis and look at the results
- Repeat

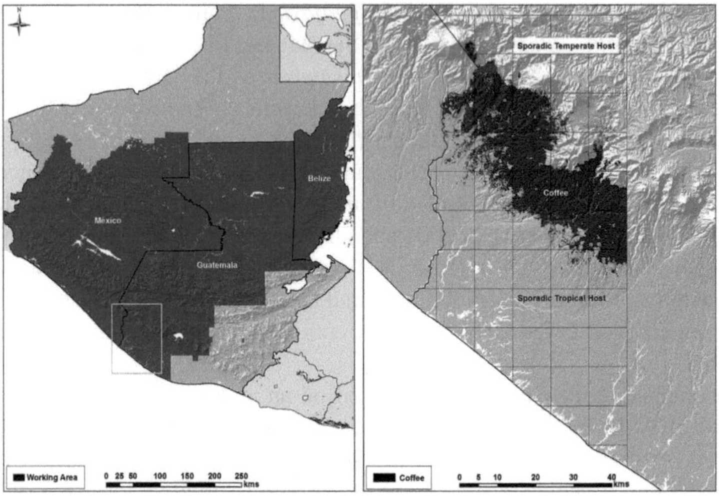

Fig. 9.1 *Left*: Map of the action area of the Regional Medfly Program in Mexico, Guatemala, and Belize with the area of interest outlined by a square. *Right*: More detailed map of the area of interest in Guatemala showing three climatic zones: Temperate, Tropical and Intermediate driven by differences in temperature, which is inversely correlated with altitude

4.2 *Framing the Question*

The questions that form the basis of the analyses are really just basic hypotheses: Is the location of the population stable or does it change over time? Is there a cyclical component? Do the population dynamics of the pest coincide with the presence and phenology of any host or set of hosts? Are captures located more often in some hosts than others and when? "Where is it?" is the basic question for tephritid fruit fly analysis, and often the answer can *be* the analysis. The detection of a single fly does not involve a great deal of analysis, and the answer to "Where is it?" is sufficient information to understand the situation and guide a response both technically and administratively. Often, however, the answer to "where is it" is not sufficient, and questions that follow may include "How long has it been there?", "How large is the infestation?", and "Is it related to something else in the area?", such as a free area, international border, airport, known infested locations, an SIT release block, hosts susceptible to export requirements, or a quarantine station.

There is no correct question for each analysis, and the basic function of the questions is to gain a better understanding of the tephritid fruit flies and the data available (Camara et al. 2004). In the early stages of analysis, different tools and procedures should be used to examine and explore the data to answer a question. A variety of different analytical methods can be tried, and the results used to guide further analysis. There are several ways to analyze any question, and each method provides a different perspective in order to reach an acceptable answer. The results of each analysis should be scrutinized and used to select an additional method or to change the parameters of a method already used.

4.3 Data Characterization

The trap information in the Medfly Program is entered and stored in ESRI ArcGIS® as point vectors. The basic characterization of the trapping data provides support for later analyses. Characterization includes data aggregation in order to explore the general trends of the information. Data aggregation is any process in which information is gathered and expressed in summary form. To illustrate, from 2004 to 2011 (an 8-year period) a total of 294,917 wild *C. capitata* was captured in the area of interest. Table 9.1 shows this number aggregated by year, along with the mean, variance, skewness, and kurtosis. The frequency distribution (Fig. 9.2) indicates that the trap results follow a negative binomial distribution typical of samples for many insect species, with most traps capturing few *C. capitata* but with a small proportion capturing large numbers. The frequency distribution is important to select the spatial and traditional analyses that can be conducted and the best way to present them (symbology).

4.3.1 Symbology: Map Language

Although symbology is not a spatial analysis, it is an essential consideration that ensures that the results of any analysis can be interpreted easily and accurately. Symbology is the design employed to communicate information on a map, i.e., the language of the map, and, like any language, the readers must be able to understand it to transmit information effectively. Combining the visual variables color, intensity, size, orientation, transparency, and fill creates the symbology (Huisman and de By 2009). A general recommendation is to use a single shape for a given phenomenon while varying the colors or fills. For example, when making a map showing the locations of different trap types, the phenomenon is "Traps", the qualitative variable is the trap type, and points of the same size, but of different colors, can be used to represent them. For quantitative and ordinal variables, the recommendations are to use the same shape and color but different sizes or color intensities (Slocum et al. 2008). To make a map using polygons to show the different densities of sterile flies released in SIT blocks, a single color with different intensity (e.g., light green to dark green) can be used to reflect low to high densities. Quantities can also be grouped into classes that suggest quality, such as a scale based on temperature (cold-blue to red-hot), to clearly demonstrate the quality and quantity of trap capture (blue as "low and good" moving to red for "high and bad"). A color language that most people have learned to understand automatically is green, yellow, orange, and red as these colors are related with the sense of danger and importance. Green is normally considered as "no or very low danger", "good", or

Table 9.1 Number of wild captures in traps within the area of interest each year from 2004 to 2011

Year	Total	Mean	Variance	Skewness	Kurtois
2004	1,742	1.59	3.93	7.20	73.15
2005	17,733	4.41	183.60	11.55	201.15
2006	3,614	3.16	50.64	6.07	46.53
2007	183,680	9.89	795.79	6.42	56.99
2008	74,236	5.81	474.23	12.67	204.35
2009	11,465	2.55	31.49	10.57	158.48
2010	2,421	1.71	5.69	8.63	114.20
2011	26	1.00	0.00	–	–

Basic statistics, such as the mean per trap with capture, variance and indices of normality are also presented

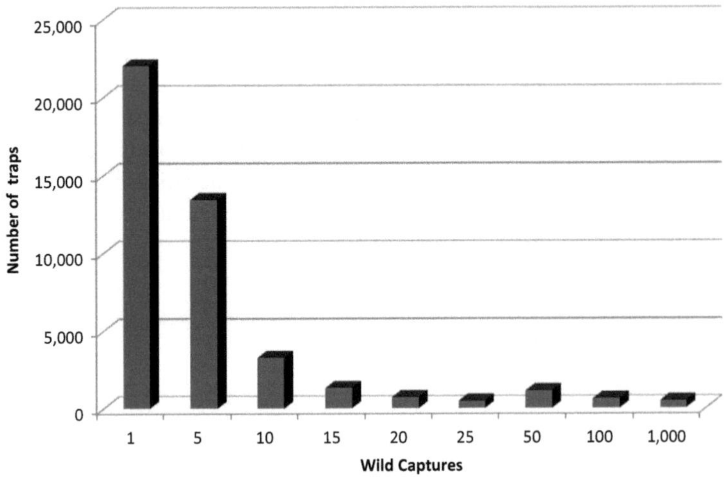

Fig. 9.2 The frequency distribution histogram of the traps with wild captures from 2004 to 2011. Each positive trap is tabulated separately; traps with no captures are not included

"very low importance" (because it is already good). Yellow is considered "low danger", "not that good" or "low importance" (because it has to improve). Orange is "medium to high danger", "poor or bad" or "high importance". Red is "very high danger", "very bad" or "very high importance". In the case of *C. capitata*, and based on International Atomic Energy Agency (IAEA) general thresholds of flies per trap per day (FTD) to define phytosanitary status, FTD captures above 1.0 (considered as infested) can be represented as red. FTD from 0.1 to 1.0 (considered appropriate for suppression) can be represented as orange. FTD higher than 0 and lower than 0.1 (considered for eradication) can be represented as yellow. An FTD of zero (free) can be represented with green.

In any case, the criteria used for categorization and symbology of the data classes used will vary depending on the purpose of the map and the audience. The rule of thumb to use when developing a symbology is to make the most important information stand out most prominently so that the map can be interpreted quickly and easily. A tephritid program should standardize and

Table 9.2 General recommendations for symbology of tephritid fruit fly trap data

Type of variable	Recommendation
Qualitative	Use same shape, different colors
Quantitative	Use same shape and color, different size or intensity of the color
Importance of classes	Danger: Green, Yellow, Orange, Red
	Temperature: Blue to Red

Qualitative can be type, purpose, or bait of trap or even species of fly captured (if an area with multiple species). Quantitative most commonly will be number of captures in a trap, but could be mean, mean per trap per day, total over a given time period, or other. Importance is a subjective interpretation of the data. There may be a large difference in importance between 1 and 2 captures of wild flies, but not much difference between 100 and 500. Likewise, important captures depend on the objective. Many wild captures is high danger, while low captures is very low danger; so the symbology should reflect this importance

publicize its symbology to enhance the understanding of the maps and avoid confusion. Table 9.2 shows the general recommendations by type of variable.

4.3.2 Capture Representation

The symbology can be designed to show the number or range of captures per trap or area when representing the trap results. Categories or classes of fly captures can be used to make the map easier to interpret, because the number of captures per trap can be quite variable (as shown in Table 9.1 and Fig. 9.2). There are many options for grouping data values based on how they are distributed. The four most common are natural breaks, quantile, equal interval, and standard deviation. Natural breaks group similar values into the same category and can be used to help identify where these values cluster. Quantiles group values so that there are near equal numbers in each category. This approach is useful for data with high variability and where categories above a certain level can be considered equally in an analysis. Equal interval groups the values into categories with the same high and low values and can be useful if the frequency distribution of the data is nearly uniform. Standard deviation defines each class in relation to the mean value and is useful to show data that are above or below a mean. In some instances, a priori classes based on program goals or international standards may be useful (e.g., IAEA recommends that an FTD > 0.1 is not suitable for eradication by SIT).

4.3.3 Animation

One drawback of maps is their static nature when presented as images, making it difficult to absorb information presented over many time periods. A time series is very helpful, but the need to remember locations while shifting from one image to another can hinder interpretation. Animation shows temporal changes and trends in a manner that makes the information more accessible. These temporal animations are much more readily interpreted as they display the fruit fly trap capture over time fluidly as a series of frames in a movie. When using animation, it is even more important that the symbology used in the maps show the most important information as clearly as possible due to the limited time each frame is displayed. Animation based on fine-scale temporal partitioning (e.g., weeks instead of months) yields a more integrated view of population behavior that may make it easier to discern patterns. Transitions between time periods can also be made easier to visualize using "morphing" or "tweening" software. There are a number of image processing software options available commercially or free that can be used to create animations; some of these are listed in the appendix.

4.4 Analysis Methods and Process

The process of selecting the method, running the analysis, and interpreting the results is iterative and not linear. Therefore, in this section we will go through several different analytical methods to try to answer questions about the trap locations and captures.

4.4.1 Vector Analysis

A good first step in spatial analysis is a simple point map of the locations of the trap data. It may be possible to get a sense of the spatial distribution, such as whether the captures are clustered, uniform, or random, simply by careful observation. The ability to identify a distribution or pattern can vary with scale, so it is useful to display the points at a variety of scales (zoom). Different areas can then be compared to see where high and low captures are located, paying particular attention to transitions between these areas. There may be a gradual trend across the area from low to high or there may be a sharper gradient or even a defined line between high and lows (Mitchell 2009).

Each dot in Fig. 9.3 represents a trap with wild fly captures from 2004 to 2011. The capture locations are clearly not uniform within the area of interest: there appear to be fewer wild captures on the tropical coast than further north in the more temperate, mountainous area. Is this perceived pattern consistent and can it be explained by factors important to fruit fly biology? One way of analyzing the capture data further is to present them broken out by year in a time series as shown in Fig. 9.4. When the information in Table 9.1 is combined with the maps in Fig. 9.4, the wild fly captures can be interpreted more meaningfully. The number of captures was different each year, but the locations or pattern of captures appears consistent for the first 4 years regardless of the number of captures. There was a 100-fold difference in the number of captures between 2004 and 2007, but the locations of the detections appear similar. Further observation may give the impression that the last 3 years showed a decrease in captures in the west and a

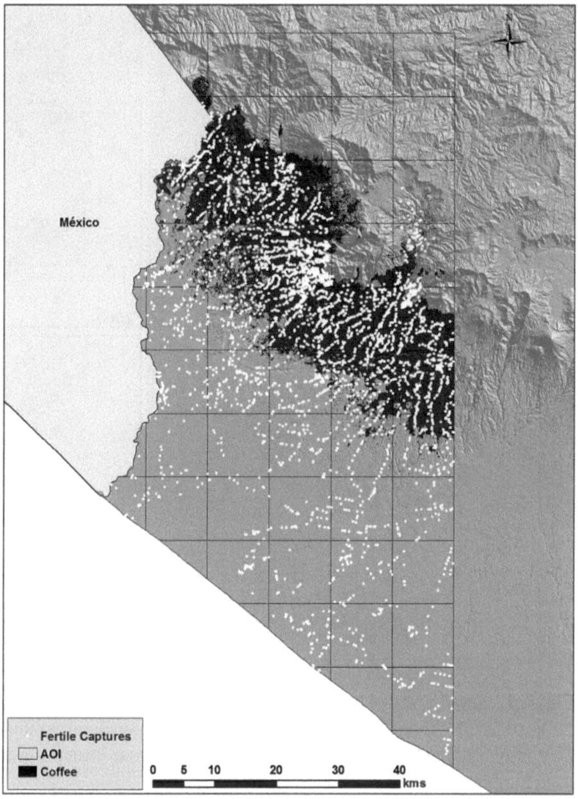

Fig. 9.3 Map of the area of interest showing the locations of wild captures as points from 2004 to 2011

correspondingly increasing concentration of captures to the east. Can these observations be validated objectively, and can they be interpreted in a meaningful way? The analysis of the locations of events is termed point pattern analysis. The objective of point pattern analysis is to determine whether the observed events exhibit a pattern as opposed to being distributed randomly (Camara et al. 2004). If a pattern is detected, a secondary objective is to determine whether the pattern is associated with proximity to another factor(s) (Bailey and Gatrell 1995).

A basic method to analyze point patterns is a measure of central tendency, such as the spatial mean, median, and standard deviation (SD). The mean location of traps is the average of their x, y locations, and the standard deviation of these values can be shown as an ellipse. In Fig. 9.4 we can also see a plot of the mean location and SD ellipse of traps with positive captures each year. This figure shows that the center was similar between 2004 and 2008, but from 2008 to 2010 there was a tendency for the number of captures to be located more to the east, and the shape and diameter of the ellipse changed. This lends support to the observation that the general spatial pattern of trap captures was spatially similar initially but changed in later years. The shape of the ellipse gives an indication of the directionality of the distribution of the fly captures and may assist in identifying the spatial trend of

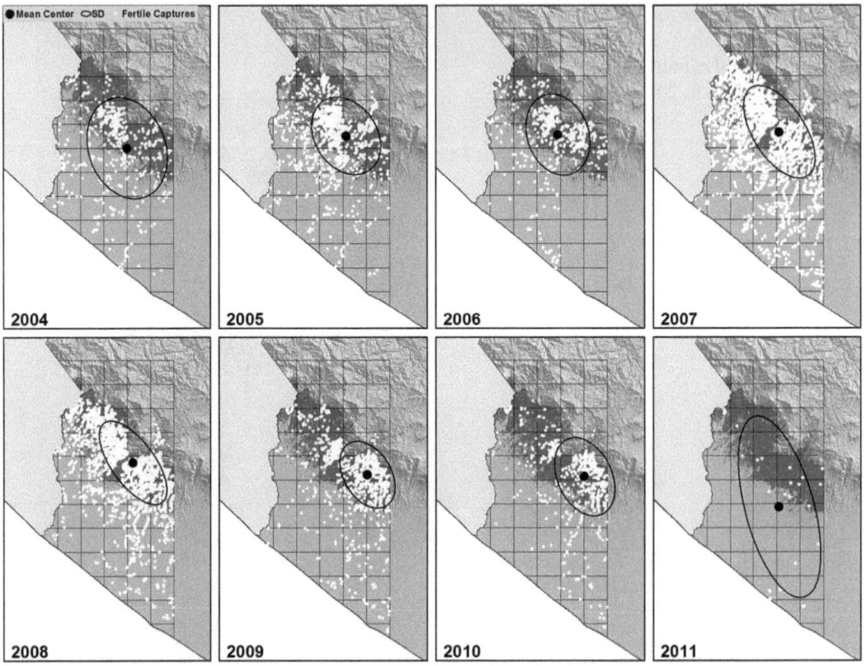

Fig. 9.4 Time series of locations of traps that captured wild *C. capitata* in each of the 8 years from 2004 to 2011. Mean Center (mean of x and y coordinates) and Standard Ellipse (one Standard Deviation of the X and Y coordinates) for each year indicating the spatial central tendency of the wild captures

medfly populations or the probability of capture in space. What are some of the characteristics of the area in this center that may be important to medfly detection? Altitude and temperature within the area as well as land use (principally host presence) may be key factors, and a map including that information along with the measures of central tendency of trap catch can show how these features coincide. One of the attributes of the trap data is the host in which the trap is placed, and we can use the GIS software to select the traps present inside the ellipse or within a given distance of the mean location and aggregate the traps by host or elevation in a table or graph. This analysis shows there are 1,274 traps inside the ellipse placed in 25 different hosts, but 75 % are located in a single host, coffee. Alternatively, a separate layer of host locations and altitude can be overlaid on the trap capture to see if any association is apparent (Fig. 9.5). The layer of hosts shows that the mean center and ellipse of captures is also located within the area of coffee production, and the altitude indicates that the majority of captures occur between 900 and 2,100 m. This provides evidence that coffee and climate (influenced by altitude) may play an important role in the presence of medfly captures. However, it also presents another question that may need to be answered, i.e., How do the captures compare to the distribution of the samples or the trap grid itself? Are the

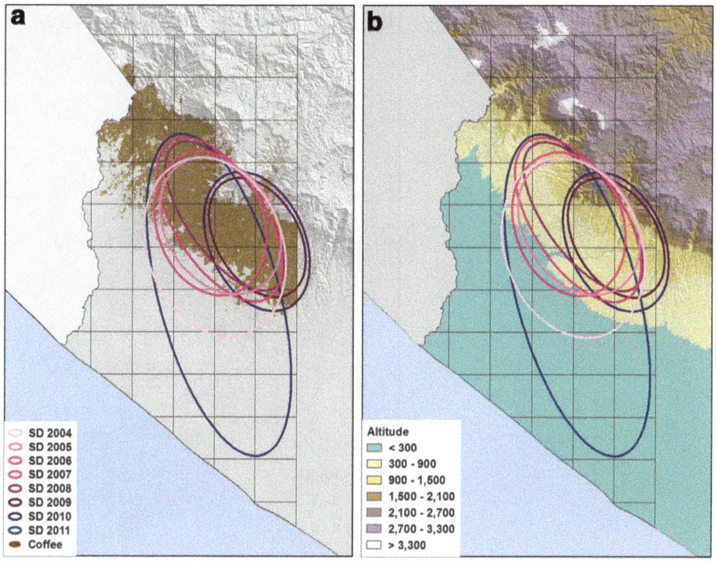

Fig. 9.5 Standard Ellipses from 2004 to 2011 overlaid on (**a**) coffee production and (**b**) elevation strata. Standard ellipses show the central tendency of the points and indicate change over time as well as a potential source of influence

samples also located mainly in coffee? A more direct question is: Are the results due to the underlying population of medflies, or do locations with more traps capture more flies? Comparing the spatial mean and SD location of all traps against the mean center of traps with positive captures could begin to answer this question. However, another point pattern method may be more useful.

Density methods group events by their number per area and show where the relative concentrations of a feature (such as traps or trap capture) are located. While it may be possible to discern different concentrations on a point map, it can be difficult to compare areas accurately where there are many events or if those events (e.g., trap capture) are repeated at the same location. A density map shows the number of features using a uniform areal unit, such as hectares or square miles, and allows the distribution of the concentrations to be compared directly. Density layers can be used to compare all trap locations to the locations of traps with positive captures, which addresses the question of whether areas with the highest number of captures are also those with the highest number of traps. Figure 9.6 presents the results of a density kernel analysis for the total number of traps and another for the traps with positive captures. The kernel method uses the number of events within a given distance, or a defined neighborhood, to determine and display the density, which is shaded darker to indicate higher density and lighter for lower density. In a GIS these two map layers can be overlaid and compared directly for different areas. Even without overlays, Fig. 9.6 clearly shows that the concentration of trap locations and trap captures are both predominantly located in the coffee production

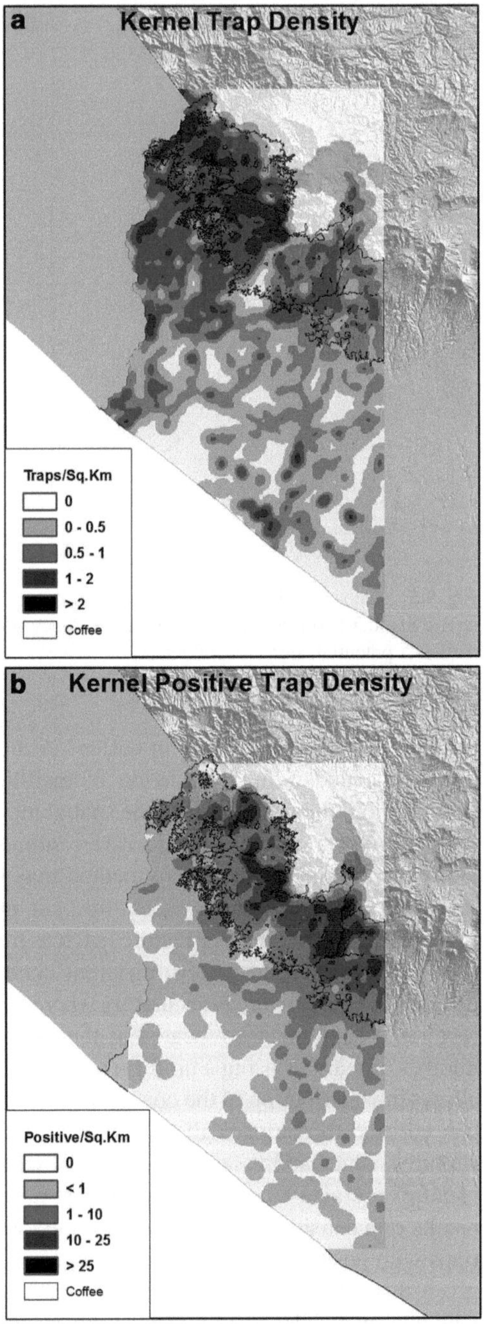

Fig. 9.6 Two Kernal Density analyses that group points by density per unit area where (**a**) displays the results based on the location of all traps in the area of interest and (**b**) shows results of wild captures. This allows the evaluation of the influence on trap distribution on the location of captures

area. However, the relationship is not direct; the location with the highest number of traps does not have the highest number of captures. So, the distribution of traps only partially explains the higher number of captures.

Point pattern analysis may or may not use all of the information available for a trap. In the previous examples, the analyses display the locations, mean, SD, and density of the captures but do not provide information about the quantity or intensity of the captures. A trap could capture 1 or 100 flies and would be represented as a single dot with equal weight in the analysis. These same analyses can also be weighted by the value of the trap capture. This would not affect the values of the trap locations themselves but would change the measures of central tendency and the trap capture density by including the number of flies in each trap in the analysis.

Another way to include the trap capture numbers in the analysis is using classifications and symbology. The search for patterns can be facilitated by using a standard classification scheme to group similar values (as discussed in Capture Representation). Figure 9.7 shows the same information as Fig. 9.4 but with the number of captures classified using six categories grouped into equal proportion of samples based on the frequency distribution (quantiles): a single fly, 2 to 5 flies, 6 to 10 flies, 10 to 100 flies, 100 to 1,000 flies, and more than 1,000 flies. The use of symbology to represent the trap captures as classes results in a clearer appreciation of the distribution of trap values and allows a more realistic interpretation of the capture data. The most highly infested area is where there is coffee production, and the trend for increased number of captures to the east over time is even more noticeable.

Displaying the trap captures as points is a useful technique, however, there are some drawbacks. While interpretation of areas with high numbers of captures is clear, areas where the trap capture is highly variable are less readily deciphered. Interpretation may be clarified by changing the symbology of the points by, for example, increasing the size of the symbols of classifications of particular interest (e.g., very high captures). However, depending on the scale, these larger points may overlap or hide others. There are many locations where high trap captures are located very close to low trap captures, and, though these areas are clearly intermediate, they are difficult, if not impossible, to interpret. Most people can hold only seven or so points at any one time in active memory, and it is unrealistic to expect a reader of a map to interpret traps with significant variation located close together objectively (Miller 1956).

Another vector method that can be used to analyze trap data is aerial in which data are aggregated within predefined polygons. The polygons can aggregate the trap data in a wide variety of ways such as mean capture, total number of flies captured, captures per trap per day, or other statistics or indices specific to a given analysis, such as the density or ratio of captures of sterile to wild males or male to female wild flies. Areal data are often collected and aggregated within polygons originally defined for other objectives. Outside the tephritid world, polygons are used because they are often the smallest spatial units in which data are collected or for which they are most meaningful (e.g., census information or cancer rates per

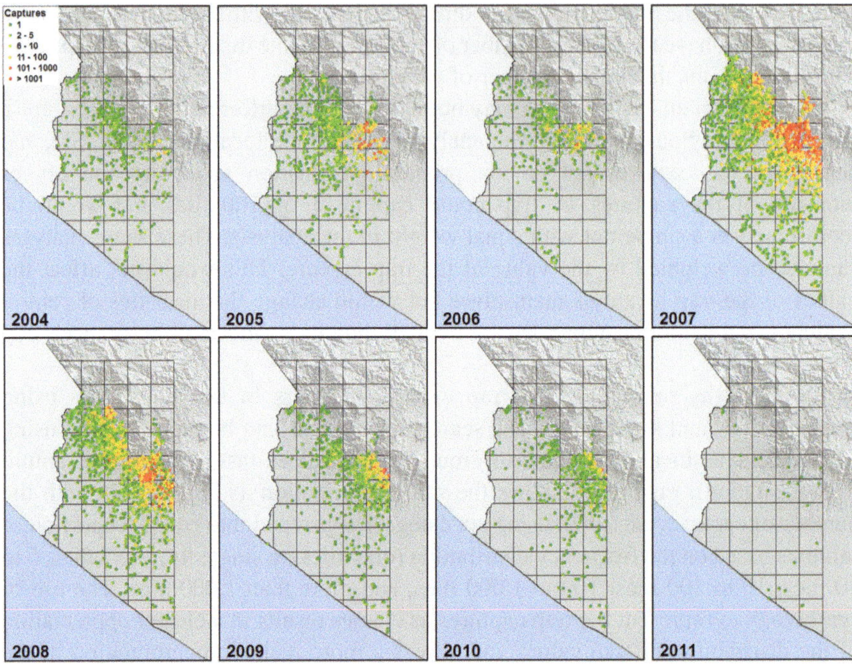

Fig. 9.7 Time series of captures presented as points as in Fig. 9.3 but using symbology to represent trap captures. Point plot of the location of wild *C. capitata* captured categorized by the quantity per trap using quantiles, with an equal number of traps in each category based on the frequency distribution in Fig. 9.2

county). Tephritid fruit fly programs use polygons that represent hosts, production zones, SIT release blocks, or phytosanitary status, and trap data stored as points can be converted or aggregated using those polygons. Keep in mind that data collected for small areas or points can be summarized and applied to larger ones, but the reverse is not possible.

For tephritid fruit fly trap analyses, data are more typically aggregated using polygons specifically designed for that purpose, such as regularly spaced quadrats of equal size in a matrix. The size chosen for the polygons is a key element in analysis, as the size of the polygon may directly affect the outcome, an influence known as the Modified Areal Unit Problem for point-based data in geography (Bailey and Gatrell 1995; Camara and Carvalho 2004). In general, larger polygons result in less variable data. Reduced variability can be beneficial but can also obscure small-scale information that the traps are intended to provide. One way to select quadrat size is to choose a target mean number of traps present per polygon. A polygon with no traps is not useful, and polygons with a single trap limit the statistics that can be presented. One general recommendation is to use a cell size that contains an average of 1.6 to 2 points (Bailey and Gatrell 1995; Rogerson 2001) or to divide the area of the map by the number of traps, multiply by two, and take the square root to give the recommended measure of one side of each polygon (Mitchell 2009). A common presentation of tephritid trap catch in a matrix of polygons is to display the total or average FTD trap catch in each, shaded or colored by classification. Depending on the size of the polygons and the variability of the data, the results may still be too inconsistent to interpret meaningfully (Bailey and Gatrell 1995). One solution, instead of increasing the size of each polygon, is to display the average for each polygon and all adjacent polygons. Thus, the value of each polygon is the average value of itself and all polygons on its border. However, this calculation is easier to conduct and involves more reasonable assumptions through the use of interpolated rasters as will be discussed later on the Raster Analysis section.

As an example of trap data displayed aerially, the Medfly Program has for many years used a 10×10 km matrix to summarize trap statistics and make management decisions. Figure 9.8 shows the yearly trap capture data aggregated using 100 km^2 polygons and the same classes for symbology as Fig. 9.7. A similar pattern as the point analysis can be seen, and perhaps made clearer, though the identification of small-scale intermediate differences is obscured. The advantage to this areal method is that it summarizes the trap information and allows numerous statistics to be calculated and displayed. Disadvantages include the arbitrary boundaries around the sample locations and the fact that the relative location of the individual traps is not taken into account.

4.4.2 Statistical Methods to Identify Patterns

One key objective for tephritid fruit fly trapping is to identify local areas of high or (especially for SIT) low captures. Locations with unusually high or low captures may be grouped into clusters and be associated with other features. Clusters occur when high or low values are located near each other and can be formed using the locations of features alone or using the location influenced by the trap catch. Based on what is known about the biology of the fly, these associations may be useful in understanding the situation and planning actions. Statistical methods can be used to assist in identifying clusters in vector data in addition to searching for patterns in the data visually by varying the symbology, scale, and areal grouping. There are two basic types of statistical methods to identify patterns: general and local. General methods evaluate whether the data have characteristics that commonly result in patterns, and local methods identify likely clusters of individual features (traps), often outliers, that potentially create patterns.

One of the most general indicators of the possible existence of a pattern is the variance to mean ratio (Cox and Lewis 1966, and discussed in relation to geostatistical methods in Midgarden et al. 1993). This index of dispersion along with other similar analyses (e.g., Taylor's Power Law) is based on the observation that samples from similar spatial distributions (uniform, clumped, or random) tend to exhibit a similar relationship between the variance and mean and do not use the locations of samples in the calculation. Others, such as the nearest neighbor index

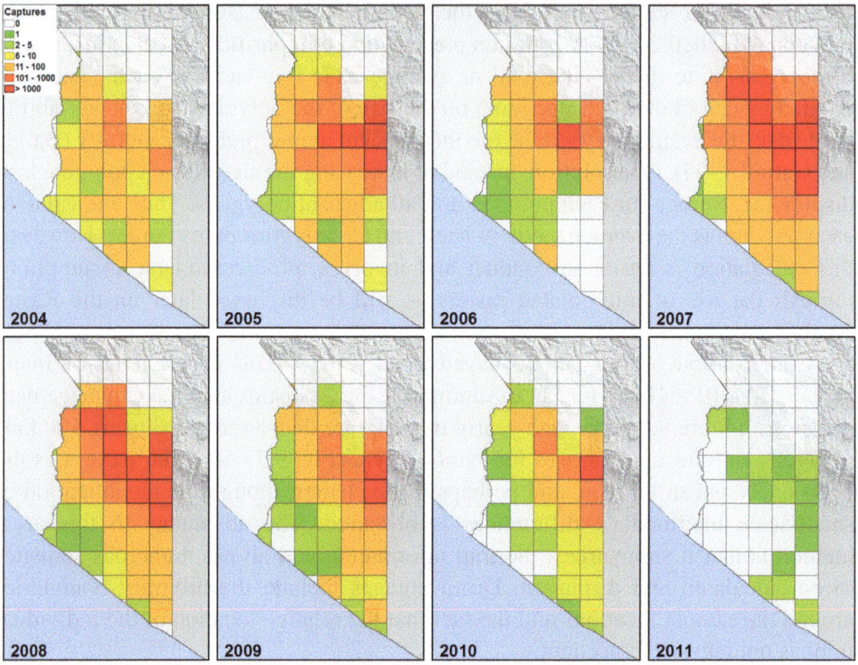

Fig. 9.8 Vector (areal polygon) plot of the location of total number wild *C. capitata* captured each year categorized by the quantity within each polygon using same categories as Fig. 9.7

(Clark and Evans 1954), use the distance among samples in the analysis and can test for significance against a null hypothesis of a random distribution. Moran's I is an index of spatial autocorrelation and uses the relative location of samples (traps) to measure how similar samples nearby are to those that are further away. Spatial autocorrelation, sometimes called spatial dependence, is based on the assumption that things that are closer together are more similar than things that are further apart, a concept so useful that it is known as the first law of geography and credited to Waldo Tobler (1970). In tephritid trap analysis, it is a frequent occurrence that a trap located nearby another trap will have a similar number of flies (e.g., a trap close to one with no flies will likely not have any flies either, while a trap close to one with many flies will likely also capture a large number of flies). The concept of spatial dependence is the basis for geostatistics, which was developed beginning in the 1960s by Matheron (1962) to assist in identifying spatial patterns in ore and petroleum sample cores to find the best location for a mine or well. Moran's I can be tested against a null hypothesis of no spatial autocorrelation and provide a significance level for similarity among nearby traps throughout the area under study. Spatially autocorrelated data are likely to have clusters and an observable pattern. In summary, these general methods do not assist in determining how the patterns might appear or where they might be located, but they do suggest whether a pattern may exist.

Local statistical methods can be used to help identify where, within an area of interest, patterns may be present. Two statistical methods used to measure local variation are Local Moran's I (Anselin 1995) and the Getis-Ord Gi (Ord and Getis 1995), which are denoted I_i and Gi, respectively. In these local analyses, if the difference in values of nearby features is less than the difference in values among all features, then those values are considered to be clustered. I_i identifies likely clusters but not whether the clusters are made up of high or low values. The Getis-Ord Gi statistic measures how high and low values (hotspots and coldspots) are distributed over an area of interest and identifies whether either or both are clustered. The local methods allow the features that may be clustered to be displayed on a map, where they can be evaluated as part of a pattern in the data.

4.4.3 Analysis Using Raster Surfaces

Another way to represent trapping data is to use continuous raster surfaces. The reinterpretation of trap data from vector to rasters is done by either deterministic or geostatistical interpolation. Spatial interpolation is a method of estimating a value at any location based on information from a limited number of sampled locations. A common deterministic interpolation is Inverse Distance Weighted (IDW), which is relatively simple to conduct using easily defined parameters. This method assumes spatial dependence and uses the distance from neighboring sample values to estimate a weighted moving average for an unsampled location: the closer a sample is to the location being estimated, the more influence it has on the value. It is calculated as the inverse of the distance ($1/D^k$), the constant k defining the weight or importance of the distance while interpolating (the higher k is, the more effect closer samples have on the estimation compared to samples that are farther away). The value calculated for each location is based on the captures from traps within a selected distance or neighborhood (e.g., 500 m, 1 mile). The closer a trap is to a given location, the more the capture value of that trap influences the estimate, traps located outside this distance would be excluded from the calculation. For example, the value of a location halfway between two traps would be calculated as the average capture of the two traps, but the value of a location closer to one of the traps than the other will consider the information from the closer trap as more important in the estimate. The values of the parameters used in the interpolation of tephritid fruit fly traps should be chosen by a specialist based on knowledge of the species in question, the characteristics of the traps and baits, the number of traps, their distribution, and the environment.

The result of IDW interpolation is a smoothed continuous surface with values located much closer together than the actual traps. The surfaces can be represented as isolines connecting similar values, analogous to elevation contours on topographic maps, or as categorized pixels commonly seen to represent temperatures on a weather report. The advantage of this method is that it uses the distance to estimate a moving average of trap values and smoothes over those values that have a high level of variation while still reflecting that variation in the resulting raster layer. The disadvantage of IDW (in fact, all interpolation methods) is that the algorithm estimating the values relies only on the relationship of the distance between traps and their capture without considering other spatial information, which can result in unrealistic values between traps where no tephritid population could possibly exist, such as a body of water. Another disadvantage frequently mentioned is that, because they are averages, the values predicted using IDW could never be more or less than any of the trap values, which limits the extension of trends to their logical conclusion.

Geostatistical interpolation methods, such as kriging (Isaaks and Srivistava 1989), are also inverse distance weighted, but use the spatial dependence measured in the data to determine the weight assigned to a given distance. The distance at which this relationship exists can be calculated using a variogram,[2] which plots the variation of samples within a given distance against a number of separation distances and measures how the data are related (correlated) with distance and modeled to use for spatial prediction.

Although it is possible to place traps in a manner that allows characterization of their spatial dependence, this is not usually a priority for most programs. Economics is more often the deciding factor in placing traps (e.g., How many traps can we afford? What area do we want to monitor?). In the presence of significant spatial autocorrelation, maps made by kriging are more accurate than deterministic methods. An additional advantage of kriging is that the spatial variation can be modeled for different cardinal directions using directional variograms. If the spatial dependence from north to south is different from east to west, for example, the interpolation algorithm can be adjusted accordingly (Isaaks and Srivastava 1989). In the absence of measurable spatial dependence, however, kriging defaults to IDW. One disadvantage in using kriging to generate raster representations of trap data on a regular basis is the time needed to determine the best model to use. The parameters and results will likely be different for each time period and make the routine use of geostatistical interpolation problematic. In addition, kriging carries assumptions about the data that are likely not valid for tephritid trapping data, such as a normal distribution and stationarity (i.e., an invariant spatial relationship exists throughout the mapped area) and may require data transformation and other remediating efforts. Despite these issues, if predictions of trap captures in unsampled areas are needed, kriging may be useful, and new methods are continually being developed to improve geostatistical methods for highly variable data, such as tephritid fruit fly traps (e.g., Krivoruchko 2012).

The advantage of both deterministic and geostatistical interpolation methods is that they objectively interpret and display the trap data. The resulting raster surfaces

[2] Bivand et al. (2008) defined variogram as a scatter plot of the average variation of pairs grouped by their separation distance. The value of one sample can be compared to the value of another sample while measuring the distance between them. With this, the variation in function of the distance can be estimated for that pair. If the process is conducted with all the pairs of samples, and the variation is plotted (putting the distance in the x-axis and the variation in the y-axis) a variogram is constructed and a regression line can be fit to use to estimate intermediate values.

make sense out of the clutter of highly variable trap results and are especially useful for displaying the results of a large number of traps over a wide area.

The following aspects should be considered when using interpolation methods to create surface raster layers:

- Distribution of the traps: Trap distribution affects the interpolation process regardless of the method used. As discussed, traps are rarely distributed uniformly (which would be ideal for interpolation) and may be placed randomly or, more likely, in clusters. This heterogeneity of trap density and environment results in an extremely variable distance between traps as well as number of flies captured. The trap data will not necessarily exhibit a similar amount of spatial dependence throughout the area of interest, complicating geostatistical methods. Likewise, it can be challenging to define parameters (maximum distance, number of neighbors, weight, etc.) using deterministic methods.
- Installed traps and serviced traps: Not all the traps are serviced on schedule, and some may be on a different schedule (e.g., monthly instead of weekly). It is important to use only traps that have been serviced and have updated information, including how long they were active for conducting interpolation for the time period of interest. A trap that is not serviced is not the same as a trap with no captures.
- Edge effect: The trap grid does not always cover the complete area of interest, especially on the borders where there may be few or no traps. The resulting interpolation in these areas may generate "extraordinary" values. Those values are due to the use of a small number of traps at the extreme locations to make estimates beyond the border and thus cause nonsense estimates to be reported. The readers of the maps need to be educated about this issue and to interpret border areas in an informed manner. This problem may be reduced by placing additional traps close to the edge of the area of interest and/or the interpolation area can be "clipped" or cut to include only the area with adequate distribution of traps.

The objective of depicting tephritid trap data as a raster surface, regardless of interpolation method, is to produce a representative summary of the trap data for a given time period. Figure 9.9 shows the same trap information as Figs. 9.6, 9.7, and 9.8 interpolated to a maximum 2 km away from a trap location using the IDW method. The same classes as Figs. 9.7 and 9.8 were used to represent the data. One of the advantages of raster surfaces is that the information is presented in a way that allows the reader to observe patterns easily and quickly but still preserves some of the local variation that overwhelms maps of point locations and can be hidden by areal depiction of the data. The spatial averaging and increased readability does not come without costs, however. The main disadvantage is the loss of reality: the values are now estimates rather than actual captures of the traps. There may be (and likely are) important reasons why the data within a particular area are so variable, e.g., spatial variability in host presence, insecticide treatments, etc. This information can be hidden in the interpolated representation of the trap data and cause the reader to see values as equally likely in all locations unless other layers are included

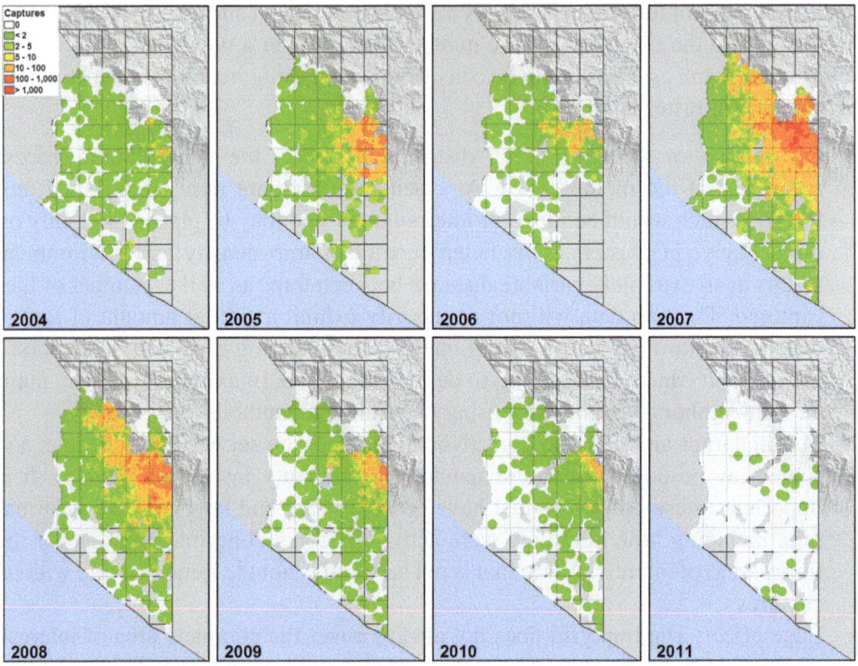

Fig. 9.9 Surface plot of the locations of total number of wild *C. capitata* captured each year categorized by the estimated trap capture at each 200 × 200 m raster, using same categories as Figs. 9.7 and 9.8. Each raster was estimated using Inverse Distance Weighted interpolation of the points presented in Fig. 9.4. The neighborhood for the interpolation was 2,000 m and the weight the inverse of the square of the distance

in the map to show key areas and more accurately reflect reality (bodies of water and other areas with no hosts, for example).

4.4.4 Raster Algebra

Raster surfaces that result from interpolation are easier to interpret than maps of points and additionally can be used to create derived layers. A layer created by the combination of the information from more than one layer is considered to be derived. Raster algebra refers to mathematical or logical operations among different raster layers. For example, if decisions would be assisted by identifying where within the area of interest population size is increasing the most rapidly and where it is decreasing, such a layer can be derived using the interpolated trap capture surfaces from two or more different time periods.

$$RI = (TW_1 - TW_0)/(TW_0)$$

Where,

RI is the rate of increase
TW_1 is the surface raster total wild flies of the most recent period
TW_0 is the total wild flies of the previous period

Let's assume that the population fluctuations from years 2006 to 2007 and from 2007 to 2008 are of interest. One way to show population increase or decrease spatially in GIS software is by using the rasters shown in Fig. 9.9. The values in each time period can be input into a "raster calculator", and the results are shown in Fig. 9.10. The areas with a decrease in the population will have negative values, shown in green shades on the map. The areas with an increase in the population will be positive values, shown progressively as yellow to red shade in the maps. With this kind of analysis, it is easy to conclude that from 2006 to 2007 the population density was increasing (more yellow and red than green in most of the map), while from 2007 to 2008 the population decreased generally (more green in most of the map), with the exception of isolated red locations (hot spots).

4.5 Evaluation of Sterile Recapture

As mentioned in the introduction, one of the uses of tephritid fruit fly traps is to monitor releases of sterile flies under an SIT program. The objectives of releasing sterile insects into the environment are: (i) to prevent the establishment of an invading tephritid pest species or (ii) to suppress and eradicate established populations. To accomplish these goals, the insects should be released in the correct locations at a determined density (number of insects per unit area). The traps serve as the main tool to monitor how well this objective is met. The two trap types used in the Medfly Program (baited with male lures [trimedlure] and synthetic food-based lures [ammonium acetate, putrescine, and trimethylamine]) capture an equal number of wild flies, however, the male lure-baited traps captures many more sterile flies (e.g., seven times more, Midgarden et al. 2004). For this reason, only the results of food-based lure-baited traps are used for analysis of sterile recapture. Once an area is selected as appropriate for sterile fly release, the objective is to distribute the insects as evenly as possible within that area, resulting, ideally, in a uniform distribution of captures.

So, the questions are: *"Are the sterile flies distributed evenly within the release blocks?"* and *"Are the locations with wild flies getting enough sterile flies?"*. We are going to target these questions one at a time, in both cases beginning by selecting and characterizing the data.

Are the sterile flies distributed evenly within the release blocks?

Sterile tephritid flies are usually released in specific areas (polygons) called blocks. Each block requires a given density of sterile flies. For this reason, it makes sense to evaluate the recapture based on these blocks rather than over the entire trapping program area. Results of sterile recapture are traditionally reported by

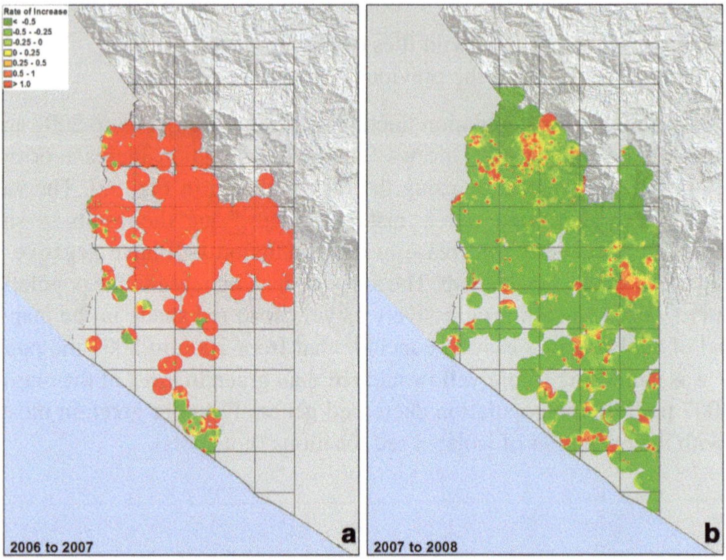

Fig. 9.10 Two examples of derived maps showing change in wild *C. capitata* capture for locations within the area of interest, each based on two interpolated surface plots shown in Fig. 9.9. (**a**) shows a situation in which the population is generally increasing. (**b**) shows a situation in which the population is generally decreasing. This example allows the identification of specific locations where the general trend is reversed and, in the case of plot on the Right, highlights locations of potential concern

aggregating the data within each block. The most common variables are: number of sterile flies, sterile to wild fly ratio, percentage of traps with sterile fly capture, and mean capture of sterile flies per trap per day. As useful as these indicators are, they do not actually show the spatial pattern of recaptures and evaluate the main objective of the release, which is the presence of an adequate number of sterile flies throughout the release block. More importantly, they do not show where this goal failed to be met (i.e., clusters of low sterile fly captures).

One might assume that the captures of released tephritid fruit flies would be more uniform than wild captures as a great deal of effort is made to release the flies uniformly throughout the blocks. The histogram of sterile captures in Fig. 9.11 shows that, although the number of traps without captures is still the most frequent category, the captures are less variable than wild captures. An interpolated map of sterile captures can identify locations where the rate of sterile recaptures is below a desired threshold. The distribution of sterile flies within a block can change dramatically from one time period to the next due to factors such as wind, temperature, altitude of release, or time of day. One way to evaluate the consistency recapture of sterile flies is to calculate a derived map averaging several time periods. In this way, locations with consistent low recapture over several trapping periods can be identified, and changes can be made to try to improve the recapture and/or release in areas identified as problematic (by possibly relocating traps to

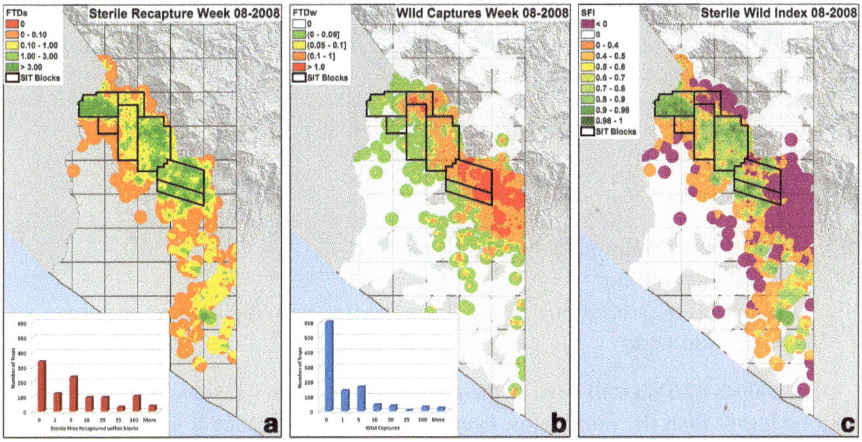

Fig. 9.11 Derived map using interpolations to show wild-sterile index (ratio) on a per trap basis. (**a**) Recapture of sterile *C. capitata* for the same week and histogram of captures. (**b**) Interpolated layer of wild *C. capitata* captures in week and histogram of captures (**c**) Sterile Wild Index using the two previous layers of capture, combined using the formula SWI = (TS − TF)/(TS + 1). Table 9.3 shows the equivalent values in SWI and sterile to fertile ratio

more appropriate locations or modifying flight lines, release altitudes, or release densities).

So, the answer to the question is that there are some areas within the blocks with good recaptures but others with low recaptures as shown in Fig. 9.11a by areas of green and yellow, respectively.

"Are the areas with wild flies getting enough sterile flies?"

The distribution of the sterile flies is telling us about the consistency of sterile releases but not whether all areas are receiving enough sterile flies in relation to wild flies. One of the traditional ways to evaluate sterile release is the sterile to wild ratio, calculated by dividing the number of sterile flies recaptured by the number of wild flies. Ideally, there is a target ratio of steriles captured to every wild fly. The IAEA (IAEA/FAO 2004) recommends a varying Sterile-to-Wild Ratio (SWR) for tephritid programs depending on the goal for a given location: 100–150:1 for eradication, 25–100:1 for suppression, and 25–50:1 for prevention.

This calculation is valid if the number of wild captures is greater than zero. If no wild flies area captured (W = 0), then it is not possible to calculate the SWR due to the zero in the denominator. If the ratio was made individually for each of thousands of traps, the frequency of W = 0 would likely be very high, resulting in many traps with no SWR. As a result, the SWR is usually calculated on an aerial basis, such as per release block. This results in an averaging of the spatial variability within what is typically a very large area (e.g., hundreds or even thousands of km^2).

Spatial analysis can use this spatial variation within release blocks. In the following example, we show an index that captures the information about sterile to wild ratio on a per trap basis and allows for interpolation. The Sterile Wild Index

(SWI) is computed as the ratio of the difference of the sterile flies recaptured and the wild flies captured in the trap to the number of sterile males in the trap using the formula:

$$\mathbf{SWI} = (\mathbf{S} - \mathbf{W})/(\mathbf{S} + \mathbf{1})$$

Where:

S = Sterile flies recaptured from the releases
W = Wild flies captured
1 = Assumption of at least one sterile fly in the trap to avoid the presence of a zero in the denominator.

The values of SWI will exist in the range from $-\infty$ to <1, since the denominator will be larger than the numerator, even when the $W = 0$ since $S + 1 > S$.

The numerator of the expression is the "excess" of sterile flies in comparison to the wild flies. The denominator of the expression indicates the sterile flies present in the traps plus one fly, which is the expected minimum. Thus, the index is a measure of the over flooding of sterile against wild flies relative to the presence of sterile flies. The index will be negative when the number of wild flies is higher than the number of sterile. The SWI will never reach the value of 1.0, but the closer the index is to 1.0, the greater the number of sterile flies in comparison with the wild flies in the trap.

Table 9.3 shows some SWI values, their interpretation, comparison with SWR, and the equivalences between the ranges of both parameters. The SWI can be calculated for a whole area of interest using raster algebra. The input layers will be two raster maps: one with the sterile fly re-captures (Fig. 9.11a) and the other with the wild fly captures (Fig. 9.11b). Figure 9.11c shows the resulting SWIs. The advantage of this index is that it preserves the spatial variation of the trap data regarding the relation between sterile and wild fly trap capture. When averaged over a block, the SWR can be affected by outliers with very high or very low SWR, so that even if the majority of traps show that the SWR is high, the average could show a poor recapture or vice versa. The SWI allows the identification of local "problem areas" within the blocks.

The answer to the question is that some areas are well covered with sterile flies in comparison to the wild captures; meanwhile localized areas within the blocks are not receiving enough sterile flies. Specific decisions to "attack" these focalized areas should be considered either to reduce wild fly numbers (e.g., aerial or ground sprays or bait stations) or increase sterile fly numbers (i.e., supplementary ground releases).

Table 9.3 The relationship between the SWI values and traditional the traditional index of sterile to wild ratio (SWR)

SWI value	Interpretation	SWR equivalent
SWI < 0	The number of wild captures is greater than the number of sterile captures. Very bad sterile to fertile ratio	0 <= SWR < 1.0
SWI = 0	The number of wild captures is equal to the number of sterile captures, including the traps with no capture	SWR = 1.0 or SWR not defined (Wild = 0)
SWI > 0	The number of wild captures is less than the number of sterile captures	SWR > 1.0
SWI > 0.98	The number of wild captures is significant smaller than the number of sterile captures	SWR ~ 100

4.6 Summary of Analysis

Trapping information can be presented in different forms, maps, tables, and graphs, and the information can be converted from points to surfaces using interpolation. The trapping information can be analyzed as points, polygons, or surfaces. Geographical statistical analyses can be conducted to improve the interpretation and identify patterns. In the case of rasters, new information can be derived by mathematical manipulation of the surfaces. In each step, the symbology of the map should provide the information in the most easily understood manner. Any analysis conducted will be justified if it: (1) provides results that improve decision-making or (2) helps detect or better explain a pattern(s).

5 Decision Making

The main objective of conducting spatial analysis is to allow program managers to make better decisions. Below, we present some instances where spatial analysis proved critical in improving control efforts.

5.1 Dispersal of C. capitata *Adults*

A significant analysis conducted in the Medfly Program in Guatemala and Mexico studied the annual cycle of increased wild captures in traps. Each year, traps detect wild flies throughout areas where no captures had occurred for the last half of the previous year. These captures occur from March to July with a peak in May, followed by the detections decreasing and finally disappearing. The longstanding explanation for this pattern was that low (undetectable) population densities were present throughout the year and that increased availability of the main host, coffee, from September-January led to population increases that reached detectable levels in March. The response from the program was to redouble area-wide control efforts to eradicate these areas using aerial bait sprays followed by high density SIT, only to have a similar pattern repeat the following year.

Using the newly available spatial database covering several successive years, the Program's technical staff began looking more carefully at the annual spatial and temporal pattern of the *C. capitata* population dynamics as reported by trap captures. Captures were plotted as point vectors over time along with layers showing parameters likely influencing population growth, such as host availability, land use, altitude, and temperature along with factors long thought to be important in influencing population dynamics, such as onset of the rainy season. The points were interpolated as raster surfaces and animated at weekly intervals over several years to observe the spatio-temporal behavior.

The analysis revealed a number of inconsistencies with the longstanding hypothesis of undetectable populations re-emerging. For example, there was a wide range of temperatures over the Program area due to dramatic differences in altitude. The effect of temperature on *C. capitata* life cycle is powerful: the higher the temperature (to the upper limit), the faster the flies develop and emerge. Therefore, the population dynamics would be expected to vary based on temperature as well as host differences. However, *C. capitata* were captured simultaneously at locations having a variety of temperatures (generation times), host presence, and phenology. A clear pattern was apparent showing a gradient in wild captures from highest at the leading edge of the trapping program, where the wild populations were uncontrolled, to lowest the further away the traps were from this highly infested area. This gradient from high to medium to low capture was consistent regardless of direction, temperature, or host. The animations clearly showed trap captures increasing in untreated areas throughout the coffee season, peaking after harvest, and then captures occurring in adjacent areas where no flies were detected previously. This capture behavior suggested a migration or movement of flies from infested areas without suppression into controlled areas.

The result of this analysis was to develop an alternate hypothesis to explain the seasonal detection of flies in normally fly-free areas based on the dispersal of adults from adjacent highly infested locations in response to reduced host presence at their origin, explaining the peak emergence of adults after coffee harvest (Midgarden and Lira 2006). The design of a control strategy around this new hypothesis led to a dramatic shift in the timing and location of control efforts to prevent or reduce the buildup of populations that result in dispersal. Perhaps more important was the ability to interpret this seasonal pattern as an expected part of the *C. capitata* spatial population dynamics and not as a failure of the control strategy. Implementation of the new strategy has allowed the program to move the front of infestation more than 100 km in 5 years.

5.2 Combine Area-Wide and Site Specific Control Methods

Area-wide programs use control methods that are not effective if applied to only a part of the infested area or region. Tephritid programs, with their large trapping grids and work areas, are emblematic of an area-wide situation. However, not all control techniques need to be applied in an area-wide manner for effective management. Site-specific control methods that operate on a small spatial scale can also be effective in some situations, such as isolated host areas. In those situations, control methods, such as ground application of bait, mass trapping, attract and kill bait stations, fruit stripping and ground release of sterile insects, can be effective. In other situations (continuous host areas), effectiveness of site-specific techniques (Fleischer et al. 1997) to control fruit fly populations are considered inefficient due to the large areas to be controlled. There is a lack of confidence in ability of the traps to sample at a fine enough scale to reliably identify locations that contain a significant portion of the population and other locations that can be safely left untreated. The potential problem is due to not treating areas that should be treated as well as treating areas that should not. GIS and spatial analysis can help to avoid this problem and ensure that both control methods (area-wide and site-specific controls) are used in an effective way.

The Mexico-Guatemala-Belize Medfly Program began using spatial analysis to identify relatively small areas for aerial insecticidal bait applications to control high population densities while leaving areas with lower population densities untreated. The SIT component, however, remains area-wide, with only generalized efforts to adjust the density of flies released over the infested areas. This strategy is modeled on site-specific or precision pest management, where samples are taken within a field with the objective of making a map of the insect population and treating only the areas above a threshold population. In this way, precision management reduces the quantity of insecticide used and the area treated. For large areas, the advantages are similar but may also reduce the negative reaction of people living or working in the areas where the Program operates.

The steps used in the Program vary somewhat from year to year but begin with the mapping of weekly wild captures. Raster surfaces are generated for each week, added together and averaged (FTD). Areas with the highest mean trap catch over the preceding 6–8 weeks are considered for bait application based on a capture threshold and the resources available. The actual locations to be treated are selected on the basis of the above information along with host presence and phenology (when fruit is susceptible to oviposition). The result has been a dramatic decrease in cost, from US$10 million to less than US$1 million while maintaining progress in eradication.

5.3 Variable Release Blocks in Valencia, Spain

SIT programs commonly adjust release densities per block based on current or historical wild detections. The Medfly Program in Valencia, Spain, uses spatial analysis to apply SIT in a site-specific manner within release blocks (Briasco et al. 2012). The program, in place in over 150,000 ha of citrus in Valencia since 2005, releases an average of 400 million sterile males every week. Both the wild and sterile populations in the fields are monitored with 742 Tephri traps and 1,027 Nadel traps, which are baited with synthetic food-based lures and trimedlure, respectively, and serviced weekly. Data from the traps are interpolated using universal kriging to generate raster surface maps of the wild captures. The spatial distribution of wild *C. capitata* varies within the release area according to changing eco-climatic conditions, in part due to the abundance of ripening hosts of the different citrus varieties and the slope and orientation of the terrain, which affects temperature. Optimally, the sterile releases should track the spatial and temporal variation in the wild population as closely as possible and even anticipate likely future distributions. The map of wild flies, along with the citrus variety, historical data, and current trends in population growth, are used to determine the release density of sterile flies to achieve a target overflooding ratio, which is then saved as a raster map of optimal release densities. Software developed for the equipment, which releases chilled, adult *C. capitata* from the airplane, transforms the flight path line into a series of 100 m sections and allocates a release rate to each of them. The release rate map is updated weekly, and based on all available information, the field manager determines the final number of flies released in each location.

The automatic system releases the flies using three sources of information:

- A georeferenced map (raster) providing target release rates over the area, including areas of exclusion to indicate where not to release
- A vector file of the planned flight path
- The GPS information during the flight

The flies are distributed according to the route length and number of hectares with suitable crops for the pest. The system permits a better distribution of the sterile insects available, reallocating insects from areas where the overflooding ratio is exceeded to areas with suboptimal ratios. The objective of this methodology is to increase the SIT efficiency by making more rational insect releases. The recorded log file of the flight provides confirmation that the release device behaves as expected during the flight operation. Furthermore, the information obtained from the monitoring grids in the fields furnishes additional confirmation that the number of sterile flies captured in the traps is correlated with the density of insects released.

5.4 Colonization and Spread of Carambola Fruit Fly in South America

The carambola fruit fly (*Bactrocera carambolae* Drew & Hancock) is a tropical tephritid indigenous to Southeast Asia, which was found in Suriname, northeastern South America, in the 1980s (van Sauers 1991; Hancock 1990). By the time the fly was discovered, it had already infested a large area in Suriname and was found in neighboring French Guiana and the State of Amapa, Brazil, as soon as traps were placed in the field in 1990 and 1993, respectively (van Sauers 1993). One of the key questions was to determine the pattern of spread of the pest and to predict where it might move in the future. The first step was to plot the current distribution of the fly, with the assumed point of origin at Paramaribo, Suriname (van Sauers-Muller 1991). If all else is equal, the point of origin should be near the center of the distribution, and the spread would be more or less equal in all directions. However, the observed pattern was quite different. To the east, south, and 100 km to the west, the fly population was associated with populated areas. This relationship with inhabited areas can be explained by host presence: in areas with few inhabitants, there are also fewer of the cultivated trees that serve as hosts. To the south, inhabited areas disappear within 100 km as did detections of the pest. However, to the west, toward the border between Suriname and Guyana, the detections fall off and disappear completely despite wide availability of hosts and voluminous trade between infested and free areas. In contrast, to the east, the fly has established in inhabited locales as far as Macapa, Brazil (da Silva et al. 2005), a linear distance of 800 km from the assumed point of origin.

What could be causing this differential pattern of spread? Obvious explanations seem unhelpful: temperature, rainfall, trade, and other factors are similar in both east and west. As part of the fly's life cycle (the pupal stage) is spent in the soil, could differences in soil composition provide an explanation? Digitized soil maps overlayed on the infested area were compelling: the pest is present in areas with sandy, well-drained soil but not in heavy, clay soil with very poor drainage (Suriname of Land and Forest Managment: Soil Survey Department – http://www.gov.sr/sr/ministerie-van-rgb/contact.aspx). Additional years of data corroborate this pattern between fly establishment and soil type. For example, the soil in the highly populated coastal area of Guyana, the country to the west of Suriname, is heavy clay and the fly has never established there despite high host density and commercial trade in fruit. However, in 2005, the Carambola fruit fly was discovered to be present in locations far from the most inhabited areas but where there is sandy, well-drained soil and hosts (Guyana Ministry of Agriculture personal communication). Likewise, detections were made in 2009 in Roraima, Brazil, on the border and along a trade route with Guyana (Brazil Ministry of Agriculture press release: http://www.agricultura.gov.br/vegetal/noticias/2012/10/mapa-envia-tecnicos-a-guiana-para-discutir-combate-a-mosca-da-carambola). With enough time, then, the pest reached areas where conditions were suitable. The correlation of establishment of carambola fruit fly with soil type/drainage and host presence is consistent with the biology of the pest and allows the identification of areas at risk to invasion of carambola fruit fly in the South America and the Caribbean Region.

6 Future Directions

Spatial analysis techniques are constantly improving, partly benefiting from the increasing ability to conduct multiple complex calculations rapidly. The data collected from tephritid fruit fly traps are highly variable in both space and time, and new analyses are being developed that may improve modeling and characterization of this type of "messy" data. As new analyses become available, they will undoubtedly increase the usefulness of spatial analysis of trap data for tephritid program managers as well as basic scientific studies and risk analysis (Castrignano et al. 2012). An obvious next step will be the further use of site-specific SIT, such as that used in Spain, while at the same time keeping within the traditional area-wide guidelines. This will involve not only the development of analytical techniques, but also the development of newly engineered equipment that will permit the accurate delivery of variable densities of sterile insects and ways to monitor their efficacy.

In addition to improving the effectiveness of large tephritid management programs, there is the opportunity to integrate spatial analysis into decisions made by single growers or groups of growers within the same (or nearby) production areas. Most often, the decision will be to use non-area-wide methods, such as bait sprays or attract and kill tools (mass trapping or bait stations). Steps have been initiated in the Middle East and Europe (Pontikakos et al. 2010; Cohen et al. 2008) to integrate spatial elements into expert systems (a computer system that emulates the decision-making ability of a human expert). Individual producers often focus on preventing losses to their crops, without paying attention to the fly population in the region at large or the effect of future host availability (i.e., other crops surrounding the crop area, abundant enough to sustain a fruit fly population, but at that moment are not yet sufficiently mature to be infested). These expert systems can assist individuals or small groups to work throughout the production system in a coordinated manner, which if attaining full compliance, can work to reduce the overall population damage, thus becoming, in effect, an area-wide strategy.

The continual development of database technologies and internet capabilities opens the possibility to share information among tephritid programs in different regions or countries. This will make standardization of the information even more important. The challenge in the future is to integrate "local" or "national" Fruit Fly Program information into a Regional or World Geographical Database that allows increased understanding of tephritid fruit flies around the world.

Addendum: Software and Relevant Web Links

ESRI's series of Arc GIS programs are arguably the most well-known and used software for GIS throughout the world, however, the cost is high for a small program or private user (between $1,500 and $15,000 depending on the extensions). Free open source software (FOSS), in general, and FOSS GIS software, in particular, have bloomed over the past decade, offering computer users the full spectrum of applications. These applications have proved to be reliable, and competitive with proprietary software. Today, a GIS technician can build a full GIS from FOSS. In the table below are some of the most useful examples for data storage, mapping, spatial analysis, and animation.

OSGeo	The umbrella organization that promotes and oversees open source geospatial software. OSGeo offers the web infrastructure to software developers, who work on Free Open Source Software for GIS (FOSS4G), and organizes activities worldwide to promote the use of open source GIS software
	http://www.osgeo.org/
Gentle Introduction	A good online tutorial covering all the basic aspects of GIS
	http://docs.qgis.org/html/en/docs/gentle_gis_introduction/index.html
QuantumGIS	The most popular general purpose open source GIS application. QGIS has a very flexible architecture for adding python based plugins, thus offering access to a wide range of GIS procedures and an interface to advanced analysis routines. Extensive documentation is available in several languages
	http://qgis.osgeo.org/
	http://docs.qgis.org/html/en/docs/user_manual/index.html
GRASS GIS	The oldest and most advanced GIS software package in the open source realm. GRASS has grown from a US military and academic project in the 1980s to a widely employed and professional application. The application offers modules covering all aspects GIS for both raster and vector spatial data layers. Many tutorials are available on-line, and a built-in help system explains the options for each operation
	http://grass.osgeo.org/
Spatialite	A light-weight, single file geospatial database, based on the highly popular sqlite database
	http://www.gaia-gis.it/gaia-sins/
	http://www.gaia-gis.it/spatialite-3.0.0-BETA/spatialite-cookbook/index.html
	http://www.bostongis.com/content_name=spatialite_tut01#19
GDAL	A library of utilities for converting between and manipulating various spatial data formats. Used "behind the scenes" by all open source GIS software
	http://www.gdal.org/
PostGIS	A high-end spatial database, built on PostgreSQL. Enables multi-user concurrent access under heavy loads, and contains a full set of spatial analysis functions
	http://postgis.refractions.net/
	http://workshops.opengeo.org/postgis-intro/
	http://www.bostongis.com/?content_name=postgis_tut01#304

(continued)

SAGA GIS	A rich collection of routines for raster data analysis. The software offers a very broad set of modules for terrain analysis, raster algebra, watershed delineation, etc. However, documentation is lacking http://www.saga-gis.org/en/index.html
R Project	A statistics programming language similar to the proprietary SAS. R contains an extensive set of spatial analysis libraries for interacting with regular GIS data, running spatial interpolations, correlations and regressions http://www.r-project.org/ http://cran.r-project.org/doc/manuals/r-release/R-intro.html http://cran.r-project.org/web/views/Spatial.html
GeoDa	An open source spatial analysis program developed by Luc Anselin. The tutorial section of the web site has a broad collection of online presentations and courses https://geodacenter.asu.edu
Animation Software	Two animation programs we have used are JASC's Animation Shop and Adobe ImageReady. Animation shop is still available by download but is no longer supported. ImageReady has been discontinued, though most features are now integrated into Photoshop. There are a number of free programs available on a number of web-based software sites (cnet, etc.), however, they have limited functionality (some only allow two frames), and it may take a few trials to find one that works well for animating a series of maps over time

References

Anselin L (1995) The local indicators of spatial association – LISA. Geogr Anal 27:93–115

Bailey TC, Gatrell AC (1995) Interactive spatial data analysis. Longman Group Limited, London

Barclay H, Matlock R, Gilchrist S, Suckling D, Reyes J, Enkerlin W, Vreysen M (2011) A conceptual model for assessing the minimum size area for an area-wide pest management program. Int J Agron 2011:Article 409328

Bivand RS, Pebesma EJ, Gomez-Rubio V (2008) Applied spatial data analysis with R. Springer, New York

Briasco M, Larrosa D, Salvador G, Montesa J, Argiles R (2012) Variable rate of aerial release of sterile insects in area-wide pest management. International Conference of Agricultural Engineering, Valencia

Burrough PA (1996) Principles of geographic information systems for land resource assessment. Clarendon, Oxford

Camara G, Carvalho M (2004) A tutorial on Spatial Analysis of Areas. [Online]. http://www.dpi.inpe.br/gilberto/tutorials/spatial_analysis/spatial_analysis_areas.pdf. [Accessed]. In: Druck S, Carvalho MS, Camara G, Monteiro AMV (eds) Analise Espacial de Dados Geograficos. Planaltina, DF, Brazil

Camara G, Monteiro A, Drucks S, Carvalho M (2004) Spatial analysis and GIS: a primer [Online]. http://www.dpi.inpe.br/gilberto/tutorials/spatial_analysis_primer.pdf. [Accessed]. In: Druck S, Carvalho MS, Camara G, Monteiro AMV (eds) Analise Espacial de Dados Geograficos. Planaltina, DF, Brazil

Castrignano A, Boccaccio L, Cohen Y, Nestel D, Kounatidis I, Papadopoulos N, De Benedetto D, Mavragani-Tsipidou P (2012) Spatio-temporal population dynamics and area-wide delineation

of *Bactrocera oleae* monitoring zones using multi-variate geostatistics. Precis Agric 13:421–441
Clark P, Evans F (1954) Distance to nearest neighbor as a measure of spatial relationship in populations. Ecology 35:445–453
Cohen Y, Cohen A, Hetzroni A, Alchanatis V, Broday D, Gazit Y, Timar D (2008) Spatial decision support system for Medfly control in citrus. Comput Electron Agric 62:107–117
Cox D, Lewis P (1966) The statistical analysis of series of events. Methuen, London
Cressie N, Wilke CK (2011) Statistics for spatio-temporal data. Wiley, Hoboken
da Silva R, Jordao A, de Sa L, de Oliveira M (2005) Mosca-da-carambola: Uma Ameaça à Fruticultura Brasileira, Circular Tecnica 31. EMBRAPA, Brasilia
Dminić I, Kozina A, Bažok R, Barčić JI (2010) Geographic information systems (GIS) and entomological research: a review. J Food Agric Environ 8:1193–1198
ESRI (2004) Understanding map projections. ESRI Press, Redlands, California
Fleischer S, Weisz R, Smilowitz Z, Midgarden D (1997) Spatial variation in Insect Populations and Site-specific Integrated Pest Management. In: Pierce FG, Sadler EJ (eds) The state of site-specific management for agriculture. American Society of Agronomy, Madison
Gould P, Wallace R (1994) Spatial structures and scientific paradoxes in the AIDS pandemic. Geogr Ann 76:105–116
Hancock D (1990) Oriental fruit fly survey and detection. FAO project report (TCP/RLA/8858), Rome
Huisman O, de By R (2009) Principles of Geographic Information Systems: an introductory textbook. International Institute for Aerospace Survey and Earth Sciences, Enschede
IAEA/FAO (International Atomic Energy Agency/Food and Agriculture Organization) (2004) Protocol for packing, shipping and release of sterile fruit flies. IAEA, Vienna
Isaaks EH, Srivastava RM (1989) Applied geostatistics. Oxford University Press, New York
Knipling EF (1960) Use of insects for their own destruction. J Econ Entomol 55:782–786
Krivoruchko K (2012) Empirical Bayesian Kriging implemented in Arc GIS Geostatistical Analyst. Arc User 15:6–10
Lira E (2010) Uso de Sistemas de Informacion Geografica. In: Montoya P, Toledo J, Hernandez E (eds) Moscas de la Fruta: Fundamentos y procedimientos para su Manejo. S y G Editores, Mexico
Matheron G (1962) Traité de Géostatistique Appliquée. Editions Technip, Paris
Midgarden D, Lira E (2006) Ecological relationship of medfly and coffee in Guatemala and Mexico. In: Sugayama R, Zucchi R, Ovrulki S, Sivinski J (eds) Fruit flies of economic importance: from basic to applied knowledge. Proceedings of the 7th international symposium on fruit flies of economic importance, BioFabrica, Bahia
Midgarden D, Youngman R, Fleischer S (1993) Spatial analysis of counts of western corn rootworm (Coleoptera: Chrysomelidae) adults on yellow sticky traps in corn: Geostatistics and Dispersion Indices. Environ Entomol 22:1124–1133
Midgarden D, Ovalle O, Epsky ND, Puche H, Kendra PE, Rendon P, Heath RR (2004) Capture of Mediterranean fruit flies (Diptera: Tephritidae) in dry traps baited with a food-based attractant and Jackson traps baited with trimedlure during sterile male release in Guatemala. J Econ Entomol 97:2137–2143
Miller GA (1956) The magical number seven, plus or minus two: some limits on our capacity for processing information. Psychol Rev 62:81–97
Mitchell A (2009) The ESRI guide to GIS analysis. ESRI Press, Redlands, California
Ord J, Getis A (1995) Local spatial autocorrelation statistics: distributional issues and an application. Geogr Anal 27:287–306
Pedigo LP (1996) Entomology and pest management. Prentice Hall, Upper Saddle River
Pontikakos CM, Tsiligiridis TA, Drougka ME (2010) Location-aware system for olive fruit fly spray control. Comput Electron Agric 70:355–368
Rogerson PA (2001) Statistical methods for geography. Sage, London

Schmitt A, Danisik M, Ayadar E, Sen E, Ulusoy I, Lovera O (2014) Identifying the volcanic eruption depicted in a Neolithic painting at Catalhoyuk, Central Anatolia, Turkey. PLoS One 9: e84711. doi:10.1371/journal.pone.0084711

Slocum T, McMaster R, Kessler F, Howard H (2008) Thematic cartography and geovisualization. Prentice Hall, Upper Saddle River

Southwood T, Henderson P (2000) Ecological methods. Wiley-Blackwell, London

Tobler W (1970) A computer movie simulating urban growth in the Detroit region. Econ Geogr 46:234–240

Utrilla P, Mazo C, Sopena MC, Martinez BM, Domingo R (2009) A palaeolithic map from 13,660 calBP: engraved stone blocks from the Late Magdalenian in Abauntz Cave (Navarra, Spain). J Hum Evol 57:99–111

van Sauers-Muller A (1991) An overview of the carambola fruit fly *Bactocera* species (Diptera: Tephritidae), found recently in Suriname. Fla Entomol 74:432–440

van Sauers-Muller A (1993) Pilot eradication project for the Carambola fruit fly in Coronie, Suriname. In: Aluja M, Liedo P (eds) Fruit flies biology and management. Springer, New York

Chapter 10
Using Molecules to Identify the Source of Fruit Fly Invasions

Norman Barr, Raul Ruiz-Arce, and Karen Armstrong

Abstract Fruit flies trapped or intercepted as part of inspection and surveillance activities could be the result of resurgent pest populations, incursions from offshore introductions, or intentional releases of sterile flies. Knowing the source of these flies can help a plant protection organization determine how best to respond to an urgent detection and which pathways pose a greater risk of future pest introductions. In this chapter, we review how biological molecules, such as DNA, proteins, and stable isotopes, have been used to estimate the geographic and population source of tephritid fruit flies. The merits and limitations of molecules as source estimators are treated by molecule origin (i.e., nuclear DNA, mitochondrial DNA, allozyme, and isotope) and technique (e.g., endonuclease digestion, DNA sequencing, genomics, biogeochemical) to provide an overview of available methods. The importance of experimental sampling, data interpretation, and data archiving are considered with reference to other insect and non-insect examples. Lastly, a case study of source estimation methods for the Mediterranean fruit fly, *Ceratitis capitata* (Wiedemann), which has been particularly well studied in this respect, is reported based on published work and ongoing studies.

Keywords Pathway analysis • Source estimation • Population assignment • Population structure • Haplotype matching • Haplotype diversity • Mitochondrial DNA • Nuclear DNA • Microsatellite DNA • Stable isotope analysis (SIA) • Next generation sequencing (NGS) • Single nucleotide polymorphism (SNP) • Restriction fragment length polymorphism (RFLP)

N. Barr (✉) • R. Ruiz-Arce
Center for Plant Health Science and Technology, Mission Laboratory, USDA-APHIS, Moore Air Base, Edinburg, TX 78541, USA
e-mail: norman.b.barr@aphis.usda.gov; raul.a.ruiz@aphis.usda.gov

K. Armstrong
Bio-Protection Research Centre, Lincoln University, Lincoln, Christchurch, New Zealand
e-mail: karen.armstrong@lincoln.ac.nz

T. Shelly et al. (eds.), *Trapping and the Detection, Control, and Regulation of Tephritid Fruit Flies*, DOI 10.1007/978-94-017-9193-9_10,

1 Introduction

The detection of pestiferous fruit flies in transported commodities, shipping vessels, and passenger baggage or within agricultural production areas that were free of the pest can raise important questions for fruit fly exclusion and eradication programs. Understanding where these flies originated, when and how they became associated with the commodity or shipment, and what pathway or transportation route was taken prior to detection can help managers identify which geographic sources might pose the greatest risk of future introduction events. By targeting the most likely sources of invasive fly populations, it is possible to direct resources for surveying, sampling, post-harvest treatment, preventive releases, and education activities along those pathways and thereby enhance the likelihood of pest exclusion. In the case of flies being detected during transport across borders, it is possible to compile detection records and evaluate trends in the data. For example, fly detections could be associated with a commodity, shipping route, foreign passenger, and specific date or season. Assuming that the detection efforts implemented by a plant protection organization are equivalent across commodities, ports, and time, these records can be used to estimate the pest propagule pressure or "introduction effort" (Ricklefs 2005; Malacrida et al. 2007), thereby indicating what pathways pose the greatest risk of invasion. Such pathways can then be targeted for specific resources or attention in order to avoid or reduce future introduction events.

Exotic fruit flies trapped within or near agricultural, urban, and natural areas pose a unique problem, because they lack shipping information to evaluate potential sources. For these flies, the pathway is less certain regarding both route and time. When an individual fly is detected outside of its known range, it is important to ascertain whether that fly is from a larger local population or is an isolated invader by increasing trapping and surveillance of the area. Information on population size and spatial distribution of an incursion revealed by the trapped samples is important for evaluating hypotheses regarding the introduction. Knowing the size and time since arrival of a founding population is also key information for developing pest management and eradication strategies. In summary, fruit fly response programs are interested in knowing the source of the introduction, the size of the founding event, and the subsequent expansion or contraction of the pest population.

Although geographic barriers and travel/trade policies limit the spread of exotic fruit flies, many species have the ability to expand their current range and establish in non-native regions (White and Elson-Harris 1994). This is problematic for economically important horticultural pest-free regions of the world that are located near infested areas. Consequently, pathway analyses of exotic species must sometimes consider the relatively small spatial scale of neighboring and contiguous countries, provinces, states, or even counties as a possible source population. A prime example of this is the pest management response to fruit fly outbreaks. During an eradication program, it is possible for a population outside of the eradication zone to re-infest the treated area. Consequently, it is very important to distinguish new introductions from failures of the eradication practices within the treatment zone (Roderick and Villablanca 1996). Also, when Sterile Insect Technique (SIT) is used to suppress outbreaks of a pest, it is crucial that the trapped flies be correctly identified as derived from either the lab source or the pest population to evaluate the effectiveness of the program (Aketarawong et al. 2011; Juan-Blasco et al. 2013). Given that the released sterile flies are the same species as the targeted population, distinguishing between the two is difficult by trap inspection alone. Therefore, knowing where flies have immediately come from would greatly assist in making potentially costly decisions about the success of an eradication campaign or determining the status of a quarantine zone as threatened but not yet compromised.

The problem of identifying a pest's source population and introduction route is therefore a significant one (Wares et al. 2005). However, the distribution and range of possible sources can be extensive for many economically important fruit flies species. For instance, the Mediterranean fruit fly, *Ceratitis capitata* (Wiedemann), is established in multiple regions of the old and new world and has many potential routes for transport (Vera et al. 2002; Liebhold et al. 2006). At a regional scale, even though the South American fruit fly, *Anastrepha fraterculus* (Wiedemann), is predominantly a new world pest, it has a wide range within Latin America (White and Elson-Harris 1994). Identifying sources and pathways requires a sound understanding of the species' genetic diversity, geographic range, ecology, and dispersal ability. As mentioned, when dealing with intercepted specimens, trend analysis of interception records is a useful way to estimate approach rates of an invasive species and narrow down likely sources. When dealing with SIT released flies, it is possible to mark the sterile populations with a tracking dye to assist in identifying the source of trapped flies as either laboratory or wild (Enkerlin et al. 1996; Hood-Nowotny et al. 2009). Unfortunately, neither trend analysis nor marking with tracking dyes is definitive for identifying the geographic origin of exotic flies. So, in addition to these tools, many research groups working closely with quarantine agencies around the world have investigated how molecular genetics and other methods of population typing can better inform what is commonly termed "pathway analysis" of flies; the premise being that the signature of intercepted flies represent a sample of the source population that is distinguishable from non-source populations.

Most of the techniques used today for this purpose are based on the analysis of an insect's DNA. These methods can utilize different protocols and instrumentation, but they are all rooted in the discipline of population genetics. Unlike DNA tools developed to distinguish two or more species according to their distinct genetic profiles (Armstrong and Ball 2005; Barr et al. 2012), the shared recent ancestry of populations and the lack of post-mating isolation mechanisms between populations present lower levels of genetic variation and often mixed genotypes within a population. To accurately detect differences in genotype frequencies between populations, it is often necessary to use tools that screen multiple genetic loci and provide sufficient resolving power. The quantitative data sets needed for this application can be large and complicated to interpret. Consequently, statistical analysis is necessary for the comparison of fly populations to estimate the source.

These statistical methods are also used to select loci and develop estimation methods by determining their diagnostic utility.

Depending on the level of genetic variation within each population and similarity between populations, these pathway methods can require substantial sampling of many reference populations in order to yield meaningful and reliable results. For example, if an oriental fruit fly (*Bactrocera dorsalis* (Hendel)) intercepted at a port is suspected to have originated from the Hawaiian Island of Oahu, because that was the source of the intercepted shipment, then a reference data base of DNA markers for Oahu flies is required to test that hypothesis. Sampling and analyzing 20–30 oriental fruit flies from a single collection site on Oahu may provide some information, but this is not adequate to determine if that represents variation on the entire island. Consequently, more extensive sampling is necessary to demonstrate that the Oahu population is sufficiently distinct from the other Hawaiian Islands to identify Oahu as the source. Unfortunately, there is no simple approach or benchmark to determine the required sample size or number of reference geographic populations without first estimating variation via preliminary studies. The scope of sampling requires consideration of the species' distribution over the island and possible barriers to random mating, such as altitude, weather and geographic distance. If variation on the island is low, then comparison of the intercept's genotype to the genetic profiles of flies from Oahu is not complicated. If the likely source of the intercept is not limited to Oahu, then additional sampling is required to develop an appropriate reference data base for all the Hawaiian Islands.

Geographic scope and scale are important concepts when examining invasion pathways. For pests with large geographic ranges, it may not be operationally feasible to sample many populations within each region. Unfortunately, reducing sampling effort to minimize the workload and cost of the study will also limit diagnostic resolution. The Mediterranean fruit fly is a good example, because it has established populations around the world (Sheppard et al. 1992; Vera et al. 2002). In such cases, rather than develop a data base with extensive sampling per country or province, samples from multiple countries have been pooled into larger geographic regions, such as the Mediterranean region and Central American region (Sheppard et al. 1992; Gasparich et al. 1997; Barr 2009). This proved useful, because genotypes found in some regions were absent in others. However, the reference collection has not been adequately sampled to estimate structure among populations within those regions.

Thus far, we have referred to pathway analysis as a means to assess where an intercepted fly originated from geographically plus the route to its final geographic destination. However, comparison of fruit fly DNA does not provide the sort of information that can directly address the latter issue. Strictly speaking, the genetic source reflects the location of the parental and older generations of flies, but this may not be the immediate geographic origin of the intercepted flies. If a fly intercepted at a port in California is the progeny of flies from Asia, the intercept will be assigned to an Asian source. If that detected fly was actually transported from Asia first to Hawaii and then to California, then the pathway would have multiple steps. The Hawaiian step in the pathway will not be detected using DNA information. For this reason, it is more appropriate to regard these DNA methods as estimators of the original genetic source and therefore as population assignment tools.

Molecular source estimation is hypothesis-driven: the proposed source of trapped flies is tested based on data and genetic models. The simplest test asks if the genotype of the intercept matches any of the genotypes sampled from a particular geographic locality. As the number of reference populations increases, it is possible to ask which source is the most likely (Cornuet et al. 1999; Piry et al. 2004). Recent advances in statistical analysis (e.g., DIY ABC program) enable evaluation of more complicated hypotheses that have genotypes of an invasion derived from multiple sources (Cornuet et al. 2008; Csilléry et al. 2010). This is needed if the invading flies have, over several generations, made multiple stops along the pathway and acquired genotypes from those intermediate populations.

The ability to define the true introduction pathway requires an integrated approach using multiple data sources, such as shipping records, host/commodity information, genotypes, and trapping records, and is beyond the scope of this chapter. Of these, genetic data are the most easily gathered, because they derive from the intercepted fly itself, although they are not always the most easily interpreted. Therefore, the remainder of this chapter is devoted largely to molecular methods for estimating the source population. First, we review some of the commonly used molecular methods for fruit flies and other invasive insects and introduce some alternative state-of-the-art approaches that have been applied to the issue of geographic origins. Then, we provide a review of source estimation technologies developed for the Mediterranean fruit fly, as a species whose broad distribution and invasive potential has led it to be intensively studied in this respect.

2 Overview of Molecular Methods for Source Estimation

Various types of molecules have been used in fruit fly population genetic studies. One of the earliest methods was protein-based allozyme or isozyme analysis, which compares the frequencies of different genetic alleles between populations. Each allele is defined by measuring the relative electrophoretic mobility of non-denatured enzymes isolated from frozen or fresh tissue (Murphy et al. 1996; Reyes and Ochando 1998; Lowe et al. 2004). Both heterozygous and homozygous states of these co-dominant alleles are used to characterize the genetic structure in populations. When allele frequencies are measured for multiple loci (i.e., different proteins representing unique genes), this method is sometimes referred to as multilocus enzyme electrophoresis (MLEE) (Gasperi et al. 2002) and provides greater resolution of genetic structure than single enzyme systems. The allozyme technique was never identified formally in the literature as a method useful for source estimation of individual insects or as a population assignment tool. Instead, several populations of flies were compared to determine which of the populations were most similar to each other and to estimate overall diversity of populations (e.g., Malacrida et al. 1998). This was typically accomplished by developing a matrix of pairwise divergence estimates among the populations and generating evolutionary trees to illustrate genetic similarity among the populations (Lowe et al. 2004). However, allozymes are not ideal markers for source estimation of trapped flies as proteins can degrade rapidly after death. Consequently, live or rapidly frozen material is required, which is difficult to manage in field and trap situations. Also, the analytical methods can only address population differences that are not adequate for assigning individuals to populations. Application of the allozyme technique eventually became less common after the widespread availability of polymerase chain reaction (PCR) technology in the early 1990s for amplification of specific DNA sequence targets. This enabled access to DNA as the genetic marker, which is not only much more stable after death, but provides greater population level resolution. The latter is made possible by additional variation at the nucleotide level that may not be revealed in proteins through transcription (i.e., in non-protein coding regions) or translation (i.e., 3rd codon variants).

Many methods are available to produce informative markers from DNA sequences in both protein coding and non-coding regions. Restriction Fragment Length Polymorphism (RFLP) is a method that uses restriction enzymes to digest DNA and produce characteristic, short, variable length stretches of DNA (e.g., Sheppard et al. 1992). A restriction enzyme is an endonuclease that can recognize a specific nucleotide sequence, typically 4–8 bp long, and cut the DNA strand at a specific nucleotide; the commonly used Type II endonucleases cut within that sequence. The cut strands are then separated according to their different size and electrophoretic mobility and visualized by staining the DNA fragments in the electrophoresis gel. If bases within the recognition site are polymorphic within a species, then DNA of some individuals will be cut by the enzyme and others will not. Consequently, differences in DNA sequences are observed as a different series of fragment lengths. Initial RFLP studies required isolation of large amounts of DNA for direct digestion of whole molecules and necessitated the pooling of samples, which is not helpful for population genetic analysis. The advent of PCR, however, facilitated RFLP analysis of individual samples by amplifying specific and more appropriate DNA regions, as well as enabling analysis of small amounts of starting DNA material (e.g., Steck et al. 1996). PCR-RFLP methods have been applied to target DNA regions in a fly's mitochondrial (Gasparich et al. 1997) and nuclear (He and Haymer 1999) genome.

The PCR-RFLP method is still used by laboratories for fruit fly analysis (e.g., Barr 2009), but the development of high throughput instrumentation for DNA sequencing in the late 1990s made it affordable to evaluate differences in DNA sequences by simply reading the complete stretch of DNA generated through PCR (Watson 1990; Puritz et al. 2012). The information content of this character-based DNA sequence data is typically greater than the frequency-based fragment data, such as RFLPs and allozymes, making it a more appealing approach for population genetic analysis (Villablanca et al. 1998; Barr 2009). DNA sequence analysis is more commonly used for mitochondrial loci than for nuclear loci, in part because of the accessibility of mitochondrial loci (see Sect. 3) and the additional costs involved in cloning alleles of nuclear loci in diploid genomes. However, a series of papers using DNA sequences of nuclear gene regions from the Mediterranean fruit fly demonstrate the value of the approach (Roderick and Villablanca 1996; Villablanca et al. 1998; Davies et al. 1999a).

A more rapid approach is to use only PCR, without the extra post-PCR modification steps as is necessary for RFLP or DNA Sequencing, and the PCR products are visualized directly by post electrophoretic separation staining. One method has been to produce genotypes as random amplified polymorphic DNAs (RAPDs), where a pair of short, random sequence, universal PCR primers produce amplicons from multiple anonymous regions in the genome (e.g., Baruffi et al. 1995; Haymer et al. 1997). Amplicons are scored based on size (i.e., electrophoretic mobility) differences, and the multi-amplicon profile often reveals population-level variation. However, the advantage of not requiring any prior knowledge of the DNA sequence of the target organism to generate these amplicons and ease of performance is offset by the difficulty of statistical analysis. This is due to difficulty distinguishing heterologous sequences with the same mobility and dominance of the markers, i.e., if sequence variation at the PCR priming site results in no amplification of one of the alleles in a heterozygote genotype then presence of a band does not distinguish it from a homozygote, effectively reducing the accuracy of the method to detect variation or differences between individuals. Therefore, this approach is not commonly used for source estimation. Alternatively, when the conventional PCR method targets a co-dominant locus in the fly genome, it is possible to estimate genetic variation (i.e., heterozygosity) of a single fly (Bonizzoni et al. 2001). The analysis of co-dominant markers is particularly important for understanding colonization events (Roderick and Villablanca 1996). Like the allozyme approach, differences in allele frequencies of populations can be used to compare similarity among sampled populations. However, if the number of co-dominant loci is sufficiently large, resolution is increased such that it becomes possible to compare the profile of a single fly to the expected profiles for characterized populations. The most commonly used co-dominant loci for source estimation studies are microsatellite DNA. Size differences at these loci, due to the different number of repeat sequences, are often very small (e.g., 2–4 base pairs) between alleles, so high resolution polyacrylamide gels or capillaries are needed to produce adequate separation. Microsatellite loci provide a powerful genotyping tool, especially with a genotype comprising several loci. However, a limiting step in their development as population markers has been the time and financial investment in their identification and validation (Lowe et al. 2004), which typically has involved use of enrichment techniques for microsatellite libraries (e.g., Shearman et al. 2006; Islam et al. 2011). Today, the development of newer lab protocols, genomic resources, and bioinformatics for screening genomes should facilitate the identification of fruit fly microsatellite DNA in the future (e.g., Abdelkrim et al. 2009; Meglécz et al. 2010; Guichoux et al. 2011). Microsatellites are important pathway analysis markers, and additional details on their use are provided below (Sect. 4).

There are many good books and reviews on the molecular techniques and gene regions used for population genetic studies, and on assignment tests to identify the source and paternity of an insect (Avise 2004; Lowe et al. 2004; Wan et al. 2004; Anne 2006). There is no one technique or genetic locus that is best suited for all fruit fly source estimations, but based on publications two types of molecular markers are predominant. These are mitochondrial DNA markers (using RFLP or DNA sequences) and microsatellite markers. We provide additional information on these marker systems and describe a few newer molecular methods being explored for fruit fly pests.

3 Mitochondrial DNA

3.1 What Is This?

Every respiring cell in animals carries two genomes, nuclear with genetic information on the chromosomes and mitochondrial with genetic information on circular, double stranded DNA found in the matrix within the inner membrane of mitochondria. Every cell has several hundred mitochondria, and every mitochondrion has several DNA molecules. The function of mitochondria is to convert food and oxygen into energy via the citric acid (Krebs) cycle. Mitochondrial DNA is therefore dedicated to this function, carrying genes encoding for key enzymes in that cycle. The mitochondrial genes comprise those coding for two ribosomal RNA's, 22 intervening tRNA's, and 13 proteins, the arrangement of which is highly conserved for many taxa (Boore 1999; Thao et al. 2004). These proteins fall into three of the five mitochondrial respiratory chain complexes: I NADH dehydrogenase, IV cytochrome oxidase, and V ATPase synthetase. Other proteins for these complexes and the other two complexes are encoded by nuclear DNA and moved into the mitochondria.

Mitochondrial DNA (mtDNA) is considered to essentially reflect the maternal lineage due to the low number of mitochondria in sperm and various mechanisms minimizing the transfer of sperm mitochondria into the egg during fertilization (Holland and Parsons 1999). Mutations in DNA sequence occur relatively frequently over inter-generational time (Brower 1994; Papadopoulou et al. 2010). Oxidative damage is thought to facilitate this due to proximity to reactive oxygen molecules arising as a by-product of the oxidative phosphorylation process. Unlike the nuclear genome, where strands are inherited from both parents, variation in mtDNA sequence is not thought to arise through recombination. Consequently, as the mtDNA molecule is essentially inherited as a single haploid unit without recombination, all loci are considered to be linked, and each unique genotype sampled from an individual is called a haplotype.

The mtDNA molecule has been relatively well studied, and a number of reviews and summaries are available on its structure and biology (Simon et al. 1994; Boore

1999; Avise 2004). Complete mitochondrial genomes are available for several species of tephritid fruit flies: *C. capitata* (Spanos et al. 2000), *Bactrocera oleae* (Rossi) (Nardi et al. 2003, 2005), *B. dorsalis* (Yu et al. 2007), *Bactrocera tryoni* (Froggatt) (Nardi et al. 2010), and *Bactrocera cucurbitae* (Coquillett) (Wu et al. 2013). Additional genomes of *B. dorsalis* complex species (*Bactrocera papayae* Drew & Hancock, *Bactrocera philippinensis* Drew & Hancock, *Bactrocera carambolae* Drew & Hancock) have been released on GenBank but are not yet published (see Nardi et al. 2005).

3.2 *Why Use It?*

The mitochondrial genome became an early and popular target for analysis of insect populations for both practical as well as biological reasons (Simon et al. 1994; Avise 2004) based on the characteristics described above. Compared to the nuclear genome, it is better understood owing to its small size (<20 kb vs. >100 mb for the nuclear genome) and simple arrangement of genes. As a tool, it is more easily analyzed with PCR amplification being more efficient due to a higher titer per cell and an abundant literature on PCR primer sequences many of which are universal (e.g., Simon et al. 2006). The lack of introns, repetitive DNA, and transposable elements also make sequence editing, alignment, and interpretation more straightforward. This uncomplicated organization was for a long time considered to be complemented by an assumed homoplasy (i.e., one haplotype within an individual due to no heterologous recombination) and a simpler maternal mode of inheritance in most species. There is, however, evidence of mitochondrial heteroplasmy (i.e., more than one haplotype mitochondrial genome present in the organism) in animals, and this can confound genetic analyses (e.g., Zhang and Hewitt 1997; Magnacca and Brown 2010). In addition, nuclear mitochondrial insertions (NUMTs) have been reported for insects (Zhang and Hewitt 1996; Bensasson et al. 2001). These NUMTS are fragments of mtDNA that have transferred to the nuclear genome and can evolve as a pseudogene (i.e., a copy of the gene that is not under normal selection pressures); these may still be amplified by the same PCR primers but represent a divergent sequence that will also confound genetic analyses.

Another advantage of mtDNA was believed to be the high mutation rate relative to single copy nuclear DNA (scnDNA) as found in higher animals. However, this does not hold true for insects, where scnDNA also generally evolves more rapidly (Zhang and Hewitt 1997). Neither is it always useful to compare mutation or substitution rates of the nuclear and mitochondrial genomes, because there is heterogeneity across both. Also, the commonly used substitution rate for the mitochondrial genome within an insect species is 2.3 % per million years (Brower 1994; Papadopoulou et al. 2010), which suggests that the majority of genotypes useful for distinguishing source populations have accumulated prior to our development of molecular techniques. Nevertheless, these are still useful where gene flow between populations is nonexistent. Genetic variation between populations can also be measured as differences in the proportions of those genotypes, the frequencies of which can change at a much faster rate than expected for the generation of new mutations.

Mitochondrial DNA also has a smaller effective population size in comparison to most nuclear DNA, given each locus is present as a single copy in the organism and essentially transferred to offspring only through the maternal line. This translates into fewer haplotypes in the population and allows lineage sorting, bottlenecks, and drift to create differences between isolated populations faster than expected for nuclear genes (Avise 2000, 2004). This effectively accelerates the formation of structure between populations necessary for source estimation. Consequently, especially for those populations isolated by geographic distance, mtDNA is a good source of population phylogeographic markers (Avise 2000).

3.3 Implementing mtDNA for Source Estimation

There are many studies that have explored the population genetics of fruit flies by using mtDNA (Prabhakar et al. 2012; Ruiz-Arce et al. 2012; Elfékih et al. 2010; Nardi et al. 2005; Gasparich et al. 1995). The use of such data for source estimation is then usually dealt with by one of two approaches: either assigning to a source population or excluding populations as potential sources.

The first approach is ideal. Here, a molecular diagnosis using mtDNA will be able to match the haplotype of a captured fly to the haplotype profile of a single source population. If source populations are relatively old and do not share gene pools (i.e., no inter-mating or significant levels of migration), it is possible for each source to have a distinct haplotype profile. In such a situation, the documented haplotypes should be "private" to a population (i.e., not shared) and source estimation relatively simple. Unfortunately, this has not been the case for fruit fly pest species. Their natural dispersal potential and human aided movement often results in some potential sources comprising young populations derived from multiple origins and gene pools. Therefore, source profiles can present considerable complexity. When there is evidence of shared haplotypes among sources, deciding which is the more likely source can become difficult or even impossible. In such cases, the inclusion of additional markers to improve resolution among populations is required to evaluate the likelihood of the sources. Where the inclusion of multiple unlinked genetic markers is possible, such as microsatellite DNA (Meixner et al. 2002; Shi et al. 2012) or nuclear coding gene regions as used for other pests (Hasan et al. 2009), there are robust statistical procedures for evaluating which is the most probable of all possible sources. Unfortunately, since mtDNA markers are linked, a similar approach using only mtDNA markers is not appropriate. Another problem with assigning a fly to a single source is that it assumes all sources have been included in the reference data base; if a possible source is not sampled, the fly cannot be assigned to it. The alternative approach to haplotype analysis is to apply an exclusion principle. Rather than assign the haplotype to a source, this exclusion principle determines which of the populations cannot be the source of the captured fly. This method was described by Barr (2009) for the Mediterranean fruit fly and works to generate a list of possible sources using mtDNA markers. Unlike the assignment approach, the un-sampled populations are then included as possible sources.

Once an analytical approach is selected, it is still necessary to use a scientifically robust procedure to evaluate similarity between the haplotype of a captured fly with those sampled from source populations. This is called haplotype matching and can be rather simple when the captured fly has a haplotype that is exactly matched and private to one geographic region. If the haplotype is present in multiple regions but absent from others, exact matching can still provide useful information. For example, if a haplotype is sampled from two regions, then these two would be possible sources, and the others could be excluded. This analysis is fairly crude and does not consider differences in the haplotype frequencies of populations. For example, it is possible that the haplotype is abundant (90 %) in one region and yet still present but rare (5 %) or undetected in another. In human forensics, several measures proposed to account for this and evaluate the probability of finding a haplotype from different sources include: the probability of two haplotypes being drawn randomly in a sample, the likelihood of drawing a genotype from a source based on its frequency, confidence intervals based on binomial or bootstrapping, and maximum match probabilities (Holland and Parsons 1999). The development of likelihood ratios for competing source hypotheses has also been proposed (Evett and Weir 1998; Holland and Parsons 1999). None of these methods have as yet been applied to tephritids.

However, matching haplotypes can become complicated when the differences between haplotypes are small. For example, if a haplotype from a fly differs from a haplotype(s) sampled in a population by one mutation, can that population be excluded as a source? One solution might be to develop conservative rules to handle data in a consistent manner. An example could be to apply a cut-off or threshold, such as, if the difference is just one base, treat the result as inconclusive. For human genetics cut-off values have been proposed. Salas et al. (2007), however, criticized this "deterministic criterion", because it does not consider the effects of variable mutation rates and heteroplasmy effects in humans. The use of cut-off values has not been examined for fruit fly source estimation studies.

In this section, we have assumed that the mtDNA reference data base was based on sufficiently large numbers of sampled populations and individuals per population. The source estimation results are only reliable if the frequencies of haplotypes in populations are accurately estimated (Pons and Petit 1995; Muirhead et al. 2008). How does a researcher know that a haplotype frequency is estimated accurately? How does a researcher know if a rare haplotype is not present in a population or has yet to be sampled from that population? These questions are crucial when making statements that exclude a population as a source.

Sheppard et al. (1992) proposed using the binomial distribution to evaluate the reliability of sampling rare haplotypes of the Mediterranean fruit fly. Barr et al. (2006) also applied this method to evaluate haplotypes of *Ceratitis* species. The method can estimate the size required to sample a rare haplotype at a particular probability but uses unrealistic assumptions about the populations, such as uniform genetic diversity among populations (Barr et al. 2006; Barr 2009). For example, if population structure exists for the species, then sampling 25 flies from one host tree may not yield the same diversity estimate as sampling 25 flies from 25 different host trees. An alternative approach is to generate saturation and rarefaction curves using genetic information. Literature developed on species richness for ecology and conservation biology applications explain how to interpolate and extrapolate species abundance for sample sizes using abundance values (Colwell et al. 2004, 2012). Using these methods, it is possible to estimate if smaller sample sizes can accurately estimate species abundance. These methods can also be applied to estimate haplotype abundance (Pilot et al. 2010; Panova et al. 2011). Based on regressions of haplotypes versus individuals or collections sampled, an asymptote in the curve suggests that the majority of haplotypes has been sampled.

For human forensics, Pereira et al. (2004) defined sampling saturation as "the point for which increments of 100 [additional samples] do not raise the number of haplotypes by more than 5 %" and used a regression to calculate the number of individuals required to detect rare haplotypes for an mtDNA data base of a Portuguese population. This saturation value must be calculated separately for different populations. The authors also observed that the required sample size increases when the number of variable sites at a locus increases. Likewise Egeland and Salas (2008) reported using frequenistic and principal component analysis to estimate sample coverage and the frequency of rare haplotypes in a data base. When there is no evidence that haplotype richness has saturated, then genotypic frequency and phylogenetic information may be needed to interpret results.

In addition to haplotype matching, mitochondrial markers can be evaluated using other population genetic analysis methods. These include estimation of summary statistics, such as haplotype diversity (Hd), migration rates (m), fixation indices between populations, such as F_{ST} and G_{ST} (Nei and Kumar 2000), and AMOVA (Analysis of Molecular Variance) tests of structure among populations (Excoffier et al. 1992). These methods are particularly important during the development of a reference data base and can help guide the selection of what regions or collections constitute a source population. For example, will each population within a country be regarded as a separate source? Or, will the country be one source and the populations sampled from that country used to estimate genetic variation?

Phylogeographic studies are also helpful in evaluating structure within species. Since the mutations separating haplotypes represent the genetic history of mitochondrial evolution, phylogenetic networks can be generated to determine similarity among haplotypes (Templeton 1998; Panchal and Beaumont 2007; Knowles and Maddison 2002). Unlike AMOVA, which requires the geographic populations be predefined, the network is generated from individuals and not populations. Once the genetic relationships among haplotypes are determined, it is then possible to map the geographic, host, or ecological information onto the network. This can help recognize "haplogroups" (clusters of haplotypes with shared ancestry) and develop rules of assignment/exclusion based on these groups rather than individual haplotypes directly.

However, it is important to recognize that the characteristic linkage of genes in mitochondrial DNA means it provides only one estimate of species and population evolutionary history (DeSalle and Giddings 1986). This history is influenced by a number of factors, both acquired and intrinsic, that result in positive and purifying selection (Meiklejohn et al. 2007). For example, fly populations can acquire endosymbiotic *Wolbachia* bacteria (Martínez et al. 2012) that cause cytoplasmic incompatibility (Zabalou et al. 2004) or fitness advantage (Sarakatsanou et al. 2011). Therefore, increasing the size of a linked data set by including additional regions of the mitochondrial genome may increase the number of variable sites but will not always enhance the confidence that the results are correct. In addition, the relatively small effective population size of this genome makes it more susceptible to factors, such as drift, that serve to reduce genetic variation. While this might be good for sorting variation, it provides a skewed history of the species.

3.4 Applications of mtDNA for Fruit Flies

Successful use of mtDNA for identifying a source population requires four things: a method of genotyping to detect sequence polymorphisms, a locus that evolves at an appropriate inter-population rate, data in a format that lends itself to robust analysis, and access to appropriate population reference material for confident interpretation of the genetic markers. While the latter probably remains the greatest limitation to any genotyping method, mitochondrial DNA has been shown to be a useful source estimator. The progress made for the Mediterranean fruit fly is used to illustrate this here, as well as inroads made with a few other pest species.

The two most commonly used methods to genotype mtDNA are RFLP and DNA sequencing. The earliest fruit fly studies on the Mediterranean fruit fly used RFLP genotyping of isolated whole mtDNA genomes. This rather cumbersome technique, based on Southern blots of two restriction enzyme digests (*Xba* I and *Eco*R V), revealed inter-population sequence variation in this genome that could be used to infer source populations of medfly infestations in California (Sheppard et al. 1992). Two New World haplotypes were detected, one in Hawaii and Venezuela and the other in Guatemala, Argentina, and California. At that time, comparison with the restriction pattern of flies collected in California was sufficient to exclude Hawaii as a source of those infestations. However, with improved access to PCR and efforts to refine the technique for higher throughput, additional haplotypes became apparent. Using a greater range of populations (McPheron et al. 1995) and restriction enzymes targeting different restriction enzyme sites (Gasparich et al. 1997), RFLP analysis of a 2.99 kb PCR product (NADH dehydrogenase 4, through to cytochrome b) containing the original two variable *EcoR*V and *Xba*I sites (Gasparich et al. 1995) showed that many populations were a mix of six haplotypes. These studies illustrated that most of the complexity (diversity of haplotypes) was in Africa, consistent with the likely origin of medfly. Although Africa could not be excluded as a source, this did not change the conclusion that Hawaii was not a likely source of the California infestations. Nevertheless, evidence of mixed haplotypes in South America makes decisions about source populations and the ability to follow invasion processes more complex and uncertain. Further details are provided in the case study section on source estimation of the Mediterranean fruit fly in California.

The PCR-RFLP method is still used by laboratories for fruit fly phylogeographic analysis largely because of the perceived simplicity, including the use of relatively basic equipment. However, its effectiveness for discriminating populations is dependent on the mtDNA gene target and the combination of restriction enzymes used in the RFLP assay. In recognizing this, Nakahara and Muraji (2010) used the A +T rich control region, which regulates the transcription and replication of mtDNA, to determine the likely source of oriental fruit fly *B. dorsalis* complex flies trapped in Japan (see Nakahara et al. 2008; Muraji et al. 2008 for details on methods and approach). Using the restriction enzymes *Dra*I and *Ssp*I, good resolution was achieved with 44 haplotypes detected among 513 individuals from 16 sites across SE Asia. Unfortunately, even with this relatively large number of 2–16 haplotypes per population, the only observed structure delineated the Philippines from the rest. A purely subjective assessment, based on a visual comparison with those haplotypes arriving in Japan, then brought Nakahara and Muraji (2010) to the conclusion that the invasion was coming from the Philippines as well as other regions in SE Asia. However, in this case, where pest species within the complex are extremely hard to distinguish genetically (Boykin et al. 2013) as well as morphologically (Mahmood 2004), the added confusion as to which species were being detected was acknowledged. The purely taxonomic distinction of tracing different species as opposed to different populations of the same species does not change the conclusions made here about source, but it could change the interpretation regarding the biosecurity impact of an invasion. Also of note, choice of this major non-coding region had been based on the outmoded general view from vertebrate studies that it evolves more rapidly than the rest of the mitochondrial genome. However, while under certain conditions the mtDNA control region could be a good population marker, it generally offers no advantage over other regions of the insect genome (Zhang and Hewitt 1997). It is neither the most variable region in terms of nucleotide substitution relative to the third codons of protein coding regions or to single-copy nuclear non-coding sequences, nor is it easily analysed. The latter being compromised due to its high A-T content, tandem repetition, heteroplasmy, and higher occurrence of repeat, inverted repeat, and palindromic (segment of nucleotides immediately followed by its reverse complement) (Arunkumar and Nagaraju 2006) regions, which can lead to possible technological difficulties with PCR amplification, RFLP, and direct sequencing (Zhang and Hewitt 1997).

In comparison to RFLP banding pattern data sets, reading the entire DNA sequence of a genetic locus provides a greater wealth of information. The RFLP method discriminates haplotypes based on a subsample of the nucleotide differences between two unique sequences. In contrast, reading every nucleotide in those two DNA strands increases the diagnostician's probability of detecting additional differences, as was recently demonstrated for the Mediterranean fruit fly (Elfékih et al. 2008, 2010; Lanzavecchia et al. 2008; Barr 2009). Nevertheless, despite its limited ability to detect unique haplotypes, the RFLP method is still used today for the Mediterranean fruit fly. Continued popularity of the method is partly due to the large RFLP reference data set generated for the pest (Gasparich et al. 1997), and it is also less expensive to perform and easier to score than is DNA sequencing. However, a direct comparison of these two methods revealed technical errors in the RFLP method (Barr 2009), highlighting the consequences of incomplete restriction digestion.

Ongoing work to build a larger DNA sequence reference data base for the Mediterranean fruit fly will eventually create a superior resource for source estimation studies of this fly (Ruiz-Arce, unpublished). In addition, because the RFLP method is based on DNA sequence information, it is possible to combine the two data sets for the same locus by simply converting the sequence information into RFLP information *in silico* (Elfékih et al. 2008). Using these methods, Barr (2009) explained the reasoning behind the exclusion principle employed for source estimation of Mediterranean fruit flies. Technical issues on how to circumscribe populations (i.e., define the geographic unit for diagnosis) and estimate haplotype sampling confidence (i.e., determine when a haplotype is not present) have not been standardized for all studies, but Barr (2009) described how this molecular resource has been applied for analysis of the pest across several laboratories.

DNA sequences of mtDNA loci have also been used to evaluate the sources of other economically important fruit flies. For example, Hu et al. (2008) analyzed *B. cucurbitae* collected from the Philippines, Thailand, and seven locations in China using sequences of the *cytochrome oxidase* I (COI) gene. The publication was not intended to determine source. Rather, it was an opportunity to test for variation in the collections and structure among regions and explore the results for evidence of movement between populations. Of the 72 flies sequenced, ten haplotypes were detected. Over 75 % of the samples shared one common haplotype, suggesting low variation. The other nine "rare" haplotypes were very similar to the common haplotype. The authors explained that the low variation and lack of pattern was informative, because it was consistent with a recent invasion model of the fly into China. The authors did not have an adequate reference database to test this hypothesis or exclude other models using a statistical framework. This is, however, a good example of how mtDNA information can be used to inform researchers on possible population movements and generate new hypotheses about invasion routes and colonization.

In 2012, three additional studies were published using mitochondrial DNA to examine the population genetics of *B. cucurbitae*. Prabhakar et al. (2012) reanalyzed the Hu et al. (2008) data with newly generated COI data from Indian collections and found low genetic variation for the species. In an independent study, Wu et al. (2012) increased the sampling effort of Hu et al. (2008) and reported evidence for expansion of the fly from western Asia into China as haplotype diversity was higher in the west (Nepal, Bangladesh, Burma, and western China) than in other locations. The third study by Jacquard et al. (2013) used concatenated mitochondrial DNA sequences, in addition to microsatellite loci, to test diversity on Reunion Island. This group identified a haplotype difference between African and

Asian populations and, based on a comparison of haplotype similarity, the African population was considered the more likely source of the island population.

In contrast to the *B. cucurbitae* results from China, haplotype diversity of the COI gene was relatively high among *B. dorsalis* collections in China (Shi et al. 2005a, b, 2010, 2012; Liu et al. 2007). However, despite the statistical detection of structure, there was a lack of strong geographic patterns that might explain the genetic associations. Examination of the likely source of flies in Yunnan, China, revealed two likely explanations: the region had experienced several colonization events (i.e., multiple sources) or the residential population is relatively old (Shi et al. 2005a, b, 2010). Although molecular studies have identified *B. dorsalis* populations (e.g., Hawaii) with strong structure (Aketarawong et al. 2007; Wan et al. 2012), source estimation will be complicated for this species because of the large number of areas sharing haplotypes (Shi et al. 2012). Trends in diversity estimates and evidence of population expansion and other demographic changes in Asia can be tested using these mtDNA markers (Wan et al. 2011; Shi et al. 2012), but they may have limited capability of selecting one source over another. Wan et al. (2012) provided support for invasion routes that started in Southeastern China. These inferred routes were based on estimates of immigration rates rather than formal tests to assign or exclude source populations.

Similarly, for the olive fly *B. oleae*, mtDNA was used together with microsatellite markers to develop broad hypotheses around historical colonization events. The haplotype similarities suggested that Africa, and not the Mediterranean area, was the origin of flies infesting cultivated olive and that the recent invasion of olive flies in the American region most likely originated from the Mediterranean area (Nardi et al. 2005) and more specifically Turkey (Dogaç et al. 2013). Shearman et al. (2010) proposed an interesting means to determine source based on interspecific differences of mitochondrial genomes. Specifically, in the context of an SIT program against the Queensland fruit fly, *B. tryoni*, these researchers wished to develop a laboratory strain with a unique DNA marker to allow reliable discrimination between released sterile males and their wild counterparts. To this end, they proposed interspecific hybridization with a closely related but distinct species *Bactrocera jarvisi* (Tryon) to create a strain of *B. tryoni* with the *B. jarvisi* mitochondrial genome and then a simple PCR test to then distinguish the marked from wild *B. tryoni*.

As can be seen in the examples given, the majority of studies of invasive fruit flies using mitochondrial DNA are still focused on understanding what structure exists within species as opposed to utilizing DNA to determine invasion routes and recent origins. Nevertheless, that information is critical to enable the latter. Importantly, the eventual development of methods for routine use will also require the generation of good reference data sets. Unrelated studies that report genetic diversity estimates for different fly populations contribute to these data sets. For example, population sequences that are available from studies on insect phylogeny (e.g., *Anastrepha* [Smith-Caldas et al. 2001], *Ceratitis* [Barr and McPheron 2006], and *Dacus* [Virgilio et al. 2009]), species diagnosis (Armstrong and Ball 2005; Barr et al. 2006; Blacket et al. 2012; Khamis et al. 2012; Frewin et al. 2013), and population structure (Boykin et al. 2006; Ruiz-Arce et al. 2012; Lim et al. 2012) contribute to our understanding genetic variation of pest species. Good population coverage in some studies has revealed limitations of markers within the mitochondrial genome for adequate resolution of sources; this is particularly true for recent and rapid radiations of species with haplotype rich populations such as *B. dorsalis* in Asia.

4 Microsatellite DNA

4.1 What Are These?

The term microsatellite refers to a particular class of repetitive DNA located in eukaryotic genomes. The size of a microsatellite locus can vary, but they are all relatively small (e.g., <500 bases) (Butler 2005; Hancock 1999). Each locus is comprised of tandem repeats of a motif sequence that is typically 1–6 bases long (Wan et al. 2004). For example, a motif could be the bases "AG" and, if repeated 50 times, would generate a microsatellite of 100 bases ("AGAGAG. . ..AGAG"). In contrast, the other two major classes of repetitive DNA, called minisatellite DNA and satellite DNA, have repeat motifs that range from 10 to 100 and 100 to 1,000 bases, respectively (Butler 2005). Because of their relatively short and less complex motifs, microsatellites are also known as Short Tandem Repeats (STRs) and Simple Sequence Repeats (SSRs). There are excellent books and reviews devoted to the topic of microsatellite evolution, analysis, and application. We recommend Goldstein and Schlötterer (1999) for a detailed account of microsatellite structure and evolution, Butler (2005) for a practical guide on microsatellite use in human forensics, and Guichoux et al. (2011) for a current review of methods analysis.

A microsatellite can be categorized based on the length of the repeat unit. If the motif consists of one base repeated multiple times, it is called a mononucleotide repeat, two bases a dinucleotide repeat, and similarly, tri-, tetra-, penta-, and hexanucleotide repeats describe motifs consisting of three, four, five, and six repeat units, respectively. Our previously described motif "AG" is an example of a dinucleotide repeat unit. Because each nucleotide position can be one of four bases (A, C, T, or G), there are more motifs than types of repeat units. The size of the repeat unit can have an impact on the utility of the microsatellite (Guichoux et al. 2011). For example, mono-nucleotide motifs are not recommended for analysis, because there is more difficulty in scoring alleles that differ by a single nucleotide. In contrast, data interpretation is often easier for microsatellites that have motifs with longer repeat units, because size, and therefore homology, is clearer, and PCR artifacts, such as stutter (a.k.a., shadow bands or DNA polymerase slippage products), are more identifiable (Walsh et al. 1996; Butler 2005, see discussion below). Consequently, it is important to consider the type of repeat units when identifying microsatellites from genomic resources or evaluating published methods.

So far, we have only described microsatellites with simple repeats. That is, each repeated unit of DNA is an exact copy of the motif; $(AG)_2$ means AGAG. If a motif includes a combination of two or more simple repeats, it is called a compound repeat. For example, the dinucleotide motif "AG" can be next to the trinucleotide motif "CTC" to create alleles with unique repeat units. Therefore, it is possible to have a repeat unit of 29 bases: $(AG)_2(CTC)_3(AG)_2(CTC)_4$. If the repetitive DNA includes a more complicated arrangement of various units and intervening DNA segments, it is called a complex repeat. Microsatellite DNA that includes an insertion of non-repetitive bases is called an interrupted microsatellite. It is also possible for a simple, compound, or complex microsatellite to have a base substitution that changes only one unit in the tandem repeat sequence (e.g., AG,AG,AG, AG,A$\underline{\text{C}}$,AG,AG).

Guichoux et al. (2011) described repeats as either perfect (i.e., simple repeats) or imperfect. The complex repeats would fall into the imperfect category. These imperfect repeats are commonly used to study population structure, but the evolution of alleles for an imperfect motif might not follow the expected models based on simple repeats. Therefore, additional care is required to confirm that the alleles are distinct (based on descent) and that statistical models of analysis match the evolution at the locus (Estoup et al. 1995, 2001).

Although microsatellite DNA can be found in mitochondrial genomes (Lunt et al. 1998), this is not the norm. The microsatellites used for animal population genetics are located in the nuclear genome. It is possible to map these loci using cytogenetic techniques and crossing experiments (Stratikopoulos et al. 2008). Not all microsatellite repeats are useful for population and source estimation studies. Each locus should be inherited as a co-dominant marker, where both alleles are fully expressed. These two alleles could be identical in size (homozygous) or distinct (heterozygous). Experiments are required to ensure that both copies can be detected, as technical issues, such as allele drop-out (preference for amplification of one allele over second allele), stuttering errors, and null alleles (failure to amplify an allele because of mismatches in the primer binding site), can prevent proper scoring (van Oosterhout et al. 2004; Carlsson 2008). Microsatellite loci with evidence of more than two copies will not behave as expected (Van't Hof et al. 2007) and should not be included in source estimation studies.

4.2 Why Do We Use Them?

When a pest species is introduced to a new region, and isolating mechanisms such as mountains, seas, or eradication zones, are present to prevent inter-mating of the new incursion and the old source population, it is possible for the populations to have different genotype frequencies. When this population structuring is strong, markers, such as mitochondrial genes, can sometimes distinguish if a fly is more likely to be from the original source population or from the introduced population. This structure has been observed for the Mediterranean fruit fly, because different mitochondrial genotypes are associated with populations in different parts of the world (Gasparich et al. 1997). However, when the structure is not strong (e.g., intermating and migration are common), additional markers are required to detect the smaller differences among populations. There are many cases where the potential source populations are genetically similar, because the populations lack barriers and/or were recently derived from the same ancestral population. Although mutations can accumulate in the genome and become diagnostic for some introduced populations, this is not a fast event for many genes. For example, even though mitochondrial markers are useful for distinguishing Hawaiian populations of the Mediterranean fruit fly from Guatemalan populations, these markers are not informative for distinguishing whether a fly originated from the less distantly distinct Guatemala or Costa Rica (Gasparich et al. 1997).

To discriminate among such populations that have small differences in genotype frequencies, it is necessary to target highly variable regions of the genome. These are more rapidly evolving, because they are less constrained than protein coding regions, such as those used in the mitochondrial genome. It is also necessary to include multiple loci to avoid bias in the estimate and increase overall resolving power. In human forensics, both attributes have been achieved by using repetitive DNA to match genotypes of suspects to population data bases. Initial forensics work used minisatellite DNA (Jeffreys et al. 1985), but by the 1990s microsatellites were the common method of choice for studying genetic variation among populations (Butler 2005). Microsatellite markers are very useful for population studies, because mutation rates are relatively high for repetitive DNA, and there are many microsatellite regions scattered throughout the nuclear genome (Lowe et al. 2004; Wan et al. 2004). This latter point is important to ensure that microsatellites are not co-evolving simply because of linkage via co-location in the genome. For example, if ten microsatellite regions are used to analyze a fly, each one should provide an independent estimate of genetic similarity to the possible source populations. If these ten are located right next to each other on a chromosome, then they simply estimate the same genetic history.

As a result of recombination events and independent assortment in a diploid nuclear genome, the microsatellite loci are suitable for many types of statistical tests based on population genetic theory. In contrast to mitochondrial DNA, the nuclear genome can be used to estimate heterozygosity within an individual. A population genetic text book will provide further details on the merits of using co-dominant markers for diploid analysis (e.g., Weir 1996; Hartl and Clark 1997). Microsatellite DNA is also useful because the relatively short size of each locus makes it compatible with conventional PCR methods. The PCR primers are designed using the non-repetitive DNA regions flanking the microsatellite DNA. If the primers amplify the alleles reliably, the PCR products of polymorphic microsatellites are observed as size differences. The sensitivity (i.e., ability to analyze minute amounts of DNA) of a microsatellite assay is also good, because the PCR step amplifies many copies of the target loci from samples with low DNA titers. This is important when working on organisms that are small and/or have degraded DNA resulting from less than optimal collecting practices (Maxwell et al. 2011).

4.3 Tephritid Microsatellites for Source Estimation

Microsatellite loci have been located and primers developed for several economically important species in the genera *Anastrepha*: *A. suspensa* (Loew) (Fritz and Schable 2004; Boykin et al. 2010) and *Anastrepha obliqua* (Macquart) (Islam et al. 2011); *Bactrocera*: *B. oleae* (Augustinos et al. 2002, 2008), *B. tryoni* (Kinnear et al. 1998), *B. dorsalis* (Dai et al. 2004; Aketarawong et al. 2006; Shearman et al. 2006; Li et al. 2007), *Bactrocera invadens* Drew, Tsuruta & White (Khamis et al. 2008), *B. cucurbitae* (Virgilio et al. 2010; Wu et al. 2011; Jacquard et al. 2013), and *Bactrocera cacuminata* (Hering) (Song et al. 2006); *Ceratitis*: *C. capitata* (Bonizzoni et al. 2000; Casey and Burnell 2001; Meixner et al. 2002; Stratikopoulos et al. 2009), *Ceratitis anonae* Graham, *Ceratitis rosa* Karsch, *and Ceratitis fasciventris* (Bezzi) (Delatte et al. 2013); and *Rhagoletis*: *Rhagoletis pomonella* (Walsh) (Velez et al. 2006), *Rhagoletis indifferens* Curran (Maxwell et al. 2009), and *Rhagoletis completa* Cresson (Chen et al. 2006). Although primers were developed to amplify a specific locus in the target species, several studies have demonstrated that microsatellite primers can amplify the microsatellite from multiple species (e.g., Baliraine et al. 2003; Velez et al. 2006; Stratikopoulos et al. 2009; Islam et al. 2011; Drosopoulou et al. 2011).

Published protocols for microsatellite loci have been used to study a wide range of biological questions. For example, does geographic population structure exist for a species (e.g., Bonizzoni et al. 2001 for *C. capitata*, Yu et al. 2001 for *B. tryoni*)? Do populations represent different taxonomic lineages (e.g., Cameron et al. 2010 for *B. tryoni*, Krosch et al. 2013 for *B. dorsalis*, Virgilio et al. 2013 for *Ceratitis* FAR complex)? What is the dispersal capacity of flies (Karsten et al. 2013), and is there evidence of hybridization, mixed parentage, or re-mating of females in the wild (Michel et al. 2007 for *R. pomonella*; Johannesen et al. 2013 for *R. cerasi* (L.); Bonizzoni et al. 2002 for *C. capitata*; Fritz et al. 2010 for *A. suspensa*; Gilchrist and Ling 2006 for *B. tryoni*)?

Although not their primary focus, studies of general population structure can reveal important information regarding the likely source of invasive populations. For example, based simply on relative genetic diversity estimates, it was possible to support Sri Lanka as the source of an expanding African *B. invadens* population (Khamis et al. 2009). Similarly, diversity estimates of the olive fly, *B. oleae*, support an expansion of populations into western European area from the east (Augustinos et al. 2005).

However, there are a few published microsatellite studies designed specifically to estimate the geographic source of invasive populations in support of pest management and eradication. Some of these studies used a combination of microsatellite and mitochondrial markers, e.g., *B. oleae* in Turkey (Dogaç et al. 2013);

B. oleae in Americas (Nardi et al. 2005); *C. capitata* in California (Meixner et al. 2002), *C. capitata* in Florida (Silva et al. 2003), *B. dorsalis* in China (Shi et al. 2012). Other studies have relied solely on microsatellite results to estimate the source population, e.g., *B. oleae* in California (Zygouridis et al. 2009), *C. capitata* in California (Bonizzoni et al. 2001), *C. capitata* in Australia (Bonizzoni et al. 2004), *B. tryoni* in southeastern Australia (Gilchrist and Meats 2010), *B. dorsalis* in Asia and Pacific (Aketarawong et al. 2007). Furthermore, in some microsatellite studies, a hypothesized resident population is included as a potential source (Sved et al. 2003; Gilchrist et al. 2004). Very localized applications also benefit from microsatellite markers when sterile flies are released to suppress pest populations to confirm that recaptured flies are from the sterilized lab source (e.g., Gilchrist et al. 2004; Aketarawong et al. 2011).

4.4 Implementing Microsatellites

As previously described, published microsatellite primers are available for many species of fruit fly and can be used to study related species. The papers that report these markers explain how the microsatellites were identified and how the PCR primers developed. Prior to implementing microsatellite markers for diagnostic programs, the markers need to be evaluated to ensure that the loci behave according to the expectations for repetitive DNA and that the protocols can be appropriately applied. This information can be acquired from the primary literature and through testing of PCR performance at a lab. Although there are many considerations involved in microsatellite implementation, we focus here on locus selection, allele verification, and data analysis.

Protocol development starts with the selection of microsatellite loci for analysis. This selection process includes evaluation of motif type, the expected range of allele sizes, and evidence of independence and neutrality of loci. This information should already be available when the markers are published. Those initial studies, however, typically focus on a specific population and may not sample the true diversity within the species. Consequently, analysis of samples from other regions of interest could reveal new alleles and detect violations of independence or neutrality at the locus.

Microsatellites with different motifs and repeat units can experience different rates of mutation and sensitivities to PCR artifacts. For example, di-nucleotide repeats have a higher propensity for slippage as the polymerase replicates the DNA (Kruglyak et al. 1998; Broquet et al. 2007). This can lead to higher levels of stutter (or shadow bands) during PCR that can complicate data interpretation and may contribute to inaccurate allele calls (Guichoux et al. 2011; van Oosterhout et al. 2004). Stutters are particularly problematic when scoring adjacent alleles (heterozygotes) for a dinucleotide repeat motif (Fig. 10.1). The downstream effects of miscalling heterozygotes and homozygotes include a bias in estimating statistics such as heterozygosity, inbreeding, null alleles, and deviation from Hardy

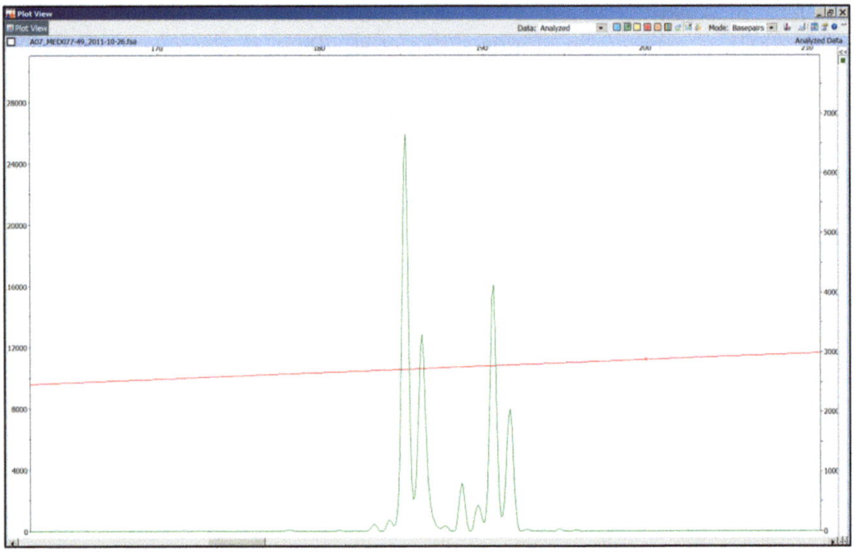

Fig. 10.1 A dinucleotide repeat motif microsatellite fragment from Medfly (*Ceratitis capitata*) analyzed with Applied Biosystems' Peak Scanner^{TM} v1.0 showing characteristic stutters

Weinberg. Technical methods are available to reduce the impact of stutter effects. These include using fusion enzymes (Fazekas et al. 2010) and decreasing the PCR denaturation temperature (Olejniczak and Krzyzosiak 2006). When these are not feasible, then the selection of markers that contain longer simple-repeat motifs is advisable (O'Reilly et al. 1996).

The use of simple repeats is often more desirable over non-simple repeats, such as compound microsatellites. A simple di-nucleotide repeat should generate a predictable pattern of alleles differing in units of two bases. Non-simple repeats units can experience more complicated patterns of mutation that result in less predictable allele sizes. Despite issues with slippage, another reason to select di-repeat motifs for analysis is their abundance. Tóth et al. (2000) evaluated microsatellites in various taxa and showed that motifs with short repeats are more common among invertebrates as compared to motifs with long repeats that are more common to vertebrates. In some invertebrates, repeats containing "CA" have been found to occur with a very high frequency (Meglécz et al. 2004; Schug et al. 1998).

Multiple peaks can also result from "non-template nucleotide addition" during PCR and not be related to the motif. For example, addition of nucleotides (e.g., poly-A's) to the end of the PCR product can result in a mixed population of DNA molecules that differ by one base. Multiple bands are visualized for such reactions, compromising the interpretation and accuracy of allele calling. To overcome this, increased the extension time and added 5′ guanine residues or "PIGtail" (GTTTCTT), or merely a guanine to the 5′ end of the reverse primer, can be employed to promote the complete adenylation of these nucleotides (Odriozola et al. 2011; Hill et al. 2009).

The lengths of motifs, as well as other factors, should be considered in the design of multiplex reactions. Multiplex PCR is the amplification of multiple loci in a single PCR reaction. With multiplex PCR, it is possible to generate products for several target loci at once and significantly reduce the cost of data generation (Guichoux et al. 2011). The interpretation of data using multiplex assays can be challenging, but the difficulties can be minimized if one takes into account various strategies. One strategy to minimize errors in interpretation of electropherograms is to use loci-distinct molecular labels. For microsatellites, the peaks observed in the electropherogram reflect coloration associated with the label for each locus. This permits the end user to distinguish between alleles from different loci that may inadvertently overlap in fragment length. The challenges increase when the loci are being analyzed with high resolution gels (and not via automated capillary instrumentation) where molecular labels are generally not used. Therefore, the use of electrophoresis in distinguishing alleles belonging to different loci when alleles overlap may be quite difficult. Taking into account the repeat motif to minimize overlap is then important in the design of a multiplex reaction. Loci that contain only dinucleotide repeat motifs can be advantageous, since the variation in allele lengths may not be as expansive for dinucleotides as it may be for tri-, tetra-, and penta-nucleotides. This strategy may help to predict the range in alleles, design a more effective PCR multiplex protocol, and reduce the chance of overlap and there by minimize difficulties in the interpretation of data.

It is important to note that the size of the PCR product, regardless of motif type, can have an impact on multiplex reactions. During PCR a competitive process is occurring where shorter fragments tend to amplify more efficiently (Frohman et al. 1988; Matz et al. 1999), and this may impact on interpretation of the PCR results. In some cases, the larger bands may appear less intense or absent. This can also happen to a sample during a normal PCR that targets a single locus. This effect is called allele drop-out and results in too few heterozygous individuals in a population, i.e., excess homozygosity.

Three statistics are typically calculated and reported for microsatellites to ensure that they are suitable for source estimation and other population genetic studies (e.g., Virgilio et al. 2010). These include tests of Hardy Weinberg Equilibrium (HWE) to assess if the locus is under selection or other violations of equilibrium (Guo and Thompson 1992; Slatkin and Excoffier 1996), linkage disequilibrium (LD) to ensure that each locus is an independent estimate of the genome (Slatkin and Excoffier 1996), and null allele presence to determine if alleles are being systematically missed (not PCR amplified and therefore not observed) resulting in excess homozygotes (Carlsson 2008). Null alleles differ from allele-drop out in that they result from mutations in the primer binding sites and not differences in allele size. Methods are available to test these measures of performance (e.g., MICRO-CHECKER, FreeNA, and ML-NullFreq). Effective screening of loci may help prevent reliance on markers that present such challenges. Information from previous studies may help in selecting the most appropriate marker, but recognizing that the repeat motif and other locus characters are important in the selection of markers is essential for developing scientifically sound methods with microsatellites (Dewoody et al. 2006).

As a result of PCR artifacts and the presence of alleles resulting from convergent or complicated evolutionary patterns, it is useful to know the DNA sequence of each unique allele in a microsatellite data set. For example, recognizing that a single allele (i.e., DNA sequence) has a propensity to appear as two forms during the scoring process would simplify data collection. Likewise, having evidence that the presence of alleles varies because of DNA differences located outside of the repeat unit (e.g., closer to the primer sites) would suggest that the marker is not evolving as a microsatellite; this would help ensure that the source estimation analysis is based on accurate allele calls (Renshaw et al. 2006). Confirming presence of the expected repeat region is important, and this quality control measure has been reported in literature on plants (Hopkins and Taylor 2011), arthropods (Lopes et al. 2009; Harr et al. 1998), fish (McDowell et al. 2002), and mammals (Brinkmeyer-Langford et al. 2012).

In addition to DNA sequence verification of the observed alleles, it may also be necessary to verify the sequence of a suspected null allele or test for evidence of multiple copies of the locus. Each microsatellite marker is supposed to represent a single locus in the nuclear genome. That locus is present on two homologous chromosomes of a diploid organism and results in either a homozygous or heterozygous genotype. However, it is possible for primers to be less specific and amplify multiple loci that generate more than the expected two PCR products (Zhang 2004; Meglécz et al. 2004). If these microsatellite copies are different, it may be possible to detect more than two alleles in the analysis. As previously mentioned for allele drop-out effects, competition during PCR can mask some of these copies. Multiple copies of a microsatellite violate the assumptions for standard analysis and is important to document. Some research has found that there is an association between mobile elements and multiple copies of microsatellites in species of Lepidoptera (Zhang 2004; Meglécz et al. 2004). These mechanisms have also been reported in other organisms (Meglécz et al. 2007; Ramsay et al. 2000). Multiple copy microsatellites have not yet been reported in Tephritidae but have been reported in other dipteran genera (Chambers et al. 2007; Wilder and Hollocher 2001).

Data collection of alleles for a microsatellite study can be a significant task in itself. A data set including just 10 loci and 100 flies would generate 2,000 data points, because there are two copies of each locus. The raw data must be scored, logged, and the various alleles named and organized. It is common practice to simply record the alleles based on the estimated size of the PCR product. This simple numeric can be easily recorded in any computational data base or spread sheet program. For example, Microsoft Excel is capable of storing and managing various fields of information in a matrix-type environment. Additionally, useful macros and other software built for microsatellite analyses have been designed to work closely with MS Excel. For example, MS Toolkit (Park 2001) and GenAlEx (Peakall and Smouse 2006, 2012) are add-in utility macros that are very useful for organizing, analyzing, and formatting. Guichoux et al. (2011) provides some examples of data management systems developed for microsatellite information.

Variation in the observed size of an allele (fragment) can result from differences in protocols, instruments, reagents, and staff. Some markers are more prone to stutters, and some fragment separation methods, such as electrophoresis, can affect the fragment scoring process (LaHood et al. 2002). Even the inclusion of a "known" fragment (samples of expected size) and other molecular standards and ladders in the assay does not remove all ambiguity. Although rounding and averaging observed values to the nearest whole number is one way for a lab to minimize slight variation in reporting, it does not solve the problem.

An alternate approach developed to deal with the problem of variation in fragment scoring is called binning (Morin et al. 2010). Binning is broadly defined as pooling different fragment sizes as one allele. This can be done by setting binning rules based upon observed variation within a lab. The different fragments can be manually compiled from the raw data, and their distributions plotted. The data can then be placed into these distributions or categories. Any alleles that vary from the expected can be further examined by either repeating the PCR or sequencing or both.

There are also automated binning algorithms in software that are available. Guichoux et al. (2011) provide a list of automated binning software. Some require formatting, but most are easily performed by PC friendly packages that are found online, such as GENEPOP (Raymond and Rousset 1995) or PGDSpider (Lischer and Excoffier 2012). As in many other methods that are used for data collection, automated binning is not error proof and must be carefully evaluated. For example, human studies have reported error rates as high as 39.62 % by automated software (Ewen et al. 2000) and even higher rates in other human studies (Weeks et al. 2002).

Binning errors may occur for different reasons. For example, GC content may influence mobility (Wenz et al. 1998). Depending on the GC content in a repeat, the binning process may place alleles somewhere between rather than within a bin. Additionally, slight variation in alleles, perhaps due to a single mutation within the repeat or at the flanking region, can also place the fragment between bins. These challenges can have serious downstream consequences on decisions of population assignment or origin of an invasion based on those conclusions. This justifies careful oversight of the analyses and performance of the markers and binning and may require numerous quality control measures in the production, interpretation, and analysis of the data.

Eventually, the scored alleles must be analyzed using software developed for population genetics (Labate 2000; Excoffier and Heckel 2006). Commonly reported statistical measures include estimates of diversity (i.e., heterozygosity) for loci and populations. Also, it is common practice to estimate the genetic differences between predefined populations using fixation indices (F_{ST}) and AMOVA (e.g., Bonizzoni et al. 2001; Virgilio et al. 2010; Dogaç et al. 2013). These values can be used to calculate whether there is significant separation of populations based on the genetic markers. They are also a critical step in defining the reference data base and determining if there is adequate structure to conduct source estimation studies.

Once matrices of these data sets are generated, it is possible to construct genetic trees (i.e., dendograms) that reflect the variation among populations (e.g., Bonizzoni et al. 2004; Virgilio et al. 2013). If an invasive population is sampled, it can be included in an analysis along with the reference data. Possible sources of the invasive population can be explored based on statistical tests (e.g., F_{ST}) or patterns in a tree. Interpreting these results requires caution, because the sampled invasive population may not represent the true variation of the invading parental population, and tree-based methods only compare relative similarity among sampled populations.

For source estimation studies, the ideal unit of analysis is the individual and not the population from which it was sampled. Methods are available to study the genetic admixture of genotypes in a fly using programs like STRUCTURE (Pritchard et al. 2000). Software has been developed based on maximum likelihood and Bayesian statistical approaches to test whether an individual's genotype is consistent with genotypes sampled from source populations (Rannala and Mountain 1997; Manel et al. 2005; Cornuet et al. 2008; Csilléry et al. 2010). These methods can be used to test which of the sampled populations is the most likely source of a captured fly (Cornuet et al. 1999; Piry et al. 2004). Other methods can ask whether the fly's genotype is inconsistent with sources, thereby eliminating (or excluding) those populations as sources (Cornuet et al. 1999; Piry et al. 2004). This exclusion method has some advantages over other analytical methods, because the decision to exclude a source is not dependent on *a priori* sampling of the other populations.

5 Additional Methods for Source Estimation

This chapter is not intended to provide details on all technologies used for population analysis but rather the most commonly used molecular techniques for fruit fly source estimation studies. However, two categories of methods deserve brief attention here because of their growing importance: next generation sequencing (NGS) and stable isotope analysis (SIA).

5.1 Next Generation Sequencing (NGS)

NGS platforms can be used for screening large genomic datasets (Schuster 2008). As mentioned, the development of multi-locus nuclear markers allows researchers to view a snapshot of variation across an insect's genome and apply sophisticated statistical methods of analysis to determine its similarity or dissimilarity with other genomic information. Microsatellites can provide a good snapshot of variation, but, for practical reasons, traditional applications of the technique in fruit flies usually screen between just 10–20 loci and only observe variation of repetitive DNA. When the entire mitochondrial and nuclear genomes are considered, the number of potential polymorphic loci increases substantially. However, these new characters are not always more informative than traditional microsatellite markers, so careful evaluation is still required in the selection of markers (Liu et al. 2005).

Mitochondrial genomes are about 15,000–16,000 bases long. In comparison, nuclear genomes are considerably larger and measured in megabases (millions of bases, Mb) rather than kilobases (thousands of bases, Kb). An initial size estimate for the *C. capitata* genome was around 540 Mb (cited by Gomulski et al. 2008), but subsequent estimates set the range between 577 and 605 Mb (Peterson et al. 2009; Tsoumani and Mathiopoulos 2011). Estimates for genome sizes of other tephritid species range from 322 Mb (*B. oleae*, Tsoumani and Mathiopoulos 2011) to 619 Mb (*B. dorsalis*, Peterson et al. 2009). The sequencing of entire genomes is becoming a more commonplace activity, and several tephritid species have been identified as subjects for an initiative to complete genomes called the Insect 5000 Genomes Project (i5k; http://arthropodgenomes.org/wiki/i5K). However, as a more practical alternative to analyzing an entire genome, some studies have targeted only those parts of the genome that are expressed as determined by the presence of mRNA. This approach involves generating an Expressed Sequence Tag (EST) library of specific tissues. These have been reported for *C. capitata* (Scolari et al. 2012), *R. pomonella* (Schwarz et al. 2009), *B. oleae* (Tsoumani et al. 2011) and *B. dorsalis* (Shen et al. 2011) and can be used to understand how genes affect traits important for invasion biology (Gomulski et al. 2012; Zheng et al. 2012).

Such genomic resources reveal many polymorphic sites that could be useful when inferring the source of a fly. One estimate of variation among individuals using single base differences is commonly called single nucleotide polymorphism (SNP) analysis (reviewed by Gibson and Muse 2002). However, these large comparative DNA sequence data sets for SNP analysis can be expensive to generate and time consuming to assemble and edit. More recently, the development of NGS instruments (Glenn 2011) has made data generation more affordable and will undoubtedly require bioinformatics to be an essential skill for future fruit fly studies. NGS instruments can be used to sequence multiple populations (or species) to identify informative markers, such as microsatellites or SNPs (Abdelkrim et al. 2009; Guichoux et al. 2011). Once variable loci in the mitochondrial and nuclear genomes have been identified and PCR primers and conditions optimized for analysis, NGS technologies can also be used to efficiently screen large numbers of samples specifically for those loci. These data sets can then serve as a reference data base for source estimation. Additional techniques, such as restriction-site-associated DNA (RAD), have already been used in conjunction with NGS to conduct population-level studies (Emerson et al. 2010; Rowe et al. 2011). These new sequencing technologies, however, do pose some challenges for molecular diagnostic applications. For example, many NGS procedures require high quality DNA isolated from well preserved specimens. Most intercepted and trapped samples will not be well preserved. The amount of DNA required to process a NGS sample is also higher in comparison to the PCR-based mitochondrial and microsatellite diagnostics previously described. These factors should be considered in selecting tools for assignment tests.

5.2 *Stable Isotope Analysis (SIA)*

Unlike the previously described methods for source estimation, SIA is not based on genetic differences. Instead, SIA uses mass spectrometers and elemental analyzers to estimate the isotope mass ratio in samples to determine the likely source (Hood-Nowotny and Knols 2007). Many elements, like carbon and hydrogen, exist in more than one form in nature. These forms, called isotopes, are characterized by having different atomic weights. For example, Carbon-13 (^{13}C) is an isotope of Carbon-12. The isotope ^{12}C is comprised of six protons and six neutrons, while the rarer isotope ^{13}C has six protons and seven neutrons. Unlike those undergoing spontaneous radioactive decay (e.g., ^{14}C), atoms used for SIA have a very long half-life as high as 10^{18} years or more. The ratio of stable isotopes present in organisms can vary because of environmental and metabolic fractionation processes but importantly also because of the composition of their dietary resources which in the case of insect herbivores are geographically associated.

Hood-Nowotny and Knols (2007) reviewed the use of SIA for arthropod studies and noted the distinction between two types of studies. The first type is called natural abundance studies, which use naturally occurring isotopes to act as the diagnostic character. For example, the photosynthetic pathways used by plant species can result in different $^{13}C/^{12}C$ ratios (Tremblay and Paquin 2007), and differences in geology, ground water and precipitation, or anthropogenic influences through industry and agricultural, can result in locality-associated variation in isotope ratios of tissues via the food web. These differences can serve as a geographic diagnostic profile. This technology has been applied to bird ecology for over 20 years (Ingler and Bearhop 2008) but only relatively recently recognized as having potential for studying the migratory patterns and natal origins of insects (Wassenaar and Hobson 1998; Ouin et al. 2011). Of note has been the promising application to biosecurity (Holder et al. 2014). Here, a multivariate approach, using ratios of the light isotopes of hydrogen together with those of the biologically inactive heavy elements strontium (Sr) and lead (Pb) and a suite of trace element concentrations, enabled individual insects (*Helicoverpa armigera* (Hübner)) to be re-assigned with significantly improved confidence to their region of origin within eastern Australia or New Zealand. Unlike previous insect applications that have had additional *a priori* knowledge of migratory patterns, and therefore geography, and concern insects with monophagous diets, this now opens the possibility of determining the source of pests that are typically geographically wide spread, non-migratory, and polyphagous. This will be particularly useful post eradication to differentiate a new independent invader from the remnant population of a failed eradication or to determine the origins of a border intercept when the pathway is not obvious.

The second use of SIA is through enrichment studies that involve the addition of exogenous chemicals to diets. The introduced chemicals with known isotope ratios are incorporated into the organism via the diet during larval development and can then be used as a marker to track the adults. Hood-Nowotny et al. (2009) demonstrated that SIA can be used to discern lab reared *C. capitata* in the field from wild flies based on a difference in diet; wild flies having used a different carbon source (C3 plants) to those reared on artificial diet using sugar cane (C4 plant). A similar study on tsetse flies (*Glossina pallidipes* Austen) demonstrated that a laboratory fly fed on an enriched diet showed measurable differences to the natural abundance of isotopes of the light elements H, N, C, and O in wild flies (Hood-Nowotny et al. 2011).

There are limitations to the technology. As for DNA methods, SIA requires adequate sampling of populations. This ensures that differences in isotope ratios are geographically or source informative and that they are reliable estimators of a population by encompassing the variation that exists within it. Besides natural individual variation occurring during accrual of isotopes through larval feeding and development, variation within a population could be compounded by subsequent adult feeding on alternate food sources. Consequently, it is important to be able to measure the isotopic compounds incorporated into the permanent tissues (e.g., wings, cuticle, mouth parts) laid down at pupation and not those involved in regular metabolic processes. This is challenging but not impossible with new laser ablation methods becoming available. Also, given that isotopic profiles are not heritable, SIA is not suitable for tracking populations that might have crossed several generations which, unless they are reared on the same diet (natural or laboratory), will have different isotope profiles. So, unlike DNA markers, SIA is not able to evaluate ancestry and historical events. Rather, it is tailored to determine immediate geographic origins, which DNA is not. Finally, it is important to note that predictions of source based on SIA can be complex as generally accepted trends in isotope composition for geographic regions (e.g., C3 plants are more common temperate regions and C4 in subtropical regions) are not absolute and the physical composition of an ecosystem can vary as can the food sources of flies within a population.

6 Sampling and Data Set Management

Techniques that can reliably and efficiently estimate genetic or chemical differences of flies are needed to conduct source estimation studies. However, these methods can only provide useful information if a well-populated reference data base exists for comparison to the query sample. In our Introduction, we described the importance of sampling fly populations. A review of studies for species invasions using mitochondrial and chloroplast markers by Muirhead et al. (2008) identified a disturbing statistic in the published literature: "the number of individuals sampled per population and the number of populations surveyed has not increased" according to publications from 1994 to 2006. Based on that review, sample sizes of <6 individuals per population was common.

Generating a genetic data set for a globally distributed pest can be a monumental task. Luckily, there are ways to facilitate the generation of genetic resources through the coordination of research projects. One obvious way to enhance sample size is for international researchers to share information. This is particularly important when a collection of flies from multiple countries or continents is required to test the introduction pathway.

Some of the molecular methods previously described are more amenable to data sharing than others. For example, DNA barcoding projects are designed to facilitate data sharing (Ratnasingham and Hebert 2007). Posting DNA sequence records onto publically accessible databases, such as GenBank (http://www.ncbi.nlm.nih.gov/genbank/), provides access to these barcodes to the entire scientific community. These projects also contribute to data sharing by using a common molecular method of analysis. Even if the protocols used to generate the data are not identical, the sequences can be easily compiled and analyzed. Other molecular projects that use DNA sequences of gene regions that are currently not classified as DNA barcodes can also use repositories, such as GenBank, to disseminate data. However, without a concerted effort to generate data of the same locus, these resources cannot be compiled and co-analyzed. The release of DNA sequences to public databases is a common requirement for publication in scientific journals. Consequently, the infrastructure for reporting and sharing DNA sequence information is good.

The common practices used by researchers to report microsatellite data in publications are not ideal for data sharing. There is not a public repository, like GenBank, designed specifically for storing results from microsatellite analyses. When microsatellites are analyzed on a capillary instrument or polyacrylamide gel, the size of the alleles are scored and eventually recorded on spread sheets or text files. These are not included in publications. Although it is possible to name alleles based on estimated size, this is not a universal practice. As a result, published microsatellite results can simply be text indicating frequencies of arbitrary allele names. Without a key matching allele name with size estimates, it is not possible to merge data sets for co-analysis. Currently, researchers must contact authors of the original publications to obtain this information. The application of binning raw allele scores using software can further complicate comparison across laboratories. One positive development to increase accessibility of microsatellite information has been the recent use of data set repositories, like DRYAD (http://datadryad.org/), by scientists. Some journals, such as Molecular Ecology, now require the submission of molecular results to an accessible repository.

The development of standard reporting practices for microsatellites would benefit the fruit fly community. The ideal approach to recording alleles would be to sequence the DNA of all unique alleles that are identified by size. Sequencing a few microsatellite alleles is common practice during the locus discovery stage, but it is not done in subsequent studies that often detect new alleles. Providing the raw data on allele sizes would also stimulate data sharing and quality assurance. At a minimum, publications could report allele sizes and include analysis files as supplementary information. Additional information, such as primer modifications, would be useful to the research community.

The lack of mechanisms for reporting microsatellite information is not unique. There is a need for transparency in reporting other types of raw research data. Kilkenny et al. (2009) surveyed information from 271 publications on general biomedical studies (molecular and non-molecular) and found only four studies (1.5 %) reported raw data, and most involved small numbers of samples. Piwowar (2011) specifically addressed gene expression data sets and suggested that, while some improvements have been made to make data sets more accessible, only 45 % of the recent gene expression studies have their data available.

Managing specimen voucher collections for data sets is also important. In many cases, it is possible to isolate DNA from a fly's leg or by using non-destructive methods of DNA extraction. The remaining fly can be retained as a specimen voucher. If this is not possible, a series voucher (a specimen that acts as a voucher of the population) should be retained. These vouchers are important for fruit fly population studies because of the large number of species complexes associated with pests and difficulty in species identification. This is especially useful if the molecular methods used for source estimation can amplify DNA of many species. Consequently, if a population from a closely related species is mistakenly included in a study of a pest, the vouchers will be invaluable to identify or confirm the mistake. Likewise, vouchers of DNA samples are important to confirm taxonomy of suspect populations and transfer the technologies to other laboratories interested in implementing the markers.

7 Case Study: The Mediterranean Fruit Fly in California

Despite its moniker, the Mediterranean fruit fly or medfly, *C. capitata*, is native to sub-Saharan Africa and thought to have evolved in eastern or southern tropical Africa (De Meyer et al. 2004; Barr 2009). The species was described by Wiedemann in 1824 using specimens with a dubious locality record; the type material was collected during a voyage in 1790–1791 from Europe to India. It is likely that the flies originated from tropical fruit brought on board the ship as provisions, but there is no way yet to confirm that hypothesis (De Meyer 2000; De Meyer et al. 2004). Its common name is the result of an early invasion history. The fly was first recorded in Spain in 1842, and some of the oldest adventive populations dating from the nineteenth century are from the Mediterranean region (Metcalf 1995; White and Elson-Harris 1994). The geographic range of the pest expanded in the late nineteenth and early twentieth centuries with first records for Australia (1893), South America (1901), and Hawaii (1907) (Metcalf 1995). Introductions of the medfly have been an issue for fruit and vegetable producing and exporting countries ever since.

The fly is regarded as a major economic pest that negatively impacts crop yields and produce marketability. There are significant costs associated with running programs to manage, exclude, or eradicate the pest and introduction events can adversely impact the environment (Messing 1993; Siebert and Cooper 1995). It is highly polyphagous with the larvae capable of using over 200 host plant species (De Meyer et al. 2002). This estimate can vary, and citations of 350–400 host species have been reported when trap catches are included (Liquido et al. 1991, 1998, 2013). Copeland et al. (2002) provides information on indigenous hosts in Africa. The Mediterranean fruit fly's host-use acceptance and tolerance to new environments contribute to its invasive behavior (Copeland et al. 2002). The pest currently has established populations in the Mediterranean regions of Europe and North Africa, Middle East, Western Australia, Central America, South America, and Hawaii. Climatic models predict that the medfly can spread even further through accidental introductions and establish in areas currently free of the pest (Vera et al. 2002; De Meyer et al. 2008). Determining the geographic source of medfly interceptions and incursions using molecular information has been an important goal over the past 20 years. Although identifying the source of these flies should greatly benefit exclusion programs, the objective is complicated because of the large number of potential geographic and host sources. It is somewhat ironic that this species, described from material lacking geographic information and having a misnomer for a common name, has been the principal subject of fruit fly source estimation studies.

The medfly was first captured in the continental United States in Florida in the late 1920s. Subsequent incursions in the 1960s were in Florida and Texas. These detections were eradicated as part of coordinated response programs (Florida in 1929–1930 and 1962–1963, Texas in 1966). Although Mediterranean fruit flies were intercepted at ports of entry, the first incursion in California dates to 1975 (Metcalf 1995). That initial incursion of 77 fly captures was reported as eradicated in 1976 by state and federal programs. From 1980 to 1994 over 1,000 flies were captured in southern California, and a debate ensued regarding the hypothesis of a resident population in California (Carey 1991, 1995, 1996, 2010; Papadopoulos et al. 2013). There were significant costs associated with these fly detections (e.g., Penrose 1993, 1996). In 1994, a quarantine area in Ventura County of 130 km^2 had an estimated cost of $50 million in agricultural loss (Batkin 1995), and annual costs due to lost markets, yield, and management of an established medfly population in California were estimated in billions of dollars (Siebert and Cooper 1995; Morse et al. 1995). The cost of eradication efforts in California have been estimated near $60 million for 1989–1990 (Sheppard et al. 1992). In response to the first detection of the medfly in Mexico in 1977, Guatemala, Mexico, and the United States agreed to cooperate on the Moscamed Regional Program to prevent establishment of the pest (Salcedo Baca et al. 2010).

Determining the high risk pathways leading to Californian outbreaks and evaluating the success of eradication practices was, and still is, crucial to fruit fly programs. Source estimation methods that can answer the question of where the Californian flies originate are of paramount importance to determine risk pathways that will greatly assist pest exclusion and eradication efforts. The information can also be used to assess the success of such efforts if flies can be shown to be new arrivals as opposed to members of a latent local population. Unfortunately, at that time, few molecular methods were available to properly evaluate the question. McPheron et al. (1995) reviewed the molecular techniques being investigated in the early 1990s for the Mediterranean fruit fly.

7.1 Protein Methods

The earliest molecular estimates of medfly gene diversity were derived using allozymes (Gasperi et al. 1991; Malacrida et al. 1992). As noted by McPheron et al. (1995), the method identified the greatest variation in sub-Saharan African populations and comparatively low variation in introduced fly populations. However, the method was not ideal for inferring sources, because it required fresh samples and large sample sizes to estimate similarity among populations and relied on statistical analyses based on clustering of populations. An ideal method would need many informative enzymatic loci and a good reference data base for comparing genetic profiles of individuals, thus enabling use of a computational tool, such as IMMANC (Rannala and Mountain 1997), that can assign an individual (rather than a population of captures) to a source. However, no such resource has been developed. The closest has been an allozyme study by Malacrida et al. (1998) on the medfly colonization process. They did not analyze Californian populations, but using 26 enzyme loci and 17 populations, the method did demonstrate significant reductions in diversity in other non-native populations, including Hawaii and South America. This led them to support the generally accepted route of colonization out of Africa.

7.2 Mitochondrial DNA

Sheppard et al. (1992) were the first to estimate genetic diversity of medfly populations using mitochondrial DNA. This was estimated using the RFLP technique, and the co-authors detected two restriction enzymes that generated variation. Unfortunately, the initial technique they used was labor-intensive, because it required southern blot analysis of RFLP-treated DNA isolates. The two restriction markers were eventually converted into a PCR-RFLP assay to generate a more economical technique for source estimation studies (McPheron et al. 1994; Gasparich et al. 1995; Steck et al. 1996). The initial database of RFLP genotypes was generated from wild populations collected in Hawaii, Nigeria, Liberia, Venezuela, and Guatemala and lab colonies from collections made in Argentina, Guatemala, and Hawaii (Sheppard et al. 1992). Each collection was represented by at least ten flies. Despite limited information (only two restriction enzymes: *Xba* I and *EcoR* V), the mitochondrial markers were able to determine that Hawaii (sampled from seven collections) was not the source of two collections made in Los Angeles in 1989 and 1991.

The mitochondrial PCR–RFLP technique was further enhanced over the next few years by adding an additional restriction enzyme (*Mnl* I) and new populations (Gasparich et al. 1997). The protocol was used by multiple laboratories, and results reported as a three-letter code (representing the three restriction patterns generated by the enzymes). This format facilitated comparisons across laboratories. A fourth enzyme (*Hae* III) identified in the earlier study (Sheppard et al. 1992) had been examined as a possible additional marker, but the utility of that enzyme for source estimation has never been formally published. It is sometimes included as a fourth letter in the depiction of a composite RFLP haplotype (e.g., San Andrés et al. 2007), but inclusion of a fourth letter does not always equate with this marker (Lanzavecchia et al. 2008). Barr (2009) provided updated protocols for the RFLP procedure and an explanation on data interpretation. Although some samples included in the Gasparich et al. (1997) study had missing data, the new reference data base supported trends found in earlier studies: sub-Saharan Africa was the most diverse area and, as for the allozyme study of Malacrida et al. (1998), the other areas had lower diversity with one or two predominant genotypes for the Mediterranean (AAA and AAB), Central America (AAA), the north-western, Andean region of South America (AAB), eastern South America (BBB), and Hawaii (BBB).

In the Gasparich et al. (1997) study, 76 flies captured in California from 1975 to 1994 were genotyped using the PCR-RFLP method. Seventy-six per cent (58/76) of the flies generated results for all three enzymes. There was no evidence to support Hawaii, Venezuela, Brazil, or Argentina as sources. The AAA genotype was present in most captures, but the Mediterranean region, Central America, and Africa could not be excluded as possible sources. Four flies collected in 1992 had the AAB genotype. Consequently, the Mediterranean region, Andean region of South America, and Africa could not be excluded as their possible source. The method does not provide resolution to one country or even one geographic region. This is due in part to the limited number of characters in the RFLP markers. Although increased sampling of populations within countries would result in better estimates of genotype frequencies using the PCR-RFLP protocol, there was obviously a need for more informative markers or methods of analysis.

Lanzavecchia et al. (2008) published an analysis of medfly genotypes from Argentinian populations using RFLP and DNA sequence information. The markers were the Gasparich et al. (1997) mitochondrial loci ND4 and ND5. The DNA sequencing technique discovered new point mutations not observed using the RFLP technique. A similar study by Elfékih et al. (2008) used DNA sequencing analysis to observe greater variation in flies in Tunisia. A subsequent publication on Tunisian populations confirmed those trends (Elfékih et al. 2010). Barr (2009) reanalyzed a subset of the Gasparich et al. (1997) collections using the ND4 and ND5 loci and a fragment of the mitochondrial COI gene. That study revealed that many of the RFLP character states (e.g., AAA, AAB, BBB) did not detect additional genetic variation present in the mitochondrial DNA loci. For example, reanalysis of flies with the AAA genotype using DNA sequencing of a fragment of the ND4-ND5 locus generated eight unique genotypes (named M03, M04, M05, M06, M07, M08, M27, and M28). Two of these genotypes (M27 and M28) are not closely related to the others. This suggests that the RFLP method underestimates diversity and can generate misleading conclusions based on non-homologous states (because of convergence in the RFLP sites). This does not invalidate previous RFLP results but does suggest that newer methods should enable more informative conclusions.

The use of DNA sequencing for mitochondrial markers clearly results in a greater number of character states for analysis. However, do these newly discovered character states (reported as genotypes) provide greater geographic resolution than the RFLP states (e.g., AAA, AAB)? This depends on whether the DNA sequence genotypes are spatially structured. If the new DNA sequence-generated genotypes (M03-M08) occur in the same localities as the RFLP genotype (AAA), then the same sources can be excluded. If some sequencing genotypes (e.g., M03) are fixed or predominant for countries within Central America, then it might be possible to exclude countries or areas of Central America as sources. The data set published by Barr (2009) does not provide adequate sample sizes to test those associations. However, it does demonstrate that the technique is more powerful for detecting differences among flies and less prone to technical error. As a result, DNA sequencing should be preferable to RFLP methods when examining differences between fly captures from incursions and building future reference databases.

McPheron et al. (1995) summarized some of the inherent problems with inferring sources of incursions using mitochondrial markers (see Sect. 3 or Avise 2004). Regardless of the assay technique (RFLP or sequencing), mitochondrial data are based on a single estimate of the fly's history. The diversity of the mitochondrial genome is especially sensitive to population dynamics because of its smaller effective population size in comparison to the nuclear genome (Davies et al. 1999a). The effects of genetic drift can be greater on species exhibiting mating bias, such as lekking behavior in medfly that further reduces effective population size (Whittier et al. 1994). As a result, the study of colonization events (e.g., estimating the size of the founding event, time since colonization, and number of source populations) can be limited when relying on one DNA marker (Roderick and Villablanca 1996). The development of additional markers in the nuclear genome was identified as an important priority for fruit fly programs during a three-day work shop of the medfly in Riverside, California in 1994 (Morse et al. 1995).

7.3 Nuclear DNA

The first nuclear DNA markers generated for the medfly used the random amplified polymorphic DNA (RAPDs) technique (Haymer and McInnis 1994; Baruffi et al. 1995; Sonvico et al. 1996; Haymer et al. 1997). The most comprehensive RAPD analysis was conducted using two populations from Hawaii, four from

Guatemala, one from Greece, and one from Argentina (Haymer et al. 1997) as a reference data set. Genetic similarity values of four fly collections made in California from 1992 to 1994 were calculated in comparison to the reference data set. The results supported evidence of multiple introductions into California, because the RAPD genotype profile of the 12 flies collected from San Jose (northern California) in 1992 was dissimilar to the genotypic profiles of the other four Californian collections. A separate collection of 12 flies from Ventura in 1994 had a genetic profile different from the other Californian populations. The remaining two Californian populations were collected from the Los Angeles area in 1992 (9 flies) and 1993 (24 flies). These had genetic profiles that were more similar, according to PCO (principle coordinate) and AMOVA analyses, to the Guatemalan populations than to the Hawaiian, Greek, or Argentinian populations. The authors explained the limitation of using these dominant RAPD markers for inferring geographic sources because of issues with accurately estimating heterozygosity and using random (anonymous) gene regions. Based on genetic clustering in a phylogenetic tree, the RAPD data exclude Hawaii as a likely source for the tested flies. Guatemala, however, could not be excluded as a source. Unfortunately, the reference data set for the study was limited and precluded exclusion of other potential sources.

Research on the use of DNA introns as markers for the study of medfly population invasions was initiated in the 1990s (Roderick and Villablanca 1996). Introns were targeted as markers, because they were located in the nuclear genomes, could be used to estimate variation within an individual, and tend to have a higher rate of nucleotide variation as they are not under selective pressures as is protein encoding DNA. To examine variation of introns between individuals, the regions of DNA were sequenced. The method, however, was cumbersome requiring additional cloning experiments to confirm alleles of heterozygous individuals.

Villablanca et al. (1998) published the first study using medfly introns. The objective of the study was to determine if the four introns selected for analysis would provide greater diversity estimates than the mitochondrial RFLP and allozyme methods previously used. Molecular markers capable of detecting greater variation are better suited for understanding colonization events. This study found that introns had greater levels of variability than the tested mitochondrial markers generations after the introduction event. A subsequent publication by these researchers specifically addressed the utility of the introns by examining the fly's invasion history in California (Davies et al. 1999a).

Citing the results of Gasparich et al. (1997) to select likely pathways, Davies et al. (1999a) focused on just two sources for the Californian incursions: Central America and eastern South America. The 1997 mitochondrial study, however, did not provide sufficient evidence to exclude regions of the Mediterranean or sub-Saharan Africa as sources. Consequently, conclusions derived from studies, such as Davies et al. (1999a), that use limited geographic representation, must be treated with caution. A total of 76 flies from 11 geographic regions (California, Hawaii, Mexico, Guatemala, Costa Rica, Ecuador, Peru, Brazil, Greece, Malawi, and Kenya) were included. The data were used to examine population structure and to conduct an assignment test for a single fly captured in Burbank, CA in 1996. The intron data were able to detect significant population structuring within the Americas. Statistical comparison of the 1996 fly to a Californian population, including 11 flies captured in California from 1992 to 1994, rejected California as the source of the 1996 fly. It should be noted that the Californian flies analyzed represent a small number of the captures in 1992 and 1994 (195 and 400 flies, respectively). Comparison of the 1996 fly's genotype to the Latin American collections failed to exclude Costa Rica, Guatemala, Mexico, and Peru as possible sources. This study presents new powerful tools for source estimation that use rigorous statistical methods (assignment tests), but the authors also stated that "baseline data from source populations" is required to conduct these studies.

He and Haymer (1999) employed an alternate technique that used RFLP methods to score intron alleles. That analysis included samples from the four Californian populations included in the RAPD study of Haymer et al. (1997) plus an additional population of 11 flies captured in Walnut Park, CA, in 1997. Structure analyses support three distinct units for the Californian populations: one from northern CA in 1992, one including the southern CA populations collected in 1992–1994, and the new 1997 population. These results echoed previous evidence for multiple introductions into California. In addition, the data again supported excluding Hawaii as a source of these introductions. Of the included regions, Guatemala could not be excluded as a source of the Californian flies captured in 1992–1994. Thus far, the studies analyzed similar geographic regions, subsamples of the same introductions, and generated very similar results.

The molecular methods published in the 1990s provided alternate ways to estimate fly diversity. Although no technique or marker was able to identify the exact source of the Californian flies, these techniques yielded important information about what sources were not likely to be contributing to the pathway. Confirming results by using multiple technologies and markers ensures that the information is valid and that exclusion and eradication programs are moving in the right direction. By the late 1990s, microsatellites were also being developed as an additional method for genotyping and would enable the implementation of more powerful statistical procedures for assigning or excluding source populations (Davies et al. 1999b). Bonizzoni et al. (2000) subsequently presented the first protocol for microsatellites markers for medfly.

The first reference data base for medfly microsatellites was generated using ten loci (70 % are imperfect/complex repeat motifs) and 242 individuals collected from seven "established" populations: Argentina, Peru, Brazil, Ecuador, El Salvador, Guatemala, and Hawaii (Bonizzoni et al. 2001). An additional 20 flies were analyzed from a native population in Kenya. This data base was used to examine potential sources for 109 flies captured in California from 1992 to 1998 (Bonizzoni et al. 2001). The Californian flies represented ten different collections, with sample sizes ranging from 5 to 25 individuals. Initial analyses of genetic distances in trees revealed that the Californian flies were more similar to Guatemalan flies than the other populations. Assignment tests generated statistically significant values for six of the 109 flies, suggesting that they were immigrants. Guatemala was the proposed source for three of the flies. The other three were assigned as immigrants from another California population. The authors point out that a collection made in Walnut Park, CA 1997 was assigned as the source of two flies collected in 1993. Hence, good sampling and careful analysis are required to understand what the tests are actually reporting. Additional exclusion tests performed on the six flies could not reject Guatemala as a source for three flies and California as source for all six flies. The remaining 103 flies were not tested using the statistical exclusion method.

The Bonizzoni et al. (2001) study supported previous work by excluding Hawaii as a likely source of the Californian flies and observing an affinity between Guatemalan and Californian flies. Although Guatemala is not excluded as a source, the authors note that many regions of Central America and the rest of the world had not been tested. The reference database did not include any samples from the Mediterranean region, so it is not possible to conclude that Central America is a more likely source than countries of southern Europe or northern Africa. Another interesting result of the analysis was that, within the Los Angeles Basin, there was evidence of homogeneity among populations over time (1992–1997). These populations also shared a "private allele (*Ccmic7*, allele 142), at a high frequency." These observations were consistent with Carey's (1991, 1996) hypothesis of an endemic population in California. Another explanation for the data is that the populations are the result of reintroductions from a common source. Based on the limited sampling of global populations, it is not possible to discriminate between the resident hypothesis proposed by Carey and a hypothesis of repeat immigration from the same source using that microsatellite data.

A year later, Meixner et al. (2002) published a second study investigating the California invasion question using two distinct microsatellite loci (one simple and one compound repeat). Unlike Bonizzoni et al. (2001), this study also included data from the previously published RFLP mitochondrial marker system (Gasparich et al. 1997). A total of 359 flies from California were examined. The RFLP mitochondrial marker was successfully scored for 329 of those flies. These RFLP data were compared among California populations to evaluate hypotheses of multiple introductions and to the global data base (Gasparich et al. 1997) to evaluate potential sources. The microsatellite loci were successfully analyzed from 293 individuals and used to evaluate the hypothesis of multiple introductions in California. The study did not report a reference database of possible source populations for the two microsatellites, precluding the use of these markers for inferring the source of flies.

The flies included in the Meixner et al. (2002) study were from collections made from 1992 to 1999 and represent the largest published Californian genetic data set to date for the pest. Based on a combined analysis of mitochondrial and microsatellite information, the authors argue that there is evidence for multiple introductions into California. Their methods of analysis do not include statistical tests of assignment or exclusion using reference datasets but rather use observed genotype frequencies within Californian populations to identify unique populations. The majority of flies had the AAA RFLP genotype (reported for California by Gasparich et al. 1997) and shared microsatellite alleles. RFLP analysis uncovered genotype AAB from flies collected in 1992, 1997, and 1998, genotype AAC from a fly collected in 1999, and the genotype BBB from flies collected in 1993 and 1998. The microsatellite alleles supported separation of these flies and populations.

Prior to this study, only the AAA genotype was documented in California collections. The fly collected in 1997 with the AAB genotype is interesting, because it has a different origin to that of the other AAB flies when microsatellite markers are considered. This fly was previously included in the He and Haymer (1999) intron DNA study and determined to be the result of a novel introduction event. Despite all previous studies indicating that Hawaii is not the source of fly captures in California, the flies reported in the 2002 study with the BBB genotype are consistent with a Hawaiian source. The main outcome of the Meixner et al. (2002) study is that introduction pathways into California are more complex than previously thought.

Meixner et al. (2002) also noted that several populations collected from the LA Basin in 1992, 1993, 1994, 1997, 1998, and 1999 shared similar genotypes (AAA) and could be evidence of a persistent population resulting from one introduction. Although it is not possible to exclude the hypothesis of re-introductions resulting from a common source, the authors noted that this is a less likely explanation. Unfortunately, that judgment on the likelihood of a resident population assumes that all pathways are equally likely. That is, if multiple introductions occurred, then the probability that they came from the same source should be low. This may not be the case for the Mediterranean fruit fly. The main objective of each genetic study was to identify unlikely sources and then infer unlikely pathways. It is reasonable to expect some pathways to be higher risk because of frequent trade or travel, differences in prevalence of pest populations, variation in inspection practices, host fruit differences, and proximity. The problem at hand is identifying those pathways.

A similar problem of assigning risk to pathways *a priori* is also associated with the selection of source populations in studies. The earlier studies of world populations using mitochondrial genotypes (Gasparich et al. 1997) included the Mediterranean region. Initial results could not exclude that region as a source of the Californian flies. However, subsequent studies include few (e.g., Haymer et al. 1997) or no (e.g., Bonizzoni et al. 2001) collections from that region as possible sources. The exclusion of this region in analyses is most likely a result of sampling limitations rather than a conscious decision to remove the region as a possible source. This exclusion, however, implies that the region is not likely or not as likely as those regions that are included. Even if the Mediterranean region were assumed a less likely source based on non-genetic information, such as geographic distance, that information is not always reliable (e.g., Rubinoff et al. 2011). Liebhold et al. (2006) give information on airline baggage pathways for the medfly, and their analysis of interception records suggests that Europe and the Middle East could be possible pathways.

7.4 Synthesis

In 2002, Gasperi et al. published an excellent review on the genetics of medfly invasions and populations. Unfortunately, the review did not include the Meixner et al. (2002) results, which increased the number of possible sources for flies in California considerably. Malacrida et al. (2007) examined the multiple source models in the context of biological invasiveness of the species. So far, the questions of how many introduction events into California occurred and the sources of those introductions have not been fully answered. Although the studies reviewed here have deepened our understanding of the complexity, no subsequent genetic studies have been published that examine the issues of invasion and colonization in California. Similar studies were conducted on other invasive medfly populations in Florida (Silva et al. 2003) and Australia (Bonizzoni et al. 2004), but the California question has come to a genetic stand still. This is somewhat surprising, given that the statistical and molecular tools to address the problem have matured over the past decade (e.g., Bohonak et al. 2001; Manel et al. 2005; Csilléry et al. 2010). One of the major obstacles to these studies seems to be an old problem: availability of biological material needed for a reference data set. During the 3-day workshop held in Riverside, CA in 1994, the need for specimens to conduct source estimation studies was stressed (McPheron et al. 1995).

To facilitate work on the Mediterranean fruit fly, the United States Department of Agriculture—Animal and Plant Health and Inspection Service (APHIS) developed a medfly Germplasm Repository (MGR) at the Otis Laboratory, Buzzards Bay, Massachusetts in 1993. Information regarding the repository and progress in molecular diagnostics for the Medfly was released in the mid-1990s as an APHIS newsletter, called "The Cooler" (curated by Douglas Prasher). Collections generated by U.S. federal, state, and academic researchers could be stored at the MGR and used to support molecular programs (e.g., Haymer et al. 1997; Bonizzoni et al. 2001). Although this repository did not house all collections used for medfly studies (e.g., Bonizzoni et al. 2004; De Meyer et al. 2002), it is an example of how a collaborative approach can be used to address the problem of sampling. Despite efforts to share samples, not all publications using these samples used a standard system to track individuals. As a result, it is not always possible to determine if material analyzed by one lab using mitochondrial markers is the same material used to analyze a nuclear marker in a second lab. The location and size of the repository has changed over the years. It is currently maintained at the APHIS Mission Laboratory in Texas and includes material from newer collections (e.g., Barr et al. 2006).

The questions of whether a resident medfly population exists in California and if this population is the source of captures within state are still unresolved (Carey 2010; Liebhold et al. 2010; Papadopoulos et al. 2013). The hypothesis of a resident population has not been rejected using molecular or non-molecular data (Bonizzoni et al. 2001; Meixner et al. 2002; Carey 2010). Similarly, the "multiple pathways hypothesis" has not been rejected (Bonizzoni et al. 2001; Meixner et al. 2002;

Liebhold et al. 2006). The molecular data show that fly captures in California are the result of multiple introductions, but this result does not eliminate the possibility that a resident population(s) also occurs in California. Carey (2010) and Liebhold et al. (2010) agree that molecular data are needed to help tease apart these complicated invasion histories. As previously described, the omission of some medfly populations from the molecular studies has resulted in errors in subsequent interpretations. For example, Carey (2010) stated that there is no evidence of flies from Europe or the Middle East despite claims by Liebhold et al. (2006) that these sources made up 83 % of origins for flies identified as *C. capitata* and intercepted in Los Angeles, CA. In fact, the predominant mitochondrial genotype of flies in California is present in flies from the Mediterranean (Meixner et al. 2002). The Mediterranean region was not included as a possible source in the majority of molecular studies, including the work of Bonizzoni et al. (2001). The Middle East is not even sampled in most published studies. As a result, it is difficult to draw firm conclusions about private alleles found in California without knowing if they are truly unique to California. They could be present in other global populations. Additional sampling is required to support those interpretations.

In addition to the need to identify geographic sources, the implementation of SIT for the medfly has added another potential source, i.e., the laboratory. Sterilized flies that are released into the wild to suppress pest populations are typically labeled with a marking dye during the rearing process to distinguish them from wild flies (e.g., Rendón et al. 2004). This dye is important, because it allows identification of released and wild flies as well as determination of overflooding (sterile male:wild male) ratios. It is possible for sterile flies to lose that tracking dye during its lifespan and appear like a wild fly (Hood-Nowotny et al. 2009). As a result, it is useful to have chemical or genetic markers that can distinguish lab reared flies from wild flies. A mitochondrial RFLP genotype (AAAA) common to lab strains used for SIT has been reported as a possible marker (e.g., San Andrés et al. 2007), but that genotype is also present in wild populations. A formal study reporting on the genotypes frequencies and diagnostic utility has not yet been published but is forthcoming (Ruiz-Arce, unpublished).

7.5 Ongoing Activities

In 2007, the APHIS lab in Mission, Texas, initiated a program to enhance medfly collections and molecular methods. To that end, the program has acquired new collections and increased the capacity for storage of biological material of cooperators. During 2009–2011, Marc De Meyer (Royal Museum for Central Africa, Belgium) coordinated a collection program for APHIS that included 32 partners from 25 countries in Africa and the Mediterranean. During this time period, De Meyer's network of collaborators provided APHIS with a total of approximately 5,500 identified flies from 97 geographic sites in 26 different countries. Additional sets of collections were provided to APHIS by Bruce A. McPheron, Pennsylvania

State University. Numerous collections of medfly were transferred to the Mission Lab in September 2010. A total of approximately 23,000 medfly specimens (whole fly and extracted DNA) were received and documented by laboratory personnel in 2011. The APHIS data base includes many California fly captures and represents a valuable historical collection because many of these specimens had been analyzed previously with PCR-RFLP methods (Sheppard et al. 1992; Gasparich et al. 1997) and included in previous studies (Meixner et al. 2002; Silva et al. 2003). These methods are still in use by Plant Protection Organizations in the US and abroad. Other collections from Hawaii and Australia were also acquired and added to the collection.

The APHIS program has also completed and published a re-evaluation of the mitochondrial RFLP method first published in 1997 and demonstrated the utility of a DNA sequencing approach (Barr 2009). To further develop the DNA sequence technique into a routine diagnostic, a reference data base is currently being generated using nearly 2,000 flies from collections made around the world. Barr (2009) reported a total of 36 unique sequence types seen in his sampling of 114 flies for a sequencing fragment of 584 bp. Due to improved methodologies and chemistries at the Mission Lab and at commercial nucleic acid sequencing facilities, it was possible to extend this fragment by 100 bases to a total of 684 bp. This increase in size resulted in an increase in the number of haplotypes for the 114 flies. There are currently 212 haplotypes for this genetic locus. A total of 130 sequences were observed to be unique to a single collection or singletons. A greater portion, 108 (83 %), of the unique haplotypes was recovered from collections from Africa where this fly has been shown to originate.

The generation of so many genotypes poses a new problem for Medfly source estimation. How should the data be managed? If multiple researchers apply this method of analysis, there is a high probability of discovering new genotypes. There needs to be a process for naming and reporting genotypes, so that redundancy is avoided in the reference data base.

Another ongoing study by the APHIS lab is a re-evaluation of 23 published primer sets developed for microsatellite analysis of the medfly (Bonizzoni et al. 2000; Stratikopoulos et al. 2009). Each microsatellite primer set was tested using approximately 30 individuals from three geographic collections to optimize PCR performance and detect allele drop out effect and null alleles. Quality control is being performed on the markers by sequencing select PCR products to confirm the expected repeat motif, re-estimating heterozygosity, and examining adherence to biological expectations of microsatellite DNA. Based on testing, 14 of the initial 23 primer sets have been selected as preferred markers. The goal of this work is to select informative markers that can be reliably analyzed by multiple labs.

A reference data base for these microsatellites is being developed using the 1,982 flies genotyped for the mitochondrial marker. This is a substantial amount of information that represents a snapshot in time in the genetic diversity for many geographic collections of medfly. These raw data are important information for the current study but may also be of benefit to future studies. For example, the associated quality control information generated for these markers can serve as an excellent reference when transferring technologies across laboratories and during the development of novel markers.

One of the advantages of having a larger data set is that the substantial amount of data generated with the numerous loci provides an excellent opportunity to more accurately evaluate molecular markers. For the above study, over 1,600 flies from 121 geographic collections (excluding California and Florida) were used. This sampling was instrumental in effectively selecting the 14 most informative loci from a total of 23 microsatellite markers. A large data set improves the confidence level in many applications that test for heterozygosity, linkage, adherence to Hardy-Weinberg, and analyses that rely on probability estimates, such as assignment tests, that are important for pathway analysis.

As has been mentioned, the data set consisting of mtDNA sequences and information from 14 microsatellite loci for 1,982 medfly samples is formidable. There are substantial data that can be analyzed from many different perspectives. For example, those data can be partitioned to look at associations to temporal, ecological, as well as geographical (broad or fine-scale) with current and new methodologies soon to be developed. These data were gathered using conventional PCR, but it may be that very soon NGS will streamline the delivery and use of microsatellites and microsatellite methods to better enable such applications. It may also be that additional sampling will improve the pathway analysis for this pest. Not all areas where the medfly is known to occur have been sampled and are represented in the ongoing APHIS study; additional collections may reveal different and important trends in the view of the genetic diversity for this pest.

8 Conclusion

Estimating the original source population of a fruit fly is an important component of pathway analysis. To that end, biological and possibly environmentally derived molecules can help determine if intercepted and field trapped fruit flies originated from a particular geographic source population. Although proteins and isotopes have been explored as biological markers, most source estimation studies of tephritids compare variation among DNA molecules. The majority of these DNA-based methods examine either DNA sequence variation in the mitochondrial genome or size differences of microsatellite DNA in the nuclear genome. For some species, such as the Mediterranean fruit fly, a number of genetic marker systems have already been published, and the procedures for source estimation explained. Although similar methods have been applied to other fly pests, these techniques have not been converted into standard operating procedures for source estimation. The vast majority of published fruit fly studies simply report and characterize variation among populations.

Interpreting molecular information for source estimation can be a complicated task. Based on statistical analysis, it is possible to determine which sampled population is most similar to a trapped fly or determine if a sampled population can be excluded as a possible source of a trapped fly. Insufficient or biased sampling of populations, however, can lead to incorrect conclusions. In addition, the evolutionary history of different molecular markers can affect how results are to be interpreted. Marker systems that incorporate multiple sources of information (i.e., different genetic loci) and rely on well-sampled data bases are expected to result in more accurate estimates.

For the relatively well-studied Mediterranean fruit fly, a synthesis of information reported by independent research programs through the 1990s and 2000s provides a better picture of its invasion history than does any single study. However, the samples selected and loci tested in these studies were not always coordinated among research programs, thereby precluding a complete synthesis of results. As the field of source estimation progresses, researchers can benefit from lessons learned in past fruit fly programs. For example, the development of national and international collaborative working groups can help generate larger and better sampled reference collections. These collections are critical to the development of diagnostic technologies based on molecular variation in both space and time. In addition, the development of better practices for reporting data sets will facilitate data sharing and enhance technology development for many important pests. These improvements will hopefully facilitate the development of source estimation tools and help restrict the spread of economically significant fruit flies.

Acknowledgements We thank Evan Braswell, Matt Ciomperlik, and John Stewart for comments on our manuscript and Silvia Barr for assistance with editing the References. K. Armstrong was supported through the Better Border Biosecurity collaboration http://b3nz.org/.

References

Abdelkrim J, Robertson BC, Stanton JL, Gemmell NJ (2009) Fast, cost-effective development of species-specific microsatellite markers by genomic sequencing. Biotechniques 46:185–192. doi:10.2144/000113084

Aketarawong N, Bonizzoni M, Malacrida AR, Gasperi G, Thanaphum S (2006) Seventeen novel microsatellite markers from an enriched library of the pest species *Bactrocera dorsalis sensu stricto*. Mol Ecol Notes 6:1138–1140

Aketarawong N, Bonizzoni M, Thanaphum S, Gomulski LM, Gasperi G, Malacrida AR, Gugliemino CR (2007) Inferences on the population structure and colonization process of the invasive oriental fruit fly, *Bactrocera dorsalis* (Hendel). Mol Ecol 16:3522–3532

Aketarawong N, Chinvinijkul S, Orankanok W, Guglielmino CR, Franz G, Malacrida AR, Thanaphum S (2011) The utility of microsatellite DNA markers for the evaluation of area-wide integrated pest management using SIT for the fruit fly, *Bactrocera dorsalis* (Hendel), control programs in Thailand. Genetica 139:129–140

Anne C (2006) Choosing the right molecular genetic markers for studying biodiversity: from molecular evolution to practical aspects. Genetica 127:101–120

Armstrong KF, Ball SL (2005) DNA barcodes for biosecurity: invasive species identification. Philos Trans R Soc B 360:1813–1823. doi:10.1098/rstb.2005.1713

Arunkumar KP, Nagaraju J (2006) Unusually long palindromes are abundant in mitochondrial control regions of insects and nematodes. PLoS One 1(1):e110. doi:10.1371/journal.pone.0000110

Augustinos AA, Stratikopoulos EE, Zacharopoulou A, Sacharopoulou A, Mathiopoulos KD (2002) Polymorphic microsatellite markers in the olive fly, *Bactrocera oleae*. Mol Ecol Notes 2:278–280

Augustinos AA, Mamuris Z, Stratikopoulos EE, D'Amelio S, Zacharopoulou A, Mathiopoulos KD (2005) Microsatellite analysis of olive fly populations in the Mediterranean indicates a westward expansion of the species. Genetica 125:231–241

Augustinos AA, Stratikopoulos EE, Drosopoulou E, Kakani EG, Mavragani-Tsipidou P, Zacharopoulou A, Mathiopoulos KD (2008) Isolation and characterization of microsatellite markers from the olive fly, *Bactrocera oleae*, and their cross-species amplification in the Tephritidae family. BMC Genomics 9:618. doi:10.1186/1471-2164-9-618

Avise JC (2000) Phylogeography: the history and formation of species. Harvard University Press, Cambridge

Avise JC (2004) Molecular markers, natural history, and evolution, 2nd edn. Sinauer Associates, Inc., Sunderland

Baliraine FN, Bonizzoni M, Osir EO, Lux SA, Mulaa FJ, Zheng L, Gomulski LM, Gasperi G, Malacrida AR (2003) Comparative analysis of microsatellite loci in four fruit fly species of the genus *Ceratitis* (Diptera: Tephritidae). Bull Entomol Res 93:1–10

Barr NB (2009) Pathway analysis of *Ceratitis capitata* (Diptera: Tephritidae) using mitochondrial DNA. J Econ Entomol 102:401–411

Barr NB, McPheron BA (2006) Molecular phylogenetics of the genus *Ceratitis* (Diptera: Tephritidae). Mol Phylogenet Evol 38:216–230

Barr NB, Copeland RS, De Meyer M, Masiga D, Kibogo HG, Billah MK, Osir E, Wharton RA, McPheron BA (2006) Molecular diagnostics of economically important *Ceratitis* fruit fly species (Diptera: Tephritidae) in Africa using PCR and RFLP analyses. Bull Entomol Res 96:505–521

Barr NB, Islam MS, De Meyer M, McPheron BA (2012) Molecular identification of *Ceratitis capitata* (Diptera: Tephritidae) using DNA sequences of the COI barcode region. Ann Entomol Soc Am 105(2):339–350

Baruffi L, Damiani G, Gugliemo CT, Gugliemo CT, Bandi C, Malacrida AR, Gasperi G (1995) Polymorphism within and between populations of *Ceratitis capitata*: comparison between RAPD and multilocus enzyme electrophoresis data. J Hered 74:425–437

Batkin T (1995) Priorities for exotic pest research from the perspective of production agriculture. In: Morse JG, Metcalf RL, Carey JR, Dowell RV (eds) The Medfly in California: defining critical research. Proceedings of meeting in Riverside, CA, 9–11 Nov 1994. College of Natural and Agricultural Sciences, University of California, Riverside

Bensasson D, Zhang D-X, Hartl DL, Hewitt GM (2001) Mitochondrial pseudogenes: evolution's misplaced witnesses. Trend Ecol Evol 16(6):314–321

Blacket MJ, Semeraro L, MALIPATIL MB (2012) Barcoding Queensland fruit flies (*Bactrocera tryoni*): impediments and improvements. Mol Ecol Resour 12:428–436. doi:10.1111/j.1755-0998.2012.03124.x

Bohonak AJ, Davies N, Villablanca FX, Roderick GK (2001) Invasion genetics of new world medflies: testing alternative colonization scenarios. Biol Invasions 3:103–111

Bonizzoni M, Malacrida AR, Guglielmino CR, Gomulski LM, Gasperi G, Zheng L (2000) Microsatellite polymorphism in the Mediterranean fruit fly, *Ceratitis capitata*. Insect Mol Biol 9:251–261

Bonizzoni M, Zheng L, Guglielmino CR, Haymer DS, Gasperi G, Gomulski LM, Malacrida AR (2001) Microsatellite analysis of medfly bioinfestations in California. Mol Ecol 10:2515–2524

Bonizzoni M, Katsoyannos BI, Marguerie R, Guglielmino CR, Gasperi G, Malacrida A, Chapman T (2002) Microsatellite analysis reveals remating by wild Mediterranean fruit fly females, *Ceratitis capitata*. Mol Ecol 11:1915–1921

Bonizzoni M, Guglielmino CR, Smallridge CJ, Gomulski M, Malacrida AR, Gasperi G (2004) On the origins of medfly invasion and expansion in Australia. Mol Ecol 13:3845–3855

Boore JL (1999) Animal mitochondrial genomes. Nucleic Acids Res 27(8):1767–1780

Boykin LM, Shatters RG Jr, Hall DG, Burns RE, Franqui RA (2006) Analysis of host preference and geographical distribution of *Anastrepha suspensa* (Diptera: Tephritidae) using phylogenetic analyses of mitochondrial cytochrome oxidase I DNA sequence data. Bull Entomol Res 96:457–469. doi:10.1079/BER2006438

Boykin LM, Shatters RG Jr, Hall DG, Dean D, Beerli P (2010) Genetic variation of *Anastrepha suspensa* (Diptera: Tephritidae) in Florida and the Caribbean using microsatellite DNA Markers. J Econ Entomol 103(6):2214–2222

Boykin LM, Schutze MK, Krosch MN, Chomic A, Chapman TA, Englezou A, Armstrong KF, Clarke AR, Hailstones D, Cameron SL (2013) Multi-gene phylogenetic analysis of south-east Asian pest members of the *Bactrocera dorsalis* species complex (Diptera: Tephritidae) does not support current taxonomy. J Appl Entomol. doi:10.1111/jen.12047

Brinkmeyer-Langford CL, Cai JJ, Gill CA, Skow LC (2012) Microsatellite variation in the equine MHC. Anim Genet 44(3):267–275

Broquet T, Ménard N, Petit E (2007) Noninvasive population genetics: a review of sample source, diet, fragment length and microsatellite motif effects on amplification success and genotyping error rates. Conserv Genet 8:249–260

Brower AVZ (1994) Rapid morphological radiation and convergence among races of the butterfly *Heliconius erato* inferred from patterns of mitochondrial DNA evolution. Proc Natl Acad Sci U S A 91:6491–6495

Butler JM (2005) Forensic DNA typing. Elsevier Academic Press, London

Cameron EC, Sved JA, Gilchrist AS (2010) Pest fruit fly (Diptera: Tephritidae) in northwestern Australia: one species or two? Bull Entomol Res 100:197–206

Carey JR (1991) Establishment of the Mediterranean fruit fly in California. Science 253:1369–1373

Carey JR (1995) The Mediterranean fruit fly invasion of southern California. In: Morse JG, Metcalf RL, Carey JR, Dowell RV (eds) The Medfly in California: defining critical research. In: Proceedings of meeting in Riverside, CA, 9–11 Nov 1994. College of Natural and Agricultural Sciences, University of California, Riverside, CA

Carey JR (1996) The incipient Mediterranean fruit fly population in California: implications for invasion biology. Ecology 77:1690–1697

Carey JR (2010) The Mediterranean fruit fly *(Ceratitis capitata).* Am Entomol 56(3):158–163

Carlsson J (2008) Effects of microsatellite null alleles on assignment testing. J Hered 99(6):616–623

Casey DG, Burnell AM (2001) The isolation of microsatellite loci in the Mediterranean fruit fly *Ceratitis capitata* (Diptera: Tephritidae) using a biotin/streptavidin enrichment technique. Mol Ecol Notes 1:120–122

Chambers EW, Meece JK, McGowan JA, Lovin DD, Hemme RR, Chadee DD, McAbee K, Brown SE, Knudson DL, Severson DW (2007) Microsatellite isolation and linkage group identification in the yellow fever mosquito *Aedes aegypti*. J Hered 98(3):202–210

Chen YH, Opp SB, Berlocher SH, Roderick GK (2006) Are bottlenecks associated with colonization? Genetic diversity and diapause variation of native and introduced *Rhagoletis completa* populations. Oecologia 149:656–667

Colwell RK, Mao CX, Chang J (2004) Interpolating, extrapolating, and comparing incidence-based species accumulation curves. Ecology 85:2717–2727

Colwell RK, Chao A, Gotelli NJ, Lin S-Y, Mao CX, Chazdon RL, Longino JT (2012) Models and estimators linking individual-based and sample-based rarefaction, extrapolation and comparison of assemblages. J Plant Ecol 5:3–21

Copeland RS, Wharton RA, Luke Q, De Meyer M (2002) Indigenous hosts of *Ceratitis capitata* (Diptera: Tephritidae) in Kenya. Ann Entomol Soc Am 95:672–694

Cornuet J-M, Piry S, Luikart G, Estoup A, Solignac M (1999) New methods employing multilocus genotypes to select or exclude populations as origins of individuals. Genetics 153:1989–2000

Cornuet J-M, Sanntos F, Beaumont MA, Robert CP, Marin J-M, Balding DJ, Guillemaud T, Estoup A (2008) Inferring population history with DIY ABC: a user-friendly approach to approximate Bayesian computation. Bioinformatics 24:2713–2719

Csilléry K, Blum MGB, Gaggiotti OE, François O (2010) Approximate Bayesian Computation (ABC) in practice. Trend Ecol Evol 25:410–418

Dai SM, Lin CC, Chang C (2004) Polymorphic microsatellite DNA markers from the oriental fruit fly *Bactrocera dorsalis* (Hendel). Mol Ecol Notes 4:629–631. doi:10.1111/j.1471-8286.2004.00732.x

Davies N, Villablanca FX, Roderick GK (1999a) Bioinvasions of the Mediterranean fruit fly *Ceratitis capitata*: source estimation using DNA sequences at multiple intron loci. Genetics 153:351–360

Davies N, Villablanca FX, Roderick GK (1999b) Determining the source of individuals: multilocus genotyping in nonequilibrium population genetics. Trend Ecol Evol 14:17–21

De Meyer M (2000) Systematic revision of the subgenus *Ceratitis* MacLeay *s. s.* (Diptera: Tephritidae). Zool J Linn Soc 128:439–467

De Meyer M, Copeland RS, Lux SA, Mansell M, Quilici S, Wharton R, White IM, Zenz NJ (2002) Annotated check list of host plants for afrotropical fruit flies (Diptera: Tephritidae) of the genus *Ceratitis*. Documentations Zoologiques, Musee Royal de l'Afrique Centrale 27:1–91

De Meyer M, Copeland RS, Wharton RA, McPheron BA (2004) On the geographical origin of the medfly, *Ceratitis capitata* (Wiedemann). In: Barnes B (ed) Proceedings of the 6th international symposium on fruit flies of economic importance. Isteg Scientific Publications, Irene, pp 45–53

De Meyer M, Robertson MP, Peterson AT, Mansell MW (2008) Ecological niches and potential distributions of Mediterranean fruit fly (*Ceratitis capitata*) and Natal fruit fly (*Ceratitis rosa*). J Biogeogr 35:270–281

Delatte H, Virgilio M, Simiand C, Quilici S, De Meyer M (2013) Isolation and characterisation of sixteen microsatellite markers cross-amplifying in a complex of three African agricultural pests (*Ceratitis rosa, C. anonae and C. fasciventris*, Diptera: Tephritidae). Dig Conserv Genet Resour. doi:10.1007/s12686-012-9722-6

DeSalle R, Giddings LV (1986) Discordance of nuclear and mitochondrial DNA phylogenies in Hawaiian *Drosophila*. Proc Natl Acad Sci U S A 83:6902–6906

Dewoody J, Nason JD, Hipkins VD (2006) Mitigating scoring errors in microsatellite data from wild populations. Mol Ecol Notes 6(4):951–957

Dogaç E, Kandemir İ, Taskin V (2013) The genetic polymorphisms and colonization process of olive fly populations in Turkey. PLoS One 8(2):e56067. doi:10.1371/journal.pone.0056067

Drosopoulou E, Augustinos AA, Nakou I, Koeppler K, Kounatidis I, Vogt H, Papadopoulos T, Bourtzis K, Mavragani-Tsipidou P (2011) Genetic and cytogenetic analysis of the American cherry fruit fly, *Rhagoletis cingulata* (Diptera: Tephritidae). Genetica 139:1449–1464

Egeland T, Salas A (2008) Estimating haplotype frequency and coverage of databases. PLoS One 3(12):e3988. doi:10.1371/journal.pone.0003988

Elfékih S, Haymer DS, Mediouni J, Hajlaoui MR, Makni M (2008) Survey of mitochondrial DNA markers in populations of *Ceratitis capitata* (Wiedemann) from Tunisia. In: Sugayama RL, Zucchi RA, Ovruski SM, Sivinski J (eds) 7th international symposium on fruit flies of economic importance: fruit flies of economic importance: from basic to applied knowledge, Salvador, Brazil, 10–15 Sept 2006, Biofábrica Moscamed, Brazil

Elfékih S, Makni M, Haymer DS (2010) Detection of novel mitochondrial haplotype variants in populations of the Mediterranean fruit fly, *Ceratitis capitata*, from Tunisia, Israel and Morocco. J Appl Entomol 134:647–651

Emerson K, Merz CR, Catchen JM, Hohenlohe PA, Cresko WA, Bradshaw WE, Holzapfel CM (2010) Resolving postglacial phylogeography using high-throughput sequencing. Proc Natl Acad Sci U S A 107(37):16196–16200

Enkerlin W, López L, Celedonio H (1996) Increased accuracy in discrimination between captured wild unmarked and released dye-marked adults in fruit fly (Diptera: Tephritidae) sterile released programs. J Econ Entomol 89(4):946–949

Estoup A, Garnery L, Solignac M, Cornuet J-M (1995) Microsatellite variation in honey bee (*Apis mellifera* L.) populations: hierarchical genetic structure and test of the infinite allele and stepwise mutation models. Genetics 140:679–695

Estoup A, Wilson IJ, Sullivan C, Cornuet J-M, Moritz C (2001) Inferring population history from microsatellite and enzyme data in serially introduced cane toads, *Bufo marinus*. Genetics 159:1671–1687

Evett IW, Weir BS (1998) Interpreting DNA evidence: statistical genetics for forensic scientists. Sinauer Associates, Sunderland

Ewen KR, Bahlo M, Treloar SA, Levinson DF, Mowry B, Barlow JW, Foote SJ (2000) Identification and analysis of error types in high-throughput genotyping. Am J Human Genet 67:727–736

Excoffier L, Heckel H (2006) Computer programs for population genetics data analysis: a survival guide. Nat Rev 7:745–758

Excoffier L, Smouse PE, Quattro JM (1992) Analysis of molecular variance inferred from metric distances among DNA haplotypes: application to human mitochondrial DNA restriction data. Genetics 131:479–491

Fazekas AJ, Steeves R, Newmaster SG (2010) Improving sequencing quality from PCR products containing long mononucleotide repeats. Biotechniques 48:277–285

Frewin A, Scott-Dupree C, Hanner R (2013) DNA barcoding for plant protection: application and summary of available data for arthropod pests. CAB Rev 8:1–13

Fritz AH, Schable N (2004) Microsatellite loci from the Caribbean fruit fly, *Anastrepha suspensa* (Diptera: Tephritidae). Mol Ecol Note 4:443–445

Fritz AH, Dhakal P, Fritz GN, Kirby LA (2010) Isolating, amplifying and quantifying sperm DNA in *Anastrepha suspensa* (Diptera: Tephritidae). Fla Entomol 93:63–72

Frohman MA, Dush MK, Martin GR (1988) Rapid production of full-length cDNAs from rare transcripts: amplification using a single gene-specific oligonucleotide primer. Proc Natl Acad Sci U S A 85:8998–9002

Gasparich GE, Sheppard WS, Han H-Y, McPheron BA, Steck GJ (1995) Analysis of mitochondrial DNA and development of PCR-based diagnostic markers for Mediterranean fruit fly (*Ceratitis capitata*) populations. Insect Mol Biol 4:61–67

Gasparich GE, Silva JG, Han H-Y, McPheron BA, Steck GJ, Sheppard WS (1997) Population genetic structure of the Mediterranean fruit fly (Diptera: Tephritidae) and implications for worldwide colonization patterns. Ann Entomol Soc Am 90:790–797

Gasperi G, Guglielmino CR, Malacrida AR, Milani R (1991) Genetic variability and gene flow in geographical populations of *Ceratitis capitata*. Heredity 67:347–356

Gasperi G, Bonizzoni M, Gomulski LM, Murelli V, Torti C, Malacrida AR, Guglielmino CR (2002) Genetic differentiation, gene flow and the origin of infestations of the medfly, *Ceratitis capitata*. Genetica 116:125–135

Gibson G, Muse SV (2002) A primer of genome science. Sinauer, Sunderland

Gilchrist AS, Ling AE (2006) DNA microsatellite analysis of naturally occurring colour intermediates between *Bactrocera tryoni* (Froggatt) and *Bactrocera neohumeralis* (Hardy) (Diptera: Tephritidae). Aust J Entomol 45:157–162

Gilchrist AS, Meats AW (2010) The genetic structure of populations of an invading pest fruit fly, *Bactrocera tryoni*, at the species climatic range limit. Heredity 105:165–172

Gilchrist AS, Sved JA, Meats A (2004) Genetic relations between outbreaks of the Queensland fruit fly, *Bactrocera tryoni* (Froggatt) (Diptera: Tephritidae), in Adelaide in 2000 and 2002. Aust J Entomol 43:157–163

Glenn TC (2011) Field guide to next-generation DNA sequencers. Mol Ecol Resour 1–11

Goldstein DB, Schlötterer C (eds) (1999) Microsatellites evolution and application. Oxford University Press, New York

Gomulski LM, Dimopoulos G, Xi Z, Soares MB, Bonaldo MF, Malacrida AR, Gasperi G (2008) Gene discovery in an invasive tephritid model pest species, the Mediterranean fruit fly, *Ceratitis capitata*. BMC Genomics 9:243. doi:10.1186/1471-2164-9-243
Gomulski LM, Dimopoulos G, Xi Z, Scolari F, Gabrieli P, Siciliano P, Clarke AR, Malacrina AR, Gasperi G (2012) Transcriptome profiling of sexual maturation and mating in the Mediterranean fruit fly, *Ceratitis capitata*. PLoS One 7(1):e30857. doi:10.1371/journal.pone.0030857
Guichoux E, Lagache L, Wagner S, Chaumeil P, Léger P, Lepais O, Lepoittevin C, Malausa T, Revardel E, Salin F, Petit RJ (2011) Current trends in microsatellite genotyping. Mol Ecol Res 11:591–611
Guo SW, Thompson EA (1992) Performing the exact test of Hardy-Weinberg proportion for multiple alleles. Biometrics 48(2):361–372
Hancock JM (1999) Microsatellites and other simple sequences: genomic context and mutational mechanisms. In: Goldstein DB, Schlötterer C (eds) Microsatellites evolution and application. Oxford University Press, New York
Harr B, Zangerl B, Brem G, Schlötterer C (1998) Conservation of locus-specific microsatellite variability across species: a comparison of two *Drosophila* sibling species, *D. melanogaster* and *D. simulans*. Mol Biol Evol 15(2):176–184
Hartl DL, Clark AG (1997) Principles of population genetics. Sinauer Associates, Sunderland
Hasan A-U, Suguria S, Ahmed SMU, Fujimotoa C, Haradaa M, Rahmanb SM, Zamane RU, Kakehi Y (2009) Molecular phylogeography of *Culex quinquefasciatus* mosquitoes in central Bangladesh. Acta Trop 112:106–114
Haymer DS, McInnis DO (1994) Resolution of populations of the Mediterranean fruit fly at the DNA level using random primers for the polymerase chain reaction. Genome 37:244–248
Haymer DS, He M, McInnis DO (1997) Genetic marker analysis of spatial and temporal relationships among existing populations and new infestations of the Mediterranean fruit fly, *Ceratitis capitata*. Heredity 79:302–309
He M, Haymer DS (1999) Genetic relationships of populations and the origins of new infestations of the Mediterranean fruit fly. Mol Ecol 8:1247–1257
Hill CR, Butler JM, Vallone PM (2009) A 26plex Autosomal STR assay to aid human identity testing. J Forensic Sci 54:1008–1015
Holder PW, Armstrong K, Van Hale R, Millet M-A, Frew R, Clough TJ, Baker J (2014) Biogeochemical markers as natal origin markers of *Helicoverpa armigera*—a model for biosecurity pests. PLoS One 9(3):e92384. doi:10.1371/journal.pone.0092384
Holland MM, Parsons TJ (1999) Mitochondrial DNA sequence analysis—validation and use for forensic casework. Forensic Sci Rev 11:22–50
Hood-Nowotny R, Knols BGJ (2007) Stable isotope methods in biological and ecological studies of arthropods. Entomol Exp Appl 124:3–16
Hood-Nowotny R, Mayr L, Islam A, Robinson A, Cáceres C (2009) Routine isotope marking for the Mediterranean fruit fly (Diptera: Tephritidae). J Econ Entomol 102(3):941–947
Hood-Nowotny R, Watzka M, Mayr L, Mekonnen S, Kapitano B, Parker A (2011) Intrinsic and synthetic stable isotope marking of tsetse flies. J Insect Sci 11:79
Hopkins SE, Taylor DL (2011) Microsatellite loci development in mycoheterotrophic *Corallorhiza maculata* (Orchidaceae) with amplification in *C. mertensiana*. Am J Bot 98(9): e253–e255
Hu J, Zhang JL, Nardi F, Zhang RJ (2008) Population genetic structure of the melon fly, *Bactrocera cucurbitae* (Diptera: Tephritidae), from China and Southeast Asia. Genetica 134:319–324. doi:10.1007/s10709-007-9239-1
Ingler R, Bearhop S (2008) Applications of stable isotope analyses to avian ecology. Ibis 150:447–461
Islam MS, Ruiz-Arce R, McPheron BA (2011) Microsatellite markers for the West Indian fruit fly (*Anastrepha obliqua*) and cross species amplification in related pest species. Conserv Genet Resour. doi:10.1007/s12686-011-9401-z

Jacquard C, Virgilio M, David P, Quilici S, De Meyer M, Delatte H (2013) Population structure of the melon fly, *Bactrocera cucurbitae*, in Reunion Island. Biol Invasions. doi:10.1007/s10530-012-0324-8

Jeffreys AJ, Wilson V, Thein SL (1985) Hypervariable 'minisatellite' regions in human DNA. Nature 314:67–73

Johannesen J, Keyghobadi N, Schuler H, Stauffer C, Vogt H (2013) Invasion genetics of American cherry fruit fly in Europe and signals of hybridization with the European cherry fruit fly. Entomol Exp Appl 147:61–72. doi:10.1111/eea.12041

Juan-Blasco M, Sabater-Muñoz B, Argilés R, Jacas JA, Castañera P, Urdaneja A (2013) Molecular tools for sterile sperm detection to monitor *Ceratitis capitata* populations under SIT programmes. Pest Manag Sci. doi:10.1002/ps.3448

Karsten M, van Vuuren BJ, Barnaud A, Terblanche JS (2013) Population genetics of *Ceratitis capitata* in South Africa: implications for dispersal and pest management. PLoS One 8(1): e54281. doi:10.1371/journal.pone.0054281

Khamis M, Karam N, Guglielmino R, Ekesi S, Masiga D, De Meyer M, Kenya EU, Malacrida AR (2008) Isolation and characterization of microsatellite markers in the newly discovered invasive fruit fly pest in Africa, *Bactrocera invadens* (Diptera: Tephritidae). Mol Ecol Res 8:1509–1511. doi:10.1111/j.1755-0998.2008.02335.x

Khamis M, Karam N, Ekesi S, De Meyer M, Bonomi A, Gomulski LM, Scolari F, Gabrieli P, Siciliano P, Masiga D, Kenya EU, Gasperi G, Malacrida AR, Guglielmino CR (2009) Uncovering the tracks of a recent and rapid invasion: the case of the fruit fly pest *Bactrocera invadens* (Diptera: Tephritidae) in Africa. Mol Ecol. doi:10.1111/j.1365-294X.2009.04391.x

Khamis FM, Masiga DK, Mohamed SA, Salifu D, de Meyer M, Ekesi S (2012) Taxonomic identity of the invasive fruit fly pest, *Bactrocera invadens*: concordance in morphometry and DNA barcoding. PLoS One 7(9):e44862. doi:10.1371/journal.pone.0044862

Kilkenny C, Parsons N, Kadyszewski E, Festing MF, Cuthill IC, Fry D, Hutton J, Altman DG (2009) Survey of the quality of experimental design, statistical analysis and reporting of research using animals. PLoS One 4(11):e782. doi:10.1371/journal.pone.0007824

Kinnear MW, Bariana HS, Sved JA, Frommer M (1998) Polymorphic microsatellite markers for population analyses of a tephitid pest species, *Bactrocera tryoni*. Mol Ecol 7:1489–1495

Knowles LL, Maddison WP (2002) Statistical phylogeography. Mol Ecol 2623–2635

Krosch MN, Schutze MK, Armstrong KF, Boontop Y, Boykin LM, Chapman TA, Englezou A, Cameron SL, Clarke AR (2013) Piecing together an integrative taxonomic puzzle: microsatellite, wing shape and aedaegus length analyses of *Bactrocera dorsalis s.l.* (Diptera: Tephritidae) find no evidence of multiple lineages in a proposed contact zone along the Thai/Malay Peninsula. Syst Entomol 38:2–13

Kruglyak S, Durrett RT, Schug MD, Aquadro CF (1998) Equilibrium distributions of microsatellite repeat length resulting from a balance between slippage events and point mutations. Proc Natl Acad Sci U S A 95:10774–10778

Labate JA (2000) Software for population genetic analyses of molecular marker data. Crop Sci 40 (6):1521–1528

LaHood ES, Moran P, Olsen J, Grant WS, Park LK (2002) Microsatellite allele ladders in two species of Pacific salmon: preparation and field-test results. Mol Ecol Notes 2:187–190

Lanzavecchia SB, Cladera JL, Faccio P, Petit Marty N, Vilardi JC, Zandomeni RO (2008) Origin and distribution of *Ceratitis capitata* mitochondrial DNA haplotypes in Argentina. Ann Entomol Soc Am 101:627–638

Li W, Yang L, Tang K, Zeng L, Liang G-W (2007) Microsatellite polymorphism of *Bactrocera dorsalis* (Hendel) populations in China. Acta Entomologica Sinica 50(12):1255–1262

Liebhold AM, Work TT, McCullough DG, McCullough DG, Cavey JF (2006) Airline baggage as a pathway for alien insect species invading the United States. Am Entomol 52:48–54

Liebhold AM, Work TT, McCullough DG, Cavey JF (2010) Response to Carey. Am Entomol 56:164–165

Lim P-E, Tan J, Suana IW, Eamsobhana P, Yong HS (2012) Distinct genetic lineages of *Bactrocera caudata* (Insecta: Tephritidae) revealed by COI and 16S DNA sequences. PLoS One 7(5):e37276. doi:10.1371/journal.pone.0037276

Liquido NJ, Shinoda LA, Cunningham RT (1991) Host plants of the Mediterranean fruit fly (Diptera: Tephritidae): an annotated world review. Misc Publ Entomol Soc Am 77:1–52

Liquido NJ, Barr PG, Cunningham RT (1998) MEDHOST: an encyclopedic bibliography of the host plants of the Mediterranean fruit fly, *Ceratitis capitata* (Wiedemann), Version 1. In: Thompson FC (ed) Fruit fly expert identification system and systematic information database. Diptera data dissemination disk 1. USDA, Washington, DC

Liquido NJ, McQuate GT, Suiter KA (2013) MEDHOST: an encyclopedic bibliography of the host plants of the Mediterranean fruit fly, *Ceratitis capitata* (Wiedemann), Version 1.1. United States Department of Agriculture, Center for Plant Health Science and Technology, Raleigh. https://www.gpdd.info/MedHost/

Lischer HE, Excoffier L (2012) PGDSpider: an automated data conversion tool for connecting population genetics and genomics programs. Bioinformatics 28(2):298–299

Liu N, Chen L, Wang S, Oh C, Zhao H (2005) Comparison of single-nucleotide polymorphisms and microsatellites in inference of population structure. BMC Genet 6(Suppl 1):S26. doi:10.1186/1471-2156-6-S1-S26

Liu J, Shi W, Hui Ye H (2007) Population genetics analysis of the origin of the oriental fruit fly, *Bactrocera dorsalis* Hendel (Diptera: Tephritidae), in northern Yunnan Province, China. Entomolog Sci 10:11–19. doi:10.1111/j.1479-8298.2006.00194.x

Lopes DM, da Silva FO, Fernandes-Salomão TM, de Oliveira Campos LA, Tavares MG (2009) A scientific note on the characterization of microsatellite loci for *Melipona mondury* (Hymenoptera: Apidae). Apidologie 41(2):138–140

Lowe A, Harris S, Ashton P (2004) Ecological genetics: design, analysis, and application. Blackwell Publishing, Malden

Lunt DH, Whipple LE, Hyman BC (1998) Mitochondrial DNA variable number tandem repeats (VNTRs): utility and problems in molecular ecology. Mol Ecol 7:1441–1455

Magnacca KN, Brown MJF (2010) Tissue segregation of mitochondrial haplotypes in heteroplasmic Hawaiian bees: implications for DNA barcoding. Mol Ecol Resour 10:60–68. doi:10.1111/j.1755-0998.2009.02724.x

Mahmood K (2004) Identification of pest species in oriental fruit fly, *Bactrocera dorsalis* (Hendel) (Diptera: Tephritidae) species complex. Pak J Zool 36:213–230

Malacrida AR, Guglielmino CR, Gasperi G, Baruffi L, Milani R (1992) Spatial and temporal differentiation in colonizing populations of *Ceratitis capitata*. Heredity 69:101–111

Malacrida AR, Marinoni F, Torti C, Gomulski LM, Sebastiani F, Bonvicini C, Gasperi G, Guglielmino CR (1998) Genetic aspects of the worldwide colonization process of *Ceratitis capitata*. J Hered 89:501–507

Malacrida AR, Gomulski LM, Bonizzoni M, Bertin S, Gasperi G, Guglielmino CR (2007) Globalization and fruitfly invasion and expansion: the medfly paradigm. Genetica 131:1–9

Manel S, Gaggiotti OE, Waples RS (2005) Assignment methods: matching biological questions with appropriate techniques. Trend Ecol Evol 20:136–142

Martínez H, Toledo J, Liedo P, Mateos M (2012) Survey of heritable endosymbionts in southern Mexico populations of the fruit fly species *Anastrepha striata* and *A. ludens*. Dig Curr Microbiol. doi:10.1007/s00284-012-0223-3

Matz M, Shagin D, Bogdanova E, Britanova O, Lukyanov S, Diatchenko L, Chenchik A (1999) Amplification of cDNA ends based on template-switching effect and step-out PCR. Nucleic Acids Res 27(6):1558–1560

Maxwell SA, Rasic G, Keyghobadi N (2009) Characterization of microsatellite loci for the western cherry fruit fly, *Rhagoletis indifferens* (Diptera: Tephritidae). Dig Perman Genet Resour Note. doi:10.1111/j.1755-0998.2009.02565.x

Maxwell SA, Thistlewood H, Keyghobadi N (2011) Analysis of single-copy, nuclear microsatellite markers from flies collected on sticky traps. J Appl Entomol 135:641–646. doi:10.1111/j.1439-0418.2010.01578.x

McDowell JR, Diaz-Jaimes P, Graves JE (2002) Isolation and characterization of seven tetranucleotide microsatellite loci from Atlantic northern Bluefin tuna *Thunnus thynnus thynnus*. Mol Ecol Notes 2:212–216

McPheron BA, Gasparich GE, Han H-Y, Steck GJ, Sheppard WS (1994) Mitochondrial DNA restriction map for the Mediterranean fruit fly, *Ceratitis capitata*. Biochem Genet 32:25–33

McPheron BA, Sheppard WS, Steck GJ (1995) Genetic research and the origin, establishment, and spread of the Mediterranean fruit fly. In: Morse JG, Metcalf RL, Carey JR, Dowell RV (eds) The Medfly in California: defining critical research. Proceedings of meeting in Riverside, CA, 9–11 Nov 1994. College of Natural and Agricultural Sciences, University of California, Riverside, CA

Meglécz E, Petenian F, Danchin E, D'Acier AC, Rasplus JY, Faure E (2004) High similarity between flanking regions of different microsatellites detected within each of two species of Lepidoptera: *Parnassius apollo* and *Euphydryas aurinia*. Mol Ecol 13:1693–1700

Meglécz E, Anderson SJ, Bourguet D, Butcher R, Caldas A, Cassel-Lundhagen A, d'Acier AC, Dawson DA, Faure N, Fauvelot C, Franck P, Harper G, Keyghobadi N, Kluetsch C, Muthulakshmi M, Nagaraju J, Patt A, Pétenian F, Silvain JF, Wilcock HR (2007) Microsatellite flanking region similarities among different loci within insect species. Insect Mol Biol 16 (2):175–185

Meglécz E, Costedoat C, Dubut V, Gilles A, Malausa T, Pech N, Martin J-F (2010) QDD: a user-friendly program to select microsatellite markers and design primers from large sequencing projects. Bioinformatics 26(3):403–404. doi:10.1093/bioinformatics/btp670

Meiklejohn CD, Montooth KL, Rand DM (2007) Positive and negative selection on the mitochondrial genome. Trend Genet 23(6):259–263. doi:10.1016/j.tig.2007.03.008

Meixner MD, McPheron BA, Silva JG, Gasparich GE, Sheppard WS (2002) The Mediterranean fruit fly in California: evidence for multiple introductions and persistent populations based on microsatellite and mitochondrial DNA variability. Mol Ecol 891–899

Messing RH (1993) Fruit fly eradication in Hawaii: environmental considerations for unique island ecosystems. In: Aluja M, Liedo P (eds) Fruit flies: biology and management. Springer, New York, pp 321–333

Metcalf RL (1995) Biography of the medfly. In: Morse JG, Metcalf RL, Carey JR, Dowell RV (eds) The Medfly in California: defining critical research. Proceedings of meeting in Riverside, CA, 9–11 Nov 1994. College of Natural and Agricultural Sciences, University of California, Riverside

Michel AP, Rull J, Aluja M, Feder JL (2007) The genetic structure of hawthorn-infesting *Rhagoletis pomonella* populations in Mexico: implications for sympatric host race formation. Mol Ecol 16:2867–2878. doi:10.1111/j.1365-294X.2007.03263.x

Morin PA, Martien KK, Archer FI, Cipriano F, Steel D, Jackson J, Taylor BL (2010) Applied conservation genetics and the need for quality control and reporting of genetic data used in fisheries and wildlife management. J Hered 101(1):1–10

Morse JG, Metcalf RL, Carey JR, Dowell RV (eds) (1995) The medfly in California: defining critical research. In: Proceedings of meeting in Riverside, CA, 9–11 Nov 1994. College of Natural and Agricultural Sciences, University of California, Riverside, CA

Muirhead JR, Gray DK, Kelly DW, Ellis SM, Heath DD, MacIsaac HJ (2008) Identifying the source of species invasions: sampling intensity vs. genetic diversity. Mol Ecol 17:1020–1035. doi:10.1111/j.1365-294X.2008.03669.x

Muraji M, Nakahara S, Ishida T, Minoura K, Miyasaki I, Kohama T (2008) The Philippines is a possible source of the *Bactrocera dorsalis* complex species (Diptera, Tephritidae) occasionally collected in the Ryukyu Islands of Japan; analyses of mitochondrial DNA. Appl Entomol Zool 43(4):609–615

Murphy RW, Sites JW Jr, Buth DG, Haufler CH (1996) Proteins: isozyme electrophoresis. In: Hillis DM, Moritz C, Mable BK (eds) Molecular systematics. Sinauer, Sunderland

Nakahara S, Muraji M (2010) PCR-RFLP analysis of the *Bactrocera dorsalis* (Tephritidae: Tephritidae) complex species collected in and around the Ryukyu Islands of Japan using the mitochondrial A-T rich control region. Res Bull Plant Protect Jpn 46:17–23

Nakahara S, Kobashigawa Y, Muraji M (2008) Genetic variations among and within populations of the oriental fruit fly, *Bactrocera dorsalis* (Diptera; Tephritidae), detected by PCR-RFLP of the mitochondrial control region. Appl Entomol Zool 43(3):457–465

Nardi F, Carapelli A, Dallai R, Frati F (2003) The mitochondrial genome of the olive fly *Bactrocera oleae*: two haplotypes from distant geographical locations. Insect Mol Biol 12 (6):605–611

Nardi F, Carapelli A, Dallai R, Roderick GK, Frati F (2005) Population structure and colonization history of the olive fly, *Bactrocera oleae* (Diptera, Tephritidae). Mol Ecol 14:2729–2738

Nardi F, Carapelli A, Boore JL, Roderick GK, Dallai R, Frati F (2010) Domestication of olive fly through a multi-regional host shift to cultivated olives: comparative dating using complete mitochondrial genomes. Mol Phylogenet Evol 57:678–686. doi:10.1016/j.ympev.2010.08.008

Nei M, Kumar S (2000) Molecular evolution and phylogenetics. Oxford University Press, Oxford

Odriozola A, Aznar JM, Celorrio D, Bravo ML, Builes JJ, Val-Bernal JF, de Pancorbo MM (2011) Development and validation of I-DNA1: a 15-Loci multiplex system for identity testing. Int J Legal Med 125:685–694. doi:10.1007/s00414-010-0539-0

Olejniczak M, Krzyzosiak WJ (2006) Genotyping of simple sequence repeat factors implicated in shadow band generation revisited. Electrophoresis 27(19):3724–3734

O'Reilly PT, Hamilton LC, McConnell SK, Wright JM (1996) Rapid analysis of genetic variation in Atlantic salmon (*Salmo salar*) by PCR multiplexing of dinucleotide and tetranucleotide microsatellites. Can J Fish Aquat Sci 53:2292–2298

Ouin A, Menozzi P, Coulon M, Hamilton AJ, Sarthou JP, Tsafack N, VialatteA PS (2011) Can deuterium stable isotope values be used to assign the geographic origin of an auxiliary hoverfly in south-western France? Rapid Commun Mass Spectrom 25:2793–2798

Panchal M, Beaumont MA (2007) The automation and evaluation of nested clade phylogeographic analysis. Evolution 61(6):1466–1480. doi:10.1111/j.1558-5646.2007.00124.x

Panova M, Blakeslee AMH, Miller AW, Mäkinen T, Ruiz GM, Johannesson K, André C (2011) Glacial history of the north Atlantic marine snail, *Littorina saxatilis*, Inferred from distribution of mitochondrial DNA lineages. PLoS One 6(3):e17511. doi:10.1371/journal.pone.0017511

Papadopoulos NT, Plant RE, Carey JR (2013) From trickle to flood: the large scale, cryptic invasion of California by tropical fruit flies. Proc R Soc B 280:20131466

Papadopoulou A, Anastasiou I, Vogler AP (2010) Revisiting the insect mitochondrial molecular clock: the mid-Aegean trench calibration. Mol Biol Evol 27(7):1659–1672. doi:10.1093/molbev/msq051

Park S (2001) Microsatellite toolkit. Department of Genetics, Trinity College, Dublin

Peakall ROD, Smouse PE (2006) GENALEX 6: genetic analysis in Excel. Population genetic software for teaching and research. Mol Ecol Notes 6(1):288–295

Peakall R, Smouse PE (2012) GenAlEx 6.5: genetic analysis in Excel. Population genetic software for teaching and research—an update. Bioinformatics 28(19):2537–2539

Penrose R (1993) The 1989/1990 Mediterranean fruit fly eradication program in California. In: Aluja M, Liedo P (eds) Fruit flies: biology and management. Springer, New York, pp 401–406

Penrose R (1996) California's 1993/1994 Mediterranean fruit fly eradication program. In: McPheron BA, Steck GJ (eds) Fruit fly pests: a world assessment of their biology and management. St. Lucie Press, Delray Beach, pp 551–554

Pereira L, Cunha C, Amorim A (2004) Predicting sampling saturation of mtDNA haplotypes: an application to an enlarged Portuguese database. Int J Legal Med 118:132–136. doi:10.1007/s00414-003-0424-1

Peterson BK, Hare EE, Iyer VN, Storage S, Conner L, Papaj DR, Kurashima R, Jang E, Eisen MB (2009) Big genomes facilitate the comparative identification of regulatory elements. Dig PLoS ONE 4(3):e4688. doi:10.1371/journal.pone.0004688

Pilot M, Branicki W, Jędrzejewski W, Goszczyński J, Jędrzejewska B, Dykyy I, Shkvyrya M, Tsingarska E (2010) Phylogeographic history of grey wolves in Europe. BMC Evol Biol 10:104

Piry S, Alapetite A, Cornuet J-M, Paetkau D, Baudouin L, Estoup A (2004) GENECLASS2: a software for genetic assignment and first-generation migrant detection. J Hered 95:536–539
Piwowar HA (2011) Who shares? Who doesn't? Factors associated with openly archiving raw research data. PLoS One 6(7):e18657. doi:10.1371/journal.pone.0018657
Pons O, Petit RJ (1995) Estimation, variance and optimal sampling of gene diversity I. Haploid locus. Theor Appl Genet 90(462):470
Prabhakar CS, Mehta PW, Sood P, Singh SK, Sharma P, Sharma PN (2012) Population genetic structure of the melon fly, *Bactrocera cucurbitae* (Coquillett) (Diptera: Tephritidae) based on mitochondrial cytochrome oxidase (COI) gene sequences. Genetica 140:83–91. doi:10.1007/s10709-012-9660-y
Pritchard JK, Stephens M, Donnelly P (2000) Inference of population structure using multilocus genotype data. Genetics 155:945–959
Puritz JB, Addison JA, Toonen RJ (2012) Next-generation phylogeography: a targeted approach for multilocus sequencing of non-model organisms. PLoS One 7:e34241. doi:10.1371/journal.pone.0034241
Ramsay L, Macaulay M, degli Ivanissevich S, MacLean K, Cardle L, Fuller J, Edwards KJ, Tuvesson S, Morgante M, Massari A, Maestri E, Marmiroli N, Sjakste T, Ganal M, Powell W, Waugh R (2000) A simple sequence repeat-based linkage map of barley. Genetics 156(4):1997–2005
Rannala B, Mountain JL (1997) Detecting immigration by using multilocus genotypes. Proc Natl Acad Sci U S A 94:9197–9201
Ratnasingham S, Hebert PDN (2007) BOLD: The Barcode of Life Data System (www.barcodinglife.org). Dig Molecular Ecol Notes. doi:10.1111/j.1471-8286.2006.01678.x
Raymond M, Rousset F (1995) GENEPOP (version 1.2): population genetics software for exact tests and ecumenicism. J Hered 86(3):248–249
Rendón P, McInnis D, Lance D, Stewart J (2004) Medfly (Diptera:Tephritidae) genetic sexing: large-scale field comparison of males-only and bisexual sterile fly releases in Guatemala. J Econ Entomol 97(5):1547–1553. doi:10.1603/0022-0493-97.5.1547
Renshaw MA, Saillant E, Bradfield SC, Gold JR (2006) Microsatellite multiplex panels for genetic studies of three species of marine fishes: red drum (*Sciaenops ocellatus*), red snapper (*Lutjanus campechanus*), and cobia (*Rachycentron canadum*). Aquaculture 253(1):731–735
Reyes A, Ochando MD (1998) Use of molecular markers for detecting the geographical origin of *Ceratitis capitata* (Dipter: Tephritidae) populations. Ann Entomol Soc Am 91:222–227
Ricklefs RE (2005) Taxon cycles. In: Sax DF, Stachowicz JJ, Gaines SD (eds) Species invasions: insights into ecology, evolution, and biogeography. Sinauer, Sunderland
Roderick GK, Villablanca FX (1996) Genetic and statistical analysis of colonization. In: McPheron BA, Steck GJ (eds) Fruit fly pests, a world assessment of the biology and management. St. Lucie Press, Delray Beach, pp 281–290
Rowe HC, Renaut S, Guggisberg A (2011) RAD in the realm of next-generation sequencing technologies. Dig Mol Ecol 1–4
Rubinoff D, Holland BS, San Jose M, Powell JA (2011) Geographic proximity not a prerequisite for invasion: Hawaii not the source of California by light brown apple moth (*Epiphyas postvittana*). PLoS One 6(1):e16361. doi:10.1371/journal.pone.0016361
Ruiz-Arce R, Barr N, Owen C, Thomas DB, McPheron B (2012) Phylogeography of *Anastrepha obliqua* inferred with mtDNA sequencing. J Econ Entomol 105(6):2147–2160
Salas A, Bandelt H-J, Macaulay V, Richards MB (2007) Phylogeographic investigations: the role of trees in forensic genetics. Forensic Sci Int 168:1–13
Salcedo Baca D, Lomeli Flores JR, Terrazas González GH, Enkerlin W (2010) Economic evaluation of the MOSCAMED regional program in Mexico (1978–2008). In: Sabater-Muñoz B, Navarro Llopis V, Urbaneja A (eds) 8th international symposium on fruit flies of economic importance, Valencia, 26 Sept–1 Oct 2010. Editorial Universitat Politècnica de València, Spain

San Andrés V, Urbaneja A, Sabater-Muñoz B, Castañera P (2007) A novel molecular approach to assess mating success of sterile *Ceratitis capitata* (Diptera: Tephritidae) males in sterile insect technique programs. J Econ Entomol 100(4):1444–1449

Sarakatsanou A, Diamantidis AD, Papanastasiou SA, Bourtzis K, Papadopoulos NT (2011) Effects of Wolbachia on fitness of the Mediterranean fruit fly (Diptera: Tephritidae). J Appl Entomol 135:554–563

Schug MD, Hutter CM, Wetterstrand KA, Gaudette MS, Mackay TF, Aquadro CF (1998) The mutation rates of di-, tri-and tetranucleotide repeats in *Drosophila melanogaster*. Mol Biol Evol 15(12):1751–1760

Schuster SC (2008) Next-generation sequencing transforms today's biology. Nat Methods 5:16–18

Schwarz D, Robertson HM, Feder JL, Varala K, Hudson ME, Ragland GJ, Hahn DA, Berlocher SH (2009) Sympatric ecological speciation meets pyrosequencing: sampling the transcriptome of the apple maggot *Rhagoletis pomonella*. BMC Genomics 10:633. doi:10.1186/1471-2164-10-633

Scolari F, Gomulski LM, Ribeiro JMC, Siciliano P, Meraldi A, Falchetto M, Bonomi A, Manni M, Gabrieli P, Malovini A, Bellazini R, Aksoy S, Gasperi G, Malacrida AR (2012) Transcriptional profiles of mating-responsive genes from testes and male accessory glands of the Mediterranean fruit fly, *Ceratitis capitata*. PLoS One 7(10):e46812. doi:10.1371/journal.pone.0046812

Shearman DCA, Gilchrist AS, Crisafulli D, Graham G, Lange C (2006) Microsatellite markers for the pest fruit fly, *Bactrocera papayae* (Diptera: Tephritidae) and other *Bactrocera* species. Mol Ecol Notes 6:4–7. doi:10.1111/j.1471-8286.2006.01024.x

Shearman DCA, Frommer M, Morrow JL, Raphael KA, Gilchrist AS (2010) Interspecific hybridization as a source of novel genetic markers for the sterile insect technique in *Bactrocera tryoni* (Diptera: Tephritidae). J Econ Entomol 103(4):1071–1079. doi:http://dx.doi.org/10.1603/EC09241

Shen G-M, Dou W, Niu J-Z, Jiang H-B, Yang W-J, Jia F-X, Hu F, Cong L, Wang J-J (2011) Transcriptome analysis of the oriental fruit fly (*Bactrocera dorsalis*). PLoS One 6(12):e29127. doi:10.1371/journal.pone.0029127

Sheppard WS, Steck GJ, McPheron BA (1992) Geographic populations of the Mediterranean fruit fly may be differentiated by mitochondrial DNA variation. Experientia 48:1010–1013

Shi W, Kerdelhué C, Ye H (2005a) Population genetics of the oriental fruit fly, *Bactrocera dorsalis* (Diptera: Tephritidae), in Yunnan (China) based on mitochondrial DNA sequences. Environ Entomol 34(4):977–983

Shi W, Kerdelhué C, Ye H (2005b) Population genetic structure of the oriental fruit fly, *Bactrocera dorsalis* (Hendel) (Diptera: Tephritidae) from Yunnan province (China) and nearby sites across the border. Genetica 138:377–385. doi:10.1007/s10709-009-9429-0

Shi W, Kerdelhué C, Ye H (2010) Population genetic structure of the oriental fruit fly, *Bactrocera dorsalis* (Hendel) (Diptera: Tephritidae) from Yunnan province (China) and nearby sites across the border. Genetica 138:377–385

Shi W, Kerdelhué C, Ye H (2012) Genetic structure and inferences on potential source areas for *Bactrocera dorsalis* (Hendel) based on mitochondrial and microsatellite markers. PLoS One 7 (5):e37083

Siebert JB, Cooper T (1995) Embargo on California produce would cause revenue job loss. Calif Agr 49:7–12

Silva JG, Meixner MD, McPheron BA, Steck GJ, Sheppard WS (2003) Recent Mediterranean fruit fly (Diptera: Tephritidae) invasions in Florida—a genetic perspective. J Econ Entomol 96:1711–1718

Simon C, Frati F, Beckenbach A, Crespi B, Liu H, Flook P (1994) Evolution, weighting, and phylogenetic utility of mitochondrial gene sequences and a compilation of conserved polymerase chain reaction primers. Annal Entomolog Soc 87(6):651–701

Simon C, Buckley TR, Frati F, Stewart JB, Beckenbach AT (2006) Incorporating molecular evolution into phylogenetic analysis, and a new compilation of conserved polymerase chain reaction primers for animal mitochondrial DNA. Annu Rev Ecol Evol Syst 37:545–579

Slatkin M, Excoffier L (1996) Testing for linkage disequilibrium in genotypic data using the expectation-maximization algorithm. Heredity 76:377–383

Smith-Caldas RB, McPheron BA, Silva JG, Zucchi RA (2001) Phylogenetic relationships among species of the *fraterculus* Group (Anastrepha: Diptera: Tephritidae) inferred from DNA sequences of mitochondrial cytochrome oxidase I. Neotrop Entomol 30(4):565–573

Song AD, Ma J, Hughes JM (2006) Polymorphic microsatellite DNA markers in *Bactrocera cacuminata* (Hering) (Diptera: Tephritidae). Mol Ecol Note 6:47–49

Sonvico A, Manso F, Quesada-Allue LA (1996) Discrimination between the immature stages of *Ceratitis capitata* and *Anastrepha fraterculus* (Diptera: Tephritidae) populations by random amplified polymorphic DNA polymerase chain reaction. J Econ Entomol 89:1208–1212

Spanos L, Koutroumbas G, Kotsyfakis M, Louis C (2000) The mitochondrial genome of the Mediterranean fruit fly, *Ceratitis capitata*. Insect Mol Biol 9:139–144

Steck GJ, Gasparich GE, Han H-Y, McPheron BA, Sheppard WS (1996) Distribution of mitochondrial haplotypes among *Ceratitis capitata* populations worldwide. In: McPheron BA, Steck GJ (eds) Fruit fly pests, a world assessment of the biology and management. St. Lucie Press, Delray Beach, pp 291–296

Stratikopoulos EE, Augustinos AA, Petalas YG, Vrahatis MN, Mintzas A, Mathiopoulos KD, Zacharopoulou A (2008) An integrated genetic and cytogenetic map for the Mediterranean fruit fly, *Ceratitis capitata*, based on microsatellite and morphological markers. Genetica 133:147–157

Stratikopoulos EE, Augustinos AA, Pavlopoulos ID, Economou KP, Mintzas A, Mathiopoulos KD, Zacharopoulou A (2009) Isolation and characterization of microsatellite markers from the Mediterranean fruit fly, *Ceratitis capitata*: cross-species amplification in other Tephritidae species reveals a varying degree of transferability. Mol Genet Genomics 282:283–306

Sved JA, Yu H, Dominiak B, Gilchrist AS (2003) Inferring modes of colonization for pest species using heterozygosity comparisons and a shared-allele test. Genetics 163:823–831

Templeton AR (1998) Nested clade analyses of phylogeographic data: testing hypothesis about gene flow and population history. Mol Ecol 7:381–397

Thao ML, Baumann L, Baumann P (2004) Organization of the mitochondrial genomes of whiteflies, aphids, and psyllids (Hemiptera, Sternorrhyncha). BMC Evol Biol 4:25. doi:10.1186/1471-2148-4-25

Tóth G, Gáspári Z, Jurka J (2000) Microsatellites in different eukaryotic genomes: survey and analysis. Genome Res 10:967–981. doi:10.1101/gr.10.7.967

Tremblay P, Paquin R (2007) Improved detection of sugar additives to maple syrup using malic acid as internal standard and in ^{13}C isotope ratio mass spectrometry (IRMS). J Agri Food Chem 55:197–203

Tsoumani KT, Mathiopoulos HD (2011) Genome size estimation with quantitative real-time PCR in two Tephritidae species: *Ceratitis capitata* and *Bactrocera oleae*. J Appl Entomol 136:626–631

Tsoumani KT, Augustinos AA, Kakani EG, Drosopoulou E, Mavragani-Tsipidou P, Mathiopoulos KD (2011) Isolation, annotation and applications of expressed sequence tags from the olive fly, *Bactrocera oleae*. Mol Genet Genomics 285:33–45

van Oosterhout C, Hutchinson WF, Wills DPM, Shipley P (2004) MICRO-CHECKER: software for identifying and correcting genotyping errors in microsatellite data. Mol Ecol Note 4:535–538

Van't Hof AE, Brakefield PM, Saccheri IJ, Zwaan BJ (2007) Evolutionary dynamics of multilocus microsatellite arrangements in the genome of the butterfly *Bicyclus anynana*, with implications for other Lepidoptera. Heredity 98:320–328

Velez S, Taylor S, Noor MAF, Lobo NF, Feder JL (2006) Isolation and characterization of microsatellite loci from the apple maggot fly *Rhagoletis pomonella* (Diptera; Tephritidae). Mol Ecol Note 6:90–92

Vera MT, Rodriguez R, Segura DF, Cladera JL, Sutherst RW (2002) Potential geographical distribution of the Mediterranean fruit fly, *Ceratitis capitata* (Diptera: Tephritidae), with emphasis on Argentina and Australia. Environ Entomol 31:1009–1022

Villablanca FX, Roderick GK, Palumbi SR (1998) Invasion genetics of the Mediterranean fruit fly: variation in multiple nuclear introns. Mol Ecol 7:547–560
Virgilio M, De Meyer M, White IM, Backeljau T (2009) African *Dacus* (Diptera: Tephritidae) molecular data and host plant associations do not corroborate morphology based classifications. Mol Phylogenet Evol 51:531–539
Virgilio M, Delatte H, Backeljau T, De Meyer M (2010) Macrogeographic population structuring in the cosmopolitan agricultural pest *Bactrocera cucurbitae* (Diptera: Tephritidae). Mol Ecol 19:1213–2724
Virgilio M, Delatte H, Quilici S, Backeljau T, De Meyer M (2013) Cryptic diversity and gene flow among three African agricultural pests: *Ceratitis rosa*, *Ceratitis fasciventris* and *Ceratitis anonae* (Diptera, Tephritidae). Mol Ecol. doi:10.1111/mec.12278
Walsh PS, Fildes NJ, Reynolds R (1996) Sequence analysis and characterization of stutter products at the tetranucleotide repeat locus vWA. Nucleic Acids Res 24:2807–2812
Wan Q-H, Wu H, Fujihara T, Fang S-G (2004) Which genetic marker for which conservation genetics issue? Electrophoresis 25:2165–2176
Wan X, Nardi F, Zhang B, Liu Y (2011) The oriental fruit fly, *Bactrocera dorsalis*, in China: origin and gradual inland range expansion associated with population growth. PLoS One 6(10): e25238. doi:10.1371/journal.pone.0025238
Wan X, Liu Y, Zhang B (2012) Invasion history of the oriental fruit fly, *Bactrocera dorsalis*, in the Pacific-Asia region: two invasion routes. PLoS One 7(5):e36176. doi:10.1371/journal.pone.0036176
Wares JP, Hughes AR, Grosberg RK (2005) Mechanisms that drive evolutionary change. In: Sax DF, Stachowicz JJ, Gaines SD (eds) Species Invasions: insights into ecology, evolution, and biogeography. Sinauer, Sunderland
Wassenaar LI, Hobson KA (1998) Natal origins of migratory monarch butterflies at wintering colonies in Mexico: new isotopic evidence. Proc Natl Acad Sci USA 95:15436–15439
Watson J (1990) The human genome project: past, present, and future. Nature 248:44–49
Weeks DE, Conley YP, Ferrell RE, Mah TS, Gorin MB (2002) A tale of two genotypes: consistency between two high-throughput genotyping centers. Genome Res 12(3):430–435
Weir BS (1996) Genetic data analysis II. Sinauer, Sunderland
Wenz H-M, Robertson JM, Menchen S, Oaks F, Demorest DM, Scheibler D, Rosenblum BB, Wike C, Gilbert DA, Efcavitch JW (1998) High-precision genotyping by denaturing capillary electrophoresis. Genome Res 8:69–80
White IM, Elson-Harris MM (1994) Fruit flies of economic significance: their identification and bionomics. CAB, Wallingford
Whittier TS, Nam FY, Shelly TE, Kaneshiro KY (1994) Male courtship success and female discrimination in the Mediterranean fruit fly (Diptera: Tephritidae). J Insect Behav 7:159–170
Wilder J, Hollocher H (2001) Mobile elements and the genesis of microsatellites in dipterans. Mol Biol Evol 18(3):384–392
Wu Y, Li Y, Ruiz-Arce R, McPheron BA, Wu J, Li Z (2011) Microsatellite markers reveal population structure and low gene flow among collections of *Bactrocera cucurbitae* (Diptera: Tephritidae) in Asia. J Econ Entomol 104:1065–1074
Wu Y, McPheron BA, Wu J-J, Li Z-H (2012) Genetic relationship of the melon fly, *Bactrocera cucurbitae* (Diptera: Tephritidae) inferred from mitochondrial DNA. Insect Sci 19:195–204. doi:10.1111/j.1744-7917.2011.01443.x
Wu P-F, Roques A, Xiong Z-P, Xu L, Huang Y-Y, Pan Y-Z (2013) The complete mitochondrial genome of the melon fly *Bactrocera cucurbitae* (Diptera: Tephritidae). Mitochondrial DNA 24:6–7
Yu H, Frommer M, Robson MK, Meats AW, Shearman DCA, Sved JA (2001) Microsatellite analysis of the Queensland fruit fly *Bactrocera tryoni* (Diptera: Tephritidae) indicates spatial structuring: implications for population control. Bull Entomol Res 91:139–147
Yu DJ, Xu L, Nardi F, Li JG, Zhang RJ (2007) The complete nucleotide sequence of the mitochondrial genome of the oriental fruit fly, *Bactrocera dorsalis* (Diptera: Tephritidae). Gene 396:66–74

Zabalou S, Riegler M, Theodorakopoulou M, Stauffer C, Savakis C, Bourtzis K (2004) *Wolbachia*-induced cytoplasmic incompatibility as a means for insect pest population control. Proc Natl Acad Sci U S A 101:15042–15045

Zhang DX (2004) Lepidopteran microsatellite DNA: redundant but promising. Trend Ecol Evol 19 (10):507–509

Zhang D-X, Hewitt GM (1996) Nuclear integrations: challenges for mitochondrial DNA marker. Tree 11(6):247–251

Zhang D-X, Hewitt GM (1997) Insect mitochondrial control region: a review of its structure, evolution and usefulness in evolutionary studies. Biochem Syst Ecol 25(2):99–120

Zheng W, Peng T, He W, Zhang H (2012) High-throughput sequencing to reveal genes involved in reproduction and development in *Bactrocera dorsalis* (Diptera: Tephritidae). PLoS One 7: e36463. doi:10.1371/journal.pone.0036463

Zygouridis NE, Augustinos AA, Zalom FG, Mathiopoulos HD (2009) Analysis of olive fly invasion in California based on microsatellite markers. Heredity 102:402–412

Chapter 11
Modeling Trapping of Fruit Flies for Detection, Suppression, or Eradication

Hugh J. Barclay and Jorge Hendrichs

Abstract Models of insect trapping in general and fruit fly trapping in particular are reviewed. These include models for detection, suppression, and eradication. Models for detection include areas of attraction for traps, probability of capture of insects in traps, and probabilistic models for declaring that a species has either not yet invaded an area or has been eradicated from the area. Dispersal is likely to have a major role in trapping, as it will bring insects into contact with more traps, and thus the probability of being trapped and killed is greater. Population dynamic models for mass-trapping insects for suppression or eradication indicate that the deployment of attract and kill devices is more effective if females are targeted, such as the use of food-based lures. If only males are targeted in an effort to deprive females of mates, the rate of trapping must be very high, otherwise the few males remaining will likely be sufficient to fertilize enough females to maintain the population. Male mating prior to being trapped is a major deterrent to the success of the male annihilation approach. If some females are also trapped, then the outcome is much more optimistic. The concurrent release of sterile insects, preferably when these are less responsive to the attractant due to pre-release exposure to a male lure such as methyl eugenol, interacts synergistically with trapping, and suppression or eradication is likely to be easier if both control methods are used simultaneously rather than sequentially. A case study of *Bactrocera dorsalis* (Hendel) is presented with relevant population equations and parameter values. Methods for calculation of the barrier width required to exclude insect pests from a protected area are presented with tentative results for medflies.

H.J. Barclay (✉)
Victoria, B. C., Canada
e-mail: hbarclay@shaw.ca

J. Hendrichs
Insect Pest Control Section, Joint FAO/IAEA Division of Nuclear Techniques in Food and Agriculture, International Atomic Energy Agency, Wagramerstrasse 5, P. O. Box 100, 1400 Vienna, Austria

T. Shelly et al. (eds.), *Trapping and the Detection, Control, and Regulation of Tephritid Fruit Flies*, DOI 10.1007/978-94-017-9193-9_11,

Keywords Modeling fruit fly trapping • Trap attraction • Detection • Suppression • Eradication • Declaring success • Barriers to invasion • *Bactrocera* • Medfly • Mass trapping • Attract and kill • Methyl eugenol • Age-structured models • Sterile releases

1 Introduction

Pest control is usually done at the population level, so the processes involved can be modeled using population dynamic models, and the results can act as hypotheses to be tested or as predictions to aid in the execution of control programs. These same kinds of results could be obtained experimentally, but usually it would require far more time and resources than does modeling. Information from the various topics addressed in this volume can be used as input to the models and to, not only obtain quantitative predictions of control types and costs, but also identify and describe interactions of control methods planned for use in Integrated Pest Management (IPM). Considerable modeling has already been done in this area, so that a modeling chapter is both justified and desirable.

Most fruit fly pest insects are anautogenous, i.e. their larval diets are very poor in amino acids, sterols, vitamins, and minerals required for sexual maturation. Hence, after emergence males and females need a relatively long sexual maturation period during which they forage for these essential nutrients. Unlike many other Diptera, tephritid females continuously produce oocytes in successive stages of development and consequently can lay eggs every day and thus also require a regular supply of these nutrients after reaching maturation (Hendrichs and Prokopy 1994). Food-based volatiles are therefore effective attractants for tephritid flies, being used for detection and monitoring purposes in traps, as well as for suppression in bait sprays, bait stations, and mass trapping (Navarro-Llopis et al. 2011; Piñero et al. 2009; Roessler 1989).

This chapter outlines methods of modeling the trapping of insects in general, and fruit flies in particular, for the purpose of detection, suppression, and eradication. Topics include (i) methods for quantifying trap attractiveness and trap efficiency, information fundamental in modeling any trapping program; (ii) probabilistic methods for declaring either that a species has not yet invaded an area or that it has been eradicated from the area, (iii) theoretical trapping models for predicting suppression or eradication of a pest from an area; (iv) assessment of types of interaction (synergism, independence, or interference) between control methods, as well as methods for obtaining this information, to better assess combinations of control methods in IPM; (v) a case study using an age structured model for suppressing or eradicating a *Bactrocera dorsalis* (Hendel) population that involves male annihilation of wild males through the use of trapping with methyl eugenol as an attractant and also the simultaneous release of sterile males; (vi) the calculation of barrier widths and trapping mortality required for pest exclusion from a delimited area, where trapping within the barrier reduces the pest population to a small proportion of its density outside the barrier; and (vii) a section outlining conclusions from such modeling efforts. In these sections we describe relevant interactions with environmental and biological factors, assumptions inherent in the models, results of the models, and methodology for obtaining the results. This treatment is not exhaustive but attempts to cover the main modeling developments in trapping of insect pests during the last few decades.

There has been considerable modeling on the structure of odor plumes, but these are not covered here as few generalities have emerged that are useful for modeling trapping. In addition, pheromone trapping models for detection and for male annihilation are not covered, as in most species of fruit flies the females do not produce long distance sex pheromones, such as occur in moths and many other insect species. Also, mating disruption is not covered, because it uses pheromone and, strictly speaking, does not involve trapping.

2 Measuring Effectiveness of Traps

This section covers four loosely related concepts and efforts to model them: (i) the idea of an attractive space within an odor plume; (ii) the definition of areas of attractiveness, but without reference to odor plumes, within which insects are stimulated to move towards the odor source; (iii) a measure of relative trap attractiveness of different types of traps; and (iv) a measure of population density obtained from trapping.

2.1 Attractive Space Within an Odor Plume

Bossert and Wilson (1963) originated the concept of active space, which is a volume of space (air) within an odor plume and within which an insect responds to the attractant and orients towards its source. These authors envisioned this space as the attractive portion of a continuous plume of odor with a well-defined boundary within which the odor was moving and becoming more dilute as it did. This neat geometric ideal has proven to be unrealistic, as turbulence in the air disturbs the plume, breaking it up into small filaments of odor, interspersed with volumes of air almost devoid of attractant, and causing the periphery to be ragged and chaotic. Wind and physical obstructions on the landscape, such as trees, hills, buildings, etc., also influence odor movement (Cardé and Willis 2008; Riffell et al. 2008) and can result in highly intermittent odor plumes even on fast time scales (Vickers 2000).

2.2 Modelling Components of Trap Attractiveness

As a result of the limited usefulness of the notion of odor plumes for trapping, the next direction taken was to define a physical space within which insects would be attracted to an attractive source but without reference to odor plumes. In this context, these shapes would be subject to physical alteration by wind, trees, and other physical influences. Although these definitions were often framed in terms of ideal shapes, they may be treated as mean values of some trapping efficiency (i.e., proportion of the existing population that is within the attractive space and that is captured by traps in a given interval of time) or of arbitrary lower limits of trapping efficiency. Various approaches have been taken to describe the maximum distance of attractiveness of traps to insects. Most of these approaches overlap, and some are almost interchangeable with appropriate parameterization. The following is a description of the major approaches.

Hartstack et al. (1971) investigated the trapping efficiency of light traps for bollworms, *Heliothis zea* (Boddie), and cabbage loopers, *Trichoplusia ni* (Hubner), using mark-recapture experiments. These authors defined two parameters: the 'effective radius' of the trap and the 'trap efficiency', the former being defined as the maximum distance from the trap at which the light was still attractive and the latter being defined as the percentage of insects caught of those that entered the effective radius of the trap. Values of effective radius and trap efficiency were estimated to be 30 m and 42.8 % for *H. zea* and 25 m and 61.4 % for *T. ni*. The data were fitted to curves and generated an equation for percentage recovery (P) of released insects

$$P = 100\,E\,R^2/(X + R)^2 \tag{11.1}$$

where E is trap efficiency, R is the effective radius of the trap, and X is the distance of the trap from the release point. In this case, trap efficiency was defined as the percentage of those insects released that were caught by the trap.

Wall and Perry (1987) distinguished three types of ranges useful for describing movement resulting in capture at traps: 'range of attraction', 'range of stimulation', and 'sampling range'.

1. The range of attraction is the maximum distance from an odor source over which insects can be shown to direct their movements to the source.
2. The range of stimulation is the maximum distance at which an attractive source can be shown to elicit any response, such as movement or wing or antennal vibrations. This range includes the range of attraction, so that the movements include those towards the odor source within the range of attraction.
3. The sampling range is the maximum distance from which insects can be shown to reach an odor source in a given time period. This includes the range of stimulation and is further extended by non-directed dispersal of insects, which may bring them into the range of attraction.

Insects within the range of stimulation, but outside the range of attraction, behave similarly to those outside the range of stimulation that are engaging in random dispersal, so it appears sufficient to consider only the ranges of sampling and attraction (Schlyter 1992).

Determining the range of attraction is difficult, as it requires distinguishing between directed and non-directed movements. Mark-recapture experiments may provide the best available evidence. However, the range of attraction will be affected by the method of dispensing of the attractant, the strength of the attractant, the release rate, and the habitat and environmental conditions and so will vary over time and space for a given species (Wall and Perry 1987). The sampling range is somewhat easier to derive. Wall and Perry (1987) were able to show that the moth, *Cydia nigricana* (Fabricius), released 500 m downwind of an attractive trap could be caught on the same afternoon, so that the sampling range was at least 500 m. By contrast, the range of attraction was shown to be about 200 m, as shown by the very quick recapture of a few male moths from that distance and that the maximum distance of interaction of traps had been shown to be about 200 m. Schlyter (1992) has provided methods for statistically analyzing the sampling range and the range of attraction, including estimates and confidence intervals. He pointed out that a simple way of estimating the range of attraction is by means of trap interference experiments, i.e., running a series of trials in which traps are placed at various distances from each other and noting the minimum distance above which there is no further increase in daily catch.

There is presently little documentation on the sampling ranges for tephritids for female-biased attractants. Kendra et al. (2010) have determined what they call the 'effective sampling range' for *Anastrepha suspensa* (Loew) to two female-targeted attractants: a torula yeast/borax solution and a two component lure consisting of ammonium acetate and putrescine. Their effective sampling range was defined as the maximum distance at which the relative trapping efficiency was ≥25 %, where the relative trapping efficiency was the percentage of females captured within each distance group. This sampling range corresponded to a distance (radius) determining the area of a circle within which 90 % of all recaptures occurred. They released and trapped both feral and mass-reared, sterile adults and found that the effective sampling range was 30 m for feral adults and 20 m for mass-reared adults. Epsky et al. (2010) determined the effective sampling range for female *Ceratitis capitata* (Wiedemann) using both contour analysis and variogram analysis. Both methods indicated an effective sampling range of about 28 m, although the contour analysis indicated that wind direction had a strong effect on sampling range, being 15 m greater upwind than downwind from the release point. They recommended that the interpretation of contours incorporate environmental variables, especially wind currents.

Barclay and Vreysen (2013) used area of attraction for the riverine tsetse (*Glossina palpalis gambiensis* (Vanderplank)) that was equivalent to Wall and Perry's range of attraction but explicitly included dispersal, effectively converting the measure to sampling range. These authors found that trapping efficiency (percentage of the population within the area of attraction that was trapped each day) increased with dispersal distance and percent of flies dispersing each day.

From the above list of measures of attractiveness of traps, it is evident that there are many formulations that could be used in modelling trapping. The area, or range, of attraction also is somewhat arbitrary as attractiveness will likely decline with distance from the trap; thus area of attraction should be defined on the basis of an area whose attraction is everywhere above some threshold or proportion of maximal attraction. Once the area of attraction is formulated, a measure of trapping success within that area is required to accurately predict daily catches. This can be estimated by determining the probability of capturing a given insect in a given time interval (e.g., one day) if the insect is within the area of attraction, and this will depend on dispersal patterns as well as attractiveness of the traps. Also, one should know the temporal rate of decay of the odor used to bait the trap.

2.3 Relative Trap Effectiveness

Byers and his Swedish colleagues (Byers et al. 1989) used a very different approach in developing a concept they call the Effective Attractive Radius (EAR) in which the catch per unit time from an attractive trap is compared with the catch from a standard, intercepting and cylindrical, but non-attractive, trap. The EAR, then, is a measure of the radius of the non-attractive trap that is equivalent in its ability to capture insects to the trap with attractant. The physical meaning of the EAR is the radius of a circular plane oriented perpendicularly to the line of flight of the approaching insects and can be regarded as the radius of a spherical volume that surrounds the attractive source. The EAR is defined as:

$$\mathrm{EAR} = [\mathrm{ATC} \times \mathrm{LCSAPT}/(\mathrm{PTC} \times \pi)]\ ½ \qquad (11.2)$$

where ATC and PTC are the catches of the attractive and passive traps, respectively, and LCSAPT is the longitudinal cross section of the passive trap. Thus, if the area of a passive cylindrical trap is A and a trap of the same dimensions contains an attractant and catches 20 times as many insects as the passive trap, then the effective area of the attractant trap is 20 times that of the passive trap, so $\mathrm{EAR} = (20\mathrm{A} / \pi)^{1/2}$. The concept was developed for bark beetles but can be generalized to other taxa. Byers et al. (1989) used passive cylindrical sticky screens hung from poles and compared the catches to those obtained by identical traps that contained attractants. The authors point out that EAR is independent of insect density, locality, and duration of test. The EAR can thus be used to compare the attractiveness of various trap types and attractants for a given species as well as comparing a given trap type containing appropriate attractants for various species. The EAR has the advantage that it is easy to measure, and it avoids the necessity of measuring or delineating the odor plume. In later publications, based on standard deviations of numbers flying at various heights, Byers (2009, 2011) calculated conversion factors for over

100 insect species that would allow the computation of EAR from existing published data that do not otherwise have sufficient information to do the computations.

2.4 *Modeling Population Density and Trap Catch*

Byers et al. (1989) noted that if captures of insects are purely passive and if the insects do not actively avoid being caught by traps, then any insect that is flying towards a cylindrical sticky screen will be captured by it. The number caught in a given time can be described as:

$$\begin{aligned}\text{Catch} = 2 \times (\text{cylinder radius}) \times (\text{trapping time}) \times (\text{insect speed}) \\ \times (\text{population density}).\end{aligned} \tag{11.3}$$

Time is part of the equation because dispersal takes time, and maximum dispersal, and hence sampling range, would be expected to increase with time interval. From this, we can calculate the population density by inverting the equation, so that

$$\text{Density} = \text{catch}/(2 \times \text{radius} \times \text{time} \times \text{speed}) \tag{11.4}$$

so that passive trapping allows an estimation of population density.

Turchin and Odendaal (1996) adopted a somewhat different approach using southern pine beetles and derived a measure termed the 'effective sampling area', which is a multiplier for converting trap catches to population density for emerging southern pine beetles when using pheromone in traps. This approach is more akin to mark-recapture, and their effective sampling area (α) is derived from

$$\mathrm{T} = \alpha\, \mathrm{B} \tag{11.5}$$

where T is the trap catch, and B is the known density of beetles per unit area, whence the estimator for α is $\alpha = \mathrm{T/B}$. Once that is established, subsequent estimates for beetle density are generated from $\mathrm{B} = \mathrm{T}/\alpha$. The units are in squared distance, but otherwise α is simply a conversion factor. Assuming the density of beetles over space is constant, calculation of α requires knowing the proportion of beetles captured by a trap that originate from a distance r from that trap. If P(r) is the proportion of captured beetles that emerged at distance r from the trap, then the total trap catch from all distances will be

$$\mathrm{T} = \int^{\infty} 2\,\pi\, \mathrm{r}\, \mathrm{P(r)}\, \mathrm{B}\, \mathrm{dr} \tag{11.6}$$

where the integral is taken from zero to infinity.

If we know P(r) by estimation from field studies, then we can estimate α from

$$\alpha = T/B = 2\,\pi \int^{\infty} r\,P(r)\,dr \qquad (11.7)$$

Turchin and Odendaal emphasized that it is not necessary to assume that all beetles at distance r from the trap are equally trappable, as the determination of r from field studies and averaging them will yield integrated (cumulative) values. Stand conditions and wind will have major effects on this lack of equality in trap capture probability. Although this example used southern pine beetles, it could be used for other species with appropriately different parameter values.

Ostrand and Anderbrant (2003) extended the notion of the effective sampling area of Turchin and Odendaal (1996) by noting that the effective sampling area, α, can be regarded as an area x probability volume. If one distributes this volume over the area within the sampling range, r_s, and uses P(r), one can estimate how much of this volume is within a certain distance of the trap. A function can then be determined for the 'cumulative proportional catch' (CPC) of insects originating from distances up to r from:

$$CPC(r) = 2\,\pi\,\alpha^{-1} \int^{r} r\,P(r)\,dr \qquad (11.8)$$

The sampling range, r_s, is then compared to a transformation of α, namely $r_\alpha = \sqrt{(\alpha/\pi)}$, and then the 'catch concentration' (CC) is

$$CC = r_\alpha / r_s \qquad (11.9)$$

If $r_\alpha << r_s$, then only a small proportion of the catch comes from near the trap. If $r_\alpha \approx r_s$, then most of the catch comes from near the trap, and immigration to the immediate vicinity is minimal.

Based on pheromone trapping data for the European sawfly, *Neodiprion sertifer* Geoffroy, the relationship between P(r) and r was obtained from the seasonal equation: P(r) = 0.198–0.0656 log(r), and from this P(r) = 0 yielded a seasonal sampling range of 1,040 m. The corresponding equation after 24 h was: P(r) = 0.144–0.0510 log(r), with a sampling range of 670 m. The effective sampling area, α, was calculated as 48,705 m^2 (=4.8705 ha) with a corresponding radius of about r_α = 125 m. The catch concentration, then, was 125/1,040 = 0.12. Calculation of the cumulative proportional catch (CPC) indicated that 50 % of the seasonal catch came from greater than 450 m, while 10 % originated more than 800 m from the trap. On the other hand, the CPC for the first 24 h yielded about 300 and 450 m for the same proportions. In total, 112 *N. sertifer* were captured during their entire flight period, so that with T = 112 and α = 48,705, the density of males was computed to be B = 112/4.8705 = 23/ha.

These forgoing considerations outline the major features that need to be included in the modeling of the effectiveness of trapping insects in general and fruit flies in particular. It appears reasonable that certain formulations may be better suited to some taxa than to others.

3 Modeling of Trapping for Detection

3.1 Probability of Capture of Insects

Capture efficiency and the probability of capture have been considered for many years. Traps/baits that contain pheromone or powerful fruit fly attractants, like methyl eugenol, will normally capture many times more insects than those that contain less powerful attractants. However, capture rates vary widely with the circumstances: Calkins et al. (1984) captured 14.4 % and 12.9 % of Caribbean fruit flies (*A. suspensa*) using a gridwork of McPhail traps at a density of 4,500 traps per km^2. Lance and Gates (1994) found an overall recovery rate of 0.6 % of released Mediterranean fruit fly males using Jackson traps baited with trimedlure at four traps per km^2 but obtained 23–27 % recovery when using trimedlure-baited yellow sticky panels at a density of 1,000 panels per km^2.

3.2 Detection

In trapping for detection, all that is required to detect the presence of a species in an area is to capture one individual. This will generally be much easier than to trap sufficient numbers to use in the estimation of population size. Also, at higher capture rates, it is more likely that at least one individual will be captured. The mathematical procedures for declaring that an invasion by a particular insect species has not yet occurred are virtually identical to those of declaring that eradication of a species from a given area is complete (see Barclay and Hargrove 2005; Barclay et al. 2005; Barclay and Humble 2009). The approaches used so far have all used probability functions in one way or another.

Kuno (1978) apparently made the first quantitative attempt to solve the problem of declaring a species to be absent from the area of interest or at minimal density, but since it was published in Japanese, it garnered little attention. Kuno subsequently (1991) published an update that generalized his first treatment for finite populations. His approach was to sample sequentially by trapping and then to make a judgment from a succession of zero captures. Using a hypergeometric probability function, which is suitable for small populations and a limited number of sampling units, he calculated the probability that the density of the remnant (or founder) population was at or above a certain low density (probability of a unit being occupied being p_0) and then based his assessment on the sequence length of zero captures. The critical length, n_0, of zero captures is given by

$$n_0 = N\left(1 - \alpha^{1/Np_0}\right) \approx \log(\alpha)/\log(1 - p_0), \text{for small } p_0 \quad (11.10)$$

where N is the total number of sampling units that might contain insects (e.g., fruits or traps), and α is the significance level for rejection. Kuno noted that the length of the sequence of zero captures increases nearly inversely with the size of the assumed p_0; p_0 is assigned to be some very small probability of pest occurrence that is deemed to be acceptable. Thus, for very small values of p_0, the required length of the sequence becomes impossibly long. For example, if $N = 50{,}000$ and the required value of $p_0 = 0.0001$, then the length of the sampling sequence yielding zero captures would need to be 30,095, although this may be achievable if routine trapping is being done.

Calkins et al. (1984) provided a table of probabilities of detecting *A. suspensa* at four population densities and at 12 McPhail trap densities using the formula

$$P = (1 - q^n) \quad (11.11)$$

where P is the probability of a single trap capturing a fly from a given population, q is the probability of not collecting a fly, and n is the number of traps. These probabilities of trapping at least one fly increase with both the population size and the number of traps.

Barclay and Hargrove (2005) provided two related methods for declaring eradication of a species from a given area. The first method deals with a spot infestation. In the area of attraction around a single trap, the probability of catching a given insect each day is σ, called the detectability, and the probability of not catching it is $1 - \sigma$. Then, if there are k insects in the area of attraction, and if catches are independent, the conditional probability of catching no insects during an activity, or sampling, period is:

$$p\left(0|k\right) = (1 - \sigma)^k \quad (11.12)$$

and the mean number caught per activity period is $k\sigma$.

For convenience, we consider k to be 1, as it is a conservative choice, because the probability of detection if only one individual is present is less than for any greater number of insects. This allows us to simplify the resulting probability to:

$$p\left(0|k = 1\right) = (1 - \sigma) \quad (11.13)$$

If the traps are used for n days and if there is one insect present in the area of attraction, then the probability of no insects being caught in the trap for n days is:

$$p\left(0|k = 1\right) = (1 - \sigma)^n \quad (11.14)$$

If we are using a rejection level of α for our test, then we can solve for the number of days required to make the assertion that there are no insects present:

$$n = \log(\alpha)/\log(1 - \sigma) \tag{11.15}$$

If large economic issues or major health issues depend on this declaration, then we may adopt $\alpha = 0.01$ or even smaller. If we use $\sigma = 0.01$ and $\alpha = 0.01$, then the number of trapping days for such a declaration for tsetse, for example, would be $n = \log(0.01)/\log(0.99) = 458$ days, a long trapping sequence. This approach does not take account of dispersal but could be modified to do so, which would greatly decrease the length of the required trapping sequence because dispersal increases the sampling range of each trap.

The second method dealt with area-wide trapping of a formerly large population. Eradication is normally attempted over large areas. Even in cases of population remnants or spot infestation, an area-wide approach is required to assure that all such spots are covered, so we considered a different approach, one involving area-wide sampling (Barclay and Hargrove 2005). We define:

A Area sampled (in km^2).
k Total surviving pests in A, randomly distributed throughout *A*.
σ The conditional probability that an insect is caught by the only trap present in that area.
s Number of traps present in all of the target area.
n Number of days for which each trap operates.
C Probability that no trap captures any insects when some are present.

If there are *s* traps in the area, operating for *n* days, and if *s* is small enough that the traps act independently, then the probability ($C(k, s, \sigma, n)$) that none of the traps catches any of the *k* insects is:

$$C(k, s, \sigma, t) = \exp(-s\, n\, \sigma\, k/A) \tag{11.16}$$

Then, the probability, p(0), of observing a sequence of n zero captures, if there are pests in the control area is:

$$p(0) = \exp(-sn\sigma\rho) \tag{11.17}$$

where $\rho = k/A$ is the population density. If the probability of a sequence of zero catches is to be below 0.01, then we require that:

$$\exp(-sn\sigma\rho) < 0.01 \tag{11.18}$$

and this yields

$$-sn\sigma\rho < \ln(0.01) = -4.605 \tag{11.19}$$

Then, if t is determined, σ is known, and ρ can be estimated:

$$s > 4.605/n\sigma\rho \tag{11.20}$$

As an example of the application of the inequality above, assume that the control area (A) is 10 km^2, the number of days for trapping (n) is 20, the number if traps (s) is 10, the conditional probability of capture (σ) on any given day of one insect that is present is 0.01, and the density of insects (ρ) is 1/km^2, is:

$$\exp(-sn\sigma\rho) = \exp(-(10)(20)(0.01)(1)) = \exp(-2.0) = 0.135 \tag{11.21}$$

which is somewhat above the normal rejection level; either more traps or more trapping days are needed. The number of traps can be calculated by (11.21) above as:

$$s > 4.605/n\sigma\rho = 4.605/(20)(0.01)(1) \approx 23 \tag{11.22}$$

This approach implicitly includes the effects of dispersal on trapping efficiency via the conditional probability, σ.

Using a very similar approach, Clift and Meats (2004) and Meats and Clift (2005) assess the probability of eradication of fruit flies in Australia by computing the probability that the fly population is below a certain density; this density can be adjusted to represent the critical density for population persistence, below which eradication would occur via the Allee effect. The number of traps and the length of trapping time can then be computed knowing the maximum acceptable density, somewhat similar to the approach of Kuno (1991). An operational protocol had been previously established in Australia (Clift and Meats 2004) that specified that eradication could be claimed three generations (based on degree-day models) plus 28 days after the last fly has been trapped. If one gets negative results, normally after three generations in some programs, one rejects the null hypothesis that flies are present. However, trapping records from an Asian Papaya Fruit Fly campaign indicated that when trapping at only one trap per 1.5 km^2, 12 weeks of successive zeros could occur when flies were still present (Clift and Meats 1997; Clift et al. 1999). In addition, if there is a residual population and it is too small to be easily trapped, then it will naturally increase when control stops, and so it becomes more detectable over time (Barclay and Hargrove 2005). In insects, such as tsetse, that have a slow rate of reproduction this requires a long period, whereas for fruit flies that have a much higher reproductive rate, any remnant population should soon become detectable (Shelly et al. 2010). The analytical methods of Meats and Clift (2005) represent a quantitative refinement of the ad hoc code of practice previously in use.

Regan et al. (2006) proposed another approach that used cost-benefit analysis to compare the costs of further trapping, after a few zero records have been obtained, with the costs of an outbreak if eradication was assumed too soon, and stochastic dynamic programming was used to determine the optimal time to stop. Rout et al. (2009a, b) extended the treatment of Regan et al. (2006) based on patterns of sightings, using methods developed by Solow (1993) in which a constant sighting rate is assumed. Their results are generally similar to those of Regan et al. (2006). However, in a majority of operational situations trapping networks would not stop and would continue to be serviced routinely to confirm the pest free status or to detect early any re-infestation after eradication or the incurrence of an exotic fruit fly pest.

Manoukis and Hoffman (2013) have constructed an agent-based simulation that models the actions of individual insects to estimate the time until extinction of populations of *C. capitata*, and their results agree well with data from seven outbreaks in California. They included ecological characteristics of *C. capitata* and include low density demographic effects.

The analytical methods presented above provide a way of objectively assessing the probability of eradication having occurred or invasion not having occurred. The common theme here is the calculation of the probability that the population is non-zero and then rejecting the non-zero hypothesis, or that the density is below some very small value, which may then result in self-extinction as a result of the mate-finding Allee effect. In many cases, these methods prescribe a very long trapping sequence in order to satisfy the probability requirements, usually because of trapping inefficiency, and shortcuts may have to be implemented in some cases.

4 Modeling Population Suppression or Eradication Using Attraction to Baits or Traps

4.1 Simple Model for Attraction and Calculation of Equilibrium

Barclay (1987a) presented a model for attraction to traps and killing insects consisting of three equations, one each for virgin females (V_t), fertilized females (F_t), and males (M_t) on day t:

$$V_{t+1} = a\,F_{t-k};\ F_{t+1} = syV_t + syF_t;\ M_{t+1} = a\,F_{t-k} + szM_t \tag{11.23}$$

This model assumes that males are polygamous, females are monogamous, and males can mate an unlimited number of times each day. The developmental period is k days and is not shown explicitly, except as a time lag in the subscript for recruitment of virgin females and males ($a\ F_{t-k}$). Parameters a and s are the recruitment to adulthood (i.e., the daily oviposition rate per adult female times the proportion of eggs that survive to adulthood) and the daily survivorship of adults, respectively, and y and z are the daily survivorship of adult females and males, respectively, as a result of trapping and killing them. Thus, there are proportions $1-y$ of females and $1-z$ of males that visit the traps each day and are killed. There is no density-dependence included, and if $y = z = 1.0$, the population increases without bound and shows asymptotically geometric increase.

Calculation of Equilibrium: A neutrally stable equilibrium, or steady state in which nothing changes, exists in this model for certain values of the control parameters, y and z, and is found by assuming that there is no change in the size of any of the population components (F, V or M) or the control effort (y and z) from time t to time $t+1$. In that case, $F_{t+1} = F_t, V_{t+1} = V_t$, and $M_{t+1} = M_t$, so that the subscripts can be dropped, and the three equations can be solved algebraically. This leads to the equilibrium: $V = \mathrm{a\,F}$; $M = \mathrm{a\,F}/(1 - \mathrm{sz})$; F is undetermined. The values of y and z for which the equilibrium is valid can then be found by solving the equilibrium values for y and z. This gives:

$$y* = 1/s(a+1) \tag{11.24}$$

These pairs of values of y and z are called critical values or a critical combination, as they separate success (population decrease) from failure (population increase) of the control program. The critical value of y is denoted y^*, but z does not appear in the condition for an equilibrium, because in this model males can mate as often as necessary to fertilize the females, so that even if $z = 0$, the females will all become fertilized by the newly emerged males. This is clearly unrealistic for many species, so this model would be too simple to describe population growth and control in these instances, and a fourth equation would be required to tally males that are too young to mate.

This equilibrium is neutrally stable, because a proportional increase or decrease in V, M, and F will not destroy the equilibrium, but the system will not return to the previous population values. However, the system is unstable with respect to changes in the value of trapping rate. An increase in y (i.e., less intensive trapping) will result in unrestricted population increase, while a decrease in y (more intensive trapping) will result in collapse and elimination of the population, provided trapping is continued until the population disappears.

It must be stressed that the unstable equilibrium is unachievable in the field. The sole purpose in calculating it is to obtain the values of y and z that will separate success from failure in the control effort. Here, the value of z is irrelevant, so that if the value of y achieved by trapping is less than y^*, then the population will decline to zero. If y is greater than y^*, the population will increase without bound.

4.2 *Males Are Limited in the Frequency of Their Mating Ability*

The model above (Eq. 11.23) assumes that males can mate any number of times in a day, and because of this the trap-related survivorship rate of males (z) does not appear in the equation for critical trapping rate, y^*. On the other hand, if males are limited in their ability to fertilize females, then the possibility exists that, because of the reduction of the male population, not all virgin females may be mated on any given day.

Barclay and Hendrichs (2014) have arrived at several cases that should be examined in models of trapping: traps attracting only males vs. both sexes, the frequency at which males are capable of mating, female monogamy vs. female polygamy, the order of mating and visiting traps each day, and whether or not trapping is successful in depriving females of mates.

If males are limited in their mating frequency, then the equations (11.23) above must be modified. In the case of traps attracting only males, we must also distinguish between the cases in which males are in insufficient numbers to fertilize all the available (i.e., virgin) females (male deficit) vs. males are in sufficient numbers for fertilization of all available receptive females (male excess). This yields a large number of equations (Barclay and Hendrichs 2014), most of which are not reproduced here. Below, we give the equations for the two cases of attraction of (i) only males and (ii) of both sexes assuming female monogamy, male deficit, attraction to traps before mating, and males are capable of only one mating per day.

4.2.1 Only Males Are Attracted to Traps

$$V_{t+1} = a\,F_{t-k} + sV_t - szM_t;\;\; M_{t+1} = a\,F_{t-k} + szM_t;\; F_{t+1} = szM_t + sF_t \quad (11.25)$$

The variables F, V, and M are as in Eq. (11.23) above, only not all the virgin females are fertilized each day, because a deficit of males is limiting the mating of females. In fact, the number that are fertilized is exactly the number of males in the population that day. The only control parameter in these equations is z, the proportion of males that do not visit the traps each day. This proportion is constant and is assumed to be independent of population density or any factor other than trap number and efficiency.

The condition for an equilibrium is: $asz = (1 - sz)\ (1{-}s)$, which is found by dropping subscripts and solving the three equations. This condition can be solved for z^*, the critical value of trap survivorship and is:

$$z* = (1 - s)/[as + s(1 - s)] \quad (11.26)$$

This set of equations has a neutrally stable equilibrium at: $V = F[a{-}(1 - s)]/(1{-}s), M = aF/(1{-}sz)$, with F being undetermined. If $z < z*$, then the population collapses and is eliminated; if $z > z*$, then the population increases without bound.

It is instructive to examine the value of z^* under different conditions of daily recruitment (a) and survivorship (s). If $a = 4$ and $s = 0.7$, then $z^* = 0.110$; if $a = 4$, $s = 0.9$, $z^* = 0.027$, if $a = 12$, $s = 0.7$, then $z^* = 0.035$, and if $a = 12$ and $s = 0.9$, then $z^* = 0.009$. In the final case, more than 99 % of males must be killed by traps each day in order to yield suppression and eventual population elimination.

4.2.2 Both Sexes Are Attracted to Traps

$$V_{t+1} = a F_{t-k} + syV_t - szM_t; \; M_{t+1} = a F_{t-k} + szM_{t;} \; F_{t+1} = szM_t + syF_t \quad (11.27)$$

These equations are the same as those in Eq. (11.25) above, except that the survivorship of females at traps, y, is included in the survivorship terms for V and F. This yields the critical male survivorship:

$$z* = (1 - sy)/[as + s(1 - sy)] \quad (11.28)$$

Using the values of recruitment and survivorship of a = 10 and s = 0.9, the values of z^* are: 0.0093 for $y = 1.0$ (no females are trapped), 0.0180 for $y = 0.9$, 0.0265 for $y = 0.8$ and 0.0353 for $y = 0.7$. It is apparent that males must be trapped in very high proportion to be of any assistance to the control program, even if some females are also being trapped (see Fig. 11.1).

If males are either not being trapped or are in excess of available females, then the equations are the same as equations (11.23) and $y^* = 0.0855$ with $a = 12$ and $s = 0.9$. That is, over 91 % of females must be trapped and killed each day at the traps in order to achieve suppression and eventually eradication for the parameter values chosen.

4.3 Factors That Might Affect Success of Male Annihilation Trapping

Barclay and Hendrichs (2014) identified the following features that need to be considered in the models for male annihilation.

- **Order of mating and trapping –** If trapping precedes mating each day, then control by trapping is difficult but possible. If mating precedes trapping, then control is impossible by trapping males alone, even if 100 % of males are trapped and killed each day (Barclay and Hendrichs 2014), as they mate first and subsequent trapping of males is futile. However, if both sexes are attracted to the traps and males are in excess, then the equations are the same as in Eq. (11.23) and control is possible, but trapping males does not contribute to control.
- **Trapping only males or both sexes –** Trapping both sexes allows suppression and eradication much more easily than trapping males alone.
- **Maximum mating frequency of males –** If males are capable of multiple matings each day, then control is much more difficult than with only one mating per day.

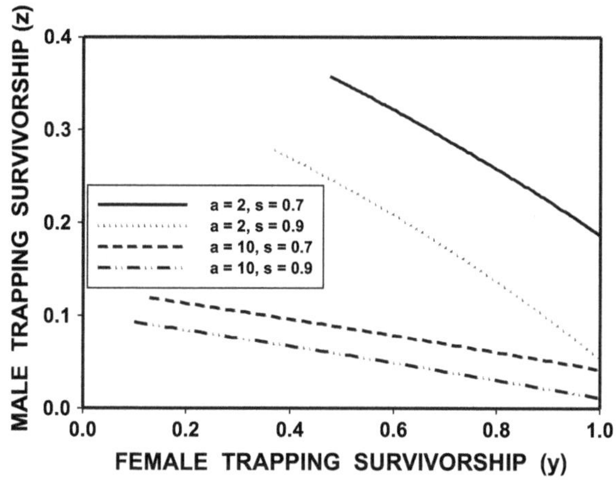

Fig. 11.1 Graphs of the critical values of trapping survivorship of males (z) and females (y) in traps baited with methyl eugenol in which both males and females are attracted and killed. The graph lines represent all possible combinations of trapping males and females that result in the equilibrium being maintained. Values of either z or y less than the critical values will result in population collapse. Values larger than the critical values result in unrestrained population increase. Parameter values for fertility (a) and survivorship (s) were: $a = 2$ or 10, $s = 0.7$ or 0.9 (Reprinted from Barclay and Hendrichs (2014; Fig. 1) with permission)

- **Females monogamous or polygamous** – This distinction makes little difference to the ease of control by trapping.

5 Interactions Among Control Methods

5.1 The Use of More Than One Pest Control Method

IPM implies the use, either sequentially or concurrently, of more than one method of pest control. Two or more pest control methods may interact in such a way that the effect of the combination is greater or less than the sum of the individual contributions of each control method. The interaction may result from differences in the density-dependent efficiency of the control methods (Barclay 1987b) and/or the degree to which increased effort in the different control methods causes increased pest mortality. As Barclay (1992) observed, increasing control effort may lead to decreased efficiency, so that two or more methods in combination would each require less killing power, and the sum of the control efforts for suppression or eradication would be less than either control method used alone.

Pest control methods can interact with each other synergistically (i.e., greater than simply additively), additively (where there is no interaction), or antagonistically (where the action of one method inhibits the action of another method).

Suckling et al. (2012) provided examples of each and added a fourth type of interaction, redundant, in which one method makes the second one irrelevant, as would be the case in the application of a broad spectrum insecticide followed by a selective insecticide. The combinations that are synergistic involve either two density-dependent (d-d) control methods (such as Sterile Insect Technique [SIT] and biological control or mating disruption) or one d-d method in combination with one density-independent (d-i) method (such as insecticides). Additive interactions might involve applications of two insecticides on different life stages, whereas the use of insecticides on a species already under control of a predator or parasitoid would usually yield an antagonistic interaction (Volterra 1926; DeBach 1974).

5.2 *Isoclines*

Constructing isoclines is a convenient way to assess interactions between two pest control methods. An isocline in this context is a line or curve that represents all combinations of pairs of values of the two control efforts that are critical, i.e., the pairs of values that will yield the equilibrium and thus separate success from failure of the control program (Fig. 11.2). Since these pairs of values yield an equilibrium, and a control program targets either suppression or eradication, the actual values of control would have to be more extreme than those specified by the isocline. Isoclines allow a quick visual inspection of the interaction of the two control methods. If the isocline curves steeply downward, then there is a positive interaction, or synergism (line S in Fig. 11.2). If the isocline is straight or curves upward, there is a negative interaction, interference or antagonism (line A of Fig. 11.2). If two control methods do not interact and are independent of each other (line I of Fig. 11.2), then the survivorships are multiplicative, and the isocline is hyperbolic and curves gently downwards (Barclay 1992). This indicates that even no interaction will produce some assistance to the control program, as it appears that a moderate amount of mortality is easy for a control method to produce, but as control effort becomes greater, the mortality produced is nonlinear so that twice the control effort yields less than twice the mortality (Barclay 1992). In what follows, we construct isoclines of the critical control efforts (e.g., number of traps, rate of release of sterile males, amount of pheromone in traps, etc.) that separate success from failure of the control program.

5.3 *Interactions Between Odor-Baited Traps and Other Control Methods*

In a tome that foreshadowed much of the later modeling of combinations of pest control methods, Knipling (1979) demonstrated that the combination of non-sex-

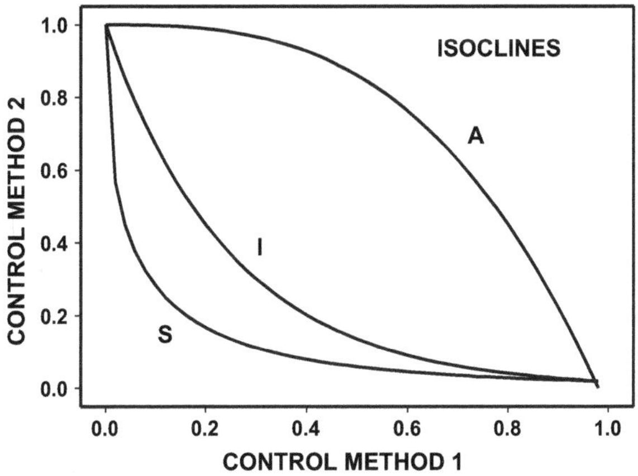

Fig. 11.2 Three hypothetical isoclines showing synergism (*line S*), independent action (*line I*) and antagonism (*line A*) between two pest control methods. Two hypothetical control methods are shown, with the critical efforts of each acting alone having been normalized so that the control efforts are shown between zero and one, one being the critical effort for each (Reprinted from Barclay and Hendrichs (2014; Fig. 5) with permission)

specific odor traps and either concurrent sterile insect releases or the use of pheromone traps was more effective than any of the three alone. This provided a quantitative argument for the integration of multiple control methods either in sequence or simultaneously. Knipling's models were very simple, but his ideas were ahead of their time.

The modeling results of Barclay (1988, 1992) and Barclay and van den Driessche (1977, 1989) reinforced Knipling's conclusions. Most of their models assumed that males were limited to one mating per day and that the non-sex-specific traps attracted and killed both sexes in equal proportions. The models of Barclay and van den Driessche (1977) investigated the generality of Volterra's (1926) result that mortality imposed on a predator–prey system would result in the decrease of the predators and an increase in the prey at equilibrium. No counter-examples to this result were found, although when only the prey were being killed the prey equilibrium remained the same and only the predators decreased. This result also applies to traps containing insecticide, especially if the attractant was also attractive to the predators or parasitoids. The veracity of this result has been amply documented by DeBach (1974). The models of Barclay (1988) combined food-baited trapping with pheromone-baited trapping in which the traps contained either insecticide or chemo-sterilant in the four combinations. He found synergism between the two trap types for both sterilants and insecticides. Barclay and van den Driessche (1989) paired sterile insect releases with either pheromone-baited trapping or non-sex-specific odor-baited trapping and found that, when pest reproductive rates are high, the best combination was sterile insect releases together with pheromone trapping in which the traps contained chemosterilant. When pest reproductive rates are low, as in tsetse, the optimal combination was sterile insect releases together with food-baited traps containing insecticide. In field trials, Navarro-Llopis et al. (2011) also found the combination of sterile releases together with traps containing chemosterilant to be useful for controlling Mediterranean fruit fly.

5.4 Models Using Both Methyl Eugenol Baits/Traps and Sterile Male Releases Simultaneously

The conventional approach to combining two or more methods of pest control is to apply them sequentially, either to avoid potential interference between them or because it is logistically easier to do so. However, to achieve full synergism between methods, it may be preferable to deploy them simultaneously. For example, Steiner et al. (1970) used methyl eugenol baits sequentially with sterile male releases against the oriental fruit fly, B. *dorsalis*, the wild males first being reduced using methyl eugenol baits and then sterile males being released in large numbers. To achieve full synergism between methods, however, it may be preferable to deploy them simultaneously. If the sterile males were not attracted in large numbers to the ME baits, then the two methods could be used simultaneously.

Simultaneously deploying male annihilation baits and sterile males appears to be a promising approach of control for those species in which males are attracted to methyl eugenol in baits. The reason is that sterile males exposed to methyl eugenol during maturation and before release are only weakly attracted to traps containing ME (Shelly 1994), so that the wild male population could be reduced by the ME baits, but the sterile male population is not reduced or only slightly so. As a consequence, wild males are largely replaced by sterile males, drastically increasing the sterile to wild male over-flooding ratio (McInnis et al. 2011; Barclay et al. 2014).

We present the equations for two cases of the use of a predominantly male attractant, methyl eugenol, used simultaneously with the release of sterile male adults. The equations for this model are for the case of female monogamy, the attraction of only wild males, and for the two cases of (i) mating occurs before trapping each day and (ii) trapping occurs before mating each day.

(i) **Mating occurs before trapping**

$$\begin{aligned} F_{i+1} &= s\,V_i[M_i/M_i+N_i] + sF_i; G_{i+1} = sV_i[N_i/M_i+N_i] + sG_i \\ V_{i+1} &= a\,F_{i-k};\; M_{i+1} = a\,F_{i-k} + szM_i;\; N_{i+1} = r + sN_i \end{aligned} \tag{11.29}$$

in which F, V, M, and z are as defined in Eq. (11.25) above, and where G and N are the number of females mated to sterile males and the number of sterile males in the population, respectively; r is the daily release rate of sterile males. Here, sterile males are not attracted to ME traps.

(ii) **Trapping occurs before mating**

$$\begin{aligned} F_{i+1} &= sV_i[zM_i/zM_i + N_i] + sF_i; \;\; G_{i+1} = sV_i[N_i/zM_i + N_i] + sG_i \\ V_{i+1} &= aF_{i-k}; \;\; M_{i+1} = aF_{i-k} + szM_i; \;\; N_{i+1} = r + sN_i \end{aligned} \tag{11.30}$$

The critical sterile release rates for these two cases are, respectively,

$$\text{(i)}\, r* = aF[as - (1 - s)]/(1 - sz) \tag{11.31}$$

$$\text{(ii)}\, r* = zaF[as - (1 - s)]/(1 - sz). \tag{11.32}$$

It is apparent that, if trapping occurs before mating, the critical value of sterile releases, r^*, is less by a factor z than if mating occurs before trapping. However, if mating takes place before trapping it does not preclude use of the control system as it does if sterile males are not released (Fig. 11.3), because control is still possible if sterile males are released and also control is still made easier by trapping as well. Also, there is considerable synergism between the two methods, as the isoclines in Fig. 11.3 show that the combination is more effective than either method alone.

5.5 The Allee Effect

The Allee Effect (Allee 1931) is the process whereby the rate of population growth declines or even becomes negative when the population declines below a given threshold, i.e., there is a change from negative to positive density-dependence as the population density declines and is exposed to demographic and environmental stochasticity at very low levels. This Allee threshold is certainly species-specific and probably habitat- and temperature-specific as well. The causes are multiple (Yamanaka and Liebhold 2009), being mainly (i) difficulty of finding mates at very low population density, (ii) decline in cooperative feeding, (iii) lack of predator satiation, and (iv) failure to overcome host defenses. The Allee Effect, in species where it exists, makes eradication easier and invasion more difficult for insect pest species.

The Allee Effect was classified into component and demographic effects by Stephens et al. (1999). Component effects are the mechanisms giving rise to an increase in individual fitness with an increase in population density, and demographic effects are reflected in an increase in population fitness (growth rate) with an increase in population density. Further, demographic effects may be strong (if growth rate becomes negative at very low population densities) or weak

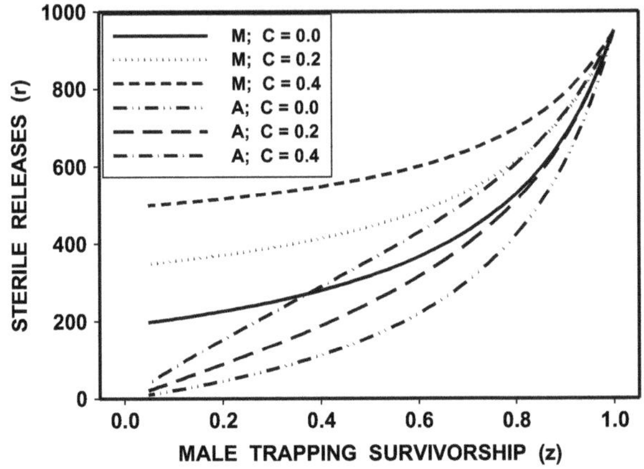

Fig. 11.3 Isoclines showing critical rates for trapping survivorship with methyl eugenol male annihilation baits and the simultaneous release of sterile males. In this case, the distinction between the order of trapping and mating is not as important as it is without the release of sterile males. The curves are very similar for high rates of sterile releases and diverge as methyl eugenol trapping becomes the dominant method in use (lower left). Here *M* refers to mating being before trapping, while *A* refers to attraction preceding mating. The parameter *C* measures the proportional attraction of females to the baits relative to that of males (Reprinted from Barclay et al. (2014; Fig. 3) with permission)

(if growth rate simply declines with density but remains positive even close to zero density).

Previously, ecologists have largely ignored the Allee Effect, as the threshold densities involved were perceived to be unrealistically low. Recently, however, conservation biologists invoked it as a possible cause of extinctions of endangered species at very low population densities. Subsequently, Kramer et al. (2009) found evidence for the Allee Effect in 63 out of 91 studies examined and across all major taxa of animals and plants. That is impressive given the difficulty of even detecting density-dependence in populations, let alone measuring it (Hassell 1978), especially at very low population density. Kramer et al. (2009) found that the dominant factor in producing Allee Effects was difficulty in finding mates, which presumably is more difficult in species with no long range pheromones than in those possessing such pheromones. In addition, Boukal and Berec (2009) found, using two-sex models, that the mate-finding Allee Effect may be more common and stronger than is implied by the single-sex models that are commonly used in population dynamics.

Although the Allee Effect is not really a pest control method, its key implication is that a population can become extinct without the need and expense of killing the last individuals. Therefore, it may combine with pest control methods to yield synergistic results in a similar way that existing biological control can be combined with control methods to achieve better results than either could alone. This possibility has been actively investigated in recent years, and the results are noteworthy

(Liebhold and Bascompte 2003; Liebhold and Tobin 2008). Berec et al. (2007) and Suckling et al. (2012) noted that density-dependent control methods will generally raise the Allee threshold and thus provide assistance to the control program. Blackwood et al. (2012) noted that pesticides in combination with mating disruption induce the Allee Effect and are more efficient in combination than either alone; the same effect is seen with pheromone trapping and predator augmentation.

6 Life Table Case Study for *Bactrocera Dorsalis* Control Using Methyl Eugenol Trapping and Sterile Male Releases

6.1 Bactrocera dorsalis

This tephritid fruit fly is one of the major pest species in the genus *Bactrocera* with a very broad host range of cultivated and wild fruits. Recent evidence showed that *Bactrocera invadens* Drew, Tsuruta, and White, *Bactrocera papayae* Drew and Hancock and *Bactrocera philippinensis* Drew and Hancock belong to the same biological species as *B. dorsalis* in spite of considerable color variation of populations throughout its distribution (Schutze et al. 2013). It is endemic to China, Southeast Asia and the Indian subcontinent and has been introduced to Africa, Hawaii, and other islands of the Indian and Pacific Oceans. Males of the species respond strongly to methyl eugenol and are pollinators of a number of plants, including wild orchids.

Virgin *Bactrocera* spp. females require protein before they become receptive to mating, and this causes a delay in the age at which they are receptive to mating (Drew and Yuval 2000). *B. dorsalis* females from laboratory strains under ideal nutritional and temperature conditions require at least six days after emergence before they are receptive (Vargas et al. 1984). In the wild, virgin females may require in some situations up to 29 days for sexual maturation, allowing much more mortality to occur than if they were receptive upon emergence (Arakaki et al. 1984).

Equations were developed in preparation for a life table treatment of *B. dorsalis*. We later use the equilibrium (in which the values of all population components remain the same) to derive critical values of z, z^*, and y, y^*, the survivorship of males and females, respectively, after daily attraction to sources baited with methyl eugenol. The recruitment and the survivorship of the pre-adult components can be reduced to two numbers: mean daily fertility (μ), called ‘mean fertile eggs per fly-day’ by Carey (1989) and pre-adult survivorship (here labelled γ), which is the proportion of eggs that develop to the adult stage emerging from pupae.

6.2 *Life Tables*

Life tables usually start with some convenient number of eggs, such as 100 or 1,000, and tabulate the numbers remaining in short time periods, such as daily or weekly. From these data, estimates, such as life expectancy, generation time, and total reproduction per generation, can be obtained.

The proportion of individuals in the original cohort at age zero (i.e., at oviposition) that are still alive at age x is l_x, or the survivorship to age x, with l_0 being 1.0. The mean number of eggs laid per female of age x during the interval x to x + 1 is m_x, the age specific fecundity. The mean daily fertility, μ, is:

$$\mu = \Sigma\, l_x h_x m_x / \Sigma\, l_x, \qquad (11.33)$$

where h_x is the hatch rate of eggs at age x as a proportion of the eggs oviposited (Carey 1989) by that age class, and the sum is taken over all ovipositing age classes. The survivorship from oviposition to age of adult emergence is termed γ. If the age at adult emergence is e, then γ is the survivorship to age e divided by the proportion alive at age zero:

$$\gamma = l_e / l_0 = l_e \qquad (11.34)$$

Life tables offer the advantages of being compact and easy to use. The disadvantages are that they are usually derived from a laboratory population that is often maintained under ideal conditions for growth and survivorship and thus overestimate the quantities γ and μ in natural settings that are then used in determining the control rates required. In particular, the change from six days needed for sexual maturation in the laboratory up to 29 days in the field would allow significant amounts of mortality to occur to the females prior to oviposition, and thus the life table would better estimate the species' resilience to control efforts. Furthermore, the reduced availability of protein to wild females in nature, which significantly restricts their egg production capacity, compounds this bias. This kind of error is fortunate for pest control managers, as it yields conservative estimates of control success in which the actual control effort required is less than that predicted from laboratory populations.

6.3 *An Age-Structured Model for* Bactrocera dorsalis *Using Only Methyl Eugenol Trapping for Control*

6.3.1 The Age-Structured Model

Barclay and Hendrichs (2014) used an age-structured model in which methyl eugenol is the attractant and identified several cases that must be considered.

Methyl eugenol attracts mainly males, because it is a requisite precursor for male pheromone production, although methyl eugenol also attracts a certain proportion of females. The following is an outline of their model, with equations given in equilibrium form without time subscripts. The equations are entirely density-independent and include age-dependent survivorship of larvae and pupae and age-dependent fecundity and survivorship of adults. To accommodate the delay in mating of females by kv days, we let the kvth term in the virgin female sequence of age classes be the one that contributes to the production of mated females (F) using $y^{kv}\ V_1\ \Pi^{\ (kv)}\ s_i$, which accommodates the natural survivorship of virgins ($\Pi^{\ (kv)}\ s_i$) and the imposed trapping survivorship (y^{kv}) in the equation for F_1 below. Here, we use the notation $\Pi^{\ (kv)}\ s_i$ to mean the product $s_1 \cdot s_2 \cdot .... \cdot s_{kv}$ of the adult female survivorships from emergence until mating and first oviposition.

At equilibrium, the numbers of each population component remains the same, so the equilibrium values of eggs are found by dropping the time subscript, *t*:

$$E_T = E_1 + E_2 + E_3 + \ldots + E_{ke} = ke\, E_1, \; = ke \sum\nolimits^{(kf)} F_i m_i h_i$$
$$= ke \sum\nolimits^{(kf)} \mu_i F_i \tag{11.35}$$

Here, *ke* is the number of days required for egg hatch, the subscript *T* denotes the total numbers for the life stage (eggs), and $\sum^{(kf)}$ denotes the sum from $i = 1$ to *kf* of the expression to the right of the summation sign, i.e., $F_i\ m_i\ h_i$, while μ_i is the product $m_i\ h_i$. The sum is taken over the *kf* mated female age classes. There is no mortality included in this sum, because the only eggs we consider are those that will hatch.

The equilibrium for larvae is:

$$L_T = L_1 + q_1 L_1 + q_1 q_2\, L_1 + \ldots + \left(\Pi^{(kl-1)} q_i\right) L_1$$
$$= \left(1 + \mathrm{q}_1 + \mathrm{q}_1 \mathrm{q}_2 + \ldots + \Pi^{(\mathrm{kl}-1)} \mathrm{q}_\mathrm{i}\right) \mathrm{E}_\mathrm{ke} \tag{11.36}$$

in which $\prod^{(kl-1)} q_i$ is the product from $i = 1$ to $kl - 1$ of the expression to the right of the product sign, i.e., q_i. The superscript *kl* is the number of larval age classes, and the q_i values are the age-specific larval survivorships and can be summed as shown when they are known. Also, $L_1 = E_{ke}$. The product, $\prod^{(kl-1)}\ q_i\, L_1$, is taken over $kl - 1$ larval age classes.

The equilibrium for pupae is similarly:

$$P_T = P_1 + u_1P_1 + u_1u_2P_1 + \ldots + \left(\Pi^{(kp-1)}u_i\right)P_1$$
$$= \left(1 + u_1 + u_1u_2 + \ldots + \Pi^{(kp-1)}u_i\right)\left(\Pi^{(kl)}q_i\right)E_{ke}. \quad (11.37)$$

since $P_1 = q_{kl}\,L_{kl}$ and also $L_{kl} = L_1 \prod^{(kl\text{-}1)} q_i$. The products, $\prod^{(kp\text{-}1)} u_i$, and $\prod^{(kl)} q_i$ are taken over *kp*-1 pupal age classes and the *kl* larval age classes. Then,

$$\gamma = \left(\Pi^{(kp)}u_i\right)\left(\Pi^{(kl)}q_i\right). \quad (11.38)$$

The equilibria for virgin females (V), mated females (*F*), and males (*M*) are similarly:

$$V_T = V_1 + V_2 + V_3 + \ldots + V_{kv} = \left(1 + s_1y + s_1s_2y^2 + s_1s_2s_3y^3 + \ldots + y^{kv-1}\Pi^{(kv-1)}s_i\right)$$
$$\left(\Pi^{(kp)}r_i\right)P_1 = \left(1 + s_1y + s_1s_2y^2 + s_1s_2s_3y^3 + \ldots + y^{kv-1}\Pi^{(kv-1)}s_i\right)\left(\Pi^{(kp)}r_i\right)$$
$$\left(\Pi^{(kl)}q_i\right)E_{ke} = \left(1 + s_1y + s_1s_2y^2 + s_1s_2s_3y^3 + \ldots + y^{kv-1}\Pi^{(kv-1)}s_i\right)\gamma\,E_{ke}$$
$$(11.39)$$

Similarly,

$$F_T = F_1 + F_2 + F_3 + \ldots + F_{kf}$$
$$= \left(1 + s_1y + s_1s_2y^2 + \ldots + y^{kf-1}\Pi^{(kf-1)}s_i\right)\left(y^{kv}\Pi^{(kv)}s_i\right)\gamma\,E_{ke} \quad (11.40)$$
$$M_T = M_1 + M_2 + M_3 + \ldots + M_{km}$$
$$= \left(1 + s_1z + s_1s_2z^2 + s_1s_2s_3z^3 + \ldots + z^{km-1}\Pi^{(km-1)}s_i\right)\gamma E_{ke} \quad (11.41)$$

Here, s_i is the adult survivorship from age i to age i + 1, y_i is the survivorship of females of age i after trapping, and z_i is the trap survivorship of males of age class i. Although there is evidence that capture probability is age-dependent (Wong et al. 1989; Shelly et al. 2008), we assume for simplicity that all adult ages have equal capture probability.

The above equations can be reduced to summary equations and solved using the two quantities: net pre-adult survivorship (γ) and the mean daily fertility (μ). The cases of male excess and male deficit are shown below.

6.3.2 Males in Excess of Receptive Females

Here, the first three equations tally the earliest age class of each of the three population components (V, F, and M), and the last three equations tally the corresponding totals over existing age classes for the three components:

$$V_1 = \gamma\mu F_T;\ \ F_1 = s_{kv}y\, V_{kv} = y^{kv}V_1\, \Pi^{(kv)}s_i;\ \ M_1 = \gamma\mu F_T \tag{11.42}$$

$$V_T = V_1\Sigma^{(kv-1)}y^j\,\Pi^{(j)}s_i;\ \ F_T = F_1\Sigma^{(kf-1)}y^j\,\Pi^{(j)}s_i;\ \ M_T = M_1\Sigma^{(km-1)}z^j\,\Pi^{(j)}s_i \tag{11.43}$$

The product $\gamma\ \mu$ takes the place of a in the models of Sect. 11.4 above. These equations yield an expression for y that is not directly solvable but can be solved by numerical methods. It is:

$$\gamma\mu\left(\Sigma^{(kf-1)}y^j\Pi^{(j)}s_i\right)\left(\Pi^{(kv)}y^js_i\right) = 1.0 \tag{11.44}$$

Note that z does not enter into this equation, as males are in excess and do not contribute to control.

6.3.3 Males Fewer than Receptive Females

$$V_1 = \gamma\mu F_T;\ \ F_1 = s_{kv}z\, M_T;\ \ M_1 = \gamma\mu F_T \tag{11.45}$$

$$\begin{aligned} V_T &= V_1\Sigma^{(kv-1)}y^j\Pi^{(j)}s_i + s_{kv}(V_{kv} - M_T)\Sigma^{\infty}\Pi^{(j)}s_iV_1; \\ F_T &= F_1\Sigma^{(kf-1)}y^j\Pi^{(j)}s_i;\ \ M_T = M_1\Sigma^{(km-1)}z^j\Pi^{(j)}s_i \end{aligned} \tag{11.46}$$

These equations yield critical values of trapping for pairs (y and z) given by:

$$(s_{kv}z)\ (\gamma\mu)\left(\Sigma^{(km-1)}z^j\Pi^{(j)}s_i\right)\left(\Sigma^{(kf-1)}y^j\,\Pi^{(j)}s_i\right) = 1.0 \tag{11.47}$$

and this must be solved numerically for z and y pairs.

6.4 Translation of Notation into Life Table Notation

We can translate the quantities in the age-structured model above into life table notation by noting that if l_x is the proportional survivorship from oviposition to age **x**, and if the preadult stages total e days, then $l_e = \gamma$. In addition, the survivorships of the adult stages are $l_{e+1}, l_{e+2}, l_{e+3}, \ldots$ etc. In the notation used above, survivorship rates of the adult stages are s_i for the survivorship from the ith adult age to the $i+1$st adult age. Then $l_{e+1} = \gamma s_1 = s_1 l_e, l_{e+2} = \gamma s_1 s_2, = s_2 l_{e+1}, l_{e+3} = \gamma s_1 s_2 s_3 = s_3 l_{e+2} \ldots$ etc., so that: $\gamma\Sigma^{(kf-1)}\Pi^{(j)}s_i = \gamma[s_1 + s_1s_2 + s_1s_2s_3 + \ldots + s_1s_2 \cdots\cdots s_{kf-1}] = l_{e+1} + l_{e+2} + l_{e+3} + \ldots + l_{e+1+kf-1}$ so that Eq. (11.44) can be written as:

$$\mu\left(\Sigma^{(kf)}y^x l_{e+x}\right)y^{kv}l_{e+kv} = \gamma, \text{ where } 1 \le x \le kf \tag{11.48}$$

and Eq. (11.47) can be written as:

$$\mu s_{kv} z\left(\Sigma^{(km)}z^x l_{e+x}\right)\left(\Sigma^{(kf)}y^x l_{e+x}\right) = \gamma \tag{11.49}$$

6.5 Parameter Values for Bactrocera dorsalis

In their laboratory study of development, Vargas et al. (1984) found that the total pre-adult mortality of eggs, larvae, and pupae of *B. dorsalis* was 37 % ($l_e = 0.63$). The immature stages had a total of about 20 days duration. Fecundity and egg hatch rate curves were given from day 26, when the first oviposition occurred by six day old mated adult females, until oviposition had virtually ceased about day 126. Barclay and Hendrichs (2014) computed the mean daily fertility (mdf) to be 8.76. This is μ, in Eq. (11.44). The period between adult female emergence and the onset of oviposition is taken as six days (=kv).

The fecundity and survivorship data of Vargas et al (1984) were used to facilitate the computation of Eq. (11.44), with male excess. The factors in Eq. (11.44) were evaluated as follows: $\mu = 8.76$, $kv = 6$, $kf = 99$, $e = 20$, $l_e = \gamma = 0.63$, and $l_{e+kv} = 0.59$. Thus, after iteratively calculating the value of the left hand side of Eq. (11.44) for a variety of values of y, $y^* = 0.724$ was obtained. If 27.6 % (i.e., 100.0 – 72.4) of females are trapped and killed every day, then the population is maintained at equilibrium. If $y < 0.724$, then the population will decline and be eliminated with continued trapping. Although the proportion of females attracted to methyl eugenol traps is usually very low (Steiner 1952), if a female attractant is also used in the ME traps, then such a combination would, in principle, be realizable, although in practice this appears unlikely.

6.6 Control of Bactrocera dorsalis *Using Methyl Eugenol Trapping and Simultaneous Sterile Male Releases*

The age-structured equations for the pre-adult and adult stages are given by Barclay et al. (2014) and are an extension of the equations for ME trapping without sterile male releases given above. Again, the effect of the pre-adult components is reduced to the two numbers: 'mean fertile eggs per fly-day' (μ) and pre-adult survivorship (γ).

Equations of the Model and Relation to Life Tables: We again use data from the study by Vargas et al. (1984) to estimate values of the control parameters z, y, and r for a population that is being suppressed by means of traps (baits) containing ME attracting both wild males (at daily trapping rate $1 - z$) and females (at daily trapping rate $1 - y$) and also for which r sterile males are being released daily. Sterile males are assumed to have been exposed to ME before release and are not to be attracted to baits. We also use simplified equations, letting $s = s_i$ for all i and estimating s by the geometric mean of the product: $s_1 \cdot s_1 \cdot \ldots \cdot s_{kf}$. In this case, we obtain an explicit solution for r*. The value of μ in the simplified model using Eq. (11.33) and a geometric survivorship curve is 11.69. We provide here only the summary equations; the full model is given by Barclay et al. (2014):

$$F_1 = V_\tau[zM_T/(zM_T + N)]; G_1 = V_\tau[N/(zM_T + N)]; V_1 = \gamma\mu F_T; \text{so } V_\tau = \gamma\mu F_T(sy)^{kv}; \\ M_1 = \gamma\mu F_T; N_1 = r \tag{11.50}$$

$$F_T = F_1/(1 - sy);\ \ G_T = G_1/(1 - sy);\ \ V_T = V_1[1.0 - (sy)^{kv}]/(1 - sy); \\ M_T = M_1/(1 - sz);\ \ N_T = r/(1 - s) \tag{11.51}$$

in which F_1, G_1, V_1, M_1 and N_1 are the values of the earliest age classes of the fertile females, sterile male-mated females, virgin females, wild males, and released sterile males, respectively, and the subscript T refers to the totals of these categories (F, G, V, M, and N). Sterile females are not released here.

Solving these for steady state, we obtain:

$$r* = z\,\gamma\,\mu\,F_T(1 - s)\left[\gamma\,\mu\,(sy)^{kv} - (1 - sy)\right]/(1 - sy)\,(1 - sz) \tag{11.52}$$

With the parameter values listed above ($\mu = 11.69$, $\gamma = 0.63$, $s = 0.961$, $kv = 6$, $s^{kv} = 0.7877$ and $F_T = 100$) for *B. dorsalis*, the values of $z*$, $y*$ and $r*$ when acting alone are: $z* = 0.00157$; $y* = 0.724$ and $r* = 118{,}274$ sterile males per day. This value of r* may seem like a huge amount compared with 100 females, showing that suppression will be difficult by SIT alone without integration with any other suppression methods. However, if we substitute $M_T = M_1/(1 - s) = \gamma\ \mu\ F_T/(1 - s)$ into the equation for $r*$ above with $y = z = 1.0$, we obtain the critical daily release rate:

$$r* = M_T\left[\gamma\,\mu\,(s)^{kv} - (1 - s)\right] = 5.77M_T \tag{11.53}$$

which is approximately 6 times the equilibrium male population (M_T is 20,515 from Eq. 11.51 above). This discrepancy between males and mated females is a result of a distortion of the configuration of population component numbers at equilibrium as compared with those in a growing population.

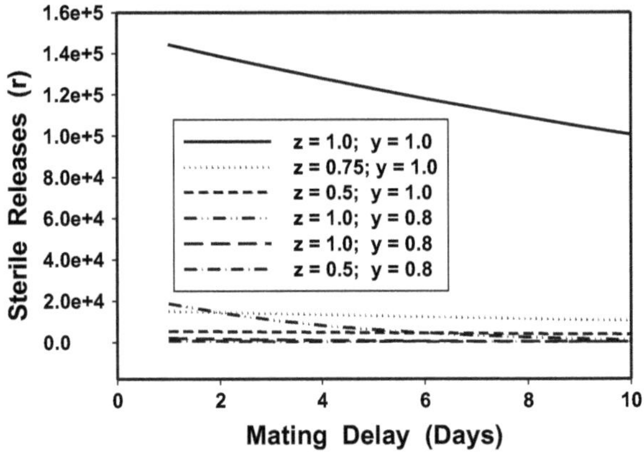

Fig. 11.4 The critical methyl eugenol-exposed sterile male release rate, r^*, for *Bactrocera dorsalis* when capture survivorship for males (z) is 1.0, 0.75 or 0.5 and that for females (y) is 1.0 or 0.8; in addition, the delay until mating is shown for values from one to ten days on the horizontal axis. The vertical axis shows values from zero to 100,000 (in exponential notation) for daily male sterile releases. If releases are made weekly rather than daily, then the release rates would need to be multiplied by seven. With even modest wild male capture rates, required sterile fly releases can be much lower than without capture. Here $F = 100$, $a = 7.67$, $s = 0.961$ (Reprinted from Barclay et al. (2014; Fig. 8) with permission)

As an indication of the synergism present in these control methods, if $z = 0.5$ and $y = 0.86$, then r^* is reduced to less than 200, i.e., less than 1 % of the value with no trapping mortality in males and females ($z = y = 1.0$). This shows that the three-way combination of attracting and killing males and some females plus the release of sterile males is very synergistic, drastically reducing the required sterile male release rate for eradication (Fig. 11.4). This value (5.77) of r^*/M_T compares well with the value (5.41) using the actual survivorship curve of *B. dorsalis* (Barclay et al. 2014).

7 Barrier Widths for Exclusion of Pests from an Area

A concern for pest managers is the minimum size of the target control area that is amenable for an area-wide integrated pest management (AW-IPM) program so that it is technically viable and economically justifiable. A conceptual mathematical model was developed that can assist with estimating the minimum area required to apply successfully a series of control tactics, including mass-trapping or baiting, as characteristic of the AW-IPM approach. The prototype model creates a basis for a decision-support tool to assess the minimum dimensions of a surrounding buffer area (in which control is also imposed) to protect a pest free area of low pest prevalence. Random dispersal would result in insects entering the buffer zone from outside the control area; these insects and their progeny would ultimately disperse across the buffer and enter the area to be protected unless they were killed within a sufficiently wide buffer. This model is sufficiently general to be applicable to a variety of insect species; it will be necessary to calibrate and validate it for any given application. Even then, it will only be a supporting tool to assist in making pest management decisions.

Dispersal is commonly modeled by means of regression (e.g., Taylor 1978), random walks (e.g., Yamamura et al. 2003), diffusion equations (e.g., Okubo 1980; Williams et al. 1992), or compartment models (e.g., Yu et al. 1996). As described below, Barclay et al. (2011) created a model that used diffusion equations to model a pest population that is diffusing across a barrier and being controlled within the barrier in an effort to protect the inner core area. Their model considers a rectangular core area, surrounded by a rectangular buffer zone. This model reflects a situation where producers wish to maintain an area (the core area) pest free or at low pest prevalence without enlarging or moving the area that contains the resource of value. The first aim of the model was to determine the minimum width of the buffer zone given the biological characteristics of the pest and the resources of the AW-IPM program. The second aim was to estimate the minimum core area that would result in a viable AW-IPM program. We are interested here only in determining the barrier width. Although their model considered the use of SIT as the control agent within the buffer, the model can be recast in terms of control using trapping mortality, as is done below.

The following simplifying assumptions were made: (i) the model assumes that the core area is already a pest free area (or an area of low pest prevalence) as a result of previous control efforts; (ii) the host density in all areas (the core area, the buffer zone, and outside the buffer zone) is assumed to be at equilibrium; (iii) there is a constant influx of pest insects from the region outside the buffer zone; and (iv) no transport of the target pest insects by wind, storms, or humans into the core area occurs. As this treatment assumes that the equilibrium has already been achieved, the barrier width calculated is that required to maintain the status quo. Dispersal describes movement of the insects across the buffer zone and will determine the width of the buffer zone, which results in the density of the invading population approaching zero at the edge of the barrier zone adjacent to the core area.

The diffusion coefficient, D, is defined in units of length squared per unit time and is usually estimated by tabulating the linear difference between the initial and final positions resulting from insect dispersal as well as the number of movements in a given time interval and then computing the means of the squared net distances travelled per unit of time. For example, if an insect takes a random walk and moves a root mean square displacement of x cm in t seconds, as a result of n individual movements over a 2-dimensional surface, then D can be estimated from these data as $D = x^2 / 4\,t$. If the n movements yield a root mean square displacement of 24 cm in 10 seconds, then the estimate of D is $D = x^2/4t = (24)\ (24)/4(10) = 14.4\ \text{cm}^2/\text{s}$.

This random walk approaches a diffusion process if the lengths of the random walks are small and the number of walks is large.

7.1 Description of the Diffusion Model

The simplest diffusion equations are described by the partial differential equation (Pielou 1969):

$$\frac{\partial u}{\partial t} = D\nabla^2 u \tag{11.54}$$

where ∇^2 is the Laplacian operator (i.e., the second partial derivatives of u with respect to x and y: $\nabla^2 = \partial^2/\partial x^2 + \partial^2/\partial y^2$) (see also Edelstein-Keshet 1988). This equation was originally developed to describe the diffusion of heat along a metal rod but has since been used for many other purposes as well, including insect dispersal. For a population of insects released simultaneously at a point, Eq. (11.54) predicts an expanding Gaussian distribution with variance 4Dt at time t:

$$\mathrm{f}(\mathrm{x},\mathrm{y},\mathrm{t}) = \frac{1}{4\pi Dt}\exp\left(-\frac{x^2+y^2}{4Dt}\right) \tag{11.55}$$

Although most insect motion is demonstrably non-random, diffusion equations have been successful at predicting longer-term patterns of insect movement (Kareiva 1983; Turchin 1998), because population level averaging occurs.

7.2 The Width of the Buffer Zone

The pest population will have a certain ambient density outside the buffer zone and will disperse from outside into the buffer zone. Because control measures are imposed within the buffer zone, the density of the pest will decrease from the outer edge of the buffer to the inner edge. The width of any buffer zone around a core area should be large enough to bring the density of the pest to zero (in case of a pest free area) or close to zero (in the case of an area of low pest prevalence) in the core area (A). The buffer zone should therefore be wide enough to prevent a gravid female insect and any of her offspring from crossing the buffer zone.

If individuals in the population outside the buffer area are reproducing and dying, as well as diffusing, then an appropriate model is:

$$\frac{\partial F(x,t)}{\partial t} = D\nabla^2 F + g(F) \tag{11.56}$$

where g is the growth function, F is the female population density, x is distance, and t is time. If g is linear and births and deaths can be separated, then:

$$\frac{\partial F(x,t)}{\partial t} = D\nabla^2 F + \beta F - \delta F \tag{11.57}$$

in which βF and δF are the instantaneous birth and death rates. We seek boundary conditions, such that at the outside of the buffer zone, F(0, t) = F_0, where F_0 is the density of insects at the outer edge of the buffer as a result of the influx of insects, and at the inside edge of the buffer, F(w, t) = a small proportion of F_0 (e.g. 10^{-6}), so that almost all the insects have been killed before reaching the inner side of the buffer. As a result of the continuous nature of the model, we can never actually achieve a zero density, but some small density below the Allee threshold that is non-viable will suffice.

If we are manipulating the death rate within the buffer by attract and kill devices that are evenly spread out to cover the whole of the buffer region, then (βF – δF) will be negative, because now δ consists of the sum of natural and imposed mortality from traps or other control sources. Since we assume that we are dealing with an equilibrium situation in which the insects have been diffusing and the buffer has been under control for a long time, the time derivative is zero, as nothing is changing over time. Only the space derivative is still non-zero. This yields the equilibrium equation:

$$D\nabla^2 F = (\delta - \beta)F \tag{11.58}$$

and this has solutions proportional to $e^{-\varphi x}$, where $\varphi^2 = (\delta - \beta)/D$. Assuming that the insect density gradient across the buffer is $F(x) = c\ e^{-\varphi x}$, the boundary conditions dictate that $c = F_0$ and that $F_0 e^{-\varphi w} = 10^{-6} F_0$. Taking logarithms, $-\varphi w = \ln(10^{-6}) = -13.8$. This leads to the minimum buffer width (w):

$$w = 13.8/\varphi = 13.8/[(\delta - \beta)/D]^{1/2} \tag{11.59}$$

to reduce the population at the inner edge of the buffer to one millionth of that at the outer edge. In this case, the diffusion coefficient is determined in the same way as it was for random walks. If a decrease down to 10^{-6} of the original density outside the buffer (F_0) is not satisfactory, then some other small fraction can be chosen, e.g., 11.5 for 10^{-5} or 16.1 for 10^{-7}. The units of w in Eq. (11.59) are in the units of D, and the units of β and δ must be the same as those of D. In demographic terms, β – δ = r, the instantaneous rate of increase. Thus, if the units of r are in terms of numbers per day, then D should also be in terms of distance^2 per day. It should be recognized that r is continuous rate of increase, although it may be stated in terms of rate of increase per generation or per year, but as with compound interest, the end result is greater than r. If the rate of increase is λ per generation, then r = ln(λ). For small values of r (e.g., daily), then $\lambda \approx r + 1$.

If trapping is to be used as the control method, then we can let τ be the trapping mortality, and it will have to be large enough that $\tau > r$ just to cancel reproductive increase. The determination of minimum buffer width proceeds in the same way as

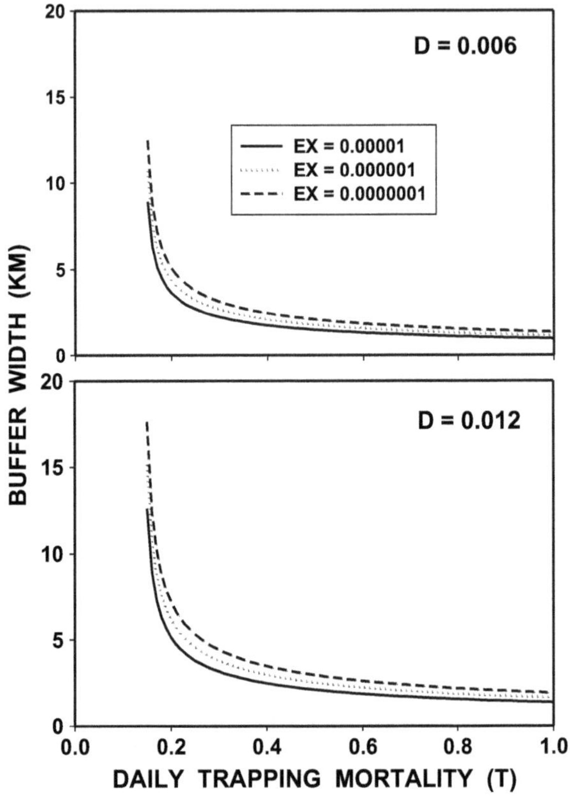

Fig. 11.5 The relationship between trapping mortality (τ) and buffer width (w) to stop the diffusion of pest insects into a protected area, shown for three values of EX, the exponent that indicates the reduction in pest density from one side of the buffer to the other, and two values of the diffusion coefficient, D

above, but with $\tau - r$ replacing $\delta - \beta$ in Eq. (11.59) above. A minimum set of parameters for the diffusion model are: the diffusion coefficient, daily birth and death rates, ambient density of the fertile population outside the buffer, percent capture rate of the traps each day, upper limit on the trap density that can be deployed, and maximum acceptable pest density in the core area. Graphs of w vs. τ are shown in Fig. 11.5 for two values of the diffusion coefficient and three values of acceptable density on the inner edge of the buffer, 10^{-5}, 10^{-6}, and 10^{-7} of the density outside the buffer.

7.3 A Case Study: The Mediterranean Fruit Fly

The Mediterranean fruit fly, *C. capitata*, was chosen as an example pest, because numerous AW-IPM programs have targeted this species successfully. These provide some practical experience against which to assess the model outputs. In addition, the Mediterranean fruit fly is relatively well studied in terms of its mobility and dispersal (Wakid and Shoukry 1976; Wong et al. 1982; Plant and

Cunningham 1991; Meats and Smallridge 2007) and ecology (McPheron and Steck 1996) so we can use values given in the literature. However, the parameter values assumed here may vary with location and are presented only to illustrate the procedure.

Fortunately, *C. capitata* has been well studied ecologically, so we can use values given in the literature. Carey (1989) has provided basic demographic information for *C. capitata* and for three other tephritids. He gives the following relevant daily values for *C. captata*: $\beta = 0.17$ and $\delta = 0.03$, and $r = 0.14$. The diffusion coefficient, D, has been estimated by Plant and Cunningham (1991) to be $D = 0.006\ km^2/day$. This leads to an equation with two unknowns, w and τ. If trapping efficiency puts an upper limit on what is possible, then we can use Eq. (11.59) directly. For example, if trapping efficiency, for the traps in total, is less that 14 % kill rate per day (100 r), then trapping mortality cannot even kill the new daily recruits; thus we need $\tau > 0.14$ Suppose $\tau = 0.20$ (i.e., 20 % of the population killed at traps every day); then the minimum buffer width is $w = 13.8/\sqrt{[(0.20–0.14)/0.006)]} = 4.36$ km. On the other hand, if physical constraints of the control area limit the buffer width, then one can invert Eq. (11.59) and calculate the required daily trap mortality as: $\tau = 190\ D/w^2 + r$, where w is the maximum allowable buffer.

Recent studies of Mediterranean fruit fly dispersal suggest that a 2 km buffer zone is a reasonable starting point for the models presented here. Meats and Smallridge (2007) studied dispersal of medfly across a grid of 3,750 surveillance traps at distances up to 10 km. They found that 90 % of released flies remained within 0.4–0.7 km of the release point. Although the required daily trapping rate to allow a 2 km buffer to be effective would be about 42 % and such a high rate might be not obtainable, perhaps a combination of trapping with some other control method, such as SIT, might allow a 2 km buffer to be sufficient. Alternatively, a wider buffer will likely be more realistic. We stress that the parameter values we have used are used for illustrative purposes only; any real control program should use data obtained from the control area.

8 Assessment and Conclusions

8.1 Importance of Assumptions

This chapter has examined modeling efforts for several different aspects of trapping fruit flies. The results of these are summarized, and an assessment is made of the likely veracity of these results. The models discussed here all have assumptions, and the degree to which they have predictive value will depend on how close the assumptions are to being met in reality and also how important the assumptions are. One way to try to get around this problem is to model a given situation with more than one type of formulation. If similar results emerge from all types of models, then the results may be robust and have more credence. Still, the results should be tested to be believed. Important processes and parameters can be varied via sensitivity analysis to assess the consequences of incorrect formulations of the processes or incorrect measurement of the parameters.

8.2 Measuring Effectiveness of Traps

The use of odor plumes in modeling trapping effectiveness has had limited success as a result of the chaotic nature of odor plumes under most natural conditions. In its place, the modeling of areas of attraction and sampling range have been more hopeful, as they depend directly on the behavior of the insects to be trapped, with insects within the range of attraction orienting towards the odor source. Along with the range of attraction there must be some measure of the likelihood of capturing an insect in the range of attraction during some specified unit of time. In addition, because odor concentration within a range of attraction is not uniform, the range of attraction will necessarily be defined in terms of some arbitrary threshold in capture efficiency. The sampling range may be the more useful of the two ranges mentioned, and a range of attraction can be converted into a sampling range by considering non-directed dispersal, which will bring insects into the area of attraction from outside it and so be susceptible to being trapped.

The formulation of the effective attractive radius (EAR) has provided an easy and quite general method of comparing the trapping effectiveness of different trap types as well as comparing (capture rates) of various species in different habitats.

8.3 Modeling of Trapping for Detection

These models have been of two closely related types: (i) models for determining the probability that the density of a founding or remnant population is below a certain low number, and (ii) models for determining the probability that any individuals at all are in the area of interest. Both types of hypotheses use probability models, and the results are likely to be fairly robust but will depend quantitatively to a certain extent on the nature of the underlying probability distribution that is used. All such models assume randomness, which may or may not be sufficiently realistic to give believable results. Most existing models of this type specify very long trapping sequences.

8.4 Modeling Population Suppression or Eradication Using Attraction to Baits or Traps

Capture of males and females is often equivalent to capture of only females because usually males can fertilize all remaining females. Thus, most trapping programs will require a female attractant, but a male attractant is not necessary unless the males have a detrimental economic or medical effect. On the other hand, the capture of males for male annihilation requires a limit to male mating frequency, otherwise a few males could in principle inseminate enough females to maintain the population. If males can mate with unlimited frequency, then trapping males does nothing to control the population, and trapping of females is necessary.

If males are limited in their mating frequency, then control by male annihilation is possible, albeit difficult because of the high proportion of males that must be trapped each day. In this case, trapping males, but not females, can only be successful if trapping occurs prior to mating. Also, control becomes much more difficult as the maximum mating frequency of males increases. If females are also trapped, then there is a synergism between the two, such that the total number trapped can be less if both sexes are trapped than if only one sex is trapped. Most of the models in this section assume that males can mate only once per day. This assumption appears to be crucial; for some tephritids, which have a short courtship period and a prolonged mating process, this assumption may be reasonable.

8.5 Interactions Among Control Methods

Pest control methods used in combination can interact positively (synergism), neutrally (independence), or negatively (antagonism), depending on how each may affect the others being used. For example, insecticide used either as a spray or in traps will generally interact negatively with existing biological control, because the insecticide also kills predators and/or parasitoids. Most other standard combinations of pest control methods will be expected to interact either positively or neutrally. It seems that density dependence of control action lends itself to positive interaction with other control methods. In addition, modeling has predicted that mortality inflicted by most pest control methods will be a sub-linear function of the control effort, such that progressive increments of control effort will yield progressively smaller increments in mortality. This phenomenon, if true, would tend to make most control methods interact positively. In addition, at low densities the Allee effect can be utilized in control programs by creating conditions that will raise the Allee threshold and/or interact positively with imposed control.

Population dynamic models using attract and kill devices for suppression or eradication indicate that their deployment is more effective if females are targeted through the use of food-based traps and baits in hot spot areas in conjunction with other methods. In addition, it appears that male annihilation through the use of methyl eugenol in traps/baits as a sole method of pest control would be very difficult if only males were attracted, because a very high level of trapping is required to deny a sufficient proportion of females the chance to mate.. If some females are also attracted, then the method offers more hope, as the required trapping level of females is much lower to yield control. In addition, if mating occurs before or simultaneously with trapping, then control by baiting males used alone is difficult, if not impossible. This method offers considerable hope, though, if integrated with other pest control methods, especially control methods in which interference does not occur or is minimal. For example, if the traps were baited with methyl eugenol and the sterile males were not attracted to them as a result of prior exposure to ME, then the two methods could be used simultaneously and would exhibit a high degree of synergism.

8.6 A Case Study for **Bactrocera dorsalis** *Using Methyl Eugenol Trapping and Sterile Male Releases*

An age-structured model is presented that is designed for use with *B. dorsalis*, but which could be used with a variety of species with little modification other than adjusting the parameter values. This model includes trapping with methyl eugenol and the simultaneous release of sterile males that have been exposed to a pre-release diet containing ME and thus are minimally attracted to the ME baits, thus increasing the sterile:fertile male ratio beyond that achievable by simply releasing sterile males. Formulae for critical values of trapping and of sterile male releases were derived and related to life table statistics.

Parameter values specific to *B. dorsalis* were obtained from the literature and used to estimate pre-adult survivorship, an adult survivorship curve, age at first oviposition, and mean daily fertility. These were then applied to the model to obtain critical values of trapping and sterile male releases to stop population growth for the two cases of male excess and male deficit. With the release of sterile males, the case of male deficit would seldom, if ever, be expected to occur. There is a strong positive interaction between trapping and the SIT, such that modest amounts of daily trapping drastically reduce the required number of sterile males to achieve population control. In addition, the simultaneous use of trapping with ME and the release of sterile males makes it feasible that control can be achieved even if mating occurs before trapping, although it is less efficient than if trapping occurs before mating.

8.7 Barrier Widths for Exclusion of Pests from an Area

A model is presented to determine the width of a buffer strip that would be required to guarantee that an insect population would not be able to cross it by means of ordinary dispersal from an uncontrolled area to an area that is intended for area-wide control. The model uses a diffusion equation and assumes that equilibrium within the buffer has already been achieved and that episodic movement of insects resulting from human transport, storms, etc. does not occur. A tradeoff occurs between buffer width and maximum possible trapping efficiency, such that the maximum trapping efficiency determines the minimum buffer width sufficient to protect the target area.

References

Allee WC (1931) Animal Aggregation. University of Chicago Press, Chicago

Arakaki N, Kuba H, Soemori H (1984) Mating behaviour of the Oriental fruit fly, *Dacus dorsalis* Hendel (Diptera: Tephritidae). Appl Entomol Zool 19:42–51

Barclay HJ (1987a) Models for pest control using food-baited traps and either insecticides or chemosterilants. Agric Ecosyst Environ 19:41–54

Barclay HJ (1987b) Combining methods of pest control: complementarity of methods and a guiding principle. Nat Resour Model 2:299–323

Barclay HJ (1988) Models for combining methods of pest control: food-baited and pheromone-baited traps containing either insecticide or sterilant. Bull Entomol Res 78:573–590

Barclay HJ (1992) Combining methods of insect pest control: partitioning mortality and predicting complementarity. Res Popul Ecol 34:91–107

Barclay HJ, Hargrove JW (2005) Probability models to facilitate a declaration of pest-free status, with special reference to tsetse (Diptera: Glossinidae). Bull Entomol Res 95:1–11

Barclay HJ, Hendrichs J (2014) Models for assessing the male annihilation of *Bactrocera* spp. with methyl-eugenol baits. Ann Entomol Soc Am 107:81–96

Barclay HJ, Humble LM (2009) Probability models to facilitate a declaration that an exotic insect species has not yet invaded an area. Biol Invasions 11:1267–1280

Barclay H, van den Driessche P (1977) Predator–prey models with added mortality. Can Entomol 109:763–768

Barclay HJ, van den Driessche P (1989) Pest control models of combinations of sterile releases and trapping. Insect Sci Appl 10:107–116

Barclay HJ, Vreysen MJB (2013) The interaction of dispersal and control methods for the riverine tsetse fly *Glossina palpalis gambiensis* (Diptera: Glossinidae): a modelling study. Popul Ecol 55:53–68

Barclay HJ, Hargrove JW, Clift A, Meats A (2005) Procedures for declaring pest-free status. In: Dyck VA, Hendrichs J, Robinson AS (eds) Sterile Insect Technique. Principles and practice in area-wide integrated pest management. Springer, Dordrecht, pp 363–386

Barclay HJ, Matlock R, Gilchrist S, Suckling DM, Reyes J, Enkerlin WR, Vreysen MJB (2011) Assessing the minimum size area for an area-wide integrated pest management program: a conceptual model involving sterile releases and a case study. Internat J Agron. doi:10.1155/2011/409328

Barclay HJ, McInnis D, Hendrichs J (2014) Modeling the area-wide integration of male annihilation and the simultaneous release of methyl-eugenol-exposed *Bactrocera* spp. sterile males. Ann Entomol Soc Am 107:97–115

Berec L, Angulo E, Courchamp F (2007) Multiple Allee effects and population management. Trends Ecol Evol 22:185–191

Blackwood JC, Berec L, Yamanaka T, Epanchin-Niell RS, Hastings A, Liebhold AM (2012) Bioeconomic synergy between tactics for insect eradication in the presence of Allee effects. Proc R Soc Ser B 279:2807–2815

Bossert WH, Wilson EO (1963) The analysis of olfactory communication among animals. J Theor Biol 5:443–469

Boukal DS, Berec L (2009) Modelling mate-finding Allee effects and population dynamics, with applications in pest control. Popul Ecol 51:445–458

Byers JA (2009) Modeling distributions of flying insects: effective attraction radius of pheromone in two and three dimensions. J Theor Biol 256:81–89

Byers JA (2011) Analysis of vertical distributions and effective flight layers of insects: three-dimensional simulations of flying insects and catch at trap heights. Environ Entomol 40:1210–1222

Byers JA, Anderbrant O, Löfqvist J (1989) Effective attraction radius: a method for comparing species attractants and determining densities of flying insects. J Chem Ecol 15:749–765

Calkins CO, Schroeder WJ, Chambers DL (1984) Probability of detecting Caribbean fruit fly, *Anastrepha suspensa* (Loew) (Diptera: Tephritidae), populations with McPhail traps. J Econ Entomol 77:198–201

Cardé RT, Willis MA (2008) Navigational strategies used by insects to find distant, wind-borne sources of odor. J Chem Ecol 34:854–866

Carey JR (1989) Demographic analysis of fruit flies. In: Robinson AS, Hooper GHS (eds) Fruit flies: their biology, natural enemies and control, vol 3B, World crop pests. Elsevier, Amsterdam, pp 253–265

Clift AD, Meats AW (1997) Using the negative binomial distribution and risk management software to simulate *Bactrocera papayae* Drew and Hancock (Diptera: Tephritidae) metapopulations in an eradication context. Proc Modsim 97:792–795

Clift A, Meats A (2004) When does zero catch in a male lure trap mean no tephritid flies in the area? In: Barnes BN (ed) Proceedings, symposium: 6th international symposium on fruit flies of economic importance, 6–10 May 2002, Stellenbosch, South Africa. Isteg Scientific Publications, Irene, South Africa, pp 183–188

Clift AD, Gleeson PJ, Meats AW (1999) Risk analysis on banana fruitfly, *Bactrocera musae*, recolonising an area near Cairns, North Queensland after area-wide insecticidal treatments had ceased. Proc Modsim 99:649–653

DeBach P (1974) Biological control by natural enemies. Cambridge University Press, London

Drew RA, Yuval B (2000) The evolution of fruit fly feeding behaviour. In: Aluja M, Norrbom AL (eds) Fruit flies (Tephritidae): phylogeny and evolution of behaviour. CRC Press LLC, Boca Raton, pp 731–750

Edelstein-Keshet L (1988) Mathematical models in biology. Random House, New York

Epsky ND, Espinoza HR, Kendra PE, Abernathy R, Midgarden D, Heath RR (2010) Effective sampling range of a synthetic protein-based attractant for *Ceratitis capitata* (Diptera: Tephritidae). J Econ Entomol 103:1886–1895

Hartstack AW, Hollingsworth JP, Ridgway RL, Hunt HH (1971) Determination of trap spacings required to control an insect population. J Econ Entomol 64:1090–1100

Hassell MP (1978) The dynamics of arthropod predator–prey systems. Princeton University Press, Princeton

Hendrichs J, Prokopy RJ (1994) Food foraging behavior of frugivorous fruit flies. In: Calkins CO, Klassen W, Liedo P (eds) Fruit flies and the sterile insect technique. CRP Press, Boca Raton, pp 37–55

Kareiva P (1983) Local movements in herbivorous insects: applying a passive diffusion model to mark-recapture field experiments. Oecologia 57:322–327

Kendra PE, Epsky ND, Heath RR (2010) Effective sampling range of food-based attractants for female *Anastrepha suspensa* (Diptera: Tephritidae). J Econ Entomol 103:533–540

Knipling EF (1979) The basic principles of insect population suppression and management. Agriculture handbook number 512. USDA/SEA, Washington, DC, USA
Kramer AM, Dennis B, Liebhold AM, Drake JM (2009) The evidence for Allee effects. Popul Ecol 51:341–354
Kuno E (1978) On the assessment of low rate of pest infestation based on successive zero-samples. Jpn J Appl Ent Zool 22:45–46 (In Japanese)
Kuno E (1991) Verifying zero-infestation in pest control: a simple sequential test based on the succession of zero-samples. Res Popul Ecol 33:29–32
Lance DR, Gates DB (1994) Sensitivity of detection trapping systems for Mediterranean fruit flies (Diptera: Tephritidae) in southern California. J Econ Entomol 87:1377–1383
Liebhold A, Bascompte J (2003) The Allee effect, stochastic dynamics and the eradication of alien species. Ecol Lett 6:133–140
Liebhold A, Tobin P (2008) Population ecology of insect invasions and their management. Ann Rev Entomol 53:387–408
Manoukis NC, Hoffman K (2013) An agent-based simulation of extirpation of *Ceratitis capitata* applied to invasions in California. J Pest Sci. doi:10.1007/s10340-013-0513-y
McInnis D, Kurashima R, Shelly T, Komatsu J, Edu J, Pahio E (2011) Prerelease exposure to methyl eugenol increases the mating competitiveness of sterile males of the oriental fruit fly (Diptera: Tephritidae) in a Hawaiian orchard. J Econ Entomol 104:1969–1978
McPheron BA, Steck GJ (eds) (1996) Fruit fly pests: a world assessment of their biology and management. St. Lucie Press, Boca Raton
Meats A, Clift AD (2005) Zero catch criteria for declaring eradication of tephritid fruit flies: the probabilities. Aust J Exp Agric 45:1335–1340
Meats A, Smallridge CJ (2007) Short- and long-range dispersal of medfly, *Ceratitis capitata* (Diptera, Tephritidae), and its invasive potential. J Appl Entomol 131:518–523
Navarro-Llopis V, Vacas S, Sanchis J, Primo J, Alfaro C (2011) Chemosterilant bait stations coupled with Sterile Insect Technique: an integrated strategy to control the Mediterranean fruit fly (Diptera: Tephritidae). J Econ Entomol 104:1647–1655
Okubo A (1980) Diffusion and ecological problems: mathematical models. Springer, New York
Ostrand F, Anderbrant O (2003) From where are insects recruited? A new model to interpret catches of attractive traps. Agric For Entomol 5:163–171
Pielou EC (1969) An introduction to mathematical ecology. Wiley, New York
Piñero JC, Mau RF, McQuate GT, Vargas RI (2009) Novel bait stations for attract-and-kill of pestiferous fruit flies. Entomol Exp Appl 133:208–2016
Plant RE, Cunningham RT (1991) Analyses of the dispersal of sterile Mediterranean fruit flies (Diptera: Tephritidae) released from a point source. Environ Entomol 20:1493–1503
Regan TJ, McCarthy MA, Baxter PWJ, Panetta FD, Possingham HP (2006) Optimal eradication: when to stop looking for an invasive plant. Ecol Lett 9:759–766
Riffell JA, Abrell L, Hildebrand JC (2008) Physical processes and real-time chemical measurement of the insect olfactory environment. J Chem Ecol 34:837–853
Roessler Y (1989) Insecticidal bait and cover sprays. In: Robinson AS, Hooper G (eds) Fruit flies, their biology, natural enemies and control, vol 3b, World crop pests. Elsevier, Amsterdam, pp 329–336
Rout TM, Salomon Y, McCarthy MA (2009a) Using sighting records to declare eradication of an invasive species. J Appl Ecol 46:110–117
Rout TM, Thompson CJ, McCarthy MA (2009b) Robust decisions for declaring eradication of invasive species. J Appl Ecol 46:782–786
Schlyter F (1992) Sampling range, attraction range and effective attraction radius: estimates of trap efficiency and communication distance in coleopteran pheromone and host attractant systems. J Appl Entomol 114:439–454
Schutze MK, Jessup A, Ul-Haq I, Vreysen MJB, Wornoayporn V, Vera MT, Clarke AR (2013) Mating compatibility among four pest members of the *Bactrocera dorsalis* fruit fly species complex (Diptera: Tephritidae). J Econ Entomol 106:695–707

Shelly TE (1994) Consumption of methyl eugenol by male *Bactrocera dorsalis* (Diptera: Tephritidae): low incidence of repeat feeding. Fla Entomol 77:201–208

Shelly TE, Edu J, Pahio E, Wee SL, Nishida R (2008) Re-examining the relationship between sexual maturation and age of response to methyl eugenol in males of the oriental fruit fly. Entomol Exp Appl 128:380–388

Shelly T, Nishimoto J, Diaz A, Leathers J, War M, Shoemaker R, Al-Zubaidy M, Joseph D (2010) Capture probability of released males of two *Bactrocera* species (Diptera: Tephritidae) in detection traps in California. J Econ Entomol 103:2042–2051

Solow AR (1993) Inferring extinction from sighting data. Ecology 74:962–964

Steiner LF (1952) Methyl eugenol as an attractant for oriental fruit fly. J Econ Entomol 45:241–248

Steiner LF, Hart W, Harris EJ, Cunningham RT, Ohinata K, Kamakahi DC (1970) Eradication of the oriental fruit fly from the Mariana Islands by the methods of male annihilation and sterile insect release. J Econ Entomol 63:131–135

Stephens PA, Sutherland WJ, Freckleton RP (1999) What is the Allee effect? Oikos 87:185–190

Suckling DM, Tobin PC, McCullough DC, Herms AA (2012) Combining tactics to exploit Allee effects for eradication of alien insect populations. J Econ Entomol 105:1–13

Taylor RAJ (1978) The relationship between density and distance of dispersing insects. Ecol Entomol 3:63–70

Turchin P (1998) Quantitative analysis of movement: measuring and modeling population redistribution in animals and plants. Sinauer Associates, New York

Turchin P, Odendaal FJ (1996) Measuring the effective sampling area of a pheromone trap for monitoring population density of southern pine beetle (Coleoptera: Scolytidae). Environ Entomol 25:582–588

Vargas RI, Miyashita D, Nishida T (1984) Life history and demographic parameters of three laboratory-reared Tephritids (Diptera: Tephritidae). Ann Entomol Soc Am 77:651–656

Vickers NJ (2000) Mechanisms of animal navigation in odor plumes. Biol Bull 198:2003–2012

Volterra V (1926) Variazione e fluttuazioni del numero d'individui in specie animali conviventi. Mem Acad Naz Lincei 2:31–113

Wakid AM, Shoukry A (1976) Dispersal and flight range of the Mediterranean fruit fly, Ceratitis capitata Wied in Egypt. J Appl Entomol 81:214–218

Wall C, Perry JN (1987) Range of attraction of moth sex-attractant sources. Entomol Exp Appl 44:5–14

Williams B, Dransfield R, Brightwell R (1992) The control of tsetse flies in relation to fly movement and trapping efficiency. J Appl Ecol 29:163–179

Wong TTY, Whitehand LC, Kobayashi RM, Ohinata K, Tanaka N, Harris EJ (1982) Mediterranean fruit fly: dispersal of wild and irradiated and untreated laboratory-reared males. Environ Entomol 11:339–343

Wong TTY, McInnis DO, Ramadan MM, Nishimoto JI (1989) Relationship of sexual maturation rate to response of oriental fruit fly strains (Diptera: Tephritidae) to methyl eugenol. J Chem Ecol 15:1399–1405

Yamamura K, Moriya S, Tanaka K (2003) Discrete random walk model to interpret the dispersal parameters of organisms. Ecol Mod 161:151–157

Yamanaka T, Liebhold AM (2009) Spatially implicit approaches to understand the manipulation of mating success for insect invasion management. Popul Ecol 51:427–444

Yu P, Habtemariam T, Oryang D, Obasa M, Nganwa D, Robnett V (1996) Stochastic model of spatial spread and control of tsetse flies (Diptera: Muscidae). Environ Entomol 25:78–84

Part IV
Attract and Kill

Chapter 12
Priorities in Formulation and Activity of Adulticidal Insecticide Bait Sprays for Fruit Flies

Robert L. Mangan

Abstract The use of adulticidal insecticide spray against fruit flies is examined as a historical development, beginning at the turn of the twentieth century and proceeding to the present. This development is considered in three phases, the first extending from the 1890s when the threats of exotic pest invasions were realized in the USA, especially California, and in Australia and focused on chemicals that were generally toxic to all animals but mainly after ingestion. After World War II, the development and recognition of synthetic organic pesticides allowed for more targeted and more toxic pesticides. A third period was initiated during the 1990s, mainly in response to social and political issues related to human exposure and impacts on environment. During all three phases there was social, economic, and political participation in establishing goals for direction of both pest management and research. Attractant baits were used during all three phases. Toxic attractant baits were developed mainly addressing control and eradication programs for *Ceratitis*, *Bactocera*, and *Anastrepha*. Although insecticides were used to control damage by established pests, programs to eradicated invasive population were associated with urgent programs.

Keywords Ingestible Bait • Insecticide • Arsenic salts • Chlorinated hydrocarbons • Carbamates • Pyrethroids • Organophospates • Tartar emetic • Residues • Resistance • Nontarget organisms • Phototoxic dyes • SureDye • Spinosad • Solbait • GF120

Fruit pests in the family Tephritidae impact the fruit industry and consumers due to direct loss of production, loss of fruit markets due to regulatory quarantines, and

R.L. Mangan (✉)
1123 Kerria Avenue, McAllen, TX 78501, USA
e-mail: r_mangan@earthlink.net

T. Shelly et al. (eds.), *Trapping and the Detection, Control, and Regulation of Tephritid Fruit Flies*, DOI 10.1007/978-94-017-9193-9_12,

undesirable exposure of people and property to pest control measures. These same impacts were recognized at the turn of the twentieth century to be important in the development of chemical control systems for fruit flies. Species that are polyphagic and have life histories that favor invasiveness require specific combinations of management practices that treat the entire population on an area-wide basis. This strategy requires treating all habitats where the pests breed, including public and residential properties where inhabitants receive no benefits from the treatment. In this chapter, I trace the history of pesticide technology and the attitudes of researchers and program managers in the development and public acceptance of effective fruit fly adult treatments applied on an area-wide basis.

Historical reviews of both the development and use of chemical based controls for fruit flies by Jeppson and Garman (1960), Klassen (1989), Roessler (1989), and Moreno and Mangan (2002) have summarized the trend in the baited insecticide technology. Studies of the biology and management of economically important fruit flies were initiated at the international level at the turn of the century (1900), coinciding with the actual and potential spread of *Ceratitis*, *Bactrocera,* and *Anastrepha* into previously fly free fruit production regions. Introductions of *Bactrocera* and *Ceratitis* into areas such as Hawaii, reports of *Anastrepha* infested fruit entering California from Mexico, and the introduction of Mediterranean fruit fly (*Ceratitis capitata* (Wiedemann)) and spread of established *Bactrocera* species in Australia resulted in surveys of pest species of these genera. New projects to develop pesticide technology to prevent invasion of established fruit fly pests (particularly in southern Europe and northern Africa) coincided with local goals to reduce losses, achieve regulatory goals by meeting quarantine standards, and preserve environmental and human health standards.

The development of baits for adult fruit fly control can be divided into three periods based on delivery systems and the insecticides. The first period, roughly covering the period from 1900 to 1944, used baits with a combination of feeding stimulants (sugar, molasses, crude brown sugar) and various substances known to be poisonous to many animals, such as arsenic salts (Anonymous 1898; Howard 1898) or plant derived insecticides (Isaac 1905; Baker et al. 1944). Use of these systems is described in reports from Mexico (Crawford 1927), Hawaii (Severin 1912; Back and Pemberton 1917; Back and Pemberton 1918a, b), and Australia (Compere 1907; Jarvis 1926a, b).

The second period extended from about 1944 to the present. Insecticides, referred to as organic insecticides at the time (Metcalf 1955), have included chlorinated hydrocarbons (DDT and dieldrin), carbamates (aldicarb), organophosphates (malathion), and pyrethroids (deltamithrin). These insecticides act mainly as contact insecticides, and the baits focus on attractant rather than phagostimulant function. Research by Jarvis (1931) in Australia for detection and control of *Bactocera tryoni* (Froggatt) (Queensland fruit fly) is credited with the development of ammonia as an attractant additive for fruit fly baits. The development of hydrolyzed protein baits combined with insecticides that killed insects on contact replaced the stomach poisons during the 1940s without the need for phagostimulants. During the 1940s, the baits were mainly hydrolyzed proteins, and pesticides were chlorinated hydrocarbons and carbamates, but arsenic salts and tartar emetics continued to be important. From the mid-1950s until present, organophosphates, particularly malathion, became the key insecticides for the contact spray baits.

A third period in fruit fly insecticide development began after widespread public concerns developed in the early 1960s concerning human and environmental exposure to pesticides. These concerns followed the publication in 1962 of Carson's book *Silent Spring*, and a growing environmental movement addressed problems about use of insecticide sprays, including damage to native wildlife and inducing secondary pest outbreaks. Hall (1964) reviewed a number of observations that portended the problems discussed by Carson (1962) of over-use, misuse, and damage to wildlife from residues and the development of resistance for the pesticide system that had worked wonders following World War II, especially in medical entomology. The requirements for area-wide sprays necessary for fruit fly eradication programs (Klassen 2005; Mangan 2005) entailed application of pesticide baits in human residential communities and public areas where non-target organisms were exposed. Pesticide systems were developed to reduce environmental impacts by more precisely targeting the pests and reducing the toxicity to non-target organisms. The impact on the target populations (the fruit flies) were usually considered adequate if they were at least equivalent to the standard hydrolyzed protein combined with malathion.

1 Insecticide in Ingestible Baits

According to United States Department of Agriculture (USDA) entomologists Back and Pemberton (1918a, b), the earliest experiments with bait sprays originated from projects in South Africa by Mally for Mediterranean fruit fly and in Italy by Berlese for olive fly (*Bactrocera oleae* (Rossi)). Severin (1912), a professor in the college of Hawaii speaking at a California growers meeting, also discussed the "Mally Bait" as a promising adulticide for Mediterranean fruit fly. He opened his report with the suggestion that Mediterranean fruit fly was accidentally introduced into Hawaii in a load of parasitized Mediterranean fruit fly pupae brought to Hawaii as part of the program against melon fly. Back and Pemberton introduced their review by refuting the claim by Severin and suggested that the Mediterranean fruit fly was introduced in fresh fruit stored on ship decks originating from recently infested Australia.

In reviews of insecticide bait, Back and Pemberton (1918a, b) suggested that the insecticide bait (2.5 lb brown sugar, 5 oz arsenate of lead, 5 gal water) could not be effective because of frequent rains, potential damage to bee populations, and the absence of large treatable orchards in the affected area. Severin disputed these claims and stated near the end of his presentation that work by H. T. Weinland, an entomologist in Hawaii supported by the state of California, was further testing the bait spray. Speaking after Severin, Weinland briefly discussed the controversy but emphasized that the introduction of Mediterranean fruit fly had little effect on the economy of the territory, since most of the economy was based on sugar and other non-fruit fly host crops. The earliest discussions of insecticide sprays for fruit flies in U.S. territories were associated with fault finding for current conditions and disagreements over approaches to address the problems.

American concern about invasive fruit flies was summarized in the 1897 Yearbook of Agriculture published (in 1898) by USDA. The report by L.O. Howard (1898) "Danger of Importing Insect Pests" (p. 529) proposed a national quarantine system and discussed the most important insect pests threatening or recently established in the U.S.A. Howard focused on three tephritid species in the discussion of "threats from the tropics", namely the Mexican orangeworm *Trypeta ludens* (now *Anastrepha ludens* (Loew)), a similar pest to peaches *Trypeta acidusa* (now *Anastrepha obliqua* (Macquart), and a pest that recently invaded Bermuda from Europe, the Bermuda peach maggot *C. capitata*. Although the main goal of this article was to establish a national quarantine law for protection from these pests, Howard also demonstrated the approaches and problems associated with gaining information about the pests and determining their entry paths. A summary of pesticides available for insect control was published by Howard in 1898 (pp. 637–640). Formulae for the various arsenic salts, carbon bisulphate, cyanic gases, kerosene, and pyrethrum as well as oils, soaps, and resins were given. Damaging effects to plants and their avoidance were discussed for most of the pesticides, however, human safety was only mentioned for pyrethrum (not poisonous to man or the higher animals), and no discussion of overall user or customer safety was given.

Studies of fruit fly pests were carried out by investigator "agents", such as Prof. C.H.T. Townsend who travelled to Mexico in 1894 (discussed in Isaac 1905) following reports of infested citrus by U.S travelers visiting Mexico. Later discovery and rearing of *A. ludens* larvae to adults from oranges sold in Maryland and Illinois originating from Morelos (imported to fill markets created by a severe freeze in the USA in 1895) prompted California to contract scientists to investigate the biology and control of *A. ludens*.

The California fruit industry also provided major funding for early evaluations of fruit flies in Mexico. John Isaac was appointed to investigate the status of the Mexican fruit fly, *A. ludens*, in 1905. Quarantine against Mexican oranges entering California was in place, and the purpose of his appointment and exploration was explained as follows:

> Our orange-growers have stood in dread of the introduction of this pest as one of the worst possible evils that could befall them. How far these fears are well grounded, and how far they are exaggerated, it is the object of the following pages to show.

Isaac's report described fruit fly pest management in Mexico, mainly in the coastal states where citrus is concentrated. The document contains discussion by Mexican scientists and officials, mainly arguing that the quarantine was biologically unnecessary and was in place as a protectionist measure and that, even if established in California, *A. ludens* would not be disastrous to the industry. Strategies for pest management were orchard management, destruction of fallen fruit, and biocontrol, including parasitoids and predators. Isaac also described the use of sweetened bait composed of extracts of *Haplophyton cimicidum* (now *Haplophyton crooksii* Benson), known as the cockroach plant. The formula, according to Isaac 1905 (p. 33), was developed by Mr. Jose Betanzos and was prepared "by boiling about two pounds of the herb, cut fine, in three gallons of water, and after the herb is thoroughly boiled two pounds of sugar is added, and the whole is strained and used as a spray." This bait spray was also discussed in Baker et al. (1944). In considering credit for first developing adulticidal baits for fruit flies, the sweetened cockroach plant bait developed by Betanzos in Mexico should also be considered with those of Mally in S. Africa and Berlese in Italy credited by Back and Pemberton (1918a, b).

Crawford (1910) took advantage of a "Pomona College Mexican Expedition" to further investigate and update Isaac's findings. His review of the Mexican program was much more negative, observing that many of the integrated efforts by the Mexican program had made little progress or had stopped. His report (Crawford 1910, p. 330) proposed an insecticide spray composed of a stock solution of 5 gal crude carbolic acid, 40 pounds whale oil soap, and 40 gal of water to be diluted one to twenty parts with water before using it as a spray. Although I could find no evidence that this formulation was ever tested, the author proposed that the formulation would provide contact toxicity to adults, repellency, and, if dripped on to the ground, mortality to larvae and pupae in the soil.

A project supported by Mexico Gulf Coast Citrus Fruit Association, the National Railways of Mexico, and the Mexican government was initiated in 1913–1914 to develop and test technologies to resolve the quarantines. The project was just beginning when the U.S.A. invasion and occupation of Veracruz ended it. Crawford (1927) published a summary of the information gathered in this project and summarized information collected during his own visits and those of Isaac as well as information gathered by Mexican scientists, particularly Manuel Fernandez Leal, Head of the Mexican Department of Agriculture and Prof. A. L. Herrera, Head of Department of Parasitology, who managed investigations and published circulars beginning in 1900.

Crawford's summary of the eastern Mexico investigations from his visit in 1913–1914 (Crawford 1927) included a discussion of poison bait adult sprays. In addition to the bait using cockroach plant extract, which is cited but not discussed, Crawford explained the use of a bait consisting of lead arsenate (1 lb), brown sugar (6 lb), and water (20 gal). Orchards were sprayed by hand or machines at approximately the same cost per tree for either application method. A table of recommended activities showed that the application of sprays and destruction of fruit depended on fruit maturity periods for oranges and grapefruit and that these activities were intended to be closely coordinated during periods when adult flies were present. According to this report, sprays were effective with infestation rates ranging from 0 to 0.2 % for fruit treated with sprays. Effects of the spray on honey bees were discounted in this report, which stated that bees could only feed on the wet toxin when freshly applied and that it dried quickly, thus minimizing the impact. Threats to humans were similarly discounted, because there was so little spray on the fruit, and washing before shipping would remove it. A combination of events, including a more strict quarantine established in California and "internecine conditions" (I assume the Mexican civil war), restricted progress developing protection systems after 1914.

The accepted technology for fruit fly eradication was challenged in Florida in a medfly outbreak that was discovered in Gainesville in April 1929 and covered about 1/3 of the state before eradication in December 1930 (Yonge 1931). This report summarized the technical, social, and financial factors in the eradication effort. Technical strategies for eradication were designed by a "team of seven" and later a "team of 5", mostly dealing with administration and justification for funding. Technical summaries of this eradication and 16 following eradications and surveys following captures from 1929 to 1988 are given in Clark and Weems (1989).

Following Mexican fruit fly outbreaks in the Rio Grande Valley in Texas during the mid-1920s, an agreement was reached between the USA and Mexico to establish a laboratory in Mexico City for studies of Mexican species of fruit flies. The history and accomplishments of this laboratory were reviewed in Shaw et al. (1970). As Shaw discussed, the activities of this laboratory, which operated from 1928 to 1968, were described in some detail in monthly reports by each scientist to the director. Periodic publications by individual scientists or, more frequently, teams of scientists were summaries of the monthly reports. The most complete summary (Baker et al. 1944) listed laboratory and field trials of pesticide and baits for adult suppression. Darby and Kapp (1934) produced an early report on the development of adult pesticides and other topics related to their study of *A. ludens*. Two sections of their report concerned adult longevity and testing toxins. Laboratory tests showed that males had a potential of 14.5 months and females a potential of 11 months adult longevity. This longevity will allow adult survival during the period when there were no host fruits in the Rio Grande Valley and suggested the importance of insecticidal baits to kill these adult populations.

Fruit fly programs in Australia were developing during this same period to focus mainly on *B. tryoni* and related species native to northern parts of Australia and *C. capitata*, which was first reported in western Australia in 1894. In the 1906 meeting of state entomologists, a proposal was made and in 1907 approved:

> That, in consequence of the increase of Fruit-fly and other pests, with the permission of New South Wales, their Entomologist be dispatched to America and Europe to inquire into the best methods of dealing with Fruit-flies and other pests; the value of parasites; to procure and dispatch same if effective.

The entomologist was W. W. Froggatt, who in 1909 published the details and conclusions from his visits to 20 countries or regions in North America and the Caribbean, Europe, Africa, and Asia.

In Sect. 1 of his report, Froggatt described the locations and activities he observed and offered evaluations and opinions about mismanagement of USA forests, abuses in area-wide programs to control white fly in California, exploitation of the guayule plant in Mexico by USA interests, and especially the exaggerated claims of successful parasite and predator introductions for fruit pest management. Campaigns against the white fly in Marysville, California, were cited as an example of state agencies requiring homeowners to destroy their gardens for control of this pest. According to Froggatt, not only was this an incursion of property owner's rights, but the program failed completely. In this report, it is clear that Froggatt had strong doubts about pest management by introduction of predators and parasites. Newspapers in Australia contained frequent anecdotal claims, such as Anonymous (1906) claiming "excellent control" of *B. tryoni* by small brown beetles (later identified as Staphylinidae). Although the claims of benefits for fruit protection from introduced predators and parasites were anecdotal and rightfully criticized by Froggatt, his evidence of their failure was also anecdotal, and it is doubtful that his criticisms affected the Australian programs. Section 2 of this report was an overall review (mostly criticism) of biological approaches to pest control through introductions of predators and parasites in a number of other countries.

Section 3 specifically addressed fruit fly programs he visited and reviewed approaches to fruit fly control in the USA, Mexico, Bermuda, Spain, Italy, South Africa, and India. The visits to California addressed other insect programs, particularly white fly and codling moth. Froggatt visited Washington D.C. and reviewed fruit fly concerns, particularly inspections and quarantines with Dr. L.-O. Howard and taxonomy of fruit flies with D.W. Coquillett, who had described many of the *Dacus* (now *Bactrocera*) species. Howard (1930) continued to follow Froggatt's career for the next 25 years, especially noting his skepticism about pest control through introduction of natural enemies. In 1935 one of Froggatt's last actions was a campaign against the introduction of the cane toad into Australia for control of two sugar cane beetle pests in 1935 and 1936 (Turvey 2010).

Following his visit to the eastern USA, Froggatt met with Prof. A. L. Herrera in Mexico City. At this time fruit fly control in Mexico was focused to alleviate quarantines against Mexican citrus for threats of infested fruit entering the USA. He further reported on the use of cockroach plant "decoction with sugar to which adult flies are attracted and rapidly killed". In the same paragraph (Froggatt 1909, p. 22) he commented on the use of a braconid wasp which parasitized 10–15 % of maggots they collected. The parasites did not seem to be "much of a check of maggots in the orange but were more common in smaller fruits." In his summary of fruit fly programs in Mexico, Froggatt (1909, p. 76) quoted Herrera stating that parasites made no difference to the pest.

The Mediterranean fruit fly had established in Bermuda in about 1865 (Back 1914) and was discussed by Froggatt, mainly as an example of needed quarantine protection for the USA. Froggatt also discussed a program by the government of Bermuda in 1907 that granted £500 for an attempt to eradicate the Mediterranean fruit fly from the island. The director of public gardens, Mr. T. I. Harris, proposed a combination of orchard cleaning and destruction of stung fruit by collecting and bagging it with a large stone and throwing it into the sea. Harris communicated to Froggatt in 1908 that the effort had greatly reduced the numbers of infested fruit the following year. Back (1914), however, reported that the program had lapsed after 1910 and restarted in 1913 but was ineffective since he readily collected infested fruit that year. Back compared the situation in Bermuda with that in Hawaii and concluded that no clean culture program can eradicate the pest if a few owners fail to clean their properties. In fact, the Mediterranean fruit fly was eradicated from Bermuda (Hilburton and Dow 1990) but only after apparent failures of control by the predator *Anolis grahami* (Gray), which was introduced in 1905, became established and common, but did not impact Mediterranean fruit fly populations, and the parasite *Opius concolor* (now *Psyttalia concolor* (Szépligeti)), which was introduced in 1926 and 1927 but did not permanently establish populations on Bermuda. The pest was eventually eradicated in 1957 using bait sprays (malathion and hydrolyzed protein), soil insecticides (dieldrin), and population monitoring by a trapping system.

In Italy, Froggatt visited laboratories near Naples and met with F. Silvestri, who was working on olive fly control and was a supporter of use of parasites. Silvestri had reared several parasites for *Dacus* (now *Bactrocera*) *oleae,* but no economically important parasite had been found. A meeting was arranged with Berlese, who developed bait sprays for *B. oleae* (which he called Dacacide), consisting of a mixture of molasses (40 %), honey (40 %), potassium arsenate (2 %), and water (18 %). This mixture was applied as a spray, but it was washed off by rain and also killed bees. Berlese also developed a system (Froggatt 1909, p. 70) consisting of a bottle hung in the branches of trees. "Into each bottle is inserted several long cotton threads, forming loose bundles hanging several feet, down which the poisoned liquid flows, and the flies find a ready resting place while they sip the poison."

In the fruit fly Sect. 3 of this report, Froggat also discussed the use of kerosene in trapping pans for Mediterranean fruit fly in Western Australia, which failed for Queensland fruit fly, and citronella oil tested in India and eventually published by F. M. Howlett (1912, 1915), for *Dacus ferrugineus* (now *Bactrocera dorsalis* (Hendel)), *D. zonatus* (now *Bactrocera zonata* (Saunders)), and *D. diversus* (now *Bactrocera diversa* (Coquillett)). After discussion with Howlett and his return to Australia, Froggatt suggested to Berlese that he test the effects of adding citronella oil to the Dacacide mixture, and Berlese's assistant later reported to Froggatt that the majority of olive flies killed were males. Severin and Severin (1914) followed the reports of attraction of kerosene to Mediterranean fruit fly by Weinland (1912) and to *Bactrocera* by Howlett with more extensive tests, but their discussion was more to describe a curiosity (that the males plunged to their own death) than a practical tool.

During this same period, although not reviewed by Froggatt, C. W. Mally, working in the Cape Town laboratory in South Africa, tested poisoned bait sprays consisting of molasses (5 gal), lead arsenate (1 lb.), and water (25 gal). His first tests were carried out in cages (1904), and later tests were conducted in the field over several years (Mally 1909). Mally concluded that the bait was effective in controlling the adult populations, since, even though the kill was delayed after feeding, the flies were incapacitated and incapable of stinging fruit. A summary recommendation on use of this poison bait, published as an anonymous review (Anonymous 1920, publication delayed due to World War I), was one of very few that discussed human and environmental safety in use of the arsenic baits.

Research programs continued in Australia following Froggatt's report. As in other regions, orchard cleanliness through destruction of fruit was mentioned as the major strategy, but reports in the early to mid-1920s discussed a series of attractants and trapping systems as means of control. Tryon, in a letter to the editor of the Brisbane Courier (1922) dated 1921, but containing observations from 1922, listed a number of issues concerning the impact of trapping baits for reducing fruit fly populations that had been recommended by the A. H. Benson, director of fruit culture. One bait discussed by Tryon was Harvey lure, a proprietary bait (formulation not given), discussed in several newspaper articles. In his letter, Tryon questioned a series of inconsistencies in reports, including identification and sex ratio of the flies killed. In reports from different experimenters, there was apparent confusion of *B. tryoni* with non-economic species and results ranging from 100 % females killed to mostly males. In addition to the Harvey lure, comparative tests had been made with eight other named lures, all of which attracted males of non-economic species.

Froggatt (1909), Benson (1906), and Tryon (1922) all expressed opinions concerning natural enemies, quarantines, and use of attractant traps and sweetened baits as pest control methods. In 1922, Hubert Jarvis was hired into a position in Stanhope, a town in southern Queensland, to develop control programs for fruit flies attacking fruit produced in the granite belt, mainly temperate pomaceous and stone fruit. In an extended series of short reports from 1922 to 1931, Jarvis described population control methods for *B. tryoni* in this region. In his first two reports, Jarvis (1922a) described the "magnet" trap and lure and the "Hall" bait which captured mostly female flies. After reporting failure to capture adults in the spring, Jarvis (1922b, c, 1923) briefly discussed a debate between H. Tryon and W. Froggatt as to whether *B. tryoni* overwintered as pupae or adults, which determines the optimal time for adulticidal sprays or baits and the effectiveness of cleaning orchards of fallen fruit over the winter. In the 1923 report, Jarvis discussed the possible role of a lure system for early season use in both detection and population control, using the Harvey lure, to be combined with orchard sanitation later in the season. Jarvis (1924) discussed further results from experiments to determine the overwintering stage(s) and dates of first interceptions of adults.

The first 1925 report (Jarvis 1925a) described the initiation of an overwintering test, various tests of host status for native fruits, biological control by parasites, and host tests of apples. In the following report, Jarvis (1925b) reported that new regulations required the removal of all fruit from the district by April 7 to prevent fruit fly over-wintering. The over-wintering experiment failed to show any survival of larvae or pupae through the winter, so in 1925 the regulation was not renewed. Additional reports in 1925 (Jarvis 1925c) included a summary of problems regulating fruit destruction and orchard sanitation and additional tests with Harvey's lure for adult population control. Jarvis (1926a, b) reported the final analysis of the overwintering test that was completed in August 1925. In the 1926 publications, Jarvis reported that no adults were produced from "several hundred-weight" (one hundred weight = 100 lb, = 45.36 kg) of infested fruit placed in cages in the fall of 1925. Jarvis (1931) reported on the formulation and testing of Jarvis lure, which was cited as the first formulation to specifically involve mixing ammonia as the attractant component of the bait. Although Jarvis specifically identified use of this bait for population control (not monitoring), the killing was done through drowning adults in the mixture, not from insecticide.

Another compound widely tested as an insecticide during the 1930s was potassium antimony tartrate, more widely known as tartar emetic. Benjamin Rush, a medical doctor and signer of the USA Declaration of Independence, cited tartar emetic as a standard treatment for a wide array of diseases (Rush 1789). The compound has been widely used for treatments of leishmaniasis and schistosomiasis but was replaced by non-antimony treatments after drugs with few side effects were discovered. Rush (1810) reviewed use of tartar emetic as a treatment for alcoholism through aversion therapy. Early work with *Anastrepha* species (probably *A. obliqua* and *Anastrepha suspensa* (Loew)) in 1932 in Key West, Florida, and the Canal Zone, Panama, and more complete tests in Key West in 1933 showed tartar emetic to be a superior insecticide compared to nicotine sulfate, lead arsenate, and copper carbonate (McAlister 1936). The spray was shown to be non-toxic to trees by the USDA Orlando laboratory in 1937 (Spencer and Osburn 1938). Tartar emetic was later tested against *Anastrepha* by Plumber (1944) in an extensive series of laboratory trials in Mexico City and later as a comparison with DDT (Plummer 1947) and in unpublished reports by Baker (1937, summarized by Baker et al. 1944). Allman (1940, 1942) and Friend (1949a, b) tested tartar emetic against *B. tryoni*, but the level of usage was not reported.

Baker et al. (1944) summarized investigations of an array of insecticides and toxic bait approaches. This report summarizes insecticide research carried out from 1930 until shortly before publication in 1943. The need for adult treatments was reinforced in the research program of the Mexico City Laboratory after Darby and Kapp (1934) showed that adult Mexican fruit flies were capable of living for extensive periods of time; females were observed to live in excess of 11 months and males in excess of 14 months. These results, and later tests by Stone (1942) with *Anastrepha serpentina* (Wiedemann), indicated that adult flies could survive through non-fruiting seasons or periods when fruit had been cleaned from orchards. Experiments were performed in the laboratory and field using inorganic compounds of copper (following tests by Miller and McBride (1931) in Florida) and nicotine. The materials (including copper carbonate, copper chloride, copper sulfate, cupric arsenite, and cupric aceto-arsenide) showed high persistence in both laboratory and field, if high humidity and rainfall did not interfere, but kill was slow; from 5 to 16 days were required to kill half the flies in cages, and high rates of fruit infestation were observed in orchard trials. A possible solution to the high humidity and rain interference was proposed for nicotine compounds. Nicotine sulfate in syrups of various concentrations (1–30 %) of molasses was shown to be much more persistent at low humidity than the copper compound but washed from the surfaces at high humidity or rain. Compounds, such as nicotine tannate, nicotine binoxalate, nicotine humate, and nicotine bitartrate, were found to be ineffective under humidity conditions known in the Rio Grande Valley.

Additional experiments were performed comparing antimony (tartar) and thallium, but thallium was rejected due to damage to vegetation. Baker et al. (1944) cited the work of Zetek (apparently unpublished) with tartar emetic and found satisfactory toxicity with complete mortality of *A. ludens* and *Anastrepha striata* Schiner in 2–3 days. Fluoroaluminate (cryolite), a compound being tested by USDA scientists against the codling moth in the USA, was also tested against *A. ludens* but found to be ineffective. Other chemicals tested included sulfur applied as dustings and extracts from chrysanthemum and pyrethrums, which gave slow mortality after ingestion and were only effective for short periods in the field. Rotenone when ingested in a 1:1000 dilution in 10 % molasses:water syrup paralyzed flies and gave high mortality for 5 days but was unstable in sunlight. Ricin, extracted from castor bean, had slow or no mortality. Extracts from pathogenic bacteria, including tetanus toxin, diphtheria, and gas gangrene, as well as toxins from rattlesnake, caused little or no mortality. Saponins from root extracts of *Saponaria* plants, Panama bark (*Quillaja saponaria* Molina), and agave gave results similar to lead arsenate. Some of the saponin preparations were effective in killing goldfish at lower pH but not fruit flies. Baker et al. (1944) cited the extensive literature reporting insecticidal use of *H. crooksii*, the cockroach plant, as discussed above, and Plummer (1938) performed tests on alcohol extracts from various stem and leaf tissues and noted syrup mixtures of molasses and water at 10–50 % that were effective in inducing paralysis and death to flies. Research carried out on extracts from this plant also suggested that the toxic factor(s) were alkaloids(s) but were not characterized further.

In addition to arsenic and other metallic inorganic compounds, tests of adult insecticide action of sodium fluosilicate were carried out in several laboratories, showed good toxicity in the laboratory but were rejected due to phytotoxicity. DeLong (1934) reviewed work on fluosilicates and other related compounds as insecticides and cited research on the walnut husk fly (*Rhagoletis completa* Cresson) in California and the natal fruit fly (*Ceratitis rosa* Karsch) in S. Africa, showing equivalent or superior activities of the fluorine compounds applied as dusts or aqueous mixtures without sweeteners in comparison with the arsenic salts. Allman (1940, 1942) reported sodium fluosilicate having only mild toxicity in peaches in Queensland, Australia, and found it to be comparable to tartar emetic, though the effectiveness of both was dependent on the sweetener used. The effects were reduced when there were alterative food sources. During the Mediterranean fruit fly eradication program in 1929–1930 in Florida, copper compounds were investigated as alternatives to lead arsenate (Miller and McBride 1931). Copper carbonate was found to be nearly as toxic to the flies as lead arsenate and was considered to have much lower phytotoxicity. Darby and Knapp (1934) extended this work to the Mexican fruit fly and concluded that the toxicity to the flies was indirect due to killing yeasts and molds.

Fruit fly programs in Japan were somewhat unique compared with those discussed above. Only one native fruit fly species, *Bactrocera* (then *Dacus*) *tsuneonis* (Miyake), has been reported from Japan. The description, biology, and economics of this species were given in Miyake (1919). The discussion of population control reviewed the author's observations of a series of predators and parasites, including a study of predation by scorpion flies, *Bittacus* spp. (Mecoptera), which he identified as significant house fly predators. He did not identify any predators of importance in fruit fly population control. Miyake discussed adult fly capture under a bounty system paid to collectors as a means of reducing damage. Flies were captured using a metal framed net (resembling a tennis racket) with a 5 ft bamboo rod. Bird lime was applied to the net to capture flies. For sampling in dense foliage, only a bird lime covered rod was used. According to Miyake, a skillful hunter could capture 130 flies a day. A purchase of 78,351 flies at Aoye village and 201,675 flies at Tsugumi Village "remarkably reduced" fruit injury level. Other measures, including attractants such as citronella oil and poison such as Mally's fruit fly remedy, were mentioned, but Miyake stated that "complicated circumstances" made these methods difficult to employ.

In the fruit fly insecticide literature from the 1880s until 1945, human safety was rarely discussed, was dismissed as unimportant, or shown to be a non-issue so long as commodities were washed. The predominant insecticides used for fruit fly control consisted of inorganic compounds, mainly of lead, mercury, arsenic, and copper (De Ong 1948). The compounds are toxic to some degree in all living organisms. The dangers of these compounds were well known, at least for their acute effects, but bioaccumulation and long term effects were not. De Ong, for example, described toxic effects of copper compounds on grazing sheep and domestic ducks but not human applicators that surely were exposed to much higher doses. However, De Ong included an appendix (pp. 298–307) with examples of warning labels that conform to Interstate Commerce Commission regulation and a listing of antidotes and their application.

Mound (2005) summarized the attitude of program managers in the 1929 Florida Mediterranean fruit fly eradication in the obituary for Wilmon Newell, Dean of the College of Agriculture and Extension Station at Gainesville:

> His "scorched earth" approach to eradicating the Mediterranean fruit fly was conducted without regard to the environment and at extreme economic losses to growers. He was highhanded because he considered eradication to be critical. Perhaps he was justified based on the numerous awards he received for meritorious service. It is easy to criticize these somewhat imperious methods—but they were successful.

A number of studies have been carried out to determine the effects of exposure to arsenic and lead arsenic on human workers, including factory production workers, orchard workers, and residents of areas exposed to high doses. In a study of factory workers in a plant in Baltimore, Maryland, USA, who were exposed to high air concentrations, Mabuchi et al. (1980) found significantly greater mortality from lung cancer and anemias compared to expected numbers. The workers had other physical symptoms as well, including keratoses and perforated nasal septum. Using data from 1938 for Wenatchee, Washington, USA, Nelson et al. (1973) and Tollestrup et al. (1995) analyzed mortality patterns among orchardists who in 1938 were regularly exposed to lead arsenate, intermediates who were retired orchardists or individuals occasionally exposed to lead arsenate, and consumers who were not directly exposed to lead arsenate but lived in the community. The source of lead arsenate was that used mainly for codling moth control. In contrast to the study of Mabuchi et al. (1980), this study did not show that orchard exposure to lead arsenate caused significantly higher toxic effects (death rate, lung cancer, anemias) in those constantly exposed and those in the intermediate or non-exposed groups.

2 Use of Synthetic Organic Insecticides

The development of insecticides with contact and vapor as well as ingestion toxicity coincided with World War II. A detailed summary of the chemistry of these insecticides is given in Metcalf (1955). Here, I summarize only information concerning the major pesticides used against fruit flies. The chlorinated hydrocarbons, although synthesized in 1874, were first developed as insecticides by Paul Hermann Muller in 1939 for which he won the Nobel Prize in 1948. DDT (1,1,1-trichloro-2,2-bis(p-chlorophenyl,)ethane: generic name dichloro diphenyl trichloroethane) was the most widespread chlorinated hydrocarbon insecticide, but aldrin and dieldrin developed in 1945 had wider use for fruit flies, especially in soil applications. The French scientist Philippe de Clermont first synthesized organophosphates in 1854, and Gerhard Schrader, working for the Bayer Corporation in Germany, recognized their activity as insecticides in 1934. Schrader's work was then diverted to human nerve gas research during World War II, and the insecticide uses were not published until 1947. The carbamate group was released as insecticides in 1958; the commercial product carbaryl, later marketed as sevin, is used on more than 200 fruits and vegetables and ranks in the top 10 of pesticides used in agriculture in the USA (in 2008) though it has been withdrawn for use in Europe. Its use against fruit flies has been mostly limited to *Rhagoletis* spp. (Aliniazee 1986), especially the western and eastern cherry fruit flies *Rhagoletis indifferens Curran* and *Rhagoletis cingulata* (Loew).

Synthetic organic insecticides were tested during the 1940s and early 1950s for the Mexican fruit fly programs. Plummer (1944) compared DDT and tartar emetic in cage tests. After dissolving the DDT in acetone then mixing into sugar solutions, he noted problems in solubility of the DDT with crystals forming in solutions with higher concentrations. Results were erratic for DDT, and tartar emetic was superior to DDT in all tests. Plummer and Shaw (1946) reported further failures in unpublished reports in 1945 for field tests of DDT spray (2 lb. DDT/100 gal water) and dust treatments. In those tests, adult populations were not reduced, and levels of fruit infestation were only slightly reduced for treated fruits (reduction 13 % for sprayed and 37 % for dusted fruit). Plummer and Shaw then presented newer results that showed higher success rates using about 3X higher concentrations of DDT for both mixed spray and dusting, and even higher rates were used in later tests. Although the reductions in trap catches and fruit infestation were significant and acceptable to the authors, clearly the use of DDT did not allow reduction in volume of pesticide use from the inorganic mixtures previously used.

Friend (1949a) tested a series of organophosphates, DDT, and tartar emetic against *B. tryoni* in laboratory and field tests carried out in New South Wales,

Australia. His baits consisted of water and sugar and gave good protection against stinging damage and infestation. The primary emphasis was to test HETP (hexaethyl tretra-phospate), one of the earliest organophospates shown by Schrader in 1942 to have insecticidal properties (Metcalf 1955). He showed HETP and parathion to be effective, have comparable persistence (about 3 weeks), and rapid knockdown, all of which were superior to tartar emetic. In a later test, Friend (1949b) showed the two organophosphates to be similar for 50 and 100 % knock-down (2.5 h and 6.5 h, respectively) to nicotine sulfate (2.3 h and 4 h, respectively) and tartar emetic (6.5 and 24 h, respectively).

Tests in Europe were also carried out with the synthetic organics for fruit fly control. Santoro (1951), working in olive orchards in the Salerno province of Italy, tested DDT, BHC (benzene hexachloride), and chlorodane in various combinations against *B. oleae*. Tests were carried out with DDT paste and mixtures of the insecticides at rates of 4 lb per 10 gal of water. Four applications were made in early fall at a time when olive fly is known to be active. Insects were collected on sheets spread under the trees, and insects killed included ants, coccinellids, scarab beetles, and *B. oleae*. Very few dead beneficial Hymenoptera were collected. Santoro reported about 40 % greater olive oil production and higher quality from trees treated with DDT or DDT-BHC mixtures. In an additional test, Santoro reported that parathion treatments of infested olives killed eggs and larvae near the surface but did not penetrate to kill larvae deeper in the fruit.

Additional tests of persistence of DDT were carried out against *Anastrepha fraterculus* (Wiedemann) in Peru, where Wills (1946) found that >60 % of flies released on DDT sprayed trees were killed up to 20 days after application, and toxicity was still evident 50 days after application. Antongiovanni (1949), however, showed that DDT and BHC were only effective against *B. oleae* for a week in Italy but attributed the short persistence to rain and high temperature, especially for dust applications.

These early reports indicated that the use of organic synthetic insecticides with fumigant and contact toxicity, in addition to being stomach poisons, did not provide greater commodity protection or reduce rates of insecticide application compared to the inorganic stomach poisons. Although some of the tests involved using sweeteners to induce consumption, the early tests did not focus on use of baits for attraction. Persistence of the synthetic chlorinated hydrocarbons (mainly DDT) was noted in some tests, but since the methods of application were identical (liquid sprays or dusts without attractants) to the inorganic insecticides, rain and other weather conditions limited persistence.

Steiner and Hinman (1952) tested a number of pesticides, including DDT, aldrin, dieldrin, dilan (a 1:1 mixture of prolan and bulan, which were chlorinated hydrocarbons placed by Metcalf (1955) in the nitroalkanes group), lindane, and the organophospates parathion and EPN. These insecticides were tested as wetable powders (dry applications) or emulsifiable formulations using xylene as the wetting agent. Results showed that these insecticides were effective for reducing infestation of Cavandish bananas against oriental fruit fly with infestation rates reduced to 57 % for treated plots vs. 94 % in untreated. Steiner and Hinman (1952) attributed instances of lower effectiveness to rain. Steiner (1952) directly compared DDT and parathion and found parathion to have faster knockdown when mixed with proteins. A mixture of parathion and a carbamate, metacide, also attracted flies from longer distances. After showing that the parathion bait was superior to DDT, Steiner (1952) tested parathion with a series of proteins, including protein hydrolysate, yeast hydrolysate, soy protein plus several sugar sources. The results showed that the sugar type did not affect kill rate and that soy was about half as effective as the protein or yeast hydrolysate. These results showed that the synthetics that had effective contact toxicity and, therefore no need to be ingested, were not improved by addition of sugars to the bait.

Gow (1954) tested a series of hydrolyzed proteins as attractants from vials or cotton wicks suspended in traps and found, in contrast to Steiner's (1952) tests with sprays, soy protein hydrolysates were generally superior to the yeast and protein hydrolysates in trapping flies. This study also showed that ammonia in trapping baits was highly repellent to the oriental fruit fly. He also concluded that specific bacteria were responsible for increasing attraction of the protein hydrolysate. Apparently, no further tests of the effects of these bacteria were carried out.

Steiner (1957) performed an extensive comparison of synthetic organic compounds and spray combinations in a series of tests carried out from 1950 to 1955 against oriental fruit fly in Hawaii. Twenty six experiments having more than 180 tests in more than 600 plots were carried out during this period. Rapid knockdown was identified as an important requirement for the bait spray to avoid fruit infestation by invading flies. Persistence was not considered important, because harvests were taken at short intervals in the mango and guava orchards. DDT was the most widely tested, but several other chlorinated hydrocarbons, including aldrin, dieldrin, and lindane, were also tested. The organophospates parathion and malathion were also tested repeatedly. Tests of baits included a number of hydrolyzed proteins with or without sugar, sugar alone, or no bait. The results of these tests were presented in a three page table giving treatment conditions, insects surviving in control plots, and percent reduction in treated plots. Although no statistical comparisons were made among the results, and the treatments only considered the oriental fruit fly, these data allowed Steiner to offer several conclusions that set the pattern for use of adulticide baits against fruit flies. Protein hydrolysate baits with either parathion or malathion bait sprays, with malathion gaining preference in later analysis, became the standard for fruit fly control.

Invasive tropical fruit fly outbreaks, before the late 1980s, were relatively rare events with no outbreaks during most years. The insecticides used for control of the western cherry fruit fly (*R. indifferens*) are given in Table 12.1 (Aliniazee 1986), which includes the insecticides, approximate number of years that they were used and reports of resistance. These patterns of usage probably follow the general pattern for fruit fly sprays in the USA. Insecticides discussed in the first section, mainly salts and plant products, were used for long periods but none were used in 1984. Malathion, diazinon, and dimethoate were the common insecticides. Resistance was only indicated for one pesticide, methoxychlore, which was rarely used.

Table 12.1 Number and type of insecticides used for the control of *Rhagoletis indifferens* in the western U.S reported in 1984

Insecticide	Mode type of application	Year introduction	Years in use (in 1984)	Current status in 1984	Resistance?
Lead arsenate	Dust	1920	30	Not used	No resistance
Rotenone	Dust	1934	20	Not used	No resistance
Lime sulphur and lead arsenate	Dust	1940	15	Not used	No resistance
DDT	Dust	1947	12	Not used	No resistance
Methoxychlore	Dust	1950	20	Rarely used	Some indication
Malathion	ULV	1970	14	Commonly used	Not documented
Parathion	Spray	1955	30	Not used	No resistance
Diazinon	Spray	1959	25	Commonly used	No resistance
Perthane	Spray	1961	23	Not used	Not documented
Sevin	Dust	1961	23	Less common	Not documented
Guthion	Spray	1965	19	Rarely used	No resistance
Dimethoate	Spray	1980	4	Commonly used	No resistance

Aerially applied bait sprays using malathion were used in the Mediterranean fruit fly outbreak in Florida in 1956. Ayers (1957) described the findings that led to this eradication campaign. Larvae were first reported from grapefruit on April 16, 1956, and their identification was verified 5 days later. The following day, fruit stripping in the area was initiated, and quarantines were established during the next week. This activity mirrored the 1929 campaign, which also started in April with larvae found in grapefruit. However, in the 1956 eradication, sprays replaced fruit stripping and tree destruction as the primary treatment, with quarantine as the primary preventative action. In addition to Ayers' (1957) fairly detailed report, Rohwer (1958), Denmark (1956a, b), Wolfenbarger (1958), and Steiner et al. (1961) commented briefly on the over-all program, including comparison of costs, treatment period, and social impact, including quarantines. It is notable that (to my knowledge) the only human death from aerial malathion application for fruit flies occurred during this eradication program. The program used a Fairchild-82A-20-FA Packet aircraft, which crashed due to engine failure and loss of control when ferrying from Miami Masters field, where it had just finished a spray application, to Boca Raton Public Airport on 8 August 1956. The accident caused deaths of two crew and three passengers (Aviation Safety Network 2009–2013).

Harris et al. (1971) performed tests to determine formulations for aerial dispersal of toxicant plus protein hydrolysate. Mortality data were collected for flies falling into trays beneath sprayed plants. The tests were replicated, and mean separation tests were performed comparing differences between the different toxicant:protein hydrolysate ratios. Although the mean separation calculation requires an analysis of variance, appropriate statistics were not given. In the first test for malathion, there were no significant differences in kill rates among the different toxicant concentrations tested (1:10; 1:20; 1:200) for oriental and Mediterranean fruit flies, but the 1:200 was inferior to the 1:10 and 1:20 treatments. In a second series of tests, again collecting dead flies under treated foliage in the field, but also examining rates of mortality for flies caged on treated foliage, concentrations of 1:4 and 1:16 had the highest mortality, 1:64 was intermediate, the lower concentrations of 1:256, 1:1230 and 1:2469 had significantly less kill than the 1:4 or 1:16 concentrations for both dead fly collections under foliage and caged foliage. Harris et al. (1971) did not clearly recommend a toxicant:bait concentration, but they did cite Lopez et al. (1969), who demonstrated effectiveness of the 1:4 concentration for the Mexican fruit fly.

The actual percent concentration of malathion:hydrolyzed protein ratios used are given in environmental assessment documents required for spray permits. The assessment document (USDA 2002) used for a 2002 Mexican fruit fly outbreak in California described the mixture in a somewhat convoluted manner: "The formulation used in the program is 0.175 lb of active ingredient per acre mixed with 9.6 fluid ounces of protein hydrolysate bait per acre, for both aerial and ground applications". My calculations of 0.175 lb = 2.8 oz and total weight = 2.8 + 9.6 = 12.4 oz indicate that the malathion concentration was 22.6 %, which is comparable to the Lopez et al. (1969) suggestion.

Technology to protect subtropical fruit and vegetable production using malathion-hydrolysed protein baits was demonstrated repeatedly after the 1956 Mediterranean fruit fly outbreak in Florida. Clark and Weems (1989) reviewed the exotic fruit fly eradication programs in Florida and listed 10 discoveries of Mediterranean fruit flies trapped in Florida between 1962 and 1989, six of which required eradication action (mainly application of sprays), and four were shown to be limited outbreaks by trapping surveys. The Mediterranean fruit fly eradication in Texas (Stevenson and McClung 1966) and the eradication Mediterranean fruit fly outbreaks in California, which began in 1975, (Hagen et al. 1981; Dawson et al. 1998) were also controlled by pesticide baits and (later) release of sterile flies.

3 Holistic Insecticidal Systems

The last section of this review refers to the approaches required to manage an area-wide program. Two components are addressed, the action components of the eradication program and the social-environmental concerns of the public in the affected area. Public concern over area-wide aerial pesticide application had, to some degree, its origin in the early 1960s following the Carson's (1962) discussion of environmental and human health damage attributed to insecticides. However, during that period (the late 1950s and 1960s) there were no fruit fly outbreaks in

California, and the Mediterranean fruit fly eradications in Florida and Texas were much smaller and were dispatched in less than a year.

The first Mediterranean fruit fly outbreak in California was reported in 1975 and was declared eliminated in 1976 after invoking a quarantine of about 100 mi^2 and application of malathion bait, all with minimal reporting or public attention. Outbreaks were detected in several counties in 1980. Hagen et al. (1981) focused mainly on biological and economic problems rather than social and political challenges and predicted future problems with continuing Mediterranean fruit fly outbreak due to multiple hosts, large areas susceptible to outbreaks which, in fact, became nearly continuous after 1987 into the mid-1990s (Penrose 1996). A history of community response to these outbreaks is summarized in Anonymous (undated a). News releases (Anderson 1982) reported that a series of biological and execution mistakes, such as confusing natural and marked florescence for sterile flies marked with fluorescent powder, believing that the pest could not over-winter in the region, and accidental release of non-irradiated fertile flies in the region were responsible for public confusion and failure to eradicate. The aerial malathion bait spray was identified in all reports as the main complaint of the public. As mentioned in the previous sections, a number of issues related to property and other damages were considered in evaluations of baited pesticide sprays, however, phytotoxicity, availability of components, and efficacy in killing the adult fruit flies were the major issues in the earlier programs. Private property contamination was sometimes considered, but public and wildlife health concerns were seldom mentioned. Steiner et al. (1961), for example, were concerned about damages caused by application of malathion-protein hydrolysate bait, but they were discussing damage to car finishes, not public health. Impact on environment, private property rights (e.g., the right not to have one's property sprayed) and human health became the major issues after the 1980s.

The second Mediterranean fruit fly eradication program in California focused on several southern areas of the state, which, after 1987, had nearly continuous outbreaks until 1995. One of the early reports (Wolfe 1987) reviewed the history of public concern and reports of threatened sabotage (Chavez and Simon 1989a, b), which became acute in California in 1993. Social and political issues related to the eradication program in the San Francisco area were summarized in the testimonies given in congressional hearings held in Washington D.C. by the Operations and Nutrition Subcommittee of the House Committee of Agriculture on May 5 1994 (United States Congress 1994). Testimony by community leaders, scientists and managers of the program, and university and commercial interests all agreed that the area-wide aerial application of the malathion bait was the crucial activity in the program subject to debate. Also included in the hearings were statements by Mr. H. Voss, Secretary of Agriculture for the State of California and proposals from the University of California for new research approaches to address the Mediterranean fruit fly problem. Statements by Mr. Voss included a contention from the Mediterranean fruit fly Science Advisory Committee:

> that only the use of sterile Medflies, preceded by pesticide applications to kill existing sexually mature adults, and aerial applications of malathion and bait, have ever eradicated any Medfly infestation.

As discussed previously here, three Mediterranean fruit fly eradication programs, Florida 1929, Florida 1955 and Texas 1956, did not use this combined approach, since the sterile insect release for fruit flies had not been developed yet. Florida 1955 and Texas 1956 relied on aerial sprays of malathion bait, fruit destruction, and quarantines, and the Florida 1929 program relied on host destruction (including trees and fruit), rigid quarantines (including vehicle inspections), and lead arsenate bait sprays with ground equipment.

The Mediterranean fruit fly program was also a social phenomenon topic in entertainment shortly after the 1987-95 eradication. The 1989 "Simpsons" cartoon show (Season 6 Episode 23) (Anonymous undated b) showed a sign for evening activities in the public park announcing "8:00 Medfly Spraying, 8:15 Springfield Pops, 8:30 Spraying: 2nd Pass". Another reference to the fruit fly outbreaks was in the character "Harry Medfly", who appeared in a single episode of Duckman (Season 2 Episode 9) (Anonymous undated c) which aired in 1995. A relevant quote "I'll be back, I'll return every fifth or sixth episode" and a clip from the next episode "Forbidden Fruit" communicated a sense of program failure to the public in California. This contrasts with the obituary presented by Mound (2005) that Newell's "draconian" approaches in Florida were accepted because they were successful.

During the 1977–1981 and 1989–1995 Mediterranean fruit fly outbreaks in California and the 1997–1998 outbreaks in Florida, the opinions of scientists were also quite diverse regarding both the progress and probable success of the eradication programs. Moreno and Mangan (2000) reviewed efforts to develop alternatives to malathion baits, specifically in response to position papers presented in the congressional hearings in 1995. New research emphasis on "novel chemical approaches" included studies of plant growth regulators as a method of maintaining resistance to infestation (Greany et al. 1983; Nguyen et al. 1992). A number of alternatives using of stomach poisons to contact and fumigant insecticides were tested. Proposed alternatives included boric acid or borax as additives to baits (Chambers et al. 1987; Enkerlin et al. 1993; Nigg and Simpson 1997; Yang et al. 2000). Insect growth regulators, including cyromasine and lufenuron (Moreno et al. 1994; Navarro-Llopis et al. 2004) and phototoxic dyes (Mangan and Moreno 1995, Moreno and Mangan 1995; Liquido et al. 1995) were developed and tested under field conditions.

USDA-ARS (Agricultural Research Service) laboratories developed a coordinated action plan that dealt with fruit fly research. The 1991 Action Plan and extensions in 1992–1993 included several lists of projects to be applied to fruit fly eradication and management programs with special emphasis on exotic tropical and subtropical species. This research plan coupled with availability of funds from state departments of agriculture, especially California and Florida, resulted in a number of programs in USDA to develop publicly acceptable insecticides to eradicate fruit fly outbreaks in the USA.

One of the first pesticide products resulting from the 1992–1993 action plans was the insecticidal bait SureDye ® marketed by PhotoDye International Inc. (Linthicum, MD, USA). This product was developed and tested under laboratory and field cage conditions at the USDA-ARS laboratory in Weslaco, Texas, and the bait formulation was tested in Hawaii and used briefly in the Moscamed Regional Program in Guatemala and Mexico. At the time this bait-toxicant system was developed, the use of photoactive dyes as insecticides was well documented in scientific literature. The first report was by Balbieri in 1928 (cited in Heitz 1995) using photoactive dyes as larvicides for mosquito control. Heitz (1995) lists mortality reports for 28 insect pest species, including flies, moths, beetles, cockroaches, and ants from the literature or tests carried out by the USDA-ARS.

In 1994–1995, Moreno (unpublished) screened 110 commercially produced dyes and found approximately 1/3 to be phototoxic when ingested by *A. ludens*. Toxicity was influenced by the cage and light conditions with plastic cages having reduced mortality compared to aluminum screen. Light intensity and wavelength strongly influenced mortality rates, so all experiments were carried out in sunlight in the mornings just after sunrise with mortality readings taken hourly for 5 h. According to laboratory notes summarized by Moreno in an unpublished annual report (1996):

> Of the dyes tested, the aminotriarylmethane, anthraquinone, disazo, hydroxytriarylmethane, monoazo, nitro, oxazene, pyrene, quinone-imine, and quinolone did not have phototoxic dye representatives. The acridines, azine, thiazole, triphenylmethane, and xanthene groups shared non- and phototoxic dye representatives; and the diphenylmethane, diphenylnaphthylmethane, rhodamine, thiazine, and triarylmethane groups had only phototoxic dye representatives. The estimated LC_{50} for the various phototoxic compounds ranged from 0.022 to > 16,000 ppm for methyl eosin and thionin, respectively. The xanthene group had the largest number of phototoxic dyes.

Lillie (1977) provided a summary of the chemical structures of biological stains and dyes, including those listed above by Moreno. Heitz (1995) discussed the phototoxic activity of dyes, especially as it occurs after ingestion in insects. Moreno and Mangan (2000, 2002) tested the Phloxine B insecticidal bait against the most important North American or invasive species of subtropical fruit flies, including *A. ludens, A. obliqua, A. suspensa, A. serpentina* and *C. capitata* under field conditions. The formulation and roles of the components are given in Table 12.2. Phloxine B's solubility in the bait was sensitive to pH, and the mixture, which was also subject to foaming, required vigorous mixing to suspend the ingredients.

Mangan and Moreno (1995) showed that the addition of the surfactant SM-9 to the bait greatly (4-7X) reduced the survival of *A. ludens* in laboratory cage tests. The original use of the surfactant/adjuvant was to aid in mixing, since phloxine B solubility is reduced at lower pH, and the Mazoferm pH was below 4.0. After observing increased fly mortality, they compared a series of 22 commercial adjuvants for insecticidal formulations and surfactants used by the food industry in laboratory cages. The most promising six products were compared in field cages (Mangan and Moreno 2001). Evaluations were made for bait consisting of the no

Table 12.2 Components and activities of the phloxine B fruit fly bait recommended for SureDye

Ingredients	% Concentration (V:V)	gm/ml/l
Phloxine B (92 %)	0.5	5.4
Mazoferm 802	70.0	700.0
Invertose	20.0	200.0
Tween 60 (as formulated)	0.01	10.0
Soybean oil	0.01	10.0
Acetic acid	0.6	6.0
Polyethylene glycol 200	2.0	20.0
Xanthan gum	0.4	4.0
Water, qs ad	4.5	45.0

The ingredients in this formulation have the following function
1. Phloxine B is the phototoxic dye
2. Mazoferm is an animal food that contains 22 % protein, minerals, and vitamins, and flies are attracted to its volatiles and readily accept it as food
3. Inverstose is an invert sugar (60 % fructose and 40 % glucose) that increases feeding in flies and is also hygroscopic, thus slowing desiccation of dye-bait drops in dry climates
4. Tween 60 is a surfactant used as an emulsificant and dispersant in medicinal products
5. Soybean oil helps to solubilize hydrophobic products and is used in cooking and the manufacturing of margarine
6. Acetic acid is used as a preservative in foods and attracts flies
7. Polyethylene glycol is used in foods and cosmetics, inhibits mold growth, and is a humectant
8. Xanthan gum is a thickener in food products, such as sauces and dressings

treatment, bait with no adjuvant, and six cages with various adjuvants in the baits (Table 12.3).

Photodye International promoted SureDye extensively from 1994 to 2000, and press releases by the USDA supported this effort (see Anonymous 1994 and Hardin 1997 for examples). This bait system was proven to be effective in field trials in Morocco (Sebbata et al. 1998), Mexico, and Guatemala as well as repeated experiments in Texas, Florida, and Hawaii (see Moreno and Mangan 2000 for references). Dowell et al. (1997) showed that SureDye was compatible with use of beneficial Hymenoptera used in citrus orchards for biological control of pests. However, other than experiments (McQuate and Peck 2000) following its use in the Moscamed Regional Program in 1997–1998, SureDye was used only briefly. Failure of SureDye to become a successful commercial product was related mainly to failure to gain registration in the USA and other major countries in South America and the European Union. Other problems we observed (Mangan, personal notes) included the high cost of the bait (at least 4X the NuLure [Miller Chemical & Fertilizer Corporation, Hanover, PA, USA]/malathion bait), especially the dye component, reluctance of the dye manufacturer to allow a profitable food and cosmetic dye to be identified as a pesticide, and lack of environmental data to support registration. One of the most contentious environmental concerns was the effect of SureDye sprays on honeybees.

Table 12.3 Summary of number of dead *Anastrepha ludens* collected for each of the adjuvant treatments on field caged mature Rio Red grapefruit trees

			Treatments						
		Control	Check	Kinetic	Latron	SM-9	Sylgard	Tergitol	Tween60
	Mean	184.50	313.63	748.25	548.88	721.13	498.63	674.38	848.13
	Std. Dev.	133.23	179.07	286.23	192.51	346.46	191.82	268.84	72.09
ANOVA-		F-ratio = 7.56, prob < 0.001, df = 7.56							
Corrected % mortality[a]			41.12	75.34	66.39	74.41	62.99	72.64	78.25
Tukey HSD Sep.		a	a	b	ab	b	ab	ab	b

Bait was SureDye with 0.5 % phloxine B. Tests were run in eight cages concurrently. A total of eight replicates was run with six adjuvants, the check (dye-bait with no adjuvant), and the control (bait, no dye, no adjuvant). The experimental schedule was to set up bait stations and release 1,500 flies into each cage at 0730 h on Monday and collect and record numbers of dead flies every 2 h from 0730 to 1530 h from Monday to Thursday and until 1130 h on Friday. Means and standard deviations were the total numbers of dead flies averaged over the eight replicates

[a]Calculated as 100*(treatment – control)/treatment

Tarshis Moreno (2001) showed that the SureDye formulation was repellent to honeybees under forced feeding laboratory conditions. This finding did little to advance registration of SureDye but formed a basis for later development of fruit fly baits.

As the bait system for SureDye was developed (Mangan and Moreno 1995; Moreno and Mangan 1995, 2002; Moreno et al. 2001), the bait was tested with an array of alternative pesticides, many of which were already registered for fruits and vegetables. As the barriers to registration and problems of cost became apparent, several other problems and solutions were identified. Mazoferm, the source of protein and some sugars in SureDye, did not have effective quality control standards, and starch grains clogged the screens and nozzles during application. A commercial product, Solulys, (Roquette Chemical and Bio-Industries) prepared by spray-drying refined Mazoferm, could be mixed in a more concentrated form with fewer problems of equipment clogging and was therefore used as a substitute. Other changes in the bait were reviewed in Moreno et al. (2001) and Moreno and Mangan (2002). In order to communicate to potential cooperators that a new bait had been developed that avoided the problems with Mazoferm, the solulys based bait was named Solbait in publications after Moreno et al. (2001). The components of this bait are given in Table 12.4. A summary of insecticides tested by Moreno and Mangan and published in 2002 are given in Table 12.5. By computing relative human toxicity, LD50 for fruit flies (mainly Mexican fruit fly) and AI concentration in the Solbait, a relative comparison of human safety is given. The spinosad based bait (later named GF120® by Dow AgroSciences, Indianapolis, IN, USA) was one of the least toxic, but a number of other candidates could be used.

The relative toxicities of malathion and spinosad to various terrestrial and aquatic organisms, as is required in an environmental impact assessment, are given in Table 12.6. These data, taken from the environmental impact assessment for use of GF120 as a treatment in the San Diego Mexican fruit fly outbreak, focused in the Valley Center region of San Diego County in 2002. The toxicities in Table 12.6 are given as LD50s in oral mg/kg animal weight for terrestrial animals and μg/L water for aquatic animals, so the rates for aquatic and terrestrial animals are not directly comparable. For the terrestrial mammals, the LD50 for spinosad ranges from about 3-5 times higher than the LD50 for malathion. For the aquatic animals, the concentration for LD50 for spinosad ranges from few hundred to several thousand times higher than LD50 for malathion. In addition to the lower toxicity of spinosad compared to malathion, GF120 contains 80 ppm AI and the NuLure-spinosad bait contains 22.6 % (226,000 ppm) AI (USDA 2002, p. 7). Thomas and Mangan (2005) further tested the effects of spinosad bait applications on populations of a number of non-target insects in orchards in south Texas and found no reductions in populations.

After GF120 was registered in the USA and major fruit producing countries, users raised a number of questions concerning dilution, persistence, methods of application, and non-target effects. Mangan et al. (2006) and Mangan (2009) studied the effects of bait dilution, spray interval, toxicant concentration, repellents for non-target beneficial insect species, and pest feeding behavior under field cage and field conditions. Results of the persistence and dilution tests (Mangan

Table 12.4 Fruit fly bait formulation (SolBait) developed as a matrix for toxins used in fruit fly control

Ingredients[a]	% Concentration, w, v/v	Gm, ml/l
Water, purified, *q.s. ad*	100.0	1,000.0
Ammonium acetate	1.0	10.0
Polyethylene glycol 200	1.0	10.0
Invert sugar	15.0	150.0
Polysorbate 60	1.0	10.0
Soybean oil	0.25	2.5
Solulys	4.4	44.0
Xanthan gum	0.4	4.0

The formula recommended in Moreno and Mangan (2002) used Spinosad at 80 ppm as the AI. Other pesticides tested with this formulation are given in Table 12.5

[a]Gustatory responses by *C. capitata* and *A. obliqua* indicated that soybean oil should be reduced to 0.25 % instead of 1 % and invert sugar to 15 % instead of 20 %, and 0.2 % methyl p-hydroxybenzoate was used as a preservative. In Dow Agroscience formulations, proxel (1, 2-benzisothiazolin- 3-one) was used in early formulations, then the mixture omitted these preservatives

et al. 2006) determined that when commercial GF120 was applied to trees at recommended dilution and rates, rates of mortality of test populations tested in field cages did not change until 14 days post-application. Dilutions could be made up to 10-fold by adding additional water to the bait, and there was no statistical loss of killing effect after material dried. At 100-fold dilution, there was a loss of killing, mostly because drops did not form on the leaves of the trees, and the material ran onto the ground. The dilution tests were important for growers wishing to use high-volume spray equipment used for cover sprays to apply the same AI/area but with higher volume.

Mangan (2009) performed further tests to determine the effects of aging on kill rates for GF120. During the development of Solbait, Mangan and Moreno (personal observation) observed a negative relationship between feeding and ammonia concentration for the Mexican fruit fly. Tarshis Moreno (2001) observed a similar relationship for sugar-ammonia mixtures for honeybees. We reasoned that, since honey bees do not have suitable mouthparts to efficiently ingest dried bait, we could minimize bee mortality in the field by increasing protein and ammonia content during the first 8 h after application. We reasoned that it would be beneficial to forfeit the fruit fly kill during the first day post-application period in order to prevent honeybee and other non-target kill. Following the commercialization of GF120, a number of publications reported the results of tests using various fruit fly species that were not killed by freshly applied GF120. Mangan (2009) discussed these results, which reported kill rates for various periods after fresh applications: "Overall mortality rates were below 10 % for 4 h, 39–43 % at 8 h, but mortality in all treatments increased to 89–93 % by 24 h, and 99 % by 48 h" for Mexican fruit flies caged with GF120 drops applied to waxed paper.

Table 12.5 Estimated LC_{50s} against female *A. ludens* for compounds tested in SolBait, projected rate of active ingredient (AI) per hectare, projected amount a person could receive, and safety index as related to standard malathion-bait sprays

Compound	Dermal LD_{50} mg kg^{-1}, rat/rabbit	LC_{50}, ppm *A. ludens*	Projected rate[a] g AI ha^{-1}	Projected[b] Mg m^{-2}, person	AI dermal index[c] μg m^{-2}, person
1 Malathion-NuLure	4,100	–	190.0	19.0	4.6342
2 Malathion in SolBait	4,100	0.44	0.176	0.176	0.04293
3 D&C Red Dye #28	396,720	41.6	16.64	1.664	0.004194
4 FD&C Red Dye #3	396,720	37.3	14.92	1.492	0.003761
5 D&C Red Dye #22	396,720	0.78	0.312	0.0312	0.000079
6 Ethyl eosin	396,720	0.017	0.0068	0.00068	0.000017
7 Methyl eosin	396,720	0.0003	0.00012	0.000012	0.000000003
8 Thiamethoxam	2,000	4.430	1.772	0.1772	0.0886
9 Imidacloprid	5,000	3.841	1.536	0.1536	0.03073
10 Spinosad	5,000	0.159	0.0636	0.0064	0.00128
11 Abamectin	2,000	0.171	0.0684	0.0068	0.00342
12 Emamectin	2,000	0.118	0.047	0.0047	0.00235
13 Milbemectin	5,000	170.4	68.16	6.816	1.3632
14 Cyromazine	5,000	5,524.0	2,209.6	220.96	44.192
15 Fipronil	2,000	2.886	1.1544	0.11544	0.05772
16 Chlorfenapyr	2,000	9.6	3.84	0.384	0.192
17 Sodium cacodylate	2,000	590.0	236.0	23.6	11.8
18 Borax	5,000	12,240.0	4,896.0	489.6	97.92
19 Boric acid	5,000	10,096.0	4,038.4	403.84	80.768
20 Indoxacarb	5,000	28.0	11.2	1.12	0.224
Sulfluramid 484	–	0.508	0.2032	0.02032	–

[a]Projected AI ha^{-1}, except malathion-NuLure, based on 100 times LC_{50} applied at 4 l per hectare
[b]Projected volume of spray m^{-2}, ≈ surface area of a person, based on 4 l per hectare, 0.4 ml
[c]Example, dermal safety index for D&C Red Dye #28 (malathion/Dye) > 1,105 times safer than malathion, spinosad > 3,620 times safer

Table 12.6 Toxicity for terrestrial (LD50s) and aquatic (LC50s) species of animals for oral and immersion doses of malathion and spinosad (United States Department of Agriculture 2002)

Acute oral LD50s for terrestrial species dosed with malathion (mg/kg)	Malathion 96-h LC50s for aquatic species (μg/L)	Acute Oral LD50s terrestrial species dosed with spinosad (mg/kg)	Spinosad 96-h LC50s for selected aquatic species (μg/L)
Mouse 720–4,060	Tadpole 200	Rat >5,000	Grass shrimp 9,760
Female rat 1,000	Rainbow trout 4.1–200	Mouse 23,100	Rainbow trout 30,000
Male rat 1,375	Bluegill 20–110	Shrew 3,400	Bluegill 5,900
Mallard 1,485	Daphnia 1–1.8	Mallard >2,000	Daphnia 92,600
Pheasant 167	Stone flies 1.1–8.8	Pheasant >2,000	Eastern oyster 295

Field trials conducted in Guatemala measured the effects of GF120 aerial sprays on honey production and bee mortality and found no effects (Spencer et al. 2003). The effects of protein and ammonia components of Solbait as repellents to nutritionally stressed (early spring) honey bees were tested in Texas bee yards (Mangan and Tarshis Moreno 2009). These experiments showed that, among bees trained to collect sugar or honey from stations, >98 % were repelled when this bait was replaced by GF120 or honey mixed with ammonium acetate and that addition of the solulys protein to the honey-ammonium acetate increased the repellence to 100 %.

Tests briefly presented in Mangan and Moreno (2004) showed that the preservative proxel could be omitted from the bait without loss of attraction. All other components of GF120 qualified for registration as "organic" under USDA and International (e.g., OMRI 2013) organic registration. In this sense, organic refers to naturally occurring or deriving from organisms as opposed to Metcalf's (1955) definition referring to organic pesticides as those composed of organic molecules. The concentrate bait contained a high enough sugar concentration to prevent fermentation, but after dilution the material rapidly (within a few hours) began to ferment. The results of the Mangan and Moreno (2004) tests led to the recommendation that proxel be omitted from GF120 (marketed as "Success" at the time), the product be registered as organic, and the product not be diluted until immediately before application.

Fruit fly programs in Australia followed a parallel path to those in the USA. Two insecticides, fenthion for pre-harvest population control and dimethoate for post-harvest disinfestation of fruit, were subject to limitation or elimination of use due to safety issues or expiration of registration. Malathion based hydrolyzed yeast baits were the major alternatives for population control. GF120 marketed in Australia as "Naturalyte fruit fly bait concentrate" based on spinosad and another product originally called Bactrogel, but later marketed as Amulet, were tested in a coordinated multi-state set of experiments, which were reported in Lloyd (2004). This report differed from the papers published by USDA authors in that formulation of the basic baits had already been completed for both Naturalyte and Bactrogel. However, methods of application (types of machine and host plant) and mixture modifications, especially with thickeners, were carried out under a number of conditions. The general discussion did not begin with consideration of public concerns about safety and environment as discussed above for the USA, but lower AI applications were mentioned in overall discussions of the results. Another interesting comparison is that in Australia the tests were carried out over a wide array of ecological conditions, involving a narrower group of fruit fly species (mainly *B. tryoni* and *C. capitata*), and with a much wider array of fruit fly hosts in tropical and subtropical conditions in comparison with North American trials.

An extensive summary of the public concerns about the use of insecticide sprays for fruit fly programs is given in the hearings carried out by the Committee of Agriculture (United States Congress 1994). A recurring theme in statements and documents from city managers and public organizations was to use technologies, such as sterile insect release, as acceptable approaches to address outbreaks and that area-wide application of malathion bait sprays were unacceptable. Penrose (1996) described the alternative to the use of insecticides as the primary response to fruit fly detection. The preventative release program (PRP) was designed to prevent reproduction after Mediterranean fruit fly introductions by having continuous weekly release of sterile flies (now sterile males) as an alternative to pesticide spray application after introductions were discovered. The logic of this approach is that, during the first generation after introduction, the emerging pests are overwhelmed by the released sterile insects and do not build population or expand the infested area to a degree that would require insecticide spray. Dowell et al. (2000) and Hendrichs et al. (2002) provided reviews of the operation and research for establishing PRP programs. A similar PRP program was established in Florida in 1998 following the eradication of the 1997–1998 outbreaks (PR Newswire 2000). The Florida program covers the areas subject to the 1998 outbreak, including areas of Dade and Broward counties near Miami and blocks in Pinellas, Hillsborough, Manatee, and Sarasota Counties on the west coast.

A similar program was developed in south Texas in 1983 (Holler et al. 1984; Nilakhe et al. 1991; Thomas et al. 1999). This program was initiated as a Mexican fruit fly management system that treated outbreaks of the flies in commercial orchards by detection of feral flies and release of sterile insects. Originally, the program released the sterile flies only during the fall, winter, and spring and only in orchards. It was assumed that, since no flies were trapped during summer, they had been destroyed by summer conditions. After gradual increases in fly detection in the mid-1990s and a serious outbreak in 1998, the Mexican fruit fly program was modified to year round releases that covered all areas containing hosts of the Mexican fruit fly. The management system for Mexican Fruit fly in the Rio Grande Valley was modified into an eradication system in the 2009–2011 protocols. The main impact of this change was a justification for use of insecticide sprays as a supporting component of the sterile insect release program.

In 2008 and amended in 2010, federal regulations (Code of Federal Regulations 2009) allowed bait spray to be an alternative to post-harvest treatment for fruit flies. Either malathion or spinosad bait at appropriate concentrations and spray rate must be applied at least 30 days prior to harvest and continued until harvest at 6–10 day intervals. This allowance alleviated some of the regulatory requirements imposed when residential areas adjacent to commercial production areas have fruit fly outbreaks. This represents a quarantine security role for the insecticide baits that can be met for regular fruit production by scheduled treatment of orchards with relatively inexpensive malathion/NuLure bait or the more costly GF120 for organic production during the period before harvest.

4 Conclusions

In reviewing the priorities for developing adulticide baits for fruit flies, it became evident to me that the typical objective of reducing losses to insect damage was not the crucial factor in setting these priorities. As trade and social awareness developed between approximately 1900 and the present, the insecticide systems for fruit flies were driven by quarantine requirements, the differing biology of the various species, especially host range, and the essential components for area-wide insecticide application. Especially important for area-wide application was a public perception of a history of misuse, or especially over-use, of insecticides in agriculture, and negative impacts on environmental and human safety. At the same time, changes in trade and human travel and migrations increased the needs for changes in the insecticide systems. During the three phases of insecticide systems I have proposed for fruit flies, the development of poisoned baits at first focused on poisons known to be toxic to all animals with a lesser effort in use of natural plant products known to be insecticidal. The second phase focused on synthetic organic compounds, some of which were found to be insecticidal but not derived from chemicals known for toxicity to other animals. However, as discussed by critics, several of these pesticides, such as the organophosphates, were developed in World War II programs as human gas weapons and others, such as the aldicarbs, have been culprits in mass human deaths due to mismanagement or sabotage during manufacturing (Broughton 2005). During this phase, social and political criticism of the programs in California and Florida (but not Texas) and, later, many other countries, became dominant forces in program operations. In addition, normal scientific discourse, which may have appeared to be disagreement among scientists, was cited in press and public communication as indicating that the programs were in disarray to a greater degree than they were. The third phase did not ignore efficacy of the baits or insecticides, however, my review of the proposed insecticides to replace organophospates suggests that improvement of pest kill rate was not the primary objective. If the toxicity (adult knock-down) and persistence were at least equivalent to the baits currently used, the more important goals were reduction in active ingredient, and overall environmental and human safety, either perceived or actual, and other characteristics, such as organic registration which increased public acceptance of the treatment. Improvements in program technology and strategies, such as preventive release programs in chronic outbreak areas in California and Florida, have largely achieved the goals listed as public concerns in the congressional hearings (United States Congress 1994) to avoid insecticide aerial sprays and use sterile insect release to control outbreaks. The environmental assessments for Mediterranean outbreaks in Pompano Beach, Florida in 2011 (USDA 2011) and Rancho Cucamonga, California in 2012 (USDA 2012) list only ground applications of GF120, referred to as "an organic formulation of spinosad bait", as pesticide treatments.

References

Aliniazee MT (1986) Management of *Rhagoletis indifferens* in western N. America. In: Cavalloro CV (ed) Fruit flies of economic importance. CRC Press, Rome, pp 197–206

Allman SL (1940) Foliage poisons for the Queensland fruit fly, (*Strumeta tryoni* Froggatt). J Aust Inst Ag Sci 6:154–160

Allman SL (1942) Foliage poison spray for Queensland fruit fly, tartar emetic and sodium fluosilicate compared. Agric Gaz N.S.W. 53:186–187

Anderson I (1982) Medfly fighters win their battle the hard way. New Sci 95, No.1324. 23 September 1982, p 815

Anonymous (1898) Preparation and use of insecticides. In: Hill GM (ed) Yearbook, U.S. Department of Agriculture, 1897, pp 637–640. http://archive.org/stream/yearbookofagricu1897unit#page/636/mode/2up

Anonymous (1906) Fruit fly parasite. Brisbane Courier, 27 Nov 1906. p 6. http://trove.nla.gov.au/ndp/del/article/19484778

Anonymous (1920) The Mally fruit fly remedy: for the prevention of maggots in fruit by the destruction of the parent flies before eggs are laid. J Dept Agric 1(April–December). Pretoria, pp 859–861

Anonymous (1994) Dye could become weapon against fruit flies. New York Times, Archives. December 18 1994. http://www.nytimes.com/1994/12/18/us/dye-could-become-weapon-against-fruit-flies.html

Anonymous (undated a) Palo Alto Pests: The 1981 Medfly Invasion. http://www.paloaltohistory.com/palo-alto-medfly-invasion.php

Anonymous (undated b) The Springfield connection. Season 6, Episode 23, 1989. http://www.tv.com/shows/the-simpsons/the-springfield-connection-1411/

Anonymous (undated c) Duckman. Season 2 Episode 9 Clip Job. First shown 8 May 1995. http://www.tv.com/shows/duckman/clip-job-62834/

Antongiovanni E (1949) Tests on the persistent effects of DDT and BHC against *D. oleae* (Gmel.). Redia, 1949 34, pp 46–66

Aviation Safety Network (2009–2013) Accident description, 8 August 1956. http://aviation-safety.net/database/record.php?id = 19560808-1. see also http://www.c82packet.com

Ayers E (1957) The Mediterranean fruit fly eradication campaign. In: Twenty-first Biennial report for the period July 1, 1954–June 30, 1956. State Plant Board, Gainesville, pp 15–23

Back EA (1914) The Mediterranean fruit fly in Bermuda. U.S. Department of Agriculture. Entomolo Bull 161, 8 p

Back EA, Pemberton CE (1917) The melon fly in Hawaii. USDA Bull 491, 58 p

Back EA, Pemberton CE (1918a) The Mediterranean fruit fly in Hawaii. USDA Bull 538, 118 p

Back EA, Pemberton CE (1918b) The Mediterranean fruit fly. USDA Bull 640, 43 p

Baker AC, Stone WE, Plummer CC, McPhail M (1944) A review of studies on the Mexican fruit fly and related Mexican species, USDA miscellaneous publication 521. U.S. Department of Agriculture, Washington, DC, p 155

Benson AH (1906) Fruit pests. Qld Agric J XVII(December) 288–289

Broughton E (2005) The Bophal disaster and its aftermath: a review. Environmental health: a global access science source 2005, 4:6. http://www.ehjournal.net/content/4/1/6

Carson R (1962) Silent spring. Houghton Mifflin Company, New York, p 297

Chambers DL, Lorraine H, Dowell RV (1987) Boron as an alternative toxicant in protein bait for Mediterranean fruit flies. In: Cavalloro R (ed) Fruit flies of economic importance. A.A. Balkema, Rotterdam, pp 501–506

Chavez S, Simon R (1989a) Mystery letter puts odd twist on Medfly crisis. Los Angeles Times December 3 1989. http://articles.latimes.com/1989-12-03/local/me-267_1_mystery-letter

Chavez S, Simon R (1989b) Months-long spraying of medfly Is ordered: infestation: officials admit underestimating the outbreak. Meantime, the malathion opposition grows. Los Angeles Times Dec 8, 1989. http://articles.latimes.com/1989-12-08/news/mn-298_1_medfly-infestation

Clark RA, Weems WV (1989) Detection, quarantine, and eradication of fruit flies invading Florida. Proc Fla State Hort Soc 102:159–164

Code of Federal Regulations (2009) 7 CFR 301.32.10 Treatments (b). http://www.lawserver.com/law/country/us/cfr/7_cfr_301-32-10

Compere G (1907) Kerosene remedy and the fruit fly. J Dept Agric. Western Australia, XV, unpaged

Crawford DL (1910) The Mexican orange maggot. Pomona J Entomol 2:321–332. Text file at: http://archive.org/stream/pomonacollegejou12pomo/pomonacollegejou12pomo_djvu.txt
Crawford DL (1927) Investigation of Mexican fruit fly (*Anastrepha ludens* Loew) in Mexico. Mon Bull Dept Agric California 16:422–445
Darby HH, Kapp EM (1934) Studies on the Mexican fruit fly, *Anastrepha ludens* (Loew). Tech Bull 444. USDA. Washington, DC, 20 p
Dawson A, Hassenpflug S, Sloan J, Yoshioka, I (1998) California agricultural trade: combating the Medfly menace. Case Study. Center for Trade and Commercial Diplomacy, Monterrey Institute of International Studies, Monterey, California. http://www.commercialdiplomacy.org/case_study/case_medfly1.htm. Viewed 17 May 2013
Delong DM (1934) The present status of Cryolite as an insecticide. Ohio J Sci 34:175–200
Denmark HA (1956a) The Mediterranean fruit fly infests Florida again: early chronology. Fla Entomol 39:85–89
Denmark HA (1956b) The Mediterranean fruit fly infests Florida again: second report. Fla Entomol 39:175–176
De Ong ER (1948) Chemistry and uses of insecticides. Reinhold Publishing, New York, 345 p
Dowell RV, Wilson W, Hennessy MK (1997) Toxicity of SureDye to beneficial insects. In: 25th annual meeting of the American Society for Photobiology, Saint. Louis, Missouri, USA, 5–10 July 1997. Photochem Photobiol 65:66s–67s
Dowell RV, Siddiqui IA, Meyer F, Spaugy EL (2000) Mediterranean fruit fly preventive release programme in Southern California. In: Tan KH (ed) Area-wide control of fruit flies and other insect pests. Penerbit Universitii Sains Malaysia, Penang, pp 369–375
Enkerlin W, Reyes J, Villalobos R (1993) Use of a mixture of boric acid, borax, hydrolyzed protein, and water to control *Anastrepha* fruit flies. In: Aluja M, Liedo P (eds) Fruit flies: biology and management. Springer, New York, pp 353–358
Friend AH (1949a) A consideration of the fruit fly problem and some details of recent control tests. Agric Gaz N.S.W. 60:35–40
Friend AH (1949b) Farther experiments on the control of the Queensland fruit fly (*Strumeta tryoni*) Agric Gaz N.S.W. 60:307–308
Froggatt WW (1909) Report on parasitic and injurious insects. In: Gullick WA, Government Printer, Sydney. 115 pp + 8 plates. http://bhl.ala.org.au/item/57287#page/5/mode/1up
Gow PL (1954) Proteinaceous baits for the oriental fruit fly. J Econ Entomol 47:153–160
Greany PD, Styer SC, Davis PL, Shaw PE, Chambers DL (1983) Biochemical resistance of citrus to fruit flies: demonstration and elucidation of resistance to the Caribbean fruit fly *Anastrepha suspensa*. Entomol Exp et Appl 34:40–50
Hagen KS, Allen WW, Tassan RL (1981) Mediterranean fruit fly: the worst may be yet to come. California Agriculture (3):5–7
Hall SA (1964) The place of insecticides. In: Farmer's world. Yearbook of Agriculture 1964. U.S. Department of Agriculture, Government Printing Office, pp 113–117
Hardin B (1997) Fruit fly dyeing continues overseas. News and Events. USDA ARS. http://www.ars.usda.gov/is/pr/1997/971208.htm
Harris EJ, Chambers DL, Steiner LF, Kamakahi DC, Komura M (1971) Mortality of tephritids attracted to guava foliage treated with either malathion or naled plus protein-hydrolyste bait. J Econ Entomol 64:1213–1216
Heitz JR (1995) Pesticidal applications of photoactivated molecules. In: Heitz JR, Downum KR (eds) Light activated pest control, ACS symposium series 616. American Chemical Society, Washington, DC, pp 1–16
Hendrichs J, Robinson AS, Cayol JP, Enkerlin W (2002) Medfly areawide sterile insect technique programmes for prevention, suppression or eradication: the importance of mating behavior studies 2002. Fla Entomol 85:1–13
Hillburton DJ, Dow RL (1990) Mediterranean fruit fly, *Ceratitis capitata*, eradicated from Bermuda. Fla Entomol 73:342–343

Holler TC, Davidson JL, Suarez A, Garcia R (1984) Release of sterile Mexican fruit flies for control of feral populations in the Rio Grande Valley of Texas and Mexico. J Rio Grande Valley Hortic Soc 37:113–121

Howard LO (1898) Danger of importing insect pests. In: Hill GM (ed) Yearbook, U.S. Department of Agriculture, 1897. pp 529–552. http://archive.org/stream/yearbookofagricu1897unit#page/529/mode/2up

Howard LO (1930) A history of applied entomology (somewhat anecdotal). Smithsonian Miscellaneous Collections. Publication 3065, 564 p. http://cluster.biodiversitylibrary.org/s/smithsonianmisce841931smit/smithsonianmisce841931smit_djvu.txt

Howlett FM (1912) VII. The effect of oil of citronella on two species of *Dacus*. Trans R Entomol Soc Lond 60:412–418

Howlett FM (1915) Chemical reactions of fruit-flies. Bull Entomol Res 6:297–305

Isaac J (1905) Report on the Mexican orange worm. (Trypeta ludens) in Mexico. Report of the commissioner. Ellwood Cooper. California State Horticulture Commission. Sacremento, CA, 48 p

Jarvis H (1922a) Fruit fly investigations. Qld Agric J 7:246–247

Jarvis H (1922b) Fruit fly investigations. Qld Agric J 18:131–133, 344–135

Jarvis H (1922c) Fruit fly investigations. Qld Agric J 17:309–312

Jarvis H (1923) Fruit fly investigations. Qld Agric J 19:1–4, 194–197, 369–171

Jarvis H (1924) Report of the entomologist (Stanthorpe District). September and October. Qld Agric J 22:435–440

Jarvis H (1925a) Fruit fly investigations. Entomologist's report (April-May 1). Qld Agric J 24:60–61

Jarvis H (1925b) The fruit fly. Report on measures of possible control. Qld Agric J 24:48–52

Jarvis H (1925c) Report of the entomologist, Stanthorpe District, Mr. H. Jarvis, April-May Qld Agric J 24:60–62

Jarvis H (1926a) Fruit fly in the Stanthorpe district. Qld Agric J 25:367–370

Jarvis H (1926b) The Queensland fruit fly (*Chaetodacus tryoni* Froggatt). Qld Agric J 26:101–104

Jarvis H (1931) Experiments with a new fruit fly lure. Qld Agric J 36:485–491

Jeppson GA, Garman GE (1960) Citrus insects and mites. Ann Rev Entomol 5:353–378

Klassen W (1989) Eradication of introduced arthropod pests: theory and historical practice. Misc Publ Entomol Soc Am 73:1–29

Klassen W (2005) Area-wide integrated pest management and the sterile insect technique. In: Dyck VA, Hendrichs J, Robinson AS (eds) Sterile insect technique principles and practice in area-wide integrated pest management. IAEA. Springer, Netherlands, pp 39–68

Lillie RD (1977) H.J. Conn's biological stains. The Williams and Wilkins Co, Baltimore

Liquido NJ, McQuate GT, Cunningham RT (1995) Light-activated toxicity of phloxine B and fluorescein in methyleugenol to oriental fruit fly, *Bactrocera dorsalis* (Hendel) (Diptera: Tephritidae), males. In: Heitz JR, Downum K (eds) Light activated pest control, ACS symposium series 616. American Chemical Society, Washington, DC, pp 107–114

Lloyd A (2004) Project AH00012, Improved protein baits for fruit fly control. Dow-Milestone 1&2 30 p. http://www.ngia.com.au/Story?Action=View&Story_id=1814

Lopez-D F, Chambers DL, Sanchez-R M, Kamasaki H (1969) Control of the Mexican fruit fly by bait sprays concentrated at discrete locations. J Econ Entomol 62:1255–1257

Mabuchi K, Lilenfeld AM, Snell LM (1980) Cancer and occupational exposure to arsenic: a study of pesticide workers. Prev Med 9:51–77

Mally CW (1909) Poisoned bait for the fruit fly. Agric J Cape Good Hope. 34 Jan–June, pp 620–633. http://babel.hathitrust.org/cgi/pt?id=coo.31924061468181;seq=632;view=1up;num=619

Mangan RL (2005) Population suppression in support of the sterile insect technique. In: Dyck VA, Hendrichs J, Robinson AS (eds) Sterile insect technique – principles and practice in area-wide integrated pest management. IACA Springer Publication, Netherlands, pp 407–425

Mangan RL (2009) Effects of bait age and prior protein feeding on cumulative time dependent mortality of *Anastrepha ludens* (Diptera: Tephritidae) exposed to GF-120 Spinosad baits. J Econ Entomol 102(3):1157–1163

Mangan RL, Moreno DS (1995) Development of phloxine B and uranine bait for control of Mexican fruit fly. In: Heitz JR, Downum K (eds) Light activated pest control. ACS symposium series 616. American Chemical Society, Washington, DC, pp 115–126

Mangan RL, Moreno DS (2001) PhotoActive dye insecticide formulations: adjuvants increase toxicityto Mexican fruit Fly (Diptera: Tephritidae). J Econ Entomol 94(1):150–156

Mangan RL, Moreno, DS (2004) Dilution and persistence of baits and safer pesticides for spray application. In: Barnes BN (ed) Proceedings of 6th international symposium on fruit flies of economic importance. Isteg Scientific Publication, Irene, South Africa, pp 305–312, 510 p

Mangan RL, Tarshis Moreno AM (2009) Honeybee foraging preferences, effects of sugars and fruit fly toxic bait components. J Econ Entomol 102(4):1472–1481

Mangan RL, Moreno DS, Thompson G (2006) Bait dilution, spinosad concentration, and efficacy of GF-120 based fruit fly sprays. Crop Prot J 25:125–133

McAlister LC (1936) Observations on the West Indian fruit fly at Key West in 1932–33. J Econ Entomol 29:440–445

McQuate GT, Peck SL (2000) Suppression of Mediterranean fruit fly populations over mountainous areas through aerial phloxine B-protein bait sprays: Regional Medfly programme in Guatemala. In: Tan KH (ed) Area-wide control of fruit flies and other insect pests. Penerbit Universitii Sains Malaysia, Penang, pp 409–418

Metcalf RL (1955) Organic insecticides, their chemistry and mode of action. Interscience Publishers Inc., New York, 392 p

Miller RL, McBride OC (1931) Experiments with copper carbonate, lead arsenate and other compounds against the Mediterranean fruit fly in Florida. J Econ Entomol 24:1119–1131

Miyake T (1919) Studies on the fruit flies of Japan, I Japanese orange-fly. Bull. Imperial Central Agric. Exp Sta. II, 2. Nishigara Tokyo. 161 p, 10 figs. http://babel.hathitrust.org/cgi/pt?id=coo.31924018311815;view=1up;seq=89

Moreno DS, Mangan RL (1995) Response of the Mexican fruit fly (Diptera: Tephritidae) to two hydrolysed proteins and incorporation of phloxine B to kill adults, Supplement. In: Heitz JR, Downum K (eds) Light activated pest control, ACS symposium series 616. American Chemical Society, Washington, DC

Moreno DS, Mangan RL (2000) Novel insecticide strategies such as phototoxic dyes in adult fruit fly control and suppression programs. In: Tan KH (ed) Area-wide control of fruit flies and other insect pests. Penerbit Universitii Sains Malaysia, Penang, pp 421–432

Moreno DS, Mangan RL (2002) A bait matrix for reduced-risk insecticides used against fruit flies (Diptera: Tephritidae). In: Hallman G, Schwalbe CP (eds) Invasive arthropods in agriculture. Science Publishers Inc., Enfield, pp 333–362

Moreno DS, Martinez AJ, Sanchez Rivello M (1994) Cyromazine effects on the reproduction of *Anastrepha ludens* (Diptera: Tephritidae) in the laboratory and in the field. J Econ Entomol 87:202–211

Moreno DS, Celedonio H, Mangan RL, Zavala JL, Montoya PG (2001) Field evaluation of a phototoxic dye, phloxine B, against three species of fruit flies (Diptera: Tephritidae). J Econ Entomol 94:1419–1427

Mound LA (2005) Florida pioneer Wilmon Newell: the past, present, and future of insect pest control. Fla Entomol 88:241–243

Navarro-Llopis V, Sanchis-Cabanes J, Ayala I, Casana-Giner V, Primo-Yufera (2004) Efficacy of lufenuron as chemosterilant against *Ceratitis capitata* in field trials. Pest Manag Sci 60:914–920

Nelson WC, Lykins MH, Mackey J, Newill VA, Finklea JF, Hammer DI (1973) Mortality among orchard workers exposed to lead arsenate spray: a cohort study. J Chron Dis 26:105–118

Nguyen R, Poucher C, Brazzel FR (1992) Seasonal occurrence of *Anastrepha suspensa* (Diptera: Tephritidae) in Indian River County Florida, 1984–1987. J Econ Entomol 85:813–820

Nigg HN, Simpson SE (1997) Use of borax toxicants to control Tephritidae fruit flies. U.S. Patent 5,698,208. December 16, 1997, 28 panels

Nilakhe SS, Worley JN, Garcia R, Davidson JL (1991) Mexican fruit fly protocol helps export Texas citrus. Subtrop Plant Sci 44:49–52

OMRI (Organic Materials Review Institute) (2013) Welcome to the Organic Materials Review Institute. https://www.omri.org/

Penrose D (1996) California's 1993/94 Mediterranean fruit fly eradication program. In: McPheron BA, Steck GJ (eds) Fruit fly pests, a world assessment of their biology and management. St. Lucie Press, Delray Beach Florida, pp 551–554, 586 p

Plummer CC (1938) Laboratory the toxicity of *Haplophyton cimicidum* A. DC. To fruit flies. USDA Circular 455. Washington, DC, 11 p. https://archive.org/details/toxicityofhaplop455plum

Plummer CC (1944) Laboratory studies on the toxicity of tartar emetic to the Mexican fruit fly. USDA Circular 697. Washington, DC, 14 p

Plummer CC (1947) DDT and the Mexican fruitfly. J Econ Entomol 37:158

Plummer CC, Shaw JG (1946) DDT and benzene hexachloride to control Mexican fruit fly. J Econ Entomol 40:483–486

PR Newswire (2000) U.S. Dept. of Agriculture: Florida's Sterile Medfly Release Program Proves Successful. Press Release Jan 24, 2000. http://www.riskworld.com/pressrel/2000/PR00a023.htm

Roessler Y (1989) Insecticidal bait and cover sprays. In: Robinson A, Hooper G (eds) Fruit flies, their biology, natural enemies and control. Elsevier, Amsterdam, pp 40–48

Rohwer GG (1958) The Mediterranean fruit fly in Florida-past present and future. Fla Entomol 41:23–25

Rush B (1789) Medical inquiries and observations. Prichard and Hall, Philadelphia, 206 p

Rush B (1810) An inquiry into effects of Ardent spirits of the human body and mind with an account of preventing and of the remedies for curing them, 8th edn, Exeter. 36 p

Santoro R (1951) Investigations with synthetic organic products against *D. oleae*, its parasites and other insects made in Ascea (Salerno) in 1950. Annali della sperimentazione agraria 5:1379–1404

Sebbata A, Leyva J, Mangan R, El Khamass M, Aboulama S, Belghiti S, Chikhi H, Humpal D, Moreno D (1998) Controle de la mouche Mediterraneenne dans les vergers d'agumes par le SureDye. Publication Societe Agricole de Service au Maroc (SASMA), 12 p

Severin HHP (1912) The introduction, methods of control, spread and migration of the Mediterranean fruit fly in the Hawaiian Islands. In: Proceedings of the 41st fruit growers convention. Santa Barbara, CA, pp 158–166

Severin HHP, Severin H (1914) Behavior of the Mediterranean fruit fly (*Ceratitis capitata* Wied.) towards kerosene. Anim Behav 4:223–227

Shaw JG, Lopez-D F, Chambers DL (1970) A review of the research done with the Mexican fruit fly and the citrus blackfly in Mexico by the entomology research division. Bull Entomol Soc Am 16:186–193

Spencer H, Osburn MR (1938) Citrus projects of the Orlando Laboratory. J Econ Entomol 31:728

Spencer JP, Ibarra J, Rendon P (2003) Effect of spinosad on honey bees (Hymenoptera: Apidae) in Guatemala. Southw Entomol 28:211–216

Steiner LF (1952) Fruit fly control in Hawaii with poison-bait sprays containing protein hydrolysates. J Econ Entomol 50:838–843

Steiner LF (1957) Field evaluation of oriental fruit fly insecticides in Hawaii. J Econ Entomol 45:16–24

Steiner LF, Hinman FG (1952) Field tests of insecticides for control of oriental fruit fly. J Econ Entomol 45:388–395

Steiner LF, Rohwer GG, Ayers EL, Christenson LD (1961) The role of attractants in the recent Mediterranean fruit fly eradication program in Florida. J Econ Entomol 54:30–35

Stevenson BC, McClung BB (1966) Mediterranean fruit fly eradication in the Lower Rio Grande Valley. Bull Entomol Soc Am 12:374

Stone A (1942) The fruit flies of the genus *Anastrepha*. U.S. Department of Agriculture miscellaneous publication no. 439, Washington, DC, 112 p

Tarshis Moreno M (2001) Detection of the phototoxic dye phloxine B in *Anastrepha ludens* (Loew) (Diptera: Tephritidae), *Apis mellifera* L. (Hymenoptera: Apidae) and honey. M.S. thesis, University of Texas, PanAmerican, Edinburg, TX

Thomas DB, Mangan RL (2005) Nontarget impact of spinosad GF-120 bait sprays for control of the Mexican fruit fly (Diptera: Tephritidae) in Texas citrus. J Econ Entomol 98:1950–1956

Thomas DB, Worley JN, Mangan RL, Vlasik RA, Davidson JL (1999) Mexican fruit fly population suppression with the sterile insect technique. Subtrop Plant Sci 51:61–71

Tollestrup K, Daling JR, Allard J (1995) Mortality in a cohort of orchard workers exposed to lead arsenate pesticide spray. Arch Environ Health Int J 50(3):221–229

Tryon H (1922) The Harvey fruit fly lure. Letters to the editor, The Brisbane Courier, QLD. 24 October 1922, pp 7–8

Turvey ND (2010) The toad's tale: a true fable of science and society. In: Weber K (ed) Cane Toads and other rogue species. Participant Media, Philadelphia, pp 3–18

United States Congress (1994) Hearing: Mediterranean fruit fly program. Subcommittee on department operations and nutrition. Commmittee on agriculture. House of Representatives. Stenholm CW, Chair. May 5 1994. http://archive.org/details/mediterraneanfru00unit

United States Department of Agriculture (2002) Mexican fruit fly cooperative eradication program, San Diego County, California, Environmental Assessment, December 2002. Animal and Plant Health Inspection Service, Riverdale MD, 30 p

United States Department of Agriculture (2011) Mediterranean fruit fly cooperative eradication program, Pompano Beach, Broward County, Florida, Environmental Assessment, February 2011. Animal and Plant Health Inspection Service, Riverdale MD, 20 p

United States Department of Agriculture (2012) Mediterranean fruit fly cooperative eradication program, Rancho Cucamonga, San Bernardino County, California, Environmental Assessment, August 2012, Animal and Plant Health Inspection Service, Riverdale MD, 16 p

Weinland HA (1912) Work on the Mediterrranean fruit fly in Hawaii. In: Proceedings of the 41st fruit growers convention. Santa Barbara, CA, pp 170–179

Wills JE (1946) Experiments with the new insecticides DDT and gammexane carried out at the agricultural experiment station of La Molina up to the end of May 1946. Boletin. estacion experimental agricola de 'la molina' (29), 33 p

Wolfe B (1987) Medfly revives a tale of techno-paranoia. Los Angeles Times. September 10, 1987. http://articles.latimes.com/1987-09-10/local/me-6834_1_medfly-invasion

Wolfenbarger DO (1958) Notes and comments on the second Mediterranean fruit fly infestation. Fla Entomol 40:29–31

Yang LK, Nigg HN, Simpson SE, Ramos LE, Cuyler NW, Barnes JI, Green CG (2000) Sodium Tetraborate effects on mortality and reproduction of *Anastrepha suspensa* (Diptera: Tephritidae). J Econ Entomol 93(5):1485–1492

Yonge PK (1931) State Plant Board of Florida, Report for the period July 1 1928–June 30 1930. Tallahassee, 112 p. http://ufdc.ufl.edu/UF00098574/0005

Chapter 13
Recent Developments and Applications of Bait Stations for Integrated Pest Management of Tephritid Fruit Flies

Jaime C. Piñero, Walther Enkerlin, and Nancy D. Epsky

Abstract The attract-and-kill approach involves the behavioral manipulation of pest insects through the integration of long-distance olfactory/visual stimuli to attract a particular target pest with a killing agent and/or a collection device. Bait stations, an element of an attract-and-kill system, can be defined as "*discrete containers of attractants and toxins, with or without a visual component, which are targeted at specific pests; these devices may or may not require service to remain active during the season, but insects that are attracted and killed, if retained, ought to be discarded and not counted*". The development of new bait station designs as well as the optimization of current ones for improved fruit fly control is currently a priority research area in several regions of the world. This chapter provides the first comprehensive review on bait stations for effective, environmentally-friendly fruit fly control. Discussion includes types of bait stations and components as well as the advantages they provide. For instance, from an environmental perspective, with use of bait stations there is no release of insecticide into the environment, consequently there is minimal or no contact between pesticides and the commodity, beneficial arthropods, and workers/applicators, thus they represent an important improvement over more toxic bait sprays. Specific examples

J.C. Piñero (✉)
Cooperative Research and Extension, Lincoln University, Allen Hall 212, 900 Chestnut St., Jefferson City, MO 65101, USA
e-mail: pineroj@lincolnu.edu

W. Enkerlin
Programa Regional Moscamed, 16 Calle 3-38 Zona 10, Ciudad de Guatemala, Guatemala
e-mail: walther.enkerlin@medfly.org.gt

N.D. Epsky
USDA-ARS, Subtropical Horticulture Research Station, 13601 Old Cutler Road, Miami, FL 33158, USA
e-mail: nancy.epsky@ars.usda.gov

T. Shelly et al. (eds.), *Trapping and the Detection, Control, and Regulation of Tephritid Fruit Flies*, DOI 10.1007/978-94-017-9193-9_13,

of bait stations that target males and females of various fruit fly species in different regions of the world are presented, thus providing a global perspective. Cost/benefit analyses are important in the development and evaluation of bait stations, and these and other factors can influence the adoption of bait station technology by growers.

Keywords Attract-and-kill • Bait station • Behavior • Attractant • Toxicant • Area-wide IPM program • Grower adoption • Protein bait • Visual/olfactory cues • Organic fruit fly control • Fruit fly suppression • Integrated Pest Management • *Anastrepha* spp. • *Bactrocera* spp. • *Rhagoletis* spp. • *Ceratitis* spp.

1 Introduction

For several decades, efforts to suppress pestiferous fruit fly (Diptera: Tephritidae) populations around the globe have relied heavily on the application of protein baits mixed with highly toxic organophosphate insecticides, such as malathion (Steiner 1955; Roessler 1989; Vargas et al. 2001; Mangan, Chap. 12, this volume). Recently, and largely in response to environmental human health concerns, negative impacts on non-target organisms, and public rejection of area-wide pesticide applications (Mangan and Moreno 2007), more environmentally-friendly methods of integrated pest management (IPM) for fruit fly control have been developed. These methods are based on a comprehensive understanding of the orientation and movement of the target fruit fly species (Prokopy et al. 2005; Dorn and Piñero 2009) as well as the numerous biotic and abiotic factors that influence fruit fly response to visual and olfactory stimuli (Prokopy and Owens 1978; Prokopy et al. 1991). One behaviorally-based approach to fruit fly management that is based on the deployment of positive visual/olfactory stimuli in association with a killing agent is termed attract-and-kill (Foster and Harris 1997; Vincent et al. 2003), and one modality of this technique for fruit fly control is the use of bait stations.

This chapter provides the first comprehensive review of bait stations as an effective and environmentally friendly approach for control of economically important fruit fly species on various continents, thereby providing a global perspective. Our review will synthesize and interpret results from field surveys, field experiments, and laboratory studies aimed at developing bait stations and assessing their effectiveness at controlling pestiferous fruit fly species. We also emphasize the need to standardize concepts and criteria associated with the use of bait stations, methodologies used to assess the efficacy, and risks associated with this approach. Critical issues affecting the efficacy of bait stations in the context of IPM of fruit flies based on published data are highlighted as well.

2 Bait Stations: Concepts and Recent History

Before attempting to define bait stations, we will describe the attract-and-kill approach, also referred to as the lure-and-kill technique by some authors (e.g., El-Sayed et al. 2009). The attract-and-kill method is a type of behavioral manipulation that combines a long-distance olfactory stimulus, with or without a visual stimulus, with some type of killing agent and/or a collection device. The more specific the stimulus is to an insect species, the greater the likelihood that a particular behavior can be manipulated successfully (Foster and Harris 1997). Various names, such as lure-and-kill, attract-and-kill, male annihilation, bait sprays, and attracticide/attracticidal, have been used to describe this method (El-Sayed et al. 2009). For the purposes of this review, the term 'attract-and-kill' will be used throughout the text given that this term has been widely accepted by fruit fly researchers around the globe.

In the context of fruit fly suppression, the attract-and-kill method can be implemented through the use of either bait sprays, bait stations, or traps (e.g., when used as part of a mass trapping system; see Chap. 15, this volume). For most bait station designs, insects are attracted and killed but not retained. If they are retained (as in the case of most trapping systems), then for a trap to be considered a bait station, the device should be serviced throughout the season, but insects captured should not be counted. Here, we propose to use the following definition of bait stations based largely on Mangan and Moreno (2007) and Epsky et al. (2012): "*Bait stations are discrete containers of attractants and toxins, with or without a visual component, which are targeted at specific pests; these devices may or may not require service to remain active during the season, but insects that are attracted and killed, if retained, ought to be discarded and not counted*". Ideally, a bait station would not need servicing and would be effective throughout the growing season, which can last up to 6 months for some crops. However, for some crops grown in tropical areas, protection needs to be provided year-round as in the case of papaya (*Carica papaya* L.) production in Hawaii. In some cases, bait stations can serve as open systems (e.g., foliage mimics) onto which bait sprays can be applied. As discussed below, a yellow bait station developed in Hawaii has been termed Papaya Leaf Mimic (e.g., Piñero et al. 2009a, b), and this design has proven to: (1) enhance the response of fruit flies to the bait, (2) protect the bait against rainfall, (3) minimize UV degradation of the insecticidal compounds, such as spinosad (as in the case of GF-120 NF Naturalyte Fruit Fly Bait; Dow AgroSciences, Indianapolis, IN, USA), (4) minimize waste of bait due to washing to the ground or undesirable areas of the target tree or plant (e.g., trunk, fruit), and (5) circumvent potential leaf phytotoxicity otherwise caused by bait spray application (e.g., Piñero et al. 2009a, b). Based on the aforementioned, the primary difference between bait stations and mass trapping for insect control is that bait stations do not retain flies via retention liquid inside the trap or sticky material on the exterior of the trap.

It is important to highlight that an IPM program involves the use of monitoring traps to assess adult population level. The quantification of fruit infestation through fruit sampling is critical to assess the effectiveness of fruit fly control systems. While monitoring of fruit fly populations is typically done as part of a fruit fly control system that involves use of bait stations, fruit sampling procedures need to be further developed and incorporated into bait station research protocols (IAEA 2007, 2009).

Bait stations have been tested for many years for all of the economically-important fruit fly genera, including *Bactrocera* Macquart, *Rhagoletis* Loew, *Ceratitis* MacLeay, *Anastrepha* Schiner, *Toxotrypana* Gerstaecker, and *Dacus* Fabricius, but this technology has received substantial attention only in recent years. The development and evaluation of attract-and-kill bait stations has been the subject of 20 research articles published from 2007 to 2013 (source: Web of Science; accessed December 09, 2013) and the report "*Development of Bait Stations for Fruit Fly Suppression in Support of SIT [Sterile Insect Technique]*" issued by the International Atomic Energy Agency in 2007 (IAEA 2007).

3 Types of Bait Stations and Components

For a bait station to suppress fruit fly populations successfully, the device and the bait should both (1) induce high levels of attraction to the source and, (2) depending on the mode of action of the toxicant, either stimulate flies to ingest a lethal dose of insecticide (e.g., spinosad; Mangan et al. 2006; Mangan and Moreno 2007; Mangan 2009) or remain in contact long enough to receive a lethal dose of insecticides that act largely upon contact (e.g., organophosphates and pyrethroids). An additional characteristic of a bait station is the presence of visual cues that are known to synergistically enhance the response of fruit flies to odor sources (Mangan and Moreno 2007; Piñero et al. 2006, 2009a; Díaz-Fleischer et al., Chap. 5, this volume). For example, Piñero et al. (2006) working with *Bactrocera cucurbitae* (Coquillett) documented that a combination of both visual and olfactory stimuli was needed to elicit high levels of female response compared to each stimulus offered alone.

Based on the type of material used for their construction, there are three general categories of bait stations (IAEA 2009): (1) long-lasting devices that can be retrieved at the end of the harvesting season, an approach suitable for commercial fruit production; (2) biodegradable devices that can remain in the field until they are degraded naturally, appropriate for suppression/eradication programs, and (3) direct, localized application to a substrate for areas where the bait needs to be protected against high rainfall or other adverse environmental conditions, such as high temperatures that can result in phytotoxicity and UV light that can degrade botanical and microbial insecticides.

According to a report issued by the International Atomic Energy Agency (2009), desirable characteristics of bait stations include (1) the ability to target and suppress female populations, (2) low cost in terms of attractant, killing agent, and the bait station material itself, (3) no trapping or retention of attracted flies, (4) long lasting attractiveness of the bait and long residual toxicity of the insecticide, resulting in reduced labor needed for servicing or replacing, (5) ease of use, disposable, and/or biodegradable, (6) high selectivity, i.e., no negative non-target effects, and (7) high effectiveness, i.e., fruit fly control using bait stations should be at least as effective as the current ground bait sprays based on insecticide/bait combination, the standard method for suppressing fruit fly populations. In the sections below, we discuss the various types of bait stations that have been developed and evaluated in the field in the last 10 years or so followed by examples of the field application of bait stations.

4 Evaluations of Bait Stations Against Pestiferous Fruit Flies

4.1 Examples of Bait Stations Targeting Male Fruit Flies

Bait stations have a long history of suppressing male fruit flies in various geographical areas. The first successful development of bait stations for pestiferous fruit flies was the combination of the powerful male-specific lure methyl eugenol (ME) with an organophosphate insecticide, which formed the basis for the male annihilation technique (MAT) (Christenson 1963). This lure, which is both an attractant and a feeding stimulant, and toxicant mixture are often presented on wooden or cardboard surfaces or applied as a gel on structures, such as telephone poles. As an example of MAT, the California Department of Food and Agriculture conducts eradication projects when evidence of an infestation of members of the *Bactrocera dorsalis* species complex is found within the state. MAT treatments are applied to an area of about 23.3 square km, and approximately 600 small gel-like "bait stations" per 1.4 square km are applied to the sides of individual utility poles and street trees on public right-of-ways.

During the implementation of the Hawaii fruit fly Area-Wide IPM (AW-IPM) program (Mau et al. 2007; Vargas et al. 2008), farmers and homeowners themselves constructed one-way traps that could be used without insecticides. Enclosing male-specific lures inside bucket traps (Fig. 13.1a, b) not only provided protection from the weather but also made the device visible, retrievable, and reusable with limited environmental contamination and exposure to humans and pets (Vargas et al. 2000). New MAT approaches include mixtures of liquid ME or cue-lure (CL) on cotton wicks inside bucket traps or solid dispensers (Vargas et al. 2010a, b, c) for suppression of both *Bactrocera dorsalis* (Hendel), and *B. cucurbitae*. More details about MAT are provided in Vargas et al. (Chap. 14, this volume).

a b

Fig. 13.1 Bucket traps used by the Hawaii Area-Wide Fruit Fly IPM program for male annihilation of *Bactrocera dorsalis* (lure: methyl eugenol), and *B. cucurbitae* (lure: cue-lure) (Vargas et al. 2008, 2010c; see also Chap. 14, this volume): (**a**) 1 l (Highland Plastic No. 36, 3.5 cm radius, 15 cm high, with holes 4 3 cm in diameter), and (**b**) 3.8 l one-way entrance traps with microcentrifuge tubes (capacity: 1.5 ml, 41.6 mm length) with the tip cut off to prevent flies from escaping from the trap (Vargas et al. 2000) (Photos: J. Piñero)

4.2 Examples of Bait Stations Targeting Female Fruit Flies

4.2.1 *Rhagoletis* spp.

One of the best examples of an attract-and-kill bait station is the so-called "attracticidal" sphere originally developed by Prokopy (1975) to control *Rhagoletis pomonella* (Walsh), a key pest of apples in eastern and central North America. The development of this device was based on a comprehensive understanding of the biology and behavior of adult *R. pomonella* males and females. Two aspects of *R. pomonella* biology and behavior that were particularly important for the development of the attracticidal sphere were: (1) wild females forage extensively for protein when immature and respond positively to compounds releasing ammonia, especially when in association with super-normal, visual mimics of foliage (e.g., yellow panels) (Prokopy 1972), and (2) mature females foraging extensively for oviposition sites respond positively to natural or synthetic host odor (Prokopy et al. 1973; Averill et al. 1988) and super-normal, visual mimics of host fruit (e.g., red-painted spheres; Prokopy 1968). The red sphere is a highly selective trap for *R. pomonella*, capturing large numbers of sexually mature adults and fewer beneficial insects (Prokopy 1975).

For several decades, odor-baited red spheres (mimicking apples) coated with Tangletrap Insect Trap Coating (Tanglefoot Company, Grand Rapids, MI, USA; Fig. 13.2a) were used for *R. pomonella* control in IPM programs (Prokopy et al. 1990). Pesticide-treated red spheres were manufactured by coating wooden spheres with a combination of latex paint and either toxicant and sucrose or toxicant and corn syrup (Duan and Prokopy 1995a). Addition of the feeding stimulant (sucrose or corn syrup) increased contact intervals and thus increased efficacy of the pesticide spheres, however, exposure to even small amounts of rainfall washed off the feeding stimulant, rendering the spheres ineffective. The spheres could be reactivated by application of an aqueous sugar solution using a household sprayer, which had to be done after every rainfall event. The spheres remained effective for at least 35 days without maintenance. Field tests found that plots treated with pesticide-treated spheres, which were baited with butyl hexanoate and ammonium acetate lures (long-distance attractants, see discussion below) and placed 5 m apart along plot borders, gave control equal to that obtained with two pesticide spray applications (Duan and Prokopy 1995b). The feeding stimulant was reactivated by dipping the spheres in aqueous sucrose solution. Effectiveness was determined both by adult fly counts on monitoring traps and fruit infestation level. Other designs evaluated have included biodegradable spheres made from a mixture containing sugar and gelatinized corn flour, which were then painted with mixtures of enamel paint, sugar, and insecticide (Fig. 13.2b; Liburd et al. 1999; Hu et al. 2000; Prokopy et al. 2003). The attracticidal sphere has been subject to more recent improvements. Wright et al. (2012) developed a reliable, maintenance-free attracticidal sphere for behavioral management of *R. pomonella*, alleviating the need for summer insecticide treatments. This new attracticidal sphere includes contoured controlled-release caps that are fixed atop visually stimulating sphere bases (Fig. 13.2c). The contoured tops provide sustained release of both insecticide and feeding stimulant under field conditions and, remarkably, the residual toxicity of the bait station lasted the entire season. In commercial orchard trials designed to evaluate the potential of these new attracticidal spheres with contoured caps for direct control of *R. pomonella*, a perimeter-based deployment provided protection comparable to plots receiving 1–2 whole-plot insecticide applications (Wright et al. 2012). Thus, the ability of this bait station to manage *R. pomonella* effectively in apple orchards without spraying insecticides has been consistently demonstrated in numerous studies spanning over nearly three decades (e.g., Prokopy et al. 1996, 2005; Bostanian et al. 1999; Bostanian and Racette 2001).

Regarding attractants used in association with the bait station for *R. pomonella* control, Zhang et al. (1999) identified a five-compound apple fruit-based blend that is highly attractive to *R. pomonella*. This blend consists of butyl butanoate (10 %), propyl hexanoate (4 %), butyl hexanoate (37 %), hexyl butanoate (44 %), and pentyl hexanoate (5 %). On average, sticky-coated red spheres deployed in association with the five-component blend capture twice as many flies as those deployed with the single component butyl hexanoate and five times as many as unbaited spheres. For a discussion on interactions between visual and olfactory stimuli and their effects on fruit fly response to bait stations and traps we refer the reader to Piñero et al. (2009a) and Díaz-Fleischer et al. (Chap. 5, this volume).

Pesticide-treated spheres have also been evaluated for the blueberry maggot, *Rhagoletis mendax* Curran. Laboratory studies were used to evaluate several neonicotinoid insectides, and imidacloprid was found to produce the highest mortality (Stelinski and Liburd 2001). Both red and green spheres, including imidacloprid-treated wooden, plastic, and biodegradable spheres, have been tested, and all bait stations significantly reduced infestation in fruit compared with

Fig. 13.2 Three versions of red spheres for control of *Rhagoletis pomonella*: (**a**) Tangletrap-coated red sphere developed by Prokopy (1975), (**b**) nonsticky, biodegradable sphere constructed with sugar (feeding stimulant) with gelatinized flour and glycerin and coated with an insecticide and red latex paint mixture as residue extending agent (Hu et al. 2000; Prokopy et al. 2003), and (**c**) attracticidal spheres consisting of a partially *round red plastic* sphere base topped with a 100 g contoured cap formulated at a ratio of 20 % wax (a 50:50 ratio of paraffin), and 80 % sucrose (granulated sugar) and spinosad (Wright et al. 2012). The *contoured tops* provide sustained release of both insecticide and feeding stimulant under field conditions (Photos: T. Leskey and J. Piñero)

untreated control plots (Stelinski and Liburd 2001; Hamill et al. 2003). Biodegradable spheres were tested with a variety of paper and plastic traps that were coated with a mixture of paint, sugar, and several insecticides (Barry et al. 2004). Laboratory and field tests found that fipronil or imidacloprid used with either plastic or biodegradable spheres baited with ammonium acetate had potential as "attract-and-kill" systems for control of *R. mendax*.

Other devices that can also be considered bait stations for *R. pomonella* control include a rectangular yellow panel sandwiched between two red 9-mm in diameter hemispheres (Kring 1970), subsequently termed a Ladd trap (Ladd Research Industries, Williston, VT, USA; Fig. 13.3) by its manufacturer. In one study (Bostanian and Racette 2001), Ladd traps were treated with cypermethrin or deltamethrin and were baited with butyl hexanoate (synthetic attractant) enveloped in semi-permeable sachets. The Ladd trap bait stations were hung on branches 1.2–1.7 m above the ground and positioned so as to be visible from outside the tree canopy. Again, since responding flies died away from the bait station, *R. pomonella* level of orchard penetration was monitored by placing four similar traps coated with Tangletrap at the four corners of the plot and another four traps near the center of the plot. Bostanian and Racette (2001) reported that Ladd traps, once deployed, were easy to handle and were maintenance free. Ladd traps provided 98.5–100 % clean fruit at harvest in McIntosh, Liberty, Royal Gala and Jonagold apple cultivars (Bostanian and Racette 2001).

Fig. 13.3 Ladd trap consisting of a rectangular yellow panel sandwiched between two red hemispheres (9 cm in diameter) (Kring 1970). In one study by Bostanian and Racette (2001), Ladd traps were treated with cypermethrin or deltamethrin and were baited with butyl hexanoate (synthetic attractant) enveloped in semi-permeable sachets. It is considered a bait station if flies killed are not counted (Photo: Anonymous)

4.2.2 *Ceratitis capitata*

Various bait stations have been developed and evaluated for *Ceratitis capitata* (Wiedemann). Hu et al. (1998) tested biodegradable spheres coated with gloss yellow latex enamel paint mixed with pesticide and sugar. In a field test, these authors compared the number of eggs laid by wild *C. capitata* females in kumquats in the presence of (1) pesticide-treated (imidacloprid or dimethoate) spheres, (2) Tangletrap-coated sugar/flour spheres, or (3) unprotected kumquats (control). They found that imidacloprid-treated sugar/flour spheres provided a significant level of protection against oviposition (equal to that provided by sticky yellow spheres), whereas dimethoate-treated spheres did not.

More recently, Heath et al. (2009) developed a wax-based matrix composed of paraffin wax, a hardener (Elvax-60; Polysciences, Inc., Warrington, PA, USA), and an emulsifier (Span 60; Croda Inc., Edison, NJ, USA) in a ratio of 16:3:1 (w/w) with yellow:green food coloring added to provide a visual cue. Corn syrup and granulated sugar were also added as feeding stimulants. Toxicants and attractants (ammonium acetate) were mixed into the wax to provide a wax matrix-based bait station. Wax matrix-based bait stations (Fig. 13.4) were formed into either plugs (2.4 cm in diameter, 2.5 cm in height) to approximate a spherical shape or into strips (2.54 cm by 7.6 cm by 4 mm thickness) that were hung horizontally to mimic leaves. A third version of the wax matrix-based bait station was developed in which insecticide, but not attractant, was mixed into the wax matrix. Instead, the wax matrix with insecticide was applied to the edges of an ammonium acetate dispenser which, along with putrescine and trimethylamine hydrochloride, make up

Biolure™ (Suterra LLC, Bend, OR, USA), a very effective lure for *C. capitata*. This bait station, termed a dipped lure bait station (Fig. 13.5), provided increased longevity of the attractant (Epsky et al. 2012). A laboratory study (Epsky et al. 2012) aimed at evaluating the efficacy of bait station strips (Fig. 13.4) with varying amounts of ammonium acetate (0, 1, 2, 3 %) and spinosad (2 %) revealed nearly complete mortality over a 9-day period and at least 78 % mortality over a 24-day period regardless of the amount of ammonium acetate present. Results from field cages indicated that bait station strips with 2 % spinosad and either 1 or 3 % ammonium acetate caused significantly greater mortality of *C. capitata* females compared to pesticide-free bait stations for 4 and 6 weeks, respectively. Overall, the concentration of ammonium acetate added to the wax matrix had little effect on the efficacy of the bait station strips in either the laboratory or field cage tests. Field tests of dipped lure bait stations with methomyl (1 %) as toxicant to provide quick knockdown were conducted in a coffee plantation in Guatemala. Bait stations were placed over funnels attached to vials containing polypropylene glycol solutions to retain dead flies. The dipped lure bait stations killed significantly more female *C. capitata* than corn cob bait stations that were impregnated with a solution of the protein bait NuLure (Miller Chemical & Fertilizer Corporation, Hanover, PA, USA) (80 %) (a hydrolyzed corn protein product) and malathion (20 %), the local bait stations commonly used by Guatemalan growers to suppress *C. capitata* (Epsky et al. 2012).

Putruele and Mouqués (2007) conducted additional field trials of the dipped lure bait station with methomyl (2 %) in Argentina. Specifically, they compared the level of control of *C. capitata* with the dipped lure bait station (ammonium acetate BioLure component only) versus that provided by the conventional ground bait spray using the hydrolysate protein NuLure with malathion and a bait spray that used a proprietary protein bait (CPH Protein, Quemar S.R.L. – Susbin, Mendoza, Argentina) combined with malathion. In an area with a low *C. capitata* population, the suppression achieved by bait stations was comparable to that obtained with bait spray application. These authors concluded that bait stations provide a useful management tool for *C. capitata* when combined with other IPM methods. Epsky et al. (2012) noted that the methomyl has a tendency to discolor the bait stations, thereby providing less visual stimulation to the flies, which did not occur in bait stations containing spinosad. Thus, wax matrix-based bait stations containing spinosad might be more effective in the long run. Additional research is underway in Guatemala to develop a method to mass produce a version of the dipped lure bait stations that will facilitate large scale testing (Heath et al. 2013).

In Portugal, Dantas and Andrade (2007), working with Jackson traps, compared the attractiveness of a Bait Station Gel comprised of proteins, sugars, and other materials present in the Solbait formulation (the basis of GF-120 Fruit Fly Bait) versus various modifications of GF-120 to adult *C. capitata*. They found that the attractiveness of the materials used in the Bait Station Gel was lower than that of GF-120. More recently, Navarro-Llopis et al. (2013) evaluated two bait stations: (1) the Magnet® MED attract-and-kill bait station (Suterra LLC, Bend, OR, USA; Casagrande 2009; Fig. 13.6), consisting of a paper envelope impregnated with

Fig. 13.4 Strip and plug bait stations developed by Heath et al. (2009) and evaluated for *Anastrepha suspensa*. Bait stations were prepared using a wax-based matrix and then molded into the corresponding shape. The bait stations are green in color to provide a visual cue in addition to the bait station shape. A variety of feeding cues, attractants and insecticides can be added to the bait matrix (Photo: N. Epsky)

Fig. 13.5 Dipped lure bait stations evaluated by Epsky et al. (2012) for the Mediterranean fruit fly, *Ceratitis capitata*. Bait stations were made by mixing a wax-based matrix with pesticide, and coating the wax matrix on two edges of combinations of ammonium acetate and trimethylamine lures (BioLure) that had been attached back-to-back (Photo: D. Midgarden)

Fig. 13.6 Magnet® MED attract-and-kill bait station consisting of a paper envelope attract-and-kill device impregnated with deltamethrin that contains two membrane dispensers (trimethylamine and ammonium acetate) as attractants, recently evaluated by Navarro-Llopis et al. (2013) for suppression of *Ceratitis capitata*. The same bait station is used with different lures to suppress *B. oleae* under the trade name Magnet OL® (Suterra LLC, Bend, OR, USA) (Photo: V. Navarro-Llopis)

deltamethrin that contains two membrane dispensers (trimethylamine and ammonium acetate) as attractants and (2) the L&K (Lure and Kill) Tube (Fig. 13.7), a prototype consisting of a yellow-colored cylinder containing a protein bait laced with cypermethrin at the bottom and with several small holes to allow the volatiles to be dispersed. Inside the tube there are two mesoporous dispensers containing ammonium acetate, trimethylamine, and methyl-pyrrolidine. In that study, bait stations were deployed at a density of 50 per ha, and plots with bait stations ranged in size from 0.8 to 1.3 ha. In each of 2 years (2010 and 2011), the efficacy of these bait stations was compared with that of the mass trapping technique and either standard treatment with insecticide bait spray (2010) or an untreated control (2011) using adult fruit fly population reduction and the fruit damage as indicators of the effectiveness of each treatment. The mass trapping device was the Moskian trap® (SanSan Agriculture Engineering, Valencia, Spain) baited with the three-component BioLure Unipack and a dichlorvos strip. Insecticide treatment consisted of foliar applications of the spinosad-containing protein bait spray Spintor Cebo® (Dow AgroSciences Ibérica, S.A., Madrid, Spain). The Magnet® MED bait station proved to be as effective in reducing populations of *C. capitata* as the mass trapping system and bait spray application, and these treatments resulted in comparatively low incidence of fruit damage in 2010. In 2011, the Magnet® MED and mass

Fig. 13.7 Lure and Kill (L&K) bait station consisting of a *yellow-colored* cylinder containing a protein bait laced with cypermethrin at the *bottom* of the cylinder, and around the bait container there are several small holes that allow attractants to be released. Two mesoporous dispensers containing ammonium acetate, trimethylamine and methyl-pyrrolidine are placed inside the tube. Evaluated by Navarro-Llopis et al. (2013) for suppression of *Ceratitis capitata* in Spain (Photo: V. Navarro-Llopis)

trapping performed equally well both in terms of reductions in female population density compared to untreated control plots.

Another device, the M3 Fruit Fly Bait Station (Fig. 13.8) produced by Quest Development CC (Brits, South Africa), was developed principally for management of *C. capitata* (Coltell Simon 2009). Its components are: (1) a clip (to secure the product to a tree branch), (2) a body (which houses and protects the foam insert), (3) a grate (that secures the foam insert to the body), and (4) a killing agent (lithium perfluorooctane sulfonate) mixed with an attractant (plant extracts and protein hydrolysate) impregnated into the foam insert. In a study conducted in grapefruit orchards in South Africa, Ware et al. (2003) evaluated the M3 Fruit Fly Bait Stations at a rate of 400 units per ha at the beginning of the season and documented excellent performance of this bait station.

In a series of studies conducted in Spain, Navarro-Llopis and collaborators (2004, 2007, 2010) evaluated the efficacy of a chemosterilant bait station system that later became commercially available as the Adress system (Syngenta Agro S.A., Madrid, Spain; Mas and Gonzalez 2009) (Fig. 13.9). The bait station consists of a yellow vertical cylinder containing the male attractant trimedlure (TML) and

Fig. 13.8 M3 Fruit Fly Bait Station developed in South Africa for management of *Ceratitis capitata* (Coltell Simon 2009) and produced by Quest. It consists of five components: (1) a clip (to secure the product to a tree branch), (2) a body (which houses the foam insert), (3) a grate (that secures the foam insert to the body), (4) insecticide, which is impregnated into the foam insert (5). The foam insert is protected from the environment by the body. The insecticide consists of a broad spectrum insecticide, and a pheromone luring agent (Photo: V. Navarro-Llopis)

the two-component female attractant *N*-methyl pyrrolidine and ammonium acetate, with slots near the bottom to emit the attractant odors. A 9-cm-diameter plate containing the gel formulation of a phagostimulant and 3 % lufenuron (the chemosterilant) is attached to the bottom of the cylinder, so that the flies can readily feed on the gel. The system is covered with a wide yellow bottomless cone to protect the gel and attractants from rain and wind. The attractants are released by three types of mesoporous dispensers (Muñoz-Pallares et al. 2001). In an area-wide trial conducted in 50,000 ha of citrus, Navarro-Llopis et al. (2011) documented a significant reduction in male and female *C. capitata* populations in areas SIT + Adress treatment versus the SIT only treatment, indicating that the SIT and the Adress system were compatible and the combination of both techniques improved the control of the *C. capitata*.

Recently, the Moscamed Regional Program has developed a novel bait station to disseminate the entomopathogenic fungus *Beauveria bassiana* (Balsamo) against *C. capitata* wild populations. Two designs of bait station have been developed. The cylindrical type is composed of a 500 ml plastic (polyethylene terephthalate) bottle (14.0 cm high × 8.5 cm diameter) with fifteen 2.5 mm holes evenly distributed on the sides, a lid containing four triangular openings of 1.5 mm on each side, and an open bottom. The basket with the TML plug is placed inside hanging from the top. The lid and the bottom are covered with tulle fabric, and the outside of the device is fully covered with a yellow plush fabric (14 cm × 22 cm) impregnated with 2 g of *B. bassiana* conidia (Fig. 13.10a). The second type is a rectangular panel composed of a galvanized panel (23 cm × 14 cm) with a basket and a TML plug inserted in a 2.5 hole in the center of the panel. This device is also covered with yellow plush

Fig. 13.9 Adress bait station (Mas and Gonzalez 2009) consisting of a *yellow vertical cylinder* containing the male attractant trimedlure and the two-component female attractant *N*-methyl pyrrolidine and ammonium acetate, with slots near the bottom to emit the attractant odors. A 9-cm-diameter plate containing the gel formulation of a phagostimulant and 3 % lufenuron is attached to the *bottom* of the cylinder, so that the flies can readily feed on the gel. The system is covered with a *wide yellow bottomless cone* to protect the gel and attractants from rain and other elements (Photo: V. Navarro-Llopis)

fabric (23 cm × 14 cm) impregnated with 2 g of *B. bassiana* conidia (Fig. 13.10b). According to Flores et al. (2013), to effectively disseminate the *B. bassiana* conidia to *C. capitata* wild population, one bait station per ha must be installed, and the conidia-treated fabric must be replaced every 15 days.

These bait station devices have been evaluated in open field tests against *C. capitata* in Guatemala. The results are promising and show over 44 % inoculated wild male flies. The dissemination of the fungus conidia is very specific as only *C. capitata* males responding to trimedlure will approach the *B. bassiana* inoculated bait station and become infected. This technology has a multiplicative effect, since inoculated wild males will infect other males and females during courtship. Because of its mode of action, this technology is considered to be species specific and environmentally friendly (Flores et al. 2013).

4.2.3 *Anastrepha* spp.

Several prototype bait stations have been tested for use against key *Anastrepha* species. Laboratory studies confirmed that imidacloprid-treated spheres were effective against *Anastrepha ludens* (Loew) (Prokopy et al. 2000) and *Anastrepha*

Fig. 13.10 Autodissemination bait station developed by the Moscamed Regional Program against wild *C. capitata*: (**a**) Cylindrical type device, composed of a 500 ml PET container (14.0 cm high × 8.5 cm diameter) with fifteen 2.5 mm holes evenly distributed on the sides, a lid containing four triangular openings of 1.5 mm on each side and an open bottom. The outside of the device is fully covered with a yellow plush fabric (14 cm × 22 cm) impregnated with 2 g of *Beauveria bassiana* conidia, (**b**) Rectangular panel-type device, composed of a galvanized panel (23 cm × 14 cm) with a basket and a TML plug inserted in a 2.5 hole in the center of the panel. This device is also covered with *yellow plush fabric* (23 cm × 14 cm) impregnated with 2 g of *B. bassiana* conidia (Photo: R. Mangan)

suspensa (Loew) (Liburd et al. 2004). Another trap type that has been evaluated for *A. suspensa* is the "Mitchell" bait station (Fig. 13.11), which consists of a badminton shuttlecock, an attractant, and a toxicant. It has been tested, not only with fruit flies (e.g., Holler et al. 2006), but also with alfalfa looper, *Autographa californica* (Speyer) (De Camelo et al. 2007). In Florida, Holler et al. (2006) evaluated the performance of "Mitchell" bait stations without attractant or baited with ammonium acetate and putrescine (two-component BioLure) against *A. suspensa* in comparison with McPhail traps baited with liquid protein bait (torula yeast/borax) and MultiLure traps (Better World Manufacturing, Fresno, CA, USA) baited with two component BioLure. In that study, bait stations were painted either green (for the field cage study) or yellow (for the field test of sterile *A. suspensa* in a 0.4 ha citrus orchard). In the field cage study, the McPhail trap attracted significantly more *A. suspensa* than any other treatment, followed by the MultiLure trap. In mixed-variety orange, *Citrus sinensis* (L.) Osbeck, plots, the numbers of released sterile *A. suspensa* adults were significantly reduced in the plot that had Mitchell bait stations compared to the untreated control.

Mangan and Moreno (2007) developed tent-shaped (Fig. 13.12) and cylindrical (Fig. 13.13) bait stations that used sponges soaked with a hydrolyzed protein and toxicant (phloxine B) for use against *A. ludens*. In preliminary field trials conducted in Mexico with bait station densities of 16–32 stations per ha, they were able to kill

Fig. 13.11 Mitchell bait station, consisting of a badminton shuttlecock (5.3 cm diameter at the base, 2.5 cm diameter at the apex, and 7.5 cm in height), an attractant, and a toxicant. It has been evaluated not only with fruit flies (e.g., *Anastrepha suspensa*; Holler et al. 2006) but also with alfalfa looper, *Autographa californica* (Speyer) (De Camelo et al. 2007) (Photo: P. Landolt)

a high number of flies but found no decrease in adults captured in monitoring traps or in fruit infestation levels. The wax matrix-based bait station developed by Heath et al. (2009) was tested against *A. suspensa* and has been evaluated in field cages in association with cyromazine, methomyl, avermectin, spinosad, SureDye (Red Dye 28-phloxine B; PhotoDye International Inc., Linthicum, MD, USA), and dimethoathe at 0.25 and 1.0 % a.i. (w/v). For insecticides tested at the 0.25 % concentration, the highest mortality of *A. suspensa* females (exposed to the toxicants for 4 h) was recorded for dimethoate (95.1 % mortality), followed by spinosad (84.1 %), SureDye (83.0 %), avermectin (81.4 %) and methomyl (74.0 %), whereas for insecticides tested at the 1.0 % level, dimethoate (99.0 % mortality), methomyl (97.8 %), and spinosad (96.1 %) were the best performing toxicants. Wax matrix-based bait stations remained effective for more than 50 days when exposed to subtropical rain conditions. These authors indicated that manufacturing the wax matrix-based bait station would cost less than 5 cents plus the cost of insecticide.

The National Program against fruit flies in Mexico recently evaluated an insecticide-free bait station termed MS2 (River Bioscience (Pty) Ltd, Port Elizabeth, South Africa; Fig. 13.14) used in combination with Cera Trap (Bioiberica, Barcelona, Spain), a bait produced by cold enzymatic hydrolysis (de los Santos-Ramos et al. 2011). In their study, de los Santos-Ramos et al. (2011) documented that the MS2 trap was more effective at trapping *A. ludens* in grapefruit orchards compared to other two trap types (MS2-Spinosad and a locally produced bait station [plastic bottle with holes baited with hydrolyzed protein]). In a subsequent study, de los Santos-Ramos et al. (2012) compared the effectiveness of various bait station designs and food-based attractants against *A. ludens* in grapefruit orchards in Veracruz, Mexico. They found that the MS2 bait station baited with Cera Trap was the most effective at capturing adult *A. ludens* compared with MS2 baited with GF-120 Fruit Fly Bait and bait stations made using a soda bottle with two 5 cm windows cut on the sides and baited with 150 ml of NuLure and malathion (9:1 ratio) and water. Bait stations were deployed at a density of 52 per ha. No fruit infestation data were recorded in that study.

Fig. 13.12 Tent-shaped bait station consisting of a sheet of sponge attached under a folded sheet of plastic (21.5 by 24.0 cm) folded in half to make 90° angle. The sheet of sponge material (19 by 20 cm and 8 mm in thickness when baited with a protein-based attractant) was stapled to the underside of the tent to line the under surface leaving a 2.0-cm overhang of the plastic on all sides. Developed by Mangan and Moreno (Mangan and Moreno 2007) and tested against *Anastrepha ludens* (Photo: R. Mangan)

Fig. 13.13 Cylindrical bait station made of a polyvinyl chloride (PVC) pipe (10.1 cm internal diameter and 10 cm in length) with a conical top and supported bait tray inserts that can be filled with protein-based bait for efficient storage and transport (Mangan and Moreno 2007) (Photo: R. Mangan)

4.2.4 *Bactrocera* spp.

Below, we describe some bait stations that have been developed for management of females of various *Bactrocera* species. In some cases, the same bait stations have been evaluated for *C. capitata* for interspecific comparisons (e.g., Piñero et al. 2011a, b).

Fig. 13.14 MS2 bait station used in combination with the food-based Cera Trap attractant. This bait station has been evaluated for suppression of *Anastrepha ludens* in citrus orchards in Mexico (de los Santos-Ramos et al. 2011) (Photo: R. Hernandez-Perez)

Umbrella Bait Station This bait station was developed by Dr. Edward Y. Cheng at the Taiwan Agricultural Research Institute and consists of a yellow funnel with a hook to hang from tree branches (Fig. 13.15), and it functions as a rain-fast device for applications of methyl eugenol or cue lure dispensers. Area-wide oriental fruit fly management in Taiwan used the combination of umbrella traps with male lures as a male annihilation tactics in fruit orchards, including guava (*Psidium guajava* L), sugar apple (*Annona squamosa* L.), wax apple (*Syzygium samarangense* [Blume] Merr. & L.M. Perry), and citrus (*Citrus* spp.), covering approx. 55,000 ha (Cheng et al. 2003). During 4 years of implementation, male annihilation alone reduced the *B. dorsalis* population in citrus and sugar apple orchards by 60–65 %, hence, foliar applications of insecticides were no longer needed for fruit fly control, and fruit harvest was increased by 20 %. The umbrella trap was also adopted by bitter gourd (*Momordica charantia* L.) growers as a protein bait station for melon fly suppression in high rain fall areas, which reduced fruit infestation from 75 % to less than 5 % within one season (E.Y. Cheng et al., personal comm.). Its performance relative to other bait stations has not been assessed in formal studies.

Fig. 13.15 Umbrella bait station developed by Dr. Edward Y. Cheng at the Taiwan Agricultural Research Institute and consisting of a *yellow funnel* with a hook to hang from tree branches. This bait station has been used for applications of methyl eugenol and cue lure dispensers for male annihilation of *Bactrocera dorsalis* and *B. cucurbitae*, respectively, in Taiwan (Photo: S. Souder)

Papaya Leaf Mimics Piñero et al. (2009a, 2010) demonstrated the potential of bait spray applications of GF-120, when used in combination with other management techniques, in achieving a low fruit fly prevalence area in papaya orchards in Hawaii. However, high levels of rainfall affect the efficacy of bait sprays. In an attempt to overcome this problem, Piñero et al. (2009a) developed a novel, visually attractive bait station for application of insecticidal baits against *B. dorsalis, B. cucurbitae*, and *C. capitata*. The bait station developed was termed the Papaya Leaf Mimic (PLM) (Fig. 13.16a, b), because it represents a supernormal visual stimulus of papaya foliage. PLMs were constructed using plant pot saucers (36 cm outer diameter, 5 cm height of the lip). A metallic shelf bracket was attached to the interior of the saucer using screws and glue. This simple design allowed for easy deployment to vertical structures, such as the trunks of papaya trees or coffee plants. To increase adherence of the protein bait, the interior area of each saucer was scraped in a circular fashion using a wire-wheel brush attached to an electric drill. With the grooves created by this brushing and using the hand-held sprayer to apply GF-120, virtually no bait dripping was observed. Subsequently, a primer was applied onto the saucer, followed by a layer of cadmium yellow medium paint. The physical structure of the PLMs has shown to endure at least 5 years of continuous weathering (J.C. Piñero et al., unpub).

PLMs have been proven to enhance the behavioral response of adult fruit flies to GF-120 and extend its attractiveness for at least 1 week (Piñero et al. 2009a). In addition, by using PLMs, waste of bait due to washing to the ground or to undesirable areas of the target tree or plant (e.g., trunk, fruit) can be avoided. PLMs are also advantageous in that they circumvent leaf phytotoxicity observed in the field, which is likely caused by one or more ingredients in the bait matrix (DeLury et al. 2009), and minimize degradation of spinosad by photolysis (Mangan et al. 2006).

For PLMs to be considered by fruit and vegetable growers as a viable alternative to foliar bait sprays, they should be cost-competitive and show good performance in

Fig. 13.16 Attract and kill bait station termed Papaya Leaf Mimic, representing a supernormal visual stimulus of papaya foliage baited with, either (**a**) sprayed GF-120 NF Naturalyte Fruit Fly Bait, or (**b**) GF-120 NF in conjunction with a proprietary rain-fast amorphous polymer matrix (APM). This simple design allows for easy deployment to vertical structures such as the trunks of papaya trees, and coffee plants. Developed by Piñero et al. (2009b), this bait station has been evaluated in various agro-ecosystems (e.g., papaya, coffee and cherry orchards) against various fruit fly species including *Bactrocera dorsalis*, *B. cucurbitae*, *Ceratitis capitata* and *Rhagoletis indifferens* (Piñero et al. 2009b, 2010, 2011a, b; D. Alston and J.C. Piñero, unpub, data) (Photo: J. Piñero)

commercial orchards. In Hawaii, Piñero et al. (2010) compared the effectiveness of GF-120 Fruit Fly Bait applied to PLMs at controlling *B. dorsalis* with that of bait sprays. Trapping using MultiLure traps baited with torula yeast and infestation data were used as indicators of the effectiveness of the two bait application methods. A key finding of that study was that GF-120 applied to PLMs performed as well as foliar bait sprays in suppressing *B. dorsalis* (trapping data) from treated plots for the first 10 weeks following the first bait spray when *B. dorsalis* populations were low. During this trapping period, infestation levels decreased 71.4 % and 63.1 % for plots with bait stations and foliar sprays, respectively, relative to control plots. For the last 3 weeks of the study when *B. dorsalis* populations were increasing rapidly, there was a decrease in the effectiveness of the bait sprays as determined by trap captures, and fruit infestation rates were, on average, 54.5 % and 45.4 % lower for plots with bait stations and foliar sprays, respectively, than control plots. Overall, substantially less GF-120 (~42 %) was applied to PLMs than in foliar applications, and this resulted in cost-savings as well as release of less insecticide into the environment. More recently, Piñero et al. (2011a) compared the response of wild *C. capitata* females to PLMs baited with GF-120 NF in conjunction with a proprietary rain-fast amorphous polymer matrix (APM) (Fig. 13.16b) to PLMs baited with the standard GF-120 NF bait spray formulation (Fig. 13.15a) under Hawaii's climatic conditions and found that APM mixed with either 50 or 75 % GF-120 applied to PLMs was attractive to female *C. capitata* for up to 3 weeks longer than the standard sprayed GF-120 NF. The performance of PLMs has also been evaluated in cherry orchards in Utah for management of Western cherry fruit fly, *Rhagoletis indifferens* Curran. In one field study, GF-120 applied onto PLMs successfully reduced infestation in cherries to 0.3 % or less using PLMs at a density of 18 and 30 per acre (D. Alston and J.C. Piñero, unpub. data). Interestingly, Piñero et al. (2009a, b) suggested the possibility of using multiple lures targeting males (e.g., cue lure and methyl eugenol) and females (e.g., protein bait sprays, host-based attractants) applied to this visually-attractive bait station.

Bait Stations for Olive Fruit Fly Unlike many of the *Bactrocera* spp., males of the olive fruit fly, *Bactrocera oleae* (Rossi), do not respond to male-specific lures, so male-targeted bait stations have not been an option for control of this pest. Instead, since the 1980s research has been ongoing to develop traps with toxicants on the exterior surface and baited with food-based attractants alone or in combination with the synthetic female-produced sex pheromone, such as 'Olive Fruit Fly Lure-Spiroketal Male' (Suterra LLC, Bend, OR, USA). Attracted by the sex pheromone, food-based attractant, and/or the yellow color, both males and females land on the toxicant-impregnated panel and receive a lethal dose of insecticide. These have often been called poison traps, and control efforts that use them are referred to as mass trapping. However, these devices are, in fact, bait stations, since flies contacting the poison traps are not retained. Nadel et al. (1989) reported on a bait station called the "Foam Fatale" device. It is made from polyurethane foam impregnated with nutrient lure and insecticide and molded into a suitable shape. Initial studies found that foam bait stations used without toxicant but with sticky material and baited with ammonium bicarbonate and pheromone captured more *B. oleae* than bait stations with ammonium bicarbonate only or McPhail traps baited with aqueous ammonium bicarbonate solution. Field longevity of the attractants added to the foam bait station was estimated to be 76 d for the ammonium bicarbonate and 103 d for the pheromone. In subsequent studies, toxicant (2,2-dichlorovinyl dimethyl phosphate [DDVP]) was mixed into the foam along with the attractants. When used with an appropriate amount of toxicant, foam bait stations weathered for 5 months were effective in killing flies that were kept in contact with the surface of the bait station for 30 s in laboratory tests.

Haniotakis et al. (1991) referenced early studies by Allen (1976) that reported that bait stations that used aqueous solutions of honey and deltamethrin sprayed at regular intervals on yellow plastic panels gave satisfactory control of *B. oleae*. Haniotakis et al. (1991) reported on large scale field tests conducted from 1984 to 1989 that tested plywood rectangles (20 cm in height by 15 cm in width by 0.4 cm in thickness) that were dipped in aqueous solutions of insecticide (deltamethrin and dichlorvos) alone or in combination with sugar (added as a feeding stimulant) or sugar plus glycerine (added for its hygroscopic properties that provided moisture as an attractant). These were baited with ammonium bicarbonate salt alone or in combination with synthetic pheromone lure. Bait station density ranged from 1:1 to 1:3 bait station/tree. Overall, the bait stations were effective in decreasing both adult counts and fruit infestation levels, although combinations of bait stations and bait sprays were needed in non-isolated orchards or in areas with high population levels. Mazomenos et al. (2002) evaluated two bait stations for control of *B. oleae* in field tests conducted from 1992 to 1996 in Greece. Effectiveness was compared with insecticide-treated plots. The first 2 years of the study were conducted with bags (described by the authors as 'paper plastic'). The bags were impregnated with deltamethrin, sugar solution, and glycerol (Vioryl A.E., Kifissia, Greece) and were filled with ammonium bicarbonate salt and baited externally with a pheromone lure. For the last 3 years of the study, they tested bait stations that were made using a cloth-covered cylindrical wire frame, with the cloth dipped in toxicant and the device baited externally with ammonium bicarbonate lures (formerly AgriSense, now Suterra UK, Pontypridd, UK) and pheromone lures. The same results were obtained with both types of bait stations. In isolated groves located in areas with low or medium population levels, bait stations deployed in every tree gave control (as indicated by capture of adult flies or by fruit infestation levels) similar to that achieved with ground bait spray application. However, combinations of bait spray application and bait station treatments were needed in olive groves that were not isolated or were in areas with high population levels to achieve levels of control equal to multiple applications of bait spray.

Broumas et al. (2002) conducted field tests from 1996 to 1999 in Greece and compared *B. oleae* control using bait stations (which they called a mass trapping method) versus bait sprays. They used Eco-Trap type bait stations (Vioryl S.A., Athens, Greece), i.e., envelopes (15 by 20 cm) of light-green-colored paper that contained ammonium bicarbonate salt, were coated with UV-light protected deltamethrin (Decis flow 2.5 %, AgrEvo Hellas, Athens, Greece), and were baited externally with a pheromone lure. Bait stations were placed in alternate trees in orchards with small to medium trees (>150 trees/ha) and in every tree in orchards with large trees. Adult fly populations and larval infestation levels were as low or lower in areas receiving bait stations than in areas receiving bait spray.

Petacchi et al. (2003) reported on studies that used EcoTrap bait stations against *B. oleae* in 15 mass trapping areas that ranged in size from 5 to 394 ha in Italy. From 1999 to 2001, bait stations were placed in alternating trees at the start of the summer and placed in all trees in the fall. Results were compared to areas that received the conventional control strategy that included 1–3 pesticide applications over the growing season. In 3 out of 4 locations, bait stations were more effective in reducing fruit infestation. Tsolakis et al. (2011) evaluated Eco-Traps alone or in combination with application of a repellant spray (copper hydroxide sprays) in field tests in Italy. Bait stations were placed in every tree in plots containing 421 and 223 trees, respectively. The combination of bait stations and two spray applications over the growing season was found to reduce infestation levels below that achieved using bait stations alone. Volakakis et al. (2012) compared the effectiveness of Eco-Traps baited with ammonium bicarbonate only with two mass trap treatments (plastic bottles baited with aqueous torula yeast/borax and Elkofon-1 traps (Phytophyl S.A., Athens, Greece) with liquid protein bait) and with untreated plots. These field tests were conducted with replicated small plots of 20 trees per plot in Greece. They found no differences in either adult fly population or fruit infestation among different treatments and the control plots and attributed the lack of effects on the small size of the plots and/or the lack of pheromone lures with the Eco-Trap.

Another bait station similar to the Eco-Trap is the Greek DakoFaka® (meaning trap for the olive fly; Viotrap Fruit, Crete, Greece) bait station (Fig. 13.17), which actually is not specific to olive fly. It integrates a visual stimulus, food attractants

Fig. 13.17 Greek DakoFaka bait station for *Bactrocera oleae* suppression, consisting of two polyethylene chambers that contain a liquid attractant in one and a solid lure in the other. The outer walls are impregnated a pyrethroid insecticide (Photo: N. Epsky)

(ammonia releasing compounds), and no pheromone. The killing agent is a pyrethroid insecticide.

5 Bait Station Applications

Bait stations can be utilized for:

(i) Fruit fly control in commercial fruit orchards.
(ii) Fruit fly control as part of an area-wide control programs in commercial fruit orchards and marginal host areas.

5.1 Commercial Fruit Orchards

In recent years, bait stations (including traps for mass trapping) have been used for fruit fly control in commercial orchards in some countries, such as Spain, Italy, and Argentina. This type of application has been triggered by the need to reduce insecticide use and its residues in fruits to comply with the current more stringent human health and environmental laws as well as with the increasing public demand, especially in the developed countries, for organic products or products with low insecticide residues.

This type of bait station application is normally limited to the area covered by the commercial fruit crop (i.e., applied on an orchard basis). In most cases, fruit fly response to bait stations (as well as traps) is considered to be weak, therefore, since bait stations are aimed at population suppression, high bait station densities are required in orchards for effective fruit fly control. Densities will depend on the following: (1) number of fruit trees per ha (a general rule is to use one bait station per tree or one every other tree), (2) fruit fly population density, (3) desired level of crop protection (i.e., economic damage threshold), and (4) relationship between costs and benefits. For example, in a mango orchard with an average of 100 trees per ha (trees planted at 10×10 m), a large fruit fly population, and an economic threshold of 1–3 %, a density of 50–100 bait stations per ha is required. In the case of a citrus orchard (e.g., orange) with an average 400 trees per ha (trees planted at 5×5 m), a large fruit fly population, and the same damage threshold as above, a density of 200–400 bait stations is required. With a lower population density and higher economic threshold, the required number of bait stations per ha could be substantially reduced.

In general, bait stations are distributed evenly in orchards. If information on the fruit fly spatial distribution within the orchard is known, bait stations may be aggregated to overlap with the fruit fly population. Moreover, if information is available on the dispersion behavior of fruit flies from areas surrounding the orchard into the orchard, bait stations may be placed around the orchard's periphery to reduce or eliminate immigrating flies (Alemany et al. 2004). The number of bait stations per ha as well as the number of treatments required may be optimized if information on fruit fly spatial and temporal distribution within and outside a commercial fruit orchard is available.

5.2 *Area-Wide Control Programs*

Bait stations have been discussed as potentially important components of area-wide IPM action programs that utilize the SIT and biological control (IAEA 2009). The Moscamed Regional Program (USA, Mexico and Guatemala) has tested protein-based bait stations as part of area-wide suppression/eradication of *C. capitata* (Programa Moscamed 1984). The suppression effect obtained for *C. capitata* populations has shown the effectiveness of bait stations, nevertheless, experimental evidence verifying their impact is lacking; therefore, in this case, we will only discuss the use of bait stations from a practical point of view.

When applied in this manner, bait stations are one of several tools of an area-wide IPM program that covers a large geographical area. Therefore, in contrast with the previous application, the area covered is not limited to the commercial fruit orchards. Because of the size of the target area, large numbers of bait stations would have to be used in order to cover the total host area at a prohibitive operational cost. Thus, in this case, the primary use of bait stations is to control populations in specific localized sites with presence of non-commercial fruit hosts, typically the main source of immigrating flies into commercial orchards. These sites may include: marginal areas with scattered fruit hosts, small rural communities with backyard hosts, and sites where a high volume of fruit is gathered to be processed, such as in coffee processing facilities – to reduce source of flies that may disperse in search of hosts. Bait stations are an important area-wide IPM tool in rural populated areas as well as in ecological sensitive areas where ground or aerial insecticide-bait sprays are restricted. Bait stations may also become a key control tool during the rainy season in areas with heavy rain, since the insecticide-bait applied by ground or by air is washed-off from the canopy of host trees. In contrast to commercial orchards, when used in specific localized situations, bait stations are used at relatively low densities raging from 5 to 50 bait stations per ha. The lower density is related to the availability of fruit hosts in marginal areas, which are normally scattered and scarce compared to the commercial orchards (Programa Moscamed 2010).

Normally, when deployed in non-commercial fruit hosts, bait stations are placed in the field using irregular patterns that follow the distribution of host trees in the target areas. In some cases, such as sites where high volumes of fruits are gathered for processing (coffee mills, packing facilities, etc.), bait stations are deployed around the processing facility. As in commercial orchards, the number of bait stations that are deployed in the field as well as the number of treatments required may be optimized if information on fruit fly spatial and temporal distribution in the target area is available.

An example of the practical use of bait stations as a tool for area-wide IPM interventions is the USA-Mexico-Guatemala Moscamed Regional Program. Since the early 1980s, the Program has used low cost protein based bait stations made of recycled cheap material. Common bait stations used are: (1) a piece of corn cob impregnated with a solution of protein bait NuLure and malathion at a 4:1 proportion, (2) a so-called "killing bag", which is a small bag made of natural fiber and filled with an absorbent material, such as wood chips, and soaked in a protein bait NuLure and malathion at a 4:1 proportion and, more recently, (3) a bait station consisting of a 600 ml plastic bottle with side openings and baited with a sponge impregnated with 250 ml of a mixture of GF-120 (a.i. spinosad) and water at a 1:4 proportion (Fig. 13.18). These bait stations have been used effectively as part of the eradication protocol in *C. capitata* localized outbreaks occurring in fruit fly-free areas (Programa Moscamed 1984). For example, in 1984, the Program used 642 "killing bags" to treat 10 *C. capitata* outbreaks that occurred in the State of Chiapas, which contributed to *C. capitata* eradication in the free areas. However, in general, these bait stations have a short life span requiring replacement every week and thus are labor intensive.

A more recent bait station used by the Moscamed Regional Program is a long lasting bait station that is called "Wax-BS" bait station (Fig. 13.19, Heath et al. 2013). It has a life span of 6–8 weeks, significantly reducing hand labor. This bait station has been used to suppress *C. capitata* populations preventatively. Ten to fifteen bait stations per ha are placed in backyard fruit hosts in rural communities that have shown recurrence of *C. capitata* presence. The bait stations

Fig. 13.18 Bait station used as a fruit fly suppression tool for area-wide IPM interventions in the USA-Mexico-Guatemala Moscamed Regional Program. It consists of a 600 ml plastic bottle with side openings baited with a sponge impregnated with 250 ml of a mixture of GF-120 Fruit Fly Bait (a.i. spinosad) and water at a 1:4 proportion (Photo: W. Enkerlin)

Fig. 13.19 Long lasting bait station termed 'Wax-BS' tested by the USA-Mexico-Guatemala Moscamed Regional Program for control of *Ceratitis capitata*. Wax-based matrix composed of paraffin wax with *yellow:green* food coloring added to provide a visual cue, corn syrup and granulated sugar as feeding stimulants, and toxicant is coated on a bait station device (Heath et al. 2013) that is baited with ammonium acetate and trimethylamine lures (BioLure) (Photo: W. Enkerlin)

are placed one life cycle (approximately 30 days) prior to the appearance of the first *C. capitata* populations in these sites according to the historical profile of *C. capitata* detections and outbreaks. Bait stations control the first incursions of *C. capitata* adults in these sites, preventing establishment of the population and the high costs associated with eradication of established *C. capitata* populations. In addition, the Program uses the long lasting bait station during the rainy season to suppress high populations in *C. capitata* reservoirs that occur in delimited localized sites. Ten to fifty bait stations per ha, depending on host availability and population density, are placed in host trees for 3 months (two treatments). Once populations are suppressed below an established threshold, populations are eradicated with the continuous release of sterile male flies. Under the conditions described, this is the most cost-effective tool for suppression of *C. capitata* reservoirs currently available in the program. One other application is the use of bait stations as part of eradication efforts in isolated *C. capitata* outbreaks that occur in *C. capitata* free areas. Fifteen to twenty five bait stations per ha, depending on the availability of hosts and magnitude of the outbreak, are placed in the first square kilometer area around the outbreak. Bait stations remain in the field for two *C. capitata* life cycles (approximately 60 days). Since bait stations last for 6–8 weeks, there is the need for one replacement. In contrast to the two treatments needed for bait stations, eight ground bait sprays would be required to span the same interval, and the cost would be substantially higher (Programa Moscamed 2012).

6 Environmental Benefits Associated with Use of Bait Stations

Several types of bait stations have been shown to be effective at managing fruit flies, but being effective is only part of the equation. As part of an IPM approach to managing fruit flies, bait stations should also be environment and grower friendly. From an environmental perspective, use of bait stations has several advantages: (1) there is no release of insecticide into the environment, (2) there is minimal or no contact between the pesticide and the commodity, (3) there is minimal or no contact between the pesticide and beneficial arthropods, and (3) there is reduced worker contact with pesticide, thus representing an important improvement in worker safety. Given the above, bait stations have been proposed as alternative treatments in areas where broadcast insecticides are not acceptable (Heath et al. 2009; Epsky et al. 2012). We anticipate that the use of bait stations will have more extensive applications, including AW-IPM programs, detection/eradication, organic farming systems, etc.

7 Cost Considerations

The acceptance of bait stations as an IPM tool to control fruit fly populations and reduce the level of injury to fruit is dependent upon various factors, including the desired level of crop protection (i.e. economic threshold; low income farmers might be happy reducing crop damage from 60 to 15 % with a cheap and easy to use bait station, others might require less than 1 % damage), cost of the bait station, lure and toxicants, bait station deployment pattern and density, and area protected. Other factors are related to fruit fly population density, host suitability for fly reproduction, and type of habitat adjacent to the orchard. Another consideration is the relative value of the crop to be protected, the added value to the crop from the perspective of using an environment friendly control tool (i.e., organic crop) and the environmental cost savings. Thus, conducting cost-benefit analyses that take into account direct and indirect costs and benefits is one of the most critical aspects that needs to be considered for the development, evaluation, and practical use of bait stations in fruit fly pest management systems. These are critical components that will determine the feasibility of any bait station adoption by growers. However, this type of information, although is very much needed, is scarce.

While some studies have reported good performance of bait stations at suppressing fruit flies, cost-benefit analyses are infrequently discussed. For example, in one study conducted in commercial grapefruit orchards in Mexico, de los Santos-Ramos et al. (2012) reported adequate suppression of *A. ludens* using the bait station MS2 baited with Cera Trap at a density of 52 bait stations per ha. However, no further account of costs was provided. Working with released, sterile adults of *A. suspensa* in an orange grove in Florida, Holler et al. (Holler et al. 2006) evaluated the ability of Mitchell stations to suppress fly abundance in areas with bait station deployment versus control areas lacking bait stations. When used at a density of about 52 bait stations per ha, bait stations suppressed a free-ranging *A. suspensa* population relative to the control plots, but no information was provided on the cost of constructing and deploying the bait stations or adequacy of the suppression achieved. In turn, Mangan and Moreno (2007) evaluated Tent Bait stations consisting of a sheet of sponge material fastened to a plastic peaked cover and baited with Mazoferm (a mixture of protein hydrolysate, sugar, adjuvants, a photoactive dye toxicant, and other additives) in a grapefruit orchard in Texas. They found that populations of released, sterile adults of *A. ludens* were reduced by 70–90 % in 4 days using a density of 50 bait stations per ha compared with control plots. These results demonstrated the potential efficacy of these stations if deployed in commercial orchards, in particular if more attractive chemical components, such as those in the BioLure formulations, would be added. However, cost considerations were not discussed.

In some instances, cost considerations have been discussed, but new technology has been developed in such a way that new cost-benefit analyses are needed. For example, Barry et al. (2004) indicated that the biodegradable attracticidal spheres evaluated by Stelinski and Liburd (2001) have not been used in Michigan blueberry orchards to control *R. mendax* because of the deployment density, lack of attractive selective lure, associated costs of products (i.e., spheres and residue extending agents), and labor requirements (i.e., monitoring and applying insecticides to spheres). However, a new sphere design that literally requires no maintenance over the entire growing season has been developed for *R. pomonella* (Prokopy et al. 2005; Wright et al. 2012), and it could be adapted to attract *R. mendax*.

As noted above, Piñero et al. (2010), working in commercial papaya orchards in Hawaii, compared the effectiveness of GF-120 applied to PLMs (a 20 % dilution; 30 PLMs per ha) or applied to the foliage (a 10 % dilution) to control *B. dorsalis*. They also conducted an analysis of costs associated with the construction and deployment of PLMs for fruit fly control. The foliar application of GF-120 required an average of 0.25 l of undiluted GF-120/ha/week, resulting in a total cost of $8.32/ha/week (based on 2008 costs). The projected cost of spraying GF-120 weekly to papaya foliage using a 10 % solution is therefore $432.60/ha/year (papayas are produced year-round in Hawaii) assuming no re-application after rainfall events and $515.80 in the hypothetical (yet conservative) situation that 10 re-applications are needed in 1 year. In contrast, application of ca. 20 ml of a 20 % solution of GF-120 to 30 PLMs/ha required an average of 0.12 l of undiluted GF-120/ha/week, for a total of $4.00/ha/week, and the projected cost of weekly bait applications to PLMs is $208.0/ha/year. The cost of materials to make one PLM was around $6.50 (for a total of $195/ha) for research purposes, an amount that can be reduced nearly by half if cheaper materials (e.g., a zip tie or Velcro) instead of shelf brackets are used for attachment to host trees. The cost of materials needed to make PLMs plus annual bait application is $403.0/ha, clearly less than the cost associated with foliar applications. It is also important to consider that foliar sprays require more equipment (e.g., backpack sprayers) and more time for application than PLMs.

Navarro-Llopis et al. (2013) indicated that, for a density of 50 devices per ha, the cost of managing *C. capitata* using bait stations is nearly half (≈USD$ 133 per ha) the amount needed to manage this pest with mass trapping (≈USD$ 266 per ha), indicating that the former method could be applied to European crops in a cost-effective way (Navarro-Llopis et al. 2013).

8 Conclusions and Future Research Needs

Fruit fly management can no longer be seen as an effort with a narrow view made by a farmer, a crop consultant, or a government agency, to control a pest without regard to the environment and society. Numerous types of bait stations have been developed, most of them in response to various needs, such as the need to protect protein-based bait, feeding stimulant, and toxicant against rainfall and UV light, and the need to avoid applications in sensitive areas, such as natural protected areas and communities in rural and suburban areas.

From our review, it seems clear that in the case of protecting crops in commercial orchards the greatest potential of bait station technology is in areas with low fruit fly population densities. This was demonstrated by Putruele and Mouqués (2007) with *C. capitata* in Argentina and by Piñero et al. (2010) with *B. dorsalis* in papaya orchards in Hawaii. Likewise, in area-wide applications the use of bait stations to suppress high density populations in reservoirs has proven effective. This emphasizes the need to develop new, more powerful and long-lasting attractants that can increase bait station effectiveness for species without effective lures and for areas and crops that harbor high fruit fly population densities. Other areas of research needed to optimize bait stations are the development of effective and environmentally-friendly killing agents and the integration of visual and olfactory cues. Bait stations have the potential of more reliable attraction of the target insect under variable environmental conditions if they are able to exploit the multiple sensory modalities that fruit flies use to locate food, mates, and oviposition sites.

In terms of procedures, the density and deployment patterns of bait stations should be optimized, and evaluation must ultimately be based on fruit infestation levels. Quantification of non-target effects to demonstrate the environmental benefits is another aspect of research that is needed. Conducting side-by-side comparisons of the various bait station types that have been developed in recent years to determine actual effectiveness against multiple fruit fly species in various geographical areas and using standardized methodologies that involve quantification of incidence of fruit infestation would provide valuable information on this environmentally-friendly technology.

Acknowledgements We are most indebted to T. Shelly and R.I. Vargas for their invitation to write this chapter after the unexpected death of Mr. Robert R. Heath. This chapter is dedicated to the memory of RRH, who made invaluable contributions in the development and applications of insect semiochemicals and other attractants for the control of agricultural insect pests.

References

Alemany A, Miranda MA, Castro D, Martin Escorza C (2004) Computer graphic simulation of Mediterranean fruit fly population density changes in a citrus orchard. In: Barnes BN (ed) Proceedings of the 6th international symposium on fruit flies of economic importance. Isteg Scientific Publications, Stellenbosch, pp 61–65

Allen WW (1976) Insecticide treated yellow boards for control of *Dacus oleae* (Gmel.). In: UNDP/SF/FAO Project GRE 69-525, Report on chemical control investigation carried out in Athens, 9 Apr–12 Dec 1976, 59 pp

Averill AL, Reissig WH, Roelofs WL (1988) Specificity of olfactory responses in the tephritid fruit fly, *Rhagoletis pomonella*. Entomol Exp Appl 47:211–222

Barry JD, Polavarapu S, Teixeira LAF (2004) Evaluation of traps and toxicants in an attract-and-kill system for *Rhagoletis mendax* (Diptera: Tephritidae). J Econ Entomol 97:2006–2014

Bostanian NJ, Racette G (2001) Attract and kill, an effective technique to manage apple maggot, *Rhagoletis pomonella* (Diptera: Tephritidae) in high density Quebec apple orchards. Phytoprotection 82:25–34

Bostanian NJ, Vincent C, Chouinard G, Racette G (1999) Managing apple maggot, *Rhagoletis pomonella*, by perimeter trapping. Phytoprotection 80:21–33

Broumas T, Haniotakis G, Liaropoulos C, Tomazou T, Ragoussis N (2002) The efficacy of an improved form of the mass-trapping method, for the control of the olive fruit fly, *Bactrocera oleae* (Gmelin) (Dipt., Tephritidae): pilot-scale feasibility studies. J Appl Entomol 126:217–223

Casagrande E (2009) Magnet® MED: "attract and kill" system for the control of the Mediterranean fruit fly (*Ceratitis capitata*). In: Proceedings, 2nd conference on pheromones, food lures, traps and biological control: alternatives for the 21st century. Murcia, Spain. http://www.feromonasmurcia.es/ingles/citrus.html

Cheng EY, Kao CH, Chiang MY, Hwang YB (2003) Modernization of oriental fruit fly control in Taiwan: the planning and execution of an area-wide control project. In: Ho CC et al (eds) Proceedings of the workshop on plant protection management for sustainable development: technology and new dimension. The Plant Protection Society of the Republic of China, Taiwan, pp 49–64

Christenson LD (1963) The male-annihilation technique in the control of fruit flies. Adv Chem Ser 41:31–35

Coltell Simon N (2009) M3 baited traps to fight *Ceratitis capitata*. In: Proceedings, 2nd conference on pheromones, food lures, traps and biological control: alternatives for the 21st century. Murcia, Spain. http://www.feromonasmurcia.es/ingles/citrus.html

Dantas L, Andrade J (2007) Evaluation of different bait stations systems for use in Mediterranean fruit fly Sterile Insect Technique (SIT) programmes. In: Development of improved attractants and their integration into fruit fly SIT, IAEA-TECDOC-1574. IAEA, Vienna, pp 67–72

De Camelo LA, Landolt PJ, Zack RS (2007) A kairomone based attract-and-kill system effective against alfalfa looper (Lepidoptera: Noctuidae). J Econ Entomol 100:366–374

de los Santos-Ramos M, Hernandez-Perez R, Cerdá-Subirachs JJ, Nieves-Ordaz F, Torres-Santillán JA, Bello-Rivera A, Leal-García DF (2011) An environmentally friendly alternative (MS2 ®-CeraTrap ®) for control of fruit flies in Mexico. J Food Agric Environ 9:926–927

de los Santos-Ramos M, Bello-Rivera A, Hernandez-Perez R, Leal-García DF (2012) Efectividad de la estación cebo MS2® y atrayente alimenticio CeraTrap® como alter nativa en la captura de moscas de la fruta en Veracruz, México. Interciencia 37:279–283

DeLury NC, Thistlewood H, Routledge R (2009) Phytotoxicity of GF-120® NF Naturalyte® fruit fly bait carrier on sweet cherry (*Prunus avium* L.) foliage. Pest Manag Sci 65:52–59

Dorn S, Piñero JC (2009) How do key tree-fruit pests detect and colonize their hosts: mechanisms and application for IPM. In: Aluja M, Leskey TC, Vincent C (eds) Biorational tree-fruit pest management. CABI Publishers, Wallingford, pp 85–109, 295 p

Duan JJ, Prokopy RJ (1995a) Development of pesticide-treated spheres for controlling apple maggot flies (Diptera: Tephritidae): pesticides and residue-extending agents. J Econ Entomol 88:117–126

Duan JJ, Prokopy RJ (1995b) Control of apple maggot flies (Diptera: Tephritidae) with pesticide-treated red spheres. J Econ Entomol 88:700–707

El-Sayed AM, Suckling DM, Byers JA, Jang EB, Wearing CH (2009) Potential of "lure and kill" in long-term pest management and eradication of invasive species. J Econ Entomol 102:815–835

Epsky ND, Midgarden D, Rendon P, Villatoro D, Heath RR (2012) Efficacy of wax matrix bait stations for Mediterranean fruit flies (Diptera: Tephritidae). J Econ Entomol 105:471–479

Flores S, Camposa S, Villaseñor A, Valleb A, Enkerlin W, Toledo J, Liedo P, Montoya P (2013) Sterile males of *Ceratitis capitata* (Diptera: Tephritidae) as disseminators of *Beauveria bassiana* conidia for IPM strategies. Biocon Sci Technol 23:1186–1198

Foster SP, Harris MO (1997) Behavioral manipulation methods for insect pest management. Annu Rev Entomol 42:123–146

Hamill JE, Liburd OE, Alm SR (2003) Comparison of biodegradable, plastic and wooden imidacloprid-treated spheres for control of *Rhagoletis mendax* (Diptera: Tephritidae) flies. Fla Entomol 86:206–210

Haniotakis G, Kozyrakis M, Fitsakis T, Antonidaki A (1991) An effective mass trapping method for the control of *Dacus oleae* (Diptera: Tephritidae). J Econ Entomol 84:564–569

Heath RR, Lavallee SL, Schnell EQ, Midgarden DG, Epsky ND (2009) Laboratory and field cage studies on female-targeted attract-and-kill bait stations for *Anastrepha suspensa* (Diptera: Tephritidae). Pest Manag Sci 65:672–677

Heath RR, Rendon P, Epsky ND (2013) Insect bait station method and apparatus. Pat Appl 20130212927

Holler T, Gillett J, Sivinski J, Moses A, Mitchell E (2006) Efficacy of the "Mitchell Station", a new bait station for the control of the Caribbean fruit fly, *Anastrepha suspensa* (Loew) (Diptera: Tephritidae). Proc Hawaiian Entomol Soc 38:111–118

Hu XP, Duan JJ, Prokopy RJ (1998) Effects of sugar/flour spheres coated with paint and insecticide on alighting female *Ceratitis capitata* (Diptera: Tephritidae) flies. Fla Entomol 81:318–325

Hu XP, Prokopy RJ, Clark JM (2000) Toxicity and residual effectiveness of insecticides on insecticide-treated spheres for controlling females of *Rhagoletis pomonella* (Diptera: Tephritidae). J Econ Entomol 93:403–411

IAEA (International Atomic Energy Agency) (2007) Development of improved attractants and their integration into fruit fly SIT management programmes, IAEA-TECDOC-1574. IAEA, Vienna

IAEA (International Atomic Energy Agency) (2009) Development of bait stations for fruit fly suppression in support of SIT. Report and recommendations of a consultants group meeting, Mazatlan, Mexico, 30 October–1 November 2008. IAEA-314.D4 08CT11588, Vienna, Austria

Kring JB (1970) Red spheres and yellow panels combined to attract apple maggot flies. J Econ Entomol 63:466–469

Liburd OE, Gut LJ, Stelinski LL, Whalon ME, McGuire MR, Wise JC, Willett JL, Hu XP, Prokopy RJ (1999) Mortality of *Rhagoletis* species encountering pesticide-treated spheres (Diptera: Tephritidae). J Econ Entomol 92:1151–1156

Liburd OE, Holler TC, Moses AL (2004) Toxicity of imidacloprid-treated spheres to Caribbean fruit fly, *Anastrepha suspensa* (Diptera: Tephritidae) and its parasitoid *Diachasinimorpha longicaudata* (Hymenoptera: Braconidae) in the laboratory. J Econ Entomol 97:525–529

Mangan RL (2009) Effects of bait age and prior protein feeding on cumulative time-dependent mortality of *Anastrepha ludens* (Diptera: Tephritidae) exposed to GF-120 spinosad baits. J Econ Entomol 102:1157–1163

Mangan RL, Moreno DS (2007) Development of bait stations for fruit fly population suppression. J Econ Entomol 100:440–450

Mangan RL, Moreno DS, Thompson GD (2006) Bait dilution, spinosad concentration, and efficacy of GF-120 based fruit fly sprays. Crop Prot 25:125–133

Mas E, Gonzalez F (2009) Adress, a long-lasting safe protection against the Mediterranean fruit fly (*Ceratitis capitata*) and an updated vision on the biology of this fly. In: Proceedings, 2nd conference on pheromones, food lures, traps and biological control: alternatives for the 21st century. Murcia, Spain. http://www.feromonasmurcia.es/ingles/citrus.html

Mau RFL, Jang EB, Vargas RI (2007) The Hawaii fruit fly area-wide fruit fly pest management programme: influence of partnership and a good education programme. In: Vreysen MJB, Robinson AS, Hendrichs J (eds) Area-wide control of insect pests: from research to field implementation. Springer, Dordrect, pp 671–683

Mazomenos BE, Pantazi-Mazomenou A, Stefanor D (2002) Attract and kill of the olive fruit fly *Bactrocera oleae* in Greece as a part of an integrated control system. IOBC-WPRS Bull 25:137–147. http://phero.net/iobc/samos/bulletin/mazomenos.pdf

Muñoz-Pallares J, Corma A, Primo J, Primo-Yufera E (2001) Zeolites as pheromone dispensers. J Agric Food Chem 49:4801–4807

Nadel H, Kraaijeveld AR, Nadel DJ (1989) Non-polluting polyurethane system for slow-release of lure and insecticide to control Tephritids: the developmental phase with field experiments using olive fruit fly, *Dacus oleae* (Gmelin). In: Cavalloro R (ed) Fruit flies of economic importance 87: proceedings of the CEC/IOBC international symposium, Rome 7–10, April 1987. A.A. Balkema, Rottendam, pp 461–485

Navarro-Llopis V, Sanchis-Cabanes J, Ayala I, Casana-Giner V, Primo-Yufera E (2004) Efficacy of lufenuron as chemosterilant against *Ceratitis capitata* in field trials. Pest Manag Sci 60:914–920

Navarro-Llopis V, Sanchis J, Primo-Millo J, Primo-Yufera E (2007) Chemosterilants as control agents of *Ceratitis capitata* (Diptera: Tephritidae) in field trials. Bull Entomol Res 97:359–368

Navarro-Llopis V, Domínguez-Ruiz J, Zarzo M, Alfaro C, Primo J (2010) Mediterranean fruit fly suppression using chemosterilants for areawide integrated pest management. Pest Manag Sci 66:511–519

Navarro-Llopis V, Vacas S, Sanchis J, Primo J, Alfaro C (2011) Chemosterilant bait stations coupled with sterile insect technique: an integrated strategy to control the Mediterranean fruit fly (Diptera: Tephritidae). J Econ Entomol 104:1647–1655

Navarro-Llopis V, Primo J, Vacas S (2013) Efficacy of attract-and-kill devices for the control of *Ceratitis capitata*. Pest Manag Sci 69:478–482

Petacchi R, Rizzi I, Guidotti D (2003) The 'lure and kill' technique in *Bactrocera oleae* (Gmel.) control: effectiveness indices and suitability of the technique in area-wide experimental trials. Int J Pest Manag 49:305–311

Piñero JC, Jácome I, Vargas RI, Prokopy RJ (2006) Response of female melon fly, *Bactrocera cucurbitae*, to host-associated visual and olfactory stimuli. Entomol Exp Appl 121:261–269

Piñero JC, Mau RFL, McQuate GT, Vargas RI (2009a) Novel bait stations for attract-and-kill of pestiferous fruit flies. Entomol Exp Appl 133:208–216

Piñero JC, Mau RFL, Vargas RI (2009b) Managing oriental fruit fly, *Bactrocera dorsalis* (Diptera: Tephritidae), through spinosad-based protein bait sprays and sanitation in papaya orchards in Hawaii. J Econ Entomol 102:1123–1132

Piñero JC, Mau RFL, Vargas RI (2010) Comparison of rain-fast bait stations versus foliar bait sprays for control of oriental fruit fly, *Bactrocera dorsalis* (Hendel), in papaya orchards in Hawaii. J Insect Sci 10:157. Available online: insectscience.org/10.157

Piñero JC, Souder SK, Gomez LE, Mau RFL, Vargas RI (2011a) Response of female Mediterranean fruit fly, *Ceratitis capitata* (Wiedemann), (Diptera: Tephritidae) to a spinosad bait/polymer matrix mixture with extended residual effect in Hawaii. J Econ Entomol 104:1856–1863

Piñero JC, Mau RFL, Vargas RI (2011b) A comparative assessment of the response of three fruit fly species (Diptera: Tephritidae) to a spinosad-based bait: Effect of ammonium acetate, female age, and protein hunger. Bull Entomol Res 101:373–381

Programa Moscamed (1984) Programa Mosca del Mediterráneo. Informe Anual 1984. DGSV-SARH. Talleres Gráficos de la Nación. México, 156p

Programa Moscamed (2010) Manual de procedimientos de estaciones cebo para su uso en el Programa regional Moscamed. MAGA, USDA, SAGARPA. Guatemala, Centro América, p 8

Programa Moscamed (2012) Memoria Anual de Labores. MAGA, USDA, SAGARPA. Guatemala, Centro América, p 36

Prokopy RJ (1968) Visual responses of apple maggot flies, *Rhagoletis pomonella*: orchard studies. Entomol Exp Appl 11:403–422

Prokopy RJ (1972) Response of apple maggot flies to rectangles of different colors and shades. Environ Entomol 1:720–726

Prokopy R (1975) Apple maggot control by sticky red spheres. J Econ Entomol 68:197–198

Prokopy RJ, Moericke V, Bush GL (1973) Attraction of apple maggot flies to odor of apples. Environ Entomol 2:743–749

Prokopy RJ, Owens ED (1978) Visual generalist with visual specialist phytophagous insects – host selection behavior and application to management. Entomol Exp Appl 24:609–620

Prokopy RJ, Christie M, Johnson SA, O'Brien MT (1990) Transitional steps toward second-stage integrated management of arthropod pests of apple in Massachusetts orchards. J Econ Entomol 83:2405–2410

Prokopy RJ, Drew RAI, Sabine BNE, Lloyd AC, Hamacek E (1991) Effect of physiological and experiential state of *Bactrocera tryoni* flies on intra-tree foraging behavior for food (bacteria) and host fruit. Oecologia 87:394–400

Prokopy RJ, Mason JL, Christie M, Wright SE (1996) Arthropod pest and natural enemy abundance under second-level versus first-level integrated pest management practices in apple orchards: a 4-year study. Agric Ecosyst Environ 57:35–47

Prokopy RJ, Jácome I, Piñero J, Guillen L, Diaz-Fleischer F, Aluja M (2000) Post-alighting responses of Mexican fruit flies (Diptera: Tephritidae) to different insecticides in paint on attractive spheres. J Appl Entomol 124:239–244

Prokopy RJ, Chandler BW, Wright SE (2003) Improved sugar delivery onto pesticide-treated spheres for controlling *Rhagoletis pomonella* (Diptera: Tephritidae). Can Entomol 135:909–918

Prokopy RJ, Jácome I, Bigurra E (2005) An index for assigning distances between odor-baited spheres on perimeter trees of orchards for control of apple maggot flies. Entomol Exp Appl 115:371–377

Putruele G, Mouqués J (2007) Evaluation of bait station against chemical control for Mediterranean fruit fly *Ceratitis capitata* Wied. in Argentina. In: Development of improved attractants and their integration into fruit fly SIT, IAEA-TECDOC-1574. IAEA, Vienna, pp 125–127

Roessler Y (1989) Insecticidal bait and cover sprays. In: Robinson AS, Hooper G (eds) Fruit flies, their biology, natural enemies and control, vol 3B. Elsevier Science Publishers, Amsterdam, pp 329–335

Steiner LF (1955) Bait sprays for fruit fly control. Proc Hawaiian Entomol Soc 5:601–607

Stelinski LL, Liburd OE (2001) Evaluation of various deployment strategies of imidacloprid-treated spheres in highbush blueberries for control of *Rhagoletis mendax* Curran (Diptera: Tephritidae). J Econ Entomol 94:905–910

Tsolakis H, Ragusa E, Tarantino P (2011) Control of *Bactrocera oleae* by low environmental impact methods: NPC methodology to evaluate the efficacy of lure-and-kill method and copper hydroxide treatments. Bull Insectol 64:1–8

Vargas RI, Stark JD, Kido MH, Ketter HM, Whitehand LC (2000) Methyl eugenol and cue-lure traps for suppression of male oriental fruit flies and melon flies (Diptera: Tephritidae) in Hawaii: effects of lure mixtures and weathering. J Econ Entomol 93:81–87

Vargas RI, Peck SL, McQuate GT, Jackson CG, Stark JD, Armstrong JW (2001) Potential for areawide integrated management of Mediterranean fruit fly (Diptera: Tephritidae) with a braconid parasitoid and a novel bait spray. J Econ Entomol 94:817–825

Vargas RI, Mau RFL, Jang EB, Faust RM, Wong L (2008) The Hawaii fruit fly area-wide pest management program. In: Koul O, Cuperus GW, Elliott NC (eds) Area wide pest management: theory to implementation. CABI Books, London, pp 300–325

Vargas RI, Mau RFL, Stark JD, Piñero JC (2010a) Evaluation of methyl eugenol and cue-lure traps with solid lure and insecticide dispensers for monitoring and male annihilation in Hawaii. J Econ Entomol 103:409–415

Vargas RI, Piñero JC, Jang EB, Mau RFL, Stark JD, Gomez L, Stoltman L, Mafra-Neto A (2010b) Response of melon fly (Diptera: Tephritidae) to weathered SPLAT-Spinosad-Cue-Lure. J Econ Entomol 103:1594–1602

Vargas RI, Shelly TE, Leblanc L, Piñero JC (2010c) Recent advances in methyl eugenol and cue-lure technologies for fruit fly detection, monitoring, and control in Hawaii. In: Litwack G (ed) Vitamins and hormones, vol 83. Academic, Burlington, pp 575–596

Vincent C, Hallman G, Panneton B, Fleurat-Lessard F (2003) Management of agricultural insects with physical control methods. Annu Rev Entomol 48:261–282

Volakakis NG, Eyre MD, Kabourakis EM (2012) Olive fly *Bactrocera oleae* (Diptera, Tephritidae) activity and fruit infestation under mass trapping in an organic table olive orchard in Crete. Greece J Sustain Agric 36:683–698

Ware T, Richards G, Daneel JH (2003) The M3 bait station: a novel method of fruit fly control. South African Fruit J 2:44–47

Wright SE, Leskey TC, Jacome I, Piñero JC, Prokopy RJ (2012) Integration of insecticidal, phagostimulatory, and visual elements of an attract and kill system for apple maggot fly (Diptera: Tephritidae). J Econ Entomol 105:1548–1556

Zhang A, Linn C, Wright S, Prokopy R, Reissig W, Roelofs W (1999) Identification of a new blend of volatiles attractive to the apple maggot, *Rhagoletis pomonella*. J Chem Ecol 25:1221–1232

Chapter 14
Male Annihilation, Past, Present, and Future

Roger I. Vargas, Luc Leblanc, Jaime C. Piñero, and Kevin M. Hoffman

Abstract We review past use of the male lures, methyl eugenol (ME), cue-lure (CL)/raspberry ketone (RK) and trimedlure (TML), for eradication and suppression of invasive fruit flies (Diptera: Tephritidae), primarily on islands. In addition, we describe the more recent application of these attractants, during the last 25 years, to Area-Wide Integrated Pest Management (AW-IPM) programs and their current use in eradication programs on the U.S mainland (i.e., California and Florida). Finally, we summarize future trends for their application, such as the use of reduced risk insecticides, new lures, lure mixtures, and new dispenser formulations.

Keywords Tephritid flies • MAT • Male lures • Methyl eugenol • Cue-lure/raspberry ketone • Trimedlure • Eradication • Suppression

---

R.I. Vargas (✉)
United States Department of Agriculture, Agricultural Research Service, Daniel K. Inouye U. S. Pacific Basin Agricultural Research Center, 64 Nowelo St., Hilo, HI 96720, USA
e-mail: roger.vargas@ars.usda.gov

L. Leblanc
Department of Plant and Environmental Protection Sciences, College of Tropical Agriculture and Human Resources, University of Hawaii, 3050 Maile Way, Room 310, Honolulu, HI 96822-2271, USA
e-mail: leblancl@ctahr.hawaii.edu

J.C. Piñero
Cooperative Research and Extension, Lincoln University of Missouri, Allen Hall 212, 900 Chestnut St., Jefferson City, MO 65102, USA
e-mail: pineroj@lincolnu.edu

K.M. Hoffman
California Department of Food and Agriculture, Pest Detection/Emergency Projects, 1220 N Street, Suite 315, Sacramento, CA 95814, USA
e-mail: kevin.hoffman@cdfa.ca.gov

T. Shelly et al. (eds.), *Trapping and the Detection, Control, and Regulation of Tephritid Fruit Flies*, DOI 10.1007/978-94-017-9193-9_14,

1 Fruit Flies and Economic Importance

True fruit flies (Diptera: Tephritidae) include over 4,000 species and some of the most economically important pests attacking soft fruits worldwide (White and Elson-Harris 1992). From an economic perspective, they (1) inflict extensive direct damage to fruits and fleshy vegetables, (2) cause quarantine restrictions on infested areas, (3) require commercial fruits to undergo protective and postharvest treatment prior to export, and (4) provide a breeding reservoir for their introduction into other parts of the world (Vargas et al. 2010c). Among the most notorious members of the family Tephritidae are the Mediterranean fruit fly, *Ceratitis capitata* (Wiedemann), oriental fruit fly, *Bactrocera dorsalis* (Hendel) and its close relatives, Queensland fruit fly, *Bactrocera tryoni* (Froggatt), peach fruit fly, *Bactrocera zonata* (Saunders), melon fly, *Bactrocera cucurbitae* (Coquillett), olive fly, *Bactrocera oleae* (Rossi), South American fruit fly, *Anastrepha fraterculus* (Wiedemann), Mexican fruit fly, *Anastrepha ludens* (Loew), West Indian fruit fly, *Anastrepa obliqua* (Macquart), and apple maggot fly, *Rhagoletis pomonella* (Walsh) (Carey and Dowell 1989; Metcalf and Metcalf 1992).

The recent alarming spread of fruit flies worldwide (e. g., *Bactrocera invadens* Drew, Tsuruta, and White into Africa and carambola fruit fly, *Bactrocera carambolae* Drew and Hancock, into South America) can be attributed to increased: (1) production of fruits and vegetables worldwide, (2) global trade of fruits and vegetables between countries, (3) movement of plants between countries, (4) movement of people between nearby countries, and (5) air travel with baggage containing infested fruits. In the USA, the largest fruit-producing state is California. California accounts for over half of the harvested fruit acreage (CDFA 2013). Eradication treatments against exotic fruit flies that are accidentally introduced from various parts of the world into California are very costly. For example, due to continuous introductions, current annual costs to exclude *C. capitata* from California total over $15 million (CDFA http://www.cdfa.ca.gov). Annual introductions of *Bactrocera* spp. often result in temporary trade restrictions and associated area-wide eradication treatments, lasting up to 9 months before movement of agricultural commodities can resume without postharvest treatments (CDFA http://www.cdfa.ca.gov).

Bactrocera is a tephritid fly genus of more than 655 species distributed primarily in tropical Asia, the south Pacific, and Australia (White and Elson-Harris 1992; Dacine Fruit Flies of the Asia-Pacific website 2012). Relatively few species have existed in Africa until the recent introduction of several major pest species (*B. cucurbitae*, *B. zonata*, *B. invadens*, and *Bactrocera latifrons* (Hendel)). At least 407 species of the male Dacini (comprised of the two major genera *Bactrocera* Macquart and *Dacus* Fabricius) are attracted to cue-lure (CL) (4-(p-acetoxyphenyl)-2-butanone)/raspberry ketone (RK) (4-(p-hydroxyphenyl)-2-butanone) and 129 species to methyl eugenol (ME) (4-allyl-1, 2-dimethoxybenzene-carboxylate) (Dacine Fruit Flies of the Asia-Pacific website 2012). Of the 86 Dacini species that are agricultural pests, 41 respond to CL/RK and 18 to ME (Dacine Fruit Flies of the Asia-Pacific website 2012). Based on this attraction, detection and monitoring traps and the suppression/eradication technique called male annihilation technique (MAT) were developed using these chemicals.

Metcalf and Metcalf (1992) reviewed the chemistry and role of plant kairomones in fruit fly ecology and their application for control. Cunningham (1989) reviewed the early development of male annihilation through the 1980s. Vargas et al. (2010c) reviewed recent advances in ME and CL technologies for detection, monitoring, and control in Hawaii, their influence on male behavior, and the environmental impact of their application. In this chapter, we review past use of ME and CL for eradication and suppression of invasive fruit flies. In addition, we describe the more recent application of these attractants, during the last 25 years, to Area-Wide Pest Integrated Management (AW-IPM) programs and their current use in eradication programs on the U.S mainland (i.e., California and Florida). Finally, we summarize future trends for their application, such as the use of reduced risk insecticides, new lures, lure mixtures, and new dispenser formulations.

2 Pheromones, Parapheromones, and Parakairomones

Pheromones are chemical compounds secreted by animals, such as insects, that mediate behavior of another animal belonging to the same species (Karlson and Butenandt 1959). As defined by Renou and Guerrero (2000), parapheromones are "*chemical compounds of anthropogenic origin, not known to exist in nature but structurally related to some natural pheromone components that in some way affect physiologically or behaviorally the insect pheromone communication system.*" These chemicals have shown a large variety of effects, and accordingly have been called pheromone mimics, synergists or else pheromone antagonists, anti-pheromones and inhibitors. Nonetheless, males of many tephritid species are strongly attracted to specific chemical compounds, which either occur naturally in plants (e.g. ME and RK) or are synthetic analogues of plant-borne substances (e.g., CL) (Cunningham 1989; Fletcher 1987). Primarily on the basis of synthetic origin, although knowing, contrary to the opinion of Renou and Guerrero (2000), "are known to exist in nature," Cunningham (1989) referred to the male lures used for detection and MAT, as "parapheromones." More about the "inappropriateness" of the term "parapheromone" is found in the first chapter of this book; however, much of the disagreement in terminology maybe discipline-based (i.e., chemist or ethologist).

Host plant odors acting as kairomones, which are chemicals produced by an organism that benefit an individual of a different species, including ME and RK, are particularly significant in the ecology of fruit flies. For instance, kairomones can naturally attract animals to plants that are scattered throughout dense tropical forests, consequently the more appropriate term than "parapheromone," "parakairomone" is used in some references (Metcalf and Metcalf 1992; Metcalf 1990) for synthetic analogues of these compounds. Some of these kairomone responses have been used for taxonomic, evolutionary, zoogeographical, and behavioral implications, such as their correlation with systematic classification based on morphological characteristics (Drew 1974; Drew and Hooper 1981). Nonetheless, male lures are used for detection and monitoring of tephritid pests and to control or suppress populations through Male Annihilation Technique (MAT) (Vargas et al. 2008a).

3 ME, CL, RK, and TML Attraction

The most commonly used tephritid male lures for detection are ME, CL/RK, and trimedlure (TML). Their chemical structures are provided in Fig. 14.1. ME is a widely distributed natural plant product that occurs in >450 plant species in 80 families found mainly in the tropics (Tan and Nishida 2012). For example, clove oil contains approximately 15 % ME and is generally recognized as a safe food additive (GRAS category) compound by the U.S. Food and Drug Administration (USEPA http://www.epa.gov). CL has not been isolated as a natural product but is rapidly hydrolyzed to form 4-(p-hydroxyphenyl)-2-butanone, rheosimin, or raspberry ketone (RK), a constituent of raspberries (*Rubus idaeus* L. and *R. strigosus* Michx.) with a raspberry-like odor (Metcalf and Metcalf 1992). RK was originally isolated from an orchid, *Dendrobium superbum* Rchb. F (Nishida et al. 1993). Nonetheless, virtually all of the lures for *B. cucurbitae* and other CL responding species are based on lure hydrolysis leading back to RK (Metcalf and Metcalf 1992), and of these, CL (the acetate of RK) has been the most widely used over the past 50 years (Oliver et al. 2002; Jang et al. 2007).

Ceratitis is a genus of about 65 species found in tropical and southern Africa with numerous pest species (White and Elson-Harris 1992). One species, *C. capitata,* has spread to almost all tropical and warm temperate areas of the world. TML (*tert*-butyl 4- and 5-chloro-*cis*- and *trans*-2-methylcyclohexane-1-carboxylate), a mixture of eight isomers, is attractive to many male *Ceratitis* species (e.g., *C. rosa* Karsch and *C. capitata*) and is commonly used in detection traps worldwide (Warthen et al. 1993).

4 MAT for Fruit Fly Eradication

MAT involves the use of a high density of various dispensers (Vargas et al. 2003) or traps (Vargas et al. 2000, 2003, 2010a) baited with a male lure combined with an insecticide, to reduce the male population of fruit flies to such a low level that eradication (Steiner and Lee 1955; Steiner et al. 1965) or suppression (UHCTAHR 2002; Vargas et al. 2010a) is achieved. More general uses of "bait stations" (Piñero et al., Chap. 13) and "mass trapping" (Navarro-Llopis et al., Chap. 15) are covered in other chapters of this book. MAT has been used with varying degrees of success to eradicate invasive *Bactrocera* species in insular and continental countries

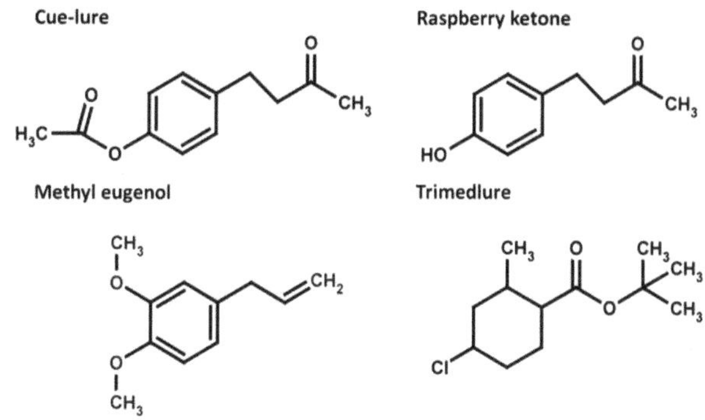

Fig. 14.1 Chemical structure of Cue-lure (CL), raspberry ketone (RK), Methyl eugenol (ME) and trimedlure (TML)

(Table 14.1). In all cases, eradication was attempted or achieved through repeated area-wide applications, by airdrops and/or manual ground applications, of bait stations (fiberboard or coconut husk blocks, cotton string or wick, or molded paper pulp) impregnated with male lures mixed with a toxicant (naled, malathion, or fipronil). MAT was frequently combined with weekly protein bait sprays in hotspots with fly breeding and, in some cases, crop sanitation as removal and destruction of ripe and fallen fruit (Piñero et al. 2009).

Following the successful initial suppression experiments in Hawaii (Steiner and Lee 1955), the first fruit fly eradication using MAT was achieved with *B. dorsalis* in the Mariana Islands (1962–1965) (Steiner et al. 1965). MAT alone was initially used on Rota; however, sterile insect releases, successful alone eradicating *B. dorsalis* from Guam, had to be augmented with MAT to achieve eradication on the other islands after 1 year of releases (Steiner et al. 1970). In an ambitious larger-scale program, the same species was eradicated from the entire Ryuku Island chain, providing a solid barrier against *B. dorsalis* introduction to the main islands of Japan (Ushio et al. 1982; Koyama et al. 1984; Nakamori et al. 1991). While the above two programs targeted long established fruit fly populations, MAT was subsequently applied to eradicate more recent invasions of *B. dorsalis* and its closely related ME-responding species. Two years of efforts over a large area were needed to eradicate *B. papayae* Drew and Hancock from Northern Queensland, because insufficient trapping surveillance failed to detect the early stage of invasion (Hancock et al. 2000; Cantrell et al. 2002). Lloyd et al. (1998) found that ME and malathion (maldison) efficacy with fiberboard blocks followed an exponential curve over 52 weeks. After 8 weeks of exposure to weather, efficacy of blocks was reduced by 50 % in comparison with a new block, and the ME content was reduced by 73 %. Malathion content of blocks did not change over 28 weeks, with a small loss over 52 weeks. Improved vigilance subsequently resulted in quick detection and eradication of *B. philippinensis* Drew and Hancock in the Northern Territory of Australia and *B. dorsalis* in Mauritius (Smith 2000; Seewooruthun et al. 2000a, b). In another example, *B. dorsalis* was nearly eradicated from French Polynesia after six MAT applications, but recovered from untreated residual pockets of breeding populations. It is believed that eight applications would have

Table 14.1 History of male annihilation programs worldwide

Island/country	Dispenser	Mixture	Application mode	Area treated	Application rate per surface area	Treatment frequency	Results	Reference
Mariana Islands (*B. dorsalis*) by USDA-ARS								
Rota (1962–63)	Fiberboard blocks	97 % ME, 3 % naled; 24 g per station	Airdrops and ground (in villages)	85 Km^2	324 stations/Km^2 (airdrop) and 104/Km^2 (ground)	Airdrops every 2 weeks and ground stations re-treated monthly	Eradication: 5 months to last fly capture	Steiner et al. (1965)
Saipan, Tinian, Agiguan (1965)	Fiberboard blocks	95 % ME, 5 % naled	Airdrops	230 Km^2	85–230 stations/Km^2	2 weeks	Eradication: 4 months to last fly capture on Saipan, 9 on Tinian	Steiner et al. (1970)
Ryuku Islands (*B. dorsalis*) by the Okinawa Prefectural Government								
Amami Group (1968–79)	Fiberboard blocks	97 % ME, 3 % naled	Airdrops and ground	1,239 Km^2	22–27.5 g lure-toxicant/ha (airdrops) and 20–202 g/ha (ground)	**Unknown**	Eradication: 4 months to last fly capture on Kikai-Jima and up to 5 years on the other islands	Ushio et al. (1982)
Okinawa Group (1977–82)	Fiberboard blocks[a]	67.5–80 % ME, 3.5–5 % naled, 15–29 % solvent; 10 g per station	Airdrops (90 % of area) and ground (10 %)	1,438 Km^2	3.3–15 g lure-toxicant/ha (airdrops) and 7.5–30 g/ha (ground)	**Unknown**	Eradication: 5 years to last fly capture	Koyoma et al. (1984)

Miyako and Yaeyama Groups (1982–85)	Fiberboard blocks	67.5–80 % ME, 3.5–5 % naled, 15–29 % solvent; 10 g per station	Airdrops and ground (residential areas)	812 Km^2	At least 2g lure-toxicant/ha in airdrops	6–17 airdrops and 1–11 ground applications per year, depending on island	Eradication: 14 months to last fly capture on Miyako and 28 months on Yaeyama	Nakamori et al. (1991)
Easter Island (*B. tryoni*) by the Chile Ministry of Agriculture								
Easter Island (1972)	Cotton string	Cue-lure and malathion	Airdrops and ground (town) + bait sprays	164 Km^2	30 stations/ha over standing vegetation	One MAT application	Eradication: 5 weeks to last fly capture	Bateman et al. (1973)
Australia (*B. papayae* in Cairns and *B. philippinensis* in Darwin) by the Departments of Primary Industries								
Cairns, Queensland (1995–97)	Fiberboard blocks	75 % methyl eugenol, 25 % malathion; 18–20 ml per station	Ground + bait sprays	**Unknown**	400–600 stations/Km^2	8 weeks	Eradication: 20 months to last fly capture	Hancock et al. (2000), Cantrell et al. (2002)
Darwin, Northern Territory (1997–98)	Fiberboard blocks	Methyl eugenol and malathion	Ground + bait sprays and crop sanitation	60 Km^2	600 stations/Km^2	6–7 weeks	Eradication: 36 days to last fly capture	Smith (2000)
Mauritius Island (*B. dorsalis*) by the Mauritius Ministry of Agriculture								
Mauritius (1996–97)	Fiberboard blocks	85 % methyl eugenol, 15 % malathion; 7 g per station	Airdrops and ground + bait sprays and crop sanitation	300 Km^2	10–14 stations/ha (ground), 3,000 stations over 50 Km^2 (airdrop)	2.5–3 months	Eradication: 11 months to last fly capture	Seewooruthun et al. 2000a; 2000b

(continued)

Table 14.1 (continued)

Island/country	Dispenser	Mixture	Application mode	Area treated	Application rate per surface area	Treatment frequency	Results	Reference
South America (*B. carambolae*) as a regional multi country collaborative project								
Brazil, French Guyana, Guyana, Surinam (1997–2002)	Fiberboard blocks	75 % methyl eugenol, 25 % malathion	Ground + bait sprays	**Unknown**	400–2,000/Km^2, depending on host density	6 weeks	Eradication in Guyana (15 months to last fly capture), Brazil, and western part of Surinam, but not achieved in eastern Surinam and French Guyana	Malavasi (2000), Malavasi et al. (2000)
French Polynesia (*B. dorsalis*) by the Territorial Rural Development Service								
Tahiti and Moorea (1997–2002)	Coconut husk blocks	75 % methyl eugenol, 25 % malathion; 16–20 ml per station	Airdrops and ground + bait sprays	1,180 Km^2	400 stations/Km^2 (ground)	8 weeks	Nearly eradicated after six MAT applications in 1997; program resumed in 1999 and fipronil-based BactroMAT ME introduced in 2000, but eradication failed	Allwood et al. (2001)
Nauru (*B. dorsalis, B. cucurbitae, B. xanthodes, B. frauenfeldi*) by the Secretariat of the Pacific Community								
Nauru (1998–2001)	Fiberboard blocks	99.6 % methyl eugenol (*B. dorsalis, B. xanthodes*) or cue-lure	Ground + bait sprays	21 Km^2	>400 stations/Km^2	8 weeks	Eradication of *B. dorsalis* (2.5 months to last fly capture), *B. cucurbitae*	Allwood et al. (2002)

		(*B. cucurbitae*, *B. frauenfeldi*), 0.4 % fipronil; 10–15 ml per station					(3 months), and *B. xanthodes* (16 months). *B. frauenfeldi* not eradicated and program abandoned after 36 months of MAT	
Cook Islands (*B. tryoni*) by the Cook Islands Ministry of Agriculture								
Rarotonga (2001–2002)	BactroMAT CL	95 % cue-lure, 5 % fipronil; 0.5 g per station	Three ground applications and one air-drop + bait sprays and crop sanitation	8 Km^2	310–360 stations/Km^2 (ground); 440/Km^2 (airdrop)	8 weeks	Eradication: 2.5 months to last fly capture	Allwood (2002)

[a]Cotton rope and rolls initially used as dispensers in Okinawa during the first year, but ineffective and replaced by fiberboard blocks

been sufficient to achieve eradication (Leblanc et al. 2013). Unsuccessful subsequent MAT attempts and reduced commitment to the waning program led to abandonment of eradication in favor of introducing parasitoids from Hawaii, which resulted in population decreases to levels more manageable through area-wide control (Vargas et al. 2007). Similarly, *B. carambolae* was eradicated from Guyana, Brazil, and most of Surinam, but the breeding populations in eastern Surinam and French Guyana proved very hard to handle. Hence, complete eradication was not achieved, and the option of introducing parasitoids from Hawaii into Brazil is being considered (Malavasi 2000; Malavasi et al. 2000). An eradication program targeting four species on the Pacific Island of Nauru, implemented by the Regional Fruit Fly Project in the Pacific, was used as training grounds for plant protection and quarantine officers from all the Pacific Islands, who subsequently improved their surveillance trap networks and formulated emergency response plans for their own countries. Two species, *B. dorsalis* and *B. cucurbitae* were quickly eradicated (Allwood et al. 2002). The Pacific fruit fly, *B. xanthodes* (Broun), proved more challenging, in part due to its relatively weak attraction to methyl eugenol compared to most other *Bactrocera* species (Allwood et al. 2002). Eradication of the CL responding *B. frauenfeldi* (Schiner) was not achieved, despite 3 years of MAT and the switching to BactroMAT units on the third year, in great part due to reduced commitment (Allwood et al. 2002).

Generally, results of MAT programs with CL have not been as spectacular as with ME and often had to be used in combination with other techniques. Nonetheless, there have been many examples of successes. A *B. cucurbitae* population was reduced by 99 % throughout 5.2 km^2 plot for over 7 months in Hawaii with CL (+naled) on fiberboard blocks (Cunningham and Steiner 1972). The same species was successfully suppressed for 5 months in Okinawa with cotton rope soaked in a solution of CL and naled (Taniguchi et al. 1988). *Bactrocera tryoni* was eradicated from Rapa Nui (Easter Island) in the southern Pacific using a combined treatment of 2 g each of CL and malathion on pieces of cotton string and spot spraying with protein–malathion bait spray in the 1970s (Bateman et al. 1973). The positive impact of the Nauru training exercise resulted in the prompt detection and eradication of *B. tryoni* from the Cook Islands and a general improvement of quarantine surveillance in the Pacific Islands (Allwood 2002).

5 Male Annihilation Technique in a Fruit Fly AWPM Suppression Program

In IPM systems where populations are large, MAT is most effective in combination with other fruit fly suppression techniques, as was demonstrated in the Hawaii fruit fly AW-IPM program where other components included: (1) field sanitation, (2) protein bait, (3) Sterile Insect Technique (SIT), and (4) biological control (Vargas et al. 2008a). This program registered new technologies for farmers and homeowners and promoted the use of safer or reduced risk fruit fly protein baits and MAT traps (Vargas et al. 2008a, 2010c) in what became popularly referred to as the 1 (sanitation), 2 (protein bait), 3 (male lure trapping) program for fruit fly control (UHCTAHR 2010; Vargas et al. 2011). MAT involved mass trapping using the male lures ME, CL/RK, and TML with an approved killing agent or, with certain traps, with no insecticide at all. The approach was most successful if applied as an "area-wide" suppression strategy by groups of farmers and homeowners. Male lure traps or other dispensers were deployed in a given area in numbers sufficient to "attract and kill" large numbers of males in the population. The number of MAT traps recommended per ha varied accordingly and ranged from 2 to 4 for ME, 4 to 8 for CL, and 10 to 20 for TML (UHCTAHR 2002) . Remaining males fertilized fewer females, and the population gradually would decline due to a shortage of males throughout the treatment area. Lowering the male population reduced chances of successful reproduction and regeneration by females. Effectiveness of MAT varied with the strength of the lure. ME is considered the most attractive male lure for *B. dorsalis* and related species; CL/RK is reasonably effective in attracting *B. cucurbitae*, *B. tryoni*, and related species; and TML is not very powerful against *C. capitata* and related species (Lance and Gates 1994; Shelly 2013). Lures, especially CL, can last from weeks to months in the field, often outlasting the insecticide included with the dispenser. Still integration with other techniques (i.e., protein bait sprays) can significantly contribute to overall suppression (Piñero et al. 2009).

Before the Hawaii AWPM program, no fruit fly lures were registered for general control of fruit flies in the U.S. They were used only on an emergency registration basis. Vargas et al. (2003) found that spinosad, although not as persistent as naled or malathion, was safer to handle and a more environmentally friendly substitute to organophosphate insecticides in ME and CL traps for use in MAT against *B. dorsalis* and *B. cucurbitae*. Eventually, solid lure and insecticide formulations replaced liquid formulations. In the Hawaii program, a solid formulation with DDVP (Hercon Vaportape TM II (DDVP), Emigsville, PA) strips in place of liquid naled was an improvement from a worker safety viewpoint. An AWPM program that included MAT with CL (+DDVP) in bucket traps in addition to protein baits, SIT, and parasitoid releases reduced a *B. cucurbitae* population to near zero for 1 year throughout a 40 km^2 area (Jang et al. 2008; Vargas et al. 2008a). Amulet CL (BASF, USA) molded paper fiber stations (Vargas et al. 2005) with CL and fipronil, similar to BactroMAT for control of *B. tryoni* in Australia (BASF, USA), was registered in Hawaii.

Placement of many traps or paper stations in the field can be time consuming and is not always ideal for eradication programs. A promising alternative involves spraying male lure (plus a carrier and a reduced risk insecticide) on to existing surfaces in the environment. Recent research, development, and registration has focused on SPLAT^{TM}-MAT-Spinosad-ME (aka STATIC^{TM}-Spinosad-ME) (Dow AgroSciences, Indianapolis, IN and ISCA Technologies, Riverside, CA) (Vargas et al. 2008b) as a replacement for the currently used Min-U-Gel-naled-ME. Min-U-Gel, an anti-splash agent, is a fine grade of attapulgite clay (anhydrous magnesium aluminum silicate) that was developed for spot applications in male annihilation programs in California for eradication of *B. dorsalis* (Chambers et al. 1974; Cunningham and Suda 1985). SPLAT^{TM} (Specialized Pheromone and Lure Application Technology) has a waxy outer coating that acts as a reservoir with time release properties, which allows the lure to last longer when applied to surfaces than Min-U-Gel. SPLAT, like Min-U-Gel, can be sprayed from small sprayers, trucks, and aircraft making the technology convenient and flexible. Current work is also focusing on research and development of a SPLAT-spinosad-CL/RK product for use against CL/RK responding species, such as *B. cucurbitae* and *B. tryoni* (Vargas et al. 2009, 2010b).

6 Recent Male Annihilation Technique Programs in California

The California Department of Food and Agriculture (CDFA) directs continuous programs to detect and eradicate invasive *Bactrocera* fruit flies (CDFA 2000a, b). From 1960 to 2012 nine different *Bactrocera* species have been detected, with B. *dorsalis* detected most frequently (126 times). There have been 140 eradication programs with 25 quarantines (Table 14.2). Most of these programs have occurred in southern California where the usual number of detection traps deployed is 5 separate Jackson Traps (JT) (with TML, ME, or CL) and 5 McPhail Traps (MPT) (Torula Yeast Solution) per mi^2 (2.59 km^2). Total number of sites in operation is approximately 25,000 for the Los Angeles area (IPRFFSP 2006) and over 30,000 for the entire state (Vargas et al. 2010a). When a pest fly (e.g., *B. dorsalis*) is detected, the number of traps in the core area (2.59 km^2), which is normally 5 JT and 5 MPT, is increased to 25 JT and 25 MPT (for illustrations and more information see CDFA 2010). The density of traps in the rest of the 209.8 km^2 (81 mi^2= 9 × 9 grid) is left unchanged. In southern California, the rest of the 209.8 km^2 in the delimitation area already has the required 5 JT and 5 MPT (CDFA 2010). During the 1st week following the find, all traps in the core mi^2 (now 25 JT + 25 MPT), and all the traps in the first surrounding buffer area (5 JT + 5 MPT per 2.59 km^2), are serviced daily for the first 7 days and once a week after that for treatment of the F_1 generation (CDFA 2010). Treatment is initiated only if more than one male fly or one female fly is captured. If new finds occur as a result of delimitation trapping, a new 2.59 km^2 area is drawn around the new finds. The density of traps within the new area is then increased to 25 traps of each trap type per mi^2 if not already at that density. Within the 2.59 km^2 new core and new buffer, daily services start again for a period of 1 week. These increases in trap densities effectively act as a concentrated MAT program around the initial fly finds.

Table 14.2 Male annihilation programs in California from 1960 to end of 2012 summarizing species introduced, initial detection, total detections, counties where detected, number of eradication events, and quarantines initiated

Name	Initial detection	Years detected	Total detections	California counties with # of detections	Eradication events	Quarantines initiated	Adult flies detected (total, range, mean ± stdev)
B. dorsalis	1960	47 of 53	126	Los Angeles (41), Orange (21), San Diego (17), San Bernardino (11), Santa Clara (11), Alameda (5), Sacramento (4), Santa Barbara (3), Riverside(3), Marin (3), San Mateo (2), Ventura (2), Contra Costa (1), Madera (1), San Joaquin (1)	104	18	~1479, 1–535, 13.69 ± 52.50
B. correcta	1986	24 of 27	63	Los Angeles (20), Orange (11), Santa Clara (9), San Diego (7), Alameda (4), Sacramento (3), Fresno (2), Riverside (2), San Mateo (3), Stanislaus (1), Ventura (1)	23	0	140, 1–10, 2.22 ± 1.87
B. zonata	1984	12 of 29	20	Los Angeles (4), Orange (4), Riverside (3), Alameda (2), Sacramento (2), Santa Clara (2), Contra Costa (1), Fresno (1), San Mateo (1)	8	2	61, 1–23, 3.05 ± 4.97
B. cucurbitae	1956	15 of 57	17	Los Angeles (12), Alameda(1), San Bernardino (1), San Diego (1), Kern (1), Fresno (1)	4	4	35, 1–8, 2.06 ± 2.19
B. scutellata	1987	6 of 26	6	Los Angeles (5), Orange (1)	0	0	16, 1–9, 2.67 ± 3.20
B. tryoni	1985	2 of 28	2	Orange (1), San Diego (1)	0	0	2, 1, 1
B. latifrons	1998	1 of 15	1	Los Angeles (1)	0	0	1, 1, 1
B. facialis	1998	1 of 15	1	Los Angeles (1)	0	0	1, 1, 1
B. albistrigata	2009	1 of 4	1	Los Angeles (1)	1	1	8, 8, 8

7 Male Annihilation Technique Treatments in California and Florida

Min-U-Gel with naled was developed as a sprayable formulation applied to telephone poles and tree trunks in California and Florida for the eradication of *B. dorsalis* (Chambers et al. 1974; Cunningham and Suda 1985). In ME eradication projects, a Min-U-Gel-naled-ME mixture is applied within a 2.41 km (1.5 mile) radius of the detection site (CDFA 2000a). Composition of the Min-U-Gel-naled-ME mixture used by CDFA is: ME (3.8 l (1 gal), 78.5 % by volume), naled (Dibrom 14 E Insecticide) (0.56 l (19 ounces), 11.7 %), Min-U-Gel 400 thickening agent (0.9–1.4 kg (1–2 lb), 9.8 %). Treatments are applied by specialized trucks as semi-liquid dollops to street trees and utility poles at 5 ml per site at a minimum height of 1.8 m (6 ft). The target density is a minimum of 600 sites per 2.59 km^2 (1 mi^2) spaced at least 15 m (50 ft) apart. Programs for CL responding flies are similar to programs for ME responding flies (CDFA 2000a, b). The typical area treated is approximately 28.5 km^2 (11 mi^2) but may be larger. Each mi^2 receives 1.1 to 3.0 l (0.3–0.8 gal) of the mixture per application. Re-treatment occurs every 2 weeks and continues for 1–3 life cycles depending on number of flies captured following the first application (CDFA 2000a). The typical duration of eradication projects ranges from 2 to 6 months. Male annihilation is conducted when one of the following detections occurs: (1) one mated female, (2) two flies found within 3 miles of each other within the same life cycle, (3) one larva, or (4) one pupa. Generally, the size of the treatment area is 23.3 km^2 or a 4.8 km radius around find site. In addition to MAT treatments, if larvae are found, plants and trees within the area around the find for a 200 m radius are treated with GF-120 NF Naturalyte Fruit Fly Bait (GF-120 Fruit Fly Bait is a mixture of the toxicant spinosad [Dow AgroSciences, Indianapolis, IN] and a protein-based feeding attractant for control of fruit fly populations [Mangan et al. 2006]). Quarantines are lifted after no fruit fly finds for 3 life cycles determined on the basis of accumulation of "degree days." A similar program has been used in Florida against *B. dorsalis* (Weems and Heppner 2012). In conclusion, MAT programs in California have prevented large outbreaks of *Bactrocera* spp. and allowed for continuous export of agricultural products during the last 60 years.

8 Future Trends

With international trade in fresh fruits and vegetables expanding and human travel increasing, the problem of fruit flies as major quarantine pests is only getting worse and has taken on added importance, triggering the need for improved early detection and control systems. It is expected that new and improved approaches will play an increasing role in development of systems approaches used in pre-harvest treatments to meet quarantine restrictions placed on exports. In the future, more environmentally friendly MAT technologies will be required since infestations continue to occur in urban areas or agricultural-urban interfaces. During the implementation of the Hawaii fruit fly AWPM program, farmers and homeowners constructed one-way traps themselves without the need for insecticides and used solid lures (Vargas et al. 2010c). Enclosing wicks inside bucket traps not only provided protection from the weather (lasting up to 20 weeks) but also made the device visible, retrievable, and reusable with limited environmental contamination and exposure to humans and pets. However, these approaches, although employed during early detections and infestations in California and Florida, can be impractical and expensive when used over large areas. Recent research has addressed replacement of organophosphate insecticides with medium to reduced risk insecticides, such as fipronil (Vargas et al. 2005) and spinosad (Vargas et al. 2008b), respectively, and development of sprayable dispensers with a reduced risk insecticide (Vargas et al. 2008b, 2009, 2010b, c). Safer treatments would allow for the application of more treatment points within an area with little environmental impact. In small scale trials in Hawaii, Vargas et al. (2008b) first demonstrated that attraction of male *B. dorsalis* to the SPLAT-MAT-Spinosad-ME (aka STATIC™-spinosad-ME) treatments equaled or outperformed Min-U-Gel-ME with naled. New MAT approaches also include mixtures of liquid ME or C-L on cotton wicks inside bucket traps (Vargas et al. 2000) or solid dispensers (Vargas et al. 2010a) for suppression of both *B. dorsalis* and *B. cucurbitae*. Current research is also focusing on a new male lure TMR formulation that combines TML, ME and RK into a single dispenser (Vargas et al. 2012). This dispenser is placed inside an escape proof Hiramoto trap (Hiramoto et al. 2006) as "attract and kill" devices without conventional insecticides for suppression and control of three groups of fruit flies (TML, RK, and ME responders) for application as part of IPM programs. Of primary interest would be suppression of *C. capitata* with TML. The attraction of a TML dispenser with a large surface area to *C. capitata* was demonstrated by Vargas et al. (2012) in a large coffee (*Coffea arabica* L) farm to be as effective as ceralure and will be tested further as a suppression device in subsequent studies. Nonetheless, this dispenser within a trap would provide a generic MAT suppression tool for a broad spectrum of economically important fruit flies. Also under research and development are safer or more powerful lures, such as ceralure (Vargas et al. 2012), RK formate (Jang et al. 2007), fluorinated ME (Khrimian et al. 2009), or ginger oil (Shelly 2013) with possible MAT applications. Other information on male lures is presented in Tan et al. (Chap. 2, this volume).

Acknowledgements We thank Steven Souder (Daniel K. Inouye U. S. Pacific Basin Agricultural Research Center, USDA-ARS, Hilo, HI) for assistance in summarizing data. IR-4 (Michael Braverman, IR-4), CDFA (Bob Dowell), Dow AgroSciences (Luis Gomez and Jim Dripps), ISCA Technologies (Agenor Mafra-Neto), and Farmatech International (Peter Cook) were instrumental in funding research on reduced risk products. Juan Mercagliano (CDFA, Van Nuys, CA) provided standard operating procedures summarized in the manuscript. The HAWPM program, a partnership among USDA, the University of Hawaii (Ron Mau, UH, Honolulu, HI), and the Hawaii Department of Agriculture (Lyle Wong, HDOA, Honolulu, HI), provided a research and development platform for rapid technology transfer to farmers and homeowners who validated these technologies and provided the impetus for further development. Opinions, findings, conclusions, or recommendations expressed in this publication are those of the authors and do not necessarily reflect the views of the USDA. Finally, we thank Nick Manoukis (USDA, ARS, DKIUSPBARC) for comments on an earlier draft of this chapter.

References

Allwood AJ (2002) Eradication of Queensland fruit fly (*Bactrocera tryoni* (Froggatt)) in Rarotonga, Cook Islands. Secretariat of the Pacific Community Report. Suva, Fiji

Allwood AJ, Vargas RI, Leblanc L, Bull R (2001) Assessment of the eradication of oriental fruit fly (*Bactrocera dorsalis* (Hendel)) in French Polynesia. Secretariat of the Pacific Community Report. Suva, Fiji

Allwood AJ, Vueti ET, Leblanc L, Bull R (2002) Eradication of introduced *Bactrocera* species (Diptera: Tephritidae) in Nauru using male annihilation and protein bait application techniques. In: Veitch CR, Clout MN (eds) Turning the Tide: the eradication of invasive species. Proceedings of the international conference on eradication of Island invasives, Auckland, 19–23 Feb 2001. IUCN Publications Services Unit, Cambridge, UK, pp 19–25

Bateman MA, Insungza V, Arreta P (1973) The eradication of Queensland fruit fly from Easter Island. FAO Plant Prot Bull 21:114

Cantrell BK, Chadwick B, Cahill A (2002) Fruit fly fighters: eradication of papaya fruit fly. CSIRO, Collingwood

Carey JR, Dowell RV (1989) Exotic fruit fly pests and California agriculture. Calif Agric 43 (3):38–40

[CDFA] California Department of Food and Agriculture. http://www.cdfa.ca.gov. Accessed 1 Apr 2014

[CDFA] California Department of Food and Agriculture (2000a) Action plan for methyl eugenol attracted fruit flies including the oriental fruit fly *Bactrocera dorsalis* (Hendel). State of California, Department of Food and Agriculture, Pest Detection and Emergency Projects, Sacramento. 11 pp

[CDFA] California Department of Food and Agriculture (2000b) Action plan for Cuelure Attracted Fruit Flies including the Melon Fly *Bactrocera cucurbitae* (Coquillett). State of California, Department of Food and Agriculture, Pest Detection and Emergency Projects, Sacramento. 10 pp

[CDFA] California Department of Food and Agriculture (2010) Insect trapping guide, 12th edn. CDFA, Sacramento

[CDFA] California Department of Food and Agriculture (2013) Fruit and nut crops. California Agriculture Statistics Review, Sacramento

Chambers DL, Ohinata K, Fujimoto M, Kashiwai S (1974) Treating tephritids with attractants to enhance their effectiveness in sterile-release programs. J Econ Entomol 65:279–282

Cunningham RT (1989) Parapheromones. In: Robinson AS, Hooper GHS (eds) Fruit flies, their biology, natural enemies and control, vol 3A, World crop pests. Elsevier, Amsterdam, pp 221–230

Cunningham RT, Steiner LF (1972) Field trial of cue-lure + naled on saturated fiberboard blocks for control of the melon fly by the male-annihilation technique. J Econ Entomol 65:505–507

Cunningham RT, Suda DY (1985) Male annihilation of the oriental fruit fly, *Dacus dorsalis* Hendel (Diptera: Tephritidae): a new thickener and extender for methyl eugenol formulations. J Econ Entomol 75:503–504

Dacine Fruit Flies of the Asia-Pacific Website (2012) University of Hawaii at Manoa. http://www.herbarium.hawaii.edu/fruitfly/index.php. Accessed 2 Apr 2014

Drew RAI (1974) The responses of fruit fly species (Diptera: Tephritidae) in the South Pacific area to male attractants. J Aust Entomol Soc 13:267–270

Drew RAI, Hooper GHS (1981) The response of fruit fly species (Diptera: Tephritidae) to various attractants. J Aust Entomol Soc 20:201–208
Fletcher BS (1987) The biology of dacine fruit flies. Annu Rev Entomol 32:115–144
Hancock DL, Osborne R, Broughton S, Gleeson PJ (2000) Eradication of Bactrocera papayae (Diptera: Tephritidae) by male annihilation and protein baiting in Queensland, Australia. In: Tan KH (ed) Area-wide control of fruit flies and other insect pests. Joint proceedings of the international conference on area-wide control of insect pests, 28 May–2 June 1998 and the fifth international symposium on fruit flies of economic importance, 1–5 June 1998, Penang, Malaysia. USM Press, Penang, pp 381–388
Hiramoto MKL, Arita-Tsutsumi L, Jang EB (2006) Test of effectiveness of newly formulated plastic matrix with methyl eugenol for monitoring *Bactrocera dorsalis* (Hendel) populations. Proc Hawaii Entomol Soc 38:103–110
[IPRFFSP] International Panel for Review of Fruit Fly Surveillance Programs (2006) Review of fruit fly surveillance programs in the United States. U.S. Dept. Agric./APHIS/PPQ/Fruit Fly Program, Riverdale
Jang EB, Casaña-Giner V, Oliver JE (2007) Field captures of wild melon fly (Diptera: Tephritidae) with an improved male attractant, raspberry ketone formate. J Econ Entomol 100:1124–1128
Jang EB, McQuate GT, McInnis DO, Harris EJ, Vargas RI, Bautista RC, Mau RFL (2008) Targeted trapping, bait-spray, sanitation, sterile-male, and parasitoid releases in an area wide integrated melon fly (Diptera: Tephritidae) control program in Hawaii. Am Entomol 54:240–250
Karlson P, Butenandt A (1959) Pheromones (ectohormones) in insects. Annu Rev Entomol 4:49–58
Khrimian A, Siderhurst MS, McQuate GT, Liquido NJ, Nagata J, Carvalho L, Guzman F, Jang EB (2009) Ring-fluorinated analog of methyl eugenol: attractiveness to and metablolism in the oriental fruit fly, *Bactrocera dorsalis* (Hendel). J Chem Ecol 35:209–218
Koyoma J, Tadashi T, Kenji T (1984) Eradication of the oriental fruit fly (Diptera: Tephritidae) from Okinawa Islands by male annihilation method. J Econ Entomol 77:468–472
Lance DR, Gates DB (1994) Sensitivity of detection trapping systems for Mediterranean fruit flies (Diptera: Tephritidae) in southern California. J Econ Entomol 77:744–749
Leblanc L, Vargas RI, Putoa R (2013) From eradication to containment: invasion of French Polynesia by *Bactrocera dorsalis* (Hendel) (Diptera: Tephritidae) and releases of two natural enemies: a 17 year case study. Proc Hawaii Entomol Soc 45:31–43
Lloyd A, Leach P, Kopittke R (1998) Effects of exposure on chemical content and efficacy of male annihilation blocks used in the eradication of *Bactrocera papayae* in North Queensland. Gen Appl Entomol 28:1–8
Malavasi A (2000) An eradication programme of the Carambola fruit fly, *Bactrocera carambolae*, in the north of South America. In: Price NS, Seewooruthun I (ed) Proceedings of the Indian Ocean commission regional fruit fly symposium, Flic en Flac, Mauritius, 5–9 June 2000, pp 131–134
Malavasi A, Sauers-Muller A van, Midgarden D, Kellman V, Didelot D, Caplong P, Ribeiro O (2000) Regional programme for the eradication of the Carambola fruit fly in South America. In: Tan KH (ed) Area-wide control of fruit flies and other insect pests. Joint proceedings of the international conference on area-wide control of insect pests, 28 May-2 June 1998 and the fifth international symposium on fruit flies of economic importance, 1–5 June 1998, Penang, Malaysia. USM Press, Penang, pp 395–399
Mangan RL, Moreno DS, Thompson GD (2006) Bait dilution, spinosad concentration, and efficacy of GF-120 based fruit fly sprays. Crop Prot 25:125–133
Metcalf RL (1990) Chemical ecology of Dacinae fruit flies (Diptera: Tephritidae). Ann Entomol Soc Am 83:1017–1030
Metcalf RL, Metcalf ER (1992) Fruit flies of the family Tephritidae. In: Metcalf RL, Metcalf ER (eds) Plant kairomones in insect ecology and control. Routledge/Chapman & Hall Inc., New York, pp 109–152

Nakamori H, Nishimura M, Kakinohana H (1991) Eradication of the oriental fruit fly, *Dacus dorsalis* Hendel (Diptera: Tephiritidae), from Miyako and Yaeyama Islands by the male annihilation method. In: Proceedings of the first international symposium on fruit flies in tropics, Kuala Lumpur, Malaysia, 14–16 March 1988, pp 220–231

Nishida R, Iwahashi O, Tan KH (1993) Accumulation of *Dendrobium superbum* (Orchidaceae) fragrance in the rectal glands by males of the melon fly, *Dacus cucurbitae*. J Chem Ecol 19:713–722

Oliver JE, Casaña-Giner V, Jang EB, McQuate GT, Carvalho L (2002) Improved attractants for melon fly, *B. cucurbitae*. In: Barnes BN (ed) Proceedings of 6th international fruit fly symposium, Stellenbosch, South Africa, pp 283–290

Piñero JC, Mau RFL, Vargas RI (2009) Managing oriental fruit fly (Diptera: Tephritidae) with spinosad-based protein bait sprays and sanitation in papaya orchards in Hawaii. J Econ Entomol 102:1123–1132

Renou M, Guerrero A (2000) Insect parapheromones in olfaction research and semiochemical-based pest control strategies. Annu Rev Entomol 45:605–630

Seewooruthun SI, Permalloo S, Sookar P, Soonnoo AR (2000a) The oriental fruit fly, *Bactrocera dorsalis* eradicated from Mauritius. In: Price NS, Seewooruthun I (eds) Proceedings of the Indian Ocean Commission regional fruit fly symposium, Flic en Flac, Mauritius, 5–9 June 2000, pp 207–210

Seewooruthun SI, Permalloo S, Gungah B, Soonnoo AR, Alleck M (2000b) Eradication of an exotic fruit fly from Mauritius. In: Tan KH (ed) Area-wide control of fruit flies and other insect pests. Joint proceedings of the international conference on area-wide control of insect pests, 28 May-2 June 1998 and the fifth international symposium on fruit flies of economic importance, 1–5 June 1998, Penang, Malaysia. USM Press, Penang, pp 389–394

Shelly TE (2013) Detection of male Mediterranean fruit flies (Diptera: Tephritidae): performance of trimedlure relative to capilure and enriched ginger root oil. Proc Hawaii Entomol Soc 45:1–7

Smith ESC (2000) Detection and eradication of the exotic fruit fly *Bactrocera philippinensis* from Australia. In: Price NS, Seewooruthun I (eds) Proceedings of the Indian Ocean Commission regional fruit fly symposium, Flic en Flac, Mauritius, 5–9 June 2000, pp 201–205

Steiner LF, Lee RKS (1955) Large area tests of a male-annihilation method for oriental fruit fly control. J Econ Entomol 48:311–317

Steiner LF, Mitchell WC, Harris EJ, Kozuma TT, Fujimoto MS (1965) Oriental fruit fly eradication by male annihilation. J Econ Entomol 58:961–964

Steiner LF, Hart WG, Harris EJ, Cunningham RT, Ohinata K, Kamakahi DC (1970) Eradication of the oriental fruit fly from the Mariana Islands by the methods of male annihilation and sterile insect release. J Econ Entomol 63:131–135

Tan KH, Nishida R (2012) Methyl eugenol: its occurrence, distribution, and role in nature, especially in relation to insect behavior and pollination. J Insect Science 12:56, available online: insectscience.org/12.5

Taniguchi M, Nakamori H, Kakinohana H, Yogi Y, Zukeyama H (1988) Suppression of male melon fly, *Dacus cucurbitae* Coquillett, (Diptera: Tephritidae) populations using cotton ropes infiltrated with a lure-toxicant. Appl Entomol Zool 32:126–128

[UHCTAHR] University of Hawaii, College of Tropical Agriculture and Human Resources (2002) Fruit Fly Male Annihilation. HAW-FLYPM. Hawaii Area-Wide Fruit Fly Integrated Pest Management. Information and Training Leaflet. Publications and Information Office, UH-CTAHR

[UHCTAHR] University of Hawaii, College of Tropical Agriculture and Human Resources. (2010) Easy as 1-2-3 Fruit Fly Suppression in Hawaii. Information and Training Leaflet. Publications and Information Office, UH-CTAHR

[USEPA] U.S. Food and Drug Administration (http://www.epa.gov)

Ushio S, Yoshioka K, Nakasu K, Waki K (1982) Eradication of the oriental fruit fly from Amami Islands by male annihilation (Diptera: Tephritidae). Jpn J Appl Entomol Zool 26:1–9

Vargas RI, Stark JD, Kido MH, Ketter HM, Whitehand LC (2000) Methyl eugenol and cue-lure traps for suppression of male oriental fruit flies and melon flies (Diptera: Tephritidae) in Hawaii: effects of lure mixtures and weathering. J Econ Entomol 93:81–87

Vargas RI, Miller N, Stark JD (2003) Field trials of Spinosad as a replacement for naled, ddvp, and malathion in methyl eugenol and cue-lure bucket traps to attract and kill male oriental fruit flies and melon flies (Diptera: Tephritidae) in Hawaii. J Econ Entomol 96:1780–1785

Vargas RI, Stark JD, Mackey B, Bull R (2005) Weathering trials of Amulet Cue-lure and methyl eugenol "attract and kill" stations with male melon flies and oriental fruit flies (Diptera: Tephritidae) in Hawaii. J Econ Entomol 98:1551–1559

Vargas RI, Leblanc L, Putoa R, Eitam A (2007) Impact of introduction of *Bactrocera dorsalis* (Diptera: Tephritidae) and classical biological control releases of *Fopius arisanus* (Hymenoptera: Braconidae) on economically important fruit flies in French Polynesia. J Econ Entomol 100:670–679

Vargas RI, Mau RFL, Jang EB, Faust RM, Wong L (2008a) The Hawaii fruit fly area-wide pest management program. In: Koul O, Cuperus GW, Elliott NC (eds) Area wide pest management: theory to implementation. CABI Books, London, pp 300–325

Vargas RI, Stark JD, Hertlein M, Mafra-Neto A, Coler R, Piñero JC (2008b) Evaluation of SPLAT with spinosad and methyl eugenol or cue-lure for "attract-and-kill" of oriental and melon fruit flies (Diptera: Tephritidae) in Hawaii. J Econ Entomol 101:750–768

Vargas RI, Piñero JC, Mau RFL, Stark JD, Hertlein M, Mafra-Neto A, Coler R, Getchell A (2009) Attraction and mortality of oriental fruit flies (Diptera: Tephritidae) to SPLAT-MAT-methyl eugenol with spinosad. Entomol Exp Appl 131:286–293

Vargas RI, Mau RFL, Stark JD, Piñero JC (2010a) Evaluation of methyl eugenol and cue-lure traps with solid lure and insecticide dispensers for monitoring and male annihilation in Hawaii. J Econ Entomol 103:409–415

Vargas RI, Piñero JC, Jang EB, Mau RFL, Stark JD, Gomez L, Stoltman L, Mafra-Neto A (2010b) Response of melon fly (Diptera: Tephritidae) to weathered SPLAT-spinosad-C-L. J Econ Entomol 103:1594–1602

Vargas RI, Shelly TE, Leblanc L, Piñero JC (2010c) Recent advances in methyl eugenol and cue-lure technologies for fruit fly detection, monitoring and control. In: Litwack G (ed) Vitamins and hormones, section: pheromones, vol 83. Academic, Burlington, pp 575–596

Vargas RI, Mau, RFL, Jang EB, Wong L (2011) Area-wide fruit fly integrated pest management program: a model system. In: Proceedings of the 8th international symposium on fruit flies of economic importance, Valencia, Spain

Vargas RI, Souder S, Mackey B, Cook PJ, Morse JG, Stark JD (2012) Field trials of solid triple lure (trimedlure, methyl eugenol, raspberry ketone, and DDVP) dispensers for detection and male annihilation of *Ceratitis capitata* (Wiedemann), *Bactrocera dorsalis* (Hendel) and *Bactrocera cucurbitae* (Coquillett) (Diptera: Tephritidae) in Hawaii. J Econ Entomol 105:1557–1565

Warthen JD, Schmidt WF, Cunningham RT, Demilo AB, Fritz GI (1993) Quantitative structure-activity relationships (QSAR) of trimedlure isomers. J Chem Ecol 19:1323–1335

Weems HV, Heppner JB (2012) Featured creatures: oriental fruit fly. University of Florida Institute of Food and Agricultural Sciences

White IM, Elson-Harris MM (1992) Fruit flies of economic significance: their identification and bionomics. CAB International, Wallingford

Chapter 15
Mass Trapping for Fruit Fly Control

Vicente Navarro-Llopis and Sandra Vacas

Abstract This chapter reviews the main methods, trapping devices, and effectiveness of the mass trapping technique along with the history of mass trapping applied to fruit flies. Since the success of mass trapping depends heavily on the type of trap and bait used, studies addressing these parameters are summarized. Based on the authors' personal experiences and the most relevant studies, we present an evaluation of the strengths and weaknesses of mass trapping and comment on the implementation strategies that may influence its effectiveness.

Keywords Dry trap • Dry attractant • Liquid attractant • Three component lure • Female attractant • DDVP • Contact insecticide • Visual cues • Trap shape • Clear lid • Lateral holes • Distribution of traps • Perimeter trapping • Trap density • Trap placement • Fruit infestation • Olipe • Tephri-trap • Moskisan • Maxitrap

1 The History of Mass Trapping for Fruit Flies

1.1 First Trap Designs and Uses

The first references to mass trapping applications for tephritid fruit flies date back to the early 1920s. In 1925, Gurney described a trapping system for Australian fruit fly pests that utilized baits of protein and fermenting sugar (cited in Epsky and Heath 1998). In the same decade, Constantino in the Portici Entomological Laboratory (Naples, Italy) described the invaginated glass jar baited with molasses and wine vinegar as a control method for fruit fly populations (Gómez-Clemente 1929). The use of mass trapping with protein and fermenting sugars was studied in Spain

V. Navarro-Llopis (✉) • S. Vacas
Instituto Agroforestal del Mediterráneo-Universidad Politécnica de Valencia, Camino de Vera s/n. Edificio 6C. 5ª planta, 46022 Valencia, Spain
e-mail: vinallo@ceqa.upv.es; sanvagon@ceqa.upv.es

T. Shelly et al. (eds.), *Trapping and the Detection, Control, and Regulation of Tephritid Fruit Flies*, DOI 10.1007/978-94-017-9193-9_15,

Picture 15.1 "Cazamoscas". Early designs of McPhail traps

between 1923 and 1925 (Gómez-Clemente 1929; Planes 1936, 1959), with the "cazamoscas" trap (Picture 15.1). This trap was a bell-shaped, invaginated, clear glass trap with a water reservoir. Newell (1936) was apparently the first to refer to this bell-shaped trap as the McPhail trap, although Dahl described this type of trap for flying insects 40 years earlier in Germany (described in Steyskal 1977). A poster (Picture 15.2) of the Valencia Exportation Association (Unión Nacional de la Exportación Agrícola, 1924) included the recommendation to use glass traps (named "mosquero Estación Fitopatológica de Valencia") against fruit flies by placing one trap per tree baited with acetic acid, peach fruit, or molasses. Moreover, the same publication mentioned the sanitation practices that should be taken, such as the removal and burial of damaged fruit treated with an insecticide. One of the first examples of mass trapping against the Mexican fruit fly (*Anastrepha ludens* (Loew)) was reported by Balock and López (1969), who used McPhail traps deployed at 110 traps/ha and baited with cotton seed hydrolysate to protect mango and citrus groves.

Later, new trap designs were developed, such as the Steiner (Steiner 1957; Picture 15.3a), Jackson (Harris et al. 1971; Picture 15.3b), and Nadel-Harris traps (Picture 15.3c). All these traps are currently used in various countries for fruit fly surveys in support of control activities and eradication campaigns (IAEA 2003). However, the Steiner, Jackson, or Nadel traps are generally baited with male lures, such as trimedlure for *Ceratitis capitata* (Wiedemann) or methyl eugenol and cue-lure for many *Bactrocera* and *Dacus* species, and very few females are captured with these lures (Nakagawa et al. 1970, 1971; Tan et al. Chap. 2, this volume). Therefore, the use of these traps baited with male lures is limited to survey purposes but not for conventional mass trapping. The Male Annihilation Technique (MAT) by lure and kill was tested in the past and had certain success against *Bactrocera dorsalis* (Hendel) and *Bactrocera cucurbitae* (Coquillett) (Steiner et al. 1965; Cunningham and Steiner 1972) but not with *C. capitata* (Avery et al. 1994). The success of MAT clearly depends on the availability of attractants powerful enough to attract almost all males in the population, since even a small number of uncaptured males may

Picture 15.2 Poster of the Valencia Exportation Association (Unión Nacional de la Exportación Agrícola, 1924) with recommendation to use "EFV trap" to achieve healthy fruit

Picture 15.3 (**a**) Steiner trap (Photo courtesy of Roger Vargas). (**b**) Jackson trap with trimedlure plug (Photo courtesy of Eric Jang). (**c**) Traditional Nadel-Harris trap (*left*) for use with liquid trimedlure (Photo courtesy of Eric Jang) and evolution of the design to use with trimedlure plug (*right*)

Table 15.1 Classification and commercially available products of attract and kill

		Types	Mode of action	Description	Example
Attract and kill	**Mass trapping**	Wet traps	Flies drown in liquid	Liquid baits (protein hydrol. or ammonium salts)	Cera Trap®, Olipe, Servatray®
				Dry attractants + water	MultiLure® baited with BioLure+water
		Dry traps	Sticky traps	Dry attractant	Delta trap
			Insecticide	Inhalation (DDVP)	Moskisan®+ BioLure®
				Contact insecticide	Decis®
	Bait Stations	Lure and infect	Contact contamination		Fungi
		Lure and kill	Contact insecticide		M3®, Vioril® Magnet® MED MAT
			Ingestion insecticide		SPLAT (Anamed®) EPALure&kill®
		Lure and sterilize	Ingestion sterilizing agent		Adress®

inseminate many females. Thus, the efficacy of MAT drops severely when the proportion of responding (and subsequently killed) males is not sufficiently high, and trapping a large number of males is not, in and of itself, a reliable indicator of fruit damage reduction. In contrast, a trapping strategy focused on females is always more efficient as female population reduction can be directly related with fruit damage reduction. For this reason, research on synthetic female attractants and the development of traps or attract and kill devices focused on females were the key points in the development of mass trapping systems. A detailed classification of all the attract and kill systems is presented in Table 15.1 following Navarro-Llopis and Vacas (2013).

1.2 Improvement of Attractants for Traps

The first attractants for capturing female fruit flies were based on food or host odors. Vinegar, soaked rice bran and fruit pulp (Gómez Clemente 1929), protein lures (McPhail 1939), ammonia (Gow 1954), or fish remains were used in the first half of the twentieth century. Ammonium soap, known as Clensel, was employed for *Bactrocera oleae* (Rossi) and *C. capitata* (Newman 1930; Bua 1934), protein-ammonium lures for *Anastrepha* species (McPhail 1939), ammonium phosphate for *B. oleae* (Bohorquez 1940), ammonium carbonate for *Bactrocera tryoni* (Froggatt) (Perkins and Hines 1934), and ammonia for *Rhagoletis pomonella* (Walsh) (Hodson 1943). Liquid protein baits replaced these early attractants and are still employed for catching a wide variety of fruit fly species. Corn protein hydrolysate was found to be very effective in capturing *C. capitata* (Gothilf and Levin 1987; Roessler 1989), whereas yeast hydrolysate was more effective for *Anastrepha* species (Heath et al. 1993). Liquid protein baits capture a higher percentage of females but also large numbers of non-target organisms, including ground invertebrates (Asquith and Messing 1992) and flying insects (Katsoyannos et al. 1999), which is a serious weakness of these attractants (as further discussed below).

Since the conception of mass trapping in the 1930s, one of the main drawbacks for its implementation was the handling of liquid attractants. McPhail type traps that use aqueous protein (food-based) baits offer the advantage of attracting both male and female fruit flies, i.e., not only males like previous described attractants (Severin and Severin 1915). However, these traps are fragile, cumbersome, and difficult to service. Liquid attractants require large amounts of water that may be difficult to transport to inaccessible places, and spillage of protein or sugar solutions on the trap may cause fungal contamination that requires trap cleaning and extra manpower. Evaporation of liquid solutions can be reduced with hygroscopic substances, such as propylene glycol, but in hot regions frequent refilling of traps may still be necessary. Although the use of borax at 8 % reduces fungus proliferation and helps preserve trapped insects (Epsky et al. 1999), these solutions may become contaminated with other microorganisms after several days in field, making it difficult to empty the trap and identify and count the captured individuals.

These difficulties motivated the development of synthetic food-based attractants. One of the most important components for fruit fly attraction, identified in protein hydrolysate emissions, is ammonia (Hodson 1948). This attraction is clearly related to the females' need to feed on protein in order to reach sex maturity and complete development of eggs (Christenson and Foote 1960). Other substances identified as female attractants were: ammonium acetate for *Rhagoletis* spp. (Reissig 1976) and putrescine, ammonium bicarbonate, pyrrolidine, and linolenic acid for *Bactrocera* spp. (Wakabayashi and Cunningham 1991), although the last compound has never been efficiently used as an attractant. The discovery of the synergistic effect of trimethylamine with ammonium acetate and putrescine (the three-component lure) (Heath et al. 1997) to attract *C. capitata* females was the key point in the development of more user-friendly mass trapping technology for the Mediterranean fruit fly. For *Rhagoletis* species, the use of ammonium carbonate and/or ammonium acetate with protein bait provided a good attraction, although the formulation of ammonium salts for long-lasting emission has not been totally optimized (Reynolds and Prokopy 1997; Yee 2007).

By contrast, the three component combination of ammonium carbonate or ammonium bicarbonate along with methylamine and putrescine has been demonstrated to be attractive for the Mexican fruit fly (Robacker and Warfield 1993; Heath et al. 1995; Robacker et al. 1996), although a two-component combination of ammonium acetate and putrescine was more attractive than the three component mixture with trimethylamine salts or protein liquid baits in field trials (Heath et al. 2004; Thomas et al. 2008). The same two components were also tested with the Caribbean fruit fly (*Anastrepha suspensa* (Loew)) and were found to be more attractive than yeast hydrolysate (Holler et al. 2006).

The use of dry traps with the previously described components (i.e., trimethylamine, ammonium acetate, and putrescine for *C. capitata*) results in easier handling of traps and therefore a great reduction in manpower costs. Furthermore, the use of synthetic food lures allows the release of specific active attractants in the correct proportions, avoiding secondary products produced in uncontrolled fermentations, and therefore increased capture efficacy with reduced attraction of non-target organisms.

The discovery of dry attractants inspired the development of new types of traps to replace the heavy and fragile glass McPhail traps. The first step in this evolution was the replacement of glass with clear plastic. All new designs incorporate a clear plastic lid to help retain flies that have entered the trap. Field trials have demonstrated the importance of a clear lid: less transparent lids reduce drastically the number of captures using the same design of trap. In 2005 the same McPhail type trap was tested with two different lids, and those with less transparent lids captured 50 % fewer *C. capitata* than those with transparent lids (Alfaro-Lassala, personal communication). As flies are photopositive, a clear lid maintains flies in the upper part of the trap and so reduces escapes. The bottom part of the trap, the container, is opaque in the majority of the new designs and is colored and shaped to provide preferred visual stimuli for the target fruit fly. For example, the most recently designed traps for *C. capitata* include rounded shapes and yellow colors (see Chap. 5, this volume, for more detailed information about color preference of fruit flies). The Tephri-Trap (Utiplast, Madrid, Spain) and MultiLure trap (Better World Manufacturing, Fresno, CA, USA) were developed in the last decade of the twentieth century and constituted improvements over previously used traps (Ros and Castillo 1994; Ros et al. 1996; IAEA 2003). However, research during the last 10 years has contributed to the further development of new trap designs with different colors, shapes, configurations, and materials that improve trapping efficacy and at the same time provide a more user-friendly control method. Navarro-Llopis et al. (2008) reported that new Moskisan® (SanSan Agriculture Engineering, Valencia, Spain) or Probodelt designs trapped between 1.5 to 2 times more *C. capitata* females than MultiLure or Tephri-Traps. Lucas-Espadas and Hermosilla-Cerón (2008a) obtained similar results in southeastern Spain, with increased performance for the Moskisan trap and the Maxitrap® (Probodelt, Amposta, Spain) relative to the Tephri-Trap.

2 Description and Classification of Traps

2.1 *Importance of Visual and Olfactory Cues*

2.1.1 Visual Cues

Color

Many studies have tried to improve trap efficacy with visual cues. These cues could be related to host-finding, mate-finding, or oviposition behaviors (Prokopy and Owens 1983), and numerous studies have examined the effect of shape, size, and color of visual stimuli on fruit fly response (see Epsky and Heath 1998 and Katsoyannos 1989 for reviews).

Although there is general consensus that the color yellow or fluorescent yellow (520–540 nm peak reflectance) is highly attractive to many fruit fly species (*Anastrepha fraterculus* (Wiedemann) (Cytrynowicz et al. 1982); *Rhagoletis cingulata* (Loew) (Frick et al. 1954); *R. pomonella* (Prokopy 1968); *Rhagoletis cerasi* L. (Prokopy and Boller 1971); *B. oleae* (Prokopy et al. 1975); *B. tryoni* (Froggatt), *Bactrocera neohumeralis* (Hardy), *Bactrocera cacuminatus* (Hering) (Hill and Hooper 1984); *A. ludens* (Robacker et al. 1990) and *C. capitata* (Prokopy and Economopoulos 1976)), there are other species that respond maximally to other colors (*A. suspensa* to orange (Greany et al. 1977, 1982); *Toxotrypana curvicauda* Gerstaeker and *Rhagoletis completa* Cresson to green (Landolt et al. 1988 and Riedl and Hislop 1985, respectively)). Most likely, the general attractiveness of tephritids to yellow occurs because this color acts as super-normal green-foliage stimuli (Prokopy 1972). However, in the last two cases cited, green may be attractive, at least in part, because females normally oviposit into green fruit.

Yellow seems to be the most attractive color for *C. capitata* as well. Prokopy and Economopoulos (1976) demonstrated that more Mediterranean fruit flies were captured on yellow rectangles than light orange, light green, red, gray, or clear rectangles. Similarly, Heath et al. (1995) reported significantly more male captures in yellow traps than in orange traps baited with ammonium acetate and putrescine. Vargas et al. (1991) demonstrated that yellow fruit-mimicking spheres were an excellent device for capturing the oriental fruit fly (*B. dorsalis*), especially females, in guava trees. The high response of *B. dorsalis* to yellow spheres and low response to dark green spheres probably indicates that the color yellow was easily identified in a dark green background. In the same way, although yellow and white spheres were the most attractive to *B. dorsalis* (compared with green or red) in the trials conducted by Cornelius et al. (1999), when red spheres were placed over a yellow panel, attraction was higher than with yellow spheres alone. These results only can be explained by the effect of the visual contrast of red spheres against a light background. Stark and Vargas (1992) also studied the effect of color in bucket traps and found that white and yellow traps caught the largest number of males when they were placed close together on stakes (compared to blue, orange, green, red or black traps), suggesting that male *B. dorsalis* respond primarily to intensity of reflected light, not hue. Hill and Hooper (1984) found the daylight fluorescent saturn yellow was the most effective color for *B. tryoni*, *B. neohumeralis*, and *B. cacuminatus* in sticky traps.

Many authors (Agee 1985; Drew et al. 2003; Sivinski et al. 2005) have noted that using a color name as a reference to describe a fly's preference is not correct, since our color vision is quite different from that of an insect and that it is more accurate to define a fly's chromatic preference based on the insect visible spectrum. For example, although yellow color has been used in *B. dorsalis* traps for a very long time, UV and green stimuli (spectra: 300–380 nm and 500–570 nm, respectively) would enhance their attractiveness and blue stimuli (380–500 nm) would diminish their attractiveness (Wu et al. 2007). Similar results were obtained by Katsoyannos and Kouloussis (2001) with *B. oleae*, in which more males were captured by colored spheres reflecting maximally between 580 and 600 nm (yellow to orange), with peak response at 590 nm, but females were mostly captured by colors reflecting maximally between 610 and 650 nm (orange to red) with peak response at 650 nm.

The scientific literature contains some reports of male-female differences regarding color preference. For example, *B. oleae* females are attracted to darker colors than those preferred by males: yellow and orange spheres trapped the greatest number of males, while red and black spheres were the most attractive for females (Katsoyannos and Kouloussis 2001). Some authors pointed out that females of some *Rhagoletis* species also prefer darker colors meanwhile males prefer lighter ones (Messina 1990; Henneman and Papaj 1999). In these works it is suggested that female response to ripeness cues is innate and corresponds to dark colors for *Rhagoletis* species; however, males develop a preference for green based on their encounter rate with females. Thus, as females seek oviposition sites, which are associated with black spherical shapes resembling mature olives for *B. oleae* or yellow-green spherical shapes resembling guava for *A. fraterculus* (Cytrynowicz et al. 1982).

The reproductive state of the fly can also affect visual responses. Virgin females of *A. suspensa* were equally attracted to same-sized (20 cm diameter) white spheres and to orange spheres, whereas white spheres were significantly less attractive for mated females than orange spheres (Sivinski 1990). Given that mated females seek out oviposition sites, the preference for host-like traps would explain the results observed. Response to trap color may also vary through the fruiting season: for *R. pomonella*, many authors (Prokopy 1972; Reissig 1975; Prokopy and Hauschild 1979; Neilson et al. 1981) reported an increased response to yellow sticky panels early in the season, whereas red sticky spheres generally were more attractive later in the season.

Nakagawa et al. (1978) suggested that the observed preference of *C. capitata* was based on color contrast against the background. Other studies also indicate that contrast against the background may be more important than the color of the trap itself. In the case of *Neoceratitis cyanescens* (Bezzi), females oriented preferentially towards an orange sphere when placed against a fluorescent yellow background as opposed to a black background (Brévault and Quilici 2007). Cornelius et al. (1999) obtained similar results with B. *dorsalis* in which red spheres attached to the center of yellow panels captured more females than any other trap tested without the panel.

Interaction of Color, Shape and Size

However, other trap parameters, particularly shape and size, may influence the attractiveness of a given color. Regarding the interaction of color and size, Nakagawa et al. (1978) found that yellow color was more effective for *C. capitata* in larger spheres (18 cm diameter) but not in the smaller ones. On the other hand, trap shape was highly important for *R. pomonella*, which preferred fluorescent yellow traps when rectangular but enamel red in the case of spherical traps (Prokopy 1968, 1972, 1973). Trap configuration also affected capture of *R. cerasi*: two crossed, sticky 15 by 20-cm yellow panels (Rebell trap) captured seven times more individuals than the most effective single sticky-coated yellow panel (14 by 23 cm) (Katsoyannos et al. 2000). *Bactrocera cucurbitae* females were particularly attracted to objects of spherical shape colored yellow, white, or orange relative to cylindrical shapes colored red, green, or black Piñero et al. (2006). The interaction of color, shape and size was reported by Russ et al. (1973) in cherry fruit fly attraction. The authors found that response of *R. cerasi* to 3-dimensional traps was superior compared with the 2-dimensional sticky rectangular boards and that a medium or large yellow surface (15×20 cm) captured significantly more flies than small boards.

Anastrepha spp. also appear to be attracted to yellow colored spherical objects. Sivinski (1990) found that 20 cm-diameter orange and green spheres were more attractive to *A. suspensa* females than white, yellow, or black spheres. However, for smaller spheres (14 cm-diameter), yellow was the preferred color. Thus, again, color attraction is apparently dependent on trap size, with yellow color preferred when traps are smaller (similar to a host fruit) and green color when traps are larger for *A. ludens* (Robacker 1992). This effect may be observed only in plots without fruit. Where fruit is present, the size effect is likely less apparent, and as a general rule in orchards with fruit, trap size should be larger than the fruit they mimic. Robacker (1992) also studied interactions between color, shape, and size in *A. ludens* and observed that yellow and green were significantly more attractive than white and that large size spherical or vertical rectangular shapes were the preferred ones versus horizontal rectangles or small spheres.

However, in considering the interaction of color, shape and size in trap design, the most straightforward solution, i.e., use the optimum state for each factor, may not yield the best performance. Russ et al. (1973), for example, found that, for traps of optimum color and size, a more attractive 3-dimensional shape did not yield significantly higher captures than a less-preferred 2-dimensional trap.

2.1.2 Interactions Between Visual and Olfactory Cues

The importance of olfactory stimuli in affecting trap effectiveness is discussed in detail in Chaps. 2, 3, 4, and 5 in this volume, and here we discuss the impact of visual and olfactory cues on trap catch.

A combination of visual and chemical cues may influence the attractiveness of traps for pest tephritid fruit flies (Cunningham 1989; Economopoulos 1989). Epsky and Heath (1998) suggested that traps for tropical tephritids, such as *C. capitata*, have relied primarily on chemical lures, while the traps for temperate tephritids, such as the apple maggot, *R. pomonella*, give more importance to visual cues. Prokopy (1968) demonstrated that olfactory cues were less important than visual cues for *R. pomonella*, as olfactory stimuli only elicited feeding-type reactions when employed in conjunction with shapes and colors that imitate foliage on which food occurs. Green et al. (1994) obtained similar results showing that odor in the absence of visual cues did not increase attraction of *R. pomonella*. However, this differentiation between temperate and tropical species was based on early findings of Prokopy (1968), when only a few efficient attractants had been described for fruit fly species. More recently, several studies (Brévault and Quilici 2007; Piñero et al. 2006) have demonstrated that particular combinations of both visual and olfactory stimuli elicit higher levels of attraction compared to each stimulus offered alone.

As a result of research in visual and chemical cues, new traps and attractants have been developed recently, and the efficacy of these systems has improved markedly. The higher efficacy of new dry attractants for *C. capitata* (ammonium acetate, trimethylamine, and putrescine), when compared with traditional protein baits, has notably increased the effectiveness of mass trapping. As an example, Ros et al. (2002) demonstrated that replacement of protein bait (9 % NuLure (Miller Chemical & Fertilizer Corporation, Hanover, PA, USA) + 3 % borax) by these new attractants allows reduction of the required density of traps for mass trapping, because captures were nearly doubled when dry attractants were used instead of NuLure. Traditional McPhail, Nadel, Jackson, or Steiner traps were white or transparent. New trap designs, with different color patterns, have improved trap efficacy. For example, the bottom part of the plastic McPhail type (IPTM; International Pheromone Systems, Cheshire, UK) designed for *C. capitata* is currently rounded and yellow reflecting the fly preferences described above.

2.2 *Classification of Traps*

Traps can be classified according to different characteristics, such as color, shape, or attractant, but we propose a classification based on two groups: wet traps with liquid attractants and dry traps.

2.2.1 Wet Traps: Invaginated and Side-Hole Traps

Wet traps have been used since the beginning of the twentieth century to reduce fruit fly populations. Initially, they were invaginated clear glass traps with a volume capacity between 100 and 500 ml, baited with acetic acid, fruit pulp, and/or molasses with water (Gómez-Clemente 1929). More recently, protein baits and ammonium salts have been employed and are still used (Broumas et al. 2002; El-Sayed et al. 2006; Lucas-Espadas and Hermosilla-Cerón 2008b). Presently, wet traps are usually made of plastic, and designs are based on the McPhail trap but with some improvements. They are usually cylindrical and have a clear plastic lid and a colored opaque container. The holes allowing the flies to enter the trap can be placed at the bottom, with an invaginated entry (as in McPhail traps), or in the walls of the cylinder (as in traditional Nadel bucket traps), or a combination of both as in the Tephri-Trap (Picture 15.4).

It is intuitively obvious that the number of traps required to treat a certain area varies with the effectiveness of the attractants. When a high density of traps is required, due to the low attractiveness of some liquid attractants, growers are constrained by cost and must use very cheap traps, such as bottles. For example, Zervas (1982) tested protein and ammonium salt liquid attractants in yellow painted sticky coated bottles against *B. oleae,* which yielded higher captures than traditional McPhail traps. Currently, one of the most commonly used protein or ammonium salts liquid traps are bottles with lateral holes, such as the Olipe trap (Ros et al. 2009) (Picture 15.5), Elkofon (Phytophyl, S.A., Athens, Greece; Eliopoulos 2007) (Picture 15.6), and Cera Trap® (Bioibérica, Barcelona, Spain) (Picture 15.7) (Lucas-Espadas and Hermosilla-Cerón 2008b; Llorens et al. 2008). At present, several companies sell bottle traps for use with new liquid attractants, such as Cera Trap® or Starce® (BIAGRO, Valencia, Spain). The Cera Trap® attractant is an enzymatic hydrolyzed protein that contains piperazindiones, developed for *C. capitata* (Sierras et al. 2006). The same attractant has been tested in Mexico with new traps (termed MS2®; River Bioscience (Pty) Ltd, Port Elizabeth, South Africa) against *A. ludens* (de los Santos-Ramos et al. 2011), resulting in an improved efficacy compared with GF-120 NF Naturalyte® fruit fly bait (Dow AgroSciences, Indianapolis, IN, USA). The use of improved attractants allows farmers to reduce the required density of traps.

The longevity of the liquid is a key component affecting the efficacy of wet traps. Given that the liquid serves both as the attractant and the killing agent, when it evaporates completely, the trap stops working. Bait 'dregs' may still attract flies, but the lack of liquid allows the flies to escape. The rate of evaporation depends on the size of the holes and the liquid composition. Evaporation can be reduced by adding a hygroscopic substance, such as propylene glycol (Thomas et al. 2001), which could even act as a synergist, increasing captures, when combined with other attractants (as demonstrated for *A. ludens*, Robacker and Czokajlo 2006). However, the addition of propylene glycol may also increase the number of non-target insects attracted (Leblanc et al. 2010a). On the other hand, small holes reduce evaporation, although a minimum size should be determined for each target species. For example, Luque-López and Pereda-Cruz (2003) tested traps with 3, 4, 5, or 7 mm-diameter holes for capturing *B. oleae* and *C. capitata* and found that hole diameter should be at least 4–

Picture 15.4 Tephri-trap® (*left*) and inside view (*right*) containing a trimedlure plug and DDVP tablet

Picture 15.5 Olipe trap with ammonium biphosphate solution

Picture 15.6 Elkofon (*left*) and new Elkofon (*right*) with ammonium salts solution (Photo courtesy of Jose Manuel Llorens)

Picture 15.7 Ceratrap® containing Ceratrap®bait (*left*) and Servatray® bottle trap (*right*) baited with protein hydrolysate

5 mm. Furthermore, 7 mm-diameter holes captured as many flies as 5 mm holes, but the number of non-target organisms increased significantly. Traps with 3 mm holes captured significantly fewer individuals of both species than traps with larger holes.

The longevity, and therefore the efficacy, of the wet trap also depends on the climatic conditions of the particular area. Cunningham et al. (1978) demonstrated that wet traps are more efficient in dry than in wet climates, probably due to fly attraction to a water source. However, this would result only if the liquid attractant is properly maintained. As an example, Olipe traps baited with 4 % ammonium biphosphate plus 25 % propylene glycol should be serviced at least every 2 months as done for *B. oleae* in Spain (unpublished results).

2.2.2 Dry Traps: Sticky Traps, Physical Traps, or Insecticide Traps

In the case of wet traps, the flies must fall into the liquid and drown to actually be trapped, an essentially passive catch mechanism. The use of a knockdown insecticide or a sticky contact surface might circumvent that problem, but use of these alternatives raises other issues. Sticky traps are traditionally employed for detection purposes because of their cheap and disposable nature. Several studies (Harris et al. 1971; Nakagawa et al. 1975) demonstrated that Jackson traps with sticky inserts or yellow panels with sticky coating perform as well as the Steiner trap, which involves fly entry and entrapment in a cylindrical tube. However, the use of these sticky traps is not always recommended for mass trapping as they lose effectiveness when they are overloaded with insects, leaves, or even dust. In these conditions, the low cost of the sticky material might be offset by labor costs as the sticky boards would require frequent servicing. As an example, the *C. capitata* Frutect® trap (RonPal Ltd., Rishpon, Israel) was designed as a red/brown sphere over a yellow sticky board baited with protein bait or the three component lure (Gazit et al. 1998). Although this trap included some effective visual cues for flies, as it combines the red spherical shape over a yellow background, Frutect did not surpass the McPhail trap in medfly captures. In order to avoid high costs in trap servicing, these traps were eventually sold with an aerosol glue that could be re-applied if the sticky surface was covered by trapped insects or dust. These re-applications were usually not necessary when traps remained in the field for less than 3 months, but in dusty areas or orchards with high pest populations, more frequent servicing is required. Furthermore, although the attractants employed could be more specific for the target species, like the three-component lure compared with diammonium phosphate for *C. capitata* (Boulahia Kheder et al. 2011), many non-target insects may come into contact with the adhesive surface and be trapped by chance or color attraction, accelerating trap saturation. Heath et al. (1996) also considered adhesive paper material as an easy way to prepare the cylindrical traps used in their work, but they also noted the capture of non-target organisms, including small vertebrates, such as certain birds and lizard species.

An alternative to retain flies without using a cumbersome wet trap or a sticky trap that can become saturated is to design a dry trap that prevents fly escape without the need for an insecticide. The one-way trap described by Tan (1985), and later tested by Hiramoto et al. (2006) and Jang (2011), is a modified bucket trap in which microcentrifuge tubes are inserted in the four lateral holes of the trap. The tip of the microcentrifuge tubes was cut off leaving a tapered tube containing a ca. 6 mm opening. These one-way entrances allowed the flies to easily enter the trap but not to escape. For *B. dorsalis* and *B. cucurbitae* a 6 mm-diameter hole has been proven to be efficient, as individuals can easily enter but not escape (Jang 2011). Uchida et al. (1996) tested the response of *C. capitata* to five different entrance hole sizes (7.1, 8.9, 12.7, 16.4, and 23 mm in diameter, respectively) and found a direct relationship between hole size and trap captures of wild flies in modified bucket traps (with clear cylindrical polystyrene vials inserted through the trap holes), suggesting an optimal value between 16.4 and 23 mm.

Another prototype proposed consists of a double funnel design that makes it very difficult for flies to escape. This trap is clear, and entry occurs at the bottom of the trap. The first invagination leads the flies to a midway chamber and the next invagination to a larger upper chamber. In this way, flies that escape from the upper chamber are retained in the chamber below, and many of them subsequently return to the upper chamber. Thus, a flow of flies from one chamber to the other helps to reduce escapes. The authors (Navarro-Llopis, unpublished data) compared the efficacy of one of these double-funnel trap prototypes with a McPhail plastic trap containing DDVP, using in both traps the three-component lure against *C. capitata*. However, efficacy of the double funnel design did not reach 50 % of that obtained with the trap containing DDVP.

Perhaps, one of the best options to improve the efficacy of trapping devices is the use of contact insecticides. Initially, most of the insecticides used were organophosphates, active by inhalation due to their high vapor pressure, such as naled or DDVP. A strip or polyvinyl chloride (PVC) tablet with 50 to 500 mg of DDVP (Navarro-Llopis et al. 2008) or 1 % of naled (1,2-dibromo-2,2-dichloroethyl dimethyl phosphate) added to the male attractant (Vargas et al. 2010) is sufficient to kill the flies that enter the trap. However, these insecticides are now banned in many countries (CEC 2007, 2012) and have been replaced with contact insecticides. Pyrethroids have largely replaced organophosphates, although other insecticides, such as methomyl (Ros et al. 2005) or fipronil (Vargas et al. 2010), have also been employed in baits. Another pyrethroid, deltamethrin, may be applied to a plastic strip (Alemany et al. 2004; Leza et al. 2008) or impregnated in the lid to increase the contact surface of walking flies with the insecticide (Tapia-Ramos et al. 2012). Other solutions have been developed, such as a retrievable disc containing cypermethrin placed in the lid of the trap just below the top (Sancho-Sánchez 2009; Boulahia Kheder et al. 2012). Traps baited with these alternative insecticides appear as effective as those baited with the traditionally used DDVP (Navarro-Llopis, unpublished results), and in some cases the average number of fruit flies captured in traps with these alternative insecticides is even higher than with DDVP, probably due to the repellency effect of this insecticide against tephritid fruit flies (Barrera et al. 2006).

The longevity of dry female attractants depends on the initial loading, the dispenser, and the weather conditions. As the main components of the dry attractants are ammonium salts (ammonium acetate, ammonium carbonate, ammonium sulfate) or amine salts (trimethylamine hydrochloride, putrescine di-hydrochloride), humidity is a key determinant of the release rate of the active substances. The use of a membrane that regulates the release of the substances and avoids hydrolysis of the salts can increase the lifetime of the attractants and provide a constant release rate. Commercially available dispensers for *B. oleae* (ammonium carbonate) are active for as long as 3 months (Broumas et al. 2002), and some *C. capitata* female attractant dispensers have been demonstrated to be effective for more than 100 days in the field (Navarro-Llopis et al. 2008). Knowledge about the functional longevity of the dispenser used is essential for mass trapping as this technique should operate effectively from the beginning of fruit susceptibility to infestation to the end of harvest. In the case of citrus crops, fruit susceptibility starts with color change from dark to light green. In deciduous orchards, such as peach, this can be even earlier, and in olives susceptibility can be operationally identified by the hardening (woodification) of the olive pit. Insecticide lifetime is, of course, as important as attractant longevity. PVC tablets with 500 mg DDVP have a lifetime over 6 months (Navarro-Llopis, unpublished results), although other insecticide dispensers should be tested to ensure insecticide activity during the entire trapping period.

As a result of recent research, dry traps are now the easiest designs for grower handling as there is no need to touch toxic compounds and servicing is less frequent due to the development of long lasting dry attractants.

2.3 *Recently Developed Traps: State of the Art*

As mentioned before, traditional glass McPhail traps have been largely replaced by plastic traps with a clear lid and a colored container, and the incorporation of lateral holes has noticeably improved the efficiency of this trap type. Perhaps one of the most efficient, commercially available traps, is the Maxitrap, which includes small tubes in the lateral holes (Picture 15.8) as one-way entrances. These small clear tubes prevent fruit flies from escaping when they walk on the inner wall of the trap, as described by Tan (1985). Other improvements with the new trap design are related to the release of attractants. In the case of the Moskisan trap (Picture 15.9), volatiles are released by three lateral chimneys, although the main fly entry is through an invaginated hole in the bottom of the trap. Another variable that has been widely studied is the diameter of the lateral holes. Almost all of the traps currently sold have a lateral hole diameter between 5 and 15 mm. Lower diameters are used in Cera Trap, meanwhile wider diameters are used in dry traps.

Recent developments by European companies have focused on the effects of trap color, shape, and size on captures of *C. capitata*. One of these new designs (Decis-Trap®; Bayer CropScience, Paterna, Spain) consists of a yellow half-sphere container (ca. 14 cm in diameter) with a clear lid coated with deltamethrin (Picture 15.10). Lateral holes are 14 ca. mm in diameter through which short tubes are inserted inward to reduce escapes. Another recent improvement in trap design involves the packaging and construction of easy-to-assemble traps intended to facilitate

Picture 15.8 Maxitrap with small tubes in the lateral holes as one-way entrances and invaginated hole in the bottom of the trap

handling and reduce volume and thus shipping costs. For example, the tubes in the lateral holes of the traps increase the efficacy of traps but prevent efficient stacking of traps. Thus, 20 traps require a box of 100 liter volume. However, the new Decis Trap allows placing five times as many traps in the same volume. As a result, this trap is cheaper, and the manpower required to place traps in the field is lower than needed for previous models.

3 Distribution, Density and Placement of Traps for Optimized Efficacy

3.1 What Is the Best Position in the Tree?

In general, the best position for traps on a tree is that volume of the canopy most frequently visited by fruit flies. There are many studies on the distribution of flies within trees but, as a general statement, their occurrence is influenced by multiple factors, including latitude, time of the day, weather, and height of the tree canopy.

In warm places like Israel, Mediterranean fruit flies moved from the upper lighted and most exterior leaves of figs, apple, and pitanga trees to the lower shaded leaves as the air warmed up at midday (Warburg and Yuval 1997; Kaspi and Yuval 1999). In the case of males, individuals occupy locations that confer suitable

Picture 15.9 Moskisan with lateral chimneys and invaginated hole in the bottom of the trap

Picture 15.10 Decis Trap showing small tubes in the lateral holes and without bottom entry

microclimates for calling and copulating as well as provide protection from predators, wind, direct sunlight, and water loss. In addition, the location of medfly leks within the tree canopy influences female visitation; females appear more likely to visit males clustered in deeply shaded sites with dense foliage than in brighter sites with sparser foliage (Shelly and Kennelly 2007). Thus, it appears that, for medfly in warm environments, traps should be positioned in fairly dense portions of the canopy and not in the outer part of the canopy. Ye et al. (2012) also studied the influence of rainfall on trap captures of *B. dorsalis* in mango orchards. Rainfall inhibited flight activity as flies typically perched on leaf undersurfaces, and consequently fewer captures were recorded on rainy days than sunny days.

There are many examples demonstrating that the optimum height for catching flies depends on tree size. In the case of tall plants, captures tend to be greater at intermediate-upper heights as reported in the following examples: 4 m height of 8 m tall mango trees for *B. dorsalis* (Ye et al. 2012), 3 m in 10 m tall guava trees (Siddiqui et al. 2003) for *Bactrocera zonata* (Saunders), 5 m height of 8 m tall mango trees for several *Anastrepha* species (Aluja et al. 1989), 5 m in field trials conducted with diverse tall vegetation (including many species of tropical and subtropical fruits and truck crops) for *C. capitata* (Holbrook and Fujimoro 1969), and 4.5 m in 4.65 m tall cherry trees for *R. cingulata* (Pelz-Stelinski et al. 2006). Optimum trapping heights for smaller sized plants with fruits near the ground tend to be at low heights, typically below the half-height of the plant. For example, traps near the ground (from 3 cm to 1.8 m) were more effective than traps placed at 3 and 5 m for *B. cucurbitae* (Holbrook and Fujimoro 1969) as this species usually attacks low or creeper plants. For small trees, from 1 to 3 m, such as peaches or citrus in the Mediterranean Basin, maximum catch is obtained around 1.5 m height as shown for *B. zonata* in peach trees (El-Gendy 2012). However, Hooper and Drew (1979) observed no effect of height on captures of *B. tryoni*, *B. neohumeralis*, *B. cacuminatus* and *Bactrocera endiandrae* (Perkins & May) with cue-lure or methyl eugenol within the range of 0.1–3.6 m above ground in citrus orchards with trees averaging 4.6 m in height or in suburban gardens. This absence of height effect could be explained by the high attractant power of male lures that can reduce the importance of other variables, such as trap height.

Furthermore, as reported by Reissig (1975), the most effective height for trap placement may depend on the type of trap being used. In an apple orchard in New York (USA), catch of *R. pomonella* was compared among three heights for three types of trap: a yellow Sectar® trap (Zoëcon Corp., Schaumburg, IL, USA) baited with a solution of 50 % ammonium acetate and 5 % yeast hydrolysate, a sticky yellow card (22 × 30 cm) with 1 g of Hy-case® (casein acid hydrolysate) and ammonium acetate mixed in the adhesive, and a sticky 8 cm diameter red sphere. The sphere and the Sectar traps were most effective when hung near the top of the tree, 3 m above ground, but the yellow card was more effective when hung near the middle of the tree, 2.1 m above ground. Reissig also included two additional variables, canopy radius and compass direction. All three traps were most effective when hung near the middle of the canopy radius, and all caught more *R. pomonella* when hung on the south side of the tree. Studies carried out in Egypt showed that traps placed in the northwest quadrant of peach trees captured the greatest number of *B. zonata*, with 2 times as many captures compared to the south and 1.7 times more than in the east (El-Gendy 2012). Collectively, these results suggest that the warmer parts of the tree, with the highest insolation, are the most preferred for flies in temperate climates, while flies tend to seek shadow and cooler areas in warmer climates.

As a general rule, in tall to medium trees, maximum trap captures are achieved by placing traps in the upper half of the tree. Exceptions to this rule would involve species in which the females oviposit in vegetables, in which case the recommended height would be near the soil. For species inhabiting warm and dry places, shadier regions of trees will be the most preferred by flies, and for colder places or seasons, sunnier areas of trees will be the best location for traps.

3.2 Distribution of Traps in the Orchards

Efficient mass trapping strategies commonly employ a high density of traps evenly distributed over the treated area, which seems intuitively the most effective way of removing a high proportion of pest individuals. However, the dynamics, and particularly seasonal trends, of fruit fly populations must be taken into account in planning the distribution of traps within the orchards for both monitoring and mass trapping strategies. The spatial and temporal distribution patterns of fruit flies are influenced by many factors but especially by environmental conditions and host fruit availability. For example, Dimou et al. (2003) documented significant heterogeneity of *B. oleae* captures within olive orchards based on particular microclimate conditions and the presence of preferred hosts. Very low captures were obtained in traps hung in wild olive trees, neighboring the orchard, as olives from managed orchards were preferred to olives from wild olive trees. The same effect was observed inside the orchard where more irrigated areas and healthy trees were clearly preferred by *B. oleae* over weaker or less irrigated trees. In areas of the orchards with suitable irrigation, olive fly populations may increase even before fruiting, probably due to aggregation of *B. oleae* adults near water sources. Moreover, fruiting levels increase in these properly irrigated areas, causing the immigration of fruit flies to these orchards. In orchards with mixed cultivars, differences in the timing of fruit maturation may also determine the distribution of fruit flies within a plot. Murphy et al. (1991), for example, found that traps located near cultivars with earlier ripening fruit captured a large number of *R. pomonella*. In a polyphagous species, like *C. capitata*, flies have been shown to track fruit ripening of different cultivars (Navarro-Llopis et al. 2007) or even different crops, from peaches in June–July to persimmon in August–September and finally citrus in October–November (Domínguez 2007).

As a general statement, fruit fly populations increase when hosts become available, and during peak populations flies may move into non-host areas, possibly in search of feeding or mating sites (Vargas et al. 1990). As growing fruit fly populations may at times also move into orchards and farms from surrounding areas, perimeter trapping could improve the efficacy of mass trapping in some cases by reducing immigration. Aluja (1996) observed that, for *Anastrepha* species, flies often invade commercial orchards from other vegetation or abandoned orchards in the vicinity. Accordingly, a perimeter distribution of traps appears a useful strategy, helping to prevent *Anastrepha* flies from invading fruit orchards. McQuate et al. (2005) described how mass trapping performed in adjacent host orchards effectively reduced the *C. capitata* population that later attacked persimmon. The treatment was based on mass deployment of traps baited with BioLure® (Suterra LLC, Bend, OR, USA) in alternate hosts, such as peach, plum, and citrus. In this trial, good suppression was achieved in persimmon, because many of the flies derived from infested alternate hosts, which completed fruiting before the persimmon season, were eliminated from the orchards before persimmon fruits became susceptible to sting damage.

Aluja et al. (1997a) observed that *T. curvicauda* individuals move back and forth between papaya groves and native vegetation and reported that most flies were captured in traps placed along the periphery of the groves (Aluja et al. 1997b). This provided evidence for the suitability of controlling *T. curvicauda* by border or perimeter trapping. Based on field trials, Cohen and Yuval (2000) suggested the potential of perimeter trapping technique for controlling *C. capitata* populations. Captures of female *C. capitata* were affected by surrounding host plants, but damage was significantly lower with a perimeter strategy (McPhail traps baited with the three component lure arranged at 10 m intervals) than in untreated plots (c.a. 3 % in a perimeter-treated persimmon plot vs. 9 % in an untreated plot at postharvest).

Border trapping has been widely employed to protect apples, intercepting a high proportion of invading *R. pomonella* flies with sticky traps (28 cm × 21.5 cm yellow panel sandwiched between the two halves of a 9 cm diameter plastic red sphere baited with butyl hexanoate) placed every 10 m around plots. Results showed performance similar to insecticide spray, although, in this case, the lack of an untreated control confounded interpretation of the results (Bostanian et al. 1999). In most cases, the number of traps required per plot depends on the length of the perimeter of the plot. Lengthened plots are most likely to receive a fruit fly invasion, and therefore higher densities of traps are required. However, in larger and regularly shaped plots the ratio of area-to-perimeter is higher, and fewer traps are required to avoid fruit fly intrusion, compared to small and irregular plots (Prokopy et al. 2005).

Several studies on trap deployment patterns have been published. Reynolds et al. (1998) used potted trees to compare the effectiveness of two trap deployment patterns (perimeter and within-orchard) to reduce fruit injury by *R. pomonella*. Results showed that both treatments significantly reduced fruit damage relative to control plots, although in both deployment patterns distances from the center to the external part of the plot were extremely short (only 6 m from inner to outer trees). More recently, Prokopy et al. (2003) demonstrated the efficacy of several trap deployments on orchard front-rows (10 or 5 m apart) and observed that more wild flies entered from adjacent woods than from adjacent open fields. Thus, the plot borders that are most prone to invasion by flies must be identified prior trap placement in order to apply the proper strategy and reduce management costs. Prokopy et al. (2005) also reported an approach to assign the distances between perimeter traps according to an index based on four environmental variables, namely plot size, prune quality, cost-competitiveness of odor-baited perimeter spheres vs. insecticidal sprays, and the nature of the bordering habitat. Good pruning refers to trees having an open canopy with a uniform distribution of light intensity that presumably facilitated the visual detection of odor-baited spheres by foraging apple maggot adults within the canopy. Poor pruning was exemplified by trees with excessive foliage that shaded the canopy interior in such a way as to decrease the likelihood that foraging apple maggot flies would find a sphere against such a dark background.

Based on this index, shorter inter-trap distances would be required in the case of orchards with highly susceptible, large and poorly pruned trees and when population pressure in the vicinity is high. Factors responsible for high pest pressure include the presence of earlier-ripening or unharvested alternate host fruits, the proximity of hedgerows or woods that provide suitable feeding or roosting sites for flies, and the absence of treatments to reduce fruit fly populations (Bostanian et al. 1999). However, the cost of mass trapping using sticky spheres is still substantially greater than the cost of applying insecticide to control apple maggot, meaning that use of these traps is not practical for large-scale, commercial application.

Moreover, the efficacy of perimeter trapping may be highly correlated to the cultivar susceptibility itself as described for *R. pomonella* (Duan and Prokopy 1995; Bostanian et al. 1999; Bostanian and Racette 2001). As an example, plots with more susceptible hosts, such as peach, require a higher density of traps (100–120 traps/ha) than citrus orchards (50 traps/ha) for *C. capitata* (unpublished results).

The isolation and size of plots subject to mass-trapping are key factors influencing the trap deployment strategy implemented. For large or isolated areas, trap distribution should be homogeneous, but for small plots or orchards surrounded by other hosts a perimeter strategy could be a proper method to reduce the likelihood of fly intrusion along the borders. Field trials conducted in Spain by the authors over large areas demonstrated that fruit fly invasion can be observed even at 1.3 km from the edge of treated plots (Navarro-Llopis et al. 2012) when chemosterilant traps are used. This value reflects the distance at which a given fruit fly population is influenced by outer populations and suggests that *C. capitata* is able to move more than 1 km seeking hosts when they are not intercepted by traps. However, the effect of fly intrusion is lower when mass trapping is applied as traps intercept a portion of the invading individuals. As a result, a noticeable fruit fly reduction might be observed in small plots as reported by Navarro-Llopis et al. (2013) in the center of 1 ha citrus plots. For 4 years, this strategy has been tested in an area of 260 ha. Application of a perimeter mass trapping and replication of the treatment during three consecutive years has allowed a reduction in trap density from 50 to 30 traps/ha without a significant loss of efficacy (unpublished results).

3.3 Required Density on Mass Trapping and Timing of Use

Arguably, the best-studied fruit fly species regarding the application of mass trapping is the Mediterranean fruit fly. Programs that have achieved successful mass trapping for this species employ trap densities in the range of 20–150 traps/ha. For example, a density of ca. 200 traps/ha resulted in efficacious control of high *C. capitata* populations in citrus orchards in Spain (Alemany et al. 2004). Subsequent studies by the same group (Leza et al. 2008) showed that efficient control can be achieved with the combined use of 50 traps/ha baited with the three-component lure and insecticide spot-bait sprays. In this case, the density of 50 traps/ha matched the maximum cost that the growers were willing to pay. Independent of this cost constraint, other studies have also identified the density of 50 traps/ha (using Maxitrap, Moskisan or IPTM traps (plastic McPhail trap)) combined with the three component attractant (ammonium acetate, trimethylamine, and putrescine) as the required density to obtain, at least, the same efficacy as bait sprays with organophosphates or spinosad (Navarro-Llopis et al. 2008, 2012). This density, however, can be insufficient to control higher populations than those studied in the Mediterranean Basin or in other more sensitive crops, such as stone fruit or figs (Hashem et al. 1986).

In olive fruit fly trials conducted in Greece, Haniotakis et al. (1991) applied 50–150 traps/ha (from 1 trap for every 3 trees to 1 trap per tree) over wide areas, combining a female attractant (50–70 g ammonium carbonate) with the sex pheromone (50 mg 1,7-dioxaspiro[5.5]undecane). Results showed a reduction in the fruit fly population and olive damage in mass trapping treated plots when compared with control plots. Broumas et al. (2002) tested densities from 75 to 100 traps/ha (depending on the tree distances) baited with a combination of ammonium acetate and pheromone. Pest population densities and fruit infestation levels were evaluated, and mass trapping obtained statistically the same efficacy as two insecticide sprays. It is important to highlight that significant differences were obtained in the third year of consecutive treatments, when fruit damage was reduced threefold in some plots with mass trapping compared to insecticide sprays. This result agrees with our unpublished data for field trials conducted in 260 ha during the last 4 years (as described above) in which the mass trapping technique had a cumulative effect when applied in consecutive years, and consequently the trap density required for effective control was reduced.

In general, the trap density used depends on several factors, such as pest density, trap design, and the attractive power of the lures. Concerning the joint effect of trap + attractant efficacy, the distance between traps might depend on the effective attraction radius (EAR). The EAR was proposed as an index of the attractive strength for a trap releasing semiochemicals (Byers et al. 1989). The EAR is the radius of the spherical volume around an unbaited trap needed to catch, merely by interception, as many dispersing insects as were actually caught on the baited trap. Therefore, EAR correlates positively with the strength of the attractant, i.e., the distance of attraction. This index is independent of insect density, locality, or duration of the test. Peck and McQuate (2000) identified 20 m as the EAR value for sticky panels baited with the commercial three-component lure (putrescine, ammonium acetate, and trimethylamine) for *C. capitata*. Theoretically, therefore, each trap will attract the flies from 1,200 m^2, and the number of required traps would be 8–9 traps per hectare. This density has been used for monitoring purposes (Peck and McQuate 2000), however, trap density requirements are higher for mass trapping. Field trials conducted in Mediterranean countries demonstrated that densities of 20 traps/ha can obtain similar results than malathion bait sprays (Mediouni et al. 2010), although fruit damage was quite high (nearly 30 %). This trap density should be increased to 40 (Boulahia Kheder et al. 2012) or 50 traps/ha (Navarro-Llopis et al. 2013) to obtain an acceptable reduction in fruit damage. Moreover, trap density depends on the efficacy of the attractants used, and therefore the impact of lure type (as well as trap type) on optimum trap density should be investigated as done recently for *C. capitata* (Gazit et al. 1998; Navarro-Llopis et al. 2008), *Anastrepha* spp. (Martínez et al. 2007a; Díaz-Fleischer et al. 2009), *B. tryoni* (Dominiak and Helen 2010), *B. dorsalis* (Cornelius et al. 2000), and *B. oleae* (Broumas and Haniotakis 1994).

Another important issue concerns the timing of trap deployment. The time at which traps are placed in the field depends on the lifespan of the attractants and the servicing costs. The main aim of mass trapping technique is to suppress fly populations; thus, the longer the attractants are active in the field, the greater suppression achievable. The best performance will be obtained when traps are operating during the whole year in tropical climates with year-round host availability or during the whole fruiting season in temperate climates, but the cost of maintenance over such long time intervals may be prohibitive. Replacement of attractants may be expensive but, in some cases, may be necessary to cover a greater window of time to gain adequate control of fruit fly populations. In this regard, wet traps using protein-based attractants or ammonium salts are the most sensitive to weather conditions, and even monthly replacement of the baits may be required. This is certainly the case for Elkofon or Olipe traps that are often used against *B. oleae* in olive crops (Navrozidis et al. 2000; Lacasta et al. 2005; Martínez et al. 2007b). As the weather is very hot and dry during summer in olive orchards, these traps require at least two bait replacements to cover the fruit ripening period, from 1 month before olive pit formation (sensitive moment) to the harvesting date (total of 6 months) (Broumas et al. 2002). By contrast, dry attractants have longer lifespans and can be used as long as 4–6 months (Colas et al. 2012; Navarro-Llopis et al. 2012). In the case of the Mediterranean fruit fly in the Mediterranean Basin, mass trapping is usually initiated in the field at least one generation before fruit ripening to 2–3 months later when harvesting is completed. Therefore, in this particular case, attractants lasting 4 months are necessary for an efficient mass trapping strategy against *C. capitata*. By contrast, in latitudes lacking pronounced seasonal variation in temperature and with continuous host availability, a continuous mass trapping of flies should be performed to obtain positive results. This is precisely the case for many Asian fruit flies (*Bactrocera* spp.), which are established in tropical latitudes with year-long host and fly occurrence (Hui 2001). In contrast, a univoltine monophagous species, such as *R. cerasi*, with a narrow window for reproduction (1 month in the coldest places) can be easily mass trapped with short lifespan attractants that could be placed in the field just before first emergence (Katsoyannos et al. 2000). In this case, the lifespan of the attractants is obviously less important.

4 Mass Trapping Strategy for Fruit Fly Control

4.1 Mass Trapping as a Control Method

4.1.1 Efficacy Studies

In recent years, mass trapping has been shown to be a very effective pest management tool against *C. capitata* in Mediterranean agrosystems. The main factor responsible for this success was the development of new female attractant dispensers using the mixture of ammonium acetate, trimethylamine, and putrescine described by Heath et al. (1997). The incorporation of these attractants has allowed the use of mass trapping in areas with high population densities and the reduction of fruit damage to commercially admissible levels.

One of the first mass trapping trials in Spain using the three-component lure was conducted in a custard apple orchard using 1 trap per tree (160 traps/ha) (Ros et al. 1999). Through the combined use of Tephri-Traps baited with protein and the three-component lure (80 traps baited with protein and 80 baited with the three-component lure per ha), mass trapping reduced fruit damage to the same or lower levels as plots that received 10 bait sprays with malathion. Sastre et al. (1999) obtained similar results in a 0.83 ha plot in a peach orchard using 125 traps/ha baited with the three-component lure. In this case, fruits from the control plots were totally damaged, whereas in the plot treated with mass trapping only the perimeter trees had fruit infestation rates greater than 0.5 % (where infestation rate is the percentage of fruits that contain one or more medfly larvae). Alemany et al. (2004) also demonstrated the efficacy of mass trapping in Spanish citrus orchards, obtaining good control level with the female three-component lure. Tests were conducted in orchards with susceptible citrus varieties (late varieties that ripen in summer with high populations of *C. capitata*) using 165 Tephri-Traps/ha. Unfortunately, this work did not include a fruit damage assessment of the control plot, and therefore the efficacy of mass trapping was estimated using historical data on fruit infestation. Nonetheless, this work did show effective reduction of the *C. capitata* population in monitoring traps during the whole trial, with a maximum reduction from 205 females/trap/day in the control plot to 2.82 females/trap/day in the mass trapping plot.

As already noted, the cost of the mass trapping may be the impediment to its implementation, and cost is highly dependent on the number of traps required to achieve adequate control. A detailed list of mass trapping strategies with their corresponding economic assessment has been compiled in Table 15.2 based on available literature.

Table 15.2 Summary of mass trapping strategies and cost valuation

Target	Country	Lure	Trap	Density (traps/ha)	Cost[a] ($/ha)	Citations
Ceratitis capitata	Spain, Tunisia	Ammonium acetate, Trimethylamine, Alkene diamine	Maxitrap, Decis, Moskisan, Tephri-trap	50	150–250	Navarro-Llopis et al. (2013), Boulahia Kheder et al. (2012), and Lucas-Espadas and Hermosilla-Cerón (2008a)
	Spain	Protein baits (Ceratrap™)	Bottle	120	175–200	Lucas-Espadas and Hermosilla-Cerón (2008b, c)
		Ammonium biphosphate 4 %		120	50–60[b]	Alfaro (2005, personal communication)
Bactrocera oleae	Greece	Ammonium carbonate + spiroacetal	Eco-Trap	75–100	40–50	Broumas et al. (2002)
	Spain	Ammonium biphosphate	Olipe	100–150	50–60	Caballero (2001)
Anastrepha spp.	México	Protein bait (Ceratrap ™)	MS2	50–100	170–190	de los Santos-Ramos et al. (2011)
Rhagoletis pomonella	USA	Fruit volatiles	Sticky Red sphere	20–50[c]	180–600	Prokopy et al. (2005)
Rhagoletis cerasi	Central Europe	Ammonium acetate	Rebell	400–2,000	2,500–7,000[d]	Remund and Boller (1983), and Katsoyannos et al. (2000)
	Switzerland	Amonium acetate and trimethylamine	Rebell	200–800	2,600–3,500	Daniel and Grunder (2012)

[a]Cost including labour in preparation and trap deployment (labour cost are taken from cited countries)
[b]Include 2 times servicing
[c]Deployed in perimeter trapping
[d]Economic assessment from Daniel and Grunder (2012)

In some valuable crops, like citrus, the current cost of mass trapping in Spanish orchards (150–250 $/ha) (Navarro-Llopis et al. 2013) is affordable, even though more expensive than five treatments with organophosphates insecticides (100 $/ha) or spinosad (130–150 $/ha) (Navarro-Llopis et al. 2008). In the same area (Valencian Community in Spain), production of sterile males and implementation costs of releasing 3,000 males/ha for 52 weeks over 150,000 ha, can be as low as 65 $/ha (Navarro-Llopis et al. 2011). In other crops with lower profit margins, the traps employed are usually cheaper but often involve more labor for preparation. For instance, olive fruit fly trapping systems are very cheap, such as Olipe trap baited with ammonium bi-phosphate at a cost of 15 $/ha (100 units/ha) (Caballero 2001), although this cost does not include preparation and deployment of the traps that may require over 20 $/ha. Moreover, these liquid traps require servicing at least one time in the field at extra cost. Other dry traps, like Eco-Trap (Vioryl SA, Afidnes, Greece), are more expensive reaching 45–50 $/ha but do not need servicing, because no liquid needs to be refilled. This cost exceeds the 35–40 $/ha of aerial insecticide treatments, although costs associated with mass trapping may decline with implementation over successive years as fruit fly population decline with successive years of mass trapping treatment and fewer traps are required for an efficient control (Broumas et al. 2002).

However, mass trapping may be economically unfeasible for other pests, such as *R. cerasi*. In this case, no specific and effective female attractants are known and therefore only visual cues are used to increase trap effectiveness in control efforts. As a result, the number of required traps is huge and the cost prohibitive. However, the development of female attractants for the control of *R. pomonella* and the optimization of a perimeter trap deployment strategy have provided a feasible cost of mass trapping of 180 $/ha in 4 ha plots versus 112 $/ha for insecticide treatments (Prokopy et al. 2005)

4.1.2 Factors Affecting Efficacy of Mass Trapping

The various factors affecting the efficacy of mass trapping can be categorized into two groups: (i) factors related to the trap and attractants, such as trap type, density and deployment in field, efficacy of attractants, and release of the compounds and (ii) factors related to the location of mass trapping, including degree of its isolation, size, pest pressure, crop, weather, and growing techniques used.

The main objective of recent studies has been the design of new traps and the use of more efficient and specific attractants that might allow reduction of the trap density required to achieve fruit protection against medfly. Navarro-Llopis et al. (2008) found that the proper combination of trap and attractants could capture more than three times as many flies as other widely employed combinations. As an example, the combination Probodelt trap with BioLure Medfly 100 captured 3 times as many flies as the Tephri-Trap with Tri-Pack® (Econex, Murcia, Spain) attractant. Both traps were very similar, and the attractant dispensers contained the same compounds. However, the use of tubes in the lateral holes of Probodelt traps that reduced fly escapes and the release of the proper proportions of each component of the attractant in BioLure relative to Tri-Pack resulted in a significantly higher efficacy. Working in grape and citrus orchards, Lucas-Espadas and Hermosillo-Cerón (2008b, c) performed direct comparisons of several trap-lure combinations to ascertain which provided the best control against *C. capitata*. As the efficacy of mass trapping was already demonstrated, this kind of field study is more easily conducted, because untreated plots are not necessary as the goal is simply to identify the most effective combination. Results of both studies showed that the combination of three-component lure dispensers with Tephri-Traps deployed at 50 traps/ha reduced fruit damage as effectively as Cera Trap attractant in bottle traps deployed at 120 units/ha.

Another important factor influencing the success of mass trapping is the isolation and size of the field trial plots. As noted above (Sects. 3.2 and 3.3), treatments should be deployed in large areas to avoid pest intrusion from untreated areas, especially in the case of fruit flies or other pests with high mobility. Several authors determined *C. capitata* dispersal ability to be approximately 1 km during their lifespan under natural conditions (Wong et al. 1982; Plant and Cunningham 1991; Navarro-Llopis et al. 2012), although other authors have demonstrated that 90 % of population displaced only 400–700 m in their entire lifetime (Meats and Smallridge 2007). However, the same authors noted that longer movements occur, and 10 % of population movements were as great as 9.5 km. For other fruit flies, dispersal distances reported included more than 200 m for *B. oleae* (Fletcher and Economopoulos 1976) and 1 km for *B. tryoni* (Meats et al. 2003). In sum, fruit flies seeking hosts may disperse 1 km or more in their lifetime, and short movements of hundreds of meters occur commonly over several days. Therefore, mass trapping should be applied in large areas or isolated plots to avoid fruit fly invasion from the surrounding area. In field trials conducted in Spain, isolated plots of medium-large size, 15 ha (Navarro-Llopis et al. 2013) and 36 ha (Lucas-Espadas and Hermosilla-Cerón 2008b), required only 50 traps/ha to reduce *C. capitata* fruit infestation rate below 0.5 %. Therefore, mass trapping can be applied in medium sized plots if external or internal factors do not reduce trap effectiveness. One potentially important internal factor is the presence of non-treated hosts inside the mass trapping area (Alemany et al. 2004). The existence of these non-managed, alternate hosts (such as fig trees in the Mediterranean area), as pest reservoirs, inside or near the growing areas could have a major influence on the effectiveness of mass trapping (Alonso and Garcia-Marí 2012).

The intensity of pest pressure is another important factor as mass trapping is limited by cost. Consequently, when high fruit fly populations are present, the number of required traps is not economically viable. Treatments with insecticides are recommended when infestation exceeds damage thresholds, although it is not clear how the damage threshold is affected by the presence of mass trapping. For example, Spain and the USA have agreed that control measures against *C. capitata* should be carried out in citrus orchards having >0.5 FTD (flies/trap/day) using Nadel traps baited with trimedlure (USDA 2002) in order to maintain fruit infestation levels below 1.5 % (Navarro-Llopis et al. 2011). However, in several field trials where mass trapping has been applied, this threshold seems excessively restrictive as fly captures over 2 FTD resulted in infestation rates below 1 % (Lucas-Espadas and Hermosilla-Cerón 2008b; Navarro-Llopis et al. 2013). This finding can be attributed to the cumulative reduction in the fruit fly population that is associated with effective mass trapping.

As higher trap densities entail higher costs, many authors (Leza et al. 2008; Martínez-Ferrer et al. 2012; Navarro-Llopis et al. 2013) suggest the combination of mass trapping and insecticide bait sprays as the most suitable strategy for growers in areas with high *C. capitata* populations. For example, Martínez-Ferrer et al. (2012) performed field trials in 3 ha groves during three consecutive years. Results showed that fruit damage in citrus was effectively reduced by applying perimeter sprays of insecticide + protein baits combined with mass trapping, indicating this may be a suitable strategy for small plots.

4.1.3 Trap Density in Mass Trapping

Sufficiently high trap density is critical for fruit fly control, and the optimum density (i.e., one that achieves acceptable control at minimum cost) may be affected by many factors. The aforementioned trap and lure/attractant efficiency is probably the main factor, but also important are the susceptibility of host plants to fruit fly attack and the characteristics of the area treated, particularly its size and isolation. Using the three-component lure, Mediouni et al. (2010) reported that a density of 20 traps/ha was sufficient to reduce *C. capitata* fruit damage significantly in mandarin orchards, compared with malathion bait sprays, although damage levels were still above the economic threshold with both treatments. Navarro-Llopis et al. (2010) also studied the effect of different trap densities on Mediterranean fruit fly populations and fruit damage. This study, conducted with very susceptible varieties of citrus (early-ripening clementines) in orchards in Spain, demonstrated that fruit damage was reduced significantly using traps baited with the attractant mixture of ammonium acetate, trimethylamine, and methylpirrolidine deployed at 50 traps/ha compared to plots that received two bait sprays with lambda-cihalotrin. However, fruit damage was effectively avoided only with higher trap densities (75 or 100 traps/ha). These results are consistent with a later study (Martínez-Ferrer et al. 2012) in which a unique dispenser with the blend of ammonium acetate, trimethylamine, and cadaverine was tested in the Probodelt trap. This study showed that a lower density of 25 traps/ha was sufficient to control damage in mid-season varieties (when pest pressure is low), whereas a higher density of traps was required for early-ripening varieties (when pest pressure is higher). Navarro-Llopis et al. (2012) found that use of 50 traps/ha reduced fruit infestation rate below 0.5 % in vulnerable citrus varieties with high *C. capitata* populations. This fruit infestation reduction was highly significant when compared with control plots without efficient attractants in which fruit damage was 5–12 times higher than in mass trapping or in plots sprayed weekly with spinosad+bait.

In general, the required density of traps depends on the power of the attractants. This density tends to be higher when using less efficient attractants as described by Lucas-Espadas and Hermosilla-Cerón (2008b) when using the liquid protein system. In this field trial, the efficacy of two commercially available dry dispensers (both with the ammonium acetate, trimethylamine, and putrescine blend) were compared with a bottle trap baited with the Cera Trap attractant, a protein hydrolysate from pig intestinal mucosa (Sierras et al. 2006) found to be an efficient attractant for fruit flies (de los Santos-Ramos et al. 2011). Results showed that, although the number of fruit fly catches was very similar with the three tested combinations of traps + lure, the number of required traps to maintain fruit damage under the economic threshold was 50 traps/ha using the dry attractants and 120 traps/ha with the Cera Trap attractant. In spite of the higher density of liquid protein baited traps, fruit damage was four times higher in these plots compared to those receiving dry attractants. However, all the tested traps were considered efficient, because fruit infestation rate was below 1 %, which is considered an acceptable economic threshold in Spain. Although Cera Trap attractant is less effective than the three-component lure for *C. capitata*, it is also attractive to *Anastrepha* species (de los Santos-Ramos et al. 2011). Meanwhile, efficacy of three-component lure in *Anastrapha* species attraction is significantly lower than the mixture of only two components (ammonium acetate and putrescine) (Holler et al. 2006; Epsky et al. 2011). Therefore, Cera Trap can be considered a good option in countries where *Ceratitis* and *Anastrepha* coexist.

In the case of species with less effective female attractants, the number of required traps should be increased. As an example, mass trapping of *B. oleae* required deployment of 100–150 traps/ha using ammonium biphosphate as the attractant (Caballero 2001), although Broumas et al. (2002) found that 75 traps/ha were enough to obtain good control when the Eco-Trap baited with ammonium bicarbonate and the pheromone of *B. oleae* was used. More recently a similar trial was conducted in Israel with Eco-Trap and yellow boards (Rimi, Petah Tikva, Israel) baited with an ammonium bicarbonate dispenser. Traps were installed in June at a density of 1 per tree, and were replaced in August. Results showed that both traps produced a significant reduction in damage of more than 50 % compared to untreated plots (Yasin et al. 2014).

4.2 Mass Trapping Combined with Other Strategies

As a general rule, all control techniques are amenable for combined implementation with mass trapping. However, mass trapping may disrupt other control efforts (especially the Sterile Insect Technique, SIT), and this possibility needs to be considered. Insecticide treatments in combination with mass trapping may represent an effective strategy for fruit fly suppression. Bait sprays are the recommended treatments as they leave fewer residues on the fruit than cover sprays, a key point when fruit is near harvesting. When fruit fly populations are high and host fruits are vulnerable to attack, this combination is probably the most efficient strategy (Broumas et al. 2002; Leza et al. 2008). In *R. pomonella*, the combined use of spherical traps coated with toxicants and bait sprays had low efficacy with high fruit fly populations or in orchards with very susceptible varieties; therefore, bait sprays were especially important in these conditions (Bostanian et al. 1999). In several integrated pest management (IPM) programs, mass trapping has been successfully combined with other control techniques. In Spain, an IPM program included bait sprays, mass trapping, sanitation, SIT, and treatment of uncontrolled hosts (mainly host trees in backyards or in neglected plots) (Primo-Millo et al. 2003). In Tunisia, the combination of mass trapping with bait sprays, sanitation techniques, chemosterilization, and even application of giberelic acid (to delay the ripening period) has been used successfully to reduce fruit damage (Boulahia Kheder et al. 2012).

Monitoring the pest is essential when mass trapping is used as it provides information about the necessity of undertaking a second treatment to gain adequate pest suppression. In this case, application of bait sprays should be conducted when the damage threshold is exceeded, and the fruit start ripening (Martínez-Ferrer et al. 2012). For this purpose, as soon as the fruit become susceptible to fruit fly puncture, a weekly visual inspection of the fruit is suggested.

In contrast to bait spraying, mass trapping may interfere with and reduce the efficacy of other control methods. Specifically, of the effectiveness of the SIT may be reduced if mass trapping is being carried out with male attractants, and the released males are more sensitive to these than wild males (Wong et al. 1982), thus, reducing the number of sterile males available to mate with wild females. On the other hand, female mass trapping will always be a complementary method to SIT as it reduces the number of females and does not alter the overflooding ratio (sterile:wild males).

Little information is available regarding the combination of classical biological control and mass trapping. An additive effect of both techniques, or even a synergistic effect, would be desirable. But, sometimes, even an additive effect has been undetectable. Hepdurgun et al. (2009) studied the combination of *B. oleae* parasitoids (*Psyttalia concolor* (Szépligeti)) and mass trapping with Eco-Traps. During 3 years of field trials, parasitoid releases at three densities were compared with the combination of parasitoid releases plus mass trapping. Results showed that the effect of parasitoid releases was not enhanced by the use of mass trapping, although both treatments resulted in a significant decrease of fruit damage when compared with untreated plots. Although supporting data may be absent, several different control methods, including mass trapping, are often included in IPM programs (Purcell 1998). For example, the combination of bait sprays, field sanitation, pheromone or bait trapping, and augmentative releases of parasitoids constitute the control techniques included in the IPM of *B. oleae* in Greece (Kapatos 1989), *B. dorsalis*, *B. cucurbitae* and *C. capitata* in Hawaii (Mau et al. 2007), and *C. capitata* and *Bactrocera* spp. in Australia (Jessup et al. 2007).

4.3 Strengths and Limitations of Mass Trapping

The main limitation for the use of mass trapping as a pest control strategy involves the cost of the system. Prohibitive costs of implementation may derive from two factors, one related to the required trap density and the other related to the biology of the fly and the climate of the treated area. It is clear that mass trapping can be effective even with high fruit fly populations when a high density of traps is used (Alemany et al. 2004), but the cost of the traps and the attractants and the cost of deployment are limiting factors. There must be a compromise between efficacy and cost, and sometimes the minimum number of traps required to reduce fruit damage to acceptable levels has a non-acceptable cost for the growers. On the other hand, the continuously warm weather in tropical areas allows for asynchronous fruiting patterns in host plants, resulting in overlapping generations in the case of polyphagous pests and the potential for continuous infestations throughout the year (Purcell 1998). In this case, mass trapping may be required during the whole year, obviously at a higher cost than temperate regions where hosts are unavailable for part of the year.

Another factor that can affect the use of mass trapping is the inadvertent capture of natural enemies. The attraction of non-target insects to fruit fly traps has been widely studied (Asquith and Messing 1992; Uchida et al. 2004; Tschorsnig et al. 2011), and several authors pointed out the importance of reducing the number of non-target insects in traps that are intended for use over long periods of time (Uchida et al. 2004). In studies on *Bactrocera* male attractants, although the higher number of non-target captures were saprophagous insects (mostly Diptera) attracted to traps baited with decaying flies (Leblanc et al. 2009), some flower-associated insects (honey bees, syrphid flies, nitidulid beetles, and endemic crambid moths) were also trapped. Even though attraction depends mainly on the attractant employed, other factors, such as trap color (Neuenschwander 1982; Howarth and Howarth 2000), could influence the trapping of non-target individuals. In this sense, the most general attractants are the most problematic for beneficial insects. When protein baits are used for mass trapping, many non-target organisms can be found in the traps, although this effect may also be observed when synthetic attractants are used (Uchida et al. 2006). Leblanc et al. (2010a) showed that BioLure (three-component lure) attracts more non-target species than torula yeast in Hawaii, especially Drosophilidae, Neriidae, and various calyptrate flies, although very few predators, parasitoids, or pollinators were attracted (Leblanc et al. 2010b). However, field trials in Tunisia showed that BioLure was very selective towards *C. capitata*, and consequently non-target insect captures did not exceed 3 % of total catch compared with 74 % of all insects captured in traps baited with biammonium phosphate (Boulahia Kheder et al. 2011). In organic Spanish citrus orchards, the percentage of predators caught in Tephri type traps baited with the three-component lure was ca. 2 % (mainly Coccinelid beetles, Neuroptera and some Diptera species from the Cecidomyiidae and Syrphidae families) (Falcó-Gari et al. 2006), although parasitoids can reach 10 % of the total catch (mainly belonging to family

Braconidae but also to the Trichogrammatidae, Aphelinidae and Pteromalidae) (Falcó-Gari et al. 2010). The harmful effect of mass trapping on non-target species seems to be quite low compared with the effect of standard insecticide sprays or even bait sprays (Michaud 2003). In fact, Purcell et al. (1994) suggested that the toxicity of traditionally applied insecticides to natural enemies may be even higher than for fruit flies.

Certain controversy surrounds the effectiveness of the mass trapping technique. El-Sayed et al. (2009) proposed that lure and kill programs will succeed only with low density target populations and in isolated areas. However, mass trapping has succeeded in the Mediterranean Basin with high summer populations of *C. capitata* in medium sized plots (from 1 ha) as previously detailed. It is intuitively obvious that, the larger the treated areas, the higher the efficacy of this method, being also more efficient as cost is reduced.

One of the strengths of mass trapping compared to other control methods is the direct observation of the efficacy. The bodies of fruit flies inside traps can be observed and even counted to confirm the reliability of the method or follow the population pattern. For other methods, such as lure and kill or insecticide application, efficacy can be measured only indirectly with monitoring traps or assessing fruit damage. However, sometimes the interaction between mass trapping and monitoring traps is misunderstood. Some growers believe that the presence of mass trapping decreases the number of captures in monitoring traps because of competition between traps used for mass trapping traps and monitoring. Consequently, they suggest reducing the threshold of captures that is acceptable in monitoring traps and starting insecticide sprays with a lower FTD index. However, the authors' experiences do not support this practice as fruit damage observed in fields treated with mass trapping is always below the damage level expected with the recorded FTD index (Navarro-Llopis et al. 2013). This finding can be attributed to the fact that traps compete with fruit, and a proportion of the females present in the field is trapped before they sting the fruit.

Mass trapping is highly compatible with other control methods included in IPM programs (especially with bait sprays). As stated by El-Sayed et al. (2006), mass trapping can be highly effective for controlling fruit fly populations, and thus it has the potential to add value to long-term pest management. In addition, mass trapping minimizes costs and risks to the environment, because traps are placed in environmental hotspots where the target pest is likely to be locally abundant and can be left in the field to provide maintenance-free protection (IAEA 2009). Other key advantages of this technique are the absence of chemical residues in fruit, unlike for insecticide sprays, safe implementation for the growers, and easy set-up and use of the traps and baits.

5 Conclusions

Mass trapping is a valuable tool for fruit fly control, especially for those species with known female attractants that have long-lasting effectiveness in the field. Unfortunately, cost is a limiting factor that should be taken into account, and mass trapping may be feasible only in high value crops. This control method was not used extensively prior to 2001 or so. Previously, mass trapping had been used in some areas as pilot projects with variable success, but the application of this control method has increased notably in the last 10 years, and it is now applied in more than 100,000 ha in Mediterranean Basin countries. However, in the same region, there is a trend to replace mass trapping with other attract and kill techniques, specifically lure and kill. The lure-and-kill approach does not require trapping the flies and has similar efficacy and strengths as mass trapping, with some additional advantages as its lower cost because a container to retain the flies is not required. This allows designing cheaper and smaller devices that are easier to handle.

In any case, mass trapping should be considered as an available pest management tool within an IPM strategy. Nowadays, its use in area-wide operational suppression programs, comprising several square kilometers as working area, is limited to Mediterranean fruit fly in some areas of Spain and, in some cases, under government financial subsidized programs.

References

Agee HR (1985) Spectral response of the compound eye of the wild and laboratory-reared apple maggot fly, *Rhagoletis pomonella*. J Agric Entomol 2:147–154

Alemany A, Miranda MA, Alonso R, Martin Escorza C (2004) Efficacy of *C. capitata* (Wied) (Diptera: Tephritidae) female mass trapping. Edge-effect and pest multiplier role of unmanaged fruit host. Bol San Veg Plagas 30:255–264

Alonso A, Garcia-Marí F (2012) Influencia de las higueras asociadas a parcelas de cítricos sobre la abundancia poblacional de la mosca de la fruta *Ceratitis capitata*. Levante Agrícola 51:34–138

Aluja M (1996) Future trends in fruit fly management. In: Aluja M, Liedo P (eds) Fruit flies biology and management. Springer, New York, pp 309–320

Aluja M, Cabrera M, Guillen J, Celedonio H, Agora F (1989) Behaviour of *Anastrepha ludens, A. obliqua* and *A. serpentina* (Diptera: Tephritidae) on a wild mango tree (*Mangifera indica*) harbouring three McPhail traps. Int J Trop Insect Sci 10:309–318

Aluja M, Jiménez A, Piñero J, Camino M, Aldana L, Valdés ME, Castrejón V, Jácome I, Dávila AB, Figueroa R (1997a) Daily activity patterns and within-field distribution of papaya fruit flies (Diptera: Tephritidae) in Morelos and Veracruz, Mexico. Ann Entomol Soc Am 90:505–520

Aluja M, Jiménez A, Camino M, Piñero J, Aldana L, Castrejón V, Valdés ME (1997b) Habitat manipulation to reduce papaya fruit fly (Diptera: Tephritidae) damage: orchard design, use of trap crops and border trapping. J Econ Entomol 90:1567–1576

Asquith A, Messing RH (1992) Attraction of Hawaiian ground litter invertebrates to protein hydrolysate bait. Environ Entomol 21:1022–1028

Avery JW, Chambers DL, Cunningham RT, Leonhardt BA (1994) Use of ceralure and trimedlure in Mediterranean fruit fly (Diptera: Tephritidae) mass-trapping tests. J Entomol Sci 29:543–556
Balock JW, López FD (1969) Trapping for control of the Mexican fruit fly in mango and citrus groves. J Econ Entomol 62:54–56
Barrera JF, Montoya P, Rojas J (2006) Bases para la aplicación de sistemas de trampas y atrayentes en manejo integrado de plagas. In: Barrera JF, Montoya P (eds) Simposio sobre trampas y atrayentes en detección, monitoreo y control de plagas de importancia económica. Sociedad Mexicana de Entomología y El Colegio de la Frontera Sur, Tapachula, pp 1–16
Bohorquez R (1940) Experiencias de lucha contra la mosca del olivo (*Dacus oleae* Gmel.) por medio de sustancias atractivas. Bol Pat Veg Ent Agric 9:188–204
Bostanian NJ, Racette G (2001) Attract and kill, an effective technique to manage apple maggot, *Rhagoletis pomonella* [Diptera: Tephritidae] in high density Quebec apple orchards. Phytoprotection 82:25–34
Bostanian NJ, Vincent C, Chouinard G, Racette G (1999) Managing apple maggot, *Rhagoletis pomonella* (Diptera: Tephritidae), by perimeter trapping. Phytoprotection 80:21–33
Boulahia Kheder S, Salleh W, Awadi N, Fezzani M, Jrad F (2011) Efficiency of different traps and baits used in mass-trapping of the Mediterranean fly *Ceratitis capitata* (Wied.) (Diptera; Tephritidae). Integrated Control in Citrus Fruit Crops. IOBC/WPRS Bull 62:215–219
Boulahia Kheder S, Trabelsi I, Aouadi N (2012) From chemicals to IPM against the Mediterranean fruit fly *Ceratitis capitata* (Diptera, Tephritidae). In: Larramendi M, Soloneski S (eds) Integrated pest management and pest control – current and future tactics. InTech. Available from: http://www.intechopen.com/books
Brévault T, Quilici S (2007) Visual response of the tomato fruit fly, *Neoceratitis cyanescens*, to colored fruit models. Entomol Exp Appl 125:45–54
Broumas T, Haniotakis G (1994) Comparative field studies of various traps and attractants of the olive fruit fly, *Bactrocera oleae*. Entomol Exp Appl 73:145–150
Broumas T, Haniotakis G, Liaropoulos C, Tomazou T, Ragoussis N (2002) The efficacy of improved form of the mass- trapping method for the control of the olive fruit fly, *Bactrocea oleae* (Gmelin) (Dipt., Tephritidae): pilot-scale feasibility studies. J Appl Entomol 126:217–223
Bua G (1934) Experiments with substances attractive to olive fly. Rev App Ent 22:365
Byers JA, Anderbrant O, Löqvist J (1989) Effective attraction radius. J Chem Ecol 15:749–765
Caballero JA (2001) Sistemas de control de la mosca del olivo (*Bactrocera oleae*) en olivar ecológico. Experiencias en "Los Pedroches". Actas X Simposium científico- técnico. Foro del olivar y el medio ambiente. Comunicación nº 12
Christenson LE, Foote RE (1960) Biology of fruit flies. Annu Rev Entomol 5:171–192
Cohen H, Yuval B (2000) Perimeter trapping strategy to reduce Mediterranean fruit fly (Diptera: Tephritidae) damage on different host species in Israel. J Econ Entomol 93:721–725
Colas C, Domínguez M, Marti S, Alfaro C (2012) Magnet MED: a new long-life, ready to use "Attract and kill" system to controlMediterranean fruit fly *Ceratitis capitata* (Wiedemann). In: Book of abstracts of the XII international citrus congress, Valencia, Spain, November 2012, p 229
CEC (Commission of the European Communities) (2007) Commission Decision 2007/387/EC of 6 June 2007 concerning the non-inclusion of dichlorvos in Annex I to Council Directive 91/414/EEC and the withdrawal of authorisations for plant protection products containing that substance
CEC (Commission of the European Communities) (2012) Commission Decision 2012/257/EU of 11 May 2012 concerning the non-inclusion of naled for product type 18 in Annex I, IA or IB to Directive 98/8/EC of the European Parliament and of the Council concerning the placing of biocidal products on the market
Cornelius ML, Duan JJ, Messing RH (1999) Visual stimuli and the response of female oriental fruit flies (Diptera: Tephritidae) to fruit mimicking traps. J Econ Entomol 92:121–129

Cornelius ML, Nergel L, Duan JJ, Messing RH (2000) Responses of female oriental fruit flies (Diptera: Tephritidae) to protein and host fruit odors in field cage and open field tests. Environ Entomol 29:14–19

Cunningham RT (1989) Population detection. In: Robinson AS, Hooper G (eds) World Crop Pests, vol 3B, Fruit flies, their biology, natural enemies and control. Elsevier Science Publishers BV, Amsterdam, pp 169–173

Cunningham RT, Steiner LF (1972) Field trial of cue-lure + naled on saturated fiberboard blocks for control of the melon fly by the Male-Annihilation Technique. J Econ Entomol 65:505–507

Cunningham RT, Nakagawa S, Suda DY, Urago T (1978) Tephritid fruit fly trapping: liquid food baits in high and low rainfall climates. J Econ Entomol 71:762–763

Cytrynowicz M, Morgante JS, de Souza HM (1982) Visual responses of South American fruit flies, *Anastrepha fraterculus*, and Mediterranean fruit flies, *Ceratitis capitata*, to colored rectangles and spheres. Environ Entomol 11:1202–1210

Daniel C, Grunder J (2012) Integrated management of European cherry fruit fly *Rhagoletis cerasi* (L.): situation in Switzerland and Europe. Insects 3:956–988

de los Santos-Ramos M, Hernandez-Perez R, Cerdá-Subirachs JJ, Nieves-Ordaz F, Torres-Santillán JA, Bello-Rivera A, Leal-García DF (2011) An environmentally friendly alternative (MS2 ®-CeraTrap ®) for control of fruit flies in Mexico. J Food Agric Environ 9:926–927

Díaz-Fleischer F, Arredondo J, Flores S, Montoya P, Aluja M (2009) There is no magic fruit fly trap: multiple biological factors influence the response of adult *Anastrepha ludens* and *Anastrepha obliqua* (Diptera: Tephritidae) individuals to MultiLure traps baited with BioLure or NuLure. J Econ Entomol 102:86–94

Dimou I, Koutsikopoulos C, Economopoulos A, Lykakis J (2003) The distribution of olive fruit fly captures with McPhail traps within an olive orchard. Phytoparasitica 31:124–131

Domínguez J (2007) Estudio de la aplicación extensiva del método quiimioesterilizante en el control de *Ceratitis capitata* (Wiedemann). Desarrollo y mejora de los sistemas de atracción. PhD dissertation, Universidad Politécnica de Valencia, Spain

Dominiak BC, Helen N (2010) Field performance of Lynfield and McPhail traps for monitoring male and female sterile *Bactrocera tryoni* (Froggatt) and wild *Dacus newmani* (Perkins). Pest Manag Sci 66:741–744

Drew RAI, Prokopy RJ, Romig MC (2003) Attraction of fruit flies of the genus *Bactrocera* to colored mimics of host fruit. Entomol Exp Appl 107:9–45

Duan JJ, Prokopy RJ (1995) Control of apple maggot flies (Diptera: Tephritidae) with pesticide-treated red spheres. J Econ Entomol 88:700–707

Economopoulos AP (1989) Use of traps based on color and/or shape. In: Robinson AS, Hooper G (eds) World crop pests, vol 3B, Fruit flies: their biology, natural enemies and control. Elsevier, Amsterdam, pp 315–327

El-Gendy IR (2012) Elevation of attraction efficiency of Jackson trap on peach fruit fly, *Bactrocera zonata* (Saunders). Int J Agric Res 7:223–230

Eliopoulos PA (2007) Evaluation of commercial traps of various designs for capturing the olive fruit fly *Bactrocera oleae* (Diptera: Tephritidae). Int J Pest Manag 53:245–252

El-Sayed AM, Suckling DM, Wearing CH, Byers JA, Jang EB, Wearing CH (2006) Potential of mass trapping for long-term pest management and eradication of invasive species. J Econ Entomol 99:1550–1564

El-Sayed AM, Suckling DM, Wearing CH, Byers JA (2009) Potential of "lure and kill" in long-term pest management and eradication of invasive species. J Econ Entomol 102:815–835

Epsky ND, Heath RR (1998) Exploiting the interactions of chemical and visual cues in behavioral control measures for pest tephritid fruit flies. Fla Entomol 81:273–282

Epsky ND, Hendrichs J, Katsoyannos BI, Vásquez LA, Ros JP, Zümreoglu A, Pereira R, Bakri A, Seewooruthun SI, Heath RR (1999) Field evaluation of female-targeted trapping systems for *Ceratitis capitata* (Diptera: Tephritidae) in seven countries. J Econ Entomol 92:156–164

Epsky ND, Kendra PE, Peña J, Heath RR (2011) Comparison of synthetic food-based lures and liquid protein baits for capture of *Anastrepha suspensa* (Diptera: Tephritidae) adults. Fla Entomol 94:180–185

Falcó-Gari JV, Verdú MJ, Bolinches JV, Cuenca F, Alfaro F (2006) Incidencia del trampeo masivo de *Ceratitis capitata* sobre *Cryptolaemus montrouziere* y otros depredadores y parasitoides en una parcela de navelina en cultivo ecológico. Levante Agrícola 390:152–157

Falcó-Gari JV, Verdú MJ, Bolinches JV (2010) Valoración cuantitativa y cualitativa de parasitoides en capturas por trampeo masivo de *Ceratitis capitata*. Phytoma España 221:18–26

Fletcher BS, Economopoulos AP (1976) Dispersal of normal and irradiated laboratory strains and wild strains of the olive fly *Dacus oleae* in an olive grove. Entomol Exp Appl 20:183–194

Frick KE, Simkover HG, Telford HS (1954) Bionomics of the cherry fruit flies in eastern Washington, Technical bulletin. Agricultural Experiment Station, 13. Washington State Institute of Agricultural Sciences, Washington, DC

Gazit Y, Rössler Y, Epsky ND, Heath RR (1998) Trapping females of the Mediterranean fruit fly (Diptera: Tephritidae) in Israel: comparison of lures and trap type. J Econ Entomol 91:1335–1359

Gómez Clemente F (1929) Experiencias de lucha contra la *Ceratitis capitata* Wied. con cazamoscas de vidrio. Bol Pat Veg Ent Agric 4:21–38

Gothilf S, Levin G (1987) Attraction of the Mediterranean fruit fly to ammonium and protein bait traps. In: Cavalloro R (ed) Proceedings: fruit flies of economic importance 87. CEC/IOBC international symposium, 7–10 April 1987, Rome, Italy. Balkema AA, Rotterdam, pp 387–391

Gow P (1954) Proteinaceous bait for the oriental fruit fly. J Econ Entomol 47:153–160

Greany PD, Agee HR, Burditt AK, Chambers DL (1977) Field studies on color preferences of the Caribbean fruit fly, *Anastrepha suspensa* (Diptera: Tephritidae). Entomol Exp Appl 21:63–70

Greany PD, Burditt AK Jr, Chambers DL (1982) Effectiveness of Jackson traps for fruit flies improved by addition of colored patterns (*Anastrepha suspensa*). Fla Entomol 65:374–375

Green TA, Prokopy RJ, Hosmer DW (1994) Distance of response to host tree models by female apple maggot flies, *Rhagoletis pomonella* (Walsh) (Diptera: Tephritidae): interaction of visual and olfactory stimuli. J Chem Ecol 20:2393–2413

Haniotakis G, Kozyrakis M, Fitsakis T, Antonidakj A (1991) An effective mass trapping method for the control of *Dacus oleae* (Diptera: Tephritidae). J Econ Entomol 84:564–569

Harris EJ, Nakagawa S, Urago DT (1971) Sticky traps for detection and survey of three tephritids. J Econ Entomol 64:62–65

Hashem AG, Harris EJ, Saafan MH (1986) Control of the Mediterranean fruit fly in Egypt with complete coverage and partial bait sprays. Ann Agric Sci 32:1813–1825

Heath RR, Epsky ND, Landolt PJ, Sivinski J (1993) Development of attractants for monitoring Caribbean fruit flies (Diptera: Tephritidae). Fla Entomol 76:233–244

Heath RR, Epsky ND, Guzman A, Dueben BD, Manukian A, Meyer WL (1995) Development of a dry plastic insect trap with food- based synthetic attractant for the Mediterranean and the Mexican fruit flies (Diptera: Tephritidae). J Econ Entomol 88:1307–1315

Heath RR, Epsky ND, Dueben BD, Meyer WL (1996) Systems to monitor and suppress Mediterranean fruit fly (Diptera: Tephritidae) populations. Fla Entomol 79:144–153

Heath RR, Epsky ND, Dueben BD, Rizzo J, Jeronimo F (1997) Adding methyl-substituted ammonia derivatives to a food-based synthetic attractant on capture of Mediterranean and Mexican fruit flies (Diptera: Tephritidae). J Econ Entomol 90:1584–1589

Heath RR, Epsky ND, Midgarden D, Katsoyannos BI (2004) Efficacy of 1,4-diaminobutane (putrescine) in a food-based synthetic attractant for capture of Mediterranean and Mexican fruit flies (Diptera: Tephritidae). J Econ Entomol 97:1126–1131

Henneman ML, Papaj DR (1999) Role of host fruit color in the behavior of the walnut fly *Rhagoletis juglandis*. Entomol Exp Appl 93:247–256

Hepdurgun B, Turanli T, Zümreoğlu A (2009) Control of the olive fruit fly, *Bactrocera oleae*, (Diptera: Tephritidae) through mass trapping and mass releases of the parasitoid *Psyttalia concolor* (Hymenoptera: Braconidae) reared on irradiated Mediterranean fruit fly. Biocontrol Sci Tech 19:211–224

Hill AR, Hooper GHS (1984) Attractiveness of various colours to Australian tephritid fruit flies in the field. Entomol Exp Appl 35:119–128

Hiramoto MK, Arita-Tsutsumi L, Jang EB (2006) Test of effectiveness of newly formulated plastic-matrix with methyl eugenol for monitoring *Bactrocera dorsalis* (Hendel) populations. J Hawaii Entomol Soc 38:103–110

Hodson AC (1943) Lures attractive to the apple maggot. J Econ Entomol 36:545–548

Hodson AC (1948) Further studies of lures attractive to the apple maggot. J Econ Entomol 41:61–66

Holbrook FR, Fujimoro MS (1969) Mediterranean fruit flies and melon flies trapped at various heights. J Econ Entomol 62:962–963

Holler T, Sivinski J, Jenkins C, Fraser S (2006) A comparison of yeast hydrolysate and synthetic food attractants for capture of *Anastrepha suspensa* (Diptera: Tephritidae). Fla Entomol 89:419–420

Hooper GHS, Drew RAI (1979) Effect of height of trap on capture of tephritid fruit flies with cuelure and methyl eugenol in different environments. Environ Entomol 8:786–788

Howarth VMC, Howarth FG (2000) Attractiveness of methyl eugenol–baited traps to Oriental Fruit Fly (Diptera: Tephritidae): effects of dosage, placement, and color. Proc Hawaiian Entomol Soc 34:167–178

Hui Y (2001) Distribution of the oriental fruit fly (Diptera: Tephritidae) in Yunnan province. Insect Sci 8:175–182

(IAEA) International Atomic Energy Agency (2003) Trapping guidelines for area-wide fruit fly programmes. IAEA, Vienna

(IAEA) International Atomic Energy Agency (2009) Development of bait stations for fruit fly suppression in support of SIT. Report and recommendations of a consultants group meeting, Mazatlán, Mexico, 30 October–1 November 2008. IAEA-314.D4 08CT11588, Vienna

Jang EB (2011) Effectiveness of plastic matrix lures and traps against *Bactrocera dorsalis* and *Bactrocera cucurbitae* in Hawaii. J Appl Entomol 135:456–466

Jessup AJ, Dominiak B, Woods B, De Lima CPF, Tomkins A, Smallridge CJ (2007) Area-wide management of fruit flies in Australia. In: Vreysen MJB, Robinson AS, Hendrichs J (eds) Area-wide control of insect pests: from research to field implementation. Springer, Dordrecht, pp 685–697

Kapatos ET (1989) Integrated pest management systems of *Dacus oleae*. In: Robinson AS, Hooper G (eds) World crop pests, vol 3A, Fruit flies: their biology, natural enemies and control. Elsevier, Amsterdam, pp 391–398

Kaspi R, Yuval B (1999) Mediterranean fruit fly leks: factors affecting male location. Funct Ecol 13:539–545

Katsoyannos BI (1989) Response to shape, size and color. In: Robinson AS, Hooper G (eds) World crop pests, vol 3A, Fruit flies: their biology, natural enemies and control. Elsevier, Amsterdam, pp 307–324

Katsoyannos BI, Kouloussis NA (2001) Captures of the olive fruit fly *Bactrocera oleae* on spheres of different colours. Entomol Exp Appl 100:165–172

Katsoyannos BI, Heath RR, Papadopoulos NT, Epsky ND, Hendrichs J (1999) Field evaluation of Mediterranean fruit fly (Diptera: Tephritidae) female selective attractants for use in monitoring programs. J Econ Entomol 92:583–589

Katsoyannos BI, Papadopoulos NT, Stavridis D (2000) Evaluation of trap types and food attractants for *Rhagoletis cerasi* (Diptera: Tephritidae). J Econ Entomol 93:1005–1010

Lacasta C, García Rojas L, Meco R (2005) The organic control of the olive fly: the efficiency of traps and food lures. In: "Congresos y Jornadas. Serie Agricultura Ecológica". Junta de Andalucía

Landolt PJ, Heath RR, Agee HR, Tumlinson JH, Calkins CO (1988) Sex pheromone-based trapping system for papaya fruit fly (Diptera: Tephritidae). J Econ Entomol 81:1163–1169

Leblanc L, Rubinoff D, Vargas RI (2009) Attraction of non-target species to fruit fly (Diptera: Tephritidae) male lures and decaying fruit flies in traps in Hawaii. Environ Entomol 38:1446–1461

Leblanc L, Vargas RI, Rubinoff D (2010a) Captures of pest fruit flies (Diptera: Tephritidae) and non-target insects in Biolure and torula yeast traps in Hawaii. Environ Entomol 39:1626–1630

Leblanc L, Vargas RI, Rubinoff D (2010b) Attraction of *Ceratitis capitata* (Diptera: Tephritidae) and endemic and introduced non-target insects to biolure bait and its individual components in Hawaii. Environ Entomol 39:989–998

Leza MM, Juan A, Capllonch M, Alemany A (2008) Female-biased mass trapping vs. bait application techniques against the Mediterranean fruit fly, *Ceratitis capitata* (Dipt., Tephritidae). J Appl Entomol 132:753–761

Llorens JM, Matamoros E, Lucas EA, Marín C, Sierras N (2008) Integrated control of Mediterranean fruit fly *Ceratitis capitata* (Wied.) by mass trapping with an enzymatic hydrolyzed protein. IOBC/WPRS Bull 38:150–156

Lucas-Espadas A, Hermosilla-Cerón A (2008a) Evaluación de la eficacia en la captura de mosca de la fruta (*Ceratitis capitata*) de varios mosqueros y cebos, en cultivos de cítricos. Levante Agrícola 390:169–178

Lucas-Espadas A, Hermosilla-Cerón A (2008b) Eficacia de Ceratrap® y otros atrayentes y mosqueros, en el control de mosca de la fruta (*Ceratitis capitata*) en cítricos. Levante Agrícola 390:159–167

Lucas-Espadas A, Hermosilla-Cerón A (2008c) Eficacia de Ceratrap y otros atrayentes y mosqueros, en el control de mosca de fruta (*Ceratitis capitata*) en uva de mesa. Agrícola Vergel 319:304–310

Luque-López E, Pereda-Cruz L (2003) La selectividad de las trampas "Olipe" (atrayente: cebos alimenticios) en la captura de la mosca del olivo Bactrocera Oleae (Gmelin). Toll Negre 2:24–33

Martínez AJ, Salinas EJ, Rendón P (2007a) Capture of *Anastrepha* species (Diptera: Tephritidae) with multilure traps and biolure attractants in Guatemala. Fla Entomol 90:258–263

Martínez E, Miranda MA, Monerris M, Alemany A (2007b) Comparative field studies of various traps and attractants for the Olive fruit fly, *Bactrocera oleae* in the Balearic Islands. In Development of improved attractants and their integration into fruit fly SIT management programmes. Joint FAO/IAEA, pp 21–25

Martínez-Ferrer MT, Campos JM, Fibla JM (2012) Field efficacy of *Ceratitis capitata* (Diptera: Tephritidae) mass trapping technique on clementine groves in Spain. J Appl Entomol 136:181–190

Mau RFL, Jang EB, Vargas RI (2007) The Hawaii area-wide fruit fly pest management programme: influence of partnerships and a good education programme. In: Vreysen MJB, Robinson AS, Hendrichs J (eds) Area-wide control of insect pests: from research to field implementation. Springer, Dordrecht, pp 671–684

McPhail M (1939) Protein lures for fruit flies. J Econ Entomol 32:758–761

McQuate GT, Sylva CD, Jang EB (2005) Mediterranean fruit fly (Dipt., Tephritidae) suppression in persimmon through bait sprays in adjacent coffee plantings. J Appl Entomol 129:110–117

Meats AW, Smallridge CJ (2007) Short-and long-range dispersal of medfly, *Ceratitis capitata* (Dipt., Tephritidae), and its invasive potential. J Appl Entomol 131:518–523

Meats AW, Duthie R, Clift AD, Dominiak BC (2003) Trials on variants of the Sterile Insect Technique (SIT) for suppression of populations of the Queensland fruit fly in small towns neighbouring a quarantine (exclusion) zone. Anim Prod Sci 43:389–395

Mediouni Ben Jemâa J, Bachrouch O, Allimi E, Dhouibi MH (2010) Field evaluation of Mediterranean fruit fly mass trapping with Tripack® as alternative to malathion bait-spraying in citrus orchards. Span J Agric Res 8:400–408

Messina FJ (1990) Components of host choice by two *Rhagoletis* species (Diptera: Tephritidae) in Utah. J Kansas Entomol Soc 63:80–87

Michaud JP (2003) Toxicity of fruit fly baits to beneficial insects in citrus. J Insect Sci 3:8. Available online: www.insectscience.org/3.8

Murphy BC, Wilson LT, Dowell RV (1991) Quantifying apple maggot (Diptera: Tephritidae) preference for apples to optimize the distribution of traps among trees. Environ Entomol 20:981–987

Nakagawa S, Farias GJ, Steiner LF (1970) Response of females Mediterranean fruit flies to male lures in the relative absence of males. J Econ Entomol 63:227–229
Nakagawa S, Chambers DL, Urago T, Cunningham RT (1971) Trap-lure combinations for survey of Mediterranean fruit flies in Hawaii. J Econ Entomol 64:1211–1213
Nakagawa S, Chambers DL, Bradshaw TI, Urago T, Harris EJ (1975) Performance of a sticky trap with trimedlure impregnated in the adhesive material. J Econ Entomol 68:817–818
Nakagawa S, Prokopy RJ, Wong TTY, Ziegler JR, Mitchell SM, Urago T, Harris EJ (1978) Visual orientation of *Ceratitis capitata* flies to fruit models. Entmol Exp Appl 24:193–198
Navarro-Llopis V, Vacas S (2013) Attract-and-kill devices for fruit fly control. TEAM Newsl 13:3–8
Navarro-Llopis V, Sanchis J, Primo-Millo J, Primo-Yúfera E (2007) Chemosterilants as control agents of *Ceratitis capitata* (Diptera: Tephritidae) in field trials. Bull Entomol Res 97:359–368
Navarro-Llopis V, Alfaro F, Domínguez J, Sanchis J, Primo J (2008) Evaluation of traps and lures for mass trapping of Mediterranean fruit fly in citrus groves. J Econ Entomol 101:126–131
Navarro-Llopis V, Sanchis J, Vacas S, Alfaro Cañamás C, Primo J (2010) Efficacy assessment of bait stations as a control method for *Ceratitis capitata* (Wiedemann) in field trials. In: Book of Abstracts of the 8th ISFFEI, Valencia, September 2010, Universitat Politècnica de València, Valencia, Spain, p 126
Navarro-Llopis V, Vacas S, Sanchis J, Primo J, Alfaro C (2011) Chemosterilant bait stations coupled with sterile insect technique: an integrated strategy to control the Mediterranean fruit fly (Diptera: Tephritidae). J Econ Entomol 104:1647–1655
Navarro-Llopis V, Vacas S, Zarzo M, Primo J (2012) Dispersal ability of *Ceratitis capitata* (Diptera: Tephritidae): edge effect in area wide treatments. J Appl Entomol. doi:10.1111/jen.12029
Navarro-Llopis V, Primo J, Vacas S (2013) Efficacy of attract-and-kill devices for the control of *Ceratitis capitata*. Pest Manag Sci 69:478–482
Navrozidis EI, Vasara E, Karamanlidou G, Salpiggidis GK, Koliais SI (2000) Biological control of *Bactocera oleae* (Diptera: Tephritidae) using a Greek *Bacillus thuringiensis* isolate. J Econ Entomol 93:1657–1661
Neilson WTA, Knowlton AD, Whitman RJ (1981) Capture of apple maggot adults on Pherocon, Rebell, and sticky sphere traps. J Econ Entomol 74:203–206
Neuenschwander P (1982) Beneficial insects caught by yellow traps used in mass-trapping of the olive fly, *Dacus oleae*. Entomol Exp Appl 32:286–296
Newell W (1936) Progress report on the Key West (Florida) fruit fly eradication project. J Econ Entomol 29:116–120
Newman LJ (1930) A new fruit fly lure: preliminary report. J Agric West Aust 7:503–504
Peck SL, McQuate GT (2000) Field tests of environmentally friendly malathion replacements to suppress wild Mediterranean fruit fly (Diptera: Tephritidae) populations. J Econ Entomol 93:280–289
Pelz-Stelinski KS, Gut LJ, Isaacs R (2006) Vertical position of traps influences captures of eastern cherry fruit fly (Diptera: Tephritidae). Fla Entomol 89:80–82
Perkins FA, Hines HJG (1934) A note on some preliminary experiments with ammonia as a lure for the Queensland fruit fly (*Chaetoducus tryoni* Froggatt). Proc R Soc Qd 45:29
Piñero JC, Jácome I, Vargas R, Prokopy RJ (2006) Response of female melon fly, *Bactrocera cucurbitae*, to host-associated visual and olfactory stimuli. Entomol Exp Appl 121:261–269
Planes S (1936) Plagas del campo. Servicio Agronómico Nacional. Estación de patología vegetal de Burjasot (Valencia). Ed. Graficas Uguina, Madrid
Planes S (1959) Estado actual de los medios de lucha contra la mosca de los frutos, *Ceratitis capitata*, y la mosca del olivo *Dacus oleae*. Bol Pat Veg Ent Agric 24:51–66
Plant RE, Cunningham RT (1991) Analyses of the dispersal of sterile Mediterranean fruit flies (Diptera: Tephritidae) released from a point source. Environ Entomol 20:1493–1503
Primo-Millo E, Argilés-Herrero R, Alfaro-Lassala F (2003) Plan de actuación contra la mosca de las frutas (*Ceratitis capitata*) en la Comunidad Valenciana. Phytoma España 153:127–130

Prokopy RJ (1968) Visual responses of apple maggot flies, *Rhagoletis pomonella* (Diptera: Tephritidae): orchard studies. Entomol Exp Appl 11:403–422

Prokopy RJ (1972) Response of apple maggot flies to rectangles of different colors and shades. Environ Entomol 1:720–726

Prokopy RJ (1973) Dark enamel sphere capture as many apple maggot flies as fluorescent spheres. Environ Entomol 2:953–954

Prokopy RJ, Boller EF (1971) Response of European cherry fruit flies to colored rectangles. J Econ Entomol 64:1444–1447

Prokopy RJ, Economopoulos AP (1976) Color responses of *Ceratitis capitata* flies. Z Angew Entomol 80:434–437

Prokopy RJ, Hauschild KI (1979) Comparative effectiveness of sticky red spheres and Pherocon AM standard traps for monitoring apple maggot flies in commercial orchards. Environ Entomol 8:696–700

Prokopy RJ, Owens ED (1983) Visual detection of plants by herbivorous insects. Annu Rev Entomol 28:337–364

Prokopy RJ, Economopoulos AP, McFadden MW (1975) Attraction of wild and laboratory-cultured *Dacus oleae* flies to small rectangles of different hues, shades, and tints. Entomol Exp Appl 18:141–152

Prokopy RJ, Mittenthal RE, Wright SE (2003) Evaluation of trap deployment patterns for behavioural control of apple maggot flies (Dipt., Tephritidae). J Appl Entomol 127:276–281

Prokopy RJ, Jácome I, Bigurra E (2005) An index for assigning distances between odor-baited spheres on perimeter trees of orchards for control of apple maggot flies. Entomol Exp Appl 115:371–377

Purcell MF (1998) Contribution of biological control to integrated pest management of tephritid fruit flies in the tropics and subtropics. Integr Pest Manag Rev 3:63–83

Purcell MF, Stark JD, Messing RH (1994) Insecticide effect on three tephritid fruit flies and associated braconid parasitoids in Hawaii. J Econ Entomol 87:1455–1462

Reissig WH (1975) Performance of apple maggot traps in various apple tree canopy positions. J Econ Entomol 68:534–538

Reissig WC (1976) Comparison of traps and lures for *Rhagoletis fausta* and *R. cingulata*. J Econ Entomol 69:639–643

Remund U, Boller E (1983) Pièges Visuels Pour la Lutte Biotechnique et Prévision Négative de la Mouche de la Cerise, Rhagoletis cerasi L. In: Cavalloro R (ed) Fruit flies of economic importance. Balkema, Rotterdam, pp 490–494

Reynolds AH, Prokopy RJ (1997) Evaluation of odor lures for use with red sticky spheres to trap apple maggot (Diptera: Tephritidae). J Econ Entomol 90:1655–1660

Reynolds AH, Kaknes AM, Prokopy RJ (1998) Evaluation of two trap deployment methods to manage the apple maggot fly (Dipt., Tephritidae). J Appl Entomol 122:255–258

Riedl H, Hislop R (1985) Visual attraction of the walnut husk fly (Diptera: Tephritidae) to color rectangles and spheres. Environ Entomol 14:810–814

Robacker DC (1992) Effect of shape and size of colored traps on attractiveness to irradiated, laboratory-strain Mexican fruit flies (Diptera: Tephritidae). Fla Entomol 75:230–241

Robacker DC, Czokajlo D (2006) Effect of propylene glycol antifreeze on captures of Mexican fruit flies (Diptera: Tephritidae) in traps baited with bioLures and AFF lures. Fla Entomol 89:286–287

Robacker DC, Warfield WC (1993) Attraction of both sexes of Mexican fruit fly, *Anastrepha ludens*, to a mixture of ammonia, methylamine, and putrescine. J Chem Ecol 19:2999–3016

Robacker DC, Moreno DS, Wolfenbarger DA (1990) Effects of trap color, height, and placement around trees on capture of Mexican fruit flies (Diptera: Tephritidae). J Econ Entomol 83:412–419

Robacker DC, Moreno DS, Demilo AB (1996) Attractiveness to Mexican fruit flies of combinations of acetic acid with ammonium/amino attractants with emphasis on effects of hunger. J Chem Ecol 22:499–511

Roessler Y (1989) Insecticidal bait and cover sprays. In: Robinson AS, Hooper G (eds) World crop pests, vol 3B, Fruit flies: their biology, natural enemies and control. Elsevier, Amsterdam, pp 329–336

Ros JP, Castillo E (1994) Valoración de diferentes mosqueros para el control de la mosca de las frutas *Ceratitis capitata* Wiedemann. Bol San Veg Plagas 20:785–791

Ros JP, Castillo E, Garijo C, Navarro L, Castillo YE (1996) Ensayos de campo con un nuevo atrayente de hembras de la mosca Mediterranea de la fruta *Ceratitis capitata* Wied. (Diptera: Tephritidae). Bol San Veg Plagas 22:151–157

Ros JP, Escobar I, Garcia-Tapia FJ, Aranda G (1999) Pilot experiment to control medfly (*Ceratitis capitata* Wied.) using mass trapping technique in a custard apple (*Anona cherimolia* Mill) orchard. Bol San Veg Plagas 25:395–404

Ros JP, Wong E, Olivero J, Castillo E (2002) Mejora de mosqueros atrayentes y sistemas de retención contra la mosca Mediterránea de la fruta *Ceratitis capitata* Wied. Cómo hacer de la técnica del trampeo masivo una buena herramienta para controlar esta plaga. Bol San Veg Plagas 28:591–1514 597

Ros JP, Wong E, Olivero J, Rubio JR, Márquez AL, Castillo E, Blas P (2005) Development of attractants and "bait stations" for their integration into mass trapping against fruit fly (*Ceratitis capitata* Wied) and olive fly (*Bactrocera oleae* Gmel). Bol San Veg Plagas 31:599–607

Ros JP, Seris E, Castillo E, Cobo A, Gonzalez-Nuñez M (2009) A new advanced step to make a useful "Mass Trapping Method" to control de olive fruit fly *Bactrocera oleae* (Rossi). Comparative study of a new attractant. Bol San Veg 35:391–400

Russ K, Boller EF, Vallo V, Haisch A, Sezer S (1973) Development and application of visual traps for monitoring and control of populations of *Rhagoletis cerasi*. Entomophaga 18:103–116

Sancho-Sánchez J (2009) Killdisc: una alternativa para asegurar la viabilidad futura de la técnica de captura masiva contra mosca de la fruta. Phytoma 213:56–57

Sastre C, Melo JC, Borreli G (1999) La captura de hembras: una posible salida en el control de la mosca de la fruta (*Ceratitis capitata*, Wied.) en melocotonero. Phytoma 113:42–46

Severin H, Severin H (1915) Kerosene traps as a means of checking up the effectiveness of a poisoned bait spray to control the Mediterranean fruit-fly (*Ceratitis capitata* Wied) with a record of beneficial insects captured in the kerosene. J Econ Entomol 8:329–338

Shelly TE, Kennelly SS (2007) Settlement patterns of Mediterranean fruit flies in the tree canopy: an experimental analysis. J Insect Behav 20:453–472

Siddiqui QH, Ahmad D, Shad Rashdi SMM, Niazi S (2003) Effect of time of the day and trap height on the catches of peach/guava fruit flies *Bactrocea zonata* (Saunders) through male annihilation technique. Asian J Plant Sci 2:228–232

Sierras N, Marin C, Carrión M, Botta A, Piñol R (2006) Biological solution for control of *Ceratitis capitata* by mass trapping. In: Proceeding of the 22nd annual meeting of the International Society of Chemical Ecology, July 2006, Barcelona, Spain, p 71

Sivinski J (1990) Colored spherical traps for capture of Caribbean fruit fly, *Anastrepha suspensa*. Fla Entomol 73:123–128

Sivinski J, Klug H, Shapiro J, Lane J, Mankin R (2005) Ultraviolet reflectance on the heads and wings of *Anastrepha suspensa* (Loew) and *Ceratitis capitata* (Wiedemann) (Diptera: Tephritidae). Studia Dipterol 11:313–322

Stark JD, Vargas RI (1992) Differential response of male oriental fruit fly (Diptera: Tephritidae) to colored traps baited with methyleugenol. J Econ Entomol 85:808–812

Steiner LF (1957) Low-cost plastic fruit fly trap. J Econ Entomol 50:508–509

Steiner LF, Mitchell WC, Harris EJ, Kozuma TT, Fujimoto MS, Kozuma TT, Fujimoto MS (1965) Oriental fruit fly eradication by male annihilation. J Econ Entomol 58:961–964

Steyskal GC (1977) History and use of the McPhail trap. Fla Entomol 60:11–16

Tan KH (1985) Description of a new attractant trap and the effect of placement height on attraction of two *Dacus* species (Diptera: Tephritidae). J Plant Prot Trop 1:17–120

Tapia-Ramos E, Acin P, Casas JI (2012) Decis® Trap: a new biorational tool to manage populations of *Ceratitis capitata* (Diptera: Tephritidae). In: Proceedings of the 2nd international symposium of TEAM. Kolymbari, Greece. Ziti Publications, Thessaloniki, p 100

Thomas DB, Holler TC, Heath RR, Salinas EJ, Moses AL (2001) Trap-lure combinations for surveillance of *Anastrepha* fruit flies (Diptera: Tephritidae). Fla Entomol 84:344–351

Thomas DB, Epsky ND, Serra CA, Hall DG, Kendra PE, Heath RR (2008) Ammonia formulations and capture of *Anastrepha* fruit flies (Diptera: Tephritidae). J Entomol Sci 43:76–85

Tschorsnig HP, Seris E, Cobo A, Cobos G, Pascual S, Ros JP, Gonzalez-Nuñez M (2011) Tachinidae (Diptera) collected in traps used for mass- trapping of *Bactrocera oleae* (Rossi) (Diptera: Tephritidae) in olive groves in Central Spain. Span J Agric Res 9:1298–1306

Uchida GK, Walsh WA, Encarnacion C, Vargas RI, Stark JD, Beardsley JW, Mcinnis DO (1996) Design and relative efficiency of Mediterranean fruit fly (Diptera: Tephritidae) bucket traps. J Econ Entomol 89:1137–1142

Uchida GK, McInnis DO, Vargas RI, Kumashiro BR, Jang E (2004) Non-target arthropods captured in cue-lure baited bucket traps at area-wide pest management implementation sites in Kamuela and Kula, Hawaiian Islands. Proc Hawaiian Entomol Soc 36:135–143

Uchida GK, Mackey BE, Vargas RI, Beardsley JW, Hardy DE, Goff ML, Stark JD (2006) Response of non-target insects to methyl eugenol, cue-lure, trimedlure, and protein bait on Kauai Island, Hawaiian Islands. Proc Hawaiian Entomol Soc 38:41–71

(USDA) U.S. Department of Agriculture (2002) Importation of clementines from Spain. Federal Register. USDA, Beltsville

Vargas RI, Stark JD, Nishida T (1990) Population dynamics, habitat preference, and seasonal distribution patterns of oriental fruit fly and melon fly (Diptera: Tephritidae) in an agricultural area. Environ Entomol 19:1820–1828

Vargas RI, Stark JD, Prokopy RJ (1991) Response of oriental fruit fly (Diptera: Tephritidae) and associated parasitoids (Hymenoptera: Braconidae) to different-color spheres. J Econ Entomol 84:1503–1507

Vargas RI, Mau RFL, Stark JD, Piñero JC, Leblanc L, Souder SK (2010) Evaluation of methyl eugenol and cue-lure traps with solid lure and insecticide dispensers for fruit fly monitoring and male annihilation in the Hawaii Area-wide Pest Management Program. J Econ Entomol 103:409–415

Wakabayashi N, Cunningham RT (1991) Four component synthetic food bait for attracting both sexes of the melon fly (Diptera: Tephritidae). J Econ Entomol 84:1672–1676

Warburg MS, Yuval B (1997) Circadian patterns of feeding and reproductive activities of Mediterranean fruit flies (Diptera: Tephritidae) on various hosts in Israel. Ann Entomol Soc Am 90:487–495

Wong TT, Whitehand LC, Kobayashi K, Tanaka N, Harris EJ (1982) Mediterranean fruit fly: dispersal of wild and irradiated and untreated laboratory-reared males. Environ Entomol 11:339–343

Wu WY, Chena YP, Yang EC (2007) Chromatic cues to trap the oriental fruit fly, *Bactrocera dorsalis*. J Insect Physiol 53:509–516

Yasin S, Rempoulakis P, Nemny-Lavy E, Levi-Zada A, Tsukada M, Papadopoulos NT, Nestel D (2014) Assessment of lure and kill and mass- trapping methods against the olive fly, *Bactrocera oleae* (Rossi), in desert-like environments in the Eastern Mediterranean. Crop Prot 57:63–70

Ye W, Li L, Sun LL, Xiao C, Dong W (2012) Daily activity and spatial distribution pattern of the oriental fruit fly, *Bactrocera dorsalis* (Diptera: Tephritidae) in mango orchard, Yuanjiang, Yunnan. Acta Ecol Sin 32:5199–5207

Yee WL (2007) Attraction, feeding, and control of *Rhagoletis pomonella* (Diptera: Tephritidae) with GF-120 and added ammonia in Washington state. Fla Entomol 90:665–673

Zervas GA (1982) A new long-life trap for olive fruit fly, *Dacus oleae* (Gmelin) (Dipt., Tephritidae) and other Diptera. Z Angew Entomol 94:522–529

Part V
Phytosanitary Programs and Regulations

Chapter 16
Integrating Tephritid Trapping into Phytosanitary Programs

D.R. Lance

Abstract Systematic deployment of attractant-baited traps has long been a mainstay of phytosanitary programs for tephritid fruit flies. Trapping arrays are used for detecting, delimiting, monitoring, and confirming eradication of tephritid populations, as well as for demonstrating minimal risk of infestation of commodities in support of trade. Designing and optimizing trapping systems to meet these various functions requires a basic understanding of the efficiency of the trapping system, including the influence of programmatic variables, such as trap density and distribution. Distance-capture functions, based on results of release-recapture tests, are discussed as one tool for assessing the sensitivity of trapping systems. As a demonstration of their use, these functions are incorporated into Monte Carlo simulations for estimating probabilities of detecting populations, including the effects of population size, time (in fly generations), and programmatic variables. Examples are provided for three key pest species: *Ceratitis capitata* (Wiedemann), *Bactrocera dorsalis* (Hendel), and *B. cucurbitae* (Coquillett). Results of these or comparable analyses can be used to predict potential ranges of population size at the time of detection or estimate the maximum expected size of a residual population after control measures are applied given a continued absence of captures. Optimal design of trapping systems for any given location and purpose requires consideration of a number of factors that may include, among other things, the frequency of introduction of the target pest(s), availability and feasibility of using various control methods (insecticides, sterile insect releases, etc.), potential of the pest population to spread, and/or costs of survey, eradication, and other program activities. In combination with this information, quantitative assessments of tephritid trapping systems can be used to ensure adequate and appropriate system sensitivity, identify opportunities to improve overall cost-effectiveness of phytosanitary programs, and provide scientific justification for program strategies and protocols.

D.R. Lance (✉)
USDA-APHIS-PPQ CPHST Otis Laboratory, 1398 West Truck Road, Buzzards Bay, MA 02542, USA
e-mail: david.r.lance@aphis.usda.gov

T. Shelly et al. (eds.), *Trapping and the Detection, Control, and Regulation of Tephritid Fruit Flies*, DOI 10.1007/978-94-017-9193-9_16,

Keywords Monte Carlo simulation • Detection sensitivity • Program trigger • Sterile insect technique • Mark-release-recapture • Distance/capture function • Cost balancing • Allee effects • Systems approach • Trap rotation • Quality control • Pest-free area • Delimitation

1 Introduction

Phytosanitary programs, including those for tephritid fruit flies, incorporate some of the world's most extensive trap-based pest survey programs, and proper function of those programs can be critical to entire industries (FAO/IAEA 2013; Meats et al. 2002; USDA-APHIS-PPQ (US Department of Agriculture, Animal and Plant Health Inspection Service, Plant protection and Quarantine) 2006a, 2011). The chapters in this book cover many aspects of trapping tephritids, from basic biology and chemical ecology to highly applied and even statutory issues. Indeed, large bodies of work have been devoted to developing effective traps and lures for tephritids as well as the use of insect traps in general as tools in pest management. Arguably, though, efforts to develop theories of sampling and strategic use of traps in phytosanitary programs have lagged behind similar efforts in farm-based pest management, where traps are used to monitor populations of established pests for purposes such as determining if captures exceed economic thresholds and timing the resulting control applications. (Gut et al. 2004). Historically, design and protocols of some fruit fly trapping programs have evolved over time based on program outcome or, for some species, extrapolation and cross-attraction (for example, the detection system for *Bactrocera dorsalis* (Hendel) works for all methyl-eugenol-responding *Bactrocera* species) rather than through a rigorous analytic process. Globalization of trade and shrinking budgets are pressing phytosanitary programs to maximize the effectiveness and efficiency of their trapping efforts, while public and regulatory insistence on accountability is requiring them to defend the scientific bases for their activities. Accordingly, the application of ecological, behavioral, and statistical theory to survey and sampling design has increased in recent years in phytosanitary programs (Barclay et al. 2005; Bogich et al. 2008; Mehta et al. 2007; Venette et al. 2002; Meats, Chap. 8, this volume).

This chapter discusses strategies for integrating tephritid trapping systems into broader phytosanitary programs. It briefly reviews roles of trapping in various phases of these programs, including efforts to exclude, detect, and mitigate effects of tephritid fly populations. Strategies are presented for evaluating the effectiveness and sensitivity of trapping systems for such tasks as detecting and delimiting populations and determining the likelihood that they have been eradicated. To those ends, Monte Carlo modeling is presented as one potentially useful tool for assessing and characterizing trapping system performance. Conceptual discussions are included on designing trapping systems that will minimize overall program costs while maintaining phytosanitary security.

In general, the mission of phytosanitary programs encompasses two broad goals: first, to protect a region or industry from the introduction and establishment of exotic plant pests that can degrade agricultural and/or natural plant resources (plant protection), and second, to ensure that a region's agricultural products are acceptable for sale to trading partners and that foreign produce can be imported safely (phytosanitary certification and other aspects of trade facilitation) (USDA-APHIS-PPQ 2010). Because actions to achieve these two goals typically include strong regulatory components and tend to cover broad geographic areas, phytosanitary programs are often primarily or wholly governmental. The goals of plant protection and trade facilitation are, of course, intertwined to the degree that keeping key pests out of a region's agricultural system will make the resulting produce more acceptable to trading partners. As such, government agencies may adopt both goals, but the two can become uneasy bedfellows if "trade facilitation" slips over into "trade promotion". Agencies can then find themselves regulating and promoting trade in the same commodities – an inherent conflict of interest – at which point there is the potential to override caution and compromise plant protection policies, regulations, and quarantines. The primary goal of most plant regulatory agencies remains plant protection, and this chapter focuses largely on using tephritid traps to protect rather than promote agricultural resources and industries.

An overarching plant protection program can be viewed as a stool with three legs: pest exclusion, pest detection, and mitigation of detected pest populations, and these functions should be inherent in the program's broadly stated goals (Lodge et al. 2006; USDA-APHIS-PPQ 2010). Efforts to exclude exotic pests – quarantines, inspections, penalties, off-shore screening programs, treating commodities to eliminate possible infestations, etc. – are necessary to slow the influx of invasive pests into an area but can never be 100% effective as stand-alone methods. Adequate plant protection also requires detection of incipient populations of those exotic invasive pests that slip through the exclusion net, followed by mitigation of the potential negative effects of the population. The "seat" of the stool – which ties the three legs together – is made up of the agency's policies and regulations. Effective analytic processes – for example, pest risk assessment as well as cost-benefit and pathway analyses – are needed to develop and adapt those policies and regulations so that the priorities and activities of the "legs" are properly targeted and coordinated (Lodge et al. 2006). If any of the components of the "stool" are ineffective, the entire endeavor may be weakened and possibly fail.

2 Functions of Insect Trapping in Phytosanitary Programs

In relation to phytosanitary programs for tephritid fruit flies, traps are perhaps most often thought of as tools for detecting populations. In areas that are considered pest-free, trapping can serve a variety of other purposes as well (Box 16.1). Theoretical aspects of detecting tephritid populations using trap arrays are covered in detail in another chapter (Meats, Chap. 8, this volume); some additional perspectives and considerations for incorporating trapping into phytosanitary programs are discussed here. Detection trapping can serve a number of functions, but in most phytosanitary programs initial detection of an incipient population is primarily a prelude to mitigation of potential negative effects of its establishment and spread (Bogich et al. 2008; Lodge et al. 2006). With tephritids, that mitigation is often eradication, but in some cases different responses may be preferred, such as initiating a biological control or pest management program, or simply containing the population (Fraser et al. 2006; Lodge et al. 2006; Myers et al. 2000). Capture of a single fly typically will trigger a programmatic response – specifically, placing a denser "delimitation" array of traps around the site of the find. For example, in areas of the United States at high risk for tephritid establishment, detection arrays of 5 or 10 traps per square mile (=2.6 km^2) are increased to 80 or 100 in the square mile surrounding a first capture, with trap densities decreasing at several 1-mile (1.6 km) intervals beyond that until the density is back at the detection-trapping level (USDA-APHIS-PPQ 2003, 2006a). As the name implies, these arrays provide program managers with information on the size and geographic extent of the populations, but their purpose is several-fold. Initially, for example, delimitation grids provide information on whether, in fact, a breeding population exists or, more functionally, if regulatory and suppression actions are warranted.

Box 16.1: Functions of Tephritid Trapping as Components of Phytosanitary Programs

- Detecting incipient populations of exotic species
- Delimitation of detected populations
- Confirmation of eradication
- Monitoring/quality control of mitigation and exclusion programs
- Demonstrating acceptable levels of risk to trading partners
- Adjunct to agricultural quarantine inspection programs
- Ensuring proper targeting of phytosanitary programs

While capture of a single target pest typically generates a programmatic response, it usually will not trigger regulatory or mitigation actions (Meats, Chap. 8, this volume). Because those latter actions can be costly, place hardships on the local agricultural industry, and may produce political fallout, program managers will be reluctant to initiate them without additional assurances that a breeding population, in fact, exists in the area of the find. "Triggers" for regulatory action vary but normally require at least one additional capture within one generation and a specified distance (often based on pest's flight range) of the initial detection (IPPC 2006). For example, in the U.S., the trigger for *Ceratitis capitata* (Wiedemann) is the capture of two flies within a 3-mile radius and within a generation (USDA-APHIS-PPQ 2003, 2013). An exception to the multiple-fly trigger is often made if the first fly caught is a mated female, as this is evidence of the existence of additional flies in the area; another exception would be in the rare instance that the initial detection is an immature stage (egg, larva, or pupa) (IPPC 2006; USDA-

APHIS-PPQ 2013). As discussed by Meats (Chap. 8, this volume), delimitation-level arrays are also often used for confirming eradication. These arrays can simply be left in place throughout eradication and post-eradication activities, but, especially in larger programs or where the Sterile Insect Technique (SIT) is used, they can be costly to maintain and may be removed and then re-deployed once control treatments are finished (Barclay et al. 2005; Vreysen 2005).

Trapping for tephritids can also support trade in a number of contexts, including maintenance of pest-free areas, monitoring areas of low pest prevalence, and as part of a systems approach to phytosanitary security (IPPC 2006, 2008, 2012). Details on trapping for trade support, and the statutory framework for those efforts, are reviewed in detail by Jang et al. (Chap. 17, this volume). While a sampling effort, such as a grid of attractant-baited traps, can never "prove" the absence of a pest, it can demonstrate that the potential incidence is low enough that the presence of a breeding population is unlikely and/or the risk of finding the pest in produce from the area is minimal (Lance and Gates 1994; Meats and Clift 2005; Venette et al. 2002).

Design of the specific trapping effort should be tied to the goal of the overall phytosanitary program (see discussion below). Pest-free areas (PFA's), for example, are by their nature large (often an entire region or country), and the programs to maintain them are managed, or at least overseen, by National Plant Protection Organizations (NPPO) (IPPC 2006). Detection trapping arrays in PFA's are typically intended to find tephritid populations in time for eradication, and the survey protocol implemented will usually satisfy trading partners that fruit or other produce grown in the area being trapped will be pest-free. With that said, both the exporting and importing parties must agree on the intensity of sampling required (FAO/IAEA 2013; IPPC 2006). The confidence of trading partners can erode quickly if the effectiveness of detection efforts is called into question (Dawson et al. 1998).

Trapping is also an important component of tephritid control projects. Most obviously, perhaps, traps are used to monitor the effectiveness of control measures on target populations (FAO/IAEA 2013). The SIT is being using increasingly in phytosanitary programs for tephritids, both for the control of known populations and in "preventative release programs" (PRP's) (Vreysen 2005). PRP's are secondary exclusion programs, primarily for Mediterranean fruit flies, *C. capitata*, at present. This technique is used in areas at high risk of fruit fly introduction and establishment, where sterile flies are released continuously at low levels in an effort to block initial establishment (Bergsten et al. 1999; Dowell et al. 1999). Trapping is used for quality control of the SIT, both to ensure that the released, sterile flies are achieving proper distribution in the field and to monitor sterile fly survival (Koyama et al. 2004; Vreysen 2005). Traps that are being used to monitor effectiveness of control, or for detection in the case of a PRP, can typically double for SIT quality control monitoring (Vreysen 2005).

3 Determining Effectiveness of Trapping Systems

When designing a trapping system, it is critical initially to consider the purpose of the system or how the capture data from the system will be used. Many of the functions listed in Box 16.1 use traps to assess, in effect, whether or not a population of a pest is present in an area. This purpose is fundamentally different from typical uses of traps in more conventional IPM systems, where the presence of the pest is accepted, and sampling is used to assess the timing of or need for management actions, such as applications of insecticide (Binns and Nyrop 1992; Gut et al. 2004; Kuno 1991; also see Daane and Johnson 2010). This difference results in a basic discrepancy in theory and design between pest detection surveys and trapping systems for most IPM purposes. For IPM, it is typically sufficient to determine if pest pressure per area or per crop unit exceeds some limit. This information can normally be gathered with a fixed maximum number of sampling points, which is (within limits) independent of the physical size of the area being sampled, and the actual numbers of samples required may be well less than the maximum if certain strategies, such as sequential sampling plans, are used (Binns and Nyrop 1992; Metcalf and Luckman 1994). In contrast, with detection trapping, the concern is not with the proportion of area or crop infested, but – especially if eradication is the goal – with the actual size of the incipient population at the time of detection. In other words, the population must be caught at a size when eradication is logistically, economically, politically, and socially feasible. This requires a specified sampling intensity (number of traps) per unit area that is more or less independent of the overall size of the area being surveyed. That is to say, the maximum number of sample points (traps) isn't fixed, but increases proportionally with the size of the managed area which, in some programs, can be thousands of square kilometers.

Design of detection trapping systems is complicated by the fact that traps are relative rather than absolute sampling tools; that is, the numbers of insects in a trap will not, absent additional information, provide an accurate estimate of the actual numbers of insects present in the surrounding area, e.g., numbers per unit area or volume or per unit of habitat (Metcalf and Luckman 1994; Southwood and Henderson 2000). For many insect pest management applications, this is immaterial as long as capture can be related to, say, subsequent levels of damage to the crop. In contrast, understanding the likelihood that a population can go unnoticed within a detection trapping system requires at least a rough estimate of the probability that all individuals in a population can escape being trapped. The probabilistic nature of detection trapping, combined with a limited understanding of trap effectiveness, can lead to controversy over such issues as the possible persistence of low-level populations of pests in areas that are "officially" considered to be pest-free (Carey 1991, 2010; Chen 2010; Lance and Gates 1994; Liebhold et al. 2010; Papadopoulos et al. 2013).

Understanding the relationship of absolute population size to the probability of detection can help managers of phytosanitary programs in tasks ranging from designing trapping arrays for ensuring timely detection to convincing trading partners that there are no undetected populations residing within a given area. The likelihood that an entire population can escape detection (i.e., no flies will be caught) is simply the product of the probabilities of not being captured for each of the insects in the population (Lance and Gates 1994; McArdle 1990). Unfortunately, it's not always a simple matter to estimate the probability that a given fly will or will not be caught in a trap. As noted by Meats (Chap. 8, this volume), distance from insect to trap is a critical factor; it is also one of the critical factors that is under program control. More specifically, the maximum insect-to-trap distance will be determined by the density and layout of the trapping array. A large number of other factors influence capture probability, and some of these are "fixed" biologically or based on what the program deems to be the best available technologies and methods, while other factors are variable and beyond programmatic control (Box 16.2) (Barry et al. 2004; Rull and Prokopy 2001). Despite these variables, a basic understanding of the distance/capture function is needed in order to design a detection trapping array of known sensitivity. "Sensitivity" in this context refers to the ability of a trapping array to detect (capture at least one member of) a population of a given size, or, alternately, the size a population must reach before the array provides some pre-set probability of catching at least one fly.

Box 16.2: Factors Affecting Probability of Capturing a Tephritid Fly in a Trap

Under program control

- Distance from insect to trap (maximum distance can be controlled)
- Protocols for trap deployment and servicing

"Fixed" programmatic factors

- Effectiveness of lure (attractant and formulation)
- Efficiency of the trap

"Fixed" biological and chemical factors of the target species

- Flight potential/propensity for appetitive flight
- Type of attractant (male lure, pheromone, food lure, etc.)
- Volatility of compound(s)
- Sensitivity of receptor system to the attractant
- Odor-seeking behavior

Uncontrolled variables

- Weather
- Habitat structure

(continued)

Box 16.2 (continued)

- Availability and distribution of requisites for adult flies (food, moisture, oviposition substrates)
- Odors and other stimuli in the environment that may compete with traps

Factors related to individual flies:

- Sex
- Age and physiological and state
- Nutritional status
- Genetic factors

The proportions of insects captured from given distances provide estimates of capture probability, but measuring those proportions requires knowledge of both the numbers and locations of insects in the area surrounding the trap (Lance and Gates 1994). Because gleaning that information from an *in situ* population would be close to impossible, researchers have typically put known numbers of marked insects in known locations (i.e., mark-release-recapture) when trying to estimate sensitivity of detection trapping systems. A variety of test designs have been used to do this. One of the more common approaches is to release insects at multiple known distances either from traps within a grid used for detection or from a single trap (Cunningham and Couey 1986; Lance and Gates 1994; Robinet et al. 2008; Shelly and Nishimoto 2011). This requires the ability to mark insects released at the different distances distinctly (e.g., with different colors) so that initial trap-to-insect distance is known for all captured individuals. Data resulting from these types of studies can be used to develop discreet models or continuous functions to describe the relation between distance from a trap and probability of capture. Another approach is to release insects at a single point at the center of a grid or concentric circles of traps, although this has limitations for assessing distance-capture probabilities due to potential interference among traps (Mastro 1980; Meats and Smallridge 2007). Trapping grids of specific densities can also be evaluated either by releasing insects at numerous random locations or, preferably, at points that are equidistant from each of the surrounding traps (i.e., worst-case scenario) (Calkins et al. 1984; Elkinton and Carde 1980; Schwalbe 1981). Although the last approach produces data with limited value for developing distance-capture functions, it is an efficient means of assessing specific trapping protocols, as the ultimate sensitivity of a system is a function of its ability to detect incipient populations that are centered relatively long distances from any trap in the system. Capture rates from those longer distances can be small – in some cases well less than 1 % (Lance et al. 1998) – and thus multiple releases of large numbers of insects may be required to develop accurate estimates.

Of course, caveats apply to the use of release-recapture data for evaluating trapping systems. For one thing, data to develop distance-capture functions are not available for most species, and extrapolation across taxa is risky. The available data indicate that capture-distance relationships can vary widely among species, even within the same genus (Shelly et al. 2010; Shelly and Edu 2010; Shelly and Nishimoto 2011; also see Meats, Chap. 8, this volume). Even within a species, distance-capture data can vary with place and test conditions (Reynolds et al. 2012; Shelly and Nishimoto 2011). A preferred option is to run tests in areas where the trapping system is actually deployed, but those areas often are not infested with the target insect. As a result, such tests may be prevented by regulatory restrictions or, when allowed, require the use of reproductively sterile insects (e.g., Lance and Gates 1994; Shelly et al. 2010). Responsiveness of flies to traps and lures can vary among different wild and mass-produced strains and can also be affected by sterilization and handling procedures (Barry et al. 2002; Lance and Gates 1994; Shelly and Edu 2009; Weldon and Meats 2010). Despite this, data obtained with good quality sterile tephritid flies will often approximate, with correction factors if warranted, expected data from wild insects when evaluating responses to male lures (Lance and Gates 1994; Shelly and Edu 2009). Differences in responses of sterile and wild flies to food lures may be more difficult to reconcile. Ovaries of sterile females typically don't develop, which affects the flies' nutritional requirements and, presumably, may limit their response to food-based attractants (Kendra et al. 2010).

Examples of variation in distance-capture data within and among species are shown in Fig. 16.1, which includes data for Mediterranean fruit fly and two *Bactrocera* species. Data for *B. dorsalis* (oriental fruit fly) and *Bactrocera cucurbitae* (Coquillett) (melon fly) were taken from reports of trials in Hawaii (Shelly and Nishimoto 2011). The *C. capitata* data from California were derived from Lance and Gates (1994). Results from an additional, unpublished study on *C. capitata* are also included; these tests were roughly similar to those of Lance and Gates (1994) in that they were conducted in an operational detection grid at 10 trimedlure-baited traps per square mile but were developed in Florida in 1998-1999 (Lance et al. 1998). The tests were run in 41 contiguous "sections" (2.6-km^2 squares) in the area of Plant City, FL, and involved releases of nearly 700,000 flies over two 18-week periods. *Anastrepha* species were not included here as their distance-capture functions are less well characterized (but see Bressan and Teles 1991; Kendra et al. 2010; Kovaleski et al. 1999).

4 Monte Carlo Simulations for Estimating Performance of Trapping Systems

A number of approaches have been used to model or estimate performance of trapping systems (e.g., Barclay et al. 2005; Barclay and Humble 2008; Calkins et al. 1984; Lance and Gates 1994; Meats, Chap. 8, this volume), and it is not the intent here to exhaustively review or rate different methods. Instead, one type of model, the Monte Carlo simulation, is presented as an example. Monte Carlo

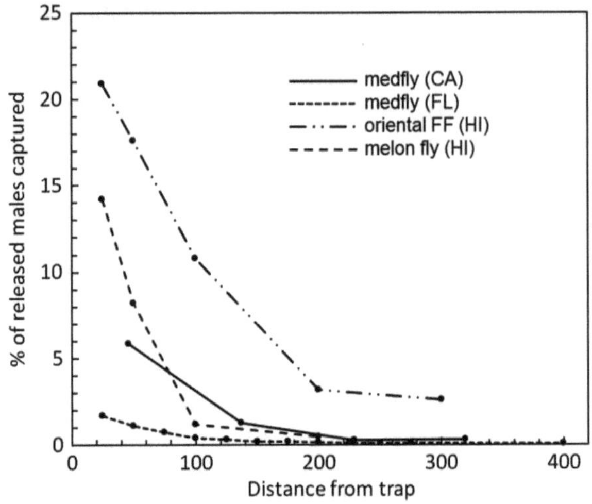

Fig. 16.1 Percentages of marked male tephritid flies captured when released at different distances (m) from survey traps. Letters in parentheses are abbreviations for names of U.S. states where trapping studies were conducted. Medfly (CA) data from Lance and Gates (1994); medfly (FL), Lance et al. (1998); *Bactrocera* spp., Shelly and Nishimoto (2011)

models can be relatively simple mathematically but, by their nature, are stochastic and iterative. The caveat of Box and Draper (1987) applies ("Essentially, all models are wrong..."), and Monte Carlo models should not be expected to generate precise predictions of trapping system performance for any specific situation. However, if properly designed and used, they may fall into the "but some are useful" category (Box and Draper 1987), as they can provide guidelines regarding the range and expected frequencies of outcomes using a specific trapping system design and aid in characterizing and visualizing the performance of trapping systems.

4.1 Description of the Models

The models presented here incorporate probabilities of capture at given distances based on release-recapture data for male tephritid flies caught in traps baited with "male lures" (Fig. 16.1). Probability of capture for a fly in the virtual population is computed based on its distance from a trap, and that probability is compared to a computer-generated random number that is sampled from a uniform distribution between 0 and 1. If the random number is less than the fly's probability of capture, the fly is considered caught; otherwise, that fly escaped capture. This procedure is repeated for every fly in the population, using a newly-generated random number for each fly, until either a fly is captured or all flies in the population have been tested. As soon as one of the flies is captured, the entire population is considered to have been detected; if none of flies are captured, the population is allowed to "grow" for another generation, and the process is repeated. As noted above, probabilities of capture can be modeled as a continuous function of probability vs. distance or discreetly simply by using proportions captured in field trials and holding them constant across applicable distance ranges (Lance and Gates 1994; Shelly and Nishimoto 2011). Here, data from release-recapture tests were fit to a modified Cauchy model (Mayer and Atzeni 1993; Meats and Smallridge 2007); specifically, $y = \alpha/\beta\pi[1 + (x/\beta)^2]$, where y is the probability of capture, and x is distance from fly to trap. Absolute differences between values of y and observed recapture rates at distances used in the field trials were weighted by the inverse of the proportion of total captures at that distance, and the sum of those values was minimized by iteratively adjusting the parameters α and β using the Solver function in Excel (Ver. 14.0, Microsoft Corporation, Redmond, WA). The weighting was done, because numeric differences between model predictions and observed recapture needed to be minimized for longer insect-to-trap distances, as relatively small numeric differences in capture probability at these distances can have a substantial influence on the ability of relatively large populations to escape detection.

The Monte Carlo model was developed in Visual Basic for Applications (VBA) using Excel macros. A run of the model consists of founding 1,000 virtual "populations", each of which is allowed to grow until detected, at which point the size and age in generations were noted. Each population was assumed to start at a unique, randomly located point within the trapping grid, and individual flies subsequently moved each generation in N-S and E-W directions. Distances moved were randomly sampled from a normal distribution with mean of zero and standard deviation set by the user. For this discussion, 25 m was used as the standard deviation in all cases, which results in a mean net displacement (per generation) of ≈ 31 m in a random direction; <1 % of individuals pupate >90 m from the site where their mother emerged. This is probably a low estimate of spread, especially for the *Bactrocera* species, although data on actual rates of spread for incipient tephritid populations are not readily available. Conservative values were chosen here to err on the side of caution and avoid over-estimating the sensitivity of the systems. The user sets values for a variety of additional parameters, including traps per unit area, whether trap locations are fixed or rotate among sites (and, if so, how frequently), maximum number of males in the first generation (30 was used here; the actual number for each population is determined randomly), and generational growth rate. In some versions of the model, population growth is stochastic and cycles annually, which can result in extinction prior to detection. Trap-site locations are based on the "quintal" system, where each 1-mi^2 (1,609 × 1,609 m) section is divided into five quintals of equal area. Trap sites are selected randomly within quintals, although there are constraints on minimum trap-to-trap distances that apply both within and between quintals. Where possible (i.e., where traps per mi^2 is evenly divisible by 5), numbers of traps per quintal is held constant.

4.2 Predicted Performance of Detection-Level Surveys

Basic output of the model predicts a number of characteristics of detection trapping systems. First, as detection trapping is probabilistic, the size of a population at first capture can vary widely just due to chance. At five traps per square mile, for example, a proportion of the virtual populations of the three species tested were,

Table 16.1 Estimated age of populations when detected in trapping grids of five traps per square mile as estimated using Monte Carlo models (populations assumed to increase 3X per generation; founders = generation 1)

	Age in generations at detection ($n = 1{,}000$)		
Species (data source)	Mean	Median	Maximum
C. capitata (CA)	3.2	3	7
C. capitata (FL)	4.2	4	10
B. dorsalis (HI)	1.8	2	5
B. cucurbitae (HI)	2.8	3	7

by chance, detected when only a few individuals were present, while others survived for a number of generations (Table 16.1) and grew quite large before a fly was caught (Fig. 16.2). Maximum size at detection and proportions of populations detected at different sizes also varied widely among species. Virtual populations of *B. dorsalis* were almost always detected before the population grew to a few hundred individuals, but a comparable degree of assurance for finding *C. capitata* didn't occur until populations were close to 10,000 based on Florida data.

The prediction of wide variation in population size at detection is in basic agreement with field experience. Detection programs will at times pick up a single insect and then catch nothing in the subsequent delimitation effort (CDFA 2008; Meats et al. 2003). To that extent, results of the model, and program experience, run counter to the concept that there is some fixed level below which detection will not occur (Carey 1991), although smaller populations are relatively less likely to be detected (Fig. 16.2). In other cases, populations that have been detected by similar trapping systems were unexpectedly large and widespread by the time the first fly was caught. This was the case in the 1997–1998 *C. capitata* infestation in central Florida, when capture of a fly in a detection trap in the Tampa area was followed by the capture of over 700 wild flies in 1997 and even greater numbers in 1998 across nine counties (Anonymous 1997). In this instance, the apparent late detection prompted an evaluation of the detection trapping system (Lance et al. 1998), and the resulting data (Fig. 16.1) indicated that the trimedlure-based detection system was indeed less sensitive in Florida than California. Despite this, estimates of the size of the actual field population in Florida, based on delimitation captures, would still have been higher than predictions of the current modeling study (Fig. 16.2). Reasons for this discrepancy remain unclear and could suggest limitations on the ability of the described method to estimate trap-system sensitivity, although other explanations, such as a large multi-locus inoculum (e.g., from commercial smuggling), are also possible.

The expected variation in population size at detection should be taken into consideration when designing detection trapping systems. More specifically, programs must be designed to ensure that available mitigation methods and resources will be effective against populations that are detected "late" and are thus relatively large when found. Tephritid eradication programs, especially those in urban areas,

Fig. 16.2 Expected frequency of tephritid populations detected by size class based on output of Monte Carlo simulations and using release-recapture data to estimate probability of capture as a function of distance from trap to emergence site of fly

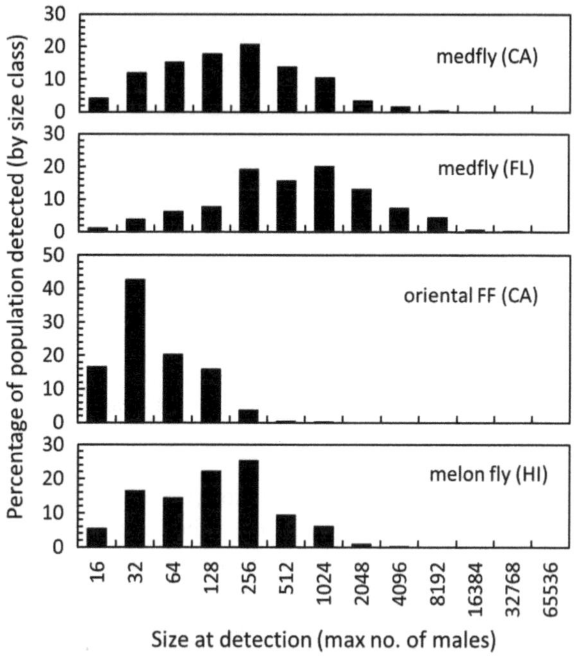

are increasingly relying on tactics for which efficacy is inversely related to the size of the target population (e.g., SIT or male annihilation). These tactics are effective at eradicating sparse populations, but their use places a premium on detecting populations while they are small and contained. The SIT, in particular, has been ineffective in some situations against large populations (Jackson and Lee 1985; Rendón et al. 2004), as ratios of sterile:wild flies have to be maintained at high enough levels to overcome the reproductive rate of the target population, any tendency of wild females to select wild males over sterile ones, and discrepancies between the local distribution of sterile and wild flies, the latter tending to be highly clumped in the field (Gavriel et al. 2012; Lance and McInnis 2005; Meats et al. 2006). Overall sterile: wild ratios as high as 100:1 or more (based on overall trap catch) have been marginally effective in some situations (Rendón et al. 2004; Vargas et al. 1994), and program managers need to consider such factors when targeting a maximum population size at detection. The primary parameter used to adjust sensitivity of a detection trapping system is trap density (Table 16.2), or, more specifically, trap spacing, which in turn sets the maximum insect-to-trap distance.

4.3 Delimitation-Level Surveys

As noted above, when a fly is initially detected, survey intensity in the immediate area is increased, often to 30–40 or more traps per km^2, to determine the size and

Table 16.2 Effects of density of trimedlure baited traps on estimated population sizes and ages (in generations) required before 10 %, 50 %, and 95 % of *C. capitata* populations are detected. Population size is the maximum number of male flies in the detected populations in the generation that detection occurred[a]

	Size (males present)			Age (generations)		
Traps per km^2	10 %	50 %	95 %	10 %	50 %	95 %
0.132[b]	224	2,187	17,496	3	6	8
0.4[b]	84	675	5,103	3	5	7
2	58	180	1,215	2	3	5
4	23	108	648	2	3	5
8	18	69	486	1	2	4

[a]Estimates are based on results of Monte Carlo simulations (1,000 repetitions each) of a detection trapping system based on mark-recapture data of Lance and Gates (1994); see text for details
[b]Estimates required substantial extrapolation of recapture rates (to longer insect-to-trap distances) from available release-recapture data

geographic extent of the infestation (USDA-APHIS-PPQ 2003; Meats, Chap. 8, this volume). A simplified version of the model above (trap locations were fixed at evenly spaced points) suggests that a grid of ca. 40 traps per km^2 (100 traps/mi^2) would be expected to catch approximately 15 % of randomly distributed male *B. dorsalis*, 5 % of *B. cucurbitae*, and 2.5 % of *C. capitata* based on California data but only 0.9 % of *C. capitata* using the Florida data. These levels of capture should be sufficient to define areas for mitigation activities, though the Florida capture rate may be marginal. In any case, treatments and elevated trapping levels should extend into "buffer" areas beyond points of capture sufficiently to account for uncertainties at the edges of the population where densities would be sparse.

Trapping at delimitation-level intensity is also used to confirm the presence of a breeding population of a pest, or, in essence, to determine if regulatory "triggers" will be met. After suppression treatments, similar trap arrays may be used to confirm eradication. For declarations of eradication, there is often a specification of trapping for a set time or number of generations, where generation time is computed using degree-day models (USDA-APHIS-PPQ 2003; Meats, Chap. 8, this volume). In U.S. programs for *C. capitata*, delimitation-level trapping continues for three generations after the last wild fly is captured before deregulation and declaration of eradication (USDA-APHIS-PPQ 2003), although this trapping has been extended to a fourth generation in some cases. The actual ability of such an array to detect a residual population within the allotted time frame is complicated by a number of factors; for example, within a fixed time period, a fast-growing population would be more likely to be identified relative to a static or slow-growing population. With that said, though, even with a generational growth rate of only 2X, the models predict that 40 traps per km^2 would find a residual *B. cucurbitae* or *C. capitata* (California data) population within 4 generations >95 % of the time and *B. dorsalis* within 4 generations on all 1,000 runs. In contrast, with the Florida *C. capitata* data, nearly a quarter of the residual populations remained undiscovered after 4 generations.

Current protocols for determining the success of an eradication effort, then, will provide varying levels of assurance that no residual population remains in the area, depending on the pest and a variety of factors that could influence pest reproduction and trap efficiency, such as weather, season, and host abundance. In the examples above, the starting conditions used in the Monte Carlo models to assess a trapping system's ability to detect residual populations following an eradication effort were the same as those for initially detecting a newly founded population. In reality, these may be unrealistic assumptions based on uncertainties regarding the expected state of any residual population following a suppression campaign and the population's subsequent growth rate.

4.4 Populations Growth Rates, Allee Effects, and Pest Detection

Growth rate of a small isolated population can have a pronounced effect on the expected size and age (in generations) at detection in either a detection or delimitation context. One way to think of this is that the likelihood of detection depends, in part, on the total number of insects in an area across both space and time. If a relatively set number of flies is needed to achieve a specific likelihood of detection, that number (summed across generations) can be achieved in a fast-growing population in a relatively few generations, whereas more generations would be required to produce that same total number of flies in slow-growing population. Total flies in the generation that was actually detected, however, will in general be lower for the slow-growing population (Table 16.3). A small population that is static in size or has a relatively constant but slow positive growth rate could potentially remain undetected in an area for enough generations to translate into several years. Consistent growth rates, however, are not a typical characteristic of tephritid flies.

Tropical and sub-tropical tephritids are multivoltine, and their population levels will fluctuate dramatically through the year based on host quality and availability, weather, and other variables (Malavasi and Morgante 1981; Papadopoulos et al. 2001a, b). The Monte Carlo models indicate that seasonal fluctuations in growth rates accelerate detection in cases where overall annual change in population size is the same (Table 16.3). In addition, those fluctuations can combine with stochastic processes to push some small populations to extinction. These models did not include Allee effects, which are factors that result in a decrease in population growth rate with decreasing population density – for example, poor reproduction due to an inability to find mates in sparse populations (Robinet et al. 2008). Allee effects could easily come into play for populations that are small enough to have moderate or low probabilities of detection in a high-density trapping grid (Liebhold and Bascompte 2003; Liebhold and Tobin 2008). While the potential state of a population following an eradication campaign is an unknown, it presumably would tend be less synchronous in time and more widely dispersed than a

Table 16.3 Effect of population growth rate on estimated population sizes and ages (in generations) required before 10 %, 50 %, and 95 % of *C. capitata* populations are detected in a grid of 2 trimedlure-baited traps per km^{2}[a]

Population growth rate[b]		Population size[c]			Age (generations)		
Per generation	Annual	10 %	50 %	95 %	10 %	50 %	95 %
1.189	2	15	36	122	2	6	15
1.414	4	20	59	264	2	5	10
2	16	24	112	640	2	4	7
4	256	29	288	1,728	2	3	5
8	4,096	48	448	4,096	2	3	4
2,4,2,0.125[d]	2	12	66	399	2	4	17
4,8,4,0.125[e]	16	25	210	1,662	2	4	8

[a]Estimates are based on results of Monte Carlo simulations (1,000 repetitions each) of a detection trapping system based on mark-recapture data of Lance and Gates (1994); see text for details
[b]Assumes four generations per year
[c]Maximum number of male flies in the generation in which detection occurred
[d]Mean growth rates for the 1st, 2nd, 3rd, and 4th generation of each year, respectively (actual rates varied around these means); 192 of 1,000 populations went extinct before being detected
[e]84 of 1,000 populations went extinct before being detected

newly founded population, and that asynchrony would tend to exacerbate Allee effects (Boukal and Berec 2009; Robinet et al. 2008). By slowing a population's growth, Allee effects could theoretically allow a residual population to linger on for some time at a level where detection probability is relatively low, even within a delimitation grid (except for cases, such as *B. dorsalis,* where capture probabilities are very high). A more likely alternative, though, might be that a population of that size would either grow within a few generations or get pushed into extinction by Allee effects in combination with seasonal and stochastic processes (Ackleha et al. 2007; Liebhold and Tobin 2008; Meats et al. 2003). For tephritids, historic trapping data suggest that this is the case (Meats and Clift 2005; Meats et al. 2003). Indeed, there is a basic assumption in invasion biology that most introductions of small numbers of insects go extinct before becoming established (Liebhold and Tobin 2008), and that assumption would certainly extend to small, isolated populations that could linger after suppression campaigns were officially ended. Accordingly, a growing body of theory suggests that it's usually not necessary to kill the last individual in order to force a population to extinction (Liebhold and Bascompte 2003; Meats and Clift 2005). Thus, while current trapping protocols for demonstrating eradication may not always provide a desired level of assurance based solely on detection of a hypothetical static or growing population, direct computation of detection probabilities likely overestimates the probability of missing a viable and sustainable population (also see Meats and Clift 2005; Meats, Chap. 8, this volume).

5 Integrating Trapping Systems into Broader Phytosanitary Programs

As noted above, the first thing to consider in designing a trapping system is its ultimate purpose. The initial discussion here assumes that a trapping array is being developed for detecting incipient populations with the goal of maintaining a pest-free zone across a broad area, such as a nation, which implies a goal of eradicating any incipient populations that are detected. Some of the basic concepts discussed will be applicable to systems designed for other purposes.

5.1 Economic Balancing

At first thought, it might seem desirable to design a system that would detect an incipient population at the earliest possible stage, thus simplifying and minimizing costs of eradication. However, unless the area being trapped is very small, such a system would likely be, in itself, prohibitively expensive. In addition, it could be considered counter-productive to detect any small populations that would otherwise go extinct rapidly without intervention. A more reasonable approach, then, could be to design a trapping system with the goal of minimizing the cost of the overall phytosanitary program for the target pest(s) (Bogich et al. 2008; Epanchin-Niell et al. 2012; Mehta et al. 2007; Pierre 2007). Specifically, greater detection trapping efforts will normally result in earlier detection of populations (when they are smaller), and money spent on detection, over time, will be inversely related to costs of the resulting eradication program. Those programmatic costs include not only the suppression measures aimed at eradication, but there may be many related expenses associated with monitoring the pest population, blocking movement of the pest out of a regulated area (typically the known infested area plus a buffer), treating and/or handling produce to keep it pest-free and thus marketable beyond the regulated area, actual loss of markets, public outreach efforts, and crop damage by the pest. Because distance from fly to trap is a critical determinant of detection sensitivity, and the number of traps in a grid increases with inverse of the square of inter-trap distance, detection sensitivity would not be expected to increase proportionally with detection system costs (Fig. 16.3). Regardless, in most systems there should be some level of trapping effort where combined costs of the detection survey and eradication is minimized (Fig. 16.4). Eradication costs in the equation need to be estimated across time, considering such factors as expected frequency of successful introductions, variation in sizes of populations at detection, and expected rate of population increase. Epanchin-Niell et al. (2012) modeled the effects of such factors on the relation between expenditures for detection and overall costs of invasive species programs, and they discuss potential strategies for optimizing the level of detection efforts.

There have been numerous introductions of major pest tephritids into many areas around the world, and the frequency of future infestations can be estimated based on

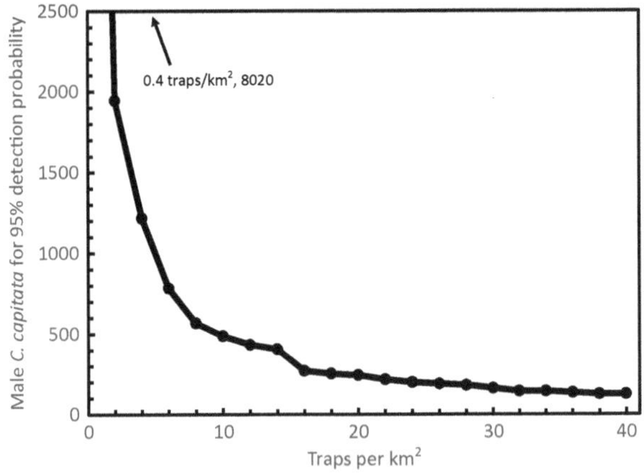

Fig. 16.3 Number of male *C. capitata* required to produce a 95 % probability of detecting a population at various trapping densities (CA data), based on results of Monte Carlo simulations. Population growth rate was assumed to be 3X per generation; trap position was static. Aside from the decrease in inter-trap travel distance (and time), survey costs should increase proportionally with traps per unit area

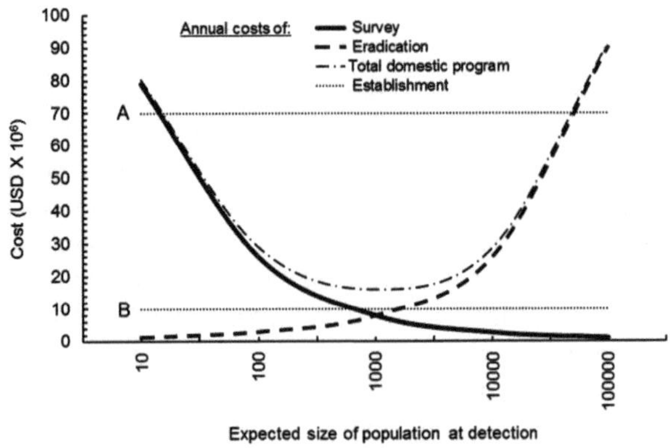

Fig. 16.4 Relation of average annual costs of operating a detection survey program and eradicating populations that are detected. This is a conceptual representation based on earlier presentations of the author; Bogich et al. (2008) and Epanchin-Niell et al. (2012) present similar graphs based on model outputs

program experience (Bergsten et al. 1999; Ha et al. 2010; Jackson and Lee 1985; Malacrida et al. 2007; Meats et al. 2003; SARDI 2001). Although there have been relatively few studies on balancing costs of detection vs. eradication as a conceptual goal (Bogich et al. 2008; Epanchin-Niell et al. 2012), trapping protocols for large-scale tephritid pest-free programs have, over the years, drifted toward minimizing overall program costs. For example, prior to 1980, trimedlure-baited traps for detecting incipient populations of *C. capitata* in were deployed at ca. 0.4 trap per km^2 in residential areas of southern California and <0.1 trap per km^2 in central California. A large, expensive, and politically contentious eradication took place in the early 1980s, the scope of which was blamed to a large extent on inefficient detection (Jackson and Lee 1985). As a result, density of detection traps in at-risk residential areas throughout the state was increased to ca. 2 traps per km^2 and subsequently to twice that level in response to ongoing increases in detections and their associated eradication costs (Dowell et al. 1999). When programs in California and Florida started treating highest-risk areas with preventative releases of sterile males, trap levels in those areas were returned to 2 traps per km^2 (Dowell et al. 1999). The reduction was in recognition of the reduced risk of new infestations in those areas and also to reduce costs associated with processing the large numbers of sterile flies caught in detection traps. A factor that may not have been considered at the time was the likely reduction in reproduction rate of incipient populations due to the presence of the sterile males, which would tend to (and has, in fact, appeared to) limit the size and spread of infestations at the time of detection. In another example, Australians opted for ca. 6 detection traps per km^2 within urban areas of their fruit fly exclusion zone, where the historically high frequency of introductions of *C. capitata* and Queensland fruit fly, *Bactrocera tryoni* (Froggatt) can be disruptive to trade (Ha et al. 2010). The high trap density has proven effective at picking up populations of those pests while they are contained in small, easily managed areas (Meats et al. 2003).

Extending cost-balancing approaches across the scope of a phytosanitary agency's programs would not be a trivial task. Ideally, costs of survey and control could also be balanced against costs of exclusion measures, such as imposing quarantines, inspections, phytosanitary treatments, and pre-clearance screening. There have been studies on relations between costs of trying to exclude pests versus managing populations of those pests (e.g., Leung et al. 2002; Olson and Roy 2005) or eradication versus other management options (Fraser et al. 2006), but those efforts have tended to consider management costs of dealing with existing pest populations rather than incorporating a detection phase that could potentially reduce subsequent management costs. A few studies have even aimed at optimizing costs and benefits of phytosanitary treatments for tephritids (Livingston et al. 2008; Livingston 2007). Currently, relations among exclusion efforts and rates of interceptions, introductions, and establishments are not clearly understood, which complicates any attempt to balance costs of exclusion against those of detection and eradication. In any case, incorporating regulatory and quarantine issues into the cost equation is beyond the scope of the present discussion.

For balancing costs of detection and eradication, calculating costs of trapping at a given level of intensity would be relatively straight-forward, as would estimating costs of an eradication effort of a given size and nature (Mumford 2005). Understanding the expected sizes of eradication projects, and their relation to survey effort, is more complicated and has greater levels of uncertainty, but the discussion here can provide a basic conceptual framework for the task. Phytosanitary agencies typically deal with a broad range of pests, and economic, environmental, and geographic aspects of risks would have to be known or estimated across the range of critical pests in order to design and prioritize the overall survey effort (USDA-APHIS-PPQ 2006b). Phytosanitary programs are designed to protect agricultural industries from the expected high costs of dealing with the pests they are excluding (Fig. 16.4, establishment cost level A). In some cases, though, projected costs associated with establishment of a species could be less than the minimum programmatic costs of survey and mitigation (Fig. 16.4; establishment cost level B), in which case survey for and subsequent mitigation of the pest would not be justified, at least strictly on economic grounds. Costs of establishment tend to be high for many exotic tephritids, where the presence of the pest adversely affects access to markets beyond the infested area or region. Export-access issues can greatly increase costs of establishment and alter the dynamics of associated cost/benefit equations (Fraser et al. 2006). Phytosanitary programs extend across years, and estimated eradication costs need be averaged across time (e.g., annualized) when balancing them against survey expenses. Introduction frequency can be estimated reasonably well for a number of major tephritid pests based on historic data. In contrast, when introductions of a pest are relatively rare, predictions of introduction frequency are problematic and rely on pest-risk and pathway analyses.

With exotic tephritids, rarely introduced species are generally not a major concern in survey design or prioritization, because efforts to balance costs of detection and eradication can be considered across most or all exotic pests in the taxon rather than on a species-by-species basis. There are two main reasons for this. First, there is a good deal of cross-attraction to a relatively small number of lures. Key pests in the genus *Ceratitis* are attracted by the "male lure" trimedlure and related compounds, such as ceralure, whereas the majority of serious pests in the genus *Bactrocera* are attracted to either methyl eugenol or cuelure (as well as a number of related compounds) (FAO/IAEA 2013). *Anastrepha* and other species that don't respond well to male lures are trapped using food-based lures (Epsky et al., Chap. 3, this volume). Depending on program focus and priorities, the food lure may be a natural product – typically torula yeast or a protein hydrolysate from corn or yeast – or a synthetic lure consisting of controlled-release devices that emit attractant compounds isolated from odors of hydrolysates. In some programs, synthetic food-based lures, rather than trimedlure, are the primary attractants used for *Ceratitis* species (FAO/IAEA 2013). The result of this cross-attraction is that survey for the majority of tropical and subtropical tephritids of concern is generally accomplished with just four types of lures, and results of some studies suggest that in some situations two or more could potentially be deployed in the same traps (Shelly et al. 2012; Vargas et al. 2012).

Another reason that survey costs for tephritids can be considered as a group is geographic coverage required for the different species. Most tephritids of regulatory concern are tropical or subtropical, so potential ranges based on climate overlap broadly among these species. Critical pest species, such as Mediterranean and oriental fruit flies, are highly polyphagous frugivores (Clarke et al. 2005; Drew and Hancock 1994; Liquido et al. 1991) and, with exceptions, their host range includes that of more specialized tephritid pests and consequently involves monitoring the same geographic areas (USDA-APHIS-PPQ 2006b).

5.2 *Additional Factors Driving the Design of Detection Systems*

In the "real world," factors other than simple economics drive goals and designs of pest detection programs (Box 16.3). Managers often tend to be risk-averse, such that a tendency to increase trapping system sensitivity could be expected if there is a real or perceived likelihood of eradication failure if populations are detected at the high end of the size distribution expected among newly detected populations. The expected spread of a population to new areas, either through the pest's own movement or human-assisted means, may also put a premium on early detection. In addition, the social and political environment within which the program operates is an increasingly important factor in the design of detection systems. In particular, public and political opposition to insecticide use, especially when applied by aircraft (Murphy 1992; Telg and Dufresne 2001), has led to increased reliance on behavioral and genetic control measures, such as the SIT and mating disruption (Aluja 1996; Bergsten et al. 1999; Brockerhoff et al. 2010; Klassen 2005; Suckling et al. 2012). The effectiveness of such measures tends to be inversely density dependent, which, as noted above, places a premium in catching the population while it is small since those same tactics will not be as effective against relatively high-density populations.

Box 16.3: Factors Affecting Optimal Design (Sensitivity) of Detection Trapping Systems Where the Program Goal Is Eradication of Detected Populations

- Cost-balancing (minimizing total costs of regulatory, survey, and mitigation components of program over time)
- Dispersal potential of the organism
- Likelihood of human-mediated spread
- Available and preferred control tactics
- Risk of eradication failure vs. population size
- Political, societal, and public relations factors
- Public health considerations, real and perceived
- Environmental consequences of eradication vs. establishment
- Pressure by affected industries and grower groups
- Funding available for detection and competing priorities.

Changes in the California trapping program for *C. capitata*, starting in the previously mentioned 1980-1982 eradication effort, illustrate how public attitudes might influence survey protocol and objectives. While part of the reason for subsequent increases in trap density has been to contain eradication costs, the resulting trapping protocols have also been designed with the goal of catching infestations before aerial application of bait sprays are required for eradication. Aerial bait sprays can be effective and economical, and they were a mainstay of the 1980's effort in California (Jackson and Lee 1985). However, their use in modern-day suburban California would likely be politically unpopular because of public concern regarding broadcast insecticide use, especially if it involves aerial applications (Anonymous 2008; Telg and Dufresne 2001).

Programmatic resources are always limited, and plant protection agencies will understandably give priority to pests that have already proven to be substantial threats. In practice, this can result in an unfortunate tendency to overlook pests that have not previously been encountered in an area when identifying targets for detection programs. Two recent examples in California, although not tephritids, are the light brown apple moth, *Epiphyas postvittana* (Walker), and European grapevine moth, *Lobesia botrana* ([Denis and Schiffermüller]). These insects are pests of regulatory concern, and the USDA has listed them as targets for detection survey since the 1980s (USDA-APHIS-PPQ 1986). Prior do their discovery, California conducted a modest survey for *E. postvittana* in 2005 and had never surveyed for *L. botrana*. The public reported the presence of the pests in 2007 (*E. postvittana*) and 2009 (*L. botrana*) (Brown et al. 2010; Gilligan et al. 2011), but both had spread to multiple counties prior to detection. The discovery of *L. botrana* spawned a large eradication program throughout much of Napa County, with smaller efforts against satellite infestations in eight other counties. Given the support of the industry and the fact that populations have been restricted almost exclusively to grape production areas, the program has progressed well to date, and eradication is anticipated in the near future. In contrast, *E. postvittana* is highly polyphagous (Wang et al. 2012), and eradication would have required repeated aerial treatments across broad residential areas. There was a brief attempt at suppressing portions of the main infestation using pheromone treatments (mating disruption), but eradication was soon dropped as a program goal for a number of reasons. These include the ongoing expansion of *E postvittana* population, an apparent lack of major damage attributed to the insect, and an organized, persistent group that produced substantial opposition to aerial spraying (Anonymous 2008). It could be argued, though, that eradication could have been successful, and with relatively little public concern, if the population had been detected while it was still restricted to a small area. For *L. botrana*, early detection could have greatly reduced the scope of the program and its resulting cost to growers and to county, state, and federal agencies.

5.3 Additional Strategies for Optimizing System Performance

Several measures may be taken to reduce the chances of missing populations until they grow unacceptably large. Increased trap density is perhaps the most effective means, but the cost of the survey program then increases proportionally. A number of other measures include:

- Rotate traps among sites on a regular basis. This can help programs in two ways. First, most of the more important tephritid pests are polyphagous, including *C. capitata* and many of the key *Bactrocera* and *Anastrepha* species. Rotating traps among sites allows traps to be kept in fruiting host plants throughout most of the year, which is generally recommended to ensure effectiveness (FAO/IAEA 2013; Jackson and Lee 1985; USDA-APHIS-PPQ 2006a). In addition, moving traps regularly – for example, approximately once per generation of the target pest – can avoid having areas that are at or near the trapping grid's maximum insect-to-trap distance for several consecutive generations and thus reduces the likelihood of a population growing relatively large prior to detection.

- Place traps to maximize capture. To the degree that effects of such variables are known, protocols should specify values for parameters such as host plant species (or hanging method if not in a host), location within the host plant and (if applicable) within the orchard or field, trap height, and aspect.
- "Piggy-back" detection trapping systems. Using the same personnel, reporting systems, and equipment for detection of multiple pests (tephritids and well as other plant pests with hosts in the same areas) improves economy.
- Improved lures and/or traps. Developing more efficient traps or lures that are more attractive and/or attract a larger proportion of the population (e.g., both sexes vs. only males) will increase the probability of catching flies present in the area (IAEA 1999, 2007). Even with highly effective traps and lures, however, detection sensitivity of a trap system will be limited if typical appetitive movement of the insect is restricted, such that the insects beyond a certain distance from a trap are unlikely to enter the active space of the attractant during their lifetimes (Weldon et al., Chap. 6, this volume).
- Optimize trap-servicing frequency. The labor and transportation for servicing traps are, in most cases, far more costly than the actual traps or lures. Within a fixed budget, a program can increase the number of traps per unit area (and thus detection sensitivity) simply by lengthening the time interval between subsequent trap checks. For example, switching from a 1- to 2-week service interval could allow a program to almost double the number of traps it maintains. There are limits to this, of course – service intervals must not exceed the field life of lures or trap-related factors affecting efficiency, such as drying or degradation of any liquid in traps. In addition, traps must be serviced frequently enough that the condition of captured insects is suitable for morphological and/or molecular identification per program requirements. Finally, service intervals should be short enough – less than a generation, for example – that populations can't build substantially between the time that the first fly is captured and traps are checked.
- Efficient data handling. Ensure that systems are in place to record trap location, trap type, and trapping results completely, effectively, and securely. To help with assessing populations following detections, negative as well as positive trap data should be maintained. Hand-held data loggers with GIS and barcoding capabilities can simplify these tasks. Many aspects of handling and analyzing trap data are covered in detail by Midgarden et al. (Chap. 9, this volume).
- Implement an effective Quality Control/Quality Assurance program. Quality control and assurance systems can be designed to ensure that traps and lures are up to specifications, trappers check traps per protocol, and captured flies move through the system and are properly identified (USDA-APHIS-PPQ 2006a). Technical specifications can be defined during procurement to ensure durability of traps and chemical purity of attractants, and chemical, bioassay, and performance testing of samples from vendors can ensure the quality of lots

delivered to program use. GIS and time/date stamping can be built into data acquisition systems to help demonstrate that trappers are visiting sites as scheduled. Seeding traps with flies can provide useful checks of the system. Periodic internal and external procedural reviews are recommended.

- Effective assessment of risk in time and space. Ideally, this is needed to balance trap effort against expected mitigation to minimize overall costs. In large, well-established programs, historic information on rates of introduction could potentially be used to fine-tune detection trapping efforts rather than, for example, trapping all residential areas at a single level. Another example, as discussed above, would be reducing trap density in high-risk areas when preventative release of sterile insects is ongoing. Again, limitations may arise. For example, detection systems in residential areas may have to be designed to catch populations at a size where a specific technique (such as sterile insects) can used for eradication with a high degree of assurance and using available capacity rather than being based strictly on minimizing costs. Also, while risk will fluctuate seasonally, gearing large trapping efforts up and down may be logistically difficult for reasons such as holding onto a crew of trained trappers. Finally, fly populations don't respect programmatic or political boundaries, and trapping efforts need to be properly coordinated across such divisions in order to be effective.

6 Conclusion

Traps for tephritid flies perform critical functions for phytosanitary programs, including detecting newly introduced populations, determining the size of populations, monitoring the progress of control operations, confirming eradication, and supporting trade (Box 16.1). An important first step in designing a trapping program is to consider the role of that trapping in the broader phytosanitary effort. With detection trapping, for example, that can include asking questions like "Will populations be detected at a point desired mitigations are possible using preferred tactics?," "What level of trapping can be expected to produce lowest overall program costs?," and "Will trading partners agree that risk of importing targeted pests from program areas is minimal?" Answering those types of questions requires an understanding of the performance and sensitivity of the trapping system as well as an appreciation of the probabilistic nature of detection trapping. This chapter, and references herein, provide a number of approaches that can hopefully be useful for developing that understanding. In addition, proper design of trapping systems requires an understanding and consideration of pest risk, in terms of potential economic and environmental damage as well as expected frequency of introduction. For tephritid flies, these risk factors can often be taken collectively across multiple species, as a relatively few trapping devices cover a wide range of species. Also, introduction of key exotic tephritids is frequent enough in many areas that future introduction rates can be estimated from historic data. Many current trapping protocols for exotic pest tephritids have evolved over time to be sufficiently sensitive to achieve program goals while operating within an acceptable budget. Regardless, rigorous quantitative assessments of these trapping systems could potentially identify opportunities to improve cost-effectiveness and provide scientific justification for program strategies and protocols in cases where performance or appropriateness is called into question.

References

Ackleha AS, Allen LJS, Carter J (2007) Establishing a beachhead: a stochastic population model with an Allee effect applied to species invasion. Theor Popul Biol 71:290–300

Aluja M (1996) Future trends in fruit fly management. In: McPheron BA, Steck GJ (eds) Fruit fly pests: a world assessment of their biology and management. St. Lucie Press, Delray Beach, pp 309–320

Anonymous (1997) A fly in the ointment. The Economist July 19, 1997. http://www.economist.com/printedition/1997-07-19

Anonymous (2008) Pheromones in a Californian storm. Int Pest Control 50:87–91

Barclay HJ, Humble L (2008) Probability models to facilitate a declaration that an exotic insect species has not yet invaded an area. Biol Invasions 11:1267–1280. doi:10.1007/s10530-008-9331-1

Barclay HJ, Hargrove JW, Clift A, Meats A (2005) Procedures for declaring pest free status. In: Dyck VA, Hendrichs J, Robinson AS (eds) Sterile Insect Technique: principles and practice in area-wide integrated pest management. Springer, Dordrecht, pp 233–268

Barry JD, Dowell RV, Morse JG (2002) Comparison of two sterile Mediterranean fruit fly (Diptera: Tephritidae) strains released in California's preventative release program. J Econ Entomol 95:936–944

Barry JD, Blessinger T, Morse JG (2004) Recapture of sterile Mediterranean fruit flies (Diptera: Tephritidae) in California's preventative release program. J Econ Entomol 97:1554–1562

Bergsten D, Lance D, Stefan M (1999) Mediterranean fruit flies and their management in the U.S.A. Pestic Outlook 10:207–212

Binns MR, Nyrop JP (1992) Sampling insect populations for the purpose of IPM decision making. Annu Rev Entomol 37:427–453

Bogich TL, Liebhold AM, Shea K (2008) To sample or eradicate? A cost minimization model for monitoring and managing an invasive species. J Appl Ecol 45:1134–1142

Boukal DS, Berec L (2009) Modelling mate-finding Allee effects and populations dynamics, with applications in pest control. Popul Ecol 51:445–458. doi:10.1007/s10144-009-0154-4

Box GEP, Draper NR (1987) Empirical model-building and response surfaces. Wiley, Hoboken, 669 p

Bressan S, Teles MDC (1991) On the recapture of marked adults of *Anastrepha* spp. (Diptera, Tephritidae) in an orchard after a one point release. Rev Bras Entomol 35:679–684

Brockerhoff EG, Liebhold AM, Richardson B, Suckling DM (2010) Eradication of invasive forest insects: concepts, methods, costs and benefits. New Zeal J For Sci 40(suppl):S117–S135

Brown JW, Epstein ME, Gilligan TM, Passoa SC, Powell JA (2010) Biology; identification, and history of the light brown apple moth, *Epiphyas postvittana*, (Walker) (Lepidoptera: Tortricidae: Archipini) in California: an example of the importance of local faunal surveys to document the establishment of exotic insects. Am Entomol 56:34–43

Calkins CO, Schroeder WJ, Chambers DL (1984) Probability of detecting Caribbean fruit fly, *Anastrepha suspensa* (Loew) (Diptera: Tephrititdae), populations with McPhail traps. J Econ Entomol 77:198–201

Carey JR (1991) Establishment of the Mediterranean fruit fly in California. Science 253:1369–1373
Carey JR (2010) The Mediterranean fruit fly (*Ceratitis capitata*). Am Entomol 56:158–163
CDFA (2008) False codling moth fact sheet. California Department of Food and Agriculture, Sacramento, CA. www.cdfa.ca.gov/plant/factsheets/FCM_FactSheet.pdf
Chen I (2010) From medfly to moth: raising a buzz of dissent. Science 327:134–136
Clarke AR, Armstrong KF, Carmichael AE, Milne JR, Raghu S, Roderick GK, Yeates DK (2005) Invasive phytophagous pests arising through a recent tropical evolutionary radiation: the *Bactrocera dorsalis* complex of fruit flies. Annu Rev Entomol 50:293–319. doi:10.1146/annurev.ento.50.071803.130428
Cunningham RT, Couey HM (1986) Mediterranean fruit fly (Diptera: Tephritidae): distance/response curves to trimedlure to measure trapping efficiency. Environ Entomol 15:71–74
Daane KM, Johnson MW (2010) Olive fruit fly: managing an ancient pest in modern times. Annu Rev Entomol 55:151–169
Dawson A, Hassenpflug S, Sloan J, Yoshioka I, Procassini A (1998) California agricultural trade: combating the medfly menace. Monterey Institute of International Studies, Monterey, California, USA. http://www.commercialdiplomacy.org/case_study/case_medfly1.htm
Dowell RV, Siddiqui IA, Meyer F, Spaugy EL (1999) Early results suggest sterile flies may protect S. California from medfly. Calif Agric 53:28–32
Drew RAI, Hancock DL (1994) The *Bactrocera dorsalis* complex of fruit flies (Diptera: Tephritidae: Dacinae) in Asia. Bull Entomol Res, Suppl Ser, pp 1–68. doi:10.1017/S1367426900000278
Elkinton JS, Carde RT (1980) Distribution, dispersal, and apparent survival of male gypsy moths as determined by capture in pheromone-baited traps. Environ Entomol 9:729–737
Epanchin-Niell RS, Haight RG, Berec L, Kean JM, Liebold AM (2012) Optimal surveillance and eradication of invasive species in heterogeneous landscapes. Ecol Lett 15:803–812
FAO/IAEA (Food and Agriculture Organization/International Atomic Energy Agency (2013) Trapping manual for area wide fruit fly Programmes. Insect Pest Control Section of the Joint FAO/IAEA Division, International Atomic Energy Agency. Vienna, Austria, 46 p. http://www-naweb.iaea.org/nafa/ipc/public/Trapping-Manual-Final-sept13.pdf
Fraser RW, Cook DC, Mumford JD, Wilby A, Waage JK (2006) Managing outbreaks of invasive species: eradication versus suppression. Int J Pest Manage 52:261–268
Gavriel S, Gazit Y, Leach A, Mumford J, Yuval B (2012) Spatial patterns of sterile Mediterranean fruit fly dispersal. Entomol Exp Appl 142:17–26. doi:10.1111/j.1570-7458.2011.01197.x
Gilligan TM, Epstein ME, Passoa SC, Powell JA, Sage OC, Brown JW (2011) Discovery of *Lobesia botrana* ([Denis & Schiffmüller]) in California: an invasive species new to North America (Lepidoptera: Tortricidae). Proc Entomol Soc Wash 113:14–30
Gut LJ, Stelinski LL, Thomson DR, Miller JR (2004) Behaviour-modifying chemicals: prospects and constraints in IPM. In: Koul O, Dhaliwal GS, Cuperus GW (eds) Integrated pest management: potential, constraints and challenges. CABI Publishing, Wallingford, pp 73–122
Ha A, Larson K, Harvey S, Fisher W, Malcolm L (2010) Benefit-cost analysis of options for managing Queensland fruit fly in Victoria. Department of Primary Industries (Victoria), Melbourne, 26 p
IAEA (International Atomic Energy Agency) (1999) Development of female medfly attractant systems for trapping and sterility assessment. In: Proceedings, final research co-ordination meeting, Joint FAO/IAEA Division of Nuclear Techniques in Food and Agriculture, Penang Malaysia, 1998. IAEA-TECDOC-1099, International Atomic Energy Agency, Vienna, Austria, 238 p. http://www-naweb.iaea.org/nafa/ipc/crp/ipc-attractans-medfly-female-TECDOC_1099.pdf
IAEA (International Atomic Energy Agency) (2007) Development of improved attractants and their integration into fruit fly SIT management programmes. In: Proceedings, final research co-ordination meeting. Joint FAO/IAEA Programme of Nuclear Techniques in Food and Agriculture, Vienna, Austria, 2005. IAEA-TECDOC-1574 International Atomic Energy

Agency, Vienna, Austria, 237 p. http://www-pub.iaea.org/MTCD/publications/PDF/te_1574_web.pdf
IPPC (International Plant Protection Convention) (2006) ISPM 26: establishment of pest free areas for fruit flies (Tephritidae). IPPC, FAO, Rome, Italy, 15 p. https://www.ippc.int/sites/default/files/documents//1361194003_ISPM_26_2006_En_2013-02-18%28WithA.pdf
IPPC (International Plant Protection Convention) (2008) ISPM 30: establishment of areas of low pest prevalence for fruit flies (Tephritidae). IPPC, FAO, Rome, Italy, 16 p. https://www.ippc.int/sites/default/files/documents//1323947574_ISPM_30_2008_En_2011-11-29_Refor.pdf
IPPC (International Plant Protection Convention) (2012) ISPM 35: systems approach for pest risk management of fruit flies (Tephritidae). IPPC, FAO, Rome, Italy, 10 p. https://www.ippc.int/sites/default/files/documents//1335957888_ISPM_35_2012_En_2012-05-02.pdf
Jackson DS, Lee BG (1985) Medfly in California 1980–1982. Bull Entomol Soc Am 31:29–37
Kendra PE, Epsky ND, Heath RR (2010) Effective sampling range of food-based attractants for female *Anastrepha suspensa* (Diptera: Tephritidae). J Econ Entomol 103:533–540. doi:10.1603/ec09313
Klassen W (2005) Area-wide integrated pest management and the sterile insect technique. In: Dyck VA, Hendrichs J, Robinson AS (eds) Sterile Insect Technique: principles and practice in area-wide integrated pest management. Springer, Dordrecht, pp 39–68
Kovaleski A, Sugayama RL, Malavasi A (1999) Movement of *Anastrepha fraterculus* from native breeding sites into apple orchards in Southern Brazil. Entomol Exp Appl 91:457–463
Koyama J, Kakinohana H, Miyatake T (2004) Eradication of the melon fly, *Bactrocera cucurbitae*, in Japan: importance of behavior, ecology, genetics, and evolution. Annu Rev Entomol 49:331–349
Kuno E (1991) Sampling and analysis of insect populations. Annu Rev Entomol 36:285–304
Lance DR, Gates DB (1994) Sensitivity of detection trapping systems for Mediterranean fruit flies (Diptera: Tephritidae) in southern California. J Econ Entomol 87:1377–1383
Lance DR, McInnis DO (2005) Biological basis of the Sterile Insect Technique. In: Dyck VA, Hendrichs J, Robinson AS (eds) Sterile Insect Technique: principles and practice in area-wide integrated pest management. Springer, Dordrecht, pp 69–94
Lance D, Sawyer A, Holler TC, Gates D, Boyer G (1998) Calibration of the medfly detection trapping grid in Florida. In: Lance DR (ed) Hawaii Plant Protection Center report of annual accomplishments. USDA-APHIS-PPQ, Waimanalo, HI, pp 16–22
Leung B, Lodge DM, Finnoff D, Shogren JF, Lewis MA, Lamberti G (2002) An ounce of prevention or a pound of cure: bioeconomic risk analysis of invasive species. Proc R Soc B Biol Sci 269:2407–2413. doi:10.1098/rspb.2002.2179
Liebhold A, Bascompte J (2003) The Allee effect, stochastic dynamics and the eradication of alien species. Ecol Lett 6:133–140
Liebhold AM, Tobin PC (2008) Population ecology of insect invasions and their management. Annu Rev Entomol 53:387–408
Liebhold AM, Work TT, McCullough DG, Cavey JF (2010) Response to Carey (2010). Am Entomol 56:164–165
Liquido NJ, Shinoda LA, Cunningham RT (1991) Host plants of the Mediterranean fruit fly (Diptera: Tephritidae): an annotated world review. Miscellaneous Publications of the Entomological Society of America 77, 52 p
Livingston MJ (2007) The Mediterranean fruit fly and the United States: is the probit 9 level of quarantine security efficient? Can J Agric Econ 55:515–526
Livingston M, Osteen C, Roberts D (2008) Regulating agricultural imports to keep out foreign pests and disease. Amber Waves 6:16–23
Lodge DM, Williams SL, MacIsaac H, Hayes K, Leung B, Reichard S, Mack RN, Moyle PB, Smith MT, Andow DA, Carlton JT, McMichael A (2006) Biological invasions: recommendations for U.S. policy and management. Position paper of the Ecological Society of America, 71 p

Malacrida AR, Gomulski LM, Bonizzoni M, Bertin S, Gasperi G, Guglielmino CR (2007) Globalization and fruitfly invasion and expansion: the medfly paradigm. Genetica 131:1–9
Malavasi A, Morgante JS (1981) Adult and larval population fluctuation of *Anastrepha fraterculus* and its relationship to host availability. Environ Entomol 10:275–278
Mastro VC (1980) Evaluation of disparlure-baited traps. In: Doane CC, McManus ML (eds) The gypsy moth: research toward integrated pest management, USDA Technical Bulletin 1584. U.S. Department of Agriculture, Washington, DC, pp. 549–554
Mayer DG, Atzeni MG (1993) Estimation of dispersal distances for *Cochliomyia hominivorax* (Diptera: Calliphoridae). Environ Entomol 22:368–374
McArdle BH (1990) When are rare species not there? Oikos 57:276–277
Meats A, Clift AD (2005) Zero catch criteria for declaring eradication of tephritid fruit flies: the probabilities. Aust J Exp Agric 45:1335–1340
Meats A, Smallridge CJ (2007) Short- and long-range dispersal of medfly, *Ceratitis capitata* (Dipt., Tephritidae), and its invasive potential. J Appl Entomol 131:518–523
Meats A, Clift AD, Perepelicia N (2002) Performance of permanent and supplementary traps for Mediterranean and Queensland fruit flies in South Australia 1975–2001: comparison of male lure and food lure traps. Gen Appl Entomol 31:53–58
Meats AW, Clift AD, Robson MK (2003) Incipient founder populations of Mediterranean and Queensland fruit flies in Australia: the relation of trap catch to infestation radius and models for quarantine radius. Aust J Exp Agric 43:397–406
Meats A, Smallridge CJ, Dominiak BC (2006) Dispersion theory and the sterile insect technique: application to two species of fruit fly. Entomol Exp Appl 119:247–254. doi:10.1111/j.1570-7458.2006.00415.x
Mehta SV, Haight RG, Homans FR, Polasky S, Venette RC (2007) Optimal detection and control strategies for invasive species management. Ecol Econ 61:237–245. doi:10.1016/j.ecolecon.2006.10.024
Metcalf RL, Luckman WH (eds) (1994) Introduction to insect pest management, 3rd edn. Wiley, New York, 661 p
Mumford JD (2005) Application of benefit/cost analysis to insect control using the Sterile Insect Technique. In: Dyck VA, Hendrichs J, Robinson AS (eds) Sterile Insect Technique: principles and practice in area-wide integrated pest management. Springer, Dordrecht, pp 69–94
Murphy SA (1992) Aerial pest eradication in Massachusetts and California and the pesticide malathion. Bos Coll Environ Aff Law Rev 19:851–884
Myers JH, Simberloff D, Kuris AM, Carey JR (2000) Eradication revisited: dealing with exotic species. Tree 15:316–320
Olson LJ, Roy S (2005) On prevention and control of an uncertain biological invasion. Rev Agric Econ 27:491–497. doi:10.1111/j.1467-9353.2005.00249.x
Papadopoulos NT, Katsoyannos BI, Carey JR, Kouloussis NA (2001a) Seasonal and annual occurrence of the Mediterranean fruit fly (Diptera: Tephritidae) in Northern Greece. Ann Entomol Soc Am 94:41–50
Papadopoulos NT, Katsoyannos BI, Kouloussis NA, Hendrichs J, Carey JR, Heath RR (2001b) Early detection and population monitoring of *Ceratitis capitata* (Diptera: Tephritidae) in a mixed-fruit orchard in Northern Greece. J Econ Entomol 94:971–978
Papadopoulos NT, Plant RE, Carey JR (2013) From trickle to flood: the large-scale, cryptic invasion of California by tropical fruit flies. Proc R Soc B 280:20131466. http://dx.doi.org/10.1098/rspb.2013.1466
Pierre RY (2007) Economic impact of a Mediterranean fruit fly outbreak in Florida. PhD dissertation, University of Florida, Food and Resource Economics, Gainesville, FL, USA, 158 p
Rendón P, McInnis D, Lance D, Stewart J (2004) Medfly (Diptera: Tephritidae) genetic sexing: large-scale field comparison of males-only and bisexual sterile fly releases in Guatemala. J Econ Entomol 97:1547–1553

Reynolds OL, Smallridge CJ, Cockington VG, Penrose LD (2012) The effect of release method and trial site on recapture rates of adult sterile Queensland fruit fly, *Bactrocera tryoni* (Froggatt) (Diptera: Tephritidae). Aust J Entomol 51:116–126. doi:10.1111/j.1440-6055.2011.00845.x

Robinet C, Lance DR, Thorpe KW, Onufrieva KS, Tobin PC, Liebhold AM (2008) Dispersion in time and space affect mating success and Allee effects in invading gypsy moth populations. J Anim Ecol 77:966–973

Rull J, Prokopy RJ (2001) Effect of apple-orchard structure on interception of *Rhagoletis pomonella* (Diptera: Tephritidae) flies by odor-baited traps. Can Entomol 133:355–363

SARDI (2001) Fruit Fly history (Parts 1–3). South Australian Research and Development Institute, Urrbrae, South Australia. http://www.sardi.sa.gov.au/pestsdiseases/horticulture/horticultural_pests/fruit_fly

Schwalbe CP (1981) Disparlure-baited traps for survey and detection. In: Doane CC, McManus ML (eds) The gypsy moth: research toward integrated pest management, USDA Technical Bulletin 1584. U. S. Department of Agriculture, Washington, DC, pp 542–549

Shelly TE, Edu J (2009) Capture of mass-reared vs. wild-like males of *Ceratitis capitata* (Dipt., Tephritidae) in trimedlure-baited traps. J Appl Entomol 133:640–646

Shelly TE, Edu J (2010) Mark-release-recapture of males of *Bactrocera cucurbitae* and *B. dorsalis* (Diptera: Tephritidae) in two residential areas of Honolulu. J Asia Pac Entomol 13:131–137

Shelly TE, Nishimoto J (2011) Additional measurements of distance-dependent capture probabilities for released males of *Bactrocera cucurbitae* and *B. dorsalis* (Diptera: Tephritidae) in Honolulu. J Asia Pac Entomol 14:271–276. doi:10.1016/j.aspen.2011.03.004

Shelly T, Nishimoto J, Diaz A, Leathers J, War M, Shoemaker R, Al-Zubaidy M, Joseph D (2010) Capture probability of released males of two *Bactrocera* species (Diptera: Tephritidae) in detection traps in California. J Econ Entomol 103:2042–2051. doi:10.1603/EC10153

Shelly TE, Nishimoto J, Kurashima R (2012) Captures of three economically important fruit fly species (Diptera: Tephritidae) in traps baited with liquid versus solid formulations of male lures in a Hawaiian coffee field. J Econ Entomol 105:1186–1193

Southwood TRE, Henderson PA (2000) Ecological methods, 3rd edn. Blackwell Science Ltd., Oxford

Suckling DM, Tobin PC, McCullough DG, Herms DA (2012) Combining tactics to exploit Allee effects for eradication of alien insect populations. J Econ Entomol 105:1–13. doi:10.1603/ec11293

Telg R, Dufresne MG (2001) Agricultural communications efforts during Florida's medfly infestations of 1997 and 1998. J Appl Commun 85:7–22

USDA-APHIS-PPQ (US Department of Agriculture, Animal and Plant Health Inspection Service, Plant Protection and Quarantine) (1986) Insects/pests not known to occur in the United States or of limited distribution ("INKTO/PNKTO"). USDA-APHIS-PPQ, Riverdale, 84 p

USDA-APHIS-PPQ (2003) Mediterranean fruit fly action plan. USDA-APHIS-PPQ, Riverdale, MD, 84 p

USDA-APHIS-PPQ (2006a) Review of fruit fly surveillance programs in the United States. USDA-APHIS-PPQ Fruit Fly Programs, Riverdale, MD, 106 p

USDA-APHIS-PPQ (2006b) Identification of areas susceptible to the establishment of fifty-three *Bactrocera* spp. (Diptera: Tephrididae: Dacinae) in the United States. USDA-APHIS-PPQ Science & Technology PERAL, Raleigh, NC. 42 p. http://www.aphis.usda.gov/plant_health/plant_pest_info/fruit_flies/downloads/bactrocera-susceptibility-analysis.pdf

USDA-APHIS-PPQ (2010) Road map to 2015: a strategic plan for plant protection & quarantine. USDA-APHIS-PPQ, Riverdale, MD, 88 p

USDA-APHIS-PPQ (2011) Exotic fruit fly strategic plan FY 2011–2015. USDA-APHIS-PPQ, Riverdale, MD, 24 p

USDA-APHIS-PPQ (2013) Cooperative fruit fly emergency response triggers & guidelines, USDA-APHIS-PPQ, Riverdale, MD, 2 p. http://www.aphis.usda.gov/plant_health/plant_pest_info/fruit_flies/downloads/FruitFlyTriggersGuidelines.pdf

Vargas RI, Walsh WA, Hsu CL, Spencer J, Mackey B, Whitehand L (1994) Effects of sterile Mediterranean fruit fly (Diptera: Tephritidae) releases on the target species, a nontarget tephritid, and a braconid parasitoid (Hymenoptera: Braconidae) in commercial coffee fields. J Econ Entomol 87:653–660

Vargas RI, Souder SK, Mackey B, Cook P, Morse JG, Stark JD (2012) Field trials of solid triple lure (trimedlure, methyl eugenol, raspberry ketone, and DDVP) dispensers for detection and male annihilation of *Ceratitis capitata, Bactrocera dorsalis,* and *Bactrocera cucurbitae* (Diptera: Tephritidae) in Hawaii. J Econ Entomol 105:1557–1565

Venette RC, Moon RD, Hutchison WD (2002) Strategies and statistics of sampling for rare individuals. Annu Rev Entomol 47:143–174

Vreysen MJB (2005) Monitoring sterile and wild insects in area-wide integrated pest management programmes. In: Dyck VA, Hendrichs J, Robinson AS (eds) Sterile Insect Technique: principles and practice in area-wide integrated pest management. Springer, Dordrecht, pp 325–361

Wang XG, Levy K, Mills NJ, Daane KM (2012) Light brown apple moth in California: a diversity of host plants and indigenous parasitoids. Environ Entomol 41:81–90. doi:10.1603/en11160

Weldon C, Meats A (2010) Dispersal of mass-reared sterile, laboratory-domesticated and wild male Queensland fruit flies. J Appl Entomol 134:16–25. doi:10.1111/j.1439-0418.2009.01436.x

Chapter 17
Trapping Related to Phytosanitary Status and Trade

Eric B. Jang, Walther Enkerlin, Charles "Ed" Miller, and Jesus Reyes-Flores

Abstract Trapping of tephritid fruit flies forms the backbone of many activites related to phytosanitary issues and regulation of trade of fresh fruits and vegetables between countries. Detection of incipient fruit fly populations can occur through a number of means, such as visual surveys, fruit cutting (to reveal the presence of eggs and larvae), collection and holding of fruits to determine if fruit flies emerge from the collected fruits, and perhaps the most commonly used method, trapping of adult flies using some combination of specific trap types, such as the McPhail, MultiLure, Jackson, etc., and semiochemical attractant, such as a food lure, pheromone, or male lure and kairomonal attractant.

Trapping for surveillance of adult fruit flies is generally a reliable method that has a long history and has been largely accepted by trading partners as a standard means to detect, delimit, and monitor tephritid fruit fly populations. Although trapping remains the most effective means for detecting early introductions of invasive exotic or native fruit fly pests into eradicated and/or pest free areas, factors, such as trap type, source and formulation of attractant, trap spacing, and frequency of trap servicing, all influence the reliability and effectiveness of the system. Over

E.B. Jang (✉)
USDA-ARS, Daniel K. Inouye, U.S. Pacific Basin Agricultural Research Center, 64 Nowelo Street, 96720 Hilo, HI, USA
e-mail: eric.jang@ars.usda.gov

W. Enkerlin
Programa Regional Moscamed, 16 Calle 3-38 Zona 10, Ciudad de Guatemala, Guatemala
e-mail: walther.enkerlin@medfly.org.gt

C. "Ed" Miller
USDA-APHIS (retired), 7939 Devilbiss Bridge Road, 21701 Frederick, MD, USA
e-mail: charles.e.miller@gmail.com

J. Reyes-Flores
Insect Pest Control Section, International Atomic Energy Agency, Wagramerstrasse 5, P.O. Box 100, A-1400 Vienna, Austria
e-mail: j.reyes-flores@iaea.org

T. Shelly et al. (eds.), *Trapping and the Detection, Control, and Regulation of Tephritid Fruit Flies*, DOI 10.1007/978-94-017-9193-9_17,

the last 30 years there has been increasing interest in the "harmonization" of fruit fly detection methods for use in regulatory programs worldwide. This has led to the development of international standards (systems approaches, areas of low prevalence, host status, pest free areas, etc), which are discussed below, that advocate the use of verifiable "surveys" to detect or confirm fruit fly presence or absence. Pest risk analysis forms the basis of nearly all trade between countries. It is apparent that trapping serves a number of roles in the establishment, verification, and subsequent maintenance of a phytosanitary condition. In this chapter we discuss the framework for international phytosanitary agreements and trapping as it relates to the regulations that support such agreements.

Keywords International standards for phytosanitary measures • Pest risk analysis • Pest free areas • Areas of low pest prevalence • Systems approach • World Trade Organization • International Plant Protection Convention • Equivalence • Survey • Official procedure • Phytosanitary status • International trade

1 Introduction

The previous chapters of this book have reviewed the current status of trapping (and attractants) for tephritid fruit flies. Here, we discuss trapping as it relates to phytosanitary issues, important for the regulatory framework applied to trade of agricultural commodities. Detection of incipient fruit fly populations can occur through a number of means, such as visual surveys, fruit cutting (to reveal the presence of eggs and larvae), collection and holding of fruits to determine if fruit flies emerge from the collected fruits, and perhaps the most commonly used method, trapping of adult flies using some combination of specific trap types, such as the McPhail, MultiLure, Jackson, etc., and semiochemical attractant, such as a food lure, pheromone, or male lure and kairomonal attractant. Trapping for surveillance of adult fruit flies is generally a reliable method that has a long history and has been largely accepted by trading partners as a standard means to detect, delimit, and monitor tephritid fruit fly populations. Although trapping remains the most effective means for detecting early introductions of invasive exotic or native fruit fly pests into eradicated and/or pest free areas, factors, such as trap type, source and formulation of attractant, trap spacing, and frequency of trap servicing, all influence the reliability and effectiveness of the system. Over the last 30 years there has been increasing interest in the "harmonization" of fruit fly detection methods for use in regulatory programs worldwide. This has led to the development of international standards, which are discussed below, that advocate the use of verifiable "surveys" to detect fruit fly presence. It is apparent that trapping serves a number of roles in the establishment, verification, and subsequent maintenance of a phytosanitary condition.

2 International Entities Developing Phytosanitary Standards Related to Trapping for Fruit Flies

2.1 *Regulatory Framework*

With a continuing increase in the level of global trade, including fruit fly host commodities, application of harmonized phytosanitary measures has become necessary to facilitate safe trade that requires compliance by exporting countries to remain competitive in international markets.

The World Trade Organization (WTO) is the international organization responsible for regulating trade of commodities and insuring that regulations are not misused to protect domestic products. WTO guidelines for applying plant health measures are contained in the multilateral Sanitary and Phytosanitary Measures Agreement (SPS Agreement). The SPS Agreement was negotiated during the 1989 Uruguay Round of the General Agreement on Tariffs and Trade (GATT) as a result of concerns over phytosanitary and sanitary measures being used as non-tariff trade barriers. The agreement was adopted with the creation of the WTO in 1995.

Under the specific provisions set out by the SPS Agreement (Articles 3.1 and 3.4), countries are encouraged to harmonize domestic SPS measures through the adoption of international standards, such as the International Standards on Phytosanitary Measures (ISPM) of the International Plant Protection Convention (IPPC).

Established in 1952, the IPPC is an international plant health agreement to which 180 signatory parties have adhered by 2013 (http://www.fao.org/fileadmin/user_upload/ legal/docs/5_004s-e.pdf). With the aim of protecting cultivated and wild plants by preventing the introduction and spread of pests, it was recognized by the 1989 Uruguay Round of the GATT as a standard setting organization for the SPS Agreement. In 1992 the IPPC Secretariat was established at the Food and Agriculture Organization (FAO) of the United Nations' headquarters in Rome and began its international standard-setting program, which was adopted by FAO the following year.

2.2 *Standard Setting Mechanism*

The Commission on Phytosanitary Measures (CPM) is the governing body of the IPPC. The members of the Commission are the contracting parties to the Convention and are responsible for implementing the work program of standards development, information exchange, and capacity building. The CPM identifies relevant phytosanitary issues affecting trade that are presented for consideration to the IPPC Secretariat by either the National Plant Protection Organizations (NPPO), Regional Plant Protection Organizations (RPPO), or by international organizations that share common interests. These phytosanitary issues form the basis of the standard setting work program.

Standard setting is managed by the Standards Committee (SC) of the IPPC, which reviews and prioritizes relevant topics presented by NPPOs and RPPOs. The SC works through expert groups (i.e., Technical Panels) for drafting the ISPMs. After being drafted by the Technical Panels, draft standards are sent out by the SC for a 90 day country consultation period. Redrafted standards are then presented to the SC for a final process of revision before submission to the CPM for adoption.

Given the importance of fruit flies in international trade, the IPPC established the Fruit Fly Technical Panel in 2004 as the first Technical Panel of the IPPC. Since then, the Fruit Fly Technical Panel has drafted the adopted standards ISPM No. 26 "Establishment of pest free areas for fruit flies (Tephritidae)" (FAO 2006a, b), ISPM No. 30 "Establishment of areas of low pest prevalence for fruit flies (Tephritidae)" (FAO 2008), and ISPM No. 35, "Systems approach for pest risk management of fruit flies (Tephritidae)" (FAO 2012a, b).

2.3 Role of Regional Plant Protection Organizations in ISPM Setting

Based on Article 3.4 of the SPS Agreement, WTO Members promote, through regional plant protection organizations (RPPOs) working within the framework of the IPPC, the development and periodic review of standards and guidelines with respect to all aspects of sanitary and phytosanitary measures (WTO 2012).

The RPPOs deal with phytosanitary issues of a transboundary nature that are more effectively controlled through regional efforts. Regional phytosanitary issues are addressed through the development and adoption of Regional Standards for Phytosanitary Measures (RSPMs).

RPPOs also facilitate the adoption and implementation of ISPMs by the member countries of its region. Additionally, RPPOs may present high profile transboundary phytosanitary issues to the IPPC that are relevant at regional or interregional level for consideration in its standard setting program. In this manner, RSPMs (Regional Standards for Phytosanitary Measures) have been used as the basis for drafting ISPMs. One example is the North American Plant Protection Organization's (NAPPO) RSPM No. 17 "Guidelines for the Establishment, Maintenance and Verification of Fruit Fly Free Areas in North America" (NAPPO 2010). This regional standard was used as a basis for the development of ISPM No. 26 "Establishment of pest free areas for fruit flies (Tephritidae)" (FAO 2006a, b).

3 International Standards Related to Fruit Fly Trapping

3.1 Trapping and Its Relation to the Terms Adopted in International Standards

As part of the standard setting process, the IPPC has developed a list of terms and definitions with specific meaning in the phytosanitary community that are included in the ISPM No. 5, "Glossary of Phytosanitary Terms" (IPPC glossary) (FAO 2013). The glossary is an internationally accepted vocabulary that has been useful in implementing ISPMs and providing clear and consistent terms facilitating the development of export/import phytosanitary protocols.

The set of terms listed in the glossary and related to phytosanitary measures is often different than terms used in scientific publications focused on the same topic. The main reason for these differences is that the terms listed in the IPPC glossary must be adopted by all the contracting parties, thus producing a vocabulary that is somehow more limited to describe a specific phytosanitary activity than the richness of terms used in science; for instance, trapping has many synonyms in the scientific literature, but it is not even used in the phytosanitary glossary.

Nevertheless, trapping as it is understood in science may be equivalent to the term survey in the IPPC glossary, which is defined as "an official procedure conducted over a defined period of time to determine the characteristics of a pest population or to determine which species occur in an area".

3.2 Current Constraints on Trapping and International Standards

Trapping is generally considered by many (but not all) in the international trading community as a primary means of survey for pest tephritid fruit flies. Terminology (as noted above) and acceptance of standard protocols are difficult to harmonize among all parties; thus, under the SPS agreement, countries are allowed to set their own level of acceptable risk. Fruit fly trapping is an example where the interest of importing countries prevail due to the perception of insufficient certainty on the effectiveness of the measure (in this case trapping as a phytosanitary measure), thus, becoming a substantial obstacle to obtaining gains from trade that should otherwise occur if unjustified barriers would not prevail in certain cases (Orden and Romano 1996). Verifiable surveys for pest risk are aligned with the fruit fly trapping guidelines developed by IPPC contracting parties as part of the ISPMs related to fruit flies. IPPC contracting parties have, however, considered these guidelines as an Appendix of the standards rather than as an Annex. According to IPPC, an Appendix is considered to be additional information and not part of a standard itself. Thus, trapping guidelines in the form of an Appendix are not considered a binding mechanism as it would be if presented as an Annex. As explained with more detail in the next paragraph, trapping has, in many cases, been an obstacle for trade for countries that have failed to comply with minimum trapping requirements. Therefore, the development of an Annex on trapping to further clarify the ISPMs on fruit flies is needed in order to provide clearer guidance to exporting/importing countries.

Roberts and Krissoff (2004) stated that, despite the potential advantages of harmonizing phytosanitary measures, its impact on horticultural trade appears to be constrained by insufficient adoption of international standards. In the phytosanitary community related to trade of fruit fly hosts, there is clearly either resistance to adopt challenging standards or impossibility of implementing such challenging standards imposing constraints in international trade. For instance, despite the availability of sound trapping technology for fruit flies (based on scientific evidence), and its extensive use by exporting and importing countries, experience in international standard setting bodies has shown that, due to the technical complexity of the issue and its direct effects on trade, member countries struggle to reach agreement on establishing harmonized procedures related to trapping. This gap leaves trapping as a key issue in the import-export bilateral negotiations of commodities in which trapping becomes a concept with economic and political meaning.

Another constraint on trapping and international standards is that trapping systems (i.e., attractants, traps and trap density) presented in bilateral negotiations are often based more on local experiences than on sound research. For example, countries tend to develop their own trapping technology based on the availability of local materials. This presents a problem for importing countries, since the technology requires validation. Moreover, the time lapse between a research outcome and implementation of this outcome on quarantine policies affecting trade usually takes several years. In some instances, the time lapse could be overcome by applying the concept of "equivalence", one of the basic principles of the IPPC, related to the recognition of alternative phytosanitary measures proposed by exporting countries as equivalent when those measures demonstrate achievement of the level of protection determined by the importing country (FAO 2006a).

These constraints highlight the urgent need for standardized trapping guidelines in order to facilitate trade.

4 Trapping and Pest Risk Analysis

A pest risk analysis (PRA) is conducted by an importing country, as a response to an official request by an exporting country, to determine the phytosanitary condition of a regulated pest in a specific part of the exporting country as per ISPM No. 11 "Pest Risk Analysis for quarantine pests, including analysis of environmental risks and living modified organisms" (FAO 2004). In some cases, PRA can be "commodity" initiated as well. A PRA goes through four basic steps: (1) identification of the risk factors, (2) characterization of the risk factors, (3) pest risk assessment (likelihood of occurrence), and (4) phytosanitary measures required to mitigate the risk. Information from pest population surveys, through the use of traps, is used throughout the four steps of a PRA.

In the initial phases of the PRA (steps 1 and 2), historical profiles of trapping data are used for an initial qualitative identification and characterization of the pest species, including its presence, relative abundance, and damage in space and time. This information is used to develop the pest risk assessment (step 3). Pest risk for quarantine pests is defined by the FAO's glossary as "the probability of introduction and spread of the pest and the magnitude of the associated potential economic consequences".

The sections below discuss the role of trapping in those ISPMs associated with pest risk management.

5 Trapping and Its Roles in Pest Risk Management

Information provided by trapping programs is fundamental to determine the phytosanitary measures required to mitigate the risk as well as to evaluate their effectiveness (step 4) in the suppression and eradication of the pest populations (FAO 1997). Trapping, being a major fruit fly surveillance tool, is the basic element used to monitor and manage the assessed pest risk. With low pest risk situations, countries may decide to reduce trap densities or to eliminate trapping from certain areas. Whereas with medium to high risk situations, they may decide to increase trap density to increase the probability for early detection.

5.1 Trapping and Fruit Fly Host Status

Consideration of the status of fruits as potential hosts for the fruit fly pest species in question is a fundamental element in PRA. If a fruit commodity bears a clear non-host status, this is sufficient as a stand-alone condition to allow the commodity to be traded without the imposition of additional risk management measures (unless other pests are present). If the fruit commodity host is deemed as natural or potential host, then pest risk management should be considered.

Unfortunately, in the initial steps of a PRA, when historical data are used for an initial assessment of the presence of the target fruit fly species infesting a commodity, some trapping records listed in the scientific literature are considered even though these may be unreliable with respect to host status (e.g., recording a fruit as a host based on the capture of the target fruit fly in a trap placed on the target fruit tree). As a consequence of this flawed trapping information, a non-host fruit may be deemed a host, resulting in the imposition of unnecessary phytosanitary measures.

In assessing the status of a commodity as a fruit fly host for quarantine purposes, a scientific determination should be carried out. There are several sources that provide guidance on determination of host status, such as the RSPM No. 30 (Guidelines for the determination and designation of host status of a fruit or vegetable for fruit flies, (NAPPO 2008) and the RSPM No. 4 ("Guidelines for the confirmation of non-host status of fruit and vegetables to Tephritid fruit flies", APPPC 2005). In these guidelines, trapping data may be used as evidence of occurrence of the target fruit fly in the area where field trials are carried out; however, additional validations are often needed to demonstrate host/non-host status (Aluja and Mangan 2008; Cowley et al. 1992).

5.2 Trapping and Pest Free Areas

Historically, fruit fly pest free areas (PFA) and postharvest probit-9 quarantine treatments have been the common pest risk management options to export commodities. Prior to the 1980s, PFA were mostly considered on a country-wide basis. Evidence for the absence of a quarantine pest was based on general surveillance, which mainly involved information from scientific and trade journals, unpublished historical data, anecdotal and museum information, and interception data. It was not common to recognize part of a country as a pest free area, and thus in many cases establishing a country-size pest free area was economically and technically unfeasible. As a result, postharvest quarantine treatments were the most common option for export.

In the 1980s, quarantine treatments based on the fumigant ethylene dibromide (EDB) were banned by the United States and other countries because of health concerns, and methyl bromide (MB) was banned or its use was restricted, because it was reported to be an ozone depleting substance. Consequently, physical postharvest probit-9 treatments continued to be developed. Furthermore, criticism of the use of probit-9 as the sole parameter to evaluate quarantine security led regulators to reinforce the application of other analytical techniques and pest risk mitigation options, such as PFA and systems approaches. Requirements for PFA became more flexible, so importing countries accepted establishing pest free areas of various sizes within a country if they were based on sound specific surveillance and exclusion measures.

Existence of effective trapping systems of adult flies based on combinations of specific trap types, such as the McPhail and Jackson, and olfactory attractants, either a food lure, pheromone, or male lure, imbedded in controlled release devices and pellets facilitated delimitation of pest free areas. This precise and easily implemented manner of conducting delimitation surveys created an opportunity for a number of exporting countries to provide to the importing countries the requested biological information for the official recognition of specific non-infested areas in such a way that establishing PFAs in part of a country became a realistic option to pest risk management.

The IPPC glossary defines PFA as "An area in which a specific pest does not occur as demonstrated by scientific evidence and in which, where appropriate, this condition is being officially maintained", and an area is defined as "an officially defined country, part of a country or all or parts of several countries". Therefore, a PFA for fruit flies can be any size but large enough to increase the practicality of establishing quarantine checkpoints to reduce the risk of being infested through the movement of infested commodities.

According to ISPM No. 4 "Requirements for the establishment of pest free areas" (FAO 1995) and ISPM No. 26 "Establishment of pest free areas for fruit flies (Tephritidae)" (FAO 2006b), a PFA can occur naturally even though, historically, sporadic invasions have taken place. Trapping is conducted to demonstrate pest freedom and, if an incursion occurs, early detection of outbreaks. PFA can also be an eradicated area, where the pest was present but subsequently eliminated and its absence from the area is being maintained. If a PFA is not physically isolated from infested areas, then a buffer zone to prevent the movement of the pest into the pest free area is necessary. Delimiting trapping to establish the boundaries of the area considered to be free, monitoring trapping to observe pest population fluctuations outside or inside the buffer zone, and detection trapping to detect invasions into the free area must be carried out. In summary, a PFA can be natural or artificially established, but in both situations surveillance is the major tool to certify pest freedom.

In order to export an agricultural commodity from a certain area, the importing country accepts and recognizes the status of the area as pest free; however, this pest free status can be lost (and regained) depending upon the occurrence of the pest, meaning that pest free status should be continually maintained in order to use it as an option to move or export commodities. In this sense, a highly sensitive trapping program is not only the major tool to establish a PFA but also the major instrument to maintain the pest free status. Without an effective trapping program, it would be impossible to engage in management of pest free areas, which depends primarily on early detection.

In addition to the requirements to establish and maintain a fruit fly free area, ISPM No. 26 includes one appendix on corrective action plans and two appendices that include guidelines related to trapping procedures and fruit sampling.

5.2.1 Examples of Trapping in Pest Free Areas

Chile

Chile has been internationally recognized as a fruit fly free country since 1995 when the Mediterranean fruit fly (medfly, *Ceratitis capitata* (Wiedemann)) was eradicated from the northernmost region of Arica bordering Peru. To gain PFA status, Chilean efforts centered on an effective surveillance program for exotic fruit flies, a strong exclusion program, and immediate implementation and successful completion of emergency procedures each time medflies were detected.

Currently, Chile's fruit fly free status is based on the National Fruit Fly Detection System operated by the Agriculture and Livestock Service of Chile (SAG), which operates 14,500 traps, placed in high risk areas for early detection. The detection of a non-native fruit fly species triggers an emergency action plan that includes highly sensitive trapping to delimit the pest incursion. SAG invests approximately US $1.5 million per year to manage the National Fruit Fly Detection System, effectively protecting the horticultural industry valued at US $3 billion per annum (SAG 2010; Enkerlin 2005; Liquido et al. 1995).

Sonora, Mexico

The first fruit fly pest free area less than the size of a country was established in 11 municipalities of the state of Sonora, Northwest Mexico. Although fruit fly monitoring was conducted as early as 1981, efforts to establish a PFA formally began in 1985 with the establishment of a comprehensive fruit fly detection network in both urban and commercial production areas. Recognition of the PFA by the United States was obtained in 1988 and in 1999 by New Zealand, Australia, the European Union, and Japan. The PFA is operated jointly by the Sonora Plant Protection Committee as well as federal and state governments. The area has been maintained fruit fly free through an effective quarantine system that includes strict checkpoints at the state international airport as well as a quarantine road checkpoint strategically located on the main interstate highway accessing the free area. The PFA is also maintained through an extensive trapping network aimed at early detection of native and non-native fruit fly incursion. The trapping network is composed of 2,400 traps strategically placed in high risk areas. The PFA approved commodities are apple, apricot, grapefruit, sweet oranges, peach, persimmon, pomegranate, grapes, and tangerine. Target pests are *C. capitata*, *Anastrepha ludens* (Loew), *Anastrepha serpentina* (Wiedemann), *Anastrepha obliqua* (Macquart) and *A. fraterculus* (Wiedemann). In 2002, citrus exports from the Sonora PFA to the United States were valued at US $10.3 million; in addition, these exports generated 2,000 jobs per year valued at US $3.2 million (Enkerlin 2005; Liquido et al. 1995).

5.3 Trapping and Pest Free Places of Production and Pest Free Production Sites

A pest free place of production (PFPoP) is a "place of production in which a specific pest does not occur as demonstrated by scientific evidence and in which, where appropriate, this condition is being officially maintained for a defined period" (FAO 1999). A pest free production site is a defined portion of a place of production that is managed as a separate unit. Therefore, a place of production can include either a single or multiple production sites. PFPoPs and PFAs are distinguished primarily by their relative size and the duration of production. The latter are much larger than a place of production and are maintained over many years without interruption. In comparison, the PFPoP may be the size of a single farm or group of farms and may be maintained for only one or few growing seasons.

PFPoP is selected as a risk mitigation measure when the biology of the pest allows an area to be kept as pest-free, (i.e., the natural spread of the pest is slow and over short distances, the pest has limited host range, the possibilities of artificial spread of the pest is limited), or the pest is temporarily absent from a certain area. Even though PFPoPs use the concept of "pest freedom", the practicality of applying the same exclusion measures of the pest free areas is questionable (i.e., quarantine checkpoints); consequently, PFPoPs are more closely related to the areas of low pest prevalence as they become major components of a systems approach.

Trapping in PFPoP plays a major role, since it is the most appropriate option to determine freedom from a pest or to detect as early as possible the occurrence of a pest. Unless a highly sensitive trapping method is used, the PFPoP as a pest risk mitigation measure will not be effective.

5.4 *Trapping and Areas of Low Pest Prevalence*

Areas of low pest prevalence (ALPP) are referenced in Article 6.2 of the SPS Agreement, Articles II, and IV.2e of the IPPC and as operational principle No.2 of the principles of plant quarantine as related to international trade of the IPPC.

An ALPP is defined by the IPPC glossary as "an area, whether all of a country, part of a country, or all or parts of several countries, as identified by the competent authorities, in which a specific pest occurs at low levels and which is subject to effective surveillance control or eradication measures" The IPPC has further detailed the ALPP in ISPMs Nos. 22 "Requirements for the establishment of areas of low pest prevalence" (FAO 2005), ISPM No. 29 "Recognition of pest free areas and areas of low pest prevalence" (FAO 2007), and ISPM No. 30 "Establishment of areas of low pest prevalence for fruit flies (Tephritidae)" (FAO 2008).

Unlike PFAs, the pest is present, albeit at low levels. Therefore, trapping sensitivity becomes a major tool to assess whether the pest population is above or below the threshold established by the exporting or importing country. An additional difference is that PFA trapping is used as a survey tool for detection, while ALPP trapping is used as a survey tool for monitoring and might also be used as a suppression measure.

Importantly, Annex 1 of ISPM No. 30 describes the use of the parameter flies per trap per day (FTD), as an index to estimate and compare the relative number of fruit fly adults in a given time and space. Moreover, Appendix 1 indicates that information about fruit fly trapping can be found in the trapping guidelines published by the IAEA (IAEA 2003). This document was reviewed and enhanced and has been made available as a fruit fly trapping manual (IAEA 2013).

In spite of their wide support as a concept and three adopted ISPMs related to them, ALPPs are not, in practice, used as a stand-alone pest risk management option for fruit flies to achieve a quarantine security level acceptable by the importing country. The main constraint is that in most fruit fly species of economic importance the quarantine security levels established by the importing countries are far above the quarantine security provided by an ALPP alone. Therefore, ALPPs have become a major component of systems approaches as pointed out in Appendix 2 of ISPM No. 30.

5.5 Trapping and Systems Approaches

As with PFAs, in the 1980s the systems approach (SA) was also seen as a new alternative to meet quarantine security of importing countries due to the banning of many post-harvest treatments based on the use of EDB and MB. Demands for new alternatives, in addition to the PFA, to replace post-harvest treatments as a means of managing the phytosanitary risk posed by fruit flies became a challenge.

The basic concept of the SA came from the realization amongst researchers and regulators that infestation of commodities by pests could be mitigated, not only by single quarantine treatments aimed at near complete mortality, but by applying a series of sequential mitigation measures (systems components), each having some role in reducing the overall pest risk in an export consignment (Jang and Moffitt 1994).

The current definition of SA in the IPPC glossary ISPM No.14 is "the integration of different risk management measures, at least two of which act independently, and which cumulatively achieve the appropriate level of protection against regulated pests" (FAO 2002). The approach had been used for many years. However, it was not until the development of ISPM No. 14 "The use of integrated measures in a systems approach for pest risk management" (FAO 2002) and the recently approved ISPM No. 35 "Systems approach for pest risk management of fruit flies (Tephritidae)" (FAO 2012) that this option has been formalized and officially accepted by the international plant protection community related to the control of fruit flies.

Based on its very nature of pest mitigation, SA is applied to export commodities from areas where the target pest occurs. Therefore, SA works best if population or infestation levels of the target pest can be easily measured. This condition provides a starting metric against which further independent actions for pest reduction can be implemented along the food chain from the field to the consumer in the importing country. A comprehensive guide on trapping related to systems approaches can be found in the guidelines for implementing systems approaches for pest risk management of fruit flies (IAEA 2010)

5.5.1 Example of Trapping in Systems Approaches

One of the most commonly used SA for the importation of fruit fly host material into the United States is the use of a pest excluding structure (greenhouse) located within an ALPP. This is commonly used for US imports of tomatoes and peppers from various locations, including Central America and the Mediterranean region.

For example, pink and ripe tomatoes are permitted to be imported into the United States from Central America with the use of a pest excluding structure (greenhouse) located within a medfly ALPP. Here, a minimum of two traps per greenhouse using protein bait must be used within the greenhouse with negative results starting 2 months before harvest through harvest. For the same period, a buffer area 500 m wide around the greenhouse must be monitored with one Jackson trap baited with TML per 10 ha with less than 0.1 flies found per trap per day.

5.6 Trapping and Post-harvest Treatments

The application of post-harvest treatments to commodities is a phytosanitary measure used to kill, remove, deactivate or make unviable the pest and therefore prevent its introduction and spread. Historically, a probit-9 efficacy level has been required for fruit flies. The pest risk mitigation is so effective that these treatments are used as a stand-alone measure to provide high quarantine security; thus, there is normally no need for any additional measure to mitigate the risk.

Following outbreaks of the Mexican fruit fly (*A. ludens*) in the Rio Grande Valley in Texas during the mid-1920s, an agreement was reached between the U.S. and Mexico to establish a laboratory in Mexico City for studies of fruit fly species present in Mexico. The history and accomplishments of this laboratory were reviewed in Shaw et al. (1970). As Shaw discussed, the activities of this laboratory, which operated from 1928 to 1968, were described in some detail in monthly reports by each scientist to the director. Periodic publications by individual scientists or, more frequently, teams of scientists were summaries of the monthly reports. The most complete summary (Baker et al. 1944) listed laboratory and field trials of pesticide and baits for adult suppression.

Since 1939, when A. C. Baker, the principal entomologist of the Division of Fruit Fly Investigation of the Bureau of Entomology and Plant Quarantine of the US Department of Agriculture (USDA) suggested the use of mortality rates as a criterion for quarantine disinfestation treatments and specifically the use of the mortality rate at probit-9 (mortality of 99.9968 % of eggs and larvae in fruit), post-harvest treatments have become the most used method to manage the risk posed by fruit flies. Since the first quarantine treatments in the 1930s (i.e., cold treatment and vapor-heat), there have been many different types of treatments approved to export fruit fly host, including application of ionizing radiation.

Even though application of these treatments has been very efficacious for fruit disinfestation, approved post-harvest treatments usually include fruit sampling. This is conducted before the fruit goes through treatment to assure low occurrence of larvae in the fruits. If this cannot be guaranteed, the fruit load is rejected. To prevent this, the importing country requests pest control in the field to assure a low infestation rate, usually measured by trapping in number of flies per trap per day.

These procedures may be claimed as obstructive by the exporting country, because they are often seen as unnecessary quarantine requirements. Nonetheless, producers also benefit from this, because it helps assure that the fruit will be free of dead larvae and thus more acceptable to the consumers.

5.6.1 Example of Trapping in Post-harvest Treatments

Guatemala exports mango to the USA from an area where medfly (*C. capitata*), Mexican fruit fly (A. *ludens*), and West Indies fruit fly (*A. obliqua*) occur at a low prevalence level. The main phytosanitary measure used to mitigate risk of moving these fruit fly pests is a postharvest treatment, namely hot water treatment. One additional phytosanitary measure required by the importing country is the use of monitoring traps to make sure that fruit fly population is kept at the established low prevalence level. An increase of population size above the set level will trigger control measures. This will help assure that harvested mangoes will not be rejected at the packing facility and that the mangoes will not contain dead larvae after the postharvest treatment. According to the USDA import protocol, Jackson traps baited with trimedlure to monitor medfly populations and MultiLure traps baited with hydrolyzed protein to monitor *Anastrepha* spp. populations are required. Traps are placed at a density of three Jackson traps per hectare and one MultiLure trap per hectare for a total of 1,600 traps in the whole mango export program. The value of mango exports to the USA has experienced a steady increase from US $2.8 million in 2005 to US $10.6 million in 2011.

6 Trapping in Fruit Fly Control Using an Area-Wide Integrated Pest Management Approach

Fruit fly trapping is a key phytosanitary activity to determine the establishment and/or maintenance of the phytosanitary condition of an area, since it provides direct information on the presence or absence of a pest in an area and thus on its phytosanitary status. Area-wide integrated pest management (AW-IPM) is a very broad and flexible concept. Furthermore, AW-IPM has been increasingly accepted for those situations where management of mobile pests at a larger landscape scale is advantageous to maximize the efficacy of management tactics. Due to the biological characteristics of fruit flies, particularly their high mobility, their surveillance and control is more effective and preferable using an area-wide approach instead of a field-by-field uncoordinated control approach.

There are basically two phytosanitary conditions that regulate the movement or export of commodities: pest infested or pest free areas. Fruit fly AW-IPM can be used to reduce the infestation level of a pest in an area, thus, modifying its phytosanitary condition to achieve the required level of pest prevalence for exports.

Accordingly, by applying AW-IPM for pest suppression, an infested area may become an area of low pest prevalence, and, if applied for eradication, the infested area may become a pest free area, although this is likely more difficult to accomplish. AW-IPM can also be used to eradicate an outbreak or incursion of the target pest occurring in a pest free area.

In some instances, ecological factors favour the establishment of a pest whose presence in the area will result in large economic losses. Therefore, AW-IPM can be applied in such an area to prevent the risk of introduction through an exclusion process, keeping the area in its pest free phytosanitary condition.

There are also circumstances where a pest is present in a limited area and threatens to spread to non-infested areas. In this case, a containment process is applied, so that AW-IPM is used in and around the infested area to prevent the spread of the pest, thus keeping the non-infested areas in their pest free phytosanitary condition.

In summary, there are four basic strategies on which AW-IPM can be used to change or maintain the phytosanitary condition of an area: suppression, eradication, exclusion, and containment. Trapping is an integral component in each of these strategies; however, trapping applications will be different for each of the four AW-IPM strategies.

Eradication programmes for newly entered pests into a previous pest free area is the most common example that can lead to the re-establishment of pest free areas. These programmes follow a simple three-step process: trapping to fully determine the occurrence and distribution of the pest, containment actions to prevent the spread of the pest, and eradication of the pest. Trapping becomes the most important tool to evaluate the efficacy of the AW-IPM approach and to verify that the desired phytosanitary condition of pest freedom has been achieved. ISPM No. 9 "Guidelines for pest eradication programmes" (FAO 1998) clearly describes the components of these programmes and the role trapping plays in eradication programmes.

As stated above, trapping applications will depend on the phytosanitary condition of the area and the AW-IPM strategy applied. Trapping applications include: (1) detection trapping to determine if the pest is present or absent in an area, (2) monitoring trapping used as an on-going survey to verify the characteristics of the pest population present in an area, (3) delimiting trapping that serves to establish the boundaries of an area considered to be infested by or free from fruit flies, and (4) verification trapping that confirms the pest status after the application of AW-IPM to eliminate an outbreak. Depending upon the type of trapping applied and the target fruit fly species, programs will likely vary in trap layouts, trap densities, trap service and inspection interval, and proportions of male specific and female biased traps. Detailed information dealing with these technical aspects is presented in the IAEA trapping guidelines (IAEA 2003) and trapping manual (IAEA 2013), mentioned in the Appendix 1 to ISPM No. 26 (FAO 2012a, b) and Appendix 1 to ISPM No. 30 (FAO 2008).

The flow chart depicted in Fig. 17.1 illustrates the relationship between different phytosanitary conditions of an area, the control process applied (i.e., AW-IPM strategies) and the type of trapping application used in each scenario to move from

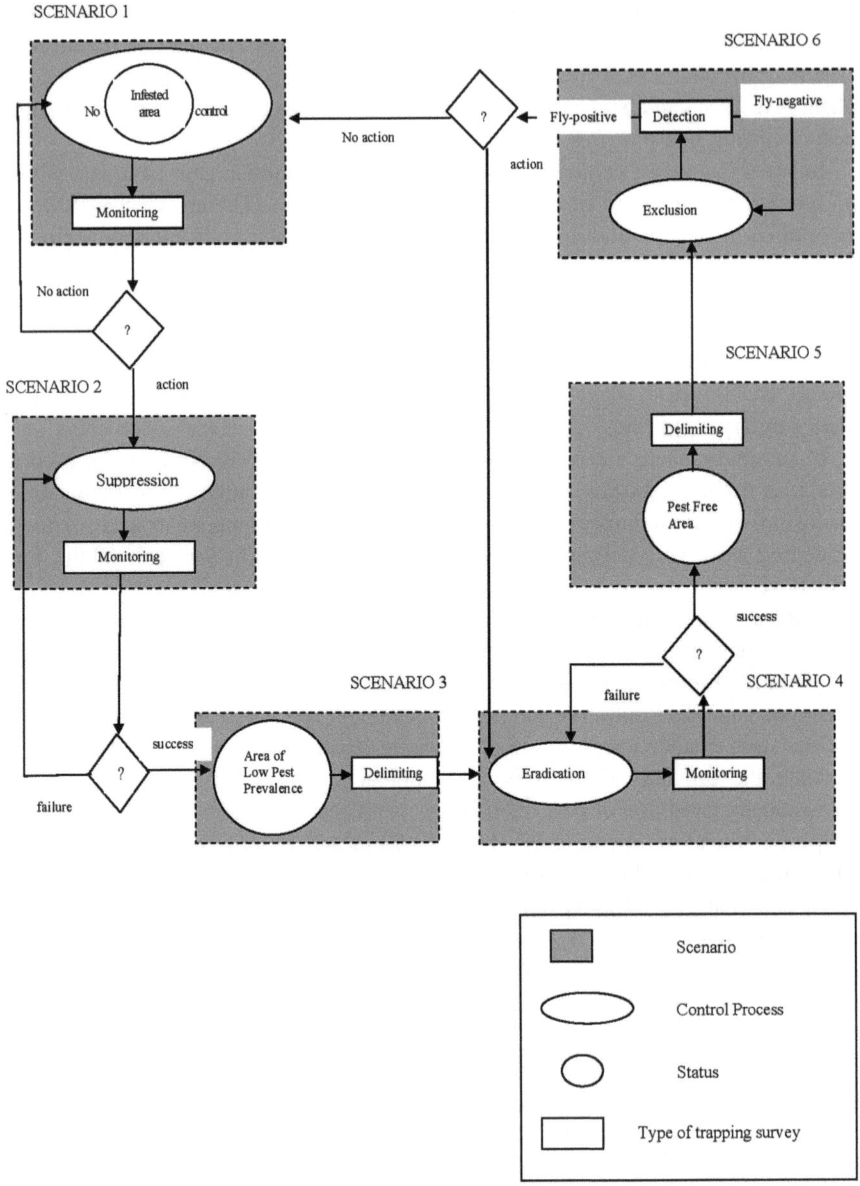

Fig. 17.1 Flow chart showing the relationship between different phytosanitary conditions of an area, the control process applied, and the type of trapping application used in each scenario (Adapted from trapping guidelines for area-wide fruit fly programmes, IAEA/FAO-TG/FFP, 2003. IAEA, Vienna)

an infested area (scenario 1) to a pest free area (scenarios 5 and 6). The chart clearly shows how trapping is an integral component of AW-IPM strategies aimed at establishing and maintaining fruit fly low prevalence and free areas.

6.1 Example of Trapping in an AW-IPM Programme

The Moscamed Regional Program in southern Mexico and Guatemala aims at stopping the northern spread of the medfly through management of a containment barrier that protects Mexico and the USA horticultural production. At the same time, the barrier front is expanding south gradually eradicating the pest from Guatemala and therefore opening new opportunities to develop its horticultural industry. The program operates a trapping network composed of 30,000 traps. The trapping network is used in the program's day to day operations as it moves forward into new infested areas to evaluate suppression and eradication measures, including, the aerial spray of the organic insecticide-bait spinosad and the massive release of sterile males. During the monitoring of the infested areas the level of the pest is assessed through the fly per trap per day index (FPD). When this index exceeds a value of 0.05 flies per trap per day, suppression actions are continued. When the value is below 0.05 flies per trap per day, suppression actions switch to the release of sterile males to achieve eradication through a competitive sterile: fertile fly ratio. The trapping network is then used to determine if eradication has been achieved and its data used for the certification of the fly free status of the area and eventual official declaration. The trapping network is also used to maintain fly free areas for delimitation of an incursion and verification of its eradication. In this way, trapping is used as a fundamental component of the pest risk management process. The regional program invests an average US $2.0 million per year in operating an extensive trapping network that protects a horticultural industry in Mexico, USA, and Guatemala valued at over US $10 billion per year (Moscamed Regional Program 2011; IICA 2009).

7 Conclusions

In this chapter, we have considered trapping primarily in the context of an official phytosanitary procedure to determine the presence or absence of the pest as well as the basis of an official verification process to confirm that the pest is either not present or present at low levels. These official procedures are complicated due to the need to set international standards agreeable to all signing parties of WTO and the development of ISPM standards under the SPS agreements makes adoption of specific definitions, even terms such as trapping, difficult. The intrinsic value of trapping for tephritid fruit flies in the context of international trade is enormous given the increasing role of agricultural trade in many countries gross domestic product (GDP). Because fruit flies can attack a wide range of host commodities, agricultural-dependent countries where the flies are not established may spend millions of dollars to detect, delimit, and eradicate new introductions. Only a few countries (mostly colder climates where the flies cannot reproduce) do not consider fruit flies a major quarantine pest. ISPM No. 6 discusses overall trapping in the context of surveillance measures. Interestingly, trapping for fruit flies preceded its importance as a regulatory concept necessary to provide adequate surveillance and arose from a need to determine pest population levels in the field. As the need for harmonization of concepts specific to phytosanitary issues became evident, trapping or its equivalence has become more and more important.

Of special note is the fact that the IPPC organized a technical panel specifically for fruit flies, which resulted in the development and adoption of three standards specifically relating fruit flies (ISPM No. 26, No. 30 and No. 35). These three ISPMs serve as a core from which regulatory approval to either import or export a commodity is determined. In the case of pest free areas (ISPM No. 26), this verification alone is often enough to certify that the pest is not present and that trade of fruit fly host materials should be allowed from these areas. For areas of low pest prevalence (ISPM No. 30) and systems approaches (ISPM No. 35), trapping is conceptually part of a process that, along with other components, may allow further movement of host materials. While specific postharvest treatments such as methyl bromide fumigation, heat and cold treatment and irradiation are experimentally designed to be effective at very high infestation levels (probit-9), many countries are increasingly relying on trapping to further insure that population levels of the pest do not pose unreasonable risk given the effectiveness of the treatment.

In conclusion, trapping plays a central role in agricultural trade of fruit fly host commodities and will continue to be an important part of the process needed to establish the required phytosanitary regulatory framework.

References

Aluja M, Mangan RL (2008) Fruit fly (Diptera: Tephritidae) host status determination: critical conceptual and methodological considerations. Ann Rev Entomol 53:473–502

APPPC (Asia and Pacific Plant Protection Commission) RSPM No. 4 (2005) Guidelines for the confirmation of non-host status of fruit and vegetables to tephritid fruit flies. APPPC, Bangkok

Baker AC, Stone WE, Plummer CC, McPhail M (1944) A review of studies on the Mexican fruit fly and related Mexican species. USDA Misc Publ 531:1–155

Cowley J, Baker RT, Harte DS (1992) Definition and determination of host status for multivoltine fruit fly (Diptera: Tephritidade) species. J Econ Entomol 85(2):312–317

Enkerlin W (2005) Sterile insect technique. In: Dyck VA, Hendrichs J, Robinson AS, IAEA (eds) Principles and practice in area-wide integrated pest management. Springer, The Netherlands, pp 651–676

FAO (Food and Agriculture Organization) (1995) Requirements for the establishment of pest free areas. ISPM No. 4, International Plant Protection Convention (IPPC). FAO, Rome, Italy

FAO (Food and Agriculture Organization) (1997) Guidelines for surveillance. ISPM No. 6, International Plant Protection Convention (IPPC). FAO, Rome, Italy

FAO (Food and Agriculture Organization) (1998) Guidelines for pest eradication programmes. ISPM No. 9, International Plant Protection Convention (IPPC). FAO, Rome, Italy
FAO (Food and Agriculture Organization) (1999) Requirements for the establishment of pest free places of production and pest free production sites. ISPM No. 10, International Plant Protection Convention (IPPC). FAO, Rome, Italy
FAO (Food and Agriculture Organization) (2002) The use of integrated measures in a systems approach for pest risk management. ISPM No. 14, International Plant Protection Convention (IPPC). FAO, Rome, Italy
FAO (Food and Agriculture Organization) (2004) Pest risk analysis for quarantine pests including analysis of environmental risk and living modified organisms. ISPM No. 11, International Plant Protection Convention (IPPC). FAO, Rome, Italy
FAO (Food and Agriculture Organization) (2005) Requirements for the establishment of areas of low pest prevalence. ISPM No. 22, International Plant Protection Convention (IPPC). FAO, Rome, Italy
FAO (Food and Agriculture Organization) (2006a) Phytosanitary principles for the protection of plants and the application of phytosanitary measures in international trade. ISPM No. 1, International Plant Protection Convention (IPPC). FAO, Rome, Italy
FAO (Food and Agriculture Organization) (2006b) Establishment of pest free areas for fruit flies (Tephritidae). ISPM No. 26, International Plant Protection Convention (IPPC). FAO, Rome, Italy
FAO (Food and Agriculture Organization) (2007) Recognition of pest free areas and areas of low pest prevalence . ISPM No. 29, International Plant Protection Convention (IPPC). FAO, Rome, Italy
FAO (Food and Agriculture Organization) (2008) Establishment of areas of low pest prevalence for fruit flies (Tephritidae). ISPM No. 30, International Plant Protection Convention (IPPC). FAO, Rome, Italy
FAO (Food and Agriculture Organization) (2012) Systems approach for pest risk management of fruit flies (Tephritidae). ISPM No. 35, International Plant Protection Convention (IPPC). FAO, Rome, Italy
FAO (Food and Agriculture Organization) (2013) Glossary of phytosanitary terms. ISPM No. 5, International Plant Protection Convention (IPPC). FAO, Rome, Italy
IICA (Instituto Interamericano de Cooperacion para la Agricultura) (2009) Evaluacion economica del Programa Moscamed en Mexico (1978–2008). In: Diznarda Salcedo Baca, Refugio Lomeli Flores J, Terrazas Gonzalezy GH. Impresora Kavers S.A, p 144. ISBN 13:978-92-9248-142-1
IAEA (International Atomic Energy Agency) (2003) FAO/IAEA trapping guidelines for area-wide fruit fly programmes, IAEA/FAO-TG/FFP, 2003. IAEA, Vienna. http://www-naweb.iaea.org/nafa/ipc/public/trapping-web.pdf
IAEA (International Atomic Energy Agency) (2010) FAO/IAEA guidelines for implementing systems approaches for pest risk management of fruit flies. Report and recommendations of the consultants group meeting organized by the Joint FAO/IAEA division of nuclear techniques in food and agriculture, Vienna, Austria, 7–11 June 2010
IAEA (International Atomic Energy Agency) (2013) FAO/IAEA trapping manual for area-wide fruit fly programmes. http://www-naweb.iaea.org/nafa/ipc/public/Trapping-Manual-Final-sept13.pdf
IPPC (International Plant Protection Convention) https://www.ippc.int
Jang E, Moffitt H (1994) Systems approaches to achieving quarantine security. In: Sharp JL, Hallman GJ (eds) Quarantine treatments for pests and food plants. Westview Press, Boulder, Colorado, USA, pp 225–236
Liquido N, Griffin R, Vick K (1995) Quarantine security for commodities: current approaches and potential strategies. In: Proceedings of joint workshops of the ARS and APHIS. Beltsville, Maryland
Moscamed Regional Program (2011) Memoria Anual de Labores. USDA-APHIS, SAGARPA-SENASICA y MAGA-VISAR. Guatemala, Centro America

NAPPO (North American Plant Protection Organization) (2010) RSPM No. 30 Guidelines for the determination and designation of host status of a fruit or vegetable for fruit flies (Diptera: Tephritidae). NAPPO, Ottawa

Orden D, Romano E (1996) The avocado dispute and other technical barriers to agricultural trade under NAFTA. Paper presented at the conference on NAFTA and agriculture: is the experiment working. San Antonio, Texas, USA

Roberts D, Krissoff B (2004) Regulatory barriers in international horticultural markets. Electronic outlook report from the economic research service. United States Department of Agriculture (USDA)

SAG (Servicio Agrícola y Ganadero) (2010) Guía para la detección de moscas de la fruta de importancia económica. Editor: Carlos Lobos Aguirre. Gobierno de Chile, p 198

Shaw JG, Lopez-D F, Chambers DL (1970) A review of research done with the Mexican fruit fly and citrus blackfly in Mexico by the Entomology Research Division. Bull Entomol Soc Am 16:186–193

WTO (World Trade Organization) The WTO agreement on the application of sanitary and phytosanitary measures (SPS agreement). http://www.wto.org

Part VI
Coda

Chapter 18
The Complexities of Knowing What It Is You Are Trapping

Anthony R. Clarke and Mark K. Schutze

Abstract The effectiveness of any trapping system is highly dependent on the ability to accurately identify the specimens collected. For many fruit fly species, accurate identification (= diagnostics) using morphological or molecular techniques is relatively straightforward and poses few technical challenges. However, nearly all genera of pest tephritids also contain groups of species where single, stand-alone tools are not sufficient for accurate identification: such groups include the *Bactrocera dorsalis* complex, the *Anastrepha fraterculus* complex and the *Ceratitis* FAR complex. Misidentification of high-impact species from such groups can have dramatic consequences and negate the benefits of an otherwise effective trapping program. To help prevent such problems, this chapter defines what is meant by a species complex and describes in detail how the correct identification of species within a complex requires the use of an integrative taxonomic approach. Integrative taxonomy uses multiple, independent lines of evidence to delimit species boundaries, and the underpinnings of this approach from both the theoretical speciation literature and the systematics/taxonomy literature are described. The strength of the integrative approach lies in the explicit testing of hypotheses and the use of multiple, independent species delimitation tools. A case is made for a core set of species delimitation tools (pre- and post-zygotic compatibility tests, multi-locus phylogenetic analysis, chemoecological studies, and morphometric and geometric morphometric analyses) to be adopted as standards by tephritologists aiming to resolve economically important species complexes. In discussing the integrative approach, emphasis is placed on the subtle but important differences between integrative and iterative taxonomy. The chapter finishes with a case study that illustrates how iterative taxonomy applied to the *B. dorsalis* species complex led to incorrect taxonomic conclusions, which has had major implications for quarantine, trade, and horticultural pest management. In contrast, an integrative approach to the

A.R. Clarke (✉) • M.K. Schutze
School of Earth, Environmental and Biological Sciences, Science and Engineering Faculty, Queensland University of Technology, GPO Box 2434, 4001 Brisbane, QLD, Australia
e-mail: a.clarke@qut.edu.au; m.schutze@qut.edu.au

T. Shelly et al. (eds.), *Trapping and the Detection, Control, and Regulation of Tephritid Fruit Flies*, DOI 10.1007/978-94-017-9193-9_18,

problem has resolved species limits in this taxonomically difficult group, meaning that robust diagnostics are now available.

Keywords Integrative taxonomy • Iterative taxonomy • Systematics • Taxonomy • Diagnostics • Cryptic species • Sibling species • Species complex • *Bactrocera dorsalis* • *Bactrocera papayae* • *Anastrepha fraterculus* • Biological species • Taxonomic species • Species delimitation • ICZN • Mate compatibility

Imagine a scenario....

As part of surveillance trapping you detect a previously unrecorded fruit fly. The same fly starts appearing in field monitoring traps and is also being reared from fruit. But, thankfully, further trapping shows the new fly is still in an area where it might be contained or eradicated. Hurray! This is what your well designed and expensively maintained trapping network is meant to do – detect a new incursion early so response can be rapid. Reading that excellent book on fruit fly trapping has been so worthwhile.

You think the fly is a well-known international pest species, but you are uncertain – it looks slightly different from photos on the internet. If this slight difference turns out to be just population level variation and it is the well-recognised pest species, then response can be prompt and highly targeted. Well-developed field controls exist and can be applied almost immediately, agreed international market access protocols exist, and these can be applied resulting in minimal trade and grower disruption, while the opportunity also exists of using SIT and other techniques to try and eradicate the fly altogether. Alternatively, if this is an entirely new pest, or one that is known but poorly researched, everything will need to start from scratch. Control treatments will need to be developed, basic biology and ecology research undertaken, market access protocols developed and negotiated, and the prospect of using SIT in the short term non-existent. In short, the fly could be disastrous, closing markets and potentially causing real food-security problems. So, you badly need to get the species confirmation correct.

To determine if your fly is, or is not, the major pest species do you:

(i) Send it to one individual who uses morphological taxonomy and professional experience to make the decision because s/he is the recognised world taxonomic authority in the group?

(ii) Use a geneticist to run a *cox1* barcode and cross-check the result against BOLD (the Barcode of Life Data System, Ratnasingham and Hebert 2007) or some similar diagnostic technique?

(iii) Send specimens to different labs with different skill sets (e.g., molecular, behavioural, taxonomic, karyotyping, etc.) and then make a decision based on the combined set of results as recommended by an integrative taxonomic approach (Schlick-Steiner et al. 2010)?

If your answer is (iii) then, unfortunately, you are unlikely to have been in charge of getting a real exotic incursion identified, because this is rarely done except in certain countries (such as the U.S.A.). If your answer is (ii) then, despite the apparent difference, it is the same as (i), as material used to populate diagnostic databases is nearly always first identified using morphological characters (Collins and Cruickshank 2013). If your answer is (i), then you are correct.

1 Trap Catch and Diagnostics

Anyone in the fruit fly community knows that the above ‘scenario’ could be replaced by many real examples. In an era when very large amounts of genomic data can be generated and analysed routinely; where international communications and networks exist and widely dispersed labs can work together promptly to seek resolution of complex biological questions; and where very sophisticated understanding on species and species delimitation theories exist, the critical call is still often made using a methodological and theoretical approach unchanged in over 200 years. Thankfully, there is an ever-growing trend amongst the taxonomic community towards integrating multiple data types in species identification (as discussed later in this chapter). However, despite this, it is worthwhile reminding ourselves of the philosophical underpinnings of how – and why – we identify species in the first place.

The chapters of this book all deal with different attributes of fruit fly trapping, with the tacit assumption being that the flies removed from the trap can be correctly identified, or to use the modern parlance – diagnosed. While diagnostics is a very different research area to trapping, the two cannot be separated, and a discussion of diagnostics in a book on trapping is clearly warranted. This is because, to be blunt, without being able to identify the flies coming out of your trapping network, then the traps are so much wasted time and energy. Diagnosis can range from the simple visual scanning of large numbers of endemic species, or it may involve a detailed genetic analysis of a single exotic detection. How accurately this is done and how easily it is achieved are common measures of diagnostics success (Barr et al. 2012; Frey et al. 2013).

Diagnostics is largely an operational discipline of placing unknown individuals within a taxonomic ‘basket’ (be it a species, genus, or some other taxonomic level) (Cranston et al. 1991). Depending on the source of the material (e.g., a border interception or an ongoing trapping program in an endemic area), diagnostics may involve one specimen every few years or hundreds of specimens per day. Diagnostics research involves finding markers, now commonly genetic, that allow an individual specimen to be identified to species level without reference to an expert taxonomist: the barcode segment of the COI gene is one well known and commonly used diagnostic tool (Blackett et al. 2012; Khamis et al. 2012; Virgilio et al. 2012; Jiang et al. 2013). As genetics has become a routine part of what we do, it is easy to forget how novel this approach is; yet the seminal paper in the area for fruit fly workers is still less than a decade old (Armstrong and Ball 2005). With the ever increasing sophistication of genetic approaches and tools, molecular diagnostic techniques can only get better in terms of accuracy, speed, and operating given databases which, while comprehensive, are often not exhaustive (Van Houdt et al. 2010; Frey et al. 2013).

While diagnostics has been promoted by fiscally-stressed managers as a way of solving declining taxonomic expertise, it is of critical importance to recognise that diagnostics is neither taxonomy nor systematics, and in this lies an inherent problem that no changes in technology will overcome. Regardless of the purpose or sophistication of the tool used, there is more to accurate diagnostics of a trap catch than matching the organism to a molecular (e.g. BOLD, http://www.boldsystems.org/) or morphological (e.g. PaDIL, http://www.padil.gov.au/) reference set. While sometimes forgotten, the quality of diagnostics is heavily dependent on the quality of its 'parent' disciplines, systematics and taxonomy, that delimited and described the species in the first place (Collins and Cruickshank 2013).

2 Systematics and Taxonomy

Although sometimes used interchangeably, diagnostics, systematics, and taxonomy are not the same things. Systematics is the science of relationships, both at higher taxonomic levels (orders, families and genera) or between individual species (Cranston et al. 1991; Cracraft 2000). Systematic approaches, such as phylogenetic analyses, can help delimit species or identify the potential existence of previously unrecognised species (O'Meara 2010; Boykin et al. 2012; Carstens et al. 2013). However, as developed towards the end of this chapter, species delimitation will commonly require more than genetic approaches alone.

Taxonomy is the description and naming of species and the placement of them within genera and higher taxa (i.e., classification). Morphology is the classical tool of taxonomy, and hence taxonomic species are traditionally described based on shared morphological traits (for individuals within a species), or morphological discontinuities (discrimination between species). The description of a new species by a taxonomist is a statement of hypothesis. The taxonomist is hypothesising that the unique markers that she or he has identified to discriminate the species accurately correspond to a real biological unit (Walter 2003). In taxa where visual cues are important in communication between individuals (e.g., many butterflies) the match between taxonomic species and biological species may be high. However, where communication is through chemical, behavioural, or other ephemeral signals (e.g., mosquitoes, many tephritids) – and hence more challenging for humans to observe, record, and analyse – the link between current taxonomic species and biological species may be poor (Paterson 1991). In such cases, multiple tools are needed to increase taxonomic accuracy, an issue we explore in subsequent sections.

2.1 Why It Is Important to Differentiate Systematics from Taxonomy from Diagnostics

It is important to clearly differentiate between systematics, taxonomy, and diagnostics, because conflation of the different concepts leads to misunderstanding of what each can achieve on its own, leading to the misapplication of different analytical approaches and potential errors (Collins and Cruickshank 2013). These problems are very well exemplified by studies of the *Bactrocera dorsalis* species complex in South-east Asia. For some 20 years, diagnostics research on flies of the *Bactrocera dorsalis* complex has been deemed to be problematic, with inconsistent molecular and morphological differentiation between *Bactrocera dorsalis* (Hendel), *Bactrocera papayae* Drew & Hancock, and *Bactrocera philippinensis* Drew & Hancock (Clarke et al. 2005), despite standard diagnostic markers, such as *cox1*, working for other *Bactrocera* species (Armstrong and Ball 2005; Blackett et al. 2012). The assumption was that the morphologically-based species limits matched biologically valid species (Clarke et al. 2005), and therefore a failure to find robust molecular diagnostics was a failure of the diagnostics to detect unique, species-level markers, which should be present. But when an integrative approach using morphology, molecular, and behavioural data was used to test the prevailing taxonomic hypothesis, it was found that *B. dorsalis*, *B. papayae*, and *B. philippinensis* most likely constitute one biological species (Schutze et al. 2012b; Boykin et al. 2013; Krosch et al. 2013): hence the 'failure' of diagnostics was not a failure at all, but a valid result. The true failure of the system was that the taxonomic species did not match the biological species.

Diagnostics only works with total accuracy if all biological species of interest are accurately delimited and named, and even then limitations can occur (Barr et al. 2012). Species delimitation, naming, and classification have traditionally been the role of taxonomists using, for the most part, morphological characters. But, the literature is increasingly demonstrating that the use of multiple tools (e.g., molecular, behavioural, physiological), integrated within a systematic framework, greatly aids accurate species delimitation: the constant recognition of new cryptic species across all animal and plant taxa is evidence of this (see further below). All three fields need to work together for a robust system. Failure of systematics and/or taxonomy will lead to a failure in diagnostics, and hence an inability to accurately identify what comes from your trap. If what you are identifying is a rare non-pest species of the rainforest or savannah, it makes little real difference, but when the identification is of a highly invasive polyphagous pest, a mistake is much more serious.

3 Biological Species Versus Taxonomic Species

Nearly all biologists recognise that species exist in nature (Dobzhansky 1941; Paterson 1985; Cracraft 2002; Walter 2003; Sites and Marshall 2004). We can look out of our windows and see a multitude of different plant and animal species at any one time. In the Tephritidae, more than 5,000 species are estimated, making it one of most species-rich of all dipteran families (Norrbom et al. 1999). Different fruit fly species vary in appearance, their basic biology and ecology, the pheromones they produce, their courtship systems, and the plants they feed upon. Yet when we sit down at a table and formally ask "Is the pest species of fruit fly invasive in Africa biologically the same as or different from the pest species in Asia?", or "Is this fly in southern Brazil the same or different as the fly in Mexico?", there is no one definition or cut-and-dry 'species test' upon which all can agree. This is not just a problem for fruit fly workers: over 150 years have passed since Darwin published his *Origin of Species*, but biologists still do not have a universally accepted species definition (Cracraft 2000). Summaries given in Mayden (1997) and de Queiroz (2007), for example, list over 20 different species theories. In such summaries it is important to realise that Mayr's well known 'Biological Species Concept', where sexual isolation is considered of critical importance in defining species (Mayr 1957), is only one species theory and a highly dated one at that.

In contrast, taxonomic species are frequently viewed in black and white terms, particularly by non-specialists and especially by regulatory agencies. This is true despite the fact that taxonomic descriptions often record intraspecific variation and recognise that subsequent research may lead to taxonomic revision. Once published, a taxonomic species (i.e., a nominal species or name) exists forever unless formally subsumed in a later taxonomic publication. Thus, taxonomically, at least, it remains unambiguous (again, particularly to non-specialists) that the invasive fly in Africa, called *Bactrocera invadens* Drew Tsuruta & White, is different from similar flies in Asia that are called *B. dorsalis* and that *Anastrepha fraterculus* (Wiedemann) is one species throughout Central and South America. The formality of naming taxonomic species is carefully defined by the International Code of Zoological Nomenclature (ICZN) (http://iczn.org/code) and, so long as the rules are followed, the species name remains available unless suppressed by a decision of the ICZN Commission. One key element of following the ICZN is the designation and lodging of a unique holotype specimen for each new species described: hence the term taxonomic species is often used interchangeably with typological species (Walter 2003). This requirement, while having many scientific benefits, can down-play the importance of intra-specific variation by focusing diagnostics onto one, dead individual organism. Information that may be critical to understanding and recognising the underlying biological species (such as behaviour, ecology, host range, or morphological variation) is lost in the designation of the type unless carefully gathered before the type is designated.

Taxonomists have always tried to match their taxonomic units with real biological units. Historically, it was considered that there should be reasonable overlap between biological and taxonomic species. For example, the famous geneticist Dobzhansky wrote: "*It is justifiable to conclude that if species separation is defined as the stage of the evolutionary divergence at which reproductive isolating mechanisms develop, the [genetical] species so delimited and the species of systematists will largely coincide*" (Dobzhansky 1941). Even 20 years ago it was still considered that, at least for taxa such as butterflies where the organisms rely on visual cues for mate recognition, and such cues can be easily detected by humans, taxonomists were more likely to align taxonomic with biological species (Paterson 1991). However, such views are now considered largely untenable. Cryptic taxa (i.e., multiple biologically unique species that previously were thought to be one species) are being routinely found across all taxonomic groups, arguably due to the increasing availability of molecular approaches and ever-more sophisticated morphological techniques. As an example, a single issue of one journal (*Biological Journal of the Linnean Society* 109 (4) [2013]) published papers dealing with cryptic species/ lineages of carnivorous plants, lycaenid butterflies, nymphalid butterflies, poecilliid fish, and squirrels.

3.1 An Example of How Easy It Is to Get It Wrong

As a postdoctoral fellow, one of the authors (A.R.C.) was employed on a large fruit fly project in Papua New Guinea (PNG) (Clarke et al. 2004). Flies were collected in a PNG-wide trapping network by skilled local scientists and subsequently sent to a specialist fruit fly laboratory in Brisbane for identification. Dedicated sorters worked daily on identification of the material, and the lab was headed by a taxonomic authority. Despite this considerable expertise, we failed to detect an incursion of banana fly (*Bactrocera musae* (Tryon)) from mainland PNG into the Gazelle Peninsular of East New Britain before it was too late to effectively respond (Mararuai et al. 2003). Why was this? It was not because the sorters had failed to detect the fly – indeed *post hoc* assessment of the daily data sheets showed we detected the fly almost immediately at the likely point of entry and could map its spread from there. Rather, because *B. musae* was not known in East New Britain and a biological incursion had not been considered, the 'odd' specimens had been set aside as possible new members of the *B. musae* species complex (Drew et al. 2011). It was only after fruit damage began to be reported that we investigated more closely. *Bactrocera musae* is now permanently established in the Gazelle, where it has become the major insect pest of bananas, the primary starch staple.

This example is used to emphasise how closely accurate diagnostics must be aligned with any trapping program, how easily the presence of cryptic complexes (see later) confuse things, and how easy it is to make mistakes in any activity. If this type of mistake can be made in a dedicated fruit fly taxonomy lab, specifically looking to identify and map the distribution of species, then it can be made anywhere. On a global scale the impact of the mistake was relatively minor, for the people of the Gazelle Peninsular it was major.

3.2 Why Is It Important to Get It Right?

While difficult to define, biologically unique species do exist, and even closely related species can have important biological differences, which impact their pest status, distribution, and ability to be controlled (Walter 2003). While some fruit fly examples follow, some of the best case studies come from outside the Tephritidae. For example, an excellent review by Garros et al. (2006) shows that among three closely related and cryptic mosquito taxa of the *Anopheles minimus* complex in South-east Asia, there is variation in human biting propensity, time of biting, ability to vector malaria, breeding habitat, and response to human manipulation of the environment. This type of study builds on older, and at the time ground breaking work, on the malaria vectoring capacity of different *Anopheles* mosquito populations in Africa (Hunt and Coetzee 1995). The recognition that multiple cryptic species often exist within a single taxonomic entity for mosquitoes has proven to be fundamental in their control and the diseases they vector (Coetzee et al. 2000).

Within the tephritids, there are many examples illustrating why accurately identifying biological species is important. In Mexico, exports of citrus were restricted due the presence of the known citrus pest *A. fraterculus*, until it was recognised that the Mexican population represents a unique species (Hernández-Ortiz et al. 2004) for which citrus are not hosts (Aluja et al. 2003). In Australia, *Bactrocera tryoni* (Froggatt) and *Bactrocera neohumeralis* (Hardy) are a closely related sibling pair that are only easily distinguished by differences in time of mating and one variable morphological trait, yet *B. tryoni* is Australia's major fruit fly pest and invasive, while *B. neohumeralis* is regarded as a minor pest of local importance and is considered non-invasive (Clarke et al. 2011). Similarly, the different members of the *Ceratitis* FAR complex (*Ceratitis fasciventris* (Bezzi), *C. anonae* Graham, and *C. rosa* Karsch) in Africa remain difficult to separate based on morphological and molecular criteria (Delatte et al. 2013), yet have different distributions, pest status, host ranges, and host preferences (Copeland et al. 2006).

4 Linking Biological Species and Taxonomic Species in the Tephritidae – A Complex Area

There are several insect groups for which it is well recognised that the alignment between taxonomic species and biological species is often poor; these include mosquitoes (Paterson 1991; Hunt and Coetzee 1995), whiteflies (Dinsdale et al. 2010; De Barro et al. 2011), and tephritids. For tephritids, the most common species issues are either the presence of multiple cryptic biological species within one taxonomic species or more taxonomic names being applied than there are biological species. Within the Tephritidae, such issues are known within most major pest genera, including *Ceratitis* (Virgilio et al. 2008), *Anastrepha* (Petit-

Marty et al. 2004; Cáceres et al. 2009), *Bactrocera* (Jamnongluk et al. 2003; Drew et al. 2011), and *Rhagoletis* (Xie et al. 2008). Why there are so many groups of closely related and morphologically similar species within the Tephritidae is unclear, but it may involve a combination of factors, including rapid radiation (Clarke et al. 2005; Krosch et al. 2012), host driven ecological speciation (Bush 1969; Feder et al. 1994), and the extensive use of non-visual courtship cues (Burk 1981).

The terminology to describe such groups of species – a 'species complex' – has different meanings within the literature, which can cause confusion. As used by several tephritid taxonomists, including E. Hardy and R.A.I. Drew, a species complex is an informal grouping of taxonomic species within a genus, where members share a suite of defining morphological characters but need not be especially morphologically cryptic with respect to each other (e.g., Drew 1989; Drew et al. 2011). In this usage, the term complex need make no inference about genetic relatedness of the different species, as morphologically similar species may be genetically divergent (Peccoud et al. 2009), while morphologically dissimilar species may be genetically close (Balvín et al. 2013). Confusion can arise, however, when the term 'species complex' is also used as a short-hand for 'cryptic species complex' and 'sibling species complex'. The former refers to a group of true cryptic (i.e., morphologically similar or identical) species, while the latter is a group of closely genetically related species. Thus, the *B. tryoni* species complex is a sibling complex of closely related taxa (Krosch et al. 2012), species within it are also all morphologically cryptic and so it also a cryptic species complex (Clarke et al. 2011), while it is also treated by Drew (1989) as a taxonomic complex. In contrast, the *B. dorsalis* species complex is a taxonomic species complex (Drew and Hancock 1994), but not all the species within it are siblings (Krosch et al. 2012), nor are the majority hard to tell apart from each other (Lawson et al. 2003). Different again is an example such as the *A. fraterculus* species complex, which has never been treated as a taxonomic complex, but is a true cryptic species complex where biologically distinct species are confounded under one nominal species (Hernández-Ortiz et al. 2004, 2012; Selivon et al. 2005; Vera et al. 2006). It is not, however, also a sibling species complex as some of the cryptic taxa (e.g., populations from the North East Andes versus those the Equatorial Andes) are distant relatives of each other and not, strictly speaking, sibling taxa (Ludeña et al. 2010). *Confused yet?*

5 Resolution of Cryptic Species Complexes

The argument being made in this chapter is that any trapping network is only as good as the subsequent diagnoses of specimens that are removed from the traps. The diagnoses are, in turn, dependent on the quality and completeness of the taxonomy of the particular target group, and the taxonomy of a group is often strengthened by a solid systematic foundation. Diagnostic approaches are about placing specimens within existing taxonomic units (Barr et al. 2012; Blackett et al. 2012; Frey et al. 2013) and thus they are reliant on good taxonomy. Taxonomy, however, should not be done without a sound systematic basis, as it is vital (particularly when dealing with economically important species) that the taxonomic species accurately reflect the underlying biological species (Paterson 1991; Walter 2003; Boykin et al. 2012).

The major problem with linking taxonomy and systematics is that, while the protocols for naming species are highly formalised and enshrined within the ICZN, no such guidelines exist for defining biological species. Indeed, as discussed earlier, no universally agreed upon biological species definition exists. This means that when difficulties arise, such as determining the species limits of economically important fruit flies, there is no simple right or wrong answer, and one individual's or one country's interpretation may be regarded as equally valid as another's.

A number of authors, including de Queiroz (2007), have articulated the reasons why delimiting biological species is difficult. He argues that every potential pair of species evolving from a shared common ancestor will be subject to different evolutionary pressures than all other pairs of evolving species. Also, as the human time-frame for studying species is a tiny fraction of the total evolutionary time-line involved, where we 'observe' different diverging populations on their respective evolutionary time-lines will vary. When these two issues are combined, it means that where one sibling pair sits, in terms of character state divergence, is very likely to differ from other sibling pairs. For example, one pair of populations isolated from each other by different ecological conditions may show ecological niche differentiation, but little or no difference in their courtship behavior; while another pair which are physically isolated, but still occur in similar habitats, may evolve mating differences while demonstrating little ecological divergence. From this basis, de Queiroz argues that to rely on a single criterion for delimiting species (e.g., species isolation, a shared mate recognition system, ecological niche differentiation, quantitative fixed differences, molecular or morphological divergence) is prone to failure as what works for one species pair may be inappropriate for another.

While de Queiroz (2007) clearly articulates the species delimitation problem, his solution is more complex. He argues that the one universally accepted fact about species is that they are separately evolving lineages, and this is how species should be defined, i.e., species are "*separately evolving (segments of) metapopulation lineages*" (de Queiroz 1998, 1999; very similar cases have also been made by Mayden (1997) and Naomi (2011)). The fact that the lineages are evolving separately means that other species properties (e.g., mate recognition, ecological niche differentiation, morphological divergence) are not the drivers of speciation, but emergent properties which may (or may not) follow from independent lineage evolution. Using these emergent species properties he then sees species delimitation as a "*methodological issue*" (de Queiroz 2005), which relies on using as many of these species properties as possible.

> ...any property that provides evidence of lineage separation is relevant to inferring the boundaries and numbers of species. Considering the properties that have previously been adopted as secondary species criteria, either the property itself (intrinsic reproductive isolation, monophyly, exclusive coalescence, diagnosability, deficits of genetic intermediates), or its converse (incompatible fertilization systems, different niches, phenetic distinquishability), provides evidence of lineage separation. Thus, all of those properties are relevant (as lines of evidence) to the problem of species delimitation. (de Queiroz 2007)

The operational side of species delimitation, using multiple lines of evidence as advocated from a theoretical basis by de Queiroz (1999, 2007) as well as Mayden (1997) and Naomi (2011), is the field now known as integrative taxonomy (Dayrat 2005). As developed fully in Schlick-Steiner et al. (2010), integrative taxonomy applies information from multiple, independent disciplines for species delimitation problems. These authors recommend input from at least three disciplines and the clear need to identify, *a priori*, the species concept being used, the delimitation criteria, and the data analysis methods. It is important that the disciplines be independent, otherwise the work becomes iterative rather than integrative. In iterative taxonomy, subsequent work supports an initial hypothesis but is not an independent test of that hypothesis (Yeates et al. 2011). For example, detailed morphometric studies that find additional evidence that previously named species are different in size and shape is iterative taxonomy (Drew et al. 2008; Schutze et al. 2012a), whereas discrete pheromone, morphological and genetic studies that independently yield the same outcome with respect to species delimitation, without species limits being *a priori* assigned, is integrative taxonomy (Tan et al. 2011; Krosch et al. 2013). It is important to note where research for diagnostic markers fits in this system. As diagnostics is about finding markers for predetermined taxa, the finding of a unique marker for a species is clearly iterative, not integrative, taxonomy. Finding a unique diagnostic marker supports the initial taxonomic hypothesis, but it does not independently strengthen the initial species hypothesis.

5.1 A Way Forward for Fruit Fly Species Complexes

For 98 % of fruit flies, classical morphological description will remain the only species delimitation tool used. Most tephritids are non-pests and remain restricted to their endemic habitats: for such flies highly detailed and costly multidisciplinary studies are not justifiable, although when they are carried out cryptic taxa will almost certainly be found (Abreu et al. 2005; Condon et al. 2008). But for the small group of important pest tephritids for which confusion exists over species delimitation, or when it is suspected that a new pest belongs to such a group, then an integrative taxonomic approach should be applied.

Both theory (Mayden 1997; de Queiroz 2007; Naomi 2011) and practice (Schlick-Steiner et al. 2010) suggest that it is largely up to the ‘user’, i.e. those interested in a particular system, to define what species delimitation criteria are most appropriate to their system. Schlick-Steiner et al. (2010) further suggest that at least three different, independent criteria should be used and that a species theory should be identified in advance. We consider identifying the species theory to be a less critical issue, as theories constantly change and nearly all are considered "the best", although all have problems (Cracraft 2000). What is more critical and needs to be recognised by fruit fly biologists are that for the truly difficult species complexes, no one particular species property (i.e., morphology vs. genetics vs. behaviour) is likely to be sufficient on its own for accurate species delimitation.

The following is a set of species delimitation criteria that should be used when attempting to determine species identities.

1. *Pre- and postzygotic mate compatibility.* Reproductive isolation (sensu Mayr 1957) and mate recognition (sensu Paterson 1985) remain a cornerstone of species definition and delimitation to the present day (The Marie Curie Speciation Network 2012). Even though biologically distinct tephritids may cross in small cages (Cruickshank et al. 2001), mate choice trials in larger flight cages have shown that mate discrimination exists between even very closely related biological species, for example, between members of the *B. dorsalis* (Schutze et al. 2013) and *A. fraterculus* (Vera et al. 2006) species complexes. While logistically complex for populations that are allopatric in the wild, mating trials can offer great insight if flies can be brought together. Caveats to any cage mating trial need to be carefully considered in interpreting the results (Walter 2003) and established protocols (FAO/IAEA/USDA 2003) should be followed to allow repeatability.
2. *Multi-gene tests.* Genetics is now a routine part of biological research, and both well established (Sites and Marshall 2004) and novel (Kubatko et al. 2011; Boykin et al. 2012; Fujita et al. 2012) analytical methods exist for delimiting species using genetic data of different types. As for any discipline, there are important caveats to genetics work. Different genes evolve at different rates, while mitochondrial and nuclear genes have totally different patterns of inheritance. It is therefore important to use multiple genes in any analysis to increase the quality of the data set. Similarly, individuals differ in their genetic makeup at both the within- and between-population levels. Without adequate sampling at the population level, it can be very difficult to determine if any genetic differences detected represent variation between species (inter-specific variation) or variation within a species (intra-specific variation) (Rittmeyer and Austin 2012).
3. *Pheromones.* Tephritid chemical ecology is complex, and putative sex pheromones are, depending on the genus, produced by rectal glands, anal glands, dermal glands and salivary glands (Nation 1981). Their specific role in sexual communication is still not clearly understood, but it is clear that analysis of pheromones can play a role in understanding the relationships of species (Carlson and Yocom 1986; Symonds et al. 2009; Tan et al. 2011). At least some of these chemicals vary with age (and obviously sex) of the specimen, and such variation needs to be understood and quantified if pheromones are to be used with confidence as a species delimitation tool (Vaníčková et al. 2012).

4. *Morphometrics and geometric morphometrics*. Morphometrics (Hernández-Ortiz et al. 2004, 2012) and geometric morphometrics (Khamis et al. 2012; Schutze et al. 2012b; Krosch et al. 2013) can be very powerful tools in detecting subtle size or shape differences between closely related taxa. However, some analyses (e.g., multivariate analysis of variance) requires *a priori* grouping of taxa, and where this done using existing taxonomy (or any existing hypothesis, such as mating 'groups'), then this should be regarded as iterative, rather than integrative, taxonomy (Drew et al. 2008; Schutze et al. 2012a). As for any discipline, morphometric analyses can be prone to errors if intra-specific variation is not well documented.

6 Closing with A Case Study – *Bactrocera papayae* and *B. dorsalis*

An unusual outbreak of maggots was found infesting papaya near Cairns in tropical north Queensland, Australia, in 1995. They were confirmed as larvae of the Asian Papaya Fruit Fly, *B. papayae,* and within 2 weeks quarantine restrictions were imposed, roadblocks established, and a major eradication campaign was underway. The response was rapid and dramatic, with over 2,600 monitoring traps laid out, more than 156,000 l of spot leaf treatment applied, and some 300 staff employed, all resulting in the total expenditure for the 42 month campaign of AU$33.5 million. When the programme was officially closed in mid-1999, *B. papayae* was declared eradicated from Australia, the campaign widely hailed a success, and the experience has become a text book example of how to intercept and eradicate an invasive fruit fly species (Cantrell et al. 2002).

Whilst *B. papayae* was first reported in Malaysia in 1991 (at the time an undescribed species known as 'Malaysian B') (Drew 1991), the events in Australia during the mid to late 1990s cemented its reputation as a pest of international significance. As *B. papayae* is recognised as an extremely close relative of the Oriental Fruit Fly (*B. dorsalis*) (Drew and Hancock 1994), and has an extremely large host range (Allwood et al. 1999), it is feared for its capacity to damage horticultural industries where it occurs (Clarke et al. 2005) or as a serious threat to countries where it is absent but has the potential to invade and establish (Plant Health Australia 2011). The threat of *B. papayae* invading countries has led to major quarantine initiatives (Tomkins 2013) and the imposition of market access restrictions. Thailand, for example, is heavily restricted in its export of fruit to markets in Taiwan, with commodities, such as rambutan, denied access, because *B. papayae* occurs in Thailand but not Taiwan (Department of International Trade Promotion, Ministry of Commerce Thailand: http://www.ditp.go.th//attachments/article/doc/51/51015335.doc). This is despite *B. dorsalis* being endemic to Taiwan; indeed, that island is the type locality of *B. dorsalis* (Hendel 1912; Drew and Hancock 1994).

When first detected in Australia, one of the key issues facing investigators was diagnosing the species (Cantrell et al. 2002). The principle challenge was to discriminate *B. papayae* from *B. dorsalis*, a species that occurs across an extremely wide native distribution from Pakistan in the west to Taiwan in the east, as well in various localities, such as Hawaii, where it is an established invasive (Drew and Hancock 1994). As developed below, the problem of diagnosing *B. papayae* from *B. dorsalis* was not limited to the Australian incursion but has been an ongoing problem for over 20 years, hampering trade, quarantine, and field control efforts for all that time. This case-study outlines the diagnostic issues, the unsuccessful approaches to solving the problem, and how an integrative taxonomy approach, as recommended above, has now resolved this complex problem.

6.1 The Diagnostic Problem

Bactrocera papayae was described in 1994 as part of an extensive taxonomic revision of the *B. dorsalis* species complex. This monograph expanded membership of the group from 16 to over 50 species and included 40 newly described taxa (Drew and Hancock 1994). As formally described in this revision, *B. papayae* is morphologically identical to the Oriental Fruit Fly with the exception of a single character: aculeus length. *Bactrocera papayae* has a longer aculeus, ranging from 1.77 to 2.12 mm, compared to a maximum of 1.6 mm in *B. dorsalis* (Drew and Hancock 1994).

Unfortunately, this character is problematic for diagnosing *B. papayae* operationally. Many quarantine interceptions occur as larvae in fruit (Plant Health Australia 2011); hence, identification is impossible unless specimens are reared to the adult stage (as was done for the 1995 Australian incursion). Further, and of greater relevance to a discussion on fruit fly trapping, the majority of individuals collected via surveillance programmes are males attracted to baited methyl eugenol traps. Given aculeus length (a *female* character) is the key diagnostic character used to separate *B. papayae* from *B. dorsalis*, lure-trapped male specimens are impossible to diagnose morphologically. The lack of diagnostic characters has meant that geographic origin of the specimen has often been the only 'character' that can be used to determine which species has been intercepted or trapped, even for specialist fruit fly taxonomists. However, this 'character' is useless for an incursive fly that appears with no known pathway.

6.2 Searching for Diagnostic Markers and Asking the Right Question

Following *B. papayae's* description in 1994, nearly 20 years of intensive diagnostic research went into the discovery of reliable characters that could be used to separate

B. papayae from *B. dorsalis*, particularly the males. This included studies of morphological and morphometric variation (Iwahashi 2001; Mahmood 2004; Drew et al. 2008), allozyme and mitochondrial barcoding research (Yong 1995; Armstrong and Ball 2005), and comparative chemical ecology work teasing apart male pheromone constituents (Fletcher and Kitching 1995; Tan and Nishida 2012). Despite tremendous efforts to find consistent markers distinguishing *B. papayae* from *B. dorsalis*, the same conclusions were almost always reached: they simply couldn't be told apart consistently. Male aedeagus length, for instance, was a logical target considering the female aculeus was different between species. Yet, while male *B. papayae* had, on average, longer aedeagi (3.02 mm) compared to *B. dorsalis* (2.73 mm), there was sufficient variation to yield extensive overlap in their ranges, rendering 'in-between' specimens unresolvable (e.g., standard deviations in aedeagus length being 0.666 mm and 0.121 mm for *B. dorsalis* and *B. papayae*, respectively) (Iwahashi 2001).

The lack of a consistent distinguishing character, or characters, poses an obvious but often overlooked question: what if the fly described in 1994 and detected in Queensland in 1995 wasn't *B. papayae* at all, but simply *B. dorsalis* by another name? Was it possible that the variation quantified (and argued over) for nearly two decades of diagnostics research represented population rather than species level variation? If *B. dorsalis* and *B. papayae* were the same biological species, and not different biological species as hypothesised by Drew and Hancock (1994), then attempts to find species-specific diagnostics were always doomed to fail. The high similarity in pheromone composition between *B. papayae* and *B. dorsalis*, coupled with their apparent ability to mate freely with each other in cage tests, led some workers to affirm they were the same biological species (Tan 2000, 2003), but a major review (Clarke et al. 2005) and a reassessment by one of *B. papayae*'s authors (Drew et al. 2008) concluded that the taxonomy was correct.

Despite the conclusions of Clarke et al. (2005) and Drew et al. (2008), robust diagnostic markers have never been identified for this sibling pair. When it is realised that exactly the same diagnostic problems apply to other major pest taxa, such as *B. philippinensis* and *B. invadens*, it is clear that not only have the earlier approaches led to unresolved debates, but the dollar and social costs of failing to find a way forward are very significant. We now recognise (see below) that Tan was right and that a focus on finding diagnostic markers meant that the right question was not asked, i.e., not "Can I separate these taxonomic species?", but "Are these taxonomic species valid biological species"? An assumption that the taxonomy was correct led to inappropriate, iterative (sensu Yeates et al. 2011) applications of genetic (Clarke et al. 2005), morphometric (Drew et al. 2011), and pheromonal (Fletcher and Kitching 1995) tools, which fuelled the debate but did not resolve the question. As argued from a theoretical basis earlier in this chapter, and seen in practice with the *B. dorsalis*/*B. papayae* system, failure to develop workable diagnostics was linked to a fundamental disconnect between systematics, taxonomy, and diagnostics. The last section of the case-study illustrates how this problem was overcome.

6.3 An Integrative Taxonomic Solution

A full integrative taxonomic study, which simultaneously used a range of independent tools as suggested earlier, has recently been applied to the Southeast Asian pest members of the *B. dorsalis* species complex, including *B. dorsalis* and *B. papayae*. The tools used have included: (i) morphological examination from traditional morphometrics of genitalic characters to fine-scale wing shape analysis using geometric morphometric techniques; (ii) molecular-based approaches using both phylogenetic and population genetic analyses of nuclear DNA sequence and microsatellite data coupled with mitochondrial DNA sequence analysis; (iii) studies of mating and post-zygotic sexual compatibility; and (iv) analysis of rectal gland (=pheromone) constituents (Schutze et al. 2012b, 2013; Tan and Nishida 2012; Boykin et al. 2013; Krosch et al. 2013; Tan et al. 2013). The outcome of this integrative study has been clear cut: variation between *B. dorsalis* and *B. papayae* is consistent with that expected at the intra-specific level, and there is no evidence these two taxonomic species are true biological species. The diagnostics issue is now greatly simplified as the markers which previously 'failed' to diagnose the 'species' can be recognised as having worked perfectly adequately: they 'failed' simply because there were no species level differences to detect.

So how has the new work differed from what has been done before, and why the confidence in the results? The key issue is not the tools (several of which are the same as earlier studies). The difference lies in the philosophical and operational approach. Critical elements of this changed approach include: (i) the involvement of multiple, independent laboratories; (ii) each of the analyses conducted independently of the others in an integrative, not iterative, framework; (iii) sampling that accounted for intra-specific variation as well as inter-specific variation; and (iv) as little reliance as was experimentally and analytically possible on the use of existing taxonomic names. But the key fundamental issue was identifying from the start that the problem was a systematic one (i.e., What is the relationship of the species/populations to each other within a biological and evolutionary framework?), not a diagnostic one (i.e., What are the appropriate markers needed to place individual specimens into the existing taxonomic framework?). This led to the design of experiments that could test very specific hypotheses relating to species delimitation (e.g., gene flow, morphological discontinuity, and mate compatibility). On their own, the results of each of these individual trials may be open to multiple interpretations, but when combined as a single body of work the consistent results are very powerful. This is the strength of the integrative taxonomic approach.

7 Conclusion

Fruit fly trapping serves many goals, but as stated earlier in this chapter a trapping network is only as good as your ability to diagnose what you remove from your trap. The diagnosis in turn is fully dependent on sound systematics and taxonomy. While the majority of fruit flies pulled from the majority of traps around the world can be easily and routinely identified, it is also clear that the family Tephritidae contains many species that cannot be so readily identified. When possible errors in diagnosis have continental scale implications, the over-arching taxonomic and systematic issues need to be promptly faced head-on and not left for 10 (*B. invadens*), 20 (*B. papayae*) or even 50+ years (*A. fraterculus*) before they are fully addressed.

Acknowledgements The paper was produced with research support through CRC National Plant Biosecurity projects 20115 and 20183: the authors would like to acknowledge the support of the Australian Government's Cooperative Research Centres Program. The authors also gratefully acknowledge the ongoing support of Dr Jorge Hendrichs and the Insect Pest Control Section of the Joint FAO/IAEA Programme Nuclear Techniques in Food and Agriculture, through the Coordinated Research Project: *Resolution of Cryptic Species Complexes of Tephritid Pests to Overcome Constraints to SIT Application and International Trade*.

References

Abreu AG, Prado PI, Norrbom AL, Solferini VN (2005) Genetic and morphological diagnosis and description of two cryptic species of flower head-infesting Tephritidae (Diptera). Insect Syst Evol 36:361–370

Allwood AJ, Chinajariyawong A, Drew RAI, Hamacek EL, Hancock DL, Hengsawad C, Jipanin JC, Jirasurat M, Kong Krong C, Kritsaneepaiboon S, Leong CTS, Vijaysegaran S (1999) Host plant records for fruit flies (Diptera: Tephritidae) in South East Asia. Raffles B Zool 7:1–92

Aluja M, Pérez-Staples D, Macías-Ordóñez R, Piñero J, McPheron B, Hernández-Ortiz V (2003) Nonhost status of *Citrus sinensis* cultivar valencia and *C. paradisi* cultivar ruby red to Mexican *Anastrepha fraterculus* (Diptera: Tephritidae). J Econ Entomol 96:1693–1703

Armstrong KF, Ball SL (2005) DNA barcodes for biosecurity: invasive species identification. Philos T Roy Soc B 360:1813–1823

Balvín O, Vilímová J, Kratochvíl L (2013) Batbugs (*Cimex pipistrelli* group, Heteroptera: Cimicidae) are morphologically, but not genetically differentiated among bat hosts. J Zool Syst Evol Res 51(4):287–295. doi:10.1111/jzs.12031

Barr NB, Islam MS, De Meyer M, McPheron BA (2012) Molecular identification of *Ceratitis capitata* (Diptera: Tephritidae) using DNA sequences of the COI barcode region. Ann Entomol Soc Am 105:339–350

Blackett MJ, Semeraro L, Malipatil MB (2012) Barcoding Queensland fruit flies (*Bactrocera tryoni*): impediments and improvements. Mol Ecol Resour 12:428–436

Boykin LM, Armstrong KF, Kubatko L, De Barro P (2012) Species delimitation and global biosecurity. Evol Bioinformatics 8:1–37

Boykin LM, Schutze MK, Krosch MN, Chomic A, Chapman TA, Englezou A, Armstrong KF, Clarke AR, Hailstones D, Cameron SL (2013) Multi-gene phylogenetic analysis of the south-east Asian pest members of the *Bactrocera dorsalis* species complex (Diptera: Tephritidae) does not support current taxonomy. J Appl Entomol. doi:10.1111/jen.12047

Burk T (1981) Signaling and sex in acalyptrate flies. Fla Entomol 64:30–43
Bush GL (1969) Sympatric host race formation and speciation in frugivorous flies of the genus *Rhagoletis* (Diptera: Tephritidae). Evolution 23:237–251
Cáceres C, Segura DF, Vera MT, Wornoayporn V, Cladera JL, Teal P, Sapountzis P, Bourtzis K, Zacharopoulou A, Robinson AS (2009) Incipient speciation revealed in *Anastrepha fraterculus* (Diptera; Tephritidae) by studies on mating compatibility, sex pheromones, hybridization, and cytology. Biol J Linn Soc 97:152–165
Cantrell B, Chadwick B, Cahill A (2002) Fruit fly fighters: eradication of the papaya fruit fly. CSIRO Publishing, Collingwood
Carlson DA, Yocom SR (1986) Cuticular hydrocarbons from six species of fruit flies. Arch Insect Biochem Physiol 3:397–412
Carstens BC, Pelletier TA, Reid NM, Satler JD (2013) How to fail at species delimitation. Mol Ecol 22:4369–4383
Clarke AR, Balagawi S, Clifford C, Drew RAI, Leblanc L, Mararuai A, McGuire D, Putulan D, Romig T, Sar S, Tenakanai D (2004) Distribution and biogeography of *Bactrocera* and *Dacus* species (Diptera: Tephritidae) in Papua New Guinea. Aust J Entomol 43:148–156
Clarke AR, Armstrong KF, Carmichael AE, Milne JR, Raghu S, Roderick GK, Yeates DK (2005) Invasive phytophagous pests arising through a recent tropical evolutionary radiation: the *Bactrocera dorsalis* complex of fruit flies. Ann Rev Entomol 50:293–319
Clarke AR, Powell KS, Weldon CW, Taylor PW (2011) The ecology of *Bactrocera tryoni* (Froggatt) (Diptera: Tephritidae): what do we know to assist pest management? Ann Appl Biol 158:26–54
Coetzee M, Craig M, Sueurb D (2000) Distribution of African malaria mosquitoes belonging to the *Anopheles gambiae* complex. Parasitol Today 16:74–77
Collins RA, Cruickshank RH (2013) The seven deadly sins of DNA barcoding. Mol Ecol Resour 13:969–975
Condon M, Adams DC, Bann D, Flaherty K, Gammons J, Johnson J, Lewis ML, Marsteller S, Scheffer SJ, Serna F, Swensen S (2008) Uncovering tropical diversity: six sympatric cryptic species of *Blepharoneura* (Diptera: Tephritidae) in flowers of *Gurania spinulosa* (Cucurbitaceae) in eastern Ecuador. Biol J Linn Soc 93:779–797
Copeland RS, Wharton RA, Luke Q, De Meyer M, Lux S, Zenz N, Macher P, Okumu M (2006) Geographic distribution, host fruit, and parasitoids of African fruit fly pests *Ceratitis anonae*, *Ceratitis cosyra*, *Ceratitis fasciventris*, and *Ceratitis rosa* (Diptera: Tephritidae) in Kenya. Ann Entomol Soc Am 99:261–278
Cracraft J (2000) Species concepts in theoretical and applied biology: a systematic debate with consequences. In: Wheeler QD, Meier R (eds) Species concepts and phylogenetic theory: a debate. Columbia University Press, New York, pp 3–14
Cracraft J (2002) The seven great questions of systematic biology: an essential foundation for conservation and the sustainable use of biodiversity. Ann Mo Bot Gard 89:127–144
Cranston PS, Gullan PJ, Taylor RW (1991) Principles and practice of systematics. In: CSIRO (ed) The insects of Australia. A textbook for students and research workers, 2nd edn. Melbourne University Press, Carlton, pp 109–124
Cruickshank L, Jessup AJ, Cruickshank DJ (2001) Interspecific crosses of *Bactrocera tryoni* (Froggatt) and *Bactrocera jarvisi* (Tryon) (Diptera: Tephritidae) in the laboratory. Aust J Entomol 40:278–280
Dayrat B (2005) Towards integrative taxonomy. Biol J Linn Soc 85:407–415
De Barro PJ, Liu S-S, Boykin LM, Dinsdale AB (2011) *Bemisia tabaci*: a statement of species status. Ann Rev Entomol 56:1–19
de Queiroz K (1998) The general lineage concept of species, species criteria, and the process of speciation: a conceptual unification and terminology recommendations. In: Howard DJ, Berlocher SH (eds) Endless forms: species and speciation. Oxford University Press, New York, pp 57–75

de Queiroz K (1999) The general lineage concept of species and the defining properties of the species category. In: Wilson RA (ed) Species: new interdisciplinary essays. MIT Press, Cambridge, pp 48–89

de Queiroz K (2005) Different species problems and their resolutions. Bioessays 27:1263–1269

de Queiroz K (2007) Species concepts and species delimitation. Syst Biol 56:879–886

Delatte H, Virgilio M, Simiand C, Quilici S, De Meyer M (2013) Isolation and characterisation of sixteen microsatellite markers cross-amplifying in a complex of three African agricultural pests (*Ceratitis rosa*, *C. anonae* and *C. fasciventris*, Diptera: Tephritidae). Conserv Genet Resour 5:31–34

Dinsdale A, Cook L, Riginos C, Buckley YM, De Barro PJ (2010) Refined global analysis of *Bemisia tabaci* (Hemiptera: Sternorrhyncha: Aleyrodoidea: Aleyrodidae) mitochondrial Cytochrome Oxidase 1 to identify species level genetic boundaries. Ann Entomol Soc Am 103:196–208

Dobzhansky T (1941) Genetics and the origin of species, 2nd edn. Columbia University Press, New York

Drew RAI (1989) The tropical fruit flies (Diptera: Tephritidae: Dacinae) of the Australasian and Oceanian regions. Mem Queensl Mus 26:1–521

Drew RAI (1991) Taxonomic studies on oriental fruit fly. In: Vijaysegaran S, Ibrahim AG (eds) Fruit flies in the tropics: proceedings of the first international symposium Malaysia March 1988. Malaysian Agricultural Research and Development Institute and Malaysian Plant Protection Society, Kuala Lumpur, Malaysia

Drew RAI, Hancock DL (1994) The *Bactrocera dorsalis* complex of fruit flies (Diptera: Tephritidae: Dacinae) in Asia. Bull Entomol Res Supplement No.2:i-iii + 1-68

Drew RAI, Raghu S, Halcoop P (2008) Bridging the morphological and biological species concepts: studies on the *Bactrocera dorsalis* (Hendel) complex (Diptera: Tephritidae: Dacinae) in South-east Asia. Biol J Linn Soc 93:217–226

Drew RAI, Ma J, Smith M, Hughes JM (2011) The taxonomy and phylogenetic relationships of species in the *Bactrocera musae* complex of fruit flies (Diptera: Tephritidae: Dacinae) in Papua New Guinea. Raffles Bull Zool 59:145–162

FAO/IAEA/USDA (Food and Agriculture Organization/International Atomic Energy Agency/ United States Department of Agriculture) (2003) Manual for product quality control and shipping procedures for sterile mass-reared tephritid fruit flies, Version 5. International Atomic Energy Agency, Vienna, Austria

Feder JL, Opp S, Wlazlo B, Reynolds K, Go W, Spisak S (1994) Host fidelity is an effective pre-mating barrier between sympatric races of the apple maggot fly. Proc Natl Acad Sci U S A 91:7990–7994

Fletcher MT, Kitching W (1995) Chemistry of fruit flies. Chem Rev 95:789–828

Frey JE, Guillén L, Frey B, Samietz J, Rull J, Aluja M (2013) Developing diagnostic SNP panels for the identification of true fruit flies (Diptera: Tephritidae) within the limits of COI-based species delimitation. BMC Evol Biol 13:106. doi:10.1186/1471-2148-1113-1106

Fujita MK, Leache AD, Burbrink FT, McGuire JA, Moritz C (2012) Coalescent-based species delimitation in an integrative taxonomy. Trends Ecol Evol 27:480–488

Garros C, Van Bortel W, Trung HD, Coosemans M, Manguin S (2006) Review of the Minimus complex of *Anopheles*, main malaria vector in Southeast Asia: from taxonomic issue to vector control strategies. Trop Med Int Health 11:102–114

Hendel F (1912) H. Sauter's Formosa-Ausbeute. Genus *Dacus* (Dipt.): Tephritidae. Supplementa Entomologica 1:13–24

Hernández-Ortiz V, Gómez-Anaya JA, Sánchez A, McPheron BA, Aluja M (2004) Morphometric analysis of Mexican and South American populations of the *Anastrepha fraterculus* complex (Diptera: Tephritidae) and recognition of a distinct Mexican morphotype. Bull Entomol Res 94:487–499

Hernández-Ortiz V, Bartolucci AF, Morales-Valles P, Frías D, Selivon D (2012) Cryptic species of the *Anastrepha fraterculus* complex (Diptera: Tephritidae): a multivariate approach for the recognition of South American morphotypes. Ann Entomol Soc Am 105:135–176

Hunt RH, Coetzee M (1995) Mosquito species concepts: their effect on the understanding of malaria transmission in Africa. In: Lambert DM, Spencer HG (eds) Speciation and the recognition concept. The John Hopkins University Press, Baltimore/London, pp 90–102

Iwahashi O (2001) Aedeagal length of the Oriental fruit fly, *Bactrocera dorsalis* (Hendel) (Diptera: Tephritidae), and its sympatric species in Thailand and the evolution of a longer and shorter aedeagus in the parapatric species of *B. dorsalis*. Appl Entomol Zool 36:289–297

Jamnongluk W, Baimai V, Kittayapong P (2003) Molecular phylogeny of tephritid fruit flies in the *Bactrocera tau* complex using the mitochondrial COI sequences. Genome 46:112–118

Jiang F, Li ZH, Deng YL, Wu JJ, Liu RS, Buahom N (2013) Rapid diagnosis of the economically important fruit fly, *Bactrocera correcta* (Diptera: Tephritidae) based on a species-specific barcoding cytochrome oxidase I marker. Bull Entomol Res 103:363–371

Khamis FM, Masiga DK, Mohamed SA, Salifu D, De Meyer M, Ekesi S (2012) Taxonomic identity of the Invasive fruit fly pest, *Bactrocera invadens*: concordance in morphometry and DNA barcoding. PLoS One 7(9):e44862. doi:10.1371/journal.pone.0044862

Krosch MN, Schutze MK, Armstrong KF, Graham GC, Yeates DK, Clarke AR (2012) A molecular phylogeny for the Tribe Dacini (Diptera: Tephritidae): systematic and biogeographic implications. Mol Phylogenet Evol 64:513–523

Krosch MN, Schutze MK, Armstrong KF, Boontop Y, Boykin LM, Chapman TA, Englezou A, Cameron SL, Clarke AR (2013) Piecing together an integrative taxonomic puzzle: microsatellite, wing shape and aedeagus length analysis of *Bactrocera dorsalis* s.l. (Diptera: Tephritidae) find no evidence of multiple lineages in a proposed contact zone along the Thai/Malay Peninsula. Syst Entomol 38:2–13

Kubatko LS, Gibbs HL, Bloomquist EW (2011) Inferring species-level phylogenies and taxonomic distinctiveness using multilocus data in *Sistrurus* rattlesnakes. Syst Biol 60:393–409

Lawson AE, McGuire DJ, Yeates DK, Drew RAI, Clarke AR (2003) DORSALIS: an interactive identification tool for fruit flies of the *Bactrocera dorsalis* complex. Griffith University, Brisbane, Multimedia CD-Rom. ISBN 0-909291-78-0

Ludeña B, Bayas R, Pintaud J-C (2010) Phylogenetic relationships of Andean-Ecuadorian populations of *Anastrepha fraterculus* (Wiedemann 1830) (Diptera: Tephritidae) inferred from COI and COII gene sequences. Annales de la Société entomologique de France (N.S.): international. J Entomol 46:344–350

Mahmood K (2004) Identification of pest species in oriental fruit fly, *Bactrocera dorsalis* (Hendel) (Diptera: Tephritidae) species complex. Pak J Zool 36:219–230

Mararuai A, Allwood A, Balagawi S, Dori F, Kalamen M, Leblanc L, Putulan D, Sar S, Schuhbeck A, Tenakanai D, Clarke A (2003) Introduction and distribution of *Bactrocera musae* (Tryon) (Diptera: Tephritidae) in East New Britain, Papua New Guinea. PNG J Agric For Fisher 45:59–65

Mayden RL (1997) A hierarchy of species concepts: the denouement in the saga of the species problem. In: Claridge MF, Dawah HA, Wilson MR (eds) Species: the units of biodiversity. Chapman & Hall, London, pp 381–424

Mayr E (1957) Species concepts and definitions. In: Mayr E (ed) The species problem. American Association for Advancement of Science, Washington, DC, pp 1–22

Naomi S-I (2011) On the integrated frameworks of species concepts: Mayden's hierarchy of species concepts and de Queiroz's unified concept of species. J Zool Syst Evol Res 49:177–184

Nation JL (1981) Sex-specific glands in tephritid fruit flies of the genera *Anastrepha*, *Ceratitis*, *Dacus* and *Rhagoletis* (Diptera: Tephritidae). Int J Insect Morphol Embryol 10:121–129

Norrbom AL, Carroll LE, Thompson FC, White IM, Freidberg M (1999) Systematic database of names. In Thompson FC (ed) Fruit fly expert identification system and systematic information database. Myia 9, pp 65–252

O'Meara BC (2010) New heuristic methods for joint species delimitation and species tree inference. Syst Biol 59:59–73

Paterson HEH (1985) The recognition concept of species. In: Vrba ES (ed) Species and speciation. Transvaal Museum, Pretoria, pp 21–29

Paterson HEH (1991) The recognition of cryptic species among economically important insects. In: Zalucki MP (ed) Heliothis: research methods and prospects. Springer, New York, pp 1–10

Peccoud J, Ollivier A, Plantegenest M, Simon J-C (2009) A continuum of genetic divergence from sympatric host races to species in the pea aphid complex. Proc Natl Acad Sci U S A 106:7495–7500

Petit-Marty N, Vera MT, Calcagno G, Cladera JL, Segura DF, Allinghi A, Rodriguero M, Cendra PG, Viscarret MM, Vilardi JC (2004) Sexual behavior and mating compatibility among four populations of *Anastrepha fraterculus* (Diptera: Tephritidae) from Argentina. Ann Entomol Soc Am 97:320–1327

Plant Health Australia (2011) The Australian handbook for the identification of fruit flies. Version 1.0. Plant Health Australia, Canberra, ACT

Ratnasingham S, Hebert PDN (2007) BOLD: the barcode of life data system http://www.barcodinglife.org. Mol Ecol Notes 7:355–364

Rittmeyer EN, Austin CC (2012) The effects of sampling on delimiting species from multi-locus sequence data. Mol Phylogenet Evol 65:451–463

Schlick-Steiner BC, Steiner FM, Seifert B, Stauffer C, Christian E, Crozier RH (2010) Integrative taxonomy: a multisource approach to exploring biodiversity. Ann Rev Entomol 55:421–438

Schutze MK, Jessup A, Clarke AR (2012a) Wing shape as a potential discriminator of morphologically similar pest taxa within the *Bactrocera dorsalis* species complex (Diptera: Tephritidae). Bull Entomol Res 102:103–111

Schutze MK, Krosch MN, Armstrong KF, Chapman TA, Englezou A, Chomič A, Cameron SL, Hailstones DL, Clarke AR (2012b) Population structure of *Bactrocera dorsalis* s.s., *B. papayae* and *B. philippinensis* (Diptera: Tephritidae) in southeast Asia: evidence for a single species hypothesis using mitochondrial DNA and wing-shape data. BMC Evol Biol 12:130. doi:10.1186/1471-2148-1112-1130

Schutze M, Jessup A, Ul-Haq I, Vreysen MJB, Wornoayporn V, Vera MT, Clarke AR (2013) Mating compatibility among four pest members of the *Bactrocera dorsalis* fruit fly species complex (Diptera: Tephritidae). J Econ Entomol 106:695–707

Selivon D, Perondini ALP, Morgante JS (2005) A genetic–morphological characterization of two cryptic species of the *Anastrepha fraterculus* complex (Diptera: Tephritidae). Ann Entomol Soc Am 98:367–381

Sites JWJ, Marshall JC (2004) Operational criteria for delimiting species. Ann Rev Ecol Evol Syst 35:199–227

Symonds MRE, Moussalli A, Elgar MA (2009) The evolution of sex pheromones in an ecologically diverse genus of flies. Biol J Linn Soc 97:594–603

Tan KH (2000) Behaviour and chemical ecology of fruit flies. In: Tan KH (ed) Area-wide control of fruit flies and other insect pests. Penerbit Universiti Sains Malaysia, Penang, pp 647–656

Tan KH (2003) Interbreeding and DNA analysis of sibling species within the *Bactrocera dorsalis* complex. Recent trends on sterile insect technique and area-wide integrated pest management – economic feasibility, control projects, farmer organization and *Bactrocera dorsalis* complex control study. Research Institute for Subtropics, Japan, pp 113–122

Tan KH, Nishida R (2012) Methyl eugenol: its occurrence, distribution, and role in nature, especially in relation to insect behavior and pollination. J Insect Sci 12. doi:10.1673/1031.1012.5601

Tan HK, Tokushima I, Ono H, Nishida R (2011) Comparison of phenylpropanoid volatiles in male rectal and pheromone gland after methyl eugenol consumptions, and molecular phylogenetic relationship of four global pest fruit fly species: *Bactrocera invadens*, *B. dorsalis*, *B. correcta* and *B. zonata*. Chemoecology 21:25–33

Tan KH, Wee S-L, Ono H, Nishida R (2013) Comparison of methyl eugenol metabolites, mitochondrial COI, and rDNA sequences of *Bactrocera philippinensis* (Diptera: Tephritidae) with those of three other major pest species within the dorsalis complex. Appl Entomol Zool 48:275–282

The Marie Curie Speciation Network (2012) Why do we need to know about speciation? Trends Ecol Evol 27:27–39

Tomkins A (2013) Review of long-term containment strategy for exotic fruit flies in Torres Strait. http://www.planthealthaustralia.com.au/wp-content/uploads/2013/05/Torres-Strait-Fruit-Fly-Strategy-Review.pdf

Van Houdt JKJ, Breman FC, Virgilio M, De Meyer M (2010) Recovering full DNA barcodes from natural history collections of Tephritid fruitflies (Tephritidae, Diptera) using mini barcodes. Mol Ecol Resour 10:459–465

Vaníčková L, Svatoš A, Kroiss J, Kaltenpoth M, Nascimento RRD, Hoskovec M, Břízová R, Kalinová B (2012) Cuticular hydrocarbons of the South American fruit fly *Anastrepha fraterculus*: variability with sex and age. J Chem Ecol 38:1133–1142

Vera MT, Cáceres C, Wornoayporn V, Islama A, Robinson AS, De La Vega MH, Hendrichs J, Cayol J-P (2006) Mating incompatibility among populations of the South American fruit fly *Anastrepha fraterculus* (Diptera: Tephritidae). Ann Entomol Soc Am 99:387–397

Virgilio M, Backelja T, Barr N, De Meyer M (2008) Molecular evaluation of nominal species in the *Ceratitis fasciventris*, *C. anonae*, *C. rosa* complex (Diptera: Tephritidae). Mol Phylogenet Evol 48:270–280

Virgilio M, Jordaens K, Breman FC, Backeljau T, De Meyer M (2012) Identifying insects with incomplete DNA barcode libraries, African fruit flies (Diptera: Tephritidae) as a test case. PLoS One 7:e31581. doi:10.1371/journal.pone.0031581

Walter GH (2003) Insect pest management and ecological research. Cambridge University Press, Cambridge

Xie X, Michel AP, Schwarz D, Rull J, Velez S, Forbes AA, Aluja M, Feder JL (2008) Radiation and divergence in the *Rhagoletis pomonella* species complex: inferences from DNA sequence data. J Evol Biol 21:900–913

Yeates DK, Seago A, Nelson L, Cameron SL, Joseph L, Trueman JWH (2011) Integrative taxonomy, or iterative taxonomy? Syst Entomol 36:209–217

Yong HS (1995) Genetic differentiation and relationships in five taxa of the *Bactrocera dorsalis* complex (Insecta: Diptera: Tephritidae). Bull Entomol Res 85:431–435

Index

A
Age-structured models, 380, 402–405, 416
Aggregation, 17, 19, 21–23, 27, 28, 31, 35, 53, 136, 155, 177, 191, 532
Allee effects, 6, 221, 254, 390, 391, 399–401, 415, 573–574
Ammonia baits, 82, 84–85, 87
Ammonium acetate, 80, 84, 85, 87, 90–94, 98, 101, 105, 107, 109, 131, 134, 135, 151, 188, 307, 383, 446, 448, 463–469, 471, 472, 483, 517–518, 522, 527, 531, 535–538, 541, 542
Anastrepha
  A. distincta, 18
  A. fraterculus, 18, 20, 78, 87, 97, 190, 224, 230, 323, 436, 494, 519, 520, 598, 616, 618, 619, 622, 627
  A. grandis, 18, 78, 87
  A. ludens, 8, 18–20, 22–24, 37, 77–82, 86, 88, 91–95, 97–99, 103, 128, 131, 133, 147, 149, 151, 153–155, 157, 158, 160, 162, 181, 182, 187, 190, 200, 201, 223, 224, 230, 233, 426, 428, 433, 442, 444, 447, 471–475, 485, 494, 518, 519, 521, 523, 598, 601, 602
  A. obliqua, 18, 20, 77, 79, 86, 94, 97, 103, 129, 131, 149, 151, 154, 155, 158–160, 162, 188, 190, 199, 201, 224, 230, 340, 426, 432, 442, 446, 494, 598, 602
  A. serpentina, 18, 20, 97, 149, 153, 162, 230, 432, 442, 598
  A. sororcula, 18–20
  A. striata, 18, 20, 77, 82, 97, 149, 224, 230, 433
  A. suspensa, 18–21, 23, 37, 77–79, 81, 86, 88, 91–96, 99, 102–107, 151, 153, 155, 157, 185, 188, 199, 206, 340, 383, 387, 388, 432, 442, 467, 471–473, 485, 518–521
Anastrepha spp., 7, 18, 20, 37, 88, 89, 91, 108, 149–151, 159, 161, 179, 187, 198, 230, 238, 240, 432, 471–474, 517, 521, 531, 532, 534, 538, 542, 567, 580, 602
Anisylacetone, 43
Area freedom, 254, 259, 260, 262, 263
Areas of low pest prevalence, 408–410, 563, 592, 599–600, 603, 606
Area-wide integrated pest management (AW-IPM) program, 281, 408, 428, 461, 481–484, 495
Area-wide pest management (AWPM), 289, 602–605
Arsenic salts, 424–426, 433
Assignment tests, 203, 206–207, 328, 348, 357, 363
Attract and kill, 134, 136, 313, 316, 411, 415, 458–460, 464, 466, 468, 474, 503, 507, 516, 546
Attractant, 5, 16, 76, 120, 147, 187, 238, 380, 424, 459, 495, 514, 563, 590
AW-IPM program. *See* Area-wide integrated pest management (AW-IPM) program
AWPM. *See* Area-wide pest management (AWPM)

B
Bactrocera
  B. carambolae, 8, 24–26, 28, 41, 224, 227, 229, 240, 315, 329, 494, 500, 502
  B. correcta, 24, 28, 29, 39, 223, 224, 227, 505
  B. cucurbitae, 8, 24, 77, 125, 146, 184, 224, 256, 329, 460, 494, 514, 567

T. Shelly et al. (eds.), *Trapping and the Detection, Control, and Regulation of Tephritid Fruit Flies*, DOI 10.1007/978-94-017-9193-9,

Bactrocera (cont.)
B. dorsalis, 8, 24, 77, 125, 151, 183, 223, 256, 324, 380, 430, 461, 494, 514, 560, 615
B. frauenfeldi, 43, 44, 500–502
B. invadens, 24–26, 39, 94, 126, 132, 179, 207, 223, 224, 227–229, 232, 340, 401, 494, 616, 625, 627
B. latifrons, 24, 46–47, 186, 191, 199, 224, 227, 494, 505
B. oleae, 30, 77, 131, 150, 184, 224, 268, 329, 425, 468, 498, 517
B. papayae, 24–26, 28, 29, 38, 41, 179, 232, 329, 401, 497, 499, 615, 623–627
B. philippinensis, 24–26, 329, 401, 497, 499, 615, 625
B. tryoni, 8, 24, 77, 128, 149, 179, 224, 256, 329, 424, 494, 517, 577, 618
B. umbrosa, 17, 24, 26, 29, 30, 38, 45
B. zonata, 24, 28–29, 38, 39, 77, 78, 94, 96, 223, 224, 227, 229, 430, 494, 505, 531
Bactrocera spp., 5–7, 9, 25, 29, 30, 32, 35, 39, 41–44, 46, 89, 125, 127, 130, 150, 189, 227–230, 238, 241, 401, 424, 429, 474–480, 494, 496, 502, 504, 506, 517, 532, 536, 543, 560, 567–569, 615
Bait station, 96, 108, 310, 313, 316, 380, 444, 457–487, 496, 497, 516
Barriers to invasion, 408–412
Behavior, 7, 16, 101, 120, 146, 177, 220, 279, 352, 414, 445, 456, 495, 519, 560, 620
Behavioral response, 21, 22, 24, 34, 103, 105, 120, 147, 158, 163, 476
Biological species, 401, 614–620, 622, 625, 626
BioLure, 80, 85, 87, 91–95, 100, 101, 108, 155, 159, 289, 465–468, 472, 483, 485, 516, 533, 539, 540, 544
Blend, 21, 23, 24, 31–35, 45, 53, 91–93, 97, 99, 120–122, 126–130, 133–136, 159, 160, 162, 463, 541, 542
Borax, 77, 85–89, 91, 92, 94, 95, 99, 100, 151, 162, 190, 193, 383, 441, 447, 472, 479, 517, 522
By-catch, 269

C
Carbamates, 424, 425, 435, 437
Ceratitis
C. capitata, 5, 32, 76, 122, 146, 180, 223, 256, 279, 323, 383, 424, 465, 494, 514, 562, 597
C. cosyra, 32, 48, 78, 92, 94, 96, 124, 126, 132, 151
C. rosa, 32, 47, 78, 92, 94, 96, 124, 224, 226, 227, 340, 433, 496
Ceratitis spp., 35, 48, 54, 124, 199, 206, 225–227, 331, 496, 578
Chemical stimuli, 16, 120, 145–163
Chlorinated hydrocarbons, 424, 425, 435–437
Clear lid, 518, 528
Color spectrum, 156
Contact insecticide, 424, 516, 527
Containment, 221, 240, 603, 605
α-Copaene, 34, 36, 47–48, 124, 161
Cost balancing, 577, 579
Cryptic species, 615, 617–623
Cue-lure (C-L), 8, 9, 16, 24, 38, 42–44, 85, 90, 93, 121, 150, 180, 188, 195, 197, 202, 207, 209, 254, 269, 461, 462, 475, 476, 478, 494–497, 499–504, 506, 507, 514, 531, 578

D
Dacus, 16, 17, 30, 35, 46, 52–53, 78, 88, 94, 95, 121, 125–128, 157, 202, 224, 230–231, 238, 336, 429, 430, 433, 460, 494, 514
Day degrees, 265, 266
DDVP. *See* 2,2-Dichlorovinyl dimethyl phosphate (DDVP)
Declaring success, 392, 396
Delimitation, 37, 38, 43, 178–179, 255, 269, 270, 504, 560, 562, 563, 570, 573, 574, 596, 605, 613–615, 620–622, 626
Density analysis, 278, 280, 289, 297–299, 305, 307, 312–315
Detection
probability, 5, 197, 238, 259, 334, 564, 574, 576
sensitivity, 5, 41, 560, 565, 566, 571, 575, 579
Diagnostics, 242, 348, 360, 613–617, 621, 625, 626
2,2-Dichlorovinyl dimethyl phosphate (DDVP), 191, 478, 503, 516, 524, 527, 528
Dietary history, 148–150
Dispersal, 39, 43, 176–209, 220, 234, 260, 261, 268, 311–312, 323, 330, 340, 382–385, 389, 390, 408–410, 412–414, 417, 438, 540, 579
Distance/capture function, 565–567
Distribution of traps, 289, 299, 305, 532–534

Dry attractant, 516, 518, 522, 527, 528, 536, 542
Dry trap, 87, 94, 95, 289, 516, 518, 522, 526–528, 539

E
EAG. *See* Electroanntenogram (EAG)
Early detection, 52, 238–240, 242, 269, 506, 507, 579, 580, 595, 597, 598
Electroanntenogram (EAG), 33, 34, 48, 49, 103–108, 122, 124, 125
Equivalence, 50, 310, 594, 606
Eradication, 37–40, 43, 52, 79, 85, 124, 125, 200, 221, 223, 226, 228, 230, 232, 233, 239–241, 254, 258–260, 262, 263, 267–269, 281, 288, 289, 292, 293, 309, 313, 322, 323, 338, 340, 348, 352, 357, 379–417, 425, 428, 433, 434, 438–441, 449, 461, 481, 482, 484, 494–506, 514, 562–564, 570, 572–580, 582, 595, 599, 603, 605, 623
Extinction, 6, 202, 234, 254, 260–262, 268–270, 391, 400, 569, 573, 574

F
Female attractant, 33, 126, 406, 415, 469, 471, 516, 517, 527, 528, 535, 537, 539, 542, 546
Floral volatile, 29
Fly age, 147
Fly responsiveness, 90, 121, 132–134, 147–157, 162, 163, 322, 458, 463, 481, 519
Food baits, 7, 9, 19, 31, 53, 149, 155, 160, 238
Food lures, 76, 132, 151, 155, 160, 162, 518, 565, 567, 578, 590, 596
Foraging behavior, 7, 152–154, 158
Fruit fly suppression, 159, 160, 483, 502, 542
Fruit infestation, 146, 432, 435, 437, 459, 463, 473, 475, 477–479, 487, 535, 537, 540–542
Fruit volatiles, 53, 90, 121, 124, 126, 128, 132, 134, 135, 151, 160, 538

G
Gene flow, 177, 178, 188, 202–206, 209, 329, 626
Geographic information systems (GIS), 278, 279, 281–288, 296, 297, 307, 313, 317, 318, 352, 581, 582
  data formats, 284, 318
Geo-reference(ing), 285–286
GF120, 445, 446, 448–450
GIS. *See* Geographic information systems (GIS)
Grower adoption, 475, 485

H
Haplotype diversity (Hd), 332, 335, 336
Haplotype matching, 331, 332
Homeostatic shifts, 147
Hormones, 107, 147, 148, 154–156
Host location, 121, 122, 135, 136, 156, 296, 581

I
ICZN. *See* International Code of Zoological Nomenclature (ICZN)
IDW. *See* Inverse distance weighted (IDW)
Incipient populations, 179, 180, 561, 562, 564, 566, 575–577
Ingestible bait, 425–435
Insecticide, 78, 134, 264, 283, 396, 423–450, 458, 495, 514, 564, 605
Insect movement, 177, 203, 208, 210, 280, 410
Integrated pest management (IPM), 43, 80, 85, 134, 162, 202, 207, 266, 281, 289, 380, 395, 408, 409, 412, 457–487, 502, 507, 535, 543, 545, 546, 564, 602–605
Integrative taxonomy, 612, 621, 623, 624, 626
International Code of Zoological Nomenclature (ICZN), 616, 620
International Plant Protection Convention (IPPC), 233, 562, 563, 591–594, 596, 599, 600, 606
International Standards for Phytosanitary Measures (ISPM), 179, 591–595, 597, 599, 600, 603, 605, 606
International trade, 254, 262, 506, 592, 594, 599, 605, 623, 627
Interpolation, 280, 303–306, 309–311, 318
Inverse distance weighted (IDW), 303–306
IPM. *See* Integrated Pest Management (IPM)
IPPC. *See* International Plant Protection Convention (IPPC)
ISPM. *See* International Standards for Phytosanitary Measures (ISPM)
Iterative taxonomy, 621

K
Kairomone, 53, 120, 121, 125, 127, 129, 135, 136, 156, 158, 159, 495–496
Kriging, 304, 314

L

Lag-phase, 221, 239, 242
Lateral holes, 523, 526, 528–530, 539
Liquid attractant, 480, 517, 522, 523, 526
Low prevalence areas, 283

M

Male annihilation technique (MAT), 8, 37, 52, 125, 176, 461, 495–504, 506, 507, 514, 516
Male lure, 7, 15–54, 76, 121, 150, 255, 289, 475, 495, 514, 565, 590
Map symbology, 291–294, 299, 311
Marking methods, 209
Mark-release-recapture (MRR), 40, 181–203, 205, 207, 209, 566
Mass trapping, 7, 15–54, 76, 108, 121, 150, 162, 255, 289, 313, 380, 459, 475, 495, 496, 513–546, 514, 565, 590
MAT. *See* Male annihilation technique (MAT)
Mate compatibility, 622, 626
Mating status, 4, 7, 104, 121, 132, 133, 148, 151, 287
Maxitrap, 518, 528, 529, 535
Mediterranean fruit fly (Medfly), 5, 32, 76, 122, 146, 180, 223, 255, 279, 323, 387, 424, 467, 494, 517, 563, 597
Methyl eugenol (ME), 7, 24, 85, 121, 150, 191, 254, 398, 449, 461, 494, 514, 538, 560, 624
Microsatellite DNA, 327, 330, 337–346, 362, 363
Mitochondrial DNA, 335, 336, 339, 353–355, 626
Modeling fruit fly trapping, 380
Molecular markers, 183, 185, 186, 202–207, 260, 328, 356, 363, 364
Monitoring, 45, 49, 92, 108, 120, 122–124, 130, 159, 162, 176, 180, 207–209, 237–239, 255, 258, 261–263, 268–270, 314, 380, 430, 432, 459, 460, 463, 472, 486, 494–496, 532, 536, 537, 543, 545, 562, 563, 575, 578, 582, 597–599, 602, 603, 605, 612, 623
Monte Carlo simulation, 186, 567–574, 576
Moskisan, 518, 528, 535, 538
MRR. *See* Mark-release-recapture (MRR)

N

Naturalization, 220–223, 234, 242
Next generation sequencing (NGS), 207, 346–348, 363
Nontarget organisms, 425, 458, 517, 518, 525, 526, 544
Nuclear DNA, 328–330, 355–359, 626
NuLure, 81, 86–96, 98–100, 132, 151, 162, 445, 447, 449, 466, 473, 482, 522
Nutrition, 82, 147

O

Odors, 8, 9, 19, 23, 35, 53, 89, 103, 119–136, 146, 150–153, 159, 160, 470, 471, 495, 566, 578
Official procedure, 593, 605
Olfactory cues, 24, 129, 130, 147, 487, 519–522
Olfactory stimuli, 32, 33, 103, 106, 130, 148, 458, 460, 463, 522
Olipe, 516, 523, 524, 526, 536, 539
Orchid, 26, 29, 43, 401, 496
Organic fruit fly control, 480
Organophosphates, 424, 425, 435–437, 450, 460, 527, 535, 539

P

Parakairomones, 495–496
Pathway analysis, 323, 324, 327, 363
Perimeter trapping, 134, 532–534
Pest free area(s), 178–179, 202, 408–410, 563, 590, 592, 596–599, 602, 603, 605, 606
Pest free zones, 254
Pest risk analysis (PRA), 230, 594–595
Phenylbutanoid, 30, 45–47
Phenylpropanoid, 41, 42, 46, 120
Pheromone, 8, 15–54, 80, 91, 100, 103, 109, 120, 124, 131, 133, 136, 146, 147, 150, 151, 153, 155, 156, 158, 160–162, 381, 385–387, 396, 397, 400, 401, 403, 478–480, 495–496, 504, 535, 542, 543, 565, 580, 590, 596, 616, 621, 622, 625, 626
Phototoxic dyes, 441–443
Physical stimuli, 156
Physiological state, 7, 8, 93, 102, 107–109, 132, 145–163
Phytosanitary status, 292, 300, 589–606
Plant odor, 9, 119–136
Population assignment, 325, 345
Population structure, 202, 204–207, 332, 337, 338, 340, 357
PRA. *See* Pest risk analysis (PRA)
Preservative, 94–95, 443, 446, 448
Probit 9, 262–264, 596, 601, 606
Program trigger, 562
Propagule pressure, 221, 236, 237, 239, 322

Protein bait(s), 76, 82–93, 95–100, 108, 127, 151, 153, 162, 424, 439, 458, 466, 468, 469, 472, 475, 476, 478, 479, 482, 497, 502, 503, 517, 518, 522, 523, 526, 538, 541, 542, 544, 601
Putrescine, 80, 90–94, 101, 105, 107, 108, 131, 155, 159, 162, 188, 307, 383, 465, 472, 517, 518, 522, 527, 535–537, 542
Pyrethroids, 134, 424, 460, 480, 527

Q
Quality control, 344, 345, 362, 445, 562, 563, 581
Quarantine, 38, 43, 52, 178–179, 187, 198, 202, 223, 233, 237, 240, 258, 259, 263, 267, 270, 283, 290, 323, 352, 423, 424, 426–429, 431, 438, 440, 441, 449, 494, 502, 504–506, 561, 562, 577, 594–602, 606, 623, 624
Quarantine regulations, 223, 233, 237, 263, 449, 494, 506, 602, 623

R
Range expansion, 202, 205, 222, 234–238
Raspberry ketone, 16, 17, 24, 36, 38, 42–44, 47, 161, 494, 496, 497
Raster, 283, 284, 301, 303–307, 310–314, 317, 318
Rectal gland, 26, 28–31, 39, 46, 47, 52, 622, 626
Remote sensing, 207–209, 269
Residues, 342, 425, 480, 542, 545
Resistance, 123, 222, 232, 425, 437, 438, 441, 594
Restriction fragment length polymorphism (RFLP), 326–328, 333–335, 353–359, 361, 362
Rhagoletis
  R. cerasi, 35, 87, 157, 231, 340, 519, 521, 536, 538, 539
  R. pomonella, 35, 80, 81, 84, 87, 98–100, 121, 132–135, 146, 150, 152, 156–160, 198, 231, 268, 340, 347, 462–464, 486, 494, 517, 519–522, 531–534, 538, 539, 543
Rhagoletis spp., 87, 121, 231–232, 238, 435, 458, 462–464, 517

S
Semiochemical(s), 103, 109, 125, 135, 160, 535, 590
Sesquiterpene, 28, 48, 124, 161
Sexual development, 148, 159
SIA. *See* Stable isotope analysis (SIA)
Sibling species, 26, 28–29, 619
Single nucleotide polymorphism (SNP), 207, 337, 347
SIT. *See* Sterile insect technique (SIT)
Solbait, 78, 87, 95, 445–448, 466
Source estimation, 325–328, 330–336, 338, 340–341, 343–349, 351–354, 357, 360, 362–364
Spatial dependence/autocorrelation, 302–305
Species complex, 25, 27, 351, 461, 615, 617, 619–624, 626
Species delimitation, 613–615, 620–622, 626
Spinosad, 150, 445–450, 459, 460, 464, 466, 468, 473, 476, 482, 483, 506, 507, 535, 539, 541, 605
Stable isotope analysis (SIA), 346, 348–349
Standard distance/ellipse, 296, 297
Sterile insect technique (SIT), 40, 108, 176, 180, 201, 240, 278, 323, 460, 470, 481, 542, 543, 563
Sterile releases, 155, 289, 309, 314, 398–400, 408
Sugar baits, 76, 79–82, 84, 86
Suppression, 7, 16, 39, 43, 97, 109, 227, 240, 292, 309, 312, 379–417, 428, 459–461, 466, 468, 469, 475, 480, 481, 483–485, 495–497, 502–504, 507, 533, 536, 542, 543, 546, 562, 572–575, 595, 599, 601, 603, 605
Suredye, 442–445, 473
Surfactant, 92, 94–95, 442, 443
Surveillance, 3, 5, 8, 9, 16, 33, 35, 38, 43, 48, 180, 237–239, 254–256, 258–259, 261, 267, 268, 270, 322, 413, 497, 502, 590, 595–597, 599, 606, 612, 624
Survey, 4, 37–39, 43–45, 52, 161, 163, 229, 238, 269, 315, 424, 428, 439, 458, 514, 560, 563, 564, 568–573, 575–580, 590, 593, 595, 596, 599, 603
Synomone, 53, 120, 136
Systematics, 614, 615, 620, 625, 627
Systems approach, 506, 563, 592, 596, 599–601, 606

T
Tartar emetic, 425, 432, 433, 435–436
Taxonomic species, 614–620, 625, 626
Taxonomy, 351, 429, 612, 614–615, 617, 619–621, 623–625, 627
Telemetry, 269–270
Tephri-trap, 314, 518, 523, 524, 537–540

Three component lure, 517–518, 526, 527, 533, 535–537, 540–542, 544
Torula yeast borax (TYB) pellets, 77, 86, 151, 162, 383, 472, 479
Toxicant, 438, 439, 445, 460–462, 465, 466, 472, 473, 478, 479, 483, 485, 486, 497, 498, 506, 543
Toxotrypana, 16, 17, 36, 53, 202
  T. curvicauda, 36, 53, 97, 108, 519, 533
Trap
  attraction, 380–384
  density, 7, 163, 180, 238, 254, 261–263, 289, 305, 412, 534–536, 538, 539, 541–542, 544, 571, 577, 579, 580, 582, 594, 595
  placement, 38, 101, 531, 533
  rotation, 180, 580
  shape, 521
  shifting, 269
Trapping program, 5, 6, 17, 37, 41, 42, 53, 147, 280, 307, 312, 380, 415, 503, 560, 579, 582, 595, 597, 613, 617
Trapping sensitivity, 599
Trimedlure, 5, 6, 8, 9, 16, 35, 36, 47–50, 52, 86, 92, 99, 100, 121, 124, 188, 195, 254, 289, 307, 314, 387, 469, 471, 496, 497, 514, 516, 524, 567, 570, 572, 574, 576, 578, 602
Trimethylamine, 80, 91–93, 97–99, 159, 307, 465–469, 483, 518, 522, 527, 535–538, 541, 542

V
Vector, 283, 284, 291, 294–303, 312, 314, 317, 618
Viable density, 574
Visual cues, 109, 121, 129, 130, 156, 209, 460, 519–522, 526, 539, 614, 617
Volatile, 18, 19, 21, 29, 32, 33, 35, 42, 44, 48, 53, 76, 83, 89–91, 94, 97, 98, 102–104, 120–136, 151, 155, 156, 159, 160, 162, 380, 443, 468, 528, 538

W
World Trade Organization (WTO), 591

Y
Yeast baits, 448

Z
Zero catch period, 262, 263

MIX
Papier aus verantwortungsvollen Quellen
Paper from responsible sources
FSC® C105338

If you have any concerns about our products,
you can contact us on
ProductSafety@springernature.com

In case Publisher is established outside the EU,
the EU authorized representative is:
Springer Nature Customer Service Center GmbH
Europaplatz 3, 69115 Heidelberg, Germany

Printed by Libri Plureos GmbH
in Hamburg, Germany